Brian Maurer
Dept. of Zoology
Brigham Young University
Provo, UT 84602

D0965908

A Complete Checklist of the
Birds of the World

A Complete Checklist of the
Birds of the World

Second Edition

Richard Howard
Alick Moore

ACADEMIC PRESS

Harcourt Brace Jovanovich, Publishers
London   San Diego   New York
Boston   Sydney   Tokyo   Toronto

Academic Press Ltd.
24–28 Oval Road
London NW1 7DX

United States edition published by
Academic Press Inc.
San Diego, CA 92101

Copyright by
© Richard Howard and Alick Moore 1980, 1984, 1991
Index of English Names © Alick Moore 1984, 1991

All Rights Reserved

No part of this book may be reproduced in any form
by photostat, microfilm or any other means
without written permission from the publishers.

This book is printed on acid-free paper.

British Library Cataloguing in Publication Data
Howard, Richard *1936–*
    A complete checklist of the birds of the world. — 2nd. ed.
    1. Birds – Lists
    I. Title   II. Moore, Alick
    598

ISBN 0–12–356910–9

Typeset by Paston Press, Loddon, Norfolk.
Printed in Great Britain by Mackays of Chatham PLC, Chatham, Kent.

Contents

Introduction

Since the revised edition of this book was published in 1984, there has been a multitude of changes necessary to bring the Checklist up to date. Many of the alterations and amendments have been brought about by the publication of the last 2 volumes of the J.L. Peters' *Checklist of the Birds of the World*, volume 8 covering the Cotingas, Manakins and Tyrant Flycatchers and volume 11 covering Old World Warblers, Flycatchers etc. Amongst many other important publications, that of the B.O.U. checklist no. 7, *The Birds of Wallacea* by the late C.M.N. White and Murray D. Bruce has added a wealth of excellent information from a previously thinly described area.

Not a single issue of an ornitholgical journal now comes out without some proposed change of name, or taxonomic status or addition to range. As bird study becomes more widespread and the methods used more sophisticated, the established order and relationships get questioned more often and more exactingly. In the not too distant future our checklists will be determined by even higher technology applied to areas of zoology which ten years ago few even recognized as relevant. Some interesting studies are currently taking place concerning the relationships of different bird groups. In particular those involving DNA hybridization and protein electrophoresis seem likely to establish quite different systematic groupings of bird families. In an article in the journal *Notornis*, R.W. Holdaway asks the question "What if Sibley and Ahlqvist are right?" He was specifically referring to the New Zealand passerine list, but the apprehension might well be expressed relevantly of the whole world list as it is established today. Drs Sibley and Ahlqvist are the leading protagonists in the DNA hybridization studies, and together with Dr B.L. Monroe Jr. are to publish a new world list which will doubtless reflect some of this technology. This may therefore be one of the last opportunities to produce an up to date usable list which bears some relationship to the historical order and form of traditional checklists, that is to say a list we can use because we know where to look and what to look for.

The argument between lumpers and splitters continues unabated. Since the first publication of this list in 1980, the names of some 75 species have disappeared due to lumping, together with about 50 generic names. Curiously only 34 generic names have been added or revived, while 24 new specific names have been erected. Another 100 or so specific names have moved from one genus to another, and the total number of species on the list is now more than 9200.

A GUIDE TO USING THIS LIST

Order of families, genera and species.

As in the previous edition, we have used the Basel sequence of families which, with our major references to the Peters' *Checklist*, continues to be the most relevant. The subfamilies where used, the genera and species are compiled from many sources. Generally we have started with the Peters' *Checklist* and then updated with more recent treatments. Generic limits have so often been a matter of individual preference as ideas of interrelationships develop, and we have tried to follow the most respected authorities in their recognition of genera and their following of the phylogenetic sequence of species within each genus. Subfamilies are only used where the authority followed deems them necessary. Many authors subscribe to tribes, but we have avoided these, as we have avoided subgenera and superspecies for the sake of simplicity. There is no question that there are groups of species sufficiently similar to be grouped together away from near-relatives yet not sufficiently dissimilar from those near-relatives to be placed within another genus. Additionally there are subspecies closer to one another in groups to be classed separately from other subspecies within the same genus. There is good validity for recognizing these sub-divisions, but their use does very much complicate a simple and clear list.

Subspecies

These are listed in common geographical order, which is to say generally from northwest to southeast, though there are naturally many places where the order of subspecies is arguable and no doubt readily acceptable if changed. Again we rely on the respected authorities for recognition of subspecies, but many of these are contentious, and those that rely on a clinal variation of size or plumage colour we have tried to avoid.

Geographical distribution

We have made several changes from the previous edition. Distribution of each species or subspecies is necessarily brief and normally adheres to the political names of the countries involved but the following should be noted.

* More use is made of the term Amazonia, covering those parts of several countries contributing to the Amazon basin.
* Rhodesia has become Zimbabwe, Ceylon is now Sri Lanka and Celebes Sulawesi.
* Central America comprises the area from Mexico to Panama inclusive.
* Indochina comprises Vietnam, Laos and Cambodia.
* Himalayas, Andes, Rocky Mountains and Ural Mountains are self evident.
* Guianas covers Guyana, Cayenne and Surinam.
* Madagascar, Borneo, New Guinea are used to denote the whole of those islands.
* Southern Australia means the southern half of that continent as distinguished from South Australia state; similarly Southern Africa indicates generally everything south of Zaire, including, of course, South Africa.

Migration areas well separated from breeding areas are indicated by the symbol >>.
N, E, S, and W are North, East, South and West and C is Central.
Smaller islands are denoted by I, island groups are Is, and where the distribution consists of a number of islands, Is is placed at the end of the list.

Extinct birds

Fossil birds are outside the scope of this book and of recent birds, those known to be extinct are omitted. Where extinction is a possibility the symbol **e?** is used.

English names

English, and sometimes colloquial names, are constantly being revised and updated. This has certainly been necessary in a number of cases and most of the revisions have more accurately described the birds in question. This process will certainly continue, possibly even to the extent of changing two of my favourites, Superciliaried Hemispingus and Guttulated Foliage-Gleaner. We have chosen not to confuse the text with more reference numbers to indicate the source of these changes, as in almost all cases they are taken from one of the references quoted for the treatment of the relevant Latin names, or from one of the main family references.

There has recently been a proposal made for a number of changes to the names of familiar European birds, so that they may be seen in a global rather than a European context. We wholeheartedly support this idea and have adopted several of the names proposed in this revision.

Acknowledgements

We are grateful for the advice and support of Dr C.M. Perrins, Jonathan Elphick and James Hancock, and would particularly like to thank Erica Moore, whose skill and determination with the word processor made a significant contribution to the production of this list.

References used in writing this introduction

J.L. Peters and successors 1931-87. *Checklist of the Birds of the World*, Vols I-XVI. Harvard University Press and Mus. Comp. Zool., Cambridge, Mass.

Mayr, E. 1969. *Principles of Systematic Zoology*. Amer. Mus. Nat. Hist., New York.

Sibley, C.G. and J.E. Ahlqvist, 1980. *Proc. Int. Orn. Congr.*, 17.125.

White, C.M.N. and M.D. Bruce, 1986. *Birds of Wallacea*. B.O.U. Checkl. 7.

Christidis, L. 1987. *Auk*, 104.556.

B.O.U. Records Comm. 1988. *Ibis*, 130 Suppl. Suggested changes to the English Names of some Western Palaearctic Birds.

Holdaway, R.W. 1988. *Notornis*, 35.63.

McKitrick, M.C. and R.M. Zink, 1988. *Condor*, 90.1.

Sibley, C.G., J.E. Ahlqvist and B.L. Monroe Jr., 1988. *Auk*, 105.409.

Arrangement of orders and families

Class AVES

Order **STRUTHIONIFORMES**

1	STRUTHIONIDAE	OSTRICHES

Order **RHEIFORMES**

2	RHEIDAE	RHEAS

Order **CASUARIIFORMES**

3	CASUARIIDAE	CASSOWARIES
4	DROMAIIDAE	EMUS

Order **APTERYGIFORMES**

5	APTERYGIDAE	KIWIS

Order **TINAMIFORMES**

6	TINAMIDAE	TINAMOUS

Order **SPHENISCIFORMES**

7	SPHENISCIDAE	PENGUINS

Order **GAVIIFORMES**

8	GAVIIDAE	DIVERS

Order **PODICIPEDIFORMES**

9	PODICIPEDIDAE	GREBES

Order **PROCELLARIIFORMES**

10	DIOMEDEIDAE	ALBATROSSES
11	PROCELLARIIDAE	PETRELS, SHEARWATERS
12	HYDROBATIDAE	STORM PETRELS
13	PELECANOIDIDAE	DIVING PETRELS

Order **PELECANIFORMES**

14	PHAETHONTIDAE	TROPIC BIRDS
15	PELECANIDAE	PELICANS
16	SULIDAE	GANNETS, BOOBIES
17	PHALACROCORACIDAE	CORMORANTS
18	ANHINGIDAE	ANHINGAS
19	FREGATIDAE	FRIGATE BIRDS

Order **CICONIIFORMES**

20	ARDEIDAE	HERONS, BITTERNS
21	SCOPIDAE	HAMMERKOP
22	BALAENICIPITIDAE	WHALE HEADED STORK
23	CICONIIDAE	STORKS
24	THRESKIORNITHIDAE	IBISES, SPOONBILLS
25	PHOENICOPTERIDAE	FLAMINGOES

Order **ANSERIFORMES**

26	ANHIMIDAE	SCREAMERS
27	ANATIDAE	DUCKS, GEESE, SWANS

Order **CAPRIMULGIFORMES**

74	STEATORNITHIDAE	OILBIRD
75	PODARGIDAE	FROGMOUTHS
76	NYCTIBIIDAE	POTOOS
77	AEGOTHELIDAE	OWLET-NIGHTJARS
78	CAPRIMULGIDAE	NIGHTJARS

Order **APODIFORMES**

79	APODIDAE	SWIFTS
80	HEMIPROCNIDAE	TREE SWIFTS
81	TROCHILIDAE	HUMMINGBIRDS

Order **COLIIFORMES**

82	COLIIDAE	MOUSEBIRDS

Order **TROGONIFORMES**

83	TROGONIDAE	TROGONS

Order **CORACIIFORMES**

84	ALCIDINIDAE	KINGFISHERS
85	TODIDAE	TODIES
86	MOMOTIDAE	MOTMOTS
87	MEROPIDAE	BEE-EATERS
88	CORACIIDAE	ROLLERS
89	BRACHYPTERACIIDAE	GROUND ROLLERS
90	LEPTOSOMATIDAE	COUROLS
91	UPUPIDAE	HOOPOES
92	PHOENICULIDAE	WOOD HOOPOES
93	BUCEROTIDAE	HORNBILLS

Order **PICIFORMES**

94	GALBULIDAE	JACAMARS
95	BUCCONIDAE	PUFFBIRDS
96	CAPITONIDAE	BARBETS
97	INDICATORIDAE	HONEYGUIDES
98	RAMPHASTIDAE	TOUCANS
99	PICIDAE	WOODPECKERS

Order **PASSERIFORMES**

100	EURYLAIMIDAE	BROADBILLS
101	DENDROCOLAPTIDAE	WOODCREEPERS
102	FURNARIIDAE	OVENBIRDS
103	FORMICARIIDAE	ANTBIRDS
104	CONOPOPHAGIDAE	GNATEATERS
105	RHINOCRYPTIDAE	TAPACULOS
106	TYRANNIDAE	TYRANT FLYCATCHERS
107	PIPRIDAE	MANAKINS
108	COTINGIDAE	COTINGAS
109	OXYRUNCIDAE	SHARPBILL
110	PHYTOTOMIDAE	PLANTCUTTERS
111	PITTIDAE	PITTAS
112	XENICIDAE	NEW ZEALAND WRENS
113	PHILEPITTIDAE	ASITIES
114	MENURIDAE	LYREBIRDS
115	ATRICHORNITHIDAE	SCRUBBIRDS
116	ALAUDIDAE	LARKS
117	HIRUNDINIDAE	SWALLOWS, MARTINS
118	MOTACILLIDAE	WAGTAILS, PIPITS
119	CAMPEPHAGIDAE	CUCKOO SHRIKES
120	PYCNONOTIDAE	BULBULS
121	IRENIDAE	LEAFBIRDS, IORAS
122	LANIIDAE	SHRIKES
123	VANGIDAE	VANGA SHRIKES

124	BOMBYCILLIDAE	WAXWINGS
125	DULIDAE	PALMCHAT
126	CINCLIDAE	DIPPERS
127	TROGLODYTIDAE	WRENS
128	MIMIDAE	MOCKING BIRDS, THRASHERS
129	PRUNELLIDAE	ACCENTORS
130	TURDIDAE	THRUSHES
131	ORTHONYCHIDAE	LOG-RUNNERS
132	TIMALIIDAE	BABBLERS
133	PANURIDAE	PARROTBILLS
134	PICATHARTIDAE	BALD CROWS
135	POLIOPTILIDAE	GNATCATCHERS
136	SYLVIIDAE	OLD WORLD WARBLERS
137	MUSCICAPIDAE	OLD WORLD FLYCATCHERS
138	PLATYSTEIRIDAE	WATTLE-EYES, PUFFBACK FLYCATCHERS
139	MALURIDAE	AUSTRALASIAN WRENS
140	ACANTHIZIDAE	THORNBILLS, FLYEATERS
141	MONARCHIDAE	MONARCHS, FANTAILS
142	EOPSALTRIDAE	AUSTRALASIAN ROBINS
143	PACHYCEPHALIDAE	WHISTLERS
144	AEGITHALIDAE	LONG-TAILED TITS
145	REMIZIDAE	PENDULINE TITS
146	PARIDAE	TITS, CHICKADEES
147	SITTIDAE	NUTHATCHES
148	CERTHIIDAE	TREECREEPERS
149	RHABDORNITHIDAE	PHILIPPINE CREEPERS
150	CLIMACTERIDAE	AUSTRALIAN CREEPERS
151	DICAEIDAE	FLOWERPECKERS
152	NECTARINIIDAE	SUNBIRDS
153	ZOSTEROPIDAE	WHITE-EYES
154	MELIPHAGIDAE	HONEYEATERS
155	EMBERIZIDAE	BUNTINGS, CARDINALS, TANAGERS
156	COEREBIDAE	BANANAQUIT
157	PARULIDAE	NEW WORLD WARBLERS
168	DREPANIDIDAE	HAWAIIAN HONEYCREEPERS
159	VIREONIDAE	VIREOS
160	ICTERIDAE	NEW WORLD BLACKBIRDS
161	FRINGILLIDAE	FINCHES
162	ESTRILDIDAE	WAXBILLS
163	PLOCEIDAE	WEAVERS, SPARROWS
164	STURNIDAE	STARLINGS
165	ORIOLIDAE	ORIOLES
166	DICRURIDAE	DRONGOS
167	CALLAEIDAE	WATTLEBIRDS
168	GRALLINIDAE	MAGPIE LARKS
169	ARTAMIDAE	WOOD SWALLOWS
170	CRACTICIDAE	BUTCHER BIRDS
171	PTILONORHYNCHIDAE	BOWERBIRDS
172	PARADISAEIDAE	BIRDS OF PARADISE
173	CORVIDAE	CROWS, JAYS

References

I have substantially revised the reference lists as reviews of various families and geographical regions have rendered a number of the previously quoted references irrelevant. At the head of each families' reference list are, in capital letters, the main works used in establishing the revised list of birds. Where a decision has been taken in an area of taxonomic contention, or where a major rearrangement has been made, a note by number refers to the work or works used in coming to that decision. This method is also used where a new species has been erected since the first revision of the list.

STRUTHIONIDAE 1

MAYR, E. 1979. *PETERS' CHECKLIST OF THE BIRDS OF THE WORLD*, Vol. 1, 2nd ed. Cambridge, Mass.
VAURIE, C. 1965. *BIRDS OF THE PALAEARCTIC FAUNA*, Vol. 2. Witherby, London.
WHITE, C.M.N. 1965. *REVISED CHECKLIST OF AFRICAN NON-PASSERINE BIRDS*. Govt. Printer, Lusaka.

RHEIDAE 2

MAYR, E. 1979. *PETERS' CHECKLIST OF THE BIRDS OF THE WORLD*, Vol. 1, 2nd ed. Cambridge, Mass.
BLAKE, E.R. 1977. *MANUAL OF NEOTROPICAL BIRDS*, Vol. 1, Chicago University Press, Chicago.

CASUARIIDAE 3

MAYR, E. 1979. *PETERS' CHECKLIST OF THE BIRDS OF THE WORLD*, Vol. 1, 2nd ed. Cambridge, Mass.
RAND, A.L. and E.T. GILLIARD, 1967. *HANDBOOK OF NEW GUINEA BIRDS*. Weidenfeld & Nicholson, London.

DROMAIIDAE 4

MAYR, E. 1979. *PETERS' CHECKLIST OF THE BIRDS OF THE WORLD*, Vol. 1, 2nd ed. Cambridge, Mass.
CONDON, H.T. 1975. *CHECKLIST OF THE BIRDS OF AUSTRALIA*, Vol. 1, R.A.O.U., Melbourne.

APTERYGIDAE 5

MAYR, E. 1979. *PETERS' CHECKLIST OF THE BIRDS OF THE WORLD*, Vol. 1, 2nd ed. Cambridge, Mass.
O.S.N.Z. 1970. *ANNOTATED CHECKLIST OF THE BIRDS OF NEW ZEALAND*. Reed, Wellington.

TINAMIDAE 6

BLAKE, E.R. 1979. *PETERS' CHECKLIST OF THE BIRDS OF THE WORLD*, Vol. 1, 2nd ed. Cambridge, Mass.
BLAKE, E.R. 1977. *MANUAL OF NEOTROPICAL BIRDS*, Vol. 1, Chicago University Press, Chicago.

Hellmayr, C.E. and B. Conover, 1942. *Field Mus. Nat. Hist. Zool*, 13.1.
Phelps, W.H. and W.H. Phelps, 1963. *Bol. Soc. Venez. Ci. N*, 12.19.24.
1 Meyer de Schauensee, R. 1982. *Guide to The Birds of South America*, (reprint). Acad. Nat. Sci., Philadelphia.

SPHENISCIDAE 7

FALLA, R.A. and J-L.MOUGIN, 1979. *PETERS' CHECKLIST OF THE BIRDS OF THE WORLD*, Vol. 1, 2nd ed. Cambridge, Mass.

Murphy, R.C. 1947. *Auk* 64.1.
Alexander, W.B. 1963. *Birds of the Ocean*, 2nd ed. Putnam, New York.
Stonehouse, B. 1970. *Ibis*, 112.1.
O.S.N.Z. 1970. *Annotated Checklist of the Birds of New Zealand*. Reed, Wellington.
1   Condon, H.T. 1975. *Checklist of the Birds of Australia*, Vol. 1, R.A.O.U., Melbourne.

GAVIIDAE 8

STORER, R.W. 1979. *PETERS' CHECKLIST OF THE BIRDS OF THE WORLD*, Vol. 1, 2nd ed. Cambridge, Mass.
VAURIE, C. 1965. *BIRDS OF THE PALAEARCTIC FAUNA*, Vol. 2, Witherby, London.

Cramp, S. *et al.* 1977. *Birds of the Western Palaearctic*, Vol. 1, 42. Oxford University Press, Oxford.

PODICIPEDIDAE 9

STORER, R.W. 1979. *PETERS' CHECKLIST OF THE BIRDS OF THE WORLD*, Vol. 1, 2nd ed. Cambridge, Mass.
STORER, R.W. 1960. *PROC. INT. ORN. CONGR.*, 12.694.
STORER, R.W. 1963. *PROC. INT. ORN. CONGR.*, 13.562.

Vaurie, C. 1965. *Birds of the Palearctic Fauna*, Vol. 2, Witherby, London.
Meyer de Schauensee, R. 1982. *Guide to the Birds of South America*, (reprint). Acad. Nat. Sci. Philadelphia.
Rumboll, L. 1974. *Comm. Mus. Arg. Cienc. Bern. Riv.*, 4(5) 33.
Condon, H.T. 1975. *Checklist of the Birds of Australia*, Vol. 1, R.A.O.U., Melbourne.
Cramp, S. *et al.* 1977. *Birds of Western Palaearctic*, Vol. 1, 65. Oxford University Press, Oxford.
1   Dickerman, R.W. 1986. *Proc. Biol. Soc. Wash.*, 99.435.

DIOMEDEIDAE 10

JOUANIN, C. and J-L. MOUGIN, 1979. *PETERS' CHECKLIST OF THE BIRDS OF THE WORLD*, Vol. 1, 2nd ed. Cambridge, Mass.
ALEXANDER, W.B. 1963. *BIRDS OF THE OCEAN*, 2nd ed. Putnam, New York.
O.S.N.Z. 1970. *ANNOTATED CHECKLIST OF THE BIRDS OF NEW ZEALAND*. Reed, Wellington.

Serventy, V. *et al.* 1971. *Handbook of Australian Seabirds*. Sydney.
Watson, G.E. *et al.* 1971. *Birds of the Antarctic and Subantarctic*. Am. Geog. Soc. New York.
Condon, H.T. 1975. *Checklist of the Birds of Australia*, Vol. 1, R.A.O.U., Melbourne.
1   Roux, J.P. *et al.* 1983. *Oiseau* R.F.U. 53.1.

PROCELLARIIDAE 11

JOUANIN, C. and J-L. MOUGIN, 1979. *PETERS' CHECKLIST OF THE BIRDS OF THE WORLD*, Vol. 1, 2nd ed. Cambridge, Mass.
ALEXANDER, W.B. 1963. *BIRDS OF THE OCEAN*, 2nd ed. Putnam, New York.
O.S.N.Z. 1970. *ANNOTATED CHECKLIST OF THE BIRDS OF NEW ZEALAND*. Reed, Wellington.
CONDON, H.T. 1975. *CHECKLIST OF THE BIRDS OF AUSTRALIA*, Vol. 1. R.A.O.U., Melbourne.

Alexander *et al.* 1965. *Ibis*, 107.401.
Vaurie, C. 1965. *Birds of the Palaearctic Fauna*, Vol. 2. Witherby, London.
1   Imber, M.J. 1985. *Ibis* 127.197.
Louette, M. and M. Herremans, 1985. *Bull. B.O.C*, 105.42.

HYDROBATIDAE, PELECANOIDIDAE, PHAETHONTIDAE, PELECANIDAE, FREGATIDAE   12, 13, 14, 15, 19.

JOUANIN, C. and J-L. MOUGIN, 1979. *PETERS' CHECKLIST OF THE BIRDS OF THE WORLD*, Vol. 1, 2nd ed. Cambridge, Mass.
ALEXANDER, W.B. 1963. *BIRDS OF THE OCEAN*, 2nd ed. Putnam, New York.
CRAMP, S. *et al.* 1977. *BIRDS OF THE WESTERN PALAEARCTIC*, Vol. 1. Oxford University Press, Oxford.

SULIDAE 16

DORST, A. and J-L. MOUGIN, 1979. *PETERS' CHECKLIST OF THE BIRDS OF THE WORLD*, Vol. 1, 2nd ed. Cambridge, Mass.
ALEXANDER, W.B. 1963. *BIRDS OF THE OCEAN*, 2nd ed. Putnam, New York.

Nelson, J.B. 1978. *The Gannet*, Berkhamstead.
1 Olson, S.L. and K.I. Warheit, 1988. *Bull. B.O.C.*, 108(1) 9.

PHALACROCORACIDAE 17

DORST, A. and J-L. MOUGIN, 1979. *PETERS' CHECKLIST OF THE BIRDS OF THE WORLD*, Vol. 1, 2nd ed. Cambridge, Mass.

Clancey, P.A. 1965. *Durban Mus. Novit.*, 7.
van Tets, G.F. 1965. *Orn. Monogr.*, 2.
Williams, J.G. 1966. *Bull B.O.C.*, 86.48.
Condon, H.T. 1975. *Checklist of the Birds of Australia*, Vol. 1. R.A.O.U., Melbourne.
Cramp, S. *et al.* 1977. *Birds of the Western Palaearctic*, Vol. 1, 199. Oxford University Press, Oxford
1 Clancey, P.A. (ed.) 1980. *Checklist of South African Birds*. S.A.O.S., Durban.

ANHINGIDAE 18

DORST, A. and J-L. MOUGIN, 1979. *PETERS' CHECKLIST OF THE BIRDS OF THE WORLD*, Vol. 1, 2nd ed. Cambridge, Mass.
VAURIE, C. 1965. *BIRDS OF THE PALAEARCTIC FAUNA*, Vol. 2. Witherby, London.
MAYR, E. and L.L. SHORT, 1970. *PUBL. NUTTALL ORN. CL.*, 9.
VOOUS, K. 1973. *IBIS*, 115.612.

ARDEIDAE 20

1 HANCOCK, J. and J. KUSHLAN, 1984. *THE HERON HANDBOOK*. Croom Helm, London.

Hancock, J. and H. Elliott, 1978. *Herons of the World*. London Editions.
Bock, W.J. 1956. *Amer. Mus. Novit.*, 1779.
Curry-Lindahl, K. 1971. *Ostrich*, Suppl. 9.
Payne, R.B. 1974. *Bull. B.O.C.*, 94.

SCOPIDAE, BALAENICIPITIDAE 21, 22

KAHL, M.P. 1979. *PETERS' CHECKLIST OF THE BIRDS OF THE WORLD*, Vol. 1, 2nd ed. Cambridge, Mass.

White, C.M.N. 1965. *Revised Checklist of African Non-Passerine Birds*. Govt. Printer, Lusaka.

CICONIIDAE 23

KAHL, M.P. 1972. *J. ZOOL. LOND.* 167.451.
KAHL, M.P. 1979. *PETERS' CHECKLIST OF THE BIRDS OF THE WORLD*, Vol. 1. 2nd ed. Cambridge, Mass.

Kahl, M.P. 1971. *Living Bird*, 10.151.
1 Wood, S.D. 1984. *J. Orn.*, 125.25.
2 Hancock, J. 1989. Pers.Comm.

THRESKIORNITHIDAE 24

STEINBACHER, J. 1979. *PETERS' CHECKLIST OF THE BIRDS OF THE WORLD*, Vol. 1, 2nd ed. Cambridge, Mass.

Amadon, D. and Woolfenden, 1953. *Amer. Mus. Novit.*, 1564.
Holyoak, D. 1970. *Bull. B.O.C.*, 90.
1 Hancock, J. 1989. Pers. Comm.

PHOENICOPTERIDAE 25

KAHL, M.P. 1979. *PETERS' CHECKLIST OF THE BIRDS OF THE WORLD*, Vol. 1, 2nd ed. Cambridge, Mass.

Sibley, C.G. *et al.* 1969. *Condor*, 71.
Sibley, C.G. and J.E. Ahlqvist, 1972. *Bull. Peabody Mus. Nat. Hist.*, 39.
Kear, J. and N. Duplaix-Hall, 1975. *Flamingoes*. Berkhamstead.

ANHIMIDAE 26

JOHNSGARD, P.A. 1979. *PETERS' CHECKLIST OF THE BIRDS OF THE WORLD*, Vol. 1, 2nd ed. Cambridge, Mass.

Meyer de Schauensee, R. 1982. *Guide to the Birds of South America*, (reprint). Acad. Nat. Sci , Philadelphia.

ANATIDAE 27

5 JOHNSGARD, P.A. 1979. *PETERS' CHECKLIST OF THE BIRDS OF THE WORLD*, Vol. 1, 2nd ed. Cambridge, Mass.
DELACOUR, J. 1954. *WATERFOWL OF THE WORLD*, 1-4. Country Life, London.

Scott, P. 1957. *Coloured Key to the Wildfowl of the World*. Wildfowl Trust, Slimbridge.
Ripley, S.D. 1964. *Bull. Peabody Mus. Nat. Hist.*, 19.
Frith, H.J. 1967. *Waterfowl of Australia*. Reed, Sydney.
1 Cramp, S. *et al.* 1977. *Birds of the Western Palaearctic*, Vol. 1. Oxford University Press, Oxford.
2 A.O.U 1983. *Checklist of North American Birds*, 6th ed. A.O.U.
3 Nowak, E. 1983. *Bonn Zool. Beitrag.*, 34.235.
4 O Myong Sok 1984. *J. Orn.*, 125.102.

CATHARTIDAE, PANDIONIDAE, SAGITTARIIDAE 28, 29, 31

STRESEMANN, E. and D. AMADON, 1979. *PETERS' CHECKLIST OF THE BIRDS OF THE WORLD*, Vol. 1, 2nd ed. Cambridge, Mass.

ACCIPITRIDAE 30

STRESEMANN, E. and D. AMADON, 1979. *PETERS' CHECKLIST OF THE BIRDS OF THE WORLD*, Vol. 1, 2nd ed. Cambridge, Mass.

1 White, C.M.N. and M.D. Bruce, 1986. *Birds of Wallacea*. B.O.U. Checkl. 7.
2 Clancey, P.A. 1987. *Bull. B.O.C.*, 107(4) 173.
3 Fry, C.H. *et al.* 1982. *Birds of Africa*, Vol.1, p.424. Academic Press, London.
4 Amadon, D. and J. Bull, 1988. *Proc.W Found.Vert.Zool.*, 3.4. 297.

FALCONIDAE 32

2 STRESEMANN, E. and D. AMADON, 1979. *PETERS' CHECKLIST OF THE BIRDS OF THE WORLD*, Vol. 1, 2nd ed. Cambridge, Mass.

1 Ellis, D.H. and C.P. Garat, 1983. *Auk*, 100.269.
5 White, C.M.N. and M.D. Bruce, 1986. *Birds of Wallacea*. B.O.U. Checkl. 7.
3 Schwartz 1972. *Condor*, 74.399.
4 Amadon, D. and J. Bull, 1988. *Proc. W. Found. Vert. Zool.*, 3.4.297.

MEGAPODIIDAE 33

PETERS, J.L. 1934. *CHECKLIST OF THE BIRDS OF THE WORLD*, Vol. 2, Cambridge, Mass.

2 Schodde, R. 1977. *CSIRO Aust. Div. Wildl. Res. Pap.*, 34.1.
1 White, C.M.N. and M.D. Bruce, 1986. *Birds of Wallacea*. B.O.U. Checkl. 7.
3 W.P.A. Journal, IX 1983.73.

CRACIDAE 34
DELACOUR, J. and D. AMADON, 1973. *CURASSOWS AND RELATED BIRDS*. Amer. Mus. Nat. Hist., New York.

Peters, J.L. 1934. *Checklist of the Birds of the World*, Vol. 2. Cambridge, Mass.
Vaurie, C. 1967. *Amer. Mus. Novit*, 2296.2299.2305 2307.
Vaurie, C. 1968. *Bull. Amer. Mus. Nat. Hist.*, 138.
Blake, E.R. 1977. *Manual of Neotropical Birds.*Chicago University Press, Chicago.

PHASIANIDAE 35

DELACOUR, J. 1977. *PHEASANTS OF THE WORLD*, 2nd ed. Spur, Hindhead.

Peters, J.L. 1934. *Checklist of the Birds of the World*, Vol. 2. Cambridge, Mass.
Hall, B.P. 1963. *Bull. Br. Mus. Nat. Hist.*, 10(2).
Wetmore, A. 1963. *Smithson Misc. Coll.*, 145(6).
Short, L.L. 1967. *Amer. Mus. Novit.*, 2289.
Farrand, J. Jr. and S.L. Olson, 1973. *Bull. B.O.C.*, 93(2).
Johnsgard, P.A. 1973. *Grouse and Quails of North America*. Nebraska University Press, Lincoln.

REFERENCES

2   Condon, H.T. 1975. *Checklist of the Birds of Australia*, Vol. 1. R.A.O.U., Melbourne.
    Davison, G.W.H. 1980. *Bull. B.O.C.*, 100(2) 141.
4   Clancey, P.A. (ed.) 1980. *Checklist of South African Birds*. S.A.O.S., Durban.
3   W.P.A. Journal, IX 1983.73.
5   A.O.U. 1983. *Checklist of North American Birds*, 6th ed. A.O.U. 6
1   White, C.M.N. and M.D. Bruce, 1986. *Birds of Wallacea*. B.O.U. Checkl. 7.
    W.P.A. 1989. Pers. Comm.

MESITORNITHIDAE   36

PETERS, J.L. 1934. *CHECKLIST OF THE BIRDS OF THE WORLD*, Vol. 2. Cambridge, Mass.

Rand, A.L. 1936. *Bull. Amer. Mus. Nat. Hist.,* 72.

TURNICIDAE   37

PETERS, J.L. 1934. *CHECKLIST OF THE BIRDS OF THE WORLD*, Vol. 2. Cambridge, Mass.

Condon, H.T. 1975. *Checklist of the Birds of Australia*, Vol. 1. R.A.O.U., Melbourne.
1   White, C.M.W. and M.D.Bruce, 1986. *Birds of Wallacea*. B.O.U. Checkl. 7.

PEDIONOMIDAE 38

PETERS, J.L. 1934. *CHECKLIST OF THE BIRDS OF THE WORLD*, Vol. 2. Cambridge, Mass.
Bock, W.J. and A.R. McEvey, 1969. *Proc. R. Soc. Vict.,* 82(2).

GRUIDAE   39

PETERS, J.L. 1934. *CHECKLIST OF THE BIRDS OF THE WORLD*, Vol. 2. Cambridge, Mass.

Vaurie, C. 1965. *Birds of the Palaearctic Fauna*, Vol. 2. Witherby, London.
Walkinshaw, L. 1973. *Cranes of the World*. Winchester, New York.
Cramp, S. *et al.* 1977. *Birds of the Western Palaearctic*, Vol. 2. Oxford University Press, Oxford.

ARAMIDAE, PSOPHIIDAE   40, 41

PETERS, J.L. 1934. *CHECKLIST OF THE BIRDS OF THE WORLD*, Vol. 2. Cambridge, Mass.

Blake, E.R. 1977. *Manual of Neotropical Birds*. Chicago University Press, Chicago.

RALLIDAE   42

RIPLEY, S.D. 1977. *RAILS OF THE WORLD*. Feheley, Toronto.

Peters, J.L. 1934. *Checklist of the Birds of the World*, Vol. 2. Cambridge, Mass.
2   Mayr, E. 1949. *Amer. Mus. Novit.*, 1417.1.
    Blake, E.R. 1977. *Manual of Neotropical Birds*, Vol. 1. Chicago University Press, Chicago.
3   Thiede, U. 1982. *Vogelwelt*, 103.143.
4   Meyer de Schauensee, R. 1982. *A Guide to the Birds of South America* (reprint). Acad. Nat. Sci., Philidelphia.
5   Schodde, R. and de Naurois, 1982. *Notornis*, 29.131.
1   Ripley, S.D. and B. Beehler, 1985. *Smithson. Contr. Zool.*, 417.

HELIORNITHIDAE, RHYNOCHETIDAE, EURYPYGIDAE, CARIAMIDAE   43, 44, 45, 46

PETERS, J.L. 1934. *CHECKLIST OF THE BIRDS OF THE WORLD*, Vol. 2, Cambridge, Mass.

Mayr, E. 1945. *Birds of the Southwest Pacific.*Macmillan, New York.
Brooke, R.K. 1954. *Ostrich*, 55.171.
White, C.M.N. 1965. *Revised Checklist of Non Passerine Birds*. Govt. Printer, Lusaka.
Blake, E.R. 1977. *Manual of Neotropical Birds*, Vol. 1. Chicago University Press, Chicago.

OTIDAE   47

1   FRY, C.H. *et al.* 1986. *BIRDS OF AFRICA*, Vol. 2. Academic Press, London.

2   Ripley, S.D. 1961. *Synopsis of the Birds of India and Pakistan*. Nat. Hist. Soc., Bombay.
    White, C.M.N. 1965. *Revised Checklist of African Non Passerine Birds*. Govt. Printer, Lusaka.
    Clancey, P.A. 1972/3. *Bokmakierie*, 24.79 & 25.10.

JACANIDAE   48

PETERS, J.L. 1934. *CHECKLIST OF THE BIRDS OF THE WORLD*, Vol. 2. Cambridge, Mass.

Wetmore 1965. *Smithson. Misc. Coll.*, 150(1).
Jehl, J.R. 1968. *Mem. San Diego Soc. Nat. Hist.*, 3.

ROSTRATULIDAE 49

PETERS, J.L. 1934. *CHECKLIST OF THE BIRDS OF THE WORLD*, Vol. 2. Cambridge, Mass.

Vaurie, C. 1965. *Birds of the Palaearctic Fauna*, Vol. 2. Witherby, London.

DROMADIDAE 50

PETERS, J.L. 1934. *CHECKLIST OF THE BIRDS OF THE WORLD*, Vol. 2. Cambridge, Mass.

Fry, C.H. *et al.* 1986. *The Birds of Africa*, Vol. 2. Academic Press, London.

HAEMATOPODIDAE 51

PETERS, J.L. 1934. *CHECKLIST OF THE BIRDS OF THE WORLD*, Vol. 2. Cambridge, Mass.

3   Vaurie, C. 1965. *Birds of the Palaearctic Fauna*, Vol. 2. Witherby, London.
1   Baker, A.J. 1975. *J. Zool. Lond.*, 175.357.
    Baker, A.J. 1977. *Bijdr. Dierkde*, 47.156.
2   Hockey, P.A.R. 1982. *Bull. B.O.C.*, 102(2) 77.

IBIDORHYNCHIDAE 52

RIPLEY, S.D. and SALIM ALI, 1980. *HANDBOOK OF THE BIRDS OF INDIA AND PAKISTAN*, Vol. 2, 2nd ed. Oxford University Press, Oxford.

Verheyen, R. 1958. *Inst. R. Sci. Nat. Belg.*, 34(18) 1

RECURVIROSTRIDAE 53

PETERS, J.L. 1934. *CHECKLIST OF THE BIRDS OF THE WORLD*, Vol. 2, Cambridge, Mass.
MAYR, E. and L.L. SHORT, 1970. *PUBL. NUTTALL ORN. SOC.*, 9.

Hamilton, R.B. 1975. *Orn. Monogr.*, 17.
Blake, E.R. 1977. *Manual of Neotropical Birds*, Vol. 1. Chicago University Press, Chicago.

BURHINIDAE 54

PETERS, J.L. 1934. *CHECKLIST OF THE BIRDS OF THE WORLD*, Vol. 2. Cambridge, Mass.

Fry, C.H. *et al.* 1986. *Birds of Africa*, Vol. 2. Academic Press, London.

GLAREOLIDAE 55

PETERS, J.L. 1934. *CHECKLIST OF THE BIRDS OF THE WORLD*, Vol. 2. Cambridge, Mass.

    Kemp A.C. and G.L.Maclean, 1973. *Ostrich*, 44.80.
    Maclean, G.L. 1978. *Bird Families of the World*, 105. Elsevier-Phaidon, Oxford.
1   Clancey, P.A. 1984. *Gerfaut*, 74.361.
2   White, C.M.N. and M.N. Bruce, 1986. *Birds of Wallacea*. B.O.U. Checkl. 7.

CHARADRIIDAE, SCOLOPACIDAE, THINOCORIDAE, CHIONIDIDAE   56, 57, 58, 59

PETERS, J.L. 1934. *CHECKLIST OF THE BIRDS OF THE WORLD*, Vol. 2. Cambridge, Mass.
BOCK, W.J. 1958. *BULL. MUS. COMP. ZOOL. HARVARD*, 118(2) 27.

    Vaurie, C. 1965. *Birds of the Palaearctic Fauna*, Vol. 2. Witherby, London.
    van Tets, G.F. *et al.* 1967. *Emu*, 67.
    Jehl, J.R. 1968. *Mem. San Diego Soc. Nat. Hist.*, 3.
4   Wolters, H.E. 1974. *Bonn Zool. Beitr.*, 25(4).
1   Parker, S.A. 1982. *S. Austr. Nat.*, 56.63.
3   Meyer de Schauensee, R. 1982. *Guide to the Birds of South America*. Acad. Nat. Sci., Philadelphia.
2   Connors, P.G. 1983. *Auk* 100.607.
    McKean, J.L. and H.A.F.Thompson, 1983. *N. Territ. Nat.*, 6.14.
    Cox, J.B. 1987. *S. Austr. Nat.*, 30.85.

STERCORARIIDAE, LARIDAE   60, 61

PETERS, J.L. 1934. *CHECKLIST OF THE BIRDS OF THE WORLD*, Vol. 2. Cambridge, Mass.
ALEXANDER, W.B. 1963. *BIRDS OF THE OCEAN*, 2nd ed. Putnam, New York.

    Moynihan, M. 1959. *Amer. Mus. Novit.*, 1928.
    Schnell, G.D. 1970. *Syst. Zool.*, 19.35.264.
    Brooke, R.K. 1978. *Durban Mus. Novit.*, 11(18) 295.
    Cramp, S. *et al.* 1983. *Birds of the Western Palaearctic*, Vol. 3. Oxford University Press, Oxford.

3 A.O.U. 1983. *Checklist of North American Birds,* 6th ed. A.O.U.
4 Harrison, P. 1983. *Seabirds, an IdentificationGuide.* Croom Helm, Kent.
1 Dubois, P. 1985. *Alauda,* 53.226.
2 Banks, R.C. 1986. *Proc. Biol. Soc. Wash,* 99.149.

RYNCHOPIDAE 62

PETERS, J.L. 1934. *CHECKLIST OF THE BIRDS OF THE WORLD,* Vol. 2. Cambridge, Mass.

Alexander, W.B. 1963. *Birds of the Ocean,* 2nd ed. Putnam, New York.

ALCIDAE 63

PETERS, J.L. 1934. *CHECKLIST OF THE BIRDS OF THE WORLD,* Vol. 2. Cambridge, Mass.

Alexander, W.B. 1963. *Birds of the Ocean,* 2nd ed. Putnam, New York.
Cramp, S. *et al.* 1985. *Birds of theWestern Palaearctic,* Vol. 4. Oxford University Press, Oxford.

PTEROCLIDIDAE 64

PETERS, J.L. 1934. *CHECKLIST OF THE BIRDS OF THE WORLD,* Vol. 2. Cambridge, Mass.

Fjeldsa, J. 1976. *Vidensk. Meddr. Dansk Naturh. Foren.,* 139.179.
1 Fry, C.H. *et al.* 1986. *Birds of Africa,* Vol. 2. Academic Press, London.

COLUMBIDAE 65

GOODWIN, D. 1970. *PIGEONS AND DOVES OF THE WORLD,* Brit. Mus. Nat. Hist., London.
PETERS, J.L. 1937. *CHECKLIST OF THE BIRDS OF THE WORLD,* Vol. 3. Cambridge, Mass.

3 Fry, C.H. *et al.* 1986. *Birds of Africa,* Vol. 2. Academic Press, London.
Buden, D.W. 1985. *Proc. Biol. Soc. Wash.,* 98.790.
2 Banks, R.C. 1986. *Auk,* 103.629.
1 White, C.M.N. and M.D. Bruce, 1986. *Birds of Wallacea,* B.O.U. Checkl. 7.

LORIIDAE, CACATUIDAE 66, 67

FORSHAW, J.W. 1973. *PARROTS OF THE WORLD,* Lansdowne, Melbourne.

Diamond, A.W. 1972. *Publ. Nuttall Orn. Cl.,* 12.
Condon, H.T. 1975. *Checklist of the Birds of Australia,* Vol. 1. R.A.O.U., Melbourne.
2 Ford, J. 1985. *Emu,* 85.163.
1 Courtney, J. 1986. *Austr. Bird Watcher,* 11.137.
White, C.M.N. and M.D. Bruce, 1986. *Birds of Wallacea.* B.O.U. Checkl. 7.

PSITTACIDAE 68

FORSHAW, J.W. 1973. *PARROTS OF THE WORLD.*Lansdowne, Melbourne.
PETERS, J.L. 1937. *CHECKLIST OF THE BIRDS OF THE WORLD,* Vol. 3. Cambridge, Mass.

Brereton, J. 1963. *Proc. XIII Int. Orn. Congr.,* 499.
Rand, A.L. and E.T. Gilliard, 1967. *Handbook of the Birds of New Guinea,* Weidenfeld & Nicholson, London.
Diamond, A.W. 1972. *Publ. Nuttall Orn. Cl.,* 12.
Holyoak, D. 1973. *Emu,* 73.
Parkes, K.C. 1976. *Bull. B.O.C.,* 96.
2 A.O.U. 1983. *Checklist of N. American Birds,* 6th ed. A.O.U.
Delgado, B. 1985. *Orn. Monogr.,* 36.17.
Taylor, R.H. *et al.* 1986. *Notornis,* 33.17.
1 White, C.M.N. and M.D. Bruce, 1986. *Birds of Wallacea.* B.O.U. Checkl. 7.
3 Rinke, D. 1989. *Bull. B.O.C.,* 185.

MUSOPHAGIDAE 69

MOREAU, R.E. 1959. *OSTRICH,* 3 (Suppl) 421 & Syst. Assoc. Bull. 3.113.

Berger, A.J. 1960. *Wilson Bull.,* 72.60.
1 Fry, C.H. *et al.* 1988. *Birds of Africa,* Vol. 3. Academic Press, London.

OPISTHOCOMIDAE 70

1 Meyer de Schauensee, R. 1982. *A Guide to the Birds of South America.* Acad. Nat. Sci., Philadelphia.

CUCULIDAE 71

PETERS, J.L. 1940. *CHECKLIST OF THE BIRDS OF THE WORLD*, Vol. 4. Cambridge, Mass.

Vaurie, C. 1965. *Birds of the Palaearctic Fauna*, Vol. 2. Witherby, London.
Dupont, J.E. 1971. *Philippine Birds*. Mus. Nat. Hist. Greenville, Del.
Sibley, C.G. and J.E. Ahlqvist, 1973. *Auk*, 90.1.
5 Ford, J. 1981. *Emu*, 81.209.
2 Parker, J. 1981. *Zool. Verh.*, 187.1.
Wells, D.R. 1982. *Bull. B.O.C.*, 102.2.62.
7 A.O.U. 1983. *Checklist of North American Birds*, 6th ed. A.O.U.
Mason, I.J. *et al.* 1984. *Emu*, 84.1.
1 Harris, A.H. and C.R. Crews, 1984. *Southwest Nat.*, 28.407.
4 Louette, M. 1986. *Bull. B.O.C.*, 106.126.
3 White, C.M.N. and M.D. Bruce, 1986. *Birds of Wallacea*. B.O.U. Checkl. 7.
6 Fry, C.H. *et al.* 1988. *Birds of Africa*, Vol. 3. Academic Press, London.

TYTONIDAE 72

PETERS, J.L. 1940. *CHECKLIST OF THE BIRDS OF THE WORLD*, Vol. 4. Cambridge, Mass.

Marshall, W.H. 1966. *Nat. Hist. Bull. Siam Soc.*, 21.
Diamond, A.W. 1972. *Publ. Nuttall Orn. Cl*, 12.
Mason, I.J. 1983. *Bull. B.O.C.*, 103(4) 123.
1 White, C.M.N. and M.D. Bruce, 1986. *Birds of Wallacea*. B.O.U. Checkl. 7.
2 A.O.U. 1983. *Checklist of North American Birds*, 6th ed. A.O.U.
3 Amadon, D. and J. Bull, 1988. *Proc. W. Found. Vert. Zool.*, 3.4.295.357.

STRIGIDAE 73

PETERS, J.L. 1940. *CHECKLIST OF THE BIRDS OF THE WORLD*, Vol. 4. Cambridge, Mass.

O'Neill, J.P. and G.R. Graves, 1977. *Auk*, 94.409.
1 Marshall, J.T. 1978. *A.O.U. Orn. Monogr.*, 25.1.
Weske, J.S. and J.W. Terborgh 1981. *Auk*, 98.1.
2 Hekstra, G.P. 1982. *Bull. Zool. Mus. Amst.*, 9(7) 50.
7 Meyer de Schauensee, R. 1982. *Guide to the Birds of South America* (reprint). Acad. Nat. Sci. Philadelphia.
9 A.O.U. 1983. *Checklist of North American Birds*, 6th ed. A.O.U.
3 Prigogine, A. 1983. *Rev. Zool. Afr.*, 97.86.
4 Prigogine, A. 1985. *Gerfaut*, 75.131.
5 Fitzpatrick, J.W. and J.P. O'Neill, 1986. *Wilson Bull.*, 98.1.
6 White, C.M.N. and M.D. Bruce, 1986. *Birds of Wallacea*. B.O.U. Checkl. 7.
8 Fry, C.H. *et al.* 1988. *Birds of Africa*, Vol. 3. Academic Press, London.
10 Marshall, J.T. and B. King, 1988. *Proc. W. Found. Vert. Zool.*, 3.4.295-357.
11 Amadon, D.and J. Bull, 1988. *Proc. W. Found. Vert. Zool.*, 3.4.295-357 .

STEATORNITHIDAE, NYCTIBIIDAE 74, 76

PETERS, J.L. 1940. *CHECKLIST OF THE BIRDS OF THE WORLD*, Vol. 4. Cambridge, Mass.

Meyer de Schauensee, R. 1982. *Guide to the Birds of South America* (reprint). Acad. Nat. Sci., Philadelphia.

PODARGIDAE, AEGOTHELIDAE 75, 77

PETERS, J.L. 1940. *CHECKLIST OF THE BIRDS OF THE WORLD*, Vol. 4. Cambridge, Mass.

Rand, A.L. and E.T. Gilliard, 1967. *Handbook to the Birds of New Guinea*. Weidenfeld & Nicholson, London.
Condon, H.T. 1975. *Checklist of the Birds of Australia*, Vol. 1. R.A.O.U., Melbourne.

CAPRIMULGIDAE 78

PETERS, J.L. 1940. *CHECKLIST OF THE BIRDS OF THE WORLD*, Vol. 4. Cambridge, Mass.

1 Mees, G.F. 1977. *Zool. Verh.*, 155.1.
Garrido, O.H. 1983. *Auk*, 100.988.
6 Fry, C.H. *et al.* 1988. *Birds of Africa*, Vol. 3. Academic Press, London.
Jackson, H.D. 1984. *Smithersia*, 1.4.
2 Dickerman, R.W. 1985. *Orn. Monogr.*, 36.356.
3 Mees, G.F. 1985. *Pr. K. Ned. Ak. Wet.*, 88(4)419.
4 White, C.M.N. and M.D. Bruce, 1986. *Birds of Wallacea*. B.O.U. Checkl. 7.
Cannell, P.F. 1986. *Abstr. XIX Int. Orn. Congr. Ottawa*, 656.
5 Dickerman, R.W. 1988. *Bull. B.O.C.*, 108(3)120.
7 A.O.U. 1983. *Checklist of North American Birds*, 6th ed. A.O.U.
8 Louette, M. 1990. *Ibis*, 132.349.

APODIDAE, HEMOPROCNIDAE 79, 80

PETERS, J.L. 1940. *CHECKLIST OF THE BIRDS OF THE WORLD*, Vol. 4. Cambridge, Mass.

Orr, R.T. 1963. *Proc. XII Int. Orn. Congr.*, 126.
Lack, D. 1956. *Ibis*, 98.34.
Brooke, R.K. 1970. *Durban Mus. Novit.*, 9(2) 13.
Brooke, R.K. 1971. *Bull. B.O.C.*, 91.76.
Brooke, R.K. 1972. *Bull. B.O.C.*, 92.53.
Diamond, A.W. 1972. *Publ. Nuttall Orn. Cl.*, 12.
Phelps, W.H. 1973. *Bol. Soc. Venez. Ci. Nat.*, 30.124.
Holyoak, D. and Thibault, 1978. *Bull. B.O.C.*, 98.
Wells, D.R. and Lord Medway, 1976. *Bull. B.O.C.*, 96.5.
6   Meyer de Schauensee, R. 1982. *Guide to the Birds of South America* (reprint). Acad. Nat. Sci., Philadelphia.
3   Salomonsen, F. 1983. *Biol. Skr. Dan. Vid. Selsk.*, 23(5) 1.
Clancey, P.A. 1983. *Bull. B.O.C.*, 103(3)80.
8   A.O.U. 1983. *Checklist of North American Birds*, 6th ed. A.O.U.
4   Somadikarta, P. 1986. *Bull. B.O.C.*, 106(1)32.
2   Pratt, H.D. 1986. *Elepaio*, 46.
5   White, C.M.N. and M.D. Bruce, 1986. *Birds of Wallacea*. B.O.U. Checkl. 7.
1   King, B. 1987. *Bull. B.O.C.*, 107(1)36.
7   Fry, C.H. *et al.* 1988. *Birds of Africa*, Vol. 3. Academic Press, London.

TROCHILIDAE 81

PETERS, J.L. 1945. *CHECKLIST OF THE BIRDS OF THE WORLD*, Vol. 5. Cambridge, Mass.

Weske, J.S. and J.W. Terborgh, 1977. *Condor*, 79.143.
Fitzpatrick, J.W. *et al.* 1979. *Wilson Bull.*, 91.177.
Graves, G.R. 1980. *Wilson Bull.*, 92.1.
Kashin, R. 1981. *Ornitologiya*, 16.167.
2   A.O.U. 1983. *Checklist of North American Birds*, 6th ed. A.O.U.
Schuchmann, K.L. 1984. *Bull. B.O.C.*, 104(1)5.
Stiles, F.G. 1985. *Orn. Monogr.*, 36.23.
Weske, J.S. 1985. *Orn. Monogr.*, 36.41.
1   Graves, G.R. 1986. *Proc. Biol. Soc. Wash.*, 99.218.

COLIIDAE 82

PETERS, J.L. 1945. *CHECKLIST OF THE BIRDS OF THE WORLD*, Vol. 5. Cambridge, Mass.
1   FRY, C.H. *et al.* 1988. *BIRDS OF AFRICA*, Vol. 3. Academic Press, London.

TROGONIDAE 83

PETERS, J.L. 1945. *CHECKLIST OF THE BIRDS OF THE WORLD*, Vol. 5. Cambridge, Mass.

Meyer de Schauensee, R. 1982. *Guide to the Birds of South America* (reprint). Acad. Nat. Sci., Philadelphia.
2   A.O.U. 1983. *Checklist of North American Birds*, 6th ed. A.O.U.
1   Fry, C.H. *et al.* 1988. *Birds of Africa*, Vol. 3. Academic Press, London.

ALCEDINIDAE 84

PETERS, J.L. 1945. *CHECKLIST OF THE BIRDS OF THE WORLD*, Vol. 5. Cambridge, Mass.
FORSHAW, J.M. 1983-1985. *KINGFISHERS AND RELATED BIRDS*, Vols 1 & 2. Lansdowne, Melbourne.

Holyoak, D. 1974. *Bull. B.O.C.*, 94(4) 147.
1   Fry, C.H. 1988. *Living Bird*, 18.113.
Cowles, G.S. 1980. *Bull. B.O.C.*, 100(4)226.
Ripley, S.D. 1983. *Bull. B.O.C.*, 103(4)145.
Clancey, P.A. 1984. *Bull. B.O.C.*, 104(2)89.
3   Fry, C.H. and R. de Naurois, 1984. *Proc. Pan Afr. Orn. Congr.*, 5.47.
2   White, C.M.N. and M.D. Bruce, 1986. *Birds of Wallacea*. B.O.U. Checkl. 7.
Clancey, P.A. 1986. *Bull. B.O.C.*, 106(3)78.
Ripley, S.D. and B.M. Beehler, 1987. *Bull B.O.C.*, 107(4)145.
4   Fry, C.H. *et al.* 1988. *Birds of Africa*, Vol. 3. Academic Press, London.

TODIDAE 85

PETERS, J.L. 1945. *CHECKLIST OF THE BIRDS OF THE WORLD*, Vol. 5. Cambridge, Mass.

MOMOTIDAE 86

PETERS, J.L. 1945. *CHECKLIST OF THE BIRDS OF THEWORLD*, Vol. 5. Cambridge, Mass.

Meyer de Schauensee, R. 1982. *Guide to the Birds of South America* (reprint). Acad. Nat. Sci., Philadelphia.

MEROPIDAE 87

FRY, C.H. 1969. *IBIS*, 111.557.
1   FRY, C.H. 1984. *THE BEE-EATERS*. T. & A.D. Poyser, Calton.
PETERS, J.L. 1945. *CHECKLIST OF THE BIRDS OF THE WORLD*, Vol. 5. Cambridge, Mass.

2   White, C.M.N. & M.D. Bruce, 1986. *Birds of Wallacea*. B.O.U. Checkl. 7.
Fry, C.H. *et al*. 1988. *Birds of Africa*, Vol. 3. Academic Press, London.

CORACIIDAE 88

PETERS, J.L. 1945. *CHECKLIST OF THE BIRDS OF THE WORLD*, Vol. 5. Cambridge, Mass.

1   White, C.M.N. and M.D. Bruce, 1986. *Birds of Wallacea*. B.O.U. Checkl. 7.
Fry, C.H. *et al*. 1988. *Birds of Africa*,Vol. 3. Academic Press, London.

BRACHYPTERACIIDAE, LEPTOSOMATIDAE, UPUPIDAE 89, 90, 91

PETERS, J.L. 1945. *CHECKLIST OF THE BIRDS OF THE WORLD*, Vol. 5. Cambridge, Mass.

PHOENICULIDAE 92

PETERS, J.L. 1945. *CHECKLIST OF THE BIRDS OF THE WORLD*, Vol. 5. Cambridge, Mass.

1   Fry, C.H. *et al*. 1988. *Birds of Africa*, Vol. 3. Academic Press, London.

BUCEROTIDAE 93

PETERS, J.L. 1945. *CHECKLIST OF THE BIRDS OF THE WORLD*, Vol. 5. Cambridge, Mass.

Ripley, S.D. 1961. *Synopsis of the Birds of India and Pakistan*. Nat. Hist. Soc., Bombay.
DuPont, J.E. 1971. *Philippine Birds*. Mus. Nat. Hist., Greenville, Del.
1   Fry, C.H. *et al*. 1988. *Birds of Africa*, Vol. 3. Academic Press, London.

GABULIDAE 94

PETERS, J.L. 1948. *CHECKLIST OF THE BIRDS OF THE WORLD*, Vol. 6. Cambridge, Mass.

Haffer, J. 1974. *Publ. Nuttall Orn. Cl.*, 14.
Meyer de Schauensee, R. 1982. *Guide to the Birds of South America* (reprint). Acad. Nat. Sci., Philadelphia.

BUCCONIDAE 95

PETERS, J.L. 1948. *CHECKLIST OF THE BIRDS OF THE WORLD*, Vol. 6. Cambridge, Mass.

Meyer de Schauensee, R. 1982. *Guide to the Birds of South America* (reprint). Acad. Nat. Sci., Philadelphia.

CAPITONIDAE 96

PETERS, J.L. 1948. *CHECKLIST OF THE BIRDS OF THE WORLD*, Vol. 6. Cambridge, Mass.

Ripley, S.D. 1961. *Synopsis of the Birds of India and Pakistan*. Nat. Hist. Soc., Bombay.
Erard, C. 1976. *Bull. B.O.C.*, 96.
Meyer de Schauensee, R. 1982. *Guide to the Birds of South America* (reprint). Acad. Nat. Sci., Philadelphia.
2   Short, L.L. and J.F.M. Horne, 1985. *Afr. Vertebr:* Syst. etc. 255, Selbstverlag, Bonn.
1   Graves, G.R. 1986. *Proc. Biol. Soc. Wash.*, 99.61.
3   Fry, C.H. *et al*. 1988. *Birds of Africa*, Vol. 3. Academic Press, London.

INDICATORIDAE 97

PETERS, J.L. 1948. *CHECKLIST OF THE BIRDS OF THE WORLD*, Vol. 6. Cambridge, Mass.
FRIEDMANN, H. 1955. *BULL. US NAT. MUS.*, 208.

1   Louette, M. 1981. *Rev. Zool. Afr.*, 95.181.
1   Colston, P.R. 1981. *Bull. B.O.C.*, 101(2)289.
2   Clancey, P.A. 1985. *Honeyguide*, 31.101.
3   Fry, C.H. *et al*. 1988. *Birds of Africa*, Vol. 3. Academic Press, London.

RAMPHASTIDAE 98

HAFFER, J. 1974. *PUBL. NUTTALL ORN. CL.*, 14.
PETERS, J.L. 1948. *CHECKLIST OF THE BIRDS OF THE WORLD*, Vol. 6. Cambridge, Mass.

Meyer de Schauensee, R. 1982. *Guide to the Birds of South America* (reprint). Acad. Nat. Sci., Philadelphia.
1 A.O.U. 1983. *Checklist of North American Birds*, 6th ed. A.O.U.

PICIDAE 99

1 SHORT, L.L. 1982. *WOODPECKERS OF THE WORLD*. Nat. Hist. Mus. Greenville, Del.

2 A.O.U. 1983. *Checklist of North American Birds*, 6th ed. A.O.U.
3 Earle, R.A. 1986. *Novors Nas. Mus. Bloemf.*, 5.79.
   Dickerman, R.W. 1987. Occ. Pap. W. Found. Vert. Zool., 3.

EURYLAIMIDAE 100

PETERS, J.L. 1951. *CHECKLIST OF THE BIRDS OF THE WORLD*, Vol. 7. Cambridge, Mass.

Ripley, S.D. 1961. *Synopsis of the Birds of India and Pakistan*. Nat. Hist. Soc., Bombay.
White, C.M.N. 1961. *Revised Checklist of African Broadbills etc.* Govt. Printer, Lusaka.
Smythies, B.E. 1981. *Birds of Borneo*. Sabah Soc., Kota Kinabalu.

DENDROCOLAPTIDAE 101

PETERS, J.L. 1951. *CHECKLIST OF THE BIRDS OF THE WORLD*, Vol. 7. Cambridge, Mass.

Feduccia, A. 1973. *Orn. Monogr.*, 13.
Meyer de Schauensee, R. 1982. *Guide to the Birds of South America* (reprint). Acad. Nat. Sci., Philadelphia.

FURNARIIDAE 102

VAURIE, C. 1971. *CLASSIFICATION OF THE OVENBIRDS*. Witherby, London.
5 PETERS, J.L. 1951. *CHECKLIST OF THE BIRDS OF THE WORLD*, Vol. 7. Cambridge, Mass.

Eisenmann, E. 1955. *Trans. Lin. Soc. N.Y.*, 7.
Koepcke, M. 1965. *Beitr. Neotrop. Fauna*, 4.
Sick, H. 1969. *Beitr. Neotrop. Fauna*, 4.
Blake, E.R. 1971. *Auk*, 88.179.
O'Neill, J.P. and T.A. Parker III, 1976. *Bull. B.O.C.*, 96.
Nores, M. and D. Yzurieta, 1979. *Ac. N. de Ci. Argent. Misc.*, 61.4.
Vaurie, C., J.S. Weske and J.W. Terborgh, 1972. *Bull. B.O.C.*, 92.
Feduccia, A. 1973. *Orn. Monogr.*, 13.
1 Meyer de Schauensee, R. 1982. *Guide to the Birds of South America* (reprint). Acad. Nat. Sci., Philadelphia.
2 Navas and Bo 1982. *Com. Mus. Argent. Cienc.*, 2.4.85.
4 Teixeira, D.M. and L.P. Gonzaga, 1983. *Bol. Mus. Para. Goeldi Zool.*,, 124.1.22.
   Oren, D.C. 1985. *Publ. Avuls. Mus. P. E. Goeldi*, 40.93.
   Graves, G.R. 1986. *Condor*, 88.120.
   Nores, M. 1986. *Hornero*, 12.262.
3 Fjeldsa, J. *et al.* 1987. *Bull. B.O.C.*, 107(3)112.
6 Fjeldsa, J, 1987. Tech Rep *Polylepis* Exp. Univ. Copenhagen (unpublished).

FORMICARIIDAE 103

PETERS, J.L. 1951. *CHECKLIST OF THE BIRDS OF THE WORLD*, Vol. 7. Cambridge, Mass.

Eisenmann, E. 1955. *Trans. Lin. Soc. N. Y.*, 7.
3 1982. *Wilson Bull.*, 94.115.
   Schulenberg, T.S. and M.D. Williams, 1982. *Wilson Bull.*, 94.105.
2 Teixeira, D.M. and L.P. Gonzaga, 1983. *Bull. B.O.C.*, 103(4)133.
4 Graves, G.R. *et al.* 1983. *Wilson Bull.*, 95.1.
5 Davis, T.J. and J.P. O'Neill 1986. *Wilson Bull.*, 98.337.
6 Graves, G.R. 1987. *Wilson Bull.*, 99.313.
1 Gonzaga, L.P. 1988. *Bull. B.O.C.*,, 108(3)132.
7 Willis, E.O. 1988. *Rev. Bras. Biol.*, 48(3)431.

CONOPOPHAGIDAE 104

PETERS, J.L. 1951. *CHECKLIST OF THE BIRDS OF THE WORLD*, Vol. 7. Cambridge, Mass.

Ames, P.L. *et al.* 1968. *Postilla*, 114.
Meyer de Schauensee, R. 1982. *Guide to the Birds of South America* (reprint). Acad. Nat. Sci., Philadelphia.

RHINOCRYPTIDAE 105

PETERS, J.L. 1951. *CHECKLIST OF THE BIRDS OF THE WORLD*, Vol. 7. Cambridge, Mass.

Meyer de Schauensee, R. 1982. *Guide to the Birds of South America* (reprint). Acad. Nat. Sci., Philadelphia.
Nores, M. 1986. *Hornero*, 12.262.

TYRANNIDAE 106

1 TRAYLOR, M.A. (ed.) 1979. *PETERS' CHECKLIST OF THE BIRDS OF THE WORLD*, Vol. 8. Cambridge, Mass.

10 Meyer de Schauensee, R. 1982. *Guide to the Birds of South America* (reprint). Acad. Nat. Sci., Philadelphia.
3 Ames, P.L. *et al.* 1968. *Postilla*, 114.
  Fitzpatrick, J.W. and J.P. O'Neill, 1979. *Auk*, 96.443.
  Schulenberg, T.S. and G.L. Graham, 1981. *Bull. B.O.C.*, 101(4)241 .
  Parkes, K.C. 1982. *Ann. Carnegie Mus.*, 51.1.
  Traylor, M.A. 1982. *Fieldiana Zool.*, 13.22.
6 A.O.U. 1983. *Checklist of North American Birds*, 6th ed. A.O.U.
  Lanyon, W.E. 1984. *Amer. Mus. Novit.*, 2797.
  Lanyon, W.E. 1985. *Orn. Monogr.*, 36.361.
5 Traylor, M.A. 1985. *Orn. Monogr.*, 36.431.
  Dickerman, R.W. 1985. *Bull. B.O.C.*, 105(2)73.
4 Lanyon, W.E. and S.M. Lanyon, 1986. *Auk*, 103.341.
  Lanyon, W.E. 1986. *Amer. Mus. Novit.*, 2846.
2 Teixeira, D.M. 1987. *Bull. B.O.C.*, 107(1)37.
  Dickerman, R.W. and W.H. Phelps Jnr. 1987. *Bol. Soc. Venez. Ci. Nat.*, 41.27.
7 Graves, G.R. 1988. *Wilson Bull.*, 100.529.
8 Lanyon, N.E. 1988. *Amer. Mus. Novit.*, 2914.
9 Lanyon, N.E. 1984. *Amer. Mus. Novit.*, 2797.

PIPRIDAE 107

2 SNOW, D.W. *et al.* 1979. *PETERS' CHECKLIST OF THE BIRDS OF THE WORLD*, Vol. 8. Cambridge, Mass.

1 Haffer, J. 1967. *Amer. Mus. Novit.*, 2294.
  Snow, D.W. 1975. *Bull. B.O.C.*, 95.20.
  Meyer de Schauensee, R. 1982. *Guide to the Birds of South America* (reprint). Acad. Nat. Sci. Philedelphia.

COTINGIDAE 108

SNOW, D.W. *et al.* 1979. *PETERS' CHECKLIST OF THE BIRDS OF THE WORLD*, Vol. 8. Cambridge, Mass.

Snow, D.W. 1973. *Breviora*, 409.
Snow, D.W. 1980. *Bull. B.O.C.*, 100(4)213.
Meyer de Schauensee, R. 1982. *Guide to the Birds of South America* (reprint). Acad. Nat. Sci. Philadelphia.
Oren, D.C. and F.C.Novaes 1985. Bull. B.O.C. 105(1)23.

OXYRUNCIDAE, PHYTOTOMIDAE 109, 110

TRAYLOR, M.A. (ed.) 1979. *PETERS' CHECKLIST OF THE BIRDS OF THE WORLD*, Vol. 8. Cambridge, Mass.

PITTIDAE 111

1 MAYR, E. 1979. *PETERS' CHECKLIST OF THE BIRDS OF THE WORLD*, Vol. 8. Cambridge, Mass.
  MORONY, J.J. Jr. *et al.* 1975. *REFERENCE LIST OF THE BIRDS OF THE WORLD*. Amer. Mus. Nat. Hist. New York.

2 White, C.M.N. and M.D. Bruce, 1986. *Birds of Wallacea*. B.O.U. Checkl. 7.

XENICIDAE 112

MAYR, E. 1979. *PETERS' CHECKLIST OF THE BIRDS OF THE WORLD*, Vol. 8. Cambridge, Mass.

O.S.N.Z. 1970. *Annotated Checklist of the Birds of New Zealand*. Reed, Wellington.

PHILEPITTIDAE 113

AMADON, D. 1979. *PETERS' CHECKLIST OF THE BIRDS OF THE WORLD*, Vol. 8. Cambridge, Mass.

MENURIDAE, ATRICHORNITHIDAE 114, 115

MAYR, E. 1979. *PETERS' CHECKLIST OF THE BIRDS OF THE WORLD*, Vol. 8. Cambridge, Mass.
SCHODDE, R. 1975. *INTERIM LIST OF AUSTRALIAN SONGBIRDS, PASSERINES.* R.A.O.U., Melbourne.

ALAUDIDAE 116

PETERS, J.L. 1960. *CHECKLIST OF THE BIRDS OF THE WORLD*, Vol. 9. Cambridge, Mass.

Ripley, S.D. 1961. *Synopsis of the Birds of India and Pakistan.* Nat. Hist. Soc. Bombay.
White, C.M.N. 1961. *Revised Checklist of African Broadbills etc.* Govt. Printer, Lusaka.
Hall, B.P. and R.E. Moreau, 1970. *Atlas of Speciation in African Passerine Birds.* Brit. Mus. Nat. Hist. London.
Winterbottom, J.M. 1972. *Ostrich*, 43.
Erard, C. and G. Jarry 1973. *Bull. B.O.C.*, 93.
Erard, C. and R. de Naurois 1973. *Bull. B.O.C.*, 93.
1   Colston, P.R. 1982. *Bull. B.O.C.*, 102.106.
2   Clancey, P.A. (ed.) 1980. *Checklist of South African Birds.* S.A.O.U., Durban.
3   Keith, G.S. 1990. *The Birds of Africa* IV. (Pers. Comm.).

HIRUNDINIDAE 117

PETERS, J.L. 1960. *CHECKLIST OF THE BIRDS OF THE WORLD*, Vol. 9. Cambridge, Mass.

Ripley, S.D. 1961. *Synopsis of the Birds of India and Pakistan.* Nat. Hist. Soc., Bombay.
White, C.M.N 1961. *Revised Checklist of African Broadbills etc.* Govt. Printer, Lusaka.
Wolters, H.E. 1971. *Bonn. Zool. Beitr.*, 22.
3   Brooke, R.K. 1974. *Durban Mus. Novit.*, 10(a).
Stepanyan, L.S. 1974. *Zool. Zh.*, 53.8.
2   Clancey, P.A. (ed.) 1980. *Checklist of South African Birds.* S.A.O.U., Durban.
Meyer de Schauensee, R. 1982. *Guide to the Birds of South America* (reprint). Acad. Nat. Sci., Philadelphia.
4   A.O.U. 1983. *Checklist of North American Birds*, 6th ed. A.O.U.
1   Fry, C.H. and D.A. Smith 1985. *Ibis*, 127.1.
5   Keith, G.S. 1990. *The Birds of Africa* IV. (Pers. Comm.).

MOTACILLIDAE 118

PETERS, J.L. 1960. *CHECKLIST OF THE BIRDS OF THE WORLD*, Vol. 9. Cambridge, Mass.

Hall, B.P. 1961. *Bull. Am. Mus. Nat. Hist.*, 7(5).
White, C.M.N. 1961. *Revised Checklist of African Broadbills etc.* Govt. Printer, Lusaka.
Hall, B.P. and R.E. Moreau 1970. *Atlas of Speciation in African Passerine Birds.* Brit Mus. Nat. Hist., London.
Colston, P.R. 1982. *Bull. B.O.C.*, 102(3)112.
3   Clancey, P.A. 1984. *Gerfaut*, 74.375.
5   Clancey, P.A. 1985. *Bull. B.O.C.*, 105.133.
1   Cooper, M.R. 1985. *Honeyguide*, 31.81.
4   Knox, A.G. 1988. *Brit. Birds*, 81.206.
7   Clancey, P.A. 1986. *Gerfaut*, 76.187.
2   Clancey, P.A. 1986. *Bull. B.O.C.*, 106(2)80.
6   Colston, P.R. 1987. *Bull. B.O.C.*, 107.92.
8   Keith, G.S. 1990. *The Birds of Africa* IV. (Pers. Comm.).

CAMPEPHAGIDAE 119

PETERS, J.L. 1960. *CHECKLIST OF THE BIRDS OF THE WORLD*, Vol. 9. Cambridge, Mass.

Galbraith, I.C.J. and E.H. Galbraith, 1962. *Bull. Br. Mus. Nat. Hist.*, 9.
Salamonsen, F. 1964. *Noona Dan Pap.*, 9.
Gilliard, E.T. and Lecroy, 1967. *Bull. Amer. Mus. Nat. Hist.*, 135(4).
Rand, A.L. and E.T. Gilliard, 1967. *Handbook of New Guinea Birds.* Weidenfeld & Nicholson, London.
Galbraith, I.C.J. 1969. *Emu*, 69.
Parkes, K.C. 1971. *Nemouria*, 4.
DuPont, J.E. 1972. *Nemouria*, 7.
Schodde, R. 1975. *Interim Checklist of Australian Songbirds, Passerines*, R.A.O.U., Melbourne.
1   Mason, I.J. and J.L. McKean, 1982. *Bull. B.O.C.*, 102(4)127.

2 White, C.M.N. and M.D. Bruce, 1986. *Birds of Wallacea.* B.O.U. Checkl. 7.
3 Erickson K.R. and P.D. Heideman, 1983. *Silliman J.*, 30.63
4 Keith, G.S. 1990. *The Birds of Africa* IV. (Pers. Comm.).

PYCNONOTIDAE   120

RAND, A.L. and H. DEIGNAN 1960. *PETERS' CHECKLIST OF THE BIRDS OF THE WORLD*, Vol. 9. Cambridge, Mass.

White, C.M.N. 1962. *Revised Checklist of African Shrikes etc.* Govt. Printer, Lusaka.
Prigogine, A. 1969. *Rev. Zool. Bot. Afr.*, 79.
Mees, G.F. 1969. *Zool.Verh.*, 102.302.
Clancey, P.A. (ed.) 1980. *Checklist of South African Birds.* S.A.O.U., Durban.
Jensen, F.P.and S.N.Stuart, 1982. *Bull. B.O.C.*, 102(3)96.
1 White, C.M.N. and M.D. Bruce, 1986. *Birds of Wallacea*, B.O.U. Checkl. 7.
2 Gatter, W. 1985. *J. Orn.*, 126.155.
3 Dee, T.J. 1986. *Endemic Birds of Madagascar.* KBP, Cambridge.
4 Keith, G.S. 1990. *The Birds of Africa* IV. (Pers.Comm.).

IRENIDAE   121

DELACOUR, J. 1960. *PETERS' CHECKLIST OF THE BIRDS OF THE WORLD*, Vol. 9. Cambridge, Mass.

Ripley, S.D. 1960. *Synopsis of the Birds of India and Pakistan*, Nat. Hist. Soc., Bombay.
DuPont, J.E. 1972. *Philippine Birds.* Mus. Nat. Hist., Greenville, Del.

LANIIDAE   122

RAND, A.L. 1960. *PETERS' CHECKLIST OF THE BIRDS OF THE WORLD*, Vol. 9. Cambridge, Mass.

White, C.M.N. 1962. *Revised Checklist of African Shrikes etc.* Govt. Printer, Lusaka.
Hall. B.P. and R.E. Moreau, 1970. *Atlas of Speciation in African Passerine Birds.* Brit. Mus. Nat. Hist., London.
3 Field, G.D. 1979. *Bull. B.O.C.*, 99.2.
1 1980. *Ibis*, 122.566.
Clancey, P.A. ed 1980. *Checklist of South African Birds.* S.A.O.U., Durban.
2 Prigogine, A. 1984. *Gerfaut*, 74.75.

VANGIDAE   123

RAND, A.L. *et al.* 1960. *PETERS' CHECKLIST OF THE BIRDS OF THE WORLD*, Vol. 9. Cambridge, Mass.

Louette, M. and M. Herremens, 1982. *Bull. B.O.C.*, 102(4)132.

BOMBYCILLIDAE, DULIDAE, CINCLIDAE   124, 125, 126

GREENWAY, J.C. 1960. *PETERS' CHECKLIST OF THE BIRDS OF THE WORLD*, Vol. 9. Cambridge, Mass.

Phillips, A.R. 1966. *Bull. B.O.C.*, 86.103.
Blake, E.R. 1972. *Birds of Mexico.* Chicago University Press, Chicago.

TROGLODYTIDAE   127

PAYNTER, R.A. and C. VAURIE 1960. *PETERS' CHECKLIST OF THE BIRDS OF THE WORLD*, Vol. 9. Cambridge, Mass.

Slud, P. 1964. *Bull. Amer. Mus. Nat. Hist.*, 128.
Phillips, A.R. 1966. *Bull. B.O.C.*, 86.103.
Blake, E.R. 1972. *Birds of Mexico.* Chicago University Press, Chicago.
Dickerman, R.W. 1975. *Amer. Mus. Novit.*, 2569.
2 Meyer de Schauensee, R. 1982. *Guide to the Birds of South America* (reprint). Acad. Nat. Sci., Philadelphia.
4 A.O.U. 1983. *Checklist of North American Birds*, 6th ed. A.O.U.
1 Parker, T.A. and J.P. O'Neill, 1985. *Orn. Monogr.*, 36.9.
3 A.O.U 1985. *Auk*, 102.680 Checkl. Suppl..

MIMIDAE   128

DAVID, J. and A.H. MILLER 1960. *PETERS' CHECKLIST OF THE BIRDS OF THE WORLD*, Vol. 9. Cambridge, Mass.

Blake, E.R. 1972. *Birds of Mexico*. Chicago University Press, Chicago.
Meyer de Schauensee, R. 1982. *Guide to the Birds of South America* (reprint). Acad. Nat. Sci., Philadelphia.
Phillips, A.R. 1961. *Anal. Inst. Biol. Mexico*, 32.333.
1 Storer, R.W. 1989. *Auk*, 106.249.

PRUNELLIDAE 129

RIPLEY, S.D. 1964. *PETERS' CHECKLIST OF THE BIRDS OF THE WORLD*, Vol. 10. Cambridge, Mass.

TURDIDAE 130

RIPLEY, S.D. 1964. PETERS' CHECKLIST OF THE BIRDS OF THE WORLD, Vol. 10. Cambridge, Mass.

Irwin, M.P.S. and P.A. Clancey, 1974. *Arnoldia*, 6.34.
Benson, C.W. and M.P.S. Irwin, 1975. *Arnoldia*, 7.1.
Prigogine, A. 1977. *Bull. B.O.C.*, 97.
Orenstein 1979. *Ibis*, 121.
12 Clancey, P.A. (ed.) 1980. *Checklist of the Birds of South Africa*. S.A.O.S., Durban.
Dickerman, R.W. 1981. *Bull. B.O.C.*, 101.(2) 285.
Jensen, F.P. and S.N. Stuart, 1982. *Bull. B.O.C.*, 102(3)97.
3 Ripley, S.D. and D. Hadden, 1982. *Yamash. Inst. Orn.,* 14.103.
6 Sluys, R. and M. van den Berg, 1982. *Ornis Scand.*, 13.123.
10 Pratt, H.D. 1982. *Living Bird*, 19.73.
13 A.O.U. 1983. *Checklist of North American Birds*, 6th ed. A.O.U.
4 A.O.U. 1983. *Checklist of North American Birds, Auk*, 99 Suppl.
2 Ford, J. 1983. *Emu*, 83.141.
Prigogine, A. 1984. *Gerfaut*, 74.181/185/383.
8 Irwin, M.P.S. and P.A. Clancey, 1985. *Honeyguide*, 31.162.
Snow, D.W. 1985. *Bull. B.O.C.*, 105.30.
Prigogine, A. 1985. *Gerfaut*, 75.285.
11 A.O.U. 1985. *Auk*, 102.680 A.O.U. Checkl. Suppl.
9 Irwin, M.P.S. and P.A. Clancey, 1986. *Bull. B.O.C.*, 106.111.
5 Tye, A. 1986. *Bull. B.O.C.*, 106.104.
1 White, C.M.N. and M.D. Bruce, 1986. *Birds of Wallacea*, B.O.U. Checkl. 7.
7 Prigogine, A. 1987. *Bull. B.O.C.*, 107(2)49
14 Collar, N.J. and P. Andrew, 1987. *World Birdwatch*, 9(4)5.
15 Keith, G.S. 1990. *The Birds of Africa* IV. (Pers. Comm.).

ORTHONYCHIDAE 131

DEIGNAN, H.G. 1964. *PETERS' CHECKLIST OF THE BIRDS OF THE WORLD*, Vol. 10. Cambridge, Mass.
RAND, A.L. and E.T. GILLIARD 1967. *HANDBOOK OF NEW GUINEA BIRDS*. Weidenfeld & Nicholson, London.

Diamond, A.W. 1969. *Amer. Mus. Novit.*, 2362.
Shodde, R. 1975. *Interim List of Australian Songbirds, Passerines*. R.A.O.U., Melbourne.
1 Ford, J. 1983. *Emu*, 83, 152.

TIMALIIDAE 132

DEIGNAN, H.G. 1964. *PETERS' CHECKLIST OF THE BIRDS OF THE WORLD*, Vol. 10. Cambridge, Mass.

Vaurie, C. 1965. *L'Oiseau*, 34.
Hoogerwerf, A. 1966. *Misc. Rep. Yamash. Inst. Orn.*, 4.
Ripley, S.D. and A. Heinrich, 1966. *Postilla*, 95.
Salim Ali and S.D. Ripley, 1971. *Birds of India and Pakistan*, Vol. 6. Oxford University Press, Oxford.
Mann, C.F. *et al.* 1978. *Bull. B.O.C.*, 98.
Cunningham-van Someren, G.R. and H. Schifter, 1981. *Bull B.O.C.*, 101(3)347.
Ash, J.S. 1981. *Bull. B.O.C.*, 101(4)399.
1 Ripley, S.D. and B.M. Beehler, 1985. *Ibis*, 127.495.
Irwin, M.P.S. 1985. *Honeyguide*, 31.99.
White, C.M.N. and M.D. Bruce, 1986. *Birds of Wallacea*, B.O.U. Checkl. 7.
3 Harrison, C.J.O. 1986. *Forktail*, 1.81.
Parkes, K.C. 1988. *Nemouria*, 30.
2 Van de Weghe, J.P. 1988. *Bull. B.O.C.*, 108(2)54.

PANURIDAE, PICATHARTIDAE 133, 134

DEIGNAN, H.G. 1964. *PETERS' CHECKLIST OF THE BIRDS OF THE WORLD*, Vol. 10. Cambridge, Mass.

POLIOPTILIDAE 135

PAYNTER, R.A. 1964. PETERS' *CHECKLISTS OF THE BIRDS OF THE WORLD*, Vol. 10. Cambridge, Mass.

Meyer de Schauensee, R. 1982. *Guide to the Birds of South America* (reprint). Acad. Nat. Sci., Philadelphia.

SYLVIIDAE 136

1 MAYR, R. and G.W. COTTRELL (ed.) 1987. *PETERS' CHECKLIST OF THE BIRDS OF THE WORLD*, Vol. 11. Cambridge, Mass.

Vaurie, C. 1959. *Birds of the Palaearctic Fauna*, Vol. 1. Witherby, London.
2 Mayr, E. 1944. *Bull. Amer. Mus. Nat. Hist.*, 83.123.
11 Williamson, K. 1967. *Identification for Ringers, Phylloscopus*, 2.2.
10 Fry, C.H. 1976. *Arnoldia*, 8.6.13.
Dowsett, R.J. and R. Sjernstedt, 1979. *Bull. B.O.C.*, 99.86.
4 Wells, D.R. 1982. *Bull. B.O.C.*, 102(2)57.
3 Orenstein, R.I. and H.D. Pratt, 1983. *Wilson Bull.*, 95.184.
Colston, P.R. and G.J. Morel, 1985. Malimbus 6.71
7 Ripley, S.D. 1985. *Bull. B.O.C.*, 105.109.
8 Watson, G.E. 1985. *Bull. B.O.C.*, 105.79.
9 White, C.M.N. and M.D. Bruce 1986. *Birds of Wallacea*, B.O.U. Checkl. 7.
5 Rozendaal, F.G. 1987. *Zool. Meded, Leiden*, 61.177.
15 Pratt, H.D. *et al.* 1987. *Field Guide to the Birds of Hawaii and the Tropical Pacific*. New York.
Parkes, K.C. 1988. *Nemouria*, 30.
12 King, B. 1989. *Bull. B.O.C.*, 109(3)162.
13 Martens, J. 1988. *J. Orn.*, 129(3)343.
14 Irwin, M.P.S. 1988. *Bull. B.O.C.*, 108(2)58.

MUSCICAPIDAE 137

1 MAYR, E. and G.W. COTTRELL, (ed.) 1987. *PETERS' CHECKLIST OF THE BIRDS OF THE WORLD*, Vol. 11. Cambridge, Mass.

2 Wells, D.R. 1982. *Bull. B.O.C.*, 102(4)148.
Wells, D.R. and C.M. Francis 1984. *Bull B.O.C.*, 104.125.
Vernon, C.J. 1985. *Honeyguide*, 31.93.
White, C.M.N. and M.D. Bruce 1986. *Birds of Wallacea*, B.O.U. Checkl. 7.

PLATYSTEIRIDAE 138

1 MAYR, E. and G.W. COTTRELL, (ed.) 1987. *PETERS' CHECKLIST OF THE BIRDS OF THE WORLD*, Vol. 11. Cambridge, Mass.
3 Erard, C. 1975. *L'Oiseau*, 45(3)235.
2 Lawson, W.J. 1984. *Bull. B.O.C.*, 104.144.

MALURIDAE, ACANTHIZIDAE, EOPSALTRIDAE 139, 140, 142

1 MAYR, E. and G.W. COTTRELL (ed.) 1987. PETERS' CHECKLIST OF THE BIRDS OF THE WORLD, Vol. 11. Cambridge, Mass.

2 Keast, A.J. 1978. *Emu*, 78.20.
Schodde, R. 1975. *Interim List of Australian Songbirds, Passerines*. R.A.O.U., Melbourne.
4 Ford, J. and R.E. Johnstone, 1983. *West Austr. Nat.*, 15.133.
3 Schodde, R. and R.G. Weatherly, 1983. *Emu*, 82.308.
6 White, C.M.N. and M.D. Bruce, 1986. *Birds of Wallacea*. B.O.U. Checkl. 7.
5 Schodde, R. 1975. *Emu*, 85.49.

MONARCHIDAE 141

1 MAYR, E. and G.W. COTTRELL (ed.) 1987. *PETERS' CHECKLIST OF THE BIRDS OF THE WORLD*, Vol. 11. Cambridge, Mass.

2 White, C.M.N. and M.D. Bruce, 1986. *Birds of Wallacea*, B.O.U. Checkl. 7.
3 Watling, R. 1988. *Bull. B.O.C.*, 108.103.
4 Erard, C. 1987. *Mem. Mus. Nat. Hist. (Ser. A) Zool.*, 138.1.

REFERENCES

PACHYCEPHALIDAE 143

MAYR, E. 1967. *PETERS' CHECKLIST OF THE BIRDS OF THE WORLD*, Vol. 12. Cambridge, Mass.

Rand, A.L. and E.T. Gilliard, 1967. *Handbook of Birds of New Guinea*. Weidenfeld and Nicholson, London.
Ford, J. 1971. *Emu*, 71.
Diamond, A.W. 1972. *Publ. Nuttall Orn. Cl.*, 12.
Schodde, R. 1975. *Interim List of Australian Songbirds, Passerines*. R.A.O.U., Melbourne.
Zuoxin, Z. 1984. *Acta Zool. Sinica*, 30.278.
1 White, C.M.N. and M.D. Bruce, 1986. *Birds of Wallacea*, B.O.U. Checkl. 7.

AEGITHALIDAE, REMIZIDAE 144, 145

SNOW, D.W. 1967. *PETERS' CHECKLIST OF THE BIRDS OF THE WORLD*, Vol. 12. Cambridge, Mass.

Blake, E.R. 1972. *Birds of Mexico*, Chicago University Press, Chicago.
Eisenmann, E. *et al.* 1973. *Auk*, 90.

PARIDAE 146

SNOW, D.W. 1967. *PETERS' CHECKLIST OF THE BIRDS OF THE WORLD*, Vol. 12. Cambridge, Mass.

Hall, B.P. and R.E. Moreau 1970. *Atlas of Speciation in African Passerine Birds*. Brit. Mus. Nat. Hist., London.
Clancey, P.A. 1972. *Durban Mus. Novit.*, 9.
Stepanyan, L.S. 1974. *By. Mosk. Ob. Isp. Priv.*, 99(6).
1 Clancey, P.A. (ed.) 1980. *Checklist of South African Birds*. S.A.O.U., Durban.

SITTIDAE 147

GREENWAY, J.C. 1967. *PETERS' CHECKLIST OF THE BIRDS OF THE WORLD*, Vol. 12. Cambridge, Mass.

Macdonald, J.D. 1969. *Emu*, 69.
Ford, J. and S. Parker 1974. *Emu*, 74.
Schodde, R. 1975. *Interim List of Australian Songbirds, Passerines*. R.A.O.U., Melbourne.
Vieilliard 1976. *Alauda*, 44.

CERTHIIDAE 148

GREENWAY, J.C. 1967. *PETERS' CHECKLIST OF THE BIRDS OF THE WORLD*, Vol. 12. Cambridge, Mass.

Thielke, G. 1962. *J. Orn.*, 103.266.
Blake, E.R. 1972. *Birds of Mexico*. Chicago University Press, Chicago.

RHABDORNITHIDAE 149

GREENWAY, J.C. 1967. *PETERS' CHECKLIST OF THE BIRDS OF THE WORLD*, Vol. 12. Cambridge, Mass.

DuPont, J.E. 1971. *Philippine Birds*. Mus. Nat. Hist., Greenville, Del.
Parkes, K.C. 1973. *Nemouria*, 11.

CLIMACTERIDAE 150

GREENWAY, J.C. 1967. *PETERS' CHECKLIST OF THE BIRDS OF THE WORLD*, Vol. 12. Cambridge, Mass.

Schodde, R. 1975. Interim List of Australian Songbirds, Passerines, R.A.O.U., Melbourne.
Boles, W.E. and N.W. Longmore, 1983. *Emu*, 83.272.

DICAEIDAE 151

SALOMONSEN, F. 1967. *PETERS' CHECKLIST OF THE BIRDS OF THE WORLD*, Vol. 12. Cambridge, Mass.

Rand, A.L. and E.T. Gilliard 1967. *Handbook of New Guinea Birds*. Weidenfeld & Nicholson London.
Rand, A.L. and Rabor 1969. *Fieldiana Zool.*, 51.
DuPont, J.E. 1971. *Philippine Birds*. Mus. Nat. Hist., Greenville, Del.
Schodde, R. 1975. *Interim List of Australian Songbirds, Passerines*. R.A.O.U., Melbourne.
2 Sheldon, F.H. 1985. *Auk*, 102.606.

1   White, C.M.N. and M.D. Bruce 1986. *Birds of Wallacea*, B.O.U. Checkl. 7.
     Parkes, K.C. 1988. *Nemouria*, 30.

NECTARINIIDAE   152

RAND, A.L. 1967. *PETERS' CHECKLIST OF THE BIRDS OF THE WORLD*, Vol. 12. Cambridge, Mass.

Clancey, P.A. 1970. *Durban Mus. Novit.*, 8.9.
Hall, B.P. and R.E. Moreau, 1970. *Atlas of Speciation in African Passerine Birds*. Brit. Mus. Nat. Hist., London.
Clancey, P.A. 1973. *Durban Mus. Novit.*, 10.
Clancey, P.A. 1975. *Durban Mus. Novit.*, 11.
5   Clancey, P.A. (ed.) 1980. *Checklist of South African Birds*. S.A.O.U., Durban.
2   Benson, C.W. and A. Prigogine 1981. *Gerfaut*, 71.47.
3   Jensen, F.P. 1983. *Ibis*, 125.447.
4   Stepanyan, L.S. 1985. *Ornitologiya*, 20.133.
1   White, C.M.N. and M.D. Bruce, 1986. *Birds of Wallacea*. B.O.U. Checkl. 7.

ZOSTEROPIDAE   153

MAYR, E. and R.E. MOREAU, 1967. PETERS' *CHECKLIST OF THE BIRDS OF THE WORLD*, Vol. 12. Cambridge, Mass.

Mees, G.F. 1957. *Zool. Verh. Rijksmus. Nat. Hist. Leiden*, 35.
Mees, G.F. 1961. *Zool. Verh. Rijksmus. Nat. Hist. Leiden*, 50.
Rand, A.L. and E. T. Gilliard, 1967. *Handbook of New Guinea Birds.* Weidenfeld & Nicholson, London.
Mees, G.F. 1969. *Zool. Verh. Rijksmus. Nat. Hist. Leiden*, 102.
2   Clancey, P.A. (ed.) 1980. *Checklist of South African Birds*. S.A.O.U., Durban.
1   White, C.M.N. and M.D. Bruce, 1986. *Birds of Wallacea*. B.O.U. Checkl. 7.

MELIPHAGIDAE   154

SALOMONSEN, F. 1967. *PETERS' CHECKLIST OF THE BIRDS OF THE WORLD*, Vol. 1. Cambridge, Mass.

Rand, A.L. and E.T. Gilliard, 1967. *Handbook of New Guinea Birds.* Weidenfeld & Nicholson, London.
Diamond, A.W. 1969. *Amer. Mus. Novit.*, 2362.
Diamond, A.W. 1972. *Publ. Nuttall Orn. Cl.*, 12.
Schodde, R. 1975. *Interim List of Australian Songbirds, Passerines*. R.A.O.U., Melbourne.
Parkes, K.C. 1980. *Bull. B.O.C.*, 100(2)143.
2   Longmore, N.W. and W.E. Boles 1983. *Emu*, 83.59.
3   Boles, W.E. and N.W. Longmore 1985. *S. Austrn. Orn.*, 29.221.
1   White, C.M.N. and M.D. Bruce 1986. *Birds of Wallacea*, B.O.U. Checkl. 7.
4   Pratt, J.D., Bruner P.L. and D.G. Berrett, 1987. *Field Guide to Birds, Hawaii.* University Press, Princeton.

EMBERIZINAE, CARDINALINAE   155 A and C

PAYNTER, R.A. 1970. *PETERS' CHECKLIST OF THE BIRDS OF THE WORLD*, Vol. 13. Cambridge, Mass.

Meyer de Schauensee, R. 1970. *Notulae Naturae*, 428.
Blake, E.R. 1972. *Birds of Mexico.* Chicago University Press, Chicago.
Matousek, B. 1973. *Zbornik Slov. Narod. Mus.*, 17(1).
Parkes, K.C. 1974. *Wilson Bull.*, 86(3).
Short, L.L. 1974. *Bull. Amer. Mus. Nat. Hist.*, 154.
Colston, P.R. 1978. *Bull. B.O.C.*, 98.
Webster, J.D. 1983. *Proc. Biol. Soc. Wash.*, 96.644.
2   A.O.U. 1983. *Checklist of North American Birds* 6th ed. A.O.U.
Oren, D.C. 1985. *Publ. Av. Mus. Para Em. Goeldi*, 40.93.
Zink, R.M. 1986. *Orn. Monogr.*, 40.
Loskot, V.M. 1986. *Tr. Zool. Inst. Ak. NAUK S.S.R.*, 150.147.
Nores, M. 1986. *Hornero*,, 12.262.
1   A.O.U. 1987. *Auk*, 104.591 Checkl. Suppl.

THRAUPINAE, CATAMBLYRHYNCHINAE, TERSINAE   155 D, B and E

STORER, R.W. 1970. *PETERS' CHECKLIST OF THE BIRDS OF THE WORLD*, Vol. 13. Cambridge, Mass.

3   ISLER, M.L. and P.R. ISLER 1987. *THE TANAGERS*. Oxford University Press, Oxford.

Vuilliumier, F. 1969. *Am. Mus. Novit.*, 2381.
2   Johnson, N.K. and A.H. Brush 1972. *Syst. Zool.*, 21.245.

REFERENCES
Graves, G.R. 1980. *Bull. B.O.C.*, 100(4)230.
5 Bock, W.J. 1985. *Orn. Monogr.*, 36.319.
4 Schulenberg, T.S. and L.C. Binford, 1985. *Wilson Bull.*, 97.413.
1 Olson, S.L. 1981. *Proc. Biol. Soc. Wash.*, 94.363.
Dickerman, R.W. 1987. *Bull. B.O.C.*, 107.42.
6 Graves, G.R. and J.S. Weske, 1987. *Wilson Bull.*, 99.1.

COEREBIDAE 156

1 Meyer de Schauensee, R.M. 1982. *A Guide to the Birds of South America* (reprint). Acad. Nat. Sci., Philadelphia.

PARULIDAE 157

LOWERY, G.H. and B.L. MONROE, 1968. *PETERS' CHECKLIST OF THE BIRDS OF THE WORLD*, Vol. 14. Cambridge, Mass.

Blake, E.R. 1972. *Birds of Mexico.* Chicago University Press, Chicago.
Kepler, C.B. and K.C. Parkes, 1972. *Auk*, 89.
Olson, S.L. 1975. *Bull. B.O.C.*, 95.
3 Meyer de Schauensee, R. 1982. *Guide to the Birds of South America* (reprint). Acad. Nat. Sci., Philadelphia.
2 A.O.U. 1983. *Checklist of North American Birds*, 6th ed. A.O.U.
Oren, D.C. 1985. *Publ. Av. Mus. Para. Em. Goeldi*, 40.93.
1 Isler, M.L. and P.R. Isler, 1987. *The Tanagers.* Oxford University Press, Oxford.

DREPANIDIDAE 158

1 A.O.U. 1983. *CHECKLIST OF NORTH AMERICAN BIRDS*, 6th ed. A.O.U.
MUNRO, G.C. 1944. *BIRDS OF HAWAII.*

2 Casey, T.L.C. and J.D. Jacobi, 1974. *Occ. Pap. Bern. P. Bishop Mus.*, 24.

VIREONIDAE 159

BLAKE, E.R. 1968. *PETERS' CHECKLIST OF THE BIRDS OF THE WORLD*, Vol. 14. Cambridge, Mass.

Blake, E.R. 1972. *Birds of Mexico.* Chicago University Press, Chicago.
Mees, G.F. 1974. *Zool. Meded Leiden*, 48(7).
1 Meyer de Schauensee, R. 1982. *Guide to the Birds of South America* (reprint). Acad. Nat. Sci., Philadelphia.
3 A.O.U. 1983. *Checklist of North American Birds*, 6th ed. A.O.U.
2 A.O.U. 1987. *Auk*, 104.591. Checkl. Suppl.

ICTERIDAE 160

BLAKE, E.R. 1968. *PETERS' CHECKLIST OF THE BIRDS OF THE WORLD*, Vol. 14. Cambridge, Mass.

Dickerman, R.W. and A. Phillips, 1970. *Condor*, 72.
Garrido, O.H. 1970. *Poeyana*, 68.
Dickerman, R.W. 1974. *Amer. Mus. Novit.*, 2538.
1 Meyer de Schauensee, R. 1982. *Guide to the Birds of South America* (reprint). Acad. Nat. Sci., Philadelphia.

FRINGILLIDAE 161

MAYR, E. 1968. *PETERS' CHECKLIST OF THE BIRDS OF THE WORLD*, Vol. 14. Cambridge, Mass.

Hall, B.P. and R. Moreau, 1970. *Atlas of Speciation in African Passerine Birds.* Brit. Mus. Nat. Hist., London.
Erard, C. 1974. *L'Oiseau*, 44(4)308.
5 Knox, A.G. 1976. *Bull. B.O.C.*, 96.
Voous, K.H. 1977. *Ibis*, 119.
4 Ash, J.S. 1979. *Ibis*, 121.
3 Clancey, P.A. (ed.) 1980. *Checklist of South African Birds.* S.A.O.S., Durban.
2 Britton, P.L. (ed.) 1980. Birds of East Africa 233. EANHS, Nairobi.
Schuchmann, K.L. and H.E. Wolters, 1982. *Bull. B.O.C.*, 101(1)12.
A.O.U. 1983. *Checklist of North American Birds*, 6th ed. A.O.U.
Phelps, W.H. Jr. and R. Avelado, 1984. *Bol. Soc. Venez. CN.*, 39(142)5.
1 Troy, D.M. 1985. *Auk*, 102.82.
Buden, D.W. 1986. *Bull. B.O.C.*, 105.106.

ESTRILDIDAE 162

MAYR, E. *et al. PETERS' CHECKLIST OF THE BIRDS OF THE WORLD*, Vol. 14. Cambridge, Mass.

GOODWIN, D. 1982. *ESTRILDID FINCHES OF THE WORLD*. Brit. Mus. Nat. Hist., London.

Hall, B.P. and R.E. Moreau, 1970. *An Atlas of Speciation in African Passerine Birds*, Brit. Mus. Nat. Hist., London.
1 Clancey, P.A. (ed.) 1980. *Checklist of South African Birds*. S.O.A.S., Durban
Cunningham-van Someren, G.R. and H. Schifter, 1981. *Bull. B.O.C.*, 101(4) 361.
Clancey, P.A. 1986. *Gerfaut*, 76.301.
White, C.M.N. and M.D. Bruce 1986. *Birds of Wallacea*. B.O.U. Checkl. 7.

PLOCEIDAE 163

TRAYLOR, M.A. *et al.* 1968/62 *PETERS' CHECKLIST OF THE BIRDS OF THE WORLD*, vols 14/15. Cambridge, Mass.

Hall, B.P. and R.E. Moreau, 1970. *An Atlas of Speciation in African Passerine Birds*, Brit. Mus. Nat. Hist., London.
4 Wolters, H.E. 1974. *Bonn Zool. Beitr.*, 25(4).
3 Clancey, P.A. (ed.) 1980. *Checklist of South African Birds*. S.A.O.S., Durban.
Louette, M. and C.W. Benson, 1982. *Bull. B.O.C.*, 102(1)24.
Franzmann, W.E. 1983. *Bull. B.O.C.*, 103(2)49.
Clancey, P.A. 1985. *Honeyguide*, 31.104.
1 Ash, J.S. 1986. *Ibis*, 128.330.
5 Baker, N.E. and E.M. 1989. Unpubl. Rpt. (Brit. Mus. Nat. Hist.).

STURNIDAE 164

AMADON, D. 1962. *PETERS' CHECKLIST OF THE BIRDS OF THE WORLD*, Vol. 15. Cambridge, Mass.

Cunningham-van Someren, G.R. 1984. *Bull. B.O.C.*, 104.120.
1 White, C.M.N. and M.D. Bruce, 1986. *Birds of Wallacea*. B.O.U. Checkl. 7.
2 Clancey, P.A. 1987. *Bull. B.O.U.*, 107(1)25.

ORIOLIDAE 165

GREENWAY, J.C. 1962. *PETERS' CHECKLIST OF THE BIRDS OF THE WORLD*, Vol. 15. Cambridge, Mass.

2 Prigogine, A. 1978. *Gerfaut*, 68.253 3.
Clancey, P.A. (ed.) 1980. *Checklist of South African Birds*. S.A.O.S., Durban.
1 White, C.M.N. and M.D. Bruce, 1986. *Birds of Wallacea*. B.O.U. Checkl. 7.

DICRURIDAE 166

VAURIE, C. 1962. *PETERS' CHECKLIST OF THE BIRDS OF THE WORLD*, Vol. 15. Cambridge, Mass.

1 White, C.M.N. and M.D. Bruce, 1986. *Birds of Wallacea*. B.O.U. Checkl. 7.

CALLAEIDAE, CRACTICIDAE 167 170

AMADON, D. 1962. *PETERS' CHECKLIST OF THE BIRDS OF THE WORLD*, Vol. 15. Cambridge, Mass.

OSNZ 1970. *Annotated Checklist of the Birds of New Zealand*. Reed, Wellington.
Schodde, R. 1975. *Interim List of Australian Songbirds, Passerines*. R.A.O.U., Melbourne.
Mees, G.F. 1983. *Emu*, 83.123.

GRALLINIDAE, ARTAMIDAE 168, 169

MAYR, E. 1962. *PETERS' CHECKLIST OF THE BIRDS OF THE WORLD*, Vol. 15. Cambridge, Mass.

Rand, A.L. and E.T. Gilliard, 1967. *Handbook of New Guinea Birds*. Weidenfeld & Nicholson, London.
Schodde, R. 1975. *Interim List of Australian Songbirds, Passerines*. R.A.O.U., Melbourne.

PTILINORHYNCHIDAE, PARADISAEIDAE 171, 172

COOPER, W.J. and J.M. FORSHAW, 1977. *BIRDS OF PARADISE AND BOWERBIRDS*. Collins, London.

REFERENCES

MAYR, E. 1962. *PETERS' CHECKLIST OF THE BIRDS OF THE WORLD*, Vol. 15. Cambridge, Mass.

Gilliard, E.T. 1969. *Birds of Paradise and Bowerbirds*. Weidenfeld & Nicholson, London.
1   Schodde, R. and J.L. McKean, 1973. *Emu*, 73.

CORVIDAE   173

VAURIE, C. and E.R. BLAKE 1962. *PETERS' CHECKLIST OF THE BIRDS OF THE WORLD*, Vol. 15. Cambridge, Mass.
GOODWIN, D. 1976. *CROWS OF THE WORLD*. Brit. Mus. Nat. Hist., London.
2   Banks, R.C. 1983. *Elepaio* 44.1.
1   White, C.M.N. and M.D, Bruce, 1986. *Birds of Wallacea*. B.O.U. Checkl. 7.

CHECKLIST

STRUTHIONIFORMES

1. STRUTHIONIDAE (OSTRICHES)

STRUTHIO
Struthio camelus (Ostrich)
  S.c.camelus
    N Africa, Sudan
  S.c.syriacus
    Syrian & Arabian deserts
  S.c.molybdophanes
    Somalia, NE Kenya
  S.c.massaicus
    E Kenya, E Tanzania
  S.c.australis
    Southern Africa

RHEIFORMES

2. RHEIDAE (RHEAS)

RHEA
Rhea americana (Greater Rhea)
  R.a.americana
    N & E Brazil
  R.a intermedia
    S Brazil, Uruguay
  R.a.nobilis
    E Paraguay
  R.a.araneipes
    E Bolivia, SE Brazil
  R.a.albescens
    N Argentina

PTEROCNEMIA
Pterocnemia pennata (Lesser Rhea)
  P.p.garleppi
    SE Peru, Bolivia, NW Argentina
  P.p.tarapacensis
    N Chile
  P.p.pennata
    S Argentina

CASUARAIIFORMES

3. CASUARIIDAE (CASSOWARIES)

CASUARIUS
Casuarius casuarius (Double-wattled Cassowary)
  C.c.casuarius
    Seram I
  C.c.bicarunculatus
    Aru Is, NW New Guinea

C.c.tricarunculatus
    Geelvink Bay, New Guinea
C.c.lateralis
    N New Guinea
C.c.sclaterii
    S New Guinea
C.c.aruensis
    Wokan I
C.c.violicollis
    Trangan I
C.c.johnsonii
    N Queensland
Casuarius bennetti (Dwarf Cassowary)
  C.b.papuanus
    NW New Guinea
  C.b.goodfellowi
    Japen I
  C.b.claudii
    NC New Guinea
  C.b.hecki
    NE New Guinea
  C.b.picticollis
    SE New Guinea
  C.b.bennetti
    New Britain
  C.b.shawmayeri
    Kratke Mts (New Britain)
Casuarius unappendiculatus (One-wattled Cassowary)
  C.u.rothschildi
    W New Guinea
  C.u.philipi
    Sepik river, New Guinea
  C.u.unappendiculatus
    Salawati I, Misol I
  C.u.occipitalis
    Japen I
  C.u.rufotinctus
    N New Guinea
  C.u.aurantiacus
    NE New Guinea

4. DROMAIIDAE (EMUS)

DROMAIUS
Dromaius novaehollandiae (Emu)
  D.n.woodwardi
    NW & Western Australia, Northern Territory
  D.n. rothschildi
    SW Australia
  D.n.novaehollandiae
    C & S Queensland to Victoria, S Australia

APTERYGIFORMES

5. APTERYGIDAE (KIWIS)

APTERYX
Apteryx australis (Brown Kiwi)
  A.a.mantelli
    S North I (New Zealand)
  A.a.novaezelandiae
    N North I (New Zealand)
  A.a. australis
    South I (New Zealand)
  A.a.lawryi
    Stewart I
Apteryx owenii (Little Spotted Kiwi)
  A.o.iredalei
    North I (New Zealand) e?
  A.o.owenii
    South I (New Zealand)
Apteryx haastii (Great Spotted Kiwi)
  W South I (New Zealand)

TINAMIFORMES

6. TINAMIDAE (TINAMOUS)

TINAMUS
Tinamus tao (Grey Tinamou)
  T.t septentrionalis
    Colombia, Venezuela, Guyana
  T.t.larensis
    C Colombia, Venezuela
  T.t.tao
    N & C Brazil
  T.t.kleei
    N Bolivia, E Peru
Tinamus solitarius (Solitary Tinamou)
  T.s.pernambucensis
    E Brazil
  T.s.solitarius
    C Brazil, Paraguay
Tinamus osgoodi (Black Tinamou)
  T.o.hershkovitzi
    S Columbia
  T.o.osgoodi
    Peru
Tinamus major (Great Tinamou)
  T.m.robustus
    EC Guatemala to N Nicaragua
  T.m.percautus
    S Mexico, Guatemala
  T.m.fuscipennis
    E Nicaragua to Panama
  T.m.brunneiventris
    C Panama
  T.m.castaneiceps
    SW Costa Rica, W Panama
  T.m.saturatus
    E Panama, N Colombia
  T.m.latifrons
    W Colombia, W Ecuador
  T.m.zuliensis
    E Colombia, W Venezuela

  T.m.major
    the Guianas, N Brazil
  T.m.olivascens
    E Brazil
  T.m.peruvianus
    E Colombia, E Ecuador, E Peru
  T.m.serratus
    S Venezuela, W Brazil
Tinamus guttatus (White-throated Tinamou)
  E Ecuador & N Bolivia to Brazil

NOTHOCERCUS
Nothocercus bonapartei (Highland Tinamou)
  N.b.frantzii
    Costa Rica, W Panama
  N.b.intercedens
    W Colombia
  N.b.discrepans
    Colombia
  N.b.bonapartei
    E Colombia, W Venezuela
  N.b.plumbeiceps
    E Ecuador
Nothocercus julius (Tawny-breasted Tinamou)
  W Venezuela & Colombia to S Peru
Nothocercus nigrocapillus (Hooded Tinamou)
  N.n.cadwaladeri
    N Peru
  N.n.nigrocapillus
    C Bolivia

CRYPTURELLUS
Crypturellus cinereus (Cinereous Tinamou)
  the Guianas to E Peru
Crypturellus berlepschi (Berlepsch's Tinamou)
  NW Colombia, N Ecuador
Crypturellus soui (Little Tinamou)
  C.s.meserythrus
    S Mexico to Honduras
  C.s.modestus
    Nicaragua to W Panama
  C.s.capnodes
    NW Panama
  C.s.poliocephalus
    W Panama
  C.s.panamensis
    Pearl Is (Panama)
  C.s.caucae
    W Colombia
  C.s.harterti
    NW Colombia, W Ecuador
  C.s.mustelinus
    NE Colombia
  C.s.caquetae
    SE Colombia
  C.s.nigriceps
    E Ecuador
  C.s.soui
    E Colombia, Venezuela, the Guianas, N Brazil

C.s.andrei
Trinidad
C.s.albigularis
E Brazil
C.s.inconspicuus
C Bolivia
Crypturellus ptaritepui (Tepui Tinamou)
SE Venezuela
Crypturellus obsoletus (Brown Tinamou)
C.o.cerviniventris
Venezuela
C.o.castaneus
E Colombia, E Ecuador, N Peru
C.o.ochraceiventris
C Peru
C.o.punensis
S Peru, W Bolivia
C.o.griseiventris
Santarem, Brazil
C.o.obsoletus
S Brazil, Paraguay, NE Argentina
C.o.traylori
E Peru
C.o.hypochracea
SW Brazil
Crypturellus undulatus (Undulated Tinamou)
C.u.manapiare
S Venezuela
C.u.simplex
S Guyana
C.u.adspersus
N Brazil
C.u.yapura
E Ecuador, E Peru, W Brazil
C.u.vermiculatus
E Brazil
C.u.undulatus
E Bolivia, SW Brazil, Paraguay
Crypturellus transfasciatus (Pale-browed Tinamou)
W Ecuador, W Peru
Crypturellus strigulosus (Brazilian Tinamou)
Brazil
Crypturellus duidae (Grey-legged Tinamou)
SE Venezuela
Crypturellus erythropus (Red-footed Tinamou)
C.e.colombianus
NC Colombia
C.e.saltuarius e?
NC Colombia
C.e.idoneus
NE Colombia, W Venezuela
C.e.cursitans
N.Colombia, NW Venezuela
C.e.spencei
Venezuela
C.e.margaritae
Margarita I
C.e.erythropus
E Venezuela, Guyana, Surinam, N Brazil
Crypturellus noctivagus (Yellow-legged Tinamou)

C.n.zabele
NE Brazil
C.n.noctivagus
SE Brazil
Crypturellus atrocapillus (Black-capped Tinamou)
C.a.atrocapillus
SE Peru
C.a.garleppi
N Bolivia
Crypturellus cinnamomeus (Thicket Tinamou)
C.c.occidentalis
W coast of Mexico
C.c.mexicanus
NE Mexico
C.c.sallaei
S Mexico
C.c.goldmani
SE Mexico, N Belize
C.c.soconuscensis
C Chiapas
C.c.cinnamomeus
El Salvador to Nicaragua
C.c.vicinior
Chiapas to C Honduras
C.c.delattrei
Nicaragua
C.c.praepes
NW Costa Rica
Crypturellus boucardi (Slaty-breasted Tinamou)
C.b.boucardi
S Mexico to N Nicaragua
C.b.costaricensis
E Honduras to Costa Rica
Crypturellus kerriae (Choco Tinamou)
Colombia
Crypturellus variegatus (Variegated Tinamou)
Amazonia, E Brazil
Crypturellus brevirostris (Rusty Tinamou)
E Peru, W Brazil, French Guiana
Crypturellus bartletti (Bartlett's Tinamou)
E Peru, N Bolivia
Crypturellus parvirostris (Small-billed Tinamou)
SE Peru to S Brazil & NE Argentina
Crypturellus casiquiare (Barred Tinamou)
E Colombia, S Venezuela
Crypturellus tataupa (Tataupa Tinamou)
C.t.inops
NW Peru
C.t.peruviana
WC Peru
C.t.lepidotus
NE Brazil
C.t.tataupa
S Brazil, E Bolivia, Paraguay

RHYNCHOTUS
Rhynchotus rufescens (Red-winged Tinamou)
R.r.catingae
C Brazil

R.r.rufescens
  SE Peru, E Bolivia to NE Brazil and
  Uruguay
R.r.pallescens
  N Argentina
R.r.maculicollis
  W & S Bolivia to W Argentina

NOTHOPROCTA
**Nothoprocta taczanowskii (Taczanowski's
  Tinamou)**
  C & S Peru
**Nothoprocta kalinowskii (Kalinowski's
  Tinamou)**
  C Peru
Nothoprocta ornata (Ornate Tinamou)
N.o.branickii
  C Peru
N.o.ornata
  SE Peru, Bolivia
N.o.rostrata
  W Argentina
Nothoprocta perdicaria (Chilian Tinamou)
N.p.perdicaria
  N & C Chile
N.p.sanborni
  W Argentina, S Chile
**Nothoprocta cinerascens (Brushland
  Tinamou)**
N.c.cinerascens
  W Argentina
N.c.parvimaculata
  La Rioja (Argentina)
Nothoprocta pentlandii (Andean Tinamou)
N.p.ambigua
  S Ecuador
N.p.oustaleti
  S Ecuador, NW Peru
N.p.niethammeri
  Central coast of Peru
N.p.fulvescens
  SE Peru
N.p.pentlandii
  Bolivia to NW Argentina
N.p.doeringi
  C Argentina
N.p.mendozae
  WC Argentina
**Nothoprocta curvirostris (Curve-billed
  Tinamou)**
N.c.curvirostris
  C Ecuador
N.c.peruviana
  C Peru

NOTHURA
Nothura boraquira (White-bellied Nothura)
  NE Brazil, E Bolivia, Paraguay
Nothura minor (Lesser Nothura)
  S Brazil
Nothura darwinii (Darwin's Nothura)
N.d.peruviana
  S Peru

N.d.agassizii
  SE Peru, W Bolivia
N.d.boliviana
  W Bolivia
N.d.salvadorii
  W Argentina
N.d.darwinii
  SC Argentina
Nothura maculosa (Spotted Nothura)
N.m.cearensis
  S Ceara (Brazil)
N.m.major
  E Brazil
N.m.paludivaga
  C Paraguay, NC Argentina
N.m.maculosa
  SE Brazil, E Paraguay, Uruguay, NE
  Argentina
N.m.pallida
  NW Argentina
N.m.annectens
  E Argentina
N.m.submontana
  SW Argentina
N.m.nigroguttata
  S Argentina
N.m.chacoensis   6.1
  Paraguay, N Argentina

TAONISCUS
Taoniscus nanus (Dwarf Tinamou)
  Paraguay, SW Brazil

EUDROMIA
Eudromia elegans (Elegant Crested-Tinamou)
E.e.intermedia
  NW Argentina
E.e.magnistriata
  NW Argentina
E.e.riojana
  NW Argentina
E.e.albida
  W Argentina
E.e.elegans
  C Argentina
E.e.multiguttata
  EC Argentina
E.e.devia
  SW Argentina
E.e.patagonica
  S Chile, S Argentina
**Eudromia formosa (Quebracho Crested-
  Tinamou)**
  N Argentina

TINAMOTIS
Tinamotis pentlandii (Puna Tinamou)
  C Peru, W Argentina, N Chile
Tinamotis ingoufi (Patagonian Tinamou)
  S Chile, S Argentina

SPHENISCIFORMES

7. SPHENISCIDAE (PENGUINS)

APTENODYTES
Aptenodytes patagonicus (King Penguin)
A.p.patagonicus
Staten, S Georgia, Falkland Is
A.p.halli
Macquarie, Kerguelen, Crozet, Marion Is
Aptenodytes forsteri (Emperor Penguin)
Antarctica

PYGOSCELIS
Pygoscelis papua (Gentoo Penguin)
P.p.papua
Falkland Is, S Georgia I
P.p.taeniata
Macquarie, Heard, Kerguelen, Marion Is
P.p.ellsworthi
S Orkney Is, Deception I, S Shetland Is
Pygoscelis adeliae (Adelie Penguin)
Antarctica, S Orkney Is, S Shetland Is
Pygoscelis antarctica (Bearded Penguin)
Antarctic Ocean, S Atlantic

EUDYPTES
Eudyptes pachyrhynchus (Victoria Penguin)
E.p.pachyrhynchus
New Zealand, Stewart I
E.p.atratus
Snares I
Eudyptes robustus (Snares I.Penguin)
Snares I
Eudyptes sclateri (Big-crested Penguin)
Auckland Is, Antipodes Is
Eudyptes chrysocome (Rockhopper Penguin)
7.1
E.c.chrysocome
Tierra del Fuego, Falkland Is
E.c.filholi
Kerguelen I & islands S of New Zealand
E.c.moseleyi
Tristan da Cunha, St Paul, Amsterdam Is
Eudyptes schlegeli (Royal Penguin)
Macquarie I
Eudyptes chrysolophus (Macaroni Penguin)
S Georgia, Kerguelen, Falkland, S Orkney,
S Shetland Is

MEGADYPTES
Megadyptes antipodes (Yellow-eyed Penguin)
South I, New Zealand & southern islands

EUDYPTULA
Eudyptula minor (Little Penguin)
E.m.novaehollandiae
Tasmania, S Australian islands
E.m.minor
New Zealand, Stewart I
E.m.chathamensis
Indian Ocean Is, Chatham I

E.m.variabilis
Wellington, New Zealand
Eudyptula albosignata (White-flippered Penguin)
E South I, New Zealand

SPHENISCUS
Spheniscus demersus (Jackass Penguin)
coast of S Africa & islands
Spheniscus humboldti (Humboldt Penguin)
W coast of S America & islands
Spheniscus magellanicus (Magellanic Penguin)
S South America & islands
Spheniscus mendiculus (Galapagos Penguin)
Galapagos Is

GAVIIFORMES

8. GAVIIDAE (DIVERS)

GAVIA
Gavia stellata (Red-throated Diver)
Holarctic, Circumpolar >> S Europe, S
China, Florida
Gavia arctica (Black-throated Diver)
G.a.arctica
N Europe, N Russia
G.a.suschkini
W Siberia to EC Asia
G.a.viridigularis
NE Siberia to W Alaska
Gavia pacifica (Pacific Diver)
Arctic of N America, Alaska >> W North
America
Gavia immer (Great Northern Diver)
G.i.immer
N North America, N Europe >>
N Mexico, Florida, North Sea
G.i.elasson
W Canada, N Dakota
Gavia adamsii (White-billed Diver)
Arctic, E Siberia, N North America >>
Norway, S Alaska

PODICIPEDIFORMES

9. PODICIPEDIDAE (GREBES)

TACHYBAPTUS
Tachybaptus ruficollis (Little Grebe)
T.r.ruficollis
Europe to Urals, NW Africa
T.r.iraquensis
Iraq, SW Iran
T.r.capensis
Caucasus to Burma, Ghana to Ethiopia &
Cape Province
T.r.poggei
E China, Hainan I, Malaysia
T.r.kunikyonis
C Riukiu Is

T.r.philippensis
Taiwan, Borneo, Philippine Is
T.r.cotabato
Mindanao I
T.r.javanicus
Java
T.r.vulcanorum
Bali I to Timor I
T r.tricolor
Sulawesi to Solomon Is
Tachybaptus novaehollandiae (Australian Dabchick)
New Guinea, Australia, New Caledonia
Tachybaptus pelzelnii (Madagascar Little Grebe)
Madagascar
Tachybaptus rufolavatus (Delacour's Little Grebe)
Madagascar

PODILYMBUS
Podilymbus podiceps (Pied-billed Grebe)
P.p.podiceps
Canada & USA >> Panama & Cuba
P.p.antillarum
Gtr. & Lesser Antilles
P.p.antarcticus
Colombia & Venezuela to S Argentina
Podilymbus gigas (Atitlan Grebe)
Lake Atitlan, Guatemala

ROLLANDIA
Rollandia rolland (White-tufted Grebe)
R.r.morrisoni
C Peru
R.r.chilensis
S Brazil & S Peru to Tierra del Fuego
R.r.rolland
Falkland Is
Rollandia micropterum (Short-winged Grebe)
Lake Titicaca, Bolivia

PODICEPS
Podiceps major (Great Grebe)
P.m.major
Amazonia to Chile
P.m.navasi
S Patagonia
Podiceps poliocephalus (Hoary-headed Grebe)
Australia, Tasmania
Podiceps rufopectus (New Zealand Dabchick)
New Zealand
Podiceps dominicus (Least Grebe)
P.d.brachypterus
S Texas to Panama
P.d.bangsi
S Baja California
P.d.dominicus
Gtr.Antilles, S Mexico, Argentina
P.d.speciosus
Colombia

P.d.eisenmanni
W Ecuador
Podiceps grisegena (Red-necked Grebe)
P.g.grisegena
Holarctic, Scandinavia to Siberia >> N Africa & Iran
P.g.holboellii
N America, NE Asia >> China, Japan, S USA
Podiceps cristatus (Great Crested Grebe)
P.c.cristatus
Europe to China, India, N Africa
P.c.infuscatus
Senegal to Ethiopia & Cape Province
P.c.australis
S Australia, Tasmania, South I, New Zealand
Podiceps auritus (Slavonian Grebe)
N America, N Europe, N Asia >> Mediterranean, E China, S USA
Podiceps nigricollis (Black-necked Grebe)
P.n.nigricollis
Europe, Russia
P.n.gurneyi
S Angola to Ethiopia & Cape Province
P.n.californicus
W Canada, W USA >> Guatemala
P.n.andinus e?
C Colombia
Podiceps occipitalis (Silvery Grebe)
P.o.juninensis
Peru, Bolivia
P.o.occipitalis
N W Argentina to Tierra del Fuego
Podiceps taczanowskii (Puna Grebe)
Lake Junin, Peru
Podiceps gallardoi (Hooded Grebe)
N W Argentina

AECHMOPHORUS   9.1
Aechmophorus occidentalis (Western Grebe)
A.o.occidentalis
W Canada, N W USA >> SW USA
A.o.ephemeralis
W Mexico
Aechmophorus clarkii (Mexican Grebe)
A.c.clarkii
N & C Mexico
A.c.transitionalis
S New Mexico

PROCELLARIIFORMES

10. DIOMEDEIDAE (ALBATROSSES)

DIOMEDEA
Diomedea exulans (Wandering Albatross)
D.e.exulans
Southern Ocean, S Georgia I
D.e.dabbenena
Tristan da Cunha, Gough I
D.e.chionoptera
S Indian Ocean, Macquarie I

Diomedea epomophora **(Royal Albatross)**
D.e.sanfordi
South I, New Zealand, Chatham I
D.e.epomophora
Auckland I to S Australia, S South America
Diomedea irrorata **(Waved Albatross)**
Peru, Galapagos Is
Diomedea albatrus **(Short-tailed Albatross)**
N Pacific, Hawaii Is, Marshall Is
Diomedea nigripes **(Black-footed Albatross)**
N Pacific, Hawaii Is, Marshall Is
Diomedea immutabilis **(Laysan Albatross)**
NW Hawaii Is
Diomedea melanophris **(Black-browed
Albatross)**
D.m.impavida
islands S of New Zealand
D.m.melanophris
S South America
Diomedea bulleri **(Buller's Albatross)**
S Pacific, Snares I
Diomedea amsterdamensis **(Amsterdam
Island Albatross)** 10.1
Amsterdam I
Diomedea cauta **(Shy Albatross)**
D.c.cauta
Bass Strait, Albatross I
D.c.eremita
Chatham I
D.c.salvini
S South America, Snares I, Bounty I
Diomedea chlororhynchos **(Yellow-nosed
Albatross)**
Tristan da Cunha, Gough I
Diomedea chrysostoma **(Grey-headed
Albatross)**
S Georgia, Marion, Crozet, Kerguelen Is

PHOEBETRIA
Phoebetria fusca **(Sooty Albatross)**
Tristan da Cunha, Gough, St Paul Is
Phoebetria palpebrata **(Light-mantled Sooty
Albatross)**
P.p.palpebrata
South Georgia I
P.p.huttoni
Kerguelen, Crozet, Heard, Antipodes,
Macquarie Is

11. PROCELLARIIDAE (PETRELS, SHEARWATERS)

MACRONECTES
Macronectes giganteus **(Giant Petrel)**
Southern Ocean
Macronectes halli **(Hall's Giant Petrel)**
Southern Ocean (N of M.giganteus)

FULMARUS
Fulmarus glacialis **(Northern Fulmar)**
F.g.glacialis
Greenland, Br Isles, Iceland, Norway

F.g.rodgersii
NE Asian Is, Bering Sea
F.g.minor
N Greenland
Fulmarus glacialoides **(Southern Fulmar)**
Antarctic, New Zealand

THALASSOICA
Thalassoica antarctica **(Antarctic Petrel)**
Antarctic

DAPTION
Daption capense **(Pintado Petrel)**
D.c.capense
Antarctica, S Georgia I
D.c.australe
Snares, Antipodes, Bounty Is

PAGODROMA
Pagodroma nivea **(Snow Petrel)**
P.n.nivea
Antarctica, S Georgia I, S Orkney Is
P.n.major
Adelie Land

LUGENSA 11.1
Lugensa brevirostris **(Kerguelen Petrel)**
Gough, Kerguelen, Marion Is

PTERODROMA 11.1
Pterodroma macroptera **(Great-winged Petrel)**
P.m.macroptera
Tristan da Cunha, Crozet, Marion,
Kerguelen Is
P.m.gouldi
SW Australia, North I, New Zealand
Pterodroma aterrima **(Mascarene Black
Petrel)** e?
Reunion I
Pterodroma lessonii **(White-headed Petrel)**
Auckland, Kerguelen, Macquarie, Antipodes
Is
Pterodroma hasitata **(Black-capped Petrel)**
Dominica, Caribbean
Pterodroma cahow **(Cahow)**
Bermuda
Pterodroma incerta **(Schlegel's Petrel)**
Tristan da Cunha, S Atlanatic
Pterodroma becki **(Solomon Is Petrel)** e?
Solomon Is
Pterodroma alba **(Phoenix Petrel)**
Christmas, Phoenix, Tonga, Marquesas Is
Pterodroma inexpectata **(Peale's Petrel)**
New Zealand, Chatham Is
Pterodroma solandri **(Solander's Petrel)**
Lord Howe, Austral Is
Pterodroma ultima **(Murphy's Petrel)**
Tuamotu, Austral Is, S Pacific
Pterodroma neglecta **(Kermadec Petrel)**
P.n.neglecta
Lord Howe, Kermadec, Tuamotu, Austral Is
P.n.juana
Juan Fernandez Is

Pterodroma magentae (Taiko)
Chatham Is
Pterodroma arminjoniana (Trinidade Petrel)
P.a.arminjoniana
Mauritius, S Trinidade I
P.a.paschae
Easter I
Pterodroma heraldica (Tonga Petrel)
Tonga, Tuamotu, Marquesas Is
Pterodroma mollis (Soft-plumaged Petrel)
Tristan da Cunha, Gough, St.Pauls,
Kerguelen Is
Pterodroma feae (Cape Verde Petrel)
Cape Verde Is
Pterodroma madeira (Madeira Petrel)
Madeira
Pterodroma baraui (Barau's Petrel)
Reunion
Pterodroma phaeopygia (Hawaiian Petrel)
P.p.phaeopygia
Galapagos
P.p.sandwichensis
Hawaiian Is
Pterodroma externa (White-necked Petrel)
P.e.externa
Masafuera I
P.e.tristani
Tristan da Cunha
Pterodroma cervicalis (Kermadec White-necked Petrel)
Kermadec I
Pterodroma cookii (Cook's Petrel)
P.c.cookii
Little Barrier I, (New Zealand)
P.c.orientalis
W South America
Pterodroma defilippiana (Juan Fernandez Petrel)
Juan Fernandez Is
Pterodroma leucoptera (White-winged Petrel)
P.l.leucoptera
Pt Stephens (N Sth Wales)
P.l.masafuerae
Masafuera I
Pterodroma brevipes (Collared Petrel)
New Caledonia, New Hebrides, Fiji Is
Pterodroma hypoleuca (Bonin Petrel)
Bonin Is, W Hawaiian Is
Pterodroma nigripennis (Black-winged Petrel)
Lord Howe, Kermadec Is
Pterodroma axillaris (Chatham Is Petrel)
Chatham Is
Pterodroma longirostris (Stejneger's Petrel)
Masafuera
Pterodroma pycrofti (Pycroft's Petrel)
New Zealand

PSEUDOBULWARIA   11.1
Pseudobulwaria macgillivrayi (Macgillivray's Petrel)
Fiji Is
Pseudobulwaria rostrata (Tahiti Petrel)

P.r.rostrata
Marquesas Is, Society Is
P.r.trouessarti
New Caledonia

HALOBAENA
Halobaena caerulea (Blue Petrel)
Kerguelen, Crozet, Falkland Is

PACHYPTILA
Pachyptila vittata (Broad-billed Prion)
P.v.vittata
Tristan da Cunha, S Georgia, Chatham Is,
SW New Zealand
P.v.macgillivrayi
Amsterdam, St Pauls Is
Pachyptila salvini (Salvin's Prion)
Marion I
Pachyptila desolata (Dove Prion)
P.d.desolata
Kerguelen I
P.d.peringueyi
S Africa
P.d.alexanderi
SW Australia
P.d.macquariensis
Macquarie I
P.d.alter
Auckland, S Orkney Is
P.d.banksi
S Georgia, S Orkney Is
Pachyptila belcheri (Slender-billed Prion)
Kerguelen I, Falkland Is
Pachyptila turtur (Fairy Prion)
SE Australia, New Zealand, Falkland Is
Pachyptila crassirostris (Thick-billed Prion)
P.c.eatoni
Kerguelen, Heard, Antipodes Is
P.c.crassirostris
Bounty I
P.c.pyramidalis
Chatham Is

BULWERIA
Bulweria bulwerii (Bulwer's Petrel)
Pacific & Atlantic Oceans
Bulweria fallax (Jouanin's Petrel)
Indian Ocean

PROCELLARIA
Procellaria cinerea (Grey Shearwater)
Tristan da Cunha, Gough, Kerguelen,
Macquarie Is
Procellaria aequinoctialis (White-chinned Petrel)
P.a.aequinoctialis
S Georgia, Falkland, Crozet, Kerguelen Is
P.a.conspicillata
Inaccessible I, Tristan da Cunha,
P.a.steadi
Auckland, Antipodes, Campbell, Macquarie
Is

Procellaria parkinsoni (Black Petrel)
New Zealand
Procellaria westlandica (Westland Petrel)
South I, New Zealand

CALONECTRIS
Calonectris leucomelas (White-faced Shearwater)
NW Pacific Ocean
Calonectris diomedea (Cory's Shearwater)
C.d.diomedea
Mediterranean Islands
C.d.borealis
Portugal, Canary Is, Madeira, Azores Is
C.d.edwardsii
Cape Verde Is
C.d.flavirostris
W Indian Ocean, Kerguelen I

PUFFINUS
Puffinus creatopus (Pink-footed Shearwater)
E Pacific, Juan Fernandez Is
Puffinus carneipes (Pale-footed Shearwater)
P.c.carneipes
SW Australia
P.c.hullianus
Lord Howe I, New Zealand
Puffinus gravis (Greater Shearwater)
Tristan da Cunha, Gough I, Falkland Is
Puffinus pacificus (Wedge-tailed Shearwater)
P.p.chlororhynchus
W Australian islands
P.p.pacificus
Kermadec I
P.p.cuneatus
Bonin Is, Hawaiian Is
P.p.royanus
E Australian islands
Puffinus bulleri (Grey-backed Shearwater)
islands off New Zealand
Puffinus griseus (Sooty Shearwater)
South I, New Zealand to Chile, Falkland Is
Puffinus tenuirostris (Short-tailed Shearwater)
SE Australia, Tasmania
Puffinus heinrothi (Heinroth's Shearwater)
New Britain
Puffinus nativitatis (Christmas Island Shearwater)
Hawaii, Christmas, Tuamotu Is
Puffinus puffinus (Manx Shearwater)
P.p.puffinus
E North Atlantic
P.p.yelkouan
E Mediterranean
P.p.mauretanicus
W Mediterranean
P.p.newelli
Hawaii Is
Puffinus gavia (Fluttering Shearwater)
islands off New Zealand
Puffinus huttoni (Hutton's Shearwater)
South I, New Zealand

Puffinus opisthomelas (Black-vented Shearwater)
W North America
Puffinus auricularis (Townsend's Shearwater)
W Mexico
Puffinus assimilis (Little Shearwater)
P.a.baroli
Azores, Madeira, Canary Is
P.a.boydi
Cape Verde Is
P.a.elegans
Tristan da Cunha, Gough I
P.a.tunneyi
islands off SW Australia
P.a.assimilis
Lord Howe, Norfolk Is
P.a.haurakiensis
New Zealand
P.a.munda
Chatham I
P.a.kermadecensis
Kermadec I
Puffinus lherminieri (Audubon's Shearwater)
P.l.bailloni
Mauritius, Reunion, Seychelles
P.l.temptator
Moheli I
P.l.bannermani
Bonin Is
P.l.nugax
New Hebrides
P.l.dichrous
Palau, Phoenix, Christmas Is
P.l.polynesiae
Samoa, Society, Marquesas, Tuamotu Is
P.l.subalaris
Galapagos Is
P.l.lherminieri
West Indies, Bahama, Bermuda Is
P.l.persicus
NW India, Iran
P.l.loyemilleri
Costa Rica to Guyana

12. HYDROBATIDAE

OCEANITES
Oceanites oceanicus (Wilson's Storm Petrel)
O.o.oceanicus
N Antarctic Is
O.o.exasperatus
S Antarctic Is
Oceanites gracilis (Elliot's Storm Petrel)
O.g.gracilis
W South America
O.g.galapagoensis
Galapagos Is

GARRODIA
Garrodia nereis (Grey-backed Storm Petrel)
S Georgia, Falkland, Kereguelen, Chatham Is

PELAGODROMA
Pelagodroma marina (White-faced Storm Petrel)(Frigate Petrel)
P.m.hypoleuca
Madeira, Canary, Cape Verde Is,
P.m.marina
Tristan da Cunha
P.m.dulciae
W & S Australia
P.m.maoriana
New Zealand, Auckland, Chatham Is
P.m.albiclunis
Kermadec Is

FREGETTA
Fregetta grallaria (White-bellied Storm Petrel)
F.g.royana
Lord Howe I
F.g.titan
Austral I
F.g.grallaria
Juan Fernandez Is
Fregetta tropica (Black-bellied Storm Petrel)
F.t.tropica
Tristan da Cunha
F.t.melanogaster
Kerguelen, Crozet Is
F.t.lineata
Samoa Is

NESOFREGETTA
Nesofregetta fuliginosa (White-throated Storm Petrel)
N.f.fuliginosa
Christmas I, Marquesas Is, Fiji Is, New Hebrides
N.f.moestissima
Samoa Is

HYDROBATES
Hydrobates pelagicus (British Storm Petrel)
N E Atlantic, Mediterranean

HALOCYPTENA
Halocyptena microsoma (Least Storm Petrel)
W Mexico to Ecuador

OCEANODROMA
Oceanodroma tethys (Galapagos Storm Petrel)
O.t.tethys
Galapagos Is
O.t.kelsalli
coast of Peru
Oceanodroma castro (Madeiran Storm Petrel)
O.c.cryptoleucura
N Pacific, Hawaiian Is
O.c.bangsi
S Pacific, Cocos Is, Galapagos Is
O.c.castro
E Atlantic islands, St Helena I
Oceanodroma leucorhoa (Leach's Storm Petrel)

O.l.leucorhoa
N Pacific & N Atlantic coasts
O.l.beali
SE Alaska to California
O.l.kaedingi
Guadelupe I, (Baja California)
Oceanodroma markhami (Sooty Storm Petrel)
O.m.markhami
W South America
O.m.owstoni
Hawaiian Is
Oceanodroma matsudairae (Matsudaira's Storm Petrel)
Volcano I
Oceanodroma tristrami (Tristram's Storm Petrel)
Japan, Laysan I, Midway I
Oceanodroma monorhis (Swinhoe's Storm Petrel)
O.m.monorhis
W North Pacific
O.m.socorroensis
E Nth Pacific, Baja California
Oceanodroma homochroa (Ashy Storm Petrel)
California, Santa Barbara I
Oceanodroma hornbyi (Ringed Storm Petrel)
W South America
Oceanodroma furcata (Fork-tailed Storm Petrel)
N Pacific Ocean
Oceanodroma melania (Black Storm Petrel)
E Pacific, Baja California

13. PELECANOIDIDAE (DIVING PETRELS)

PELECANOIDES
Pelecanoides garnotii (Peruvian Diving Petrel)
coasts of Peru & Chile
Pelecanoides magellani (Magellan Diving Petrel)
S Chile, Cape Horn
Pelecanoides georgicus (Georgian Diving Petrel)
P.g.georgicus
S Georgia I
P.g.novus
Macquarie I
Pelecanoides urinatrix (Common Diving Petrel)
P.u.berard
Falkland Is
P.u.coppingeri
S Chile
P.u.dacunhae
Tristan da Cunha
P.u.elizabethae
Gough I
P.u.urinatrix
SE Australia, Tasmania, New Zealand
P.u.chathamensis
Chatham I, Antipodes Is, Snares I

P.u.exsul
Crozet, Marion, Heard Is, Kerguelen, Auckland Is

PELECANIFORMES

14. PHAETHONTIDAE (TROPIC BIRDS)

PHAETHON
Phaethon aethereus (Red-billed Tropic Bird)
P.a.limatus
Tower I, Galapagos Is
P.a.mesonauta
Daphne I, Lesser Antilles to Cape Verde Is
P.a.aethereus
Fernando Noronha I, St Helena I, Ascension I
P.a.indicus
Persian Gulf, Gulf of Aden
Phaethon rubricauda (Red-tailed Tropic Bird)
P.r.rubricauda
Mauritius I, Aldabra I
P.r.westralis
Christmas I, Cocos Keeling Is, NW Australia
P.r.roseotincta
Lord Howe I, Norfolk I, Kermadec Is
P.r.melanorhynchos
Society, Palmerston Is
P.r.rothschildi
Bonin, Hawaiian Is
Phaethon lepturus (White-tailed Tropic Bird)
P.l.catesbyi
West Indies, Bahama, Bermuda I
P.l.ascensionis
Fernando Noronha, Ascension Is, Gulf of Guinea
P.l.lepturus
Mascarene, Seychelles, Andaman Is
P.l.fulvus
Christmas I, (Java)
P.l.dorotheae
SW Pacific islands

15. PELECANIDAE (PELICANS)

PELECANUS
Pelecanus onocrotalus (Eastern White Pelican)
S Europe, Africa, C Asia
Pelecanus roseus (White Pelican)
China, Persian Gulf
Pelecanus rufescens (Pink-backed Pelican)
C & S Africa
Pelecanus philippensis (Grey Pelican)
S Asia, Iran to Philippine Is
Pelecanus crispus (Dalmatian Pelican)
SE Europe to China
Pelecanus conspicillatus (Australian Pelican)
Australia, Tanimbar Is, New Guinea
Pelecanus erythrorhynchos (American White Pelican)
N & C America, West Indies
Pelecanus occidentalis (Brown Pelican)

P.o.occidentalis
West Indies
P.o.carolinensis
coasts of S Carolina to Venezuela
P.o.californicus
coast of California, W Mexico
P.o.murphyi
coasts of W Colombia & Ecuador
P.o.urinator
Galapagos Is
P.o.thagus
coasts of Peru & Chile

16. SULIDAE (GANNETS, BOOBIES)

MORUS
Morus bassanus (Northern Gannet)
E Canada, Iceland, Br.Isles
Morus capensis (Cape Gannet)
Southern Africa
Morus serrator (Australian Gannet)
M.s.serrator
S Australian coast, Tasmania, Bass Str
M.s.rex
North I, New Zealand

PAPASULA 16.1
Papasula abbotti (Abbott's Booby)
Christmas I, (Java)

SULA
Sula nebouxii (Blue-footed Booby)
S.n.nebouxii
California to Peru
S.n.excisa
Galapagos Is
Sula variegata (Peruvian Booby)
coasts of Peru & Chile
Sula dactylatra (Blue-faced Booby)
S.d.dactylatra
Caribbean, Bahama Is, Ascension I
S.d.melanops
W Indian Ocean, Seychelles
S.d.californica
W Mexico Islands
S.d.granti
Galapagos Is
S.d.bedouti
Christmas I, (Java), Lesser Sunda Is
S.d.personata
N Australian & Pacific islands
Sula sula (Red-footed Booby)
S.s.sula
Caribbean Is & S Trinidade I, (Brazil)
S.s.rubripes
Indian Ocean & Pacific islands
S.s.websteri
Galapagos Is
Sula leucogaster (Brown Booby)
S.l.leucogaster
Caribbean & Atlantic islands
S.l.brewsteri
coast of California & W Mexico

S.l.etesiaca
C American & Colombian islands
S.l.plotus
Indian & W Pacific Ocean islands & Java
S.l.nesiotes
Clipperton I

17. PHALACROCORACIDAE (CORMORANTS)

PHALACROCORAX

Phalacrocorax auritus (Double-crested Cormorant)
P.a.cincinatus
Alaska to Oregon
P.a.albociliatus
California to W Mexico
P.a.auritus
C & E Canada, E USA
P.a.floridanus
S USA to Bahama Is & Honduras
Phalacrocorax olivaceus (Olivaceous Cormorant)
P.o.mexicanus
S USA to Nicaragua & Cuba
P.o.olivaceus
Panama to Patagonia
P.o.chancho
Sonora(NW Mexico)
P.o.hornensis
Tierra del Fuego
Phalacrocorax sulcirostris (Little Black Cormorant)
P.s.territori
Malaysia to New Guinea & N Australia
P.s.sulcirostris
E & S Australia, Tasmania
P.s.purpuragula
New Zealand
Phalacrocorax carbo (Great Cormorant)
P.c.carbo
E Canada to Br Isles
P.c.maroccanus
N Africa
P.c.lugubris
NE Africa
P.c.sinensis
C Europe to India & China
P.c.hanedae
Japan
P.c.novaehollandiae
Australia & Tasmania
P.c.steadi
New Zealand, Chatham I
Phalacrocorax lucidus (White-breasted Cormorant)
Cape Verde Is, Senegal to E Africa
Phalacrocorax fuscicollis (Indian Cormorant)
India, Sri Lanka, Burma
Phalacrocorax capensis (Cape Cormorant)
coasts of South Africa

Phalacrocorax nigrogularis (Socotra Cormorant)
S Red Sea, Persian Gulf
Phalacrocorax neglectus (Bank Cormorant)
coasts of S Africa
Phalacrocorax capillatus (Japanese Cormorant)
NE Asia to China, Japan
Phalacrocorax penicillatus (Brandt's Cormorant)
S Alaska to NW Mexico
Phalacrocorax aristotelis (Shag)
P.a.aristotelis
Iceland & Lapland to Portugal
P.a.desmarestii
C Mediterranean
P.a.riggenbachi
coast of Morocco
Phalacrocorax pelagicus (Pelagic Cormorant)
P.p.pelagicus
N Pacific islands
P.p.resplendens
British Columbia to Mexico
Phalacrocorax urile (Red-faced Cormorant)
Bering Sea to Taiwan
Phalacrocorax magellanicus (Magellan Cormorant)
Tierra del Fuego, Falkland Is
Phalacrocorax bougainvillei (Guanay Cormorant)
coasts of Peru & Chile
Phalacrocorax featherstoni (Chatham Is Cormorant)
Chatham I
Phalacrocorax varius (Pied Cormorant)
Australia, Tasmania, New Zealand
Phalacrocorax fuscescens (Black-faced Cormorant)
S Australia, Tasmania
Phalacrocorax carunculatus (Rough-faced Cormorant)
P.c.carunculatus
South I, New Zealand
P.c.chalconotus
Stewart I
P.c.onslowi
Chatham I
Phalacrocorax campbelli (Campbell Is Cormorant)
P.c.campbelli
Campbell I
P.c.colensoi
Auckland I
P.c.ranfurlyi
Bounty I
Phalacrocorax verrucosus (Kerguelen Cormorant)
Kerguelen, Marion Is
Phalacrocorax gaimardi (Red-legged Cormorant)
P.g.gaimardi
coast of Peru & Chile

P.g.cirriger
  coast of S Argentina
Phalacrocorax punctatus (Spotted Cormorant)
  P.p.punctatus
    New Zealand
  P.p.oliveri
    Stewart I
Phalacrocorax atriceps (Blue-eyed Cormorant)
  P.a.atriceps
    S South America
  P.a.nivalis
    Heard I
  P.a.gaini
    Antarctica
Phalacrocorax georgianus (S Georgia Cormorant)
  S Georgia I
Phalacrocorax albiventer (King Cormorant)
  P.a.albiventer
    Patagonia, Falkland Is
  P.a.melanogenis
    Crozet Is
  P.a.purpurascens
    Macquarie I
Phalacrocorax melanoleucos (Little Pied Cormorant
  P.m.melvillensis
    Malaysia to N Australia
  P.m.melanoleucos
    S Australia, Tasmania
  P.m.brevirostris
    New Zealand
  P.m.brevicauda
    Rennell I
Phalacrocorax africanus (Reed Cormorant)
  P.a.africanus
    Senegal to Egypt & Cape Province
  P.a.pictilis
    Madagascar
Phalacrocorax coronatus (Crowned Cormorant)   17.1
    Namibia
Phalacrocorax niger (Javanese Cormorant)
    India to Java & Borneo
Phalacrocorax pygmeus (Pygmy Cormorant)
    C Europe, N Africa to C Asia

NANNOPTERUM
Nannopterum harrisi (Flightless Cormorant)
    Galapagos Is

18. ANHINGIDAE (ANHINGAS)

ANHINGA
Anhinga rufa (African Darter)
  A.r.rufa
    Senegal to Cape Province
  A.r.vulsini
    Madagascar
  A.r.chantrei
    Tigris & Euphrates rivers

A.r.papua
    New Guinea
Anhinga melanogaster (Indian Darter)
    India to Philippine Is, Sulawesi
Anhinga novaehollandiae (Australian Darter)
    New Guinea, Australia
Anhinga anhinga (American Darter) (Anhinga)
  A.a.anhinga
    Brazil, Argentina, E Bolivia
  A.a.leucogaster
    SE USA to Colombia

19. FREGATIDAE (FRIGATE BIRDS)

FREGATA
Fregata aquila (Ascension Frigate Bird)
    Ascension I
Fregata andrewsi (Christmas I Frigate Bird)
    E Indian Ocean, Christmas I
Fregata magnificens (Magnificent Frigate Bird)
  F.m.magnificens
    Galapagos Is
  F.m.rothschildi
    SE USA, C America, W.Indies
  F.m.lowei
    Cape Verde Is
Fregata minor (Great Frigate Bird)
  F.m.aldabrensis
    Seychelles, Aldabra I
  F.m.minor
    Christmas I, Cocos Keeling Is
  F.m.peninsulae
    Raine I (Queensland)
  F.m.palmerstoni
    C & S Pacific
  F.m.strumosa
    Hawaiian Is
  F.m.ridgwayi
    Galapagos Is
  F.m.nicolli
    S Trinidad I (Brazil)
Fregata ariel (Lesser Frigate Bird)
  F.a.iredalei
    W Indian Ocean, Aldabra I
  F.a.ariel
    Philippine Is to N Australia, S Pacific
  F.a.trinitatis
    S Trinidade I (Brazil)

CICONIIFORMES

20. ARDEIDAE (HERONS, BITTERNS)   20.1

SYRIGMA
Syrigma sibilatrix (Whistling Heron)
  S.s.fostersmithi
    NW Venezuela, NE Colombia
  S.s.sibilatrix
    S Brazil, Paraguay, Uruguay

PILHERODIUS
Pilherodius pileatus (Capped Heron)
  E Panama to E Peru & S Brazil

ARDEA
Ardea cinerea (Grey Heron)
  A.c.cinerea
    Europe to E Asia, W China, Africa
  A.c.jouyi
    Korea, Japan to Sumatra & Java
  A.c.monicae
    Banc D'Arguin, Mauretania
  A.c.firasa
    Madagascar, Aldabra, Comoro Is
Ardea herodias (Great Blue Heron)
  A.h.fannini
    W Nth America
  A.h.herodias
    NE Nth America >> C America
  A.h.wardi
    SE USA
  A.h.cognata
    Galapagos Is
  A.h.occidentalis
    SE USA, Cuba, Jamaica
Ardea cocoi (Cocoi Heron)
  S.America
Ardea pacifica (White-necked Heron)
  Australia, Tasmania
Ardea melanocephala (Black-headed Heron)
  Gambia to Sudan & Cape Province
Ardea humbloti (Malagasy Heron)
  E Madagascar, Aldabra I
Ardea imperialis (Imperial Heron)
  Sikkim to C Burma
Ardea sumatrana (Dusky-grey Heron)
  A.s.sumatrana
    Burma, Malaysia to Philippine Is, New
    Guinea
  A.s.mathewsae
    N Australia
Ardea goliath (Goliath Heron)
  Senegal to Sudan & Cape Province
Ardea purpurea (Purple Heron)
  A.p.purpurea
    S Europe to Iran, Africa
  A.p.madagascariensis
    Madagascar
  A.p.manilensis
    India, China, Gtr Sunda Is

EGRETTA
Egretta alba (Great Egret)
  E.a.alba
    SE Europe, N Asia >> N Africa, India, S
    China
  E.a.modestus
    India to Japan & Australia
  E.a.melanorhynchos
    Senegal to Sudan & Cape Province
  E.a.egretta
    S USA to Patagonia
Egretta rufescens (Reddish Egret)

  E.r.rufescens
    S USA, Mexico, Cuba, Jamaica, Hispaniola
  E.r.dickeyi
    N W Mexico, Baja California
Egretta picata (Pied Heron)
  Sulawesi to New Guinea, N Australia
Egretta vinaceigula (Slaty Egret)
  Transvaal
Egretta ardesiaca (Black Heron)
  Senegal to Sudan & Natal
Egretta tricolor (Louisiana Heron)
  E.t.ruficollis
    SE USA to N W South America
  E.t.tricolor
    Cayenne, Surinam, NE Brazil
Egretta intermedia (Intermediate Egret)
  E.i.brachyrhyncha
    Sudan to Cape Province
  E.i.intermedia
    S. India to Japan, Gtr Sunda Is
  E.i.plumifera
    S.Moluccas, New Guinea, Australia
Egretta novaehollandiae (White-faced Egret)
  E.n.novaehollandiae
    Lombok I to Timor I, Australia, New Zealand
  E.n.parryi
    NW Australia
Egretta caerulea (Little Blue Heron)
  S USA to C Sth America
Egretta thula (Snowy Egret)
  W & SE USA to N Chile, N Argentina
Egretta garzetta (Little Egret)
  E.g.garzetta
    S Europe to Japan & Africa
  E.g.gularis
    W Africa
  E.g.schistacea
    NE Africa, India
  E.g.dimorpha
    Madagascar, Pemba I, Aldabra I
  E.g.nigripes
    Sunda I, Philippine Is to
    New Guinea, Australia
  E.g.immaculata
    N & E Australia
Egretta eulophotes (Swinhoe's Egret)
  S & C China, Taiwan, Sulawesi
Egretta sacra (Reef Heron)
  E.s.sacra
    SE Asia to Australia, New Zealand
  E.s.albolineata
    New Caledonia

BUBULCUS
Bubulcus ibis (Cattle Egret)
  B.i.ibis
    Spain to Iran, N & C Africa, E Nth America,
    C & N Sth America
  B.i.coromandus
    India to S Japan, Philippine Is, Moluccas
  B.i.seychellarum
    Seychelles

ARDEOLA
Ardeola ralloides (Squacco Heron)
 S Europe to Iran & Africa
Ardeola grayii (Indian Pond Heron)
 Iran to India, Burma, SriLanka
Ardeola bacchus (Chinese Pond Heron)
 China to Malaysia & Borneo
Ardeola speciosa (Javanese Pond Heron)
 A.s.continentalis
  C Thailand, S Indochina
 A.s.speciosa
  W Borneo, Sumatra, Java, Sunda Is
Ardeola idae (Malagasy Pond Heron)
 Aldabra, Madagascar >> E Africa
Ardeola rufiventris (Rufous-bellied Heron)
 S Angola to Tanzania & Cape Province

BUTORIDES
Butorides striatus (Green-backed{Striated}Heron)
 B.s.anthonyi
  SW,USA, W Mexico
 B.s.frazari
  S Baja California
 B.s.virescens
  E Nth America, W Indies >> Panama
 B.s.bahamensis
  Bahama Is
 B.s.striatus
  E Panama to N Argentina, Bolivia
 B.s.sundevalli
  Galapagos Is
 B.s.brevipes
  Somalia, Red Sea coast
 B.s.atricapillus
  Senegal to Sudan & Cape Province
 B.s.rutenbergi
  Madagascar
 B.s.rhizophorae
  Comoro Is
 B.s.degens
  Seychelles
 B.s.crawfordi
  Assumption, Aldabra Is
 B.s.albolimbatus
  Maldive Is
 B.s.chloriceps
  India, Sri Lanka
 B.s.spodiogaster
  Andaman Is, Nicobar Is, W Sumatran Is
 B.s.carcinophilus
  Taiwan, Sulawesi, Philippine Is
 B.s.amurensis
  NE Asia >> Philippines Is, Sunda Is
 B.s.steini
  Lesser Sunda Is
 B.s.actophilus
  China, N Indochina >> Indonesia
 B.s.javanicus
  E India to Philippine Is, Sunda Is
 B.s.moluccarum
  S Moluccas

 B.s.solomonensis
  Solomon Is
 B.s.papuensis
  NW New Guinea
 B.s idenburgi
  interior of N New Guinea
 B.s.littleri
  S New Guinea, N Queensland
 B.s.macrorhynchus
  S New Guinea, E Australia
 B.s.cinerea
  coast of W Australia
 B.s.stagnatilis
  N & NW Australia
 B.s.patruelis
  Tahiti Is
 B.s.rogersi
  W C Australia

AGAMIA
Agamia agami (Chestnut-bellied Heron)
 E Mexico to Brazil & N Bolivia

NYCTICORAX
Nycticorax violaceus (Yellow-crowned Night Heron)
 N.v.violaceus
  EC USA, E Central America
 N.v.bancrofti
  W Baja California, W Indies
 N.v.gravirostris
  Socorro I
 N.v.caliginis
  Panama to Peru
 N.v.cayennensis
  S Colombia to Peru & S Brazil
 N.v.pauper
  Galapagos Is
Nycticorax nycticorax (Black-crowned Night Heron)
 N.n.nycticorax
  Holland to Japan, Sunda Is, Africa
 N.n.hoactli
  SE Canada to Argentina, Hawaii Is
 N.n.obscurus
  S Peru to Tierra del Fuego
 N.n.falklandicus
  Falkland Is
Nycticorax caledonicus (Nankeen Night Heron)
 N.c.manillensis
  Philippine Is, N Borneo
 N.c.pelewensis
  Palau Is
 N.c.mandibularis
  Solomon Is, Bismarck Arch.
 N.c.caledonicus
  New Caledonia
 N.c.hilli
  Moluccas, Sulawesi, New Guinea, Australia
Nycticorax leuconotus (White-backed Night Heron)
  Senegal to Sudan & Natal

GORSACHIUS

Gorsachius magnificus (Magnificent Night Heron) e?
SE China, Hainan I

Gorsachius goisagi (Japanese Night Heron)
E China, Japan, Philippine Is

Gorsachius melanolophus (Malaysian Night Heron)
India to SE Asia >> Gtr.Sunda Is,
Philippine Is

COCHLEARIUS

Cochlearius cochlearius (Boat-billed Heron)
C.c.zeledoni
W Mexico
C.c.phillipsi
S Mexico, Belize
C.c.ridgwayi
Guatemala, Honduras
C.c.panamensis
S Costa Rica, Panama
C.c.cochlearius
N & C South America

TIGRISOMA

Tigrisoma mexicanum (Bare-throated Tiger Heron)
W Mexico to E Panama, NW Colombia

Tigrisoma fasciatum (Fasciated Tiger Heron)
T.f.fasciatum
SE Brazil, N Argentina
T.f.salmoni
Colombia to Bolivia
T.f.pallescens
NW Argentina

Tigrisoma lineatum (Rufescent Tiger Heron)
T.l.lineatum
Honduras to NW South America
T.l.marmoratum
Brazil, Paraguay, N Argentina

ZONERODIUS

Zonerodius heliosylus (Forest Bittern)
New Guinea, Aru Is

TIGRIORNIS

Tigriornis leucolophus (White-crested Tiger Heron)
Sierra Leone to SW & NE Zaire

ZEBRILUS

Zebrilus undulatus (Zigzag Heron)
the Guianas to C Brazil

IXOBRYCHUS

Ixobrychus involucris (Stripe-backed Bittern)
S Brazil to Patagonia

Ixobrychus exilis (Least Bittern)
I.e.exilis
Nth America to W Indies >> Brazil
I.e.pullus
S Sonora

I.e.bogotensis
C Colombia
I.e.erythromelas
Trinidad, the Guianas to Paraguay
I.e.peruvianus
coast of Peru

Ixobrychus minutus (Little Bittern)
I.m.minutus
C & S Europe to C Asia and NW India >>
Africa
I.m.payesii
Senegal to Aden >> Cape Province
I.m.podiceps
Madagascar
I.m.dubius
E & SW Australia
I.m.novaezelandiae
New Zealand

Ixobrychus sinensis (Chinese Little Bittern)
Manchuria, Japan >> India, Sunda Is,
Philippine Is

Ixobrychus eurhythmus (Schrenk's Bittern)
E Asia >> Malaysia, Sunda Is, Philippine Is

Ixobrychus cinnamomeus (Cinnamon Bittern)
India to China, Philippines Is & Sulawesi

Ixobrychus sturmii (African Dwarf Bittern)
Senegal to Sudan & Cape Province

Ixobrychus flavicollis (Black Bittern)
I.f.flavicollis
C China to India, Malaysia, Sulawesi
I.f.australis
Moluccas, Timor, New Guinea, Australia
I.f.woodfordi
Solomon Is

BOTAURUS

Botaurus pinnatus (Pinnated Bittern)
B.p.pinnatus
Colombia to SE Brazil
B.p.caribaeus
SE Mexico, Belize

Botaurus lentiginosus (American Bittern)
Canada to SW & NE USA >> C America

Botaurus stellaris (Eurasian Bittern)
B.s.stellaris
Europe to E Asia >> N & C Africa
B.s.capensis
C Botswana to Natal & Cape Province

Botaurus poiciloptilus (Australian Bittern)
Australia, Tasmania, New Zealand

21. SCOPIDAE (HAMMERKOP)

SCOPUS

Scopus umbretta (Hammerkop)
S.u.umbretta
Senegal to Nigeria
S.u.minor
coast from Sierra Leone to Nigeria
S.u.bannermanni
Cameroun to Aden & Cape Province

22. BALAENICIPITIDAE (WHALE-HEADED STORK)

BALAENICEPS
Balaeniceps rex (Whale-headed Stork)
Sudan to Zambia

23. CICONIIDAE (STORKS)

MYCTERIA
Mycteria americana (American Wood Ibis)
SE USA to C South America
Mycteria cinerea (Milky Stork)
Malaysia, Sumatra, Java
Mycteria ibis (Yellow-billed Stork)
Senegal to Sudan & Cape Province
Mycteria leucocephala (Painted Stork)
India to SW China, Indochina

ANASTOMUS
Anastomus oscitans (Asian Open-bill Stork)
India to Indochina
Anastomus lamelligerus (African Open-bill Stork)
A.l.lamelligerus
Senegal to Sudan & Zimbabwe
A.l.madagascariensis
Madagascar

CICONIA
Ciconia nigra (Black Stork)
Europe to N China >> Africa, India
Ciconia abdimii (Abdim's Stork)
Ethiopia to Angola & Transvaal
Ciconia episcopus (Woolly-necked Stork)
C.e.microscelis
Senegal to Sudan & Cape Province
C.e.episcopus
India, Sri Lanka, Burma
C.e.neglecta
Malaysia to Philippine Is, Sunda Is
Ciconia stormi (Storm's Stork)   23.2
Borneo
Ciconia maguari (Maguari Stork)
the Guianas to Chile & S Argentina
Ciconia ciconia (White Stork)
C.c.ciconia
Europe, N Africa >> S Africa
C.c.asiatica
C Asia >> India
Ciconia boyciana (Oriental White Stork) 23.2
NE Asia, Japan

EPHIPPIORHYNCHUS
Ephippiorhynchus asiaticus (Black-necked Stork)
E.a.asiaticus
India to Malaysia & Indochina
E.a.australis
New Guinea, N & E Australia
Ephippiorhynchus senegalensis (Saddle-bill Stork)
Senegal to Sudan & Transvaal

Ephippiorhynchus mycteria (Jabiru)   23.1
S Mexico to C Argentina

LEPTOPTILOS
Leptoptilos javanicus (Lesser Adjutant Stork)
C India to S China
Leptoptilos dubius (Greater Adjutant Stork)
e?
India to Indochina, Gtr Sunda Is
Leptoptilos crumeniferus (Marabou Stork)
Senegal to Sudan & Transvaal

24. THRESKIORNITHIDAE (IBISES & SPOONBILLS)

THRESKIORNITHINAE

THRESKIORNIS
Threskiornis aethiopicus (Sacred Ibis)
T.a.aethiopicus
Africa, Saudi Arabia
T.a.abbotti
Aldabra I
T.a.bernieri
Madagascar
Threskiornis melanocephalus (Oriental Ibis)
India to China, Japan
Threskiornis molucca (Australian White Ibis)
T.m.molucca
W Papuan Is, Kei Is, New Guinea
T.m.strictipennis
Australia
Threskiornis spinicollis (Straw-necked Ibis)
Australia, Tasmania

PSEUDIBIS
Pseudibis papillosa (Black Ibis)   24.1
P.p.papillosa
N India
P.p.davisoni
E Burma to S Indochina

THAUMATIBIS
Thaumatibis gigantea (Giant Ibis)
S Thailand, S Indochina

GERONTICUS
Geronticus eremita (Hermit Ibis)(Waldrapp)
N Africa, Ethiopia
Geronticus calvus (Bald Ibis)
S Africa

NIPPONIA
Nipponia nippon (Japanese Crested Ibis)
NE Asia, Japan

LAMPRIBIS
Lampribis olivacea (Olive Ibis)
L.o.olivacea
Sierra Leone, Liberia
L.o.cupreipennis
S Cameroun to W Zaire

CICONIIFORMES

L.o.rothschildi
 Principé I e?
L.o.bocagei e?
 Sao Thome I
L.o.akleyorum
 Kenya
Lampribis rara (Spot-breasted Ibis)
 Liberia to E Zaire & Angola

HAGEDASHIA
Hagedashia hagedash (Hadada Ibis)
 H.h.brevirostris
 Gambia to Zaire
 H.h.nilotica
 Ethiopia to Uganda
 H.h.erlangeri
 Somalia to Malawi
 H.h.hagedash
 Southern Africa

BOSTRYCHIA
Bostrychia carunculata (Wattled Ibis)
 Ethiopia

HARPIPRION
Harpiprion caerulescens (Plumbeous Ibis)
 C Brazil to N Argentina

THERISTICUS
Theristicus caudatus (Buff-necked Ibis)
 T.c.caudatus
 E Panama to French Guina
 T.c.hyperorious
 C South America
Theristicus melanopis (Black-faced Ibis)
 T.m.melanopis
 S Chile, S Argentina
 T.m.branickii
 Ecuador, Peru, N Bolivia

CERCIBIS
Cercibis oxycerca (Sharp-tailed Ibis)
 SE Colombia to Surinam, N Brazil

MESEMBRINIBIS
Mesembrinibis cayennensis (Green Ibis)
 Panama to NE Argentina

PHIMOSUS
Phimosus infuscatus (Bare-faced Ibis)
 P.i.berlepschi
 N South America
 P.i.nudifrons
 C & S Brazil
 P.i.infuscatus
 Paraguay, Uruguay, NE Argentina

EUDOCIMUS
Eudocimus albus (White Ibis)
 S USA to N South America
Eudocimus ruber (Scarlet Ibis)
 N South America, Trinidad

PLEGADIS
Plegadis falcinellus (Glossy Ibis)
 P.f.falcinellus
 S Europe, Asia, Africa, Central America
 P.f.peregrinus
 Philippine Is, Sulawesi, Java to Australia
Plegadis chihi (White-faced Ibis)
 NW USA to SC South America
Plegadis ridgwayi (Puna Ibis)
 Peru, Bolivia

LOPHOTIBIS
Lophotibis cristata (Crested Wood Ibis)
 L.c.cristata
 E Madagascar
 L.c.urschi
 W Madagascar

PLATALEINAE

PLATALEA
Platalea leucorodia (White Spoonbill)
 P.l.leucorodia
 Holland, S Europe to Asia Minor >> Africa
 P.l.major
 C Asia to Japan, Egypt, India, Taiwan
 P.l.balsaci
 Mauretania
 P.l.archeri
 Red Sea coasts, Somalia
Platalea minor (Black-faced Spoonbill)
 S Japan to S China, Taiwan
Platalea alba (African Spoonbill)
 Gambia to Sudan, Cape Province
Platalea regia (Royal Spoonbill)
 Australia to New Guinea, Sulawesi, New
  Zealand
Platalea flavipes (Yellow-billed Spoonbill)
 Australia

AJAIA
Ajaia ajaja (Roseate Spoonbill)
 S USA to C Argentina & Chile

25. PHOENICOPTERIDAE (FLAMINGOES)

PHOENICOPTERUS
Phoenicopterus ruber (Greater Flamingo)
 P.r.ruber
 Central & South America, West Indies
 P.r.roseus
 S Europe, C Asia, NW India, S Africa
Phoenicopterus chilensis (Chilean Flamingo)
 Peru, Uruguay to Tierra del Fuego

PHOENICONAIAS
Phoeniconaias minor (Lesser Flamingo)
 S & E Africa, NW India, Madagascar

PHOENICOPARRUS
Phoenicoparrus andinus (Andean Flamingo)
 Chile & NW Argentina

Phoenicoparrus jamesi (James'Flamingo)
S Peru, N Chile, NW Argentina

ANSERIFORMES

26. ANHIMIDAE (SCREAMERS)

ANHIMA
Anhima cornuta (Horned Screamer)
N South America

CHAUNA
Chauna torquata (Crested Screamer)
Paraguay, S Brazil, N & E Argentina, Bolivia
Chauna chavaria (Northern Screamer)
N Colombia, N Venezuela

27. ANATIDAE (DUCKS, GEESE, SWANS)

ANSERANATINAE

ANSERANAS
Anseranas semipalmata (Magpie Goose)
S New Guinea, N Australia

ANSERINAE

DENDROCYGNA
Dendrocygna guttata (Spotted Whistling Duck)
Mindanao I to Sulawesi & New Guinea
Dendrocygna eytoni (Plumed Whistling Duck)
Australia, Tasmania
Dendrocygna bicolor (Fulvous Whistling Duck)
D.b.helva
S USA, N Mexico
D.b.bicolor
N South America, E Africa, India
Dendrocygna arcuata (Wandering Whistling Duck)
D.a.arcuata
Sumatra to Philippine Is, Timor I, Moluccas
D.a australis
Australia, New Guinea
D.a.pygmaea
New Britain, Fiji Is
Dendrocygna javanica (Indian Whistling Duck)
India to Java & Indochina
Dendrocygna viduata (White-faced Whistling Duck)
S America, C Africa
Dendrocygna arborea (Black-billed{W Indian}Whistling Duck)
West Indies
Dendrocygna autumnalis (Red-billed Whistling Duck)
D.a.autumnalis
SE Texas to Panama
D.a.discolor
E Panama to Ecuador & N Argentina

CYGNUS
Cygnus olor (Mute Swan)
Europe to C Asia >> N Africa, India
Cygnus atratus (Black Swan)
Australia, Tasmania
Cygnus melanocoryphus (Black-necked Swan)
S South America, Falkland Is
Cygnus cygnus (Whooper Swan)
C.c.cygnus
N Europe, N Asia >> C Europe, C Asia, China
C.c.islandicus
S Greenland, Iceland
Cygnus buccinator (Trumpeter Swan) 27.2
N Canada, N USA >> USA
Cygnus columbianus (Tundra Swan) 27.2
C.c.columbianus
N Canada >> coasts of USA
C.c.bewickii
N Russia, N Siberia >> N Europe, C Asia
C.c.jankowskii
NE Asia >> China, Japan

COSCOROBA
Coscoroba coscoroba (Coscoroba Swan)
S South America

ANSER
Anser cygnoides (Swan Goose)
NC Asia & Siberia >> China
Anser fabalis (Bean Goose)
A.f.fabalis
Lapland to Ural Mts >> W & E Europe
A.f.johanseni
W Siberia
A.f.middendorffi
E Siberia >> E China & Japan
A.f.rossicus
N Russia, NW Siberia >> WC Europe, C Asia
A.f.serrirostris
N Siberia >> China, Japan
Anser brachyrhynchus (Pink-footed Goose)
27.1
Greenland, Iceland, Spitzbergen >> W Europe
Anser albifrons (White-fronted Goose)
A.a.albifrons
N Europe, N Asia >> W Europe, S Russia, India, China
A.a.frontalis
E Siberia, N Canada >> W USA, China, Japan
A.a.flavirostris
NW Greenland >> British Isles
A.a.gambeli
NW Canada >> C California
A.a.elgasi
Alaska
Anser erythropus (Lesser White-fronted Goose)
N Russia, N Asia >> SE Europe & China

ANSERIFORMES

Anser anser (Greylag Goose)
A.a.anser
  N Europe, N Asia >> NW Africa
A.a.rubirostris
  C & E Asia >> NW India, China
Anser indicus (Bar-headed Goose)
  C Asia >> N India, N Burma
Anser caerulescens (Snow Goose)
A.c.caerulescens
  N Canada >> S USA
A.c.atlanticus
  N Greenland >> NE USA
Anser rossi (Ross's Goose)
  N Canada >> N California
Anser canagicus (Emperor Goose)
  NE Siberia, NW Alaska >> Aleutian Is

BRANTA
Branta sandvicensis (Hawaiian Goose)
  Hawaii
Branta canadensis (Canada Goose)
B.c.leucopareia
  Aleutian Is >> Japan & W USA
B.c.minima
  W Alaska >> W USA
B.c.occidentalis
  Gulf of Alaska
B.c.fulva
  S Alaska, W British Columbia
B.c.taverneri
  N Canada >> SW USA, Mexico
B.c.parvipes
  Interior Canada >> SC USA
B.c.moffitti
  SW Canada >> NW USA
B.c.hutchinsii
  NC Canada >> Texas, Mexico
B.c.interior
  C & E Canada >> E USA
B.c.canadensis
  (Introduced Europe, New Zealand)
  E Canada >> E USA
Branta leucopsis (Barnacle Goose)
  NE Greenland, Spitzbergen to W Europe
Branta bernicla (Brent Goose)
B.b.bernicla
  N Europe, NW Asia >> W Europe
B.b.hrota
  E Canada, Greenland >> NE USA, NW
  Europe
B.b.nigricans
  NE Canada >> New Jersey
B.b.orientalis
  E Siberia, W Canada >> E Asia, W USA
Branta ruficollis (Red-breasted Goose)
  N Siberia >> S Russia

CEREOPSIS
Cereopsis novaehollandiae (Cereopsis Goose)
  Islands off S Australia

STICTONETTA
Stictonetta naevosa (Freckled Duck)
  Southern Australia, Tasmania

ANATINAE

CYANOCHEN
Cyanochen cyanopterus (Blue-winged Goose)
  Ethiopia

CHLOEPHAGA
Chloephaga melanoptera (Andean Goose)
  Peru to Tierra del Fuego
Chloephaga picta (Magellan Goose)
C.p.picta
  S Chile, S Argentina
C.p.leucoptera
  Falkland Is
Chloephaga hybrida (Kelp Goose)
C.h.hybrida
  S Chile
C.h.malvinarum
  Falkland Is
Chloephaga poliocephala (Ashy-headed Goose)
  S Chile, Argentina
Chloephaga rubidiceps (Ruddy-headed Goose)
  Tierra del Fuego, Falkland Is

NEOCHEN
Neochen jubatus (Orinoco Goose)
  Orinoco & Amazon basins

ALOPOCHEN
Alopochen aegyptiacus (Egyptian Goose)
  Africa (introduced UK)

TADORNA
Tadorna ferruginea (Ruddy Shelduck)
  SE Europe & C Asia >> India & S China
Tadorna cana (South African Shelduck)
  Transvaal, Cape Province
Tadorna variegata (Paradise Shelduck)
  New Zealand
Tadorna tadornoides (Australian Shelduck)
  Southern Australia, Tasmania
Tadorna tadorna (Common Shelduck)
  W Europe to E Asia >> N Africa, India, S
  China
Tadorna radjah (Radjah Shelduck)
T.r.radjah
  S Moluccas, Aru Is, New Guinea
T.r.rufitergum
  N & E Australia
Tadorna cristata (Crested Shelduck)   27.3.4
  NE Asia >> S Korea, S Japan

TACHYERES
Tachyeres patachonicus (Flying Steamer Duck)
  S South America

Tachyeres pteneres (Flightless Steamer Duck)
S South America

Tachyeres brachypterus (Falkland Is Flightless Steamer Duck)
Falkland Is

Tachyeres leucocephalus (White-headed Flightless Steamer Duck)
S Argentina

PLECTROPTERUS
Plectropterus gambensis (Spur-winged Goose)
P.g.gambensis
Gambia to Sudan & Zimbabwe
P.g.niger
Southern Africa

CAIRINA
Cairina moschata (Muscovy Duck)
Mexico to Peru & Uruguay
Cairina scutulata (White-winged Wood Duck)
Assam & Thailand to Sumatra & Java

SARKIDIORNIS
Sarkidiornis melanotos (Comb Duck)
S.m.melanotos
Africa, India, SE China
S.m.sylvicola
Colombia
S.m.carunculatus
Venezuela to N Argentina

PTERONETTA
Pteronetta hartlaubii (Hartlaub's Duck)
P.h.hartlaubii
Liberia to E Zaire
P.h.albifrons
EC Africa

NETTAPUS
Nettapus pulchellus (Green Pygmy Goose)
S Moluccas, S New Guinea, N Australia
Nettapus coromandelianus (Cotton Teal)
N.c.coromandelianus
India to S China & NW Indonesia
N.c.albipennis
NE Australia
Nettapus auritus (African Pygmy Goose)
Gambia to Kenya & Cape Province

CALLONETTA
Callonetta leucophrys (Ringed Teal)
C South America

AIX
Aix sponsa (Wood Duck)
S Canada, USA, Cuba
Aix galericulata (Mandarin Duck)
NE Asia, E China, Japan

CHENONETTA
Chenonetta jubata (Maned Goose)
Australia, Tasmania

AMAZONETTA
Amazonetta brasiliensis (Brazilian Teal)
A.b.brasiliensis
Venezuela to N Argentina
A.b.ipecutiri
C & S Argentina

HYMENOLAIMUS
Hymenolaimus malacorhynchos (Mountain Duck)
New Zealand

MERGANETTA
Merganetta armata (Torrent Duck)
M.a.colombiana
W Venezuela, Colombia, Ecuador
M.a.leucogenis
C & S Ecuador, Peru
M.a.turneri
S Peru
M.a.garleppi
Bolivia
M.a.berlepschi
NW Argentina
M.a.armata
C Chile, W Argentina
M.a.fraenata
SC Chile

ANAS
Anas waigiuensis (Salvadori's Duck)
New Guinea, Waigeu I
Anas sparsa (African Black Duck)
A.s.maclatchyi
Gabon to E Guinea
A.s.leucostigma
Sudan & Ethiopia to Tanzania
A.s.sparsa
Angola, Malawi to Cape Province
Anas penelope (European Wigeon)
N Europe, N Asia >> N Africa, India, Japan
Anas americana (American Wigeon)
W Canada, USA >> Costa Rica, W Indies
Anas sibilatrix (Chiloe Wigeon)
S South America
Anas falcata (Falcated Teal)
NE Asia >> E & S China
Anas strepera (Gadwall)
Europe, Asia, W North America >> N Africa, India, China, Mexico
Anas formosa (Baikal Teal)
NE Asia >> China, Japan
Anas crecca (Green-winged Teal)
A.c.crecca
Europe, Asia >> N Africa, India, China
A.c.nimia
Aleutian Is
A.c.carolinensis
N Canada >> S USA, Central America, W Indies

21

Anas flavirostris (Chilean Teal)
A.f.flavirostris
C Chile.NW Argentina to Tierra del Fuego
A.f.andium
C & S Colombia, Ecuador
A.f.altipetens
W Venezuela, E Colombia
A.f.oxyptera
N Peru to N Chile, Argentina
Anas capensis (Cape Teal)
S Ethiopia to Botswana, Cape Province
Anas gibberifrons (Grey Teal)
A.g.gibberifrons
Java to Sulawesi, Timor I, Wetar I
A.g.remissa
Rennell I
A.g.gracilis
New Guinea, Australia, New Zealand
A.g.albogularis
Andaman Is.Cocos Is
Anas bernieri (Madagascar Teal)
W Madagascar
Anas castanea (Chestnut-breasted Teal)
Southern Australia, Tasmania
Anas aucklandica (New Zealand Teal)
A.a.aucklandica
Auckland I
A.a.nesiotis
Campbell I
A.a chlorotis
New Zealand
Anas platyrhynchos (Mallard)
A.p.platyrhynchos
Europe, Asia, N America >> N Africa,
India, Mexico
A.p.conboschas
Greenland
A.p.wyvilliana
Hawaiian Is
A.p.laysanensis
Laysan I
A p.fulvigula
SE USA
A.p.diazi
N & C Mexico
A p.maculosa
Mexico
Anas rubripes (North American Black Duck)
NE North America to SE USA
Anas melleri (Meller's Duck)
E Madagascar
Anas undulata (African Yellow-bill)
A.u.ruppelli
Ethiopia, Sudan
A.u.undulata
Angola to Uganda, Cape Province
Anas poecilorhyncha (Spotbill Duck)
A.p.poecilorhyncha
India, Sri Lanka
A.p.haringtoni
Burma, SW China
A.p.zonorhyncha
NE Asia, China

Anas superciliosa (Pacific Black Duck) 27.5
A.s.pelewensis
Palau, Solomon, Fiji Is, N New Guinea
A.s.rogersi
Sunda Is, Sulawesi, S New Guinea,
Australia
A.s.superciliosa
New Zealand & Is
Anas luzonica (Philippine Duck)
Philippine Is
Anas specularis (Bronze-winged Duck)
S Chile, S Argentina
Anas specularioides (Crested Duck)
A.s.alticola
C Peru, Bolivia, N Chile
A.s.specularioides
C & S Chile, WC Argentina
Anas acuta (Northern Pintail)
A.a.acuta
N Europe, Asia, N America >> Africa,
China, C America
A.a.eatoni
Kerguelen I
A.a.drygalskii
Crozet Is
Anas georgica (Georgian Teal)
A.g.niceforoi
EC Colombia
A.g.spinicauda
S Colombia to Tierra del Fuego
A.g.georgica
S Georgia I
Anas bahamensis (Bahama Pintail)
A.b.bahamensis
Bahama Is, Gtr Antilles, N South America
A.b.rubrirostris
S Brazil to C Peru
A.b.galapagensis
Galapagos Is
Anas erythrorhyncha (Red-billed Pintail)
S & E Africa
Anas versicolor (Versicolor Teal)
A.v.versicolor
Bolivia to C Chile, C Argentina
A.v.fretensis
S Chile, S Argentina
A.v.puna
Highlands of C Peru to N Chile
Anas punctata (Hottentot Teal)
Uganda to Cape Province, Madagascar
Anas querquedula (Garganey)
W Europe to Japan >> Africa, India,
Indonesia
Anas discors (Blue-winged Teal)
A.d.discors
S Canada, C USA >> Central America & N
South America
A.d.orphna
SE Canada, E USA >> W Indies, South
America
Anas cyanoptera (Cinnamon Teal)
A.c.septentrionalium
W North America to N South America

A.c.tropica
Lowlands of Colombia
A.c.borreroi
Highlands of Colombia
A.c.orinomus
Peru, Bolivia, N Chile
A.c.cyanoptera
S Brazil to Tierra del Fuego
Anas platalea (Argentine Shoveller)
Peru, S Brazil to Tierra del Fuego
Anas smithii (Cape Shoveller)
Angola, Transvaal to Cape Province
Anas rhynchotis (Australian Shoveller)
A.r.rhynchotis
Australia, Tasmania
A.r.variegata
New Zealand
Anas clypeata (Northern Shoveller)
Europe, Asia N America >> E Africa, India, China, Mexico

MALACORHYNCHUS
Malacorhynchus membranaceus (Pink-eared Duck)
Australia, Tasmania

MARMARONETTA
Marmaronetta angustirostris (Marbled Teal)
S Spain to NW India

RHODONESSA
Rhodonessa caryophyllacea (Pink-headed Duck)
NE & E India e?

NETTA
Netta rufina (Red-crested Pochard)
E Europe, C Asia >> N Africa, India
Netta erythrophthalma (Southern Pochard)
N.e.brunnea
Angola to Ethiopia & Cape Province
N.e.erythrophthalma
W South America
Netta peposaca (Rosybill)
C Chile, N & C Argentina, Paraguay, Bolivia

AYTHYA
Aythya valisineria (Canvasback)
WC Canada, WC USA >> S USA, Mexico
Aythya ferina (Common Pochard)
W Europe to C Asia >> N Africa, India, S China
Aythya americana (Redhead)
W Canada, NW USA >> NW Mexico
Aythya collaris (Ring-necked Duck)
W Canada, NW USA >> S USA, Central America, W Indies
Aythya australis (Australian White-eyed Duck)
A.a.australis
Australia, Tasmania, New Zealand
A.a.extima
New Hebrides

A.a.papuana
W New Guinea
Aythya baeri (Baer's Pochard)
NE Asia >> China, Burma, Japan
Aythya nyroca (Ferruginous Duck)
S & E Europe to C Asia >> NE Africa, Iran, Burma
Aythya innotata (Madagascar Pochard) e?
N & E Madagascar
Aythya novaeseelandiae (New Zealand Scaup)
New Zealand, Auckland I
Aythya fuligula (Tufted Duck)
Europe, Asia >> India, S China, Philippine Is
Aythya marila (Greater Scaup)
A.m.marila
N Europe, Asia >> W & S Europe, NW India
A.m.mariloides
Bering Is >> China, Korea, Japan
A.m.nearctica
N & W Canada >> W & E USA, W Indies
Aythya affinis (Lesser Scaup)
W & C Canada, W USA >> S USA, Panama, W Indies

SOMATERIA
Somateria mollissima (Common Eider)
S.m.v-nigra
Alaska, NE Asia >> Aleutian Is
S.m.borealis
N Canada, W Greenland >> Maine
S.m.dresseri
NE Canada >> Newfoundland
S.m.mollissima
Iceland to Novaya Zemlya >> NW Europe
S.m.sedentaria
Hudson Bay
S.m.faroeensis
Faroe Is
Somateria spectabilis (King Eider)
N Europe, N Asia, N Canada
Somateria fischeri (Spectacled Eider)
NE Siberia, Alaska >> Aleutian Is

POLYSTICTA
Polysticta stelleri (Steller's Eider)
NE Siberia, Alaska >> Aleutian Is

HISTRIONICUS
Histrionicus histrionicus (Harlequin Duck)
H.h.histrionicus
Iceland, Greenland, N Labrador
H.h.pacificus
E Siberia, Alaska, W USA to Japan, California

CLANGULA
Clangula hyemalis (Long-tailed Duck)
N Europe, Asia, N America >> W Europe, Japan, S USA

MELANITTA
Melanitta nigra (Black Scoter)
  M.n.nigra
    N Europe, N Asia >> W Europe to Black
    Sea
  M.n.americana
    NE Asia, W Alaska >> China, Gt Lakes,
    California
Melanitta perspicillata (Surf Scoter)
  N Canada, NE Siberia to coasts of USA
Melanitta fusca (Velvet Scoter)
  M.f.fusca
    N Europe, NW Asia to W Europe, SW Asia
  M.f.stejnegeri
    C & E Asia >> China & Japan
  M.f.dixoni
    W Alaska >> W USA
  M.f.deglandi
    NW Canada, NW USA to Gt Lakes, E USA

BUCEPHALA
Bucephala albeola (Bufflehead)
  N & W Canada >> USA
Bucephala islandica (Barrow's Goldeneye)
  SC Alaska to Iceland >> coasts of USA
Bucephala clangula (Common Goldeneye)
  B.c.clangula
    N Europe, N Asia >> S Europe, India,
    Japan
  B.c.americana
    Canada >> California, S Carolina

MERGUS
Mergus cucullatus (Hooded Merganser)
  SC Canada to SE USA >> S USA
Mergus albellus (Smew)
  N Europe, N Asia >> N India, China, Japan
Mergus octosetaceus (Brazilian Merganser)
  S Brazil, E Paraguay, NE Argentina
Mergus serrator (Red-breasted Merganser)
  M.s.serrator
    N Europe, Asia, N America >> China,
    Mexico
  M.s.schioeleri
    Greenland
Mergus squamatus (Chinese Merganser)
  NE Asia >> China
Mergus merganser (Goosander)
  M.m.merganser
    Iceland to NE Asia >> Mediterranean &
    China
  M.m.orientalis
    C Asia to Himalayas >> Assam & Japan
  M.m.americanus
    Canada, W USA >> E & S USA

HETERONETTA
Heteronetta atricapilla (Black-headed Duck)
  C Chile, Paraguay, N & C Argentina

OXYURA
Oxyura dominica (Masked Duck)
  Gtr Antilles, N & C Sth America

Oxyura jamaicensis (Ruddy Duck)
  O.j.rubida
    W Canada, W USA >> Mexico, E & S USA
  O.j.jamaicensis
    W Indies
  O.j.andina
    Colombia
  O.j.ferruginea
    Peru, Bolivia
Oxyura leucocephala (White-headed Duck)
  Mediterranean to C Asia to Egypt, N India
Oxyura maccoa (Maccoa Duck)
  S Ethiopia to Cape Province
Oxyura vittata (Argentine Lake Duck)
  N Chile, S Brazil to Tierra del Fuego
Oxyura australis (Australian Blue-billed Duck)
  S Australia, Tasmania

BIZIURA
Biziura lobata (Musk Duck)
  S Australia, Tasmania

THALASSORNIS
**Thalassornis leuconotus (White-backed
Duck)**
  T.l.leuconotus
    E Cameroun to S Ethiopia & Cape Province
  T.l.insularis
    Madagascar

FALCONIFORMES

28. CATHARTIDAE (NEW WORLD VULTURES)

CATHARTES
Cathartes aura (Turkey Vulture)
  C.a.aura
    S Canada to Costa Rica & Cuba
  C.a.septentrionalis
    E North America
  C.a.ruficollis
    Panama to N Argentina, Trinidad
  C.a.jota
    Colombia to Patagonia, Falkland Is
**Cathartes burrovianus (Lesser Yellow-headed
Vulture)**
  E Mexico to N Argentina
**Cathartes melambrotus (Greater Yellow-
headed Vulture)**
  C South America

CORAGYPS
Coragyps atratus (American Black Vulture)
  C.a.atratus
    W & S USA, N Mexico
  C.a.brasiliensis
    C Mexico to Peru & Brazil

C.a.foetens
Ecuador to C Chile & Patagonia

SARCORHAMPHUS
Sarcorhamphus papa (King Vulture)
C Mexico to N Argentina, Trinidad

GYMNOGYPS
Gymnogyps californianus (Californian Condor)
S California

VULTUR
Vultur gryphus (Andean Condor)
Andes from W Venezuela to Tierra del Fuego

29. PANDIONIDAE (OSPREY)

PANDION
Pandion haliaetus (Osprey)
P.h.haliaetus
Europe, Asia >> S Africa, India, Sunda Is
P.h.carolinensis
N America >> C South America
P.h.ridgwayi
Bahama Is, E Belize
P.h.melvillensis
Philippine Is to Sumatra & C Australia
P.h.microhaliaetus
New Caledonia, N Australia
P.h.cristatus
S Australia, Tasmania

30. ACCIPITRIDAE (HAWKS, EAGLES)

AVICEDA
Aviceda cuculoides (African Cuckoo Hawk)
A.c.cuculoides
Gambia to N Zaire
A.c.verreauxi
Angola to Uganda & Cape Province
A.c.batesi
Guinea to Cameroun
A.c.emini
NE Zaire
Aviceda madagascariensis (Madagascar Cuckoo Hawk)
Madagascar
Aviceda jerdoni (Jerdon's Baza)
A.j.jerdoni
E Himalayas, N India
A.j.ceylonensis
S India, Sri Lanka
A.j.borneensis
Borneo
A.j.magnirostris
Philippine Is
A.j.celebensis
Sulawesi, Banggai I, Sula Is
Aviceda subcristata (Crested Baza)
A.s.timorlaoensis
Lombok I to Timor I & Babar I

A.s.rufa
Obi Is, N Moluccas
A.s.stresemanni
Buru I
A.s.reinwardtii
Seram I, Ambon I
A.s.pallida
Kei Is
A.s.obscura
Biak I
A.s.waigeuensis
Waigeu I
A.s.stenozona
W New Guinea, Aru Is, Misol I
A.s.megala
E New Guinea, Fergusson I, Goodenough I
A.s.bismarckii
New Britain, New Ireland, New Hanover
A.s.coultasi
Admiralty Is, Manus I
A.s.gurneyi
San Cristobal, Ugi, Santa Anna, Malaita, Guadalcanal Is
A.s.robusta
Choiseul I, Ysabel I
A.s.proxima
Bougainville I, Shortland I
A.s.njikena
NW Australia
A.s.subcristata
NE & E Australia
Aviceda leuphotes (Black Baza)
A.l.leuphotes
Himalayas to SW India
A.l.burmana
Burma, Malaysia & Indochina
A.l.syama
Nepal, S China >> Indochina
A.l.wolfei
Szechwan
A.l.andamanica
Andaman Is

LEPTODON
Leptodon cayanensis (Grey-headed Kite)
EC Mexico to N Argentina, Trinidad

CHONDROHIERAX
Chondrohierax uncinatus (Hook-billed Kite)
C.u.uncinatus
S Mexico to N Argentina, Trinidad
C.u.aquilonis
Mexico
C.u.mirus
Grenada
C.u.wilsonii
E Cuba

HENICOPERNIS
Henicopernis longicauda (Long-tailed Honey Buzzard)
H.l.longicauda
New Guinea

H.l.minimus
islands off W New Guinea
H.l.fraterculus
Japen I
Henicopernis infuscata (Black Honey Buzzard)
New Britain

PERNIS
Pernis apivorus (Western Honey Buzzard)
Europe, N Asia >> Africa
Pernis ptilorhynchus (Oriental Honey Buzzard)
P.p.orientalis
E Siberia >> Burma, China
P.p.ruficollis
India, Burma, SW China
P.p.torquatus
Malaysia, Thailand, Sumatra, Borneo
P.p.ptilorynchus
Java
P.p.palawanensis
Palawan I
P.p.philippensis
Philippine Is
Pernis celebensis (Barred Honey Buzzard)
P.c.celebensis
Sulawesi
P.c.steerei
Philippine Is

ELANOIDES
Elanoides forficatus (Swallow-tailed Kite)
E.f.forficatus
S USA, N Mexico
E.f.yetapa
S Mexico to N Argentina

MACHAERHAMPHUS
Machaerhamphus alcinus (Bat Hawk)
M.a.alcinus
Malaysia, Sumatra, Borneo
M.a.papuanus
SE New Guinea
M.a.anderssoni
Gambia to Somalia & Natal

GAMPSONYX
Gampsonyx swainsonii (Pearl Kite)
G.s.leonae
W Nicaragua, N South America
G.s.swainsonii
C Brazil to N Argentina
G.s.magnus
S Ecuador, N Peru

ELANUS
Elanus leucurus (White-tailed Kite)
E.l.majusculus
S USA & E Mexico
E.l.leucurus
N South America to C Chile
Elanus caeruleus (Black-shouldered Kite)

E.c.caeruleus
Africa, S Asia
E.c.sumatranus
Sumatra
E.c.hypoleucus
Java to Philippine Is & Sulawesi
E.c.wahgiensis
E & C New Guinea
Elanus notatus (Australian Black-shouldered Kite)
Australia
Elanus scriptus (Letter-winged Kite)
C Australia

CHELICTINIA
Chelictinia riocourii (Scissor-tailed Kite)
Senegal to Somalia & N Kenya

ROSTRHAMUS
Rostrhamus sociabilis (Everglade Kite)
R.s.plumbeus
Florida
R.s.levis
Cuba, Isle of Pines
R.s.major
E Mexico, Guatemala
R.s.sociabilis
Nicaragua to Argentina
Rostrhamus hamatus (Slender-billed Kite)
E Panama to Peru & E Brazil

HARPAGUS
Harpagus bidentatus (Double-toothed Kite)
H.b.fasciatus
S Mexico to W Colombia & Ecuador
H.b.bidentatus
E Bolivia to E Colombia & E Brazil
Harpagus diodon (Rufous-thighed Kite)
the Guianas to Paraguay & N Argentina

ICTINIA
Ictinia plumbea (Plumbeous Kite)
EC Mexico to Paraguay & N Argentina
Ictinia mississippiensis (Mississippi Kite)
S USA to C South America

LOPHOICTINIA
Lophoictinia isura (Square-tailed Kite)
Australia

HAMIROSTRA
Hamirostra melanosternon (Black-breasted Buzzard Kite)
N & C Australia

MILVUS
Milvus migrans (Black Kite)
M.m.migrans
Europe, Middle East, W Asia
M.m.tenebrosus
Cape Verde Is, Madeira
M.m.arabicus
Egypt

M.m.aegyptius
  N Africa, Somalia, S Yemen
M.m.parasitus
  Senegal to Sudan & Cape Province
M.m.lineatus
  C & E Asia, Japan to Himalayas
M.m.govinda
  India to Indochina & Malaysia
M.m.formosanus
  Hainan I, Taiwan
M.m.affinis
  Lombok I to Timor I, Sulawesi, New Guinea, Australia
Milvus milvus (Red Kite)
M.m.milvus
  Europe, Asia Minor, NW Africa, Canary Is
M.m.fasciicauda
  Cape Verde Is

HALIASTUR
Haliastur sphenurus (Whistling Kite)
  E New Guinea, New Caledonia, Australia
Haliastur indus (Brahminy Kite)
H.i.indus
  India, Sri Lanka to S China, Indochina
H.i.intermedius
  Malaysia, Philippine Is, Borneo, Indonesia
H.i.girrenera
  Australia, New Guinea, Bismarck Arch.
H.i.flavirostris
  Solomon Is

HALIAEETUS
Haliaeetus leucogaster (White-bellied Sea Eagle)
  India to China & Australia
Haliaeetus sanfordi (Sanford's Sea Eagle)
  Solomon Is
Haliaeetus vocifer (African Fish Eagle)
  Senegal to Ethiopia & Cape Province
Haliaeetus vociferoides (Madagascar Fish Eagle)
  Madagascar
Haliaeetus leucoryphus (Pallas' Sea Eagle)
  C Asia to Iraq, N India & Burma
Haliaeetus leucocephalus (American Bald Eagle)
H.l.alascensis
  Alaska, W Canada
H.l.leucocephalus
  C & S USA
Haliaeetus albicilla (White-tailed Sea Eagle)
H.a.albicilla
  Europe, N Asia, Japan to India, China
H.a.groenlandicus
  Greenland
Haliaeetus pelagicus (Steller's Sea Eagle)
H.p.pelagicus
  NE Asia, N China, Japan
H.p.niger
  Korea

ICHTHYOPHAGA
Ichthyophaga humilis (Lesser Fishing Eagle)   30.1
I.h.plumbea
  Himalayas to N Vietnam, Hainan
I.h.humilis
  Thailand to Malaysia, Sumatra to Sulawesi
Ichthyophaga ichthyaetus (Grey-headed Fishing Eagle)
  India to Borneo & Philippine Is

AEGYPIUS
Aegypius monachus (Cinereous Vulture)
  Spain to C Asia >> N Africa, India & China
Aegypius tracheliotus (Lappet-faced Vulture)
                                                              30.4
  Israel & NW Sahara to Ethiopia & Cape Province
Aegypius occipitalis (White-headed Vulture)
  Senegal to Sudan & Cape Province
Aegypius calvus (Red-headed Vulture)   30.4
  India to Laos & SW China

NECROSYRTES
Necrosyrtes monachus (Hooded Vulture)
N.m.monachus
  Senegal to Sudan
N.m.pileatus
  E Sudan & Angola to Cape Province

GYPS
Gyps fulvus (Griffon Vulture)
G.f.fulvus
  S Europe, N Africa to C Asia
G.f.fulvescens
  Afghanistan, NW India
Gyps indicus (Long-billed Griffon)
G.i.indicus
  N & S India to Indochina
G.i.nudiceps
  NE India
Gyps himalayensis (Himalayan Griffon)
  C Asia to N India
Gyps rueppelli (Ruppell's Griffon)
G.r.rueppelli
  Egypt, Senegal to Uganda & Tanzania
G.r.erlangeri
  Ethiopia, Somalia
Gyps coprotheres (Cape Vulture)
  SW Zimbabwe, South Africa
Gyps bengalensis (Oriental White-backed Vulture)
  India to Indochina
Gyps africanus (African White-backed Vulture)
  Senegal to Sudan & Transvaal

NEOPHRON
Neophron percnopterus (Egyptian Vulture)
N.p.percnopterus
  S.Europe, Middle East, Africa
N.p.ginginianus
  Himalayas to S India

FALCONIFORMES

Melierax canorus (Pale Chanting-Goshawk)
  South Africa
Melierax poliopterus (Somali Chanting-
  Goshawk)   30.4
  Somalia to Tanzania

MICRONISUS   30.2
Micronisus gabar (Gabar Goshawk)
  M.g.niger
    N Nigeria to S Yemen
  M.g.aequatorius
    C Africa
  M.g.gabar
    Southern Africa

POLYBOROIDES
Polyboroides typus (African Harrier Hawk)
  P.t.typus
    Sudan to Angola & Cape Province
  P.t.pectoralis
    Gambia to Gabon & Zaire
Polyboroides radiatus (Madagascar Harrier
Hawk)
    Madagascar

KAUPIFALCO
Kaupifalco monogrammicus (Lizard Buzzard)
  K.m.monogrammicus
    Senegal to Ethiopia & Kenya
  K.m.meridionalis
    Angola to Tanzania & Natal

BUTASTUR
Butastur rufipennis (Grasshopper Buzzard-
Eagle)
    Senegal to Ethiopia & Tanzania
Butastur liventer (Rufous-winged Buzzard)
    S Burma to Java & Sula Is
Butastur teesa (White-eyed Buzzard)
    India, Burma
Butastur indicus (Grey-faced Buzzard Eagle)
    NE Asia, Japan, Philippine Is

CIRCUS
Circus assimilis (Spotted Harrier)
    Sulawesi, Timor I, New Guinea, WN & E
      Australia
Circus maurus (Black Harrier)
    Natal & Cape Province
Circus cyaneus (Hen Harrier)(Marsh Hawk)
  C.c.cyaneus
    Europe, Asia, N Africa
  C.c.hudsonius
    N & C America
Circus cinereus (Cinereous Harrier)
    W & S Sth America, Falkland Is
Circus macrourus (Pallid Harrier)
    E Europe & C Asia to Africa & India
Circus melanoleucos (Pied Harrier)
    E Siberia to India & Indochina
Circus pygargus (Montague's Harrier)
    W Europe & EC Asia to Africa, China

Circus aeruginosus (Western Marsh Harrier)
  C.a.aeruginosus
    Europe, Israel, C Asia
  C.a.harterti
    N Africa
Circus spilonotus (Eastern Marsh Harrier)
                                              30.4
  C.s.spilonotus
    E Asia to Japan, Philippine Is, Borneo
  C.s.spilothorax
    W New Guinea
Circus approximans (Pacific Marsh Harrier)
                                              30.4
    S.New Guinea, Australia, New Zealand,
      Pacific islands
Circus ranivorus (African Marsh Harrier)
    Angola & Kenya to Cape Province
Circus maillardi (Malagasy Marsh Harrier)
                                              30.4
    Madagascar, Reunion, Comoro Is
Circus buffoni (Long-winged Harrier)
    Colombia, Trinidad, Guianas to C Argentina

ACCIPITER   30.4
Accipiter poliogaster (Grey-bellied Goshawk)
    N & C Sth America
Accipiter trivirgatus (Asian Crested
Goshawk)
  A.t.peninsulae
    SW India
  A.t.layardi
    Sri Lanka
  A.t.indicus
    NE India to S China, Malaysia
  A.t.trivirgatus
    Sumatra
  A.t.niasensis
    Nias I
  A.t.javanicus
    Java
  A.t.microstictus
    Borneo
  A.t.palawanus
    Palawan, Calamian Is
  A.t.castroi
    Polillo Is
  A.t.extimus
    Negros, Samar, Leyte, Mindanao
Accipiter griseiceps (Sulawesi Crested
Goshawk)   30.1
    Sulawesi, Muna, Buton
Accipiter tachiro (African Goshawk)
  A.t.macroscelides
    Sierra Leone to W Cameroun
  A.t.lopezi
    Fernando Po
  A.t.tousenelii
    Cameroun, Gabon, N & W Zaire
  A.t.canescens
    E Zaire
  A.t.unduliventer
    E & S Ethiopia

A.t.croizati
  SW Ethiopia
A.t.sparsimfasciatus
  Uganda & S Zaire to Somalia & Tanzania
A.t.tachiro
  S Angola & Mozambique to Cape Province
Accipiter castanilus (Chestnut-bellied Sparrow Hawk)
  A.c.castanilus
    Nigeria W Zaire
  A.c.beniensis
    E Zaire
Accipiter brevipes (Levant Sparrow Hawk)
  Balkans to S Russia & Egypt
Accipiter badius (Shikra)
  A.b.sphenurus
    Gambia to Ethiopia & Tanzania
  A.b.polyzonoides
    S Tanzania & S Zaire to Cape Province
  A.b.cenchroides
    Transcaucasia, N Iran to C Asia
  A.b.dussumieri
    Himalayas, N India
  A.b.badius
    S India, Sri Lanka
  A.b.poliopsis
    Assam to Indochina, Taiwan
Accipiter butleri (Nicobar Shikra)
  A.b.butleri
    Car Nicobar I
  A.b.obsoletus
    S.Nicobar Is
Accipiter soloensis (Grey Frog Hawk)
  S & NE China to Indonesia
Accipiter francesii (Frances' Sparrow Hawk)
  A.f.francesii
    Madagascar
  A.f.griveaudi
    Gd Comoro I
  A.f.pusillus
    Anjouan
  A.f.brutus
    Mayotte
Accipiter trinotatus (Spot-tailed Sparrow Hawk)   30.1
  Sulawesi, Muna, Buton
Accipiter fasciatus (Australian Goshawk)
  A.f.natalis
    Christmas I
  A.f.wallacei
    Lombok to Wetar, Damar, Moa I
  A.f.tjendanae
    Sumba I
  A.f.stresemanni
    Djampea I, Tukangbesi Is
  A.f.savu
    Savu I
  A.f.hellmayri
    Alor I, Samao, Timor
  A.f.buruensis
    Buru I
  A.f.dogwa
    S New Guinea

A.f.polycryptus
  E New Guinea
A.f.vigilax
  New Caledonia, New Hebrides
A.f.didimus
  N coast of Australia >> Moluccas
A.f.fasciatus
  Rennell I, Australia, Tasmania
Accipiter novaehollandiae (White Goshawk)
                                        30.1
A.n.misoriensis
  Biak I
A.n.hiogaster
  Seram, Ambon
A.n.albiventris
  Kei Is
A.n.pallidiceps
  Buru
A.n.matthiae
  S Matthias I
A.n.manusi
  Admiralty Is
A.n.lavongai
  New Hanover
A.n.rubianae
  New Georgia, Rendova, Vellalavella I
A.n.rufoschistaceus
  Ysabel, Choiseul I
A.n.bougainvillei
  Faure I, Bougainville I
A.n.malaitae
  Malaita I
A.n.pulchellus
  Guadalcanal I
A.n.sylvestris
  Sumbawa, Flores, Pantar, Alor I
A.n.dampieri
  Rook I
A.n.lihirensis
  Lihir is, Tanga I
A.n.misulae
  Louisiade Arch.
A.n.leucosomus
  New Guinea & islands
A.n.polionotus
  Babar, Damar, Timorlaut, Banda Is
A.n.pallidimas
  d'Entrecasteaux Arch.
A.n.novaehollandiae
  N & E Australia, Tasmania
A.n.mortyi
  Morotai I
A.n.griseogularis
  C Moluccas
A.n.obiensis
  Obi I
Accipiter melanochlamys (Black-mantled Goshawk)
  New Guinea
Accipiter albogularis (Pied Goshawk)
  A.a.woodfordi
    Bougainville, Guadalcanal, Malaita, Choiseul Is

A.a.gilvus
New Georgia, Rendova, Vellalavella Is
A.a.albogularis
San Cristobal, Ugi, Santa Anna Is
A.a.eichorni
Feni I
A.a.sharpei
Vanikoro I, Utupua I
Accipiter rufitorques (Fiji Goshawk)
Fiji Is
Accipiter haplochrous (New Caledonia Sparrow Hawk)
New Caledonia
Accipiter henicogrammus (Gray's Goshawk)
Batjan, Halmahera, Morotai Is
Accipiter luteoschistaceus (Blue & Grey Sparrow Hawk)
New Britain
Accipiter imitator (Imitator Sparrow Hawk)
Choiseul I, Ysabel I
Accipiter poliocephalus (New Guinea Grey-headed Goshawk)
New Guinea & islands
Accipiter princeps (New Britain Grey-headed Goshawk)
New Britain
Accipiter superciliosus (Tiny Sparrow Hawk)
A.s.fontanieri
SE Nicaragua to W Colombia, Ecuador
A.s.superciliosus
E Peru to Venezuela, Guianas, N Argentina
Accipiter collaris (Semi-collared Sparrow Hawk)
W Venezuela, E & S Colombia, Ecuador
Accipiter erythropus (Red-thighed Sparrow Hawk)
A.e.erythropus
Gambia to Togo
A.e.zenkeri
Cameroun to S Angola & Uganda
Accipiter minullus (African Little Sparrow Hawk)
Ethiopia to Angola & Cape Province
Accipiter gularis (Japanese Sparrow Hawk)
NE Asia, Japan to S China & Philippine Is
Accipiter virgatus (Besra Sparrow Hawk)
A.v.affinis
W Himalayas to W & S China, Indonesia
A.v.besra
S India, Sri Lanka, Andaman Is
A.v.confusus
Philippine Is
A.v.quagga
Leyte
A.v.rufotibialis
N Borneo
A.v.vanbemmeli
Sumatra
A.v.virgatus
Java

A.v.quinquefasciatus
Flores
Accipiter nanus (Dwarf Sparrow Hawk)   30.1
Sulawesi
Accipiter cirrhocephalus (Australian Collared Sparrow Hawk)
A.c.papuanus
New Guinea, Aru, Waigeu, Japen Is
A.c.rosselianus
Rossel I, Louisiades Arch.
A.c.quaesitandus
N Australia
A.c.cirrhocephalus
S Australia, Tasmania
Accipiter brachyurus (New Britain Collared Sparrow Hawk)
New Britain
Accipiter erythrauchen (Grey Moluccan Collared Sparrow Hawk)
A.e.erythrauchen
Batjan, Halmahera, Morotai, Obi Is
A.e.ceramensis
Seram, Buru Is
Accipiter rhodogaster (Vinous-breasted Sparrow Hawk)   30.1
A.r.rhodogaster
Sulawesi, Buton, Muna Is
A.r.sulaensis
Peleng, Sula Is
Accipiter madagascariensis (Madagascar Sparrow Hawk)
Madagascar
Accipiter ovampensis (Ovampo Sparrow Hawk)
Ghana & Ethiopia to E Transvaal
Accipiter nisus (Northern Sparrow Hawk)
A.n.nisus
Europe to C Russia & Iran
A.n.punicus
N Africa
A.n.granti
Madeira, Canary Is
A.n.wolterstorffi
Corsica, Sardinia
A.n.nisosimilis
N Iran to Manchuria & Japan
A.n.melaschistos
Himalayas to W China
Accipiter rufiventris (Rufous-breasted Sparrow Hawk)
A.r.rufiventris
C Zaire to Kenya & Cape Province
A.r.perspicillaris
Ethiopia
Accipiter striatus (Sharp-shinned Hawk)
A.s.perobscurus
Queen Charlotte Is
A.s.velox
Canada, USA
A.s.suttoni
N Mexico
A.s.madrensis
SW Mexico

A.s.chionogaster
S Mexico, Guatemala to Nicaragua
A.s.fringilloides
Cuba
A.s.striatus
Hispaniola
A.s.venator
Puerto Rico
A.s.ventralis
W Venezuela & Colombia to W Bolivia
A.s.erythronemius
E Bolivia & S Brazil to Uruguay & N
Argentina
Accipiter cooperii (Cooper's Hawk)
S Canada to N Central America
Accipiter gundlachii (Gundlach's Hawk)
Cuba
Accipiter bicolor (Bicoloured Sparrow Hawk)
A.b.bicolor
S Mexico to E Bolivia
A.b.fidens
SW Mexico
A.b.pileatus
S Brazil
A.b.guttifer
S Bolivia, Paraguay, N Argentina
A.b.chilensis
Andes of Chile, Argentina to Tierra del
Fuego
**Accipiter melanoleucus (Great Sparrow
Hawk)**
A.m.melanoleucus
C African Rep. & Ethiopia to Cape Province
A.m.temminckii
Ghana to Gabon, Cape Verde Is
Accipiter henstii (Henst's Goshawk)
Madagascar
Accipiter gentilis (Northern Goshawk)
A.g.gentilis
Europe, SW Asia, Morocco
A.g.buteoides
N Scandinavia, N Russia
A.g.albidus
NE Siberia
A.g.arrigonii
Corsica, Sardinia
A.g.schvedowi
SE Russia to W China
A.g.fujiyamae
Japan
A.g.atricapillus
N America
A.g.laingi
Brit.Columbian islands
A.g.apache
SE USA, NW Mexico
Accipiter meyerianus (Meyer's Goshawk)
Moluccas, New Britain, Solomon Is
**Accipiter buergersi (Chestnut-shouldered
Goshawk)**
E New Guinea
Accipiter radiatus (Red Goshawk)
NC Australia

Accipiter doriae (Doria's Goshawk)
New Guinea

UROTRIORCHIS
**Urotriorchis macrourus (African Long-tailed
Hawk)**   30.4
Ghana to C Zaire

GERANOSPIZA
Geranospiza caerulescens (Crane Hawk)
G.c.livens
NW Mexico
G.c.nigra
Mexico to C Panama
G.c.balzarensis
E Panama to NW Peru
G.c.caerulescens
E Ecuador & Colombia to Guianas, N Brazil
G.c.gracilis
NE Brazil
G.c.flexipes
S Brazil & Bolivia to N Argentina

LEUCOPTERNIS
**Leucopternis schistacea (Slate-coloured
Hawk)**
Amazonia
Leucopternis plumbea (Plumbeous Hawk)
E Panama to NW Peru
Leucopternis princeps (Barred Hawk)
Costa Rica to N Ecuador
Leucopternis melanops (Black-faced Hawk)
N Amazonia
Leucopternis kuhli (White-browed Hawk)
S Amazonia
**Leucopternis lacernulata (White-necked
Hawk)**
E & S Brazil
**Leucopternis semiplumbea (Semi-plumbeous
Hawk)**
Honduras to NW Ecuador
Leucopternis albicollis (White Hawk)
L.a.ghiesbreghti
S Mexico to Nicaragua
L.a.costaricensis
Honduras to NW Colombia
L.a.williaminae
NW Colombia to W Venezuela
L.a.albicollis
Trinidad, C Venezuela & the Guianas to
Bolivia
**Leucopternis occidentalis (Grey-backed
Hawk)**
W Ecuador
Leucopternis polionota (Mantled Hawk)
E & S Brazil to N Argentina

ASTURINA
Asturina nitida (Grey Hawk)
A.n.plagiata
S USA to NW Costa Rica
A.n.costaricensis
SW Costa Rica to W Ecuador

A.n.nitida
Trinidad, N Amazonia
A.n.pallida
S Brazil, E Bolivia to NC Argentina

BUTEOGALLUS
Buteogallus anthracinus (Common Black Hawk)
B.a.anthracinus
SW USA to NW Guyana, St Vincent I
B.a.utilensis
Bay I, Honduras
B.a.gundlachii
Cuba, Isle of Pines
Buteogallus subtilis (Mangrove Black Hawk)
Coast from S Mexico to NW Peru
Buteogallus aequinoctialis (Rufous Crab Hawk)
Venezuela to SE Brazil
Buteogallus urubitinga (Great Black Hawk)
B.u.ridgwayi
N Mexico to Panama
B.u.urubitinga
N & C Sth America, Trinidad
Buteogallus meridionalis (Savannah Hawk)
E Panama to C Argentina

HARPYHALIAETUS
Harpyhaliaetus solitarius (Black Solitary Eagle)
NW Mexico to Venezuela & Peru
Harpyhaliaetus coronatus (Crowned Solitary Eagle)
S Brazil to E Bolivia & C Argentina

BUSARELLUS
Busarellus nigricollis (Black-collared Hawk)
B.n.nigricollis
Mexico to N Argentina
B.n.leucocephalus
Paraguay, NC Argentina

GERANOAETUS
Geranoaetus melanoleucus (Black-chested Buzzard Eagle)
G.m.australis
Venezuela, W Sth America, Tierra del Fuego
G.m.melanoleucus
S Brazil, Uruguay, Paraguay, N Argentina

PARABUTEO
Parabuteo unicinctus (Harris' Hawk)
P.u.harrisi
S Texas to N Peru
P.u.superior
SE California, W Mexico
P.u.unicinctus
S America

BUTEO
Buteo magnirostris (Roadside Hawk)
B.m.griseocauda
Mexico to W Peru
B.m.conspectus
SE Mexico, N Belize
B.m.gracilis
Cozumel, Holbox Is
B.m.sinushonduri
Bonacca I, Ruatan I (Honduras)
B.m.petulans
SW Costa Rica, SW Panama
B.m.alius
Pearl Is (Panama)
B.m.magnirostris
S America (N of R Amazon)
B.m.occiduus
E Peru, W Brazil
B.m.saturatus
Bolivia to W Argentina
B.m.nattereri
NE Brazil
B.m.magniplumis
S Brazil, NW Argentina
B.m.pucherani
E Argentina, Paraguay
Buteo leucorrhous (White-rumped Hawk)
Venezuela to Bolivia & NE Argentina
Buteo ridgwayi (Ridgway's Hawk)
Hispaniola
Buteo lineatus (Red-shouldered Hawk)
B.l.lineatus
E Nth America
B.l.alleni
Florida to E Texas
B.l.extimus
S Florida & Keys
B.l.texanus
SC Texas to C Mexico
B.l.elegans
S Oregon to Baja California
Buteo platypterus (Broad-winged Hawk)
B.p.platypterus
SE Canada, E USA to Peru & Brazil
B.p.cubanensis
Cuba
B.p.brunnescens
Puerto Rico
B.p.insulicola
Antigua I
B.p.rivierei
Dominica, Martinique, St Lucia Is
B.p.antillarum
Barbados, St Vincent, Grenada, Tobago Is
Buteo brachyurus (Short-tailed Hawk)
B.b.fuliginosus
S Florida to Panama
B.b.brachyurus
S America (Below 7000')
Buteo albigula (White-throated Hawk)
Andes from Colombia to C Chile
Buteo swainsonii (Swainson's Hawk)
W Canada & W USA to C Argentina

Buteo galapagoensis (Galapagos Hawk)
Galapagos Is
Buteo albicaudatus (White-tailed Hawk)
B.a.hypospodius
Texas to N Colombia, W Venezuela
B.a.colonus
E Colombia to Surinam
B.a.albicaudatus
S Brazil to C Argentina
Buteo polyosoma (Red-backed Hawk)
B.p.polyosoma
NW Colombia to Tierra del Fuego, Falkland Is
B.p.exsul
Masafuera I
Buteo poecilochrous (Variable Hawk)
SW Colombia to N Chile
Buteo albonotatus (Zone-tailed Hawk)
SW USA to N Sth America
Buteo solitarius (Hawaiian Hawk)
Hawaii Is
Buteo ventralis (Rufous-tailed Hawk)
S Chile, S Argentina
Buteo jamaicensis (Red-tailed Hawk)
B.j.borealis
E Nth America to N Mexico
B.j.calurus
W Nth America >> C America
B.j.harlani
N Brit.Columbia, N Alberta >> S USA
B.j.alascensis
SE Alaska
B.j.kriderii
SC Canada, NC USA >> S USA
B.j.fuertesi
SW USA, NW Mexico
B.j.umbrinus
S Florida, Bahama Is
B.j.hadropus
N to SC Mexico
B.j.socorroensis
Socorro I
B.j.fumosus
Tres Marias Is
B.j.solitudinis
Cuba, Isle of Pines
B.j.jamaicensis
Jamaica, Hispaniola, Puerto Rico
B.j.kemsiesi
S Mexico to Nicaragua
B.j.costaricensis
Costa Rica, W Panama
Buteo buteo (Eurasian Buzzard)
B.b.buteo
W & S Europe, Atlantic islands
B.b.vulpinus
N & E Europe, C Asia, E & S Africa
B.b.menetriesi
Caucasus & Elburz Mts
B.b.japonicus
Transbaikalia & Tibet to Japan, Indochina
B.b.toyoshimae
Bonin Is, Izu Is

Buteo oreophilus (African Mountain Buzzard)
B.o.oreophilus
S Ethiopia to C Tanzania
B.o.trizonatus
S Natal, E Cape Province
Buteo brachypterus (Madagascar Buzzard)
Madagascar
Buteo lagopus (Rough-legged Buzzard)
B.l.lagopus
Europe, C Asia
B.l.menzbieri
NE Asia
B.l.kamtschatkensis
Kamchatka, N Kurile Is
B.l.sanctijohannis
Canada, N USA
Buteo rufinus (Long-legged Buzzard)
B.r.rufinus
C Europe to C Asia
B.r.cirtensis
Morocco to Egypt
Buteo hemilasius (Upland Buzzard)
C & E Asia
Buteo regalis (Ferruginous Hawk)
SW Canada, WC USA
Buteo auguralis (African Red-tailed Buzzard)
Sierra Leone to Ethiopia & Angola
Buteo augur (Augur Buzzard)   30.4
B.a.archeri
Somalia
B.a.augur
S Ethiopia to Angola & Mozambique
Buteo rufofuscus (Jackal Buzzard)
Namibia, South Africa

MORPHNUS
Morphnus guianensis (Guiana Crested Eagle)
Honduras to N Paraguay & Argentina

HARPIA
Harpia harpyja (Harpy Eagle)
S Mexico to E Bolivia & N Argentine

HARPYOPSIS
Harpyopsis novaeguineae (New Guinea Harpy Eagle)
New Guinea

PITHECOPHAGA
Pithecophaga jefferyi (Philippine Eagle)
Luzon, Mindanao Is

ICTINAETUS
Ictinaetus malayensis (Indian Black Eagle)
I.m.perniger
India, Sri Lanka
I.m.malayensis
Burma to S China & Moluccas

AQUILA
Aquila pomarina (Lesser Spotted Eagle)
A.p.pomarina
C & E Europe, Caucasus, Transcaucasia

A.p.hastata
India, N Burma
Aquila clanga (Greater Spotted Eagle)
E Europe to E Asia, & NE Africa to S China
Aquila rapax (Tawny Eagle)
A.r.orientalis
E Europe, C Asia to C Africa
A.r.nipalensis
EC Asia & India
A.r.vindhiana
Baluchistan, India, N Burma
A.r.belisarius
Morocco to Nigeria & Ethiopia
A.r.rapax
SC & Sth Africa
Aquila heliaca (Imperial Eagle)
A.h.adalberti
Spain
A.h.heliaca
Greece to C Siberia >> NE Africa, India
Aquila gurneyi (Gurney's Eagle)
N Moluccas, New Guinea
Aquila chrysaetos (Golden Eagle)
A.c.chrysaetos
Alps, N Europe, W Asia
A.c.daphanea
Russian Turkestan to SW China
A.c.japonica
Korea, C Japan
A.c.canadensis
NE Siberia, N Mongolia, Canada, W USA
A.c.homeyeri
Spain, N Africa
Aquila audax (Wedge-tailed Eagle)
A.a.audax
S New Guinea
A.a.fleayi
Tasmania
Aquila verreauxii (Verreaux's Eagle)
Ethiopia, Sudan to Cape Province

HIERAAETUS
Hieraaetus wahlbergi (Wahlberg's Eagle) 30.4
Gambia to Ethiopia & N Cape Province
Hieraaetus fasciatus (Bonelli's Eagle)
H.f.fasciatus
S Europe, N Africa to India & China
H.f.renschi
Lesser Sunda Is
Hieraaetus spilogaster (African Hawk Eagle)
30.4
Gambia to Ethiopia & Cape Province
Hieraaetus pennatus (Booted Eagle)
H.p.pennatus
S Europe to N Africa & Caucasus
H.p.harterti
SW & C Asia
Hieraaetus morphnoides (Little Eagle)
H.m.morphnoides
Australia
H.m.weiskei
New Guinea

Hieraaetus ayresii (Ayres' Hawk Eagle) 30.3
Nigeria to Ethiopia & Cape Province
Hieraaetus kienerii (Chestnut-bellied Hawk Eagle)
H.k.kienerii
S Himalayas, W India, Sri Lanka
H.k.formosus
Burma to Philippine Is, Sulawesi & Java
Hieraaetus bellicosus (Martial Eagle) 30.4
Senegal to Somalia & Cape Province

SPIZASTUR
Spizastur melanoleucus (Black & White Hawk Eagle)
E & S Mexico to Paraguay & NE Argentina

SPIZAETUS
Spizaetus occipitalis (Long-crested Eagle)
30.4
Senegal to Ethiopia & Cape Province
Spizaetus africanus (Cassin's Hawk Eagle)
Togo to C Zaire & Uganda
Spizaetus cirrhatus (Crested Hawk Eagle)
S.c.cirrhatus
India
S.c.ceylanensis
Sri Lanka
S.c.limnaeetus
NE India to Mindanao, Borneo & Java
S.c.andamanensis
Andaman Is
S.c.vanheurni
Simalur I
S.c.floris
Sumbawa, Flores Is
Spizaetus nipalensis (Hodgson's Hawk Eagle)
S.n.nipalensis
W India, Himalayas to SE China
S.n.kelaarti
Sri Lanka
S.n.orientalis
Japan
Spizaetus bartelsi (Java Hawk Eagle)
W Java
Spizaetus lanceolatus (Sulawesi Hawk Eagle)
Sula I & Sulawesi
Spizaetus philippensis (Philippine Hawk Eagle)
Philippine Is, Palawan
Spizaetus alboniger (Blyth's Hawk Eagle)
S Burma to Sumatra & Borneo
Spizaetus nanus (Wallace's Hawk Eagle)
S.n.nanus
Malaysia, Borneo, Sumatra
S.n.stresemanni
Nias I
Spizaetus tyrannus (Black Hawk Eagle)
S.t.serus
C Mexico to E Peru
S.t.tyrannus
E & S Brazil
Spizaetus ornatus (Ornate Hawk Eagle)

S.o.vicarius
Mexico to Colombia, W Ecuador
S.o.ornatus
C Colombia to Guianas & N Argentina
Spizaetus coronatus (Crowned Eagle) 30.4
Guinea to Ethiopia & Cape Province
Spizaetus isidori (Black & Chestnut Eagle)
30.4
W Venezuela to Bolivia & NW Argentina

31. SAGITTARIIDAE (SECRETARY BIRD)

SAGITTARIUS
Sagittarius serpentarius (Secretary Bird)
Senegal to Somalia & Cape Province

32. FALCONIDAE (FALCONS, CARACARAS)

DAPTRIUS
Daptrius ater (Yellow-throated Caracara)
Amazonia
Daptrius americanus (Red-throated Caracara)
S Mexico to C Peru & S Brazil, Bolivia

PHALCOBOENUS
Phalcoboenus carunculatus (Carunculated Caracara)
Andes of Ecuador & SW Colombia
Phalcoboenus megalopterus (Mountain Caracara)
Andes of Peru to N Chile
Phalcoboenus albogularis (White-throated Caracara)
S Chile, S Argentina
Phalcoboenus australis (Forster's Caracara)
Falkland Is, Cape Horn islands

POLYBORUS
Polyborus plancus (Common Caracara)
P.p.auduboni
S USA to W Panama, Cuba
P.p.pallidus
Tres Marias Is
P.p.cheriway
E Panama & S America (N of R Amazon)
P.p.plancus
S Sth America, Falkland Is

MILVAGO
Milvago chimango (Chimango)
M.c.chimango
Paraguay & Uruguay to S Argentina
M.c.temucoensis
S Chile, Tierra del Fuego
Milvago chimachima (Yellow-headed Caracara)
Panama to E Bolivia & N Argentina

HERPETOTHERES
Herpetotheres cachinnans (Laughing Falcon)
H.c.cachinnans
NW Mexico to N Argentina

H.c.fulvescens
W Panama to N Peru

MICRASTUR
Micrastur ruficollis (Barred Forest Falcon)
M.r.guerilla
Mexico to Nicaragua
M.r.interstes
Costa Rica to W Colombia & Ecuador
M.r.zonothorax
N Venezuela, E Colombia
M.r.pelzelni
W Brazil, E Peru
M.r ruficollis
E Brazil, Paraguay, N Argentina
M.r.olrogi
NW Argentina
Micrastur gilvicollis (Lined Forest Falcon)
32.3.4
M.g.gilvicollis
S Venezuela, Guianas, Amazonia
M.g.plumbeus
W Colombia, NW Ecuador
Micrastur mirandollei (Slaty-backed Forest Falcon)
E Costa Rica to E Peru & SE Brazil
Micrastur semitorquatus (Collared Forest Falcon)
M.s.naso
Mexico to NW Peru
M.s.semitorquatus
E Colombia & N Peru to Brazil & N Argentina
Micrastur buckleyi (Traylor's Forest Falcon)
E Ecuador, NE Peru

SPIZIAPTERYX
Spiziapteryx circumcinctus (Spot-winged Falconet)
N & W Argentina

POLIHIERAX
Polihierax semitorquatus (African Pygmy Falcon)
P.s.castanotus
E Zaire & Ethiopia to C Tanzania
P.s.semitorquatus
S Angola, Namibia, Botswana, W Sth Africa
Polihierax insignis (White-rumped Pygmy Falcon)
P.i.insignis
N Burma
P.i.cinereiceps
S Burma, Thailand, N Indochina
P.i.harmandi
S Laos, S Vietnam

MICROHIERAX
Microhierax caerulescens (Collared Falconet)
M.c.caerulescens
Himalayas, N India
M.c.burmanicus
Burma to Indochina

***Microhierax fringillarius* (Black-thighed Falconet)**
  S Burma to Bali I & Sumatra
***Microhierax latifrons* (Bornean Falconet)**
  NW Borneo
***Microhierax erythrogenys* (Philippine Falconet)**
  Philippine Is
***Microhierax melanoleucus* (Pied Falconet)**
  Assam, SE China, N Indochina

FALCO
***Falco naumanni* (Lesser Kestrel)**
  S Europe to China >> S Africa
***Falco rupicoloides* (Greater Kestrel)**
  F.r.fieldi
    Somalia, N Kenya
  F.r.arthuri
    C & S Kenya, N Tanzania
  F.r.rupicoloides
    S Tanzania & Angola to S Transvaal
***Falco alopex* (Fox Kestrel)**
  Ghana to Sudan & W Kenya
***Falco sparverius* (American Kestrel)**
  F.s.sparverius
    Alaska & Canada to S Mexico
  F.s.paulus
    SE USA
  F.s.peninsularis
    Baja California, NW Mexico
  F.s.tropicalis
    S Mexico, Guatemala, N Honduras
  F.s.nicaraguensis
    NW Honduras, Nicaragua
  F.s.sparverioides
    S Bahamas Is, Cuba
  F.s.dominicensis
    Hispaniola
  F.s.caribaearum
    Puerto Rico, Virgin Is, Lesser Antilles
  F.s.brevipennis
    Netherlands West Indies
  F.s.isabellinus
    the Guianas, E Venezuela, N Brazil
  F.s.ochraceus
    E Colombia, NW Venezuela
  F.s.aequatorialis
    NW Colombia, N Ecuador
  F.s.peruvianus
    SW Ecuador, Peru, N Chile
  F.s.cinnamominus
    SE Peru to Paraguay & Tierra del Fuego
  F.s.fernandensis
    Masatierra I, Juan Fernandez Is
  F.s.cearae
    S Brazil
***Falco tinnunculus* (Common Kestrel)**
  F.t.tinnunculus
    Europe to NE Asia >> C Africa & India
  F.t.canariensis
    Madiera
  F.t dacotiae
    Lanzarote, E Canary Is

F.t.neglectus
  N Cape Verde Is
F.t.alexandri
  S Cape Verde Is
F.t.rupicolaeformis
  Egypt, S Yemen
F.t.archeri
  Socotra I, Somalia, NE Kenya
F.t.rufescens
  Guinea & N Angola to Ethiopia & Tanzania
F.t.rupicolus
  C Angola & S Tanzania to Cape Province
F.t.interstinctus
  Himalayas to Japan & Philippine Is
F.t.objurgatus
  S India
***Falco newtoni* (Madagascar Kestrel)**
  F.n.newtoni
    Madagascar
  F.n.aldabranus
    Aldabra, Anjouan Is
***Falco punctatus* (Mauritius Kestrel)**
  Mauritius
***Falco araea* (Seychelles Kestrel)**
  Seychelles
***Falco moluccensis* (Moluccan Kestrel)**   32.5
  F.m.moluccensis
    Ambon, Seram, Buru, N Moluccan Is
  F.m.microbalia
    Java, Lesser Sunda Is, Sulawesi, Timor
***Falco cenchroides* (Australian Kestrel)**
  F.c.baru
    C New Guinea
  F.c.cenchroides
    Australia, Tasmania
***Falco ardosiaceus* (Grey Kestrel)**
  Senegal to Ethiopia & S Tanzania
***Falco dickinsoni* (Dickinson's Kestrel)**
  Angola to Tanzania & Natal
***Falco zoniventris* (Madagascar Banded Kestrel)**
  Madagascar
***Falco vespertinus* (Western Red-footed Falcon)**
  C Europe to C Asia >> W & SW Africa
***Falco amurensis* (Eastern Red-footed Falcon)**   32.2
  E Siberia, N China >> E & S Asia, SE Africa
***Falco chicquera* (Red-headed Falcon)**
  F.c.chicquera
    Pakistan, India
  F.c.ruficollis
    Gambia to Ethiopia >> Zambia
  F.c.horsbrughi
    Zimbabwe, S Africa
***Falco columbarius* (Merlin)**
  F.c.subaesalon
    Iceland to Britain & Belgium
  F.c.aesalon
    Europe to N Russia & W Siberia
  F.c.insignis
    E Siberia to Japan, Indochina, India

F.c.pallidus
Transcaucasia to SC Asia
F.c.lymani
E Altai, Tien Shan to W China
F.c.columbarius
Alaska to Newfoundland & N Sth America
F.c.suckleyi
W British Columbia to N California
F.c.richardsonii
SW & C Canada to WC USA
Falco berigora (Brown Falcon)
F.b.novaeguineae
New Guinea, Dampier I
F.b.berigora
humid parts of Australia
F.b.centralia
dry interior of Australia
F.b.tasmanica
Tasmania
Falco novaezeelandiae (New Zealand Falcon)
New Zealand
Falco subbuteo (Northern Hobby)
F.s.subbuteo
Europe to Japan >> Africa & India
F.s.jugurtha
E Mediterranean to S USSR & N India
F.s.streichi
C & S China, Laos
Falco cuvieri (African Hobby)
Ghana to Ethiopia & Cape Province
Falco severus (Oriental Hobby)
E Himalayas to Philippine Is & New Guinea
Falco longipennis (Australian Hobby)
F.l.murchisonianus
dry N Australia
F.l.hanieli
N Australia >> Lesser Sunda Is, New
Guinea
F.l.longipennis
SE & SW Australia, Tasmania
Falco eleonorae (Eleonora's Falcon)
Canary Is, Mediterranean Is >>
Madagascar
Falco concolor (Sooty Falcon)
NE Africa >> Madagascar
Falco rufigularis (Bat Falcon)
F.r.petrophilus
W Mexico
F.r.rufigularis
C Mexico, all E Sth America to N Argentina
Falco femoralis (Aplomado Falcon)
F.f.septentrionalis
SW USA & Mexico
F.f.pichinchae
Andes from Colombia to Chile
F.f.femoralis
C & S America
Falco hypoleucos (Grey Falcon)
NC & Western Australia
Falco subniger (Black Falcon)
Australia
Falco biarmicus (Lanner Falcon)

F.b.feldeggii
SE Europe, Asia Minor
F.b.erlangeri
NW Africa
F.b.tanypterus
NE Africa, Arabia, Iraq
F.b.abyssinicus
Ghana to N Zaire, Uganda, Ethiopia
F.b.biarmicus
E Zaire & Kenya to Angola & Cape Province
Falco mexicanus (Prairie Falcon)
SW Canada, W USA, NW Mexico
Falco jugger (Lagger Falcon)
Baluchistan, Himalayas, N & C India
Falco cherrug (Saker Falcon)
F.c.cyanopus
C Europe, W Russia
F.c.cherrug
C Asia, NW Mongolia to N Africa, N India
F.c.milvipes
SC Asia
F.c.altaicus
mountains of C Asia
Falco rusticolus (Gyr Falcon)
Arctic Europe, Asia, N America
Falco deiroleucus (Orange-breasted Falcon)
C America to N Argentina
Falco fasciinucha (Taita Falcon)
S Ethiopia to Zambia & Malawi
Falco peregrinus (Peregrine Falcon)
F.p.pealei
coast of W Canada, W USA
F.p.anatum
NC & S America
F.p.cassini
S Chile, Tierra del Fuego, Falkland Is   32.1
F.p.peregrinus
Europe to N Russia & Caucasus
F.p.calidus
N Russia, N Siberia to Southern Africa &
New Guinea
F.p.japonensis
E Siberia, Japan, Taiwan
F.p.brookei
Mediterranean, Asia Minor
F.p.babylonicus
Iraq to Mongolia, N India
F.p.peregrinator
India, Sri Lanka to S China
F.p.minor
Ghana to Ethiopia >> Cape Province
F.p.submelanogenys
SW Australia
F.p.macropus
Australia (except SW)
F.p.madens
Cape Verde Is
F.p.radama
Madagascar, Comoro Is
F.p.furuitii
Volcano I

F.p.ernesti
Indonesia, Philippine Is, New Guinea, Sulawesi
F.p.nesiotes
New Hebrides, Loyalty Is, New Caledonia
Falco pelegrinoides (Barbary Falcon) 32.4
N Africa, Middle East

GALLIFORMES

33. MEGAPODIAE

MEGAPODIUS 33.1.2.
Megapodius nicobariensis (Nicobar Scrub Fowl)
M.n.nicobariensis
N.Nicobar Is
M.n.abbotti
Gt & Little Nicobar
Megapodius cumingii (Philippine Scrub Fowl)
M.c.pusillus
Philippine Is
M.c.tabon
Mindanao
M.c.cumingii
Palawan, Balabac Is, Sulawesi
M.c.sanghirensis
Sangihe, Talaut
Megapodius bernsteinii (Sula Scrub Fowl)
Sula Is, Banggai
Megapodius affinis (New Guinea Scrub Fowl)
N New Guinea
Megapodius reinwardt (Orange-footed Scrub Fowl)
M.r.perrufus
Peleng
M.r.tenimberensis
Tanimbar Is
M.r.aruensis
Aru Is
M.r.duperryi
S & W New Guinea
M.r.reinwardt
Lesser Sunda Is
M.r.forstenii
S Moluccas, Buru
M.r.macgillivrayi
Louisiade Arch., d'Entrecasteaux Arch.
M.r.tumulus
N Northern Territory
M.r.yorki
N Queensland
M.r.castanonotus
N & C Queensland
Megapodius eremita (Bismarck Scrub Fowl)
M.e.eremita
Admiralty Is, Bismarck Arch.
M.e.brenchleyi
Solomon Is
Megapodius freycinet (Dusky Scrub Fowl)
M.f.freycinet
N Moluccas, W Papuan Is

M.f.geelvinkianus
Biak, Numfor, Japen, Meosnum Is
Megapodius layardi (Banks Is Scrub Fowl)
New Hebrides, Banks Is
Megapodius laperouse (Marianas Scrub Fowl)
M.l.senex
Palau Is
M.l.laperouse
Marianas Is
Megapodius pritchardii (Polynesian Scrub Fowl)
Friendly Is
Megapodius wallacei (Moluccas Scrub Fowl)
33.3
Moluccas, Misol I

LEIPOA
Leipoa ocellata (Mallee Fowl)
Southern Australia

ALECTURA
Alectura lathami (Brush Turkey)
A.l.purpureicollis
N Queensland
A.l.lathami
C & S Queensland, New South Wales

TALEGALLA
Talegalla cuvieri (Red-billed Brush Turkey)
NW New Guinea, Salawati I, Misol I
Talegalla fuscirostris (Black-billed Brush Turkey)
T.f.fuscirostris
S & E New Guinea
T.f.occidentis
Aru Is, SW New Guinea
Talegalla jobiensis (Brown-collared Brush Turkey)
T.j.jobiensis
Japen I, N New Guinea
T.j.longicaudus
SW New Guinea

AEPYPODIUS
Aepypodius arfakianus (Wattled Brush Turkey)
A.a.arfakianus
NW New Guinea
A.a.misoliensis
Misol I
Aepypodius bruijnii (Bruijn's Brush Turkey)
Waigeu I

MACROCEPHALON
Macrocephalon maleo (Maleo Fowl)
Sulawesi

34. CRACIDAE (CURASSOWS, GUANS)

ORTALIS
Ortalis vetula (Plain Chachalaca)
O.v.mccallii
S Texas, SE Mexico

O.v.vetula
  E Mexico to C Nicaragua
O.v.pallidiventris
  N Yucatan
O.v.deschauenseei
  Utilia I (Honduras)
Ortalis cinereiceps (Grey-headed Chachalaca)
  SE Honduras to NW Colombia
Ortalis garrula (Chestnut-winged Chachalaca)
O.g.mira
  Panama
O.g.chocoensis
  NW Colombia
O.g.garrula
  N Colombia
Ortalis ruficauda (Rufous-vented Chachalaca)
O.r.ruficrissa
  N Colombia, NW Venezuela
O.r.lamprophonia
  Guajira peninsula
O.r.baliolus
  L Maracaibo
O.r.ruficauda
  Venezuela, Tobago I, Lesser Antilles
Ortalis erythroptera (Rufous-headed Chachalaca)
  W Ecuador, NW Peru
Ortalis poliocephala (West Mexican Chachalaca)
O.p.wagleri
  N W Mexico
O.p.lajuelae
  C Mexico
O.p.poliocephala
  W Mexico
Ortalis canicollis (Chaco Chachalaca)
O.c.canicollis
  E Bolivia, W Paraguay, N Argentina
O.c.pantanalensis
  SW Brazil
Ortalis leucogastra (White-bellied Chachalaca)
  SE Mexico to NW Nicaragua
Ortalis motmot (Variable Chachalaca)
O.m.motmot
  S Venezuela, the Guianas, N Brazil
O.m.ruficeps
  N C Brazil
O.m.superciliaris
  NE Brazil
O.m.araucuan
  E Brazil
O.m.squamata
  SE Brazil
O.m.caucae
  N Colombia
O.m.colombiana
  SC Colombia, NW Upper Amazonia
O.m.guttata
  W Upper Amazonia
O.m.subaffinis
  NE & E Bolivia, W Brazil

PENELOPE
Penelope argyrotis (Band-tailed Guan)
P.a.mesaeus
  W Venezuela, N Colombia
P.a.albicauda
  W Venezuela, NE Colombia
P.a.colombiana
  Santa Marta Mts
P.a.argyrotis
  NW Colombia, N Venezuela
P.a.olivaceiceps
  N Venezuela
Penelope barbata (Bearded Guan)
  S Ecuador, NW Peru
Penelope montagnii (Andean Guan)
P.m.montagnii
  N & C Colombia
P.m.atrogularis
  S Colombia, WC Ecuador
P.m.brooki
  S Colombia, EC Ecuador
P.m.plumosa
  C Peru
P.m.sclateri
  C Bolivia, NW Argentina
Penelope ortoni (Baudo Guan)
  W Colombia.W Ecuador
Penelope marail (Marail Guan)
P.m.jacupeba
  SE Venezuela, N Brazil
P.m.marail
  E Venezuela, the Guianas
Penelope superciliaris (Rusty-margined Guan)
P.s.superciliaris
  NC & E Brazil
P.s.jacupemba
  C & S Brazil
P.s.major
  S Brazil, E Paraguay, NE Argentina
Penelope dabbenei (Red-faced Guan)
  S Bolivia, NW Argentina
Penelope obscura (Dusky-legged Guan)
P.o.bronzina
  E Brazil
P.o.obscura
  S Brazil, E Paraguay, Uruguay, NE Argentina
P.o.bridgesi
  C Bolivia, NW Argentina
Penelope jacquacu (Spix's Guan)
P.j.granti
  E Venezuela, Guyana
P.j.orienticola
  NW Brazil, SE Venezuela
P.j.jacquacu
  Upper Amazonia
P.j.speciosa
  C & E Bolivia
Penelope albipennis (White-winged Guan)
  NW Peru
Penelope perspicax (Cauca Guan)
  W & C Colombia

Penelope purpurascens (Crested Guan)
  P.p.purpurascens
    NE & NW Mexico to Honduras, Nicaragua
  P.p.aequatorialis
    S Honduras to NW Colombia
  P.p.brunnescens
    N Colombia, E Venezuela
Penelope jacucaca (White-browed Guan)
  NE Brazil
Penelope ochrogaster (Chestnut-bellied Guan)
  SC Brazil
Penelope pileata (White-crested Guan)
  NC Brazil

ABURRIA
Aburria pipile (Blue-throated Piping Guan)
  A.p.pipile
    Trinidad
  A.p.cumanensis
    the Guianas to C Colombia, NW Peru, W Brazil
  A.p.grayi
    E Bolivia, NE Paraguay, SW Brazil
Aburria cujubi (Red-throated Piping Guan)
  A.c.cujubi
    NC Brazil
  A.c.nattereri
    S & W Amazonia
Aburria jacutinga (Black-fronted Piping Guan)
  SE Brazil, SE Paraguay
Aburria aburri (Wattled Guan)
  N Colombia, E Venezuela to SC Peru

CHAMAEPETES
Chamaepetes unicolor (Black Guan)
  Costa Rica, Panama
Chamaepetes goudotii (Sickle-winged Guan)
  C.g.goudotii
    C & W Colombia
  C.g.sanctaemarthae
    Santa Marta Mts
  C.g.fagani
    SW Colombia, W Ecuador
  C.g.tschudii
    EC Ecuador, N Peru
  C.g.rufiventris
    EC Peru

PENELOPINA
Penelopina nigra (Highland Guan)
  S Mexico to N Nicaragua

OREOPHASIS
Oreophasis derbianus (Horned Guan)
  SE Mexico, SW Guatemala

NOTHOCRAX
Nothocrax urumutum (Nocturnal Curassow)
  Upper Amazonia

CRAX
Crax tomentosa (Lesser Razor-billed Curassow)
  SE Colombia to Guyana, NW Brazil
Crax salvini (Salvin's Curassow)
  SW Colombia, E Ecuador, NE Peru
Crax mitu (Razor-billed Curassow)
  C.m.tuberosa
    S Amazonia
  C.m.mitu
    E Brazil e?
Crax pauxi (Helmeted Curassow)
  C.p.pauxi
    NC to W Venezuela
  C.p.gilliardi
    Venezuela-Colombia border
Crax unicornis (Horned Curassow)
  C.u.unicornis
    NE Bolivia
  C.u.koepckeae
    E Peru
Crax rubra (Great Curassow)
  C.r.rubra
    E Mexico to W Colombia & W Ecuador
  C.r.griscomi
    Cozumel I
Crax alberti (Blue-billed Curassow)
  N Colombia
Crax daubentoni (Yellow-knobbed Curassow)
  N Venezuela
Crax alector (Black Curassow)
  E Colombia to the Guianas, N Brazil
Crax globulosa (Wattled Curassow)
  W Upper Amazonia
Crax fasciolata (Bare-faced Curassow)
  C.f.fasciolata
    C & SW Brazil, Paraguay
  C.f.pinima
    NE Brazil
  C.f.grayi
    E Bolivia
Crax blumenbachii (Red-billed Curassow)
  SE Brazil

35. PHASIANIDAE (PHEASANTS, GROUSE)

MELAGRIDINAE

MELEAGRIS
Meleagris gallopavo (Common Turkey)
  M.g.silvestris
    SE USA
  M.g.osceola
    S Florida
  M.g.intermedia
    N Texas to NE Mexico
  M.g.onusta
    NW Mexico
  M.g.mexicana
    NC Mexico
  M.g.merriami
    SW USA, NW Mexico

M.g.gallopavo
WC Mexico

Agriocharis ocellata (Ocellated Turkey)
SE Mexico, Belize, Guatemala

TETRAONINAE

Dendragapus falcipennis (Siberian Spruce Grouse)
NE Asia, Sakhalin
Dendragapus canadensis (Spruce Grouse)
D.c.osgoodi
Alaska
D.c.atratus
S Alaska
D.c.canadensis
C Alberta to Labrador
D.c.torridus
Nova Scotia
D.c.canace
SE Canada, N & NE USA
Dendragapus obscurus (Blue Grouse) 35.5
D.o.sitkensis
SE Alaska
D.o.fuliginosus
S Yukon to NW California
D.o.sierrae
Oregon to NC California
D.o.howardi
C California
D.o.richardsonii
N British Columbia to SE Idaho
D.o.pallidus
SE British Columbia to NE Oregon
D.o.obscurus
Utah to New Mexico

Lagopus lagopus (Willow/Red Grouse)
L.l.scoticus
Scotland, Wales, N England
L.l.hibernicus
Outer Hebrides, Ireland
L.l.lagopus
Circumpolar, N Europe, N Asia, N Canada
L.l.birulai
N Siberian islands
L.l.leucopterus
islands N of N America
L.l.rossicus
European Russia
L.l.maior
SE Russia
L.l.brevirostris
S Siberia
L.l.kozlowae
C Asia, N Mongolia

L.l.alexandrae
S & SE Alaska islands, NW British Columbia
L.l.alleni
Newfoundland
Lagopus mutus (Rock Ptarmigan)
L.m.hyperboreus
Spitzbergen, Franz Josef Land
L.m.mutus
N Scandinavia, N Russia
L.m.millaisi
Scotland
L.m.helveticus
Alps
L.m.pyrenaicus
Pyrenees
L.m.komensis
N Ural Mts
L.m.pleskei
N Siberia
L.m.macrorhynchus
Tarbagatai
L.m.nadezdae
Altai, C Asia
L.m.transbaicalicus
SE Siberia
L.m.ridgwayi
Commander Is
L.m.kurilensis
N & C Kurile Is
L.m.japonicus
Hondo I, Japan
L.m.evermanni
Attu I (Aleutian Is)
L.m.townsendi
Kiska I (Aleutian Is)
L.m.sanfordi
Tanaga I (Aleutian Is)
L.m.chamberlaini
Adak I (Aleutian Is)
L.m.atkhensis
Atka I (Aleutian Is)
L.m.nelsoni
Unimak I, Unalaska I to S Alaska
L.m.gabrielsoni
C Alaska
L.m.dixoni
Glacier Bay islands
L.m.rupestris
N Nth America
L.m.saturatus
W Greenland
L.m.welchi
Newfoundland
L.m.reinhardi
SW Greenland
L.m.captus
E Greenland
L.m.islandorum
Iceland
Lagopus leucurus (White-tailed Ptarmigan)
L.l.peninsularis
C Alaska, Yukon

L.l.leucurus
N British Columbia to Vancouver I
L.l.rainierensis
Mt Rainier, C & S Washington
L.l.saxatilis
Vancouver I
L.l.altipetens
Rocky Mts from Montana to New Mexico

Tetrao mlokosiewiczi (Caucasian Black Grouse)
Caucasus
Tetrao tetrix (Black Grouse)
T.t.britannicus
Scotland, N England
T.t.tetrix
Scandinavia & France to N Siberia
T.t.viridanus
SE Russia, SW Siberia
T.t.tschusii
S Siberia
T.t.baikalensis
N Mongolia, W Manchuria
T.t.mongolicus
C Tien Shan, W Altai
T.t.ussuriensis
N Manchuria, NE Korea
Tetrao parvirostris (Black-billed Capercaillie)
T.p.turensis
NC Siberia
T.p.janensis
NE Siberia
T.p.parvirostris
E Siberia, Sakhalin
T.p.macrurus
C Asia, N Mongolia
T.p kamschaticus
Kamchatka
Tetrao urogallus (Western Capercaillie)
T.u.aquitanicus
Pyrenees & N Spain
T.u.urogallus
Scandinavia, Scotland
T.u.major
C Europe, W Russia
T.u.lugens
Finland, NW Russia
T.u.pleskei
N Russia
T.u.volgensis
C Russia
T.u.uralensis
C Ural Mts
T.u.grisescens
S Ural Mts
T.u.kureikensis
Lower Yenisei valley
T.u.taczanowskii
C Siberia, NW Mongolia

Bonasa sewerzowi (Severtzov's Hazel Grouse)
B.s.sewerzowi
Kansu
B.s.secunda
W Szechwan
Bonasa bonasia (Hazel Grouse)
B.b.bonasia
Scandinavia to Ural Mts
B.b.griseonota
N Sweden
B.b.rupestris
C Germany to Alps & Bulgaria
B.b.horicei
E Carpathians
B.b.volgensis
C Poland to C Russia
B.b.sibiricus
C Siberia, N Mongolia
B.b.kolymensis
E Siberia
B.b.amurensis
Korea, S Amur, N Manchuria
B.b.vicinitas
Sakhalin, Hokkaido
Bonasa umbellus (Ruffed Grouse)
B.u.yukonensis
Alaska, NW Canada
B.u.umbelloides
S British Columbia to Manitoba & N Colorado
B.u.castaneus
Mt.Olympus , Washington
B.u.affinis
British Columbia, Oregon
B.u.obscura
N Ontario
B.u.sabini
coast of British Columbia to California
B.u.brunnescens
Vancouver I
B.u.togata
NC & NE USA
B.u.medianus
Minnesota
B.u.phaios
Idaho
B.u.incanus
Utah
B.u.monticola
W Virginia
B.u.umbellus
EC USA
B.u.thayeri
Nova Scotia

Centrocercus urophasianus (Sage Grouse)
S British Columbia to E California & Nebraska

GALLIFORMES

TYMPANUCHUS
Tympanuchus phasianellus (Sharp-tailed Grouse)
T.p.kennicottii
NW Canada
T.p.phasianellus
EC Canada
T.p columbianus
British Columbia to N California & Utah
T.p.jamesi
EC Colorado
T.p.campestris
C Canada to Wisconsin
Tympanuchus cupido (Prairie Chicken)
T.c.pinnatus
SC Canada to NE Texas
T.c.attwateri
coast of Texas & SW Louisiana
Tympanuchus pallidicinctus (Lesser Prairie Chicken)   35.5
Kansas to New Mexico

ODONTOPHORINAE

DENDRORTYX
Dendrortyx barbatus (Bearded Tree Quail)
Vera Cruz (Mexico)
Dendrortyx macroura (Long-tailed Tree Quail)
D.m.macroura
Vera Cruz
D.m.griseipectus
Morales (Mexico)
D.m.diversus
NW Jalisco
D.m.striatus
Michoacan & Colima (Mexico)
D.m.oaxacae
E Oaxaca (Mexico)
Dendrortyx leucophrys (Buffy-crowned Tree Quail)
D.l.leucophrys
SE Mexico, Guatemala
D.l.nicaraguae
Honduras, Nicaragua
D.l.hypospodius
Costa Rica

OREORTYX
Oreortyx picta (Mountain Quail)
O.p.palmeri
coast from SW Washington to California
O.p.picta
Columbia River to California
O.p.russelli
California
O.p.confinis
Baja California

CALLIPEPLA
Callipepla squamata (Scaled Quail)
C.s.pallida
SW USA, NW Mexico

C.s.squamata
NC & C Mexico
C.s.castanogastris
S Texas, NE Mexico
C.s.hargravei
New Mexico

LOPHORTYX
Lophortyx californica (California Quail)
L.c.brunnescens
coast from SW Oregon to C California
L.c.canfieldae
EC California
L.c.decolorata
Baja California
L.c.californica
E Oregon to Baja California
L.c.catalinensis
Santa Catalina I (Los Coronados Is)
L.c.achrustera
S Baja California
Lophortyx gambelii (Gambel's Quail)
L.g.gambelii
SW USA, NW Mexico
L.g.sana
W Colorado
L.g.friedmann
coast of NW Mexico
L.g.fulvipectus
SW Sonora
L.g.pembertoni
Tiburon I
Lophortyx douglasii (Elegant Quail)
L.d.bensoni
Sonora
L.d.douglasii
Sinaloa, Jalisco
L.d.languens
C Chihuahua
L.d.impedita
Nayarit
L.d.teres
NW Jalisco

PHILORTYX
Philortyx fasciatus (Banded Quail)
Colima, Guerrero, Puebla (SW Mexico)

COLINUS
Colinus virginianus (Northern Bobwhite)
C.v.virginianus
C & E USA
C.v.floridanus
Florida, Bahama Is
C.v.cubanensis
Cuba
C.v.ridgwayi
N & SC Sonora
C.v.texanus
NE Mexico
C.v.maculatus
C Mexico

C.v.aridus
NC Mexico
C.v.graysoni
WC Mexico
C.v.nigripectus
SC Mexico
C.v.pectoralis
C Vera Cruz
C.v.godmani
E Vera Cruz
C.v.minor
NE Chiapas
C.v.insignis
NW Guatemala, W Chiapas
C.v.salvini
S Chiapas
C.v.coyolcos
S Mexico
C.v.thayeri
NE Oaxaca
C.v.atriceps
W Oaxaca
C.v.nelsoni
C Chiapas
Colinus nigrogularis (Black-throated Quail)
C.n.caboti
Campeche (Mexico)
C.n.persiccus
N Yucatan
C.n.nigrogularis
Yucatan
C.n.segoviensis
Guatemala, Honduras
Colinus cristatus (Crested Bobwhite)
C.c.incanus
Guatemala
C.c.hypoleucus
W Guatemala, W El Salvador
C.c.leucopogon
El Salvador   35.3
C.c.leylandi
Honduras
C.c.sclateri
W Honduras, W Nicaragua
C.c.dickeyi
SW Nicaragua, W Costa Rica
C.c.panamensis
W Panama
C.c.decoratus
N Colombia
C.c.bogotensis
C Colombia
C.c.badius
C Colombia
C.c.leucotis
S Colombia
C.c.parvicristatus
EC Colombia
C.c.littoralis
NE Colombia
C.c.cristatus
E Colombia, W Venezuela, Aruba I,
Curacao I

C.c.horvathi
Venezuela
C.c.sonnini
E Venezuela, the Guianas, N Brazil
C.c.mocquerysi
NW Venezuela, Margarita I
C.c.mariae
W Panama
C.c.continentis
NW Venezuela
C.c.barnesi
NC Venezuela

Odontophorus gujanensis (Marbled Wood Quail)
O.g.castigatus
SW Costa Rica, NW Panama
O.g.marmoratus
N Colombia, E Panama
O.g.polionotus
NW Venezuela
O.g.gujanensis
E Venezuela, the Guianas
O.g.medius
Mt Duida (S Venezuela)
O.g.buckleyi
SE Colombia, E Ecuador, W Brazil
O.g.rufogularis
NE Peru
O.g.pachyrhynchus
E Peru
O.g.simonsi
NW Bolivia
Odontophorus capueira (Spot-winged Wood Quail)
E Brazil, Paraguay
Odontophorus erythrops (Rufous-fronted Wood Quail)
O.e.verecundus
Honduras
O.e.melanotis
Nicaragua, N & E Costa Rica
O.e.coloratus
W Panama
O.e.erythrops
E Ecuador
O.e.parambae
W Colombia, NW Ecuador
Odontophorus atrifrons (Black-fronted Wood Quail)
O.a.atrifrons
Santa Marta Mts (Colombia)
O.a.variegatus
E Colombia
O.a.navai
W Venezuela
Odontophorus melanonotus (Black-backed Wood Quail)
W Ecuador
Odontophorus hyperythrus (Chestnut Wood Quail)
Andes of Colombia

Odontophorus speciosus (Rufous-breasted Wood Quail)
  O.s.soederstroemii
    C Ecuador
  O.s.speciosus
    E Ecuador, E Peru
  O.s.loricatus
    C Bolivia
Odontophorus strophium (Gorgeted Wood Quail)
  C Colombia
Odontophorus dialeucos (Tacarcuna Wood Quail)   35.5
  Panama
Odontophorus colombianus (Venezuela Wood Quail)
  N Venezuela
Odontophorus leucolaemus (White-throated Wood Quail)
  N Costa Rica, W Panama
Odontophorus balliviani (Stripe-faced Wood Quail)
  Peru, Bolivia
Odontophorus stellatus (Starred Wood Quail)
  E Ecuador, E Peru
Odontophorus guttatus (Spotted Wood Quail)
  S Mexico to W Panama

DACTYLORTYX
Dactylortyx thoracicus (Singing Quail)
  D.t.thoracicus
    E coast of Mexico
  D.t.sharpei
    Yucatan peninsula
  D.t.devius
    W Mexico
  D.t.lineolatus
    S Mexico
  D.t.chiapensis
    W Guatemala, C Chiapas
  D.t.fuscus
    E Chiapas to S Honduras
  D.t.salvadoranus
    C El Salvador
  D.t.taylori
    N El Salvador, S Honduras
  D.t.colophonus
    W Guatemala
  D.t.rufescens
    Honduras
  D.t.conoveri
    EC Honduras

CYRTONYX
Cyrtonyx montezumae (Montezuma's Quail)
  C.m.mearnsi
    S USA, NW Mexico
  C.m.montezumae
    N & C Mexico
  C.m.merriami
    Mt Orizaba (Vera Cruz, Mexico)
Cyrtonyx sallei (Salle's Quail)
  SW Mexico

Cyrtonyx ocellatus (Ocellated Quail)
  C.o.ocellatus
    SW Mexico
  C.o.differens
    W Honduras, N Nicaragua

RHYNCHORTYX
Rhynchortyx cinctus (Tawny-faced Quail)
  R.c.pudibundus
    N Honduras
  R.c.cinctus
    Nicaragua to SE Panama
  R.c.hypopius
    NE Panama
  R.c.australis
    W Colombia, NW Ecuador

PHASIANINAE

LERWA
Lerwa lerwa (Snow Partridge)
  Afghanistan, Himalayas, W China

AMMOPERDIX
Ammoperdix griseogularis (See See Partridge)
  A.g.peraticus
    W Afghanistan
  A.g.griseogularis
    S Russia, Iran to NW India
Ammoperdix heyi (Sand Partridge)
  A.h.heyi
    River Jordan to Sinai
  A.h.nicolli
    N Egypt
  A.h.cholmleyi
    River Nile to Red Sea
  A.h.intermedia
    S Arabia

TETRAOGALLUS
Tetraogallus caucasicus (Caucasian Snowcock)
  Caucasus
Tetraogallus caspius (Caspian Snowcock)
  T.c.caspius
    Taurus Mts to N Iran
  T.c.semenowtianschanskii
    Zagros Mts(Iran)
Tetraogallus tibetanus (Tibetan Snowcock)
  T.t.tibetanus
    Pamir Mts (W Tibet)
  T.t.tschimenensis
    N Tibet
  T.t.centralis
    NE & C Tibet
  T.t.przewalskii
    E Tibet, W Kansu
  T.t.henrici
    W China
  T.t.aquilonifer
    S Tibet, Sikkim
Tetraogallus altaicus (Altai Snowcock)

T.a.altaicus
  Altai Mts, Sajan Mts
T.a.orientalis
  NW Mongolia
Tetraogallus himalayensis (Himalayan Snowcock)
  T.h.sewerzowi
    SE Turkestan
  T.h.himalayensis
    W Himalayas, E Afghanistan
  T.h.bendi
    NW Afghansitan
  T.h.grombczewskii
    W Kwenlun Mts
  T.h.koslowi
    Humboldt & S Kokonor Mts

TETRAOPHASIS
Tetraophasis obscurus (Verreaux's Monal Partridge)
  NE Tibet, W China
Tetraophasis szechenyii (Szechenyi's Monal Partridge)
  E Tibet, SW China

ALECTORIS
Alectoris graeca (Rock Partridge)
  A.g.saxatilis
    Alps
  A.g.orlandoi
    Appenine Mts
  A.g.graeca
    SE Europe
  A.g.scotti
    Crete
  A.g.sinaica
    Syria to Sinai
  A.g.daghestanica
    N Caucasus
  A.g.caucasica
    S Caucasus
  A.g.werae
    SW Iran
  A.g.koroviakovi
    E & S Iran
  A.g.shestoperovi
    S Transcaspia
  A.g.subpallida
    Kyzylkum Mts
  A.g.falki
    W & C Tien Shan Mts
  A.g.dzungarica
    Tarbagatai Mts
  A.g.fallax
    E Tien Shan Mts
  A.g.pallida
    S Chinese Turkestan
  A.g.pallescens
    N India
  A.g.obscurata
    W Tannu Ola Mts
Alectoris chukar (Chukar Partridge)

A.c.kleini
  E Greece
A.c.cypriotes
  Cyclades, Asia Minor
A.c.kurdestanica
  S Kurdistan
A.c.chukar
  Himalayas
A.c.potanini
  W Mongolia
A.c.pubescens
  S Manchuria, N China
Alectoris magna (Przewalski's Rock Partridge)
  E Tibet, W Kansu
Alectoris philbyi (Philby's Rock Partridge)
  SW Arabia
Alectoris barbara (Barbary Partridge)
  A.b.barbara
    N Morocco, N Algeria, N Tunisia
  A.b.theresae
    S Morocco
  A.b.koenigi
    Canary Is
  A.b.spatzi
    S Algeria, S Tunisia
  A.b.barbata
    Libya
Alectoris rufa (Red-legged Partridge)
  A.r.rufa
    SC Europe
  A.r.hispanica
    NW Spain & N Portugal
  A.r.intercedens
    S Spain
  A.r.corsa
    Corsica
  A.r.australis
    Gran Canaria I
Alectoris melanocephala (Arabian Chukar)
  A.m.melanocephala
    SW Arabia
  A.m.guichardi
    E Hadhramaut

ANUROPHASIS
Anurophasis monorthonyx (Snow Mountain Quail)
  Oranje Mts (New Guinea)

FRANCOLINUS
Francolinus francolinus (Black Partridge)
  F.f.francolinus
    Cyprus to Caucasia & N Iran
  F.f.billypaynei
    Syria
  F.f.arabistanicus
    S Iraq, W Iran
  F.f.bogdanovi
    S Iran
  F.f.henrici
    Pakistan

F.f.asiae
N India
F.f.melanonotus
E Himalayas, Assam
Francolinus pictus (Painted Partridge)
F.p.pallidus
NC India
F.p.pictus
S India
F.p watsoni
Sri Lanka
Francolinus pintadeanus (Chinese Francolin)
F.p.phayrei
NE India to S China, Indochina
F.p.pintadeanus
SE China
Francolinus afer (Red-necked Spurfowl)
F.a.nyanzae
Uganda, W Kenya, W Tanzania
F.a.harterti
Rwanda, Burundi
F.a.cranchii
N Angola, N Zambia, W Tanzania
F.a.leucoparaeus
E Kenya, N Tanzania
F.a.bohmi
W Tanzania
F.a.itigi
C Tanzania
F.a.intercedens
SE Zaire, S Tanzania, N Zambia
F.a.castaneiventer
E Cape Province
F.a.loangwae
NE Zambia
F.a.benguellensis
W Angola
F.a.punctulatus
C Angola
F.a.afer
S Angola, Namibia
F.a.humboldtii
S Malawi, W Mozambique
F.a.swynnertoni
Zimbabwe, S Mozambique
F.a.lehmanni
E Transvaal, SE Cape Province, S Natal
F.a.notatus
S Cape Province
Francolinus swainsonii (Swainson's Francolin)
F.s.gilli
N Namibia, N Botswana, W Zambia
F.s.damarensis
N Namibia
F.s.lundazi
N Zimbabwe, S Mozambique
F.s.swainsonii
S Botswana, Transvaal, S Mozambique
Francolinus rufopictus (Painted Francolin)
SE Lake Victoria
Francolinus leucoscepus (Yellow-necked Francolin)

F.l.leucoscepus
E Ethiopia, N Somalia
F.l.muhamedbenabdullah
S Somalia, N Kenya
F.l.infuscatus
NE Uganda, S Ethiopa to N Tanzania
Francolinus erckelii (Erckel's Francolin)
F.e.erckelii
Ethiopia
F.e.pentoni
NE Sudan
Francolinus ochropectus (Pale-bellied Francolin)
Somalia
Francolinus castaneicollis (Chestnut-naped Francolin)
F.c.ogoensis
Somalia
F.c.castaneicollis
E Ethiopia
F.c.bottegi
S Ethiopia
F.c.kaffanus
W Ethiopia
F.c.gofanus
SW Ethiopia
F.c.atrifrons
S Ethiopia
Francolinus jacksoni (Jackson's Francolin)
F.j.jacksoni
Higher Aberdare Mts
F.j.pollenorum
Mt Kenya
F.j.gurae
Lower Aberdare Mts
Francolinus nobilis (Handsome Francolin)
F.n.nobilis
E Zaire, SW Uganda
F.n.chapini
Ruwenzori Mts (E Zaire)
Francolinus camerunensis (Cameroun Mountain Francolin)
Cameroun Mts
Francolinus swierstrai (Swierstra's Francolin)
S Angola
Francolinus ahantensis (Ahanta Francolin)
F.a.hopkinsoni
Gambia, Guinea
F.a.ahantensis
Guinea to Nigeria
Francolinus squamatus (Scaly Francolin)
F.s.squamatus
S Nigeria to N Zaire
F.s.schuetti
Angola to Ethiopia, W Kenya
F.s.zappeyi
Uganda, W Kenya
F.s.tetraoninus
W Ethiopia
F.s.maranensis
S Kenya
F.s.usambarae
Usambara Mts (Tanzania)

F.s.uzungwensis
Uzungwe Mts (Tanzania)
F.s.doni
Vipya plateau (W Malawi)
Francolinus griseostriatus (Grey-striped Francolin)
N Angola
Francolinus bicalcaratus (Double-spurred Francolin)
F.b.ayesha
W Morocco
F.b.bicalcaratus
Senegal to Niger & N Nigeria
F.b.thornei
Sierra Leone to Benin
F.b.adamauae
N Nigeria, Cameroun
F.b.ogilvie-granti
Cameroun
Francolinus icterorhynchus (Yellow-billed Francolin)
F.i.icterorhynchus
C African Republic to SW Sudan
F.i.dybowskii
NE Zaire, W Uganda
F.i.ugandensis
C Uganda
Francolinus clappertoni (Clapperton's Francolin)
F.c.clappertoni
Mali to W Sudan
F.c.heuglini
SW Sudan
F.c.cavei
SE Sudan
F.c.gedgii
Mt Elgon
F.c.sharpii
E Ethiopia
F.c.testis
Ethiopia
F.c.nigrosquamatus
Ethiopia
Francolinus hildebrandti (Hildebrandt's Francolin)
F.h.helleri
N Kenya
F.h.altumi
W Kenya
F.h.hildebrandti
E Kenya to NE Zambia & W Malawi
F.h.fischeri
C Tanzania
F.h.grotei
SE Tanzania
F.h.johnstoni
S Tanzania, E Zambia, S Malawi, Mozambique
Francolinus natalensis (Natal Francolin)
F.n.neavei
NE Zambia, W Mozambique
F.n.natalensis
S Zambia to Natal

F.n.thamnobium
N Cape Province, Zimbabwe to C Mozambique
Francolinus hartlaubi (Hartlaub's Francolin)
F.h.hartlaubi
Angola
F.h.bradfieldi
N Namibia
F.h.crypticus
Onguato area, C Namibia
Francolinus harwoodi (Harwood's Francolin)
S Ethiopia
Francolinus adspersus (Red-billed Francolin)
F.a.adspersus
S Angola, Namibia, Botswana
F.a.kalahari
NE Namibia, SW Zambia, NW Botswana
Francolinus capensis (Cape Francolin)
Cape Province
Francolinus sephaena (Crested Francolin)
F.s.somaliensis
Somalia
F.s.grantii
Ethiopia to C Tanzania
F.s.zuluensis
S Mozambique
F.s.thompsoni
N Namibia to NW Zimbabwe
F.s.sephaena
E Zimbabwe to Mozambique & N Natal
F.s.zambesiae
Namibia to S Malawi
Francolinus streptophorus (Ring-necked Francolin)
Cameroun, W Kenya, NW Tanzania
Francolinus psilolaemus (Montane Francolin)
C Ethiopia
Francolinus rovuma (Kirk's Francolin) 35.4
E Mozambique, SE Tanzania
Francolinus shelleyi (Shelley's Francolin)
F.s.elgonensis
Kenya
F.s.theresae
Mt Kenya
F.s.shelleyi
Uganda to Natal
F.s.trothae
W Tanzania
F.s.whytei
SE Zaire, Zambia, N Malawi
F.s.sequestris
Natal
F.s.canidorsalis
SE Zimbabwe, N Transvaal, W Mozambique
Francolinus africanus (Greywing Francolin)
F.a.gutteralis
N Ethiopia
F.a.eritreae
NE Ethiopia
F.a.lorti
Somalia

F.a.ellenbecki
  C Ethiopia
F.a.archeri
  SC Ethiopia
F.a.friedmanni
  S Ethiopia
F.a.uluensis
  C Kenya
F.a.macarthuri
  SE Kenya
F.a.proximus
  Lesotho, W Natal, SE Transvaal
F.a.africanus
  S Africa
Francolinus levalliantoides (Archer's Greywing Francolin)
F.l.jugularis
  S Angola
F.l.pallidor
  Etosha Pan, N Namibia
F.l.wattii
  C Namibia
F.l.levalliantoides
  Botswana, Orange Free State, Transvaal
Francolinus levaillantii (Red-winged Francolin)
F.l.kikuyuensis
  Uganda, Kenya
F.l.crawshayi
  N Malawi
F.l.benguellensis
  S Angola
F.l.clayi
  W Zambia
F.l.levaillantii
  Transvaal, Natal, NE Cape Province
Francolinus finschi (Finsch's Francolin)
  SW Zaire, Angola
Francolinus coqui (Coqui's Francolin)
F.c.buckleyi
  Ghana to S Nigeria
F.c.spinetorum
  Mali to Nigeria
F.c.maharao
  S Ethiopia
F.c.angolensis
  Gabon to Angola & Zambia
F.c.ruahdae
  S Uganda
F.c.hubbardi
  W Kenya
F.c.thikae
  C Kenya
F.c.coqui
  Kenya to Botswana & Natal
F.c.kasaicus
  C Zaire
F.c.vernayi
  Botswana
F.c.hoeschianus
  N Namibia
F.c.campbelli
  S Mozambique, Natal

Francolinus albogularis (White-throated Francolin)
F.a.albogularis
  Senegal, Gambia
F.a.buckleyi
  Ghana to Cameroun
F.a.dewittei
  SE Zaire
F.a.meinertzhageni
  E Angola, NW Zambia
Francolinus schlegelii (Schlegel's Francolin)
  S Central African Republic to SW Sudan
Francolinus lathami (Latham's Francolin)
F.l.lathami
  Sierra Leone to Gabon & NW Zaire
F.l.schubotzi
  NE Zaire, Uganda
Francolinus nahani (Nahan's Forest Francolin)
  NE Zaire
Francolinus pondicerianus (Indian Grey Francolin)
F.p.mecranensis
  S Iran to Pakistan
F.p.interpositus
  N India
F.p.pondicerianus
  S India
F.p.ceylonensis
  Sri Lanka
Francolinus gularis (Swamp Partridge)
  NE India, Assam

PERDIX
Perdix perdix (Grey Partridge)
P.p.perdix
  British Isles, W & C Europe
P.p.armoricana
  NW France
P.p.sphagnetorum
  NE Holland, NW Germany
P.p.hispaniensis
  Pyrenees, N Spain
P.p.italica
  Italy
P.p.lucida
  NE Europe
P.p.robusta
  NW Russia
P.p.arenicola
  WC Russia
P.p.furvescens
  SW Russia, N Iran
P.p.canescens
  Transcaucasia to NW Iran
Perdix dauuricae (Daurian Partridge)
P.d.dauuricae
  EC Asia, Mongolia, N China
P.d.castaneothorax
  S Manchuria
P.d.turcomana
  E Turkestan, E Tien Shan

P.d.przewalskii
  E Nanshans, NW China
P.d.suschkini
  C Amur, Ussuriland
Perdix hodgsoniae (Tibetan Partridge)
P.h.koslowi
  W Nanshans, S Kokonor Mts
P.h.sifanica
  E Nanshans, SE Tibet, W China
P.h.caraganae
  E Ladakh, W Himalayas
P.h.hodgsoniae
  Tibet, E Himalayas, Assam

RHIZOTHERA
Rhizothera longirostris (Long-billed Wood Partridge)
R.l.longirostris
  Malaysia, Sumatra, W Borneo
R.l.dulitensis
  N Borneo

MARGAROPERDIX
Margaroperdix madagarensis (Madagascar Partridge)
  Madagascar

MELANOPERDIX
Melanoperdix nigra (Black Wood Partridge)
M.n.nigra
  Malaysia, Sumatra
M.n.borneensis
  Borneo

COTURNIX   35.1
Coturnix coturnix (Common Quail)
C.c.coturnix
  Europe, W Asia to C Africa, India
C.c.ussuriensis
  NE Asia
C.c.conturbans
  Azores
C.c.confisa
  Madeira, Canary Is
C.c.inopinata
  Cape Verde Is
C.c.africana
  Sthn Africa, Madagascar
C.c.erlangeri
  Ethiopia to E Zimbabwe, W Mozambique
Coturnix japonica (Japanese Quail)
  Sakhalin, Japan to Indonesia
Coturnix coromandelica (Rain Quail)
  India, Sri Lanka, Burma
Coturnix delegorguei (Harlequin Quail)
C.d.delegorguei
  Senegal to Ethiopia & S Africa
C.d.histrionica
  Sao Thome
C.d.arabica
  S Arabia
Coturnix pectoralis (Pectoral Quail)
  Australia, Tasmania

Coturnix australis (Brown Quail)
C.a.raaltenii
  Flores, Timor
C.a.pallidior
  Sumba, Savu
C.a.saturatior
  N New Guinea
C.a.lamonti
  C New Guinea
C.a.dogura
  S New Guinea
C.a.mafulu
  SE New Guinea
C.a.castaneus
  Lesser Sunda Is
C.a.plumbeus
  SE New Guinea
C.a.cervinus
  NW Australia
C.a.queenslandicus
  N Queensland
C.a.australis
  SW Australia & Queensland to Victoria
Coturnix ypsilophora (Tasmanian Brown Quail)   35.2
  SE Australia, Tasmania
Coturnix chinensis (Indian Blue Quail)   35.3
C.c.chinensis
  India to Malaysia & Indochina
C.c.adansonii
  Ethiopia to Sierra Leone & S Africa   35.3
C.c.trinkutensis
  Nicobar Is
C.c.palmeri
  Sumatra, Java
C.c.lineata
  Philippines, Borneo, Sualwesi & to Timor
C.c.lepida
  Bismarck Arch.
C.c.papuensis
  SE New Guinea
C.c.australis
  Queensland to Victoria
C.c.colletti
  Nthn Territory

PERDICULA
Perdicula asiatica (Jungle Bush Quail)
  Himalayas, N & C India
Perdicula argoondah (Rock Bush Quail)
  SE India
Perdicula erythrorhyncha (Painted Bush Quail)
P.e.erythrorhyncha
  SW India
P.e.blewitti
  C India
Perdicula manipurensis (Manipur Bush Quail)
P.m.inglisi
  N Assam
P.m.manipurensis
  S Assam

ARBOROPHILA

Arborophila torqueola (Common Hill Partridge)
  A.t.millardi
    NW India
  A.t.torqueola
    N & E India, S Tibet
  A.t.batemani
    N Burma
  A.t.griseata
    W N Vietnam
Arborophila rufogularis (Rufous-throated Hill Partridge)
  A.r.rufogularis
    N India
  A.r.intermedia
    Assam to NW Burma
  A.r.tickelli
    S Burma to SW Laos
  A.r.euroa
    S China, N Laos
  A.r.guttata
    C Vietnam
  A.r.annamensis
    S Vietnam
Arborophila atrogularis (White-cheeked Hill Partridge)
    Assam, N Burma
Arborophila crudigularis (White-throated Hill Partridge)
    Taiwan
Arborophila mandellii (Red-breasted Hill Partridge)
    Sikkim to E Assam
Arborophila brunneopectus (Brown-breasted Hill Partridge)
  A.b.brunneopectus
    E Assam & S Yunnan to S Thailand
  A.b.henrici
    N & C Vietnam
  A.b.albigula
    S Vietnam
Arborophila rufipectus (Boulton's Hill Partridge)
    W Szechwan
Arborophila gingica (Rickett's Hill Partridge)
    SE China
Arborophila davidi (David's Tree Partridge)
    S Vietnam
Arborophila cambodiana (Chestnut-headed Tree Partridge)
  A.c.cambodiana
    Cambodia
  A.c.diversa
    SE Thailand
Arborophila orientalis (Sumatran Hill Partridge)
  A.o.campbelli
    Malaysia
  A.o.rolli
    NW Sumatra
  A.o.sumatrana
    C Sumatra

  A.o.orientalis
    E Java
Arborophila javanica (Chestnut-bellied Tree Partridge)
  A.j.javanica
    W Java
  A.j.bartelsi
    C Java
  A.j.lawuana
    C Java
Arborophila rubrirostris (Red-billed Tree Partridge)
    Sumatra
Arborophila hyperythra (Red-breasted Tree Partridge)
    NW Borneo
Arborophila ardens (Hainan Hill Partridge)
    Hainan I

TROPICOPERDIX

Tropicoperdix charltonii (Chestnut-breasted Tree Partridge)
  T.c.charltonii
    S Thailand, Malaysia
  T.c.atjehensis
    N Sumatra
  T.c.tonkonensis
    N Vietnam
  T.c.graydoni
    Borneo
Tropicoperdix chloropus (Green-legged Hill Partridge)
  T.c.chloropus
    N Burma to W & S Thailand
  T.c.olivacea
    Laos, Cambodia
  T.c.cognacqi
    S Vietnam
Tropicoperdix merlini (Annamese Hill Partridge)
  T.m.merlini
    C Vietnam
  T.m.vivida
    EC Vietnam

CALOPERDIX

Caloperdix oculea (Ferruginous Wood Partridge)
  C.o.oculea
    S Thailand, Malaysia
  C.o.sumatrana
    Sumatra
  C.o.borneensis
    Borneo

HAEMATORTYX

Haematortyx sanguiniceps (Crimson-headed Wood Partridge)
    N Borneo

ROLLULUS

Rollulus roulroul (Crested Wood Partridge)
    S Thailand, Malaysia, Sumatra, Borneo

PTILOPACHUS
Ptilopachus petrosus (Stone Partridge)
  P.p.petrosus
    Gambia to Cameroun
  P.p.saturatior
    NC Cameroun
  P.p.brehmi
    Lake Chad to Sudan
  P.p.major
    N Ethiopia
  P.p.florentiae
    S Sudan to NE Zaire, Uganda & Kenya

BAMBUSICOLA
Bambusicola fytchii (Bamboo Partridge)
  B.f.fytchii
    W China, Burma, N Vietnam
  B.f.hopkinsoni
    Assam, S Burma
Bambusicola thoracica (Chinese Bamboo Partridge)
  B.t.thoracica
    China
  B.t.sonorivox
    Taiwan

GALLOPERDIX
Galloperdix spadicea (Red Spurfowl)
  G.s.spadicea
    W Nepal to SE India
  G.s.caurina
    Rajputana
  G.s.stewarti
    C & S Travancore
Galloperdix lunulata (Painted Spurfowl)
  India
Galloperdix bicalcarata (Ceylon Spurfowl)
  Sri Lanka

OPHRYSIA
Ophrysia superciliosa (Himalayan Mountain Quail)
  NW Himalayas

ITHAGINIS
Ithaginis cruentus (Blood Pheasant)
  I.c.cruentus
    Nepal, Sikkim, W Burma
  I.c.affinis
    Sikkim
  I.c.tibetanus
    E Bhutan, SE Tibet
  I.c.kuseri
    E Assam to Yunnan
  I.c.holoptilus
    Yunnan
  I.c.rocki
    W Yunnan
  I.c.clarkei
    W Yunnan
  I.c.annae
    NW Szechwan

  I.c.beicki
    N Kansu
  I.c.marionae
    NE Burma
  I.c.geoffroyi
    SE Tibet to W Szechwan
  I.c.berezowskii
    S Kansu, N Szechwan
  I.c.sinensis
    Shensi
  I.c.michaelis
    W Kansu

TRAGOPAN
Tragopan melanocephalus (Western Tragopan)
  NW Himalayas
Tragopan satyra (Satyr Tragopan)
  C & E Himalayas
Tragopan blythii (Blyth's Tragopan)
  T.b.molesworthi
    SE Tibet
  T.b.blythii
    Assam, NW Burma
Tragopan temminckii (Temminck's Tragopan)
  SE Tibet, W China, N Vietnam
Tragopan caboti (Cabot's Tragopan)
  SE China

PUCRASIA
Pucrasia macrolopha (Koklass Pheasant)
  P.m.castanea
    Afghanistan, N Pakistan
  P.m.biddulphi
    N Kashmir, Ladakh
  P.m.macrolopha
    W Himalayas
  P.m.bethelae
    Punjab
  P.m.nipalensis
    W Nepal
  P.m.meyeri
    SE Tibet, W Yunnan
  P.m.ruficollis
    Kansu, W Shansi
  P.m.xanthospila
    SE Mongolia
  P.m.joretiana
    W Anhwei
  P.m.darwini
    Hupeh & E China

LOPHOPHORUS
Lophophorus impeyanus (Himalayan Monal Pheasant)
  Himalayas
Lophophorus sclateri (Sclater's Monal Pheasant)
  L.s.sclateri
    E Assam, N Burma, W Yunnan
  L.s.orientalis
    Myitkyina, Burma

Lophophorus lhuysii (Chinese Monal
Pheasant)
    W & NW Szechwan

GALLUS
Gallus gallus (Red Jungle-fowl)
    G.g.murghi
      Kashmir to Assam & C India
    G.g.gallus
      S Indochina, Thailand, Sumatra
    G.g.spadiceus
      SW Yunnan, N Indochina, Burma
    G.g.jabouillei
      N Vietnam
    G.g.bankiva
      Java
Gallus lafayettii (Ceylon Jungle-fowl)
    Sri Lanka
Gallus sonneratii (Grey Jungle-fowl)
    W & S India
Gallus varius (Green Jungle-fowl)
    Java to Sumba & Flores

LOPHURA
Lophura leucomelana (Kalij Pheasant)
    L.l.hamiltonii
      W Himalayas
    L.l.moffitti
      C Bhutan
    L.l.leucomelana
      Nepal
    L.l.melanotus
      E Nepal
    L.l.lathami
      E Bhutan, Assam, W Burma
    L.l.williamsi
      N Burma
    L.l.oatesi
      S Burma
    L.l.lineatus
      E Burma, Thailand
    L.l.crawfurdi
      S & W Thailand
Lophura nycthemera (Silver Pheasant)
    L.n.occidentalis
      NW Yunnan, NE Burma
    L.n.rufipes
      SW Yunnan, N Burma
    L.n.jonesi
      SW Yunnan, C Thailand, C Burma
    L.n.ripponi
      S Burma
    L.n.beaulieui
      SE Yunnan, N Indochina
    L.n.omeiensis
      Szechwan
    L.n.fokiensis
      NW Fokien
    L.n.rongjianensis
      Kweichow
    L.n.nycthemera
      S China, NE Vietnam
    L.n.berliozi

      WC Indochina
    L.n.beli
      C Vietnam
    L.n.annamensis
      S Vietnam
    L.n.lewisi
      Cambodia
    L.n.engelbachi
      S Laos
    L.n.whiteheadi
      Hainan I
Lophura imperialis (Imperial Pheasant)
    NC Indochina
Lophura edwardsi (Edwards' Pheasant)
    C Vietnam
Lophura haitiensis (Vo Quy's Pheasant)   35.6
    NC Vietnam
Lophura swinhoii (Swinhoe's Pheasant)
    Taiwan
Lophura inornata (Salvadori's Pheasant)
    L.i.hoogewerfi
      N Sumatra
    L.i.inornata
      W Sumatra
Lophura erythrophthalma (Crestless Fireback
Pheasant)
    S Malaysia, N Borneo, NE Sumatra
Lophura ignita (Crested Fireback Pheasant)
    L.i.rufa
      S Thailand, Malaysia, N & C Sumatra
    L.i.macartneyi
      SE Sumatra
    L.i.nobilis
      N Borneo
    L.i.ignita
      S Borneo
Lophura diardi (Siamese Fireback Pheasant)
    Burma, Thailand, C & S Indochina
Lophura bulweri (Bulwer's Pheasant)
    Borneo

CROSSOPTILON
Crossoptilon crossoptilon (White Eared-
Pheasant)
    C.c.dolani
      S Kokonor
    C.c.crossoptilon
      C Szechwan, NW Yunnan
    C.c.lichiangense
      NW Yunnan
    C.c.drouynii
      SE Tibet
    C.c.harmani
      SE Tibet
Crossoptilon mantchuricum (Brown Eared-
Pheasant)
    NE China
Crossoptilon auritum (Blue Eared-Pheasant)
    W China

CATREUS
Catreus wallichii (Cheer Pheasant)
    Himalayas

SYRMATICUS
Syrmaticus ellioti (Elliot's Pheasant)
  SE China
Syrmaticus humiae (Mrs Hume's Pheasant)
  S.h.humiae
    N Burma
  S.h.burmanicus
    SW Yunnan, NE Burma
Syrmaticus mikado (Mikado Pheasant)
  Taiwan
Syrmaticus soemmerringi (Copper Pheasant)
  S.s.scintillans
    Honshu I
  S.s.subrufus
    W Honshu I
  S.s.intermedius
    SW Honshu, Shikoku Is
  S.s.soemmerringi
    N & C Kyushu I
  S.s.ijimae
    SE Kyushu I
Syrmaticus reevesii (Reeves' Pheasant)
  N & C China

PHASIANUS
Phasianus colchicus (Common Pheasant)
  P.c.septentrionalis
    N Caucasus, W Caspian Sea
  P.c.colchicus
    Transcaucasia, E & SE Black Sea
  P.c.talischensis
    SW & S Caspian Sea
  P.c.persicus
    SW Transcaspia
  P.c.principalis
    S Turkestan, N Afghanistan
  P.c.chrysomelas
    Russian Turkestan
  P.c.zarudnyi
    Russian Turkestan
  P.c.bianchii
    Pamir, Hindukush Mts
  P.c.zerafschanicus
    Samarkand
  P.c.bergii
    Aral Sea
  P.c.turcestanicus
    Russian Turkestan
  P.c.mongolicus
    NE Russian Turkestan, S Dzungaria
  P.c.shawii
    Chinese Turkestan
  P.c.tarimensis
    E Chinese Turkestan
  P.c.vlangalii
    E Zaidam
  P.c.satscheuensis
    W Kansu
  P.c.edzinensis
    C Gobi
  P.c.sohokotensis
    Soho-khoto Oasis

P.c.alaschanicus
  C Alaschan Mts
P.c.hagenbecki
  W Mongolia
P.c.pallasi
  SE Siberia, C Manchuria
P.c.karpowi
  S Manchuria, Korea
P.c.kiangsuensis
  SE Mongolia, N Shansi, N Shensi
P.c.strauchi
  Kansu, C & S Shensi, NE Szechwan
P.c.suehschanensis
  NW Szechwan
P.c.elegans
  SW Szechwan to N Burma
P.c.rothschildi
  SE Yunnan, N Vietnam
P.c.decollatus
  WC China
P.c.torquatus
  E China
P.c.takatsukasae
  NE Vietnam
P.c.formosanus
  Taiwan
Phasianus versicolor (Green Pheasant)
  P.v.robustipes
    Sado I, NW Honshu
  P.v.versicolor
    Kyushu, E & S Honshu, Shikoku
  P.v.tanensis
    S peninsulas & islands off Honshu

CHRYSOLOPHUS
Chrysolophus pictus (Golden Pheasant)
  C China
Chrysolophus amherstiae (Lady Amherst's Pheasant)
  SE Tibet, SW China, N Burma

POLYPLECTRON
Polyplectron chalcurum (Sumatran Peacock-Pheasant)
  P.c.scutulatum
    N Sumatra
  P.c.chalcurum
    S Sumatra
Polyplectron inopinatum (Rothschild's Peacock-Pheasant)
  C Malaysia
Polyplectron germaini (Germain's Peacock-Pheasant)
  S Vietnam
Polyplectron bicalcaratum (Burmese Peacock-Pheasant)
  P.b.bakeri
    Sikkim to E Assam
  P.b.bicalcaratum
    C & S Burma to Laos
  P.b.bailyi
    N Thailand

P.b.ghigii
  N Vietnam
P.b.katsumatae
  Hainan I
Polyplectron malacense (Malay Peacock-Pheasant)
P.m.malacense
  SW Thailand, Malaysia, Sumatra
P.m.schleiermacheri
  Borneo
Polyplectron emphanum (Palawan Peacock-Pheasant)
  Palawan I

RHEINARTIA
Rheinartia ocellata (Crested Argus)
R.o.ocellata
  C Vietnam
R.o.nigrescens
  C Malaysia

ARGUSIANUS
Argusianus argus (Great Argus Pheasant)
A.a.grayi
  C Borneo
A.a.argus
  SW Thailand, Malaysia, Sumatra

PAVO
Pavo cristatus (Common Peafowl)
  India, Sri Lanka
Pavo muticus (Green Peafowl)
P.m.spicifer
  SE Assam, W Burma
P.m.imperator
  E Burma, Thailand, Indochina
P.m.muticus
  Java, Malaysia

AFROPAVO
Afropavo congensis (Congo Peafowl)
  Ituri Forest, EC Zaire

NUMIDINAE

PHASIDUS
Phasidus niger (Black Guineafowl)
  S Cameroun to C Zaire

AGELASTES
Agelastes meleagrides (White-breasted Guineafowl)
  Liberia, Ghana

NUMIDA
Numida meleagris (Helmeted Guineafowl)
N.m.sabyi
  W Morocco
N.m.galeata
  Senegal to Air & Cameroun
N.m.bannermanni
  Cape Verde Is

N.m.marchei
  Gabon, Central African Republic
N.m.strasseni
  E Cameroun, N Central African Republic
N.m.meleagris
  Chad, Sudan, N Ethiopia, SW Arabia
N.m.somaliensis
  SE Ethiopia, Somalia N Kenya
N.m.major
  NE Zaire, W Uganda
N.m.toruensis
  E Zaire, W Uganda
N.m.intermedia
  SW Uganda
N.m.mitrata
  S Kenya to Zimbabwe, Madagascar
N.m.macroceras
  W Kenya
N.m.reichenowi
  SW Kenya, NW Tanzania
N.m.uhehensis
  SC Tanzania
N.m.callewaerte
  N Angola to SC Zaire
N.m.marungensis
  S Zaire, Zambia
N.m.maxima
  S Angola
N.m.rikwae
  SW Tanzania
N.m.papillosa
  S Angola
N.m.coronata
  Transvaal, Natal, E Cape Province

GUTTERA
Guttera plumifera (Plumed Guineafowl)
G.p.plumifera
  Cameroun, Gabon, N Angola
G.p.schubotzi
  N Zaire
Guttera edouardi (Crested Guineafowl)
G.e.verreauxi
  Guinea to Togo
G.e.sclateri
  W Cameroun
G.e.schoutedeni
  S Zaire
G.e.chapini
  S Angola
G.e.sethsmithi
  E Zaire, NW Tanzania
G.e.suahelica
  C Tanzania
G.e.barbata
  SW Tanzania, W Mozambique
G.e.kathleenae
  W Zambia
G.e.edouardi
  S Malawi, E Transvaal, Natal
G.e.symonsi
  C Natal

G.e.pucherani
SE Somalia, E Kenya, NE Tanzania,
Zanzibar   35.3

ACRYLLIUM
Acryllium vulturinum (Vulturine Guineafowl)
E Uganda, S Somalia, E Kenya, NE
Tanzania

GRUIFORMES

36. MESITORNITHIDAE (MESITES)

MESITORNIS
Mesitornis variegata (White-breasted Mesite)
E Madagascar
Mesitornis unicolor (Brown Mesite)
E Madagascar

MONIAS
Monias benschi (Bensch's Monia)
SW Madagascar

37. TURNICIDAE (BUTTON QUAILS)

TURNIX
Turnix sylvatica (Little Button-Quail)   37.1
T.s.sylvatica
S Iberia, NW Africa
T.s.lepurana
Senegal to Sudan, W Cape Province
T.s.dussumier
India, Burma
T.s.mikado
Thailand, S China, N Indochina, Taiwan
T.s.davidi
S Indonesia
T.s.bartelsorum
Java
T.s.whiteheadi
Luzon
T.s.celestinoi
Bohol
T.s.nigrorum
Negros, Umod
T.s.alleni
E Africa to E Cape Province
Turnix maculosa (Red-backed Button-Quail)
37.1
T.m.masaaki
Mindanao
T.m.suluensis
Sulu Arch
T.m.kinneari
Peleng
T.m.beccarii
Sulawesi
T.m.maculosa
Lesser Sunda Is
T.m.saturata
New Britain, Duke of York I
T.m.furva
New Guinea

T.m.giluwensis
C New Guinea
T.m.savuensis
Savu Is
T.m.sumbana
Sumba I
T.m.floresiana
Flores
T.m.horsbrughi
S New Guinea
T.m.salamonis
Guadalcanal, Bismarck Arch.
T.m.pseutes
NW Australia
Turnix everetti (Sumba Button-Quail)
Sumba I
Turnix worcesteri (Worcester's Button-Quail)
Luzon I
Turnix hottentotta (Hottentot Button-Quail)
T.h.nana
Ghana to Uganda & SE Cape Province
T.h.luciana
C Kenya
T.h.hottentotta
SW Cape Province
Turnix tanki (Yellow-legged Button-Quail)
T.t.tanki
India, Nicobar & Andaman Is
T.t.blanfordii
Manchuria to Burma, S China, Indochina
Turnix suscitator (Bustard Quail)
T.s.plumbipes
Nepal to N Burma
T.s.bengalensis
NE India
T.s.taigoor
India
T.s.leggei
Sri Lanka
T.s.blakistoni
S China, N Indochina
T.s.rostrata
Taiwan
T.s.pallescens
SC Burma
T.s.thai
C Thailand
T.s.interrumpens
S Burma, S Thailand
T.s.atrogularis
Malaysia, N Sumatra
T.s.machetes
C Sumatra
T.s.suscitator
SE Sumatra, Java, Bali I
T.s.kuiperi
Billiton I
T.s.okinavensis
Okinawa I
T.s.fasciata
Palawan, N & W Philippine Is
T.s.nigrescens
Negros, Cebu Is

T.s.rufilata
  Sulawesi
T.s.powelli
  Lesser Sunda islands
Turnix nigricollis (Madagascar Button-Quail)
  Madagascar
Turnix ocellata (Spotted Button-Quail)
  Luzon I
Turnix melanogaster (Black-breasted Button-Quail)
  Queensland, New South Wales
Turnix varia (Painted Button-Quail)
T.v.scintillans
  Houtman Abrolhos Is
T.v.varia
  Australia
T.v.novaecaledoniae
  New Caledonia
Turnix castanota (Chestnut-backed Button-Quail)
T.c.castanota
  NW Australia, Nthn Territory, Melville I
T.c.olivii
  N Queensland
Turnix pyrrhothorax (Red-chested Button-Quail)
  N, E & SE Australia
Turnix velox (Little Quail)
  Australia

ORTYXELOS
Ortyxelos meiffrenii (Quail Plover)
  Senegal to C Sudan, N Kenya

38. PEDIONOMIDAE (PLAINS WANDERER)

PEDIONOMUS
Pedionomus torquatus (Plains Wanderer)
  New South Wales, Victoria, Sth Australia

39. GRUIDAE (CRANES)

GRUINAE

GRUS
Grus grus (Common Crane)
G.g.grus
  N & E Europe, W Russia >> NE Africa
G.g.lilfordi
  C & E Asia, China, N India
Grus nigricollis (Black-necked Crane)
  C Asia to Assam, S China
Grus monacha (Hooded Crane)
  SE Siberia, N China, Japan
Grus canadensis (Sandhill Crane)
G.c.canadensis
  E Siberia, NW Canada to NW USA
G.c.tabida
  SW Canada, W USA, N Mexico
G.c.pratensis
  SE USA
G.c.pulla
  Gulf coast, S USA

G.c.nesiotes
  Isle of Pines, W Cuba
Grus japonensis (Manchurian Crane)
  Manchuria >> E China
Grus americana (Whooping Crane)
  N Saskatchewan >> SE Texas
Grus vipio (Japanese White-naped Crane)
  NW Mongolia >> E China
Grus antigone (Sarus Crane)
G.a.antigone
  N India
G.a.sharpii
  E Assam, Burma to S Indochina
Grus rubicunda (Brolga)
  S New Guinea, N, E & S Australia
Grus leucogeranus (Great White Crane)
  SE Russia, Siberia >> NW India & China

BUGERANUS
Bugeranus carunculatus (Wattled Crane)
  Somalia to Angola & S Africa

ANTHROPOIDES
Anthropoides virgo (Demoiselle Crane)
  SE Europe, NE Africa to C Asia & China
Anthropoides paradisea (Stanley Crane)
  Sthn Africa

BALEARICINAE

BALEARICA
Balearica pavonina (Crowned Crane)
B.p.pavonina
  Senegal to Chad & N Zaire
B.p.ceciliae
  Sudan, Ethiopia
B.p.gibbericeps
  E Zaire, Uganda, Kenya, N Tanzania
Balearica regulorum (South African Crowned Crane)
  Sthn Africa

40. ARAMIDAE (LIMPKIN)

ARAMUS
Aramus guarauna (Limpkin)
A.g.pictus
  SE USA
A.g.elucus
  Hispaniola, Puerto Rico
A.g.dolosus
  S Mexico to Panama
A.g.guarauna
  N Sth America to Paraguay & Argentina

41. PSOPHIIDAE (TRUMPETERS)

PSOPHIA
Psophia crepitans (Common Trumpeter)
P.c.crepitans
  S Venezuela, the Guianas, NE Brazil
P.c.napensis
  N Upper Amazonia
Psophia leucoptera (White-winged Trumpeter)

P.l.leucoptera
E Peru, N Bolivia, W Brazil
P.l.ochroptera
NW Brazil
Psophia viridis (Green-winged Trumpeter)
P.v.viridis
N Brazil
P.v.dextralis
NC Brazil
P.v.interjecta
C Brazil
P.v.obscura
NE Brazil

42. RALLIDAE (RAILS, COOTS)

HIMANTORNIS
Himantornis haematopus (Nkulengu Rail)
H.h.haematopus
Liberia to S Cameroun
H.h.petiti
Gabon to N Angola
H.h.whitesidei
C Zaire

CANIRALLUS
Canirallus oculeus (Grey-throated Rail)
C.o.oculeus
Liberia to Nigeria
C.o.batesi
S Cameroun to C Zaire
Canirallus kioloides (Madagascar Grey-throated Rail)
C.k.berliozi
NW Madagascar
C.k.kioloides
E Madagascar
Canirallus cuvieri (White-throated Rail)
C.c.cuvieri
Madagascar, Mauritius I
C.c.aldabranus
Aldabra I

EULABEORNIS
Eulabeornis castaneoventris (Chestnut-bellied Rail)
E.c.sharpei
Aru Is
E.c.castaneoventris
coast of N & NE Australia
Eulabeornis plumbeiventris (Bare-eyed Rail)
E.p.plumbeiventris
N Moluccas, New Ireland, N New Guinea
E.p.hoeveni
S New Guinea, Aru Is
Eulabeornis rosenbergii (Bald-faced Rail)
N & C Sulawesi
Eulabeornis calopterus (Red-winged Wood Rail)
E Ecuador, E Peru
Eulabeornis saracura (Slaty-breasted Wood Rail)
SE Brazil, Paraguay

Eulabeornis ypecaha (Giant Wood Rail)
E Brazil to C Argentina
Eulabeornis wolfi (Brown Wood Rail)
Colombia to SW Ecuador
Eulabeornis mangle (Little Wood Rail)
E Brazil
Eulabeornis cajaneus (Grey-necked Wood Rail)
E.c.mexicanus
S Mexico
E.c.vanrossemi
Guatemala, El Salvador
E.c.albiventris
Yucatan to Belize
E.c.pacificus
Honduras, Nicaragua
E.c.plumbeicollis
NE Costa Rica
E.c.cajaneus
Costa Rica to Paraguay & N Argentina
E.c.latens
San Miguel, Pearl Is (Panama)
E.c.morrisoni
Pearl Is (Panama)
Eulabeornis axillaris (Rufous-necked Wood Rail)
S Mexico to Guyana & Ecuador
Eulabeornis concolor (Uniform Crake)
E.c.guatemalensis
S Mexico to Ecuador
E.c.castaneus
NE Sth America
(E.c.concolor
extinct)

RALLUS
Rallus plateni (Platen's Rail)
Sulawesi
Rallus wallacii (Wallace's Rail)
Halmahera I
Rallus insignis (New Britain Rail)
New Britain
Rallus lafresnayanus (New Caledonian Wood Rail)
New Caledonia e?
Rallus sylvestris (Lord Howe Wood Rail)
Lord Howe I
Rallus poecilopterus (Barred Wing Rail)
R.p.poecilopterus
Taveuni, Viti Levu Is e?
R.p.woodfordi
Guadalcanal I
R.p.immaculatus
Ysabel I
R.p.tertius
Bougainville I
Rallus sanguinolentus (Plumbeous Rail)
R.s.simonsi
NW Peru, N Chile
R.s.tschudii
C Peru
R.s.zelebori
Rio de Janeiro, Brazil

R.s.sanguinolentus
 S Brazil, W Argentina, Bolivia
R.s.landbecki
 C Chile
R.s.luridus
 Tierra del Fuego
Rallus nigricans (Blackish Rail)
R.n.nigricans -
 E Ecuador to E Brazil and E Argentina
R.n.caucae
 Cauca valley, Colombia
Rallus maculatus (Spotted Rail)
R.m.insolitus
 S Mexico to Costa Rica
R.m.maculatus
 Venezuela to C Argentina, Trinidad, Cuba
Rallus philippensis (Buff-banded Rail)
R.p.philippensis
 Mindoro, Luzon, Batan Is
R.p.xerophilus
 Banda Sea islands
R.p.wilkinsoni
 S Flores I
R.p.andrewsi
 Cocos Keeling Is
R.p.admiralitatis
 Admiralty Is
R.p.praedo
 Skoki I
R.p.lesouefi
 New Hanover
R.p.meyeri
 Witu I, New Britain
R.p.anchoretae
 Anchorite I
R.p.pelewensis
 Palau Is
R.p.christophori
 E Solomon Is
R.p.randi
 Mt Wilhelm (New Guinea)
R.p.lacustris
 Sentani Lake (W New Guinea)
R.p.reductus
 NE New Guinea
R.p.wahgiensis
 SE New Guinea
R.p.mellori
 Australia
R.p.assimilis
 New Zealand
R.p.tounelieri
 SW Pacific
R.p.norfolkensis
 Norfolk I
R.p.swindellsi
 New Caledonia
R.p.sethsmithi
 New Hebrides, Fiji Is
R.p.goodsoni
 Samoa Is
R.p.ecaudatus
 Tonga

Rallus striatus (Blue-breasted Banded Rail)
R.s.gularis
 Java
R.s.albiventer
 India, Burma, N Malaysia
R.s.obscurior
 Andaman, Nicobar Is
R.s.jouyi
 SE China
R.s.taiwanus
 Taiwan
R.s.striatus
 Borneo, Philippine Is, Sulawesi
R.s.paratermus
 Samar I
Rallus torquatus (Barred Rail)
R.t.torquatus
 Philippine Is
R.t.celebensis
 Sulawesi
R.t.sulcirostris
 Peling, Sula Is
R.t.kuehni
 Tukang Besi Is
R.t.limarius
 Salawati I, NW New Guinea
Rallus owstoni (Guam Rail)
 Guam I
Rallus pectoralis (Slate-breasted Rail)
R.p.exsul
 Flores I
R.p.mayri
 NW New Guinea
R.p.insulsus
 Hertzog Mts (New Guinea)
R.p.captus
 Mt Hagen (New Guinea)
R.p.alberti
 SC New Guinea
R.p.pectoralis
 SW, S & E Australia
R.p.brachipus
 Tasmania
R.p.muelleri
 Adams I, S New Zealand
R.p.mirificus
 Luzon I
Rallus caerulescens (Kaffir Rail)
 N Angola to Ethiopa & Cape Province
Rallus madagascariensis (Madagascar Rail)
 E Madagascar
Rallus aquaticus (Water Rail)
R.a.aquaticus
 W Europe to W Siberia, NW Africa
R.a.hibernans
 Iceland
R.a.korejewi
 Turkey to NW India
R.a.indicus
 Siberia to Japan, N India & China
Rallus semiplumbeous (Bogota Rail)
 Colombia, Ecuador, Peru

42.5

Rallus longirostris (Clapper Rail)
R.l.obsoletus
N California
R.l.levipes
S California
R.l.yumanensis
SW USA, N Mexico
R.l.beldingi
S Mexico to N Colombia
R.l.tenuirostris
SC Mexico
R.l.pallidus
Yucatan
R.l.grossi
Quintana Roo
R.l.belizensis
Belize
R.l.elegans
C & E USA
R.l.crepitans
E Coast USA, Connecticut to N Carolina
R.l.waynei
S Carolina to Florida
R.l.saturatus
coast of Texas, Alabama
R.l.scotti
W Florida
R.l.insularum
Florida Keys
R.l.coryi
Bahama Is
R.l.ramsdeni
Cuba
R.l.leucophaeus
Isle of Pines
R.l.caribaeus
Cuba, Jamaica, Hispaniola, Puerto Rico, Antigua
R.l.margaritae
N Venezuela
R.l.dillonripleyi
Venezuela
R.l.phelpsi
Colombia, Venezuela
R.l.pelodramus
Trinidad
R.l.longirostris
Coast of the Guianas
R.l.crassirostris
Coast of Brazil
R.l.cypereti
W Ecuador, N Peru
Rallus wetmorei (Plain-flanked Rail)
N Venezuela
Rallus limicola (Virginia Rail)
R.l.limicola
USA, N Mexico
R.l.friedmanni
C Mexico
R.l.aequatorialis
Colombia, Ecuador, Peru
Rallus antarcticus (Austral Rail)   42.4
C & S Chile, S Argentina

Rallus okinawae (Okinawa Rail)   42.1.3
Okinawa I

ATLANTISIA
Atlantisia rogersi (Inaccessible Island Rail)
Inaccessible I

GALLIRALLUS
Gallirallus australis (Weka Rail)
G.a.greyi
North I, New Zealand
G.a.australis
N & W South I, New Zealand
G.a.hectori
E South I, New Zealand
G.a.scotti
Stewart I

ROUGETIUS
Rougetius rougetii (Rouget's Rail)
N Ethiopia

CYANOLIMNAS
Cyanolimnas cerverai (Zapata Rail)
S Cuba

RALLINA
Rallina rubra (New Guinea Chestnut Rail)
R.r.rubra
NW New Guinea
R.r.telefolminensis
WC New Guinea
R.r.klossi
C & SW New Guinea
Rallina leucospila (White-striped Chestnut Rail)
NW & W New Guinea
Rallina forbesi (Forbes' Chestnut Rail)
R.f.forbesi
SE New Guinea
R.f.parva
Adelbert Mts, New Guinea
R.f.dryas
Huon peninsula, New Guinea
R.f.steini
C New Guinea
Rallina mayri (Mayr's Chestnut Rail)
R.m.mayri
Cyclops Mts (New Guinea)
R.m.carmichaeli
NE New Guinea
Rallina castaneiceps (Chestnut-headed Crake)
R.c.coccineipes
SW Colombia, NE Ecuador
R.c.castaneiceps
E Ecuador, N Peru
Rallina tricolor (Red-necked Crake)
R.t.tricolor
N Queensland, New Guinea, Damar, Tanimbar I
R.t.convicta
New Hanover, New Ireland

Rallina canningi (Andaman Banded Crake)
Andaman Is
Rallina fasciata (Red-legged Crake)
S Burma to Philippine Is, Moluccas, Java
Rallina eurizonoides (Banded Crake)
R.e.amauroptera
India >> Sri Lanka
R.e.telmatophila
Burma to Indochina & Java
R.e.sepiaria
Riukiu Is
R.e.formosana
Taiwan
R.e.eurizonoides
Philippine Is
R.e.alvarezi
Philippine Is
R.e.minahasa
Sula Is, Sulawesi
Rallina paykullii (Band-bellied Crake)
NE Asia, China >> Borneo, Java

COTURNICOPS
Coturnicops rufa (Red-chested Crake)
C.r.bonapartii
Sierra Leone to Gabon
C.r.elizabethae
NE Zaire, Uganda, N Kenya
C.r.rufa
Angola, S Africa
Coturnicops pulchra (White-spotted Crake)
C.p.pulchra
Sierra Leone to NW Cameroun
C.p.zenkeri
S Cameroun
C.p.batesi
S Cameroun
C.p.centralis
SE Cameroun to N Kenya, N Angola
Coturnicops lugens (African Chestnut-headed Crake)
C.l.lugens
Angola to NE Zaire, Tanzania
C.l.lynesi
NE Zambia
Coturnicops boehmi (Streaky-breasted Crake)
Guinea to N Kenya & Malawi
Coturnicops elegans (Buff-spotted Crake)
C.e.reichenovi
Liberia to Angola & Uganda
C.e.elegans
Somalia to E Cape Province
Coturnicops affinis (Chestnut-tailed Crake)
C.a.antonii
Sudan to Zambia & Malawi
C.a.affinis
Zimbabwe to Natal, E Cape Province
Coturnicops insularis (Madagascar Crake)
Madagascar
Coturnicops watersi (Waters'Crake)
Madagascar

Coturnicops ayresi (White-winged Crake)
Ethiopia & E Sth Africa
Coturnicops schomburgkii (Ocellated Crake)
C.s.schomburgkii
Venezuela, Guyana, French Guiana
C.s.chapmani
S Brazil
Coturnicops notata (Darwin's Rail)
Guyana, Uruguay to S Argentina
Coturnicops noveboracensis (Yellow Rail)
C.n.noveboracensis
E Canada to SW USA
C.n.goldmani
Lerma, Mexico
C.n.exquisitus
Siberia to Japan, China

LATERALLUS
Laterallus fasciatus (Black-banded Crake)
SE Colombia, NE Peru, NW Brazil
Laterallus levraudi (Rusty-flanked Crake)
N Venezuela
Laterallus ruber (Ruddy Crake)
C Mexico to N Nicaragua
Laterallus viridis (Russet-crowned Crake)
L.v.brunnescens
C Colombia
L.v.viridis
E Peru, the Guianas, Brazil
Laterallus exilis (Grey-breasted Crake)
Peru to the Guianas, Trinidad
Laterallus spilonotus (Galapagos Rail)
Galapagos Is
Laterallus melanophaius (Rufous-sided Crake)
L.m.oenops
E Colombia, E Ecuador
L.m.melanophaius
Guyana to C Argentina
Laterallus albigularis (White-throated Crake)
L.a.cinereiceps
Nicaragua to W Panama
L.a.albigularis
SW Costa Rica to W Ecuador
L.a.cerdaleus
E Colombia
Laterallus leucopyrrhus (Red & White Crake)
S Brazil, Paraguay, Uruguay
Laterallus jamaicensis (American Black Crake)
L.j.jamaicensis
C & E USA, Jamaica, Cuba
L.j.coturniculus
S California
L.j.murivagans
W Peru
L.j.tuerosi
C Peru
L.j.salinasi
C Chile
Laterallus xenopterus (Rufous-faced Crake)
Paraguay

CREX
Crex crex (Corn Crake)
  Europe, N Africa to C Asia

PORZANA
Porzana egregia (African Crake)
  Gambia to E Ethiopia & Natal
Porzana flavirostra (African Black Crake)
  Senegal to Sudan & Cape Province
Porzana olivieri (Olivier's Rail)
  W Madagascar
Porzana flaviventer (Yellow-breasted Crake)
  P.f.gossii
    Cuba, Jamaica
  P.f.hendersoni
    Hispaniola, Puerto Rico
  P.f.woodi
    El Salvador
  P.f.bangsi
    NC Colombia
  P.f.flaviventer
    SW Colombia to French Guiana
Porzana cinerea (White-browed Rail)
  P.c.cinerea
    Malaysia, Sumatra to Sumbawa I, Moluccas
  P.c.ocularis
    Philippine Is, Sulawesi
  P.c.micronesiae
    Guam, Yap, Truk Is
  P.c.leucophrys
    Bismarck Arch, New Guinea, N Australia
  P.c.meeki
    St Matthias I
  P.c.tannensis
    New Caledonia, New Hebrides, Fiji Is,
    Samoa Is
Porzana spiloptera (Dot-winged Crake)
  Uruguay, NE Argentina
Porzana albicollis (White-throated Crake)
  P.a.olivacea
    N Colombia, the Guianas, Venezuela
  P.a.albicollis
    E Brazil, E Bolivia, Paraguay, NW Argentina
Porzana marginalis (Striped Crake)
  Irregular distribution throughout Africa
Porzana erythrops (Paint-billed Crake)
  P.e.olivascens
    Venezuela, the Guianas to NW Argentina
  P.e.erythrops
    Peru, Brazil, N Argentina
Porzana columbiana (Colombian Crake)
  P.c.ripleyi
    S Panama, NW Colombia
  P.c.columbiana
    N Colombia, NW Ecuador
Porzana tabuensis (Sooty Crake)
  P.t.tabuensis
    Fiji, New Caledonia, Samoa, Tonga,
    Marquesas Is
  P.t.edwardi
    EC New Guinea
  P.t.richardsoni
    C New Guinea

  P.t.plumbea
    Chatham I, S Australia, Tasmania
Porzana atra (Henderson Island Crake)
  Henderson I
Porzana parva (Little Crake)
  E & S Europe to W India
Porzana pusilla (Baillon's Crake)
  P.p.intermedia
    W Europe to Iran & N Africa
  P.p.pusilla
    C Asia to India & China
  P.p.obscura
    Uganda to Angola & Cape Province
  P.p.mira
    Borneo, Malaysia, Sumatra
  P.p.mayri
    New Guinea
  P.p.palustris
    Australia, Tasmania
  P.p.affinis
    New Zealand
Porzana fluminea (Australian Spotted Crake)
  SW to E Australia, Tasmania
Porzana porzana (Spotted Crake)
  W Europe, N Africato C Asia, India
Porzana carolina (Sora Crake)
  Canada to Venezuela, Peru, W Indies
Porzana fusca (Ruddy-breasted Crake)
  P.f.fusca
    N India to Philippine Is, Flores I, Sulawesi
  P.f.erythrothorax
    Japan, China
  P.f.phaeopyga
    Riukiu Is
  P.f.zeylonica
    SW India, Sri Lanka

AMAURORNIS
Amaurornis olivaceus (Rufous-tailed Moorhen)
  A.o.moluccanus
    N & E New Guinea, Moluccas
  A.o.olivaceus
    Philippine Is
  A.o.nigrifrons
    Bismarck Archipelago, Solomon Is
  A.o.ultimus
    Gower I
  A.o.ruficrissus
    SE New Guinea, Nthn Territory, N Queensland
Amaurornis isabellinus (Sulawesi Water Hen)
  N & SE Sulawesi
Amaurornis ineptus (New Guinea Flightless Rail)
  A.e.ineptus
    N coast & SW coast, New Guinea
  A.e.pallidus
    S New Guinea
Amaurornis akool (Brown Crake)
  A.a.akool
    N India

A.a.coccineipes
SE China, NE Indochina
***Amaurornis bicolor* (Elwes' Crake)**
Nepal to W China, N Indochina
***Amaurornis phoenicurus* (White-breasted Water Hen)**
A.p.phoenicurus
Philippine Is, Indochina to S India, Sri Lanka
A.p.insularis
Andaman, Nicobar Is
A.p.leucomelanus
Lesser Sunda Is, Sulawesi

GALLICREX
***Gallicrex cinerea* (Water Cock)**
India to Japan, Philippine Is, Sulawesi

GALLINULA
***Gallinula ventralis* (Black-tailed Native Hen)**
Australia
***Gallinula mortierii* (Tasmanian Native Hen)**
Tasmania
***Gallinula sylvestris* (San Cristobal Mountain Rail)**
San Cristobal I
***Gallinula pacifica* (Samoan Wood Rail)**
Samoa e?
***Gallinula nesiotis* (Gough Is Coot)**
Gough Is
***Gallinula tenebrosa* (Dusky Moorhen)**
G.t.frontata
SE Borneo, S Moluccas, S New Guinea, Sulawesi
G.t.neumanni
N New Guinea
G.t.tenebrosa
Australia
***Gallinula chloropus* (Moorhen)(Common Gallinule)**
G.c.correiana
Azores Is
G.c.chloropus
Europe, N Africa, Mid East, Russia
G.c.indica
India to Japan, Taiwan, Malaysia
G.c.pyrrhorrhoa
Madagascar, Reunion, Mauritius Is
G.c.orientalis
Africa (S of Sahara) & S Malaysia to Philippine Is
G.c.guami
Mariana Is
G.c.sandvicensis
Hawaii Is
G.c.cachinnans
USA, Bermuda, Galapagos Is
G.c.cerceris
Gtr & Lesser Antilles
G.c.pauxilla
N & W Colombia, W Ecuador, NW Peru
G.c.garmani
Peru, Bolivia, N Chile, NW Argentina

G.c.galeata
the Guianas, Uruguay, N Argentina, Trinidad
***Gallinula angulata* (Lesser Moorhen)**
Senegal to Sudan, Cape Province
***Gallinula melanops* (Spot-flanked Gallinule)**
G.m.bogotensis
C Colombia
G.m.melanops
E Brazil to Paraguay, N Argentina
G.m.crassirostris
C Chile
***Gallinula flavirostris* (Azure Gallinule)**
the Guianas, N & C Brazil, Paraguay, Bolivia
***Gallinula alleni* (Allen's Gallinule)**
Senegal to Sudan, Cape Province
***Gallinula martinica* (American Purple Gallinule)**
SE USA to Argentina, West Indies

PORPHYRIO
***Porphyrio porphyrio* (Purple Swamphen)**
P.p.porphyrio
SW Europe, NW Africa
P.p.madagascariensis
E & S Africa, Madagascar
P.p.seistanicus
E Turkey, E Iran
P.p.poliocephalus
Iraq to Thailand, Andaman, Nicobar Is
P.p.viridis
Burma, Malaysia, S China, Indochina
P.p.indicus
Sumatra to Bali, Borneo, Sulawesi
P.p.melanopterus
Timor, Moluccas, New Guinea
P.p.bellus
SW Australia
P.p.chathamensis
Chatham I
P.p.pulverulentus
Philippine Is
P.p.pelewensis
Palau Is
P.p.samoensis
W Pacific islands, Timor, Moluccas, New Guinea   42.2
***Porphyrio mantelli* (Takahe)**
P.m.mantelli
North I, New Zealand
P.m.hochstetteri
SE South I, New Zealand

FULICA
***Fulica armillata* (Red-gartered Coot)**
Paraguay & S Brazil to Cape Horn
***Fulica leucoptera* (White-winged Coot)**
Bolivia & S Brazil to Cape Horn
***Fulica rufifrons* (Red-fronted Coot)**
N Chile & Uruguay to S Argentina
***Fulica gigantea* (Giant Coot)**
Peru, Bolivia, N Chile

Fulica cornuta (Horned Coot)
  Bolivia, N Chile, NW Argentina
Fulica caribaea (Caribbean Coot)
  W Indies, Trinidad, NE Venezuela
Fulica americana (American Coot)
  F.a.alai
    Hawaii Is
  F.a.americana
    Canada to Nicaragua, W Indies
  F.a.colombiana
    C Colombia to N Ecuador
  F.a.ardesiaca
    C Peru to N Chile
  F.a.atrura
    S Colombia, Ecuador, NW Peru
Fulica atra (Black Coot)
  F.a.atra
    Europe to SE Asia, N Africa
  F.a.lugubris
    Java
  F.a.novaeguineae
    NW & C New Guinea
  F.a.australis
    Buru I, Australia, Tasmania
Fulica cristata (Red-knobbed Coot)
  S Spain, Ethiopia to Cape Province,
  Madagascar

43. HELIORNITHIDAE (SUNGREBES)

PODICA
Podica senegalensis (Peters' Finfoot)
  P.s.senegalensis
    Senegal to N Zaire
  P.s.camerunensis
    Cameroun, Gabon to C Zaire
  P.s.albipectus
    W Angola
  P.s.petersii
    N Kenya to Cape Province

HELIOPAIS
Heliopais personata (Masked Finfoot)
  NE India to Malaysia, Sumatra

HELIORNIS
Heliornis fulica (American Finfoot)(Sungrebe)
  S Mexico to Paraguay & NE Argentina

44. RHYNOCHETIDAE (KAGU)

RHYNOCHETOS
Rhynochetos jubatus (Kagu)
  New Caledonia

45. EURYPYGIDAE (SUNBITTERNS)

EURYPYGA
Eurypyga helias (Sun-Bittern)
  E.h.major
    Guatemala to Colombia, E Ecuador
  E.h.meridionalis
    SC Peru

E.h.helias
  Upper Amazonia, the Guianas, N Brazil

46 CARIAMIDAE (SERIEMAS)

CARIAMA
Cariama cristata (Red-legged Seriema)
  C Brazil to NW Argentina & Paraguay

CHUNGA
Chunga burmeisteri (Black-legged Seriema)
  NW Argentina

47. OTIDAE (BUSTARDS)   47.1

TETRAX
Tetrax tetrax (Little Bustard)
  T.t.tetrax
    S France, Iberia, N Africa
  T.t.orientalis
    E Europe to Afghanistan >> Syria, NW
    India

NEOTIS
Neotis denhami (Denham's Bustard)
  N.d.denhami
    Senegal to Sudan, Ethiopia, N Zaire
  N.d.jacksoni
    E Africa, S Zaire, Zambia, Angola,
    Botswana
  N.d.stanleyi
    S Africa
Neotis ludwigii (Ludwig's Bustard)
  SW Angola, Namibia, interior Cape
  Province
Neotis nuba (Nubian Bustard)
  N.n.nuba
    Sudan
  N.n.agaze
    Mauretania to Chad
Neotis heuglinii (Heuglin's Bustard)
  E & S Ethiopia, N Somalia, N Kenya

CHLAMYDOTIS
Chlamydotis undulata (Houbara Bustard)
  C.u.fuertaventurae
    Canary Is
  C.u.undulata
    NW & N Africa
  C.u.macqueenii
    Turkestan, C Asia to Sinai, Baluchistan >>
    NW India, Egypt

ARDEOTIS
Ardeotis arabs (Arabian Bustard)
  A.a.lynesi
    NW Morocco e?
  A.a.stieberi
    Senegal to NE Sudan
  A.a.butleri
    S Sudan, Kenya
  A.a.arabs
    Ethiopia, N Somalia, Arabia

Ardeotis kori **(Kori Bustard)**
  A.k.struthiunculus
    NE Africa
  A.k.kori
    Sthn Africa

CHORIOTIS
Choriotis nigriceps **(Great Indian Bustard)**47.2
  C India
Choriotis australis **(Australian Bustard)**
  S New Guinea, Australia

OTIS
Otis tarda **(Great Bustard)**
  O.t.tarda
    C & S Europe, W Asia
  O.t.korejewi
    Turkestan to W China >> Iran
  O.t.dybowskii
    Altai to Manchuria >> China

EUPODOTIS
Eupodotis ruficrista **(Crested Bustard)**
  E.r.savilei
    Senegal to Sudan
  E.r.gindiana
    Eastern Africa
  E.r.ruficrista
    SW Angola to Mozambique, N Natal
Eupodotis afra **(Black Bustard)**
  E.a.kalaharica
    N & C Botswana
  E.a.etoschae
    Namibia, W Botswana
  E.a.afraoides
    N & C Cape Province, Or.Free St., W Transvaal
  E.a.afra
    SW Cape Province
Eupodotis vigorsii **(Vigors' Bustard)**
  E.v.namaqua
    S Namibia, NW Cape Province
  E.v.vigorsii
    W & S Cape Province
Eupodotis rueppellii **(Ruppell's Bustard)**
  E.r.rueppellii
    N Namibia, SW Angola
  E.r.fitzsimonsi
    C Namibia
Eupodotis humilis **(Little Brown Bustard)**
  E Ethiopia, N & W Somalia
Eupodotis caerulescens **(Blue Bustard)**
  E Cape Province to S Transvaal, C Natal
Eupodotis senegalensis **(White-bellied Bustard)**
  E.s.senegalensis
    Senegal to NW Ethiopia
  E.s.canicollis
    E Ethiopia, Somalia to C Tanzania
  E.s.erlangeri
    S Kenya, W Tanzania
  E.s.mackenziei
    W Zambia, E Angola, S Zaire

  E.s.barrowii
    Botswana, S Africa
Eupodotis melanogaster **(Black-bellied Bustard)**
  E.m.melanogaster
    Senegal to Ethiopia, Zambia, Angola
  E.m.notophila
    SE Africa
Eupodotis hartlaubii **(Hartlaub's Bustard)**
  E Sudan, S Ethiopia, S Kenya, N Tanzania

HOUBAROPSIS 47.2
Houbaropsis bengalensis **(Bengal Florican)**
  H.b.bengalensis
    Himalayas, N India
  H.b.blandini
    Cambodia

SYPHEOTIDES 47.2
Sypheotides indica **(Lesser Florican)**
  India

CHARADRIIFORMES

48. JACANIDAE (JACANAS)

MICROPARRA
Microparra capensis **(Smaller Jacana)**
  Sudan to Natal & Cape Province

ACTOPHILORNIS
Actophilornis africana **(African Jacana)**
  Senegal to Sudan & Cape Province
Actophilornis albinucha **(Madagascar Jacana)**
  Madagascar

IREDIPARRA
Irediparra gallinacea **(Comb-crested Jacana)**
  Borneo to Australia & New Guinea

HYDROPHASIANUS
Hydrophasianus chirurgus **(Pheasant-tailed Jacana)**
  India to Philippine Is, Taiwan & Java

METOPIDIUS
Metopidius indicus **(Bronze-winged Jacana)**
  India to Cambodia, Java, Sumatra

JACANA
Jacana spinosa **(Northern Jacana)**
  J.s.gymnostoma
    S Mexico
  J.s.violacea
    Cuba, Jamaica, Hispaniola
  J.s.spinosa
    Guatemala to W Panama
Jacana jacana **(Wattled Jacana)**
  J.j.hypomelaena
    E Panama, N Colombia
  J.j.melanopygia
    W Colombia, W Venezuela

J.j.intermedia
 N Venezuela
J.j.jacana
 Trinidad, the Guianas to E Bolivia & N Argentina
J.j.scapularis
 W Ecuador
J.j.peruviana
 E Peru

49. ROSTRATULIDAE (PAINTED-SNIPE)

ROSTRATULA
Rostratula benghalensis (Painted-Snipe)
R.b.benghalensis
 Africa, S Asia to Java & Philippine Is
R.b.australis
 Australia & Tasmania

NYCTICRYPHES
Nycticryphes semicollaris (South American Painted-Snipe)
 C Chile to N Argentina & Uruguay

50. DROMADIDAE (CRAB PLOVER)

DROMAS
Dromas ardeola (Crab Plover)
 E Africa, Indian Ocean, Andaman Is

51. HAEMATOPODIDAE (OYSTERCATCHERS)

HAEMATOPUS
Haematopus ostralegus (Palaearctic Oystercatcher)
H.o.malacophaga
 Iceland, Faroe Is
H.o.occidentalis
 British Isles
H.o.ostralegus
 Europe, Asia Minor, N Africa
H.o.longipes
 Russia, Siberia
H.o.osculans
 NE Asia, China, Japan
Haematopus meadewaldoi (Canary Is Oystercatcher) 51.2
 E Canary Is e?
Haematopus palliatus (American Oystercatcher) 51.3
H.p.frazari
 W Mexico
H.p.palliatus
 E Coast, N & C America, W Indies
H.p.prattii
 Bahamas
H.p.galapagensis
 Galapagos
H.p.pitanay
 W coast S America
H.p.durnfordi
 E coast S America

Haematopus longirostris (Pied Oystercatcher) 51.1.3
H.l.longirostris
 Aru Is, New Guinea, Australia
H.l.finschi
 South I, New Zealand
Haematopus bachmani (American Black Oystercatcher)
 W coast of N America
Haematopus moquini (African Black Oystercatcher)
 Gabon to Natal
Haematopus unicolor (Variable Oystercatcher)
 New Zealand
Haematopus chathamensis (Chatham Island Oystercatcher) 51.1
 Chatham I
Haematopus leucopodus (Magellanic Oystercatcher)
 S America, Falkland Is
Haematopus ater (Blackish Oystercatcher)
 S America, Falkland Is
Haematopus fuliginosus (Sooty Oystercatcher)
H.f.fuliginosus
 coast of Australia
H.f.ophthalmicus
 coast of N Australia

52. IBIDORHYNCHIDAE (IBIS BILL)

IBIDORHYNCHA
Ibidorhyncha struthersii (Ibis Bill)
 C Asia, Himalayas, N India

53. RECURVIROSTRIDAE (AVOCETS, STILTS)

HIMANTOPUS
Himantopus himantopus (Black-winged Stilt)
 E Europe to China, India, C Africa
Himantopus leucocephalus (Australian Stilt)
 Philippine Is to Java & Australia
Himantopus melanurus (Black-tailed Stilt)
 Peru to C Argentina, C Chile
Himantopus mexicanus (Black-necked Stilt)
 USA to N S America, W Indies
Himantopus ceylonensis (Srk Lanka Stilt)
 Sri Lanka
Himantopus knudseni (Hawaiian Stilt)
 Hawaii
Himantopus novaezelandiae (New Zealand Stilt)
 New Zealand
Himantopus meridionalis (South African Stilt)
 S Africa

CLADORHYNCHUS
Cladorhynchus leucocephalus (Banded Stilt)
 Australia

RECURVIROSTRA
Recurvirostra avosetta (Pied Avocet)
  Europe to China, India, S Africa
Recurvirostra americana (American Avocet)
  W USA to Guatemala
Recurvirostra novaehollandiae (Red-necked Avocet)
  Australia, Tasmania
Recurvirostra andina (Andean Avocet)
  S Peru to N Chile, NW Argentina

54. BURHINIDAE (STONE-CURLEWS)

BURHINUS
Burhinus oedicnemus (Stone-Curlew)
  B.o.distinctus
    W Canary Is
  B.o.insularum
    E Canary Is
  B.o.jordansi
    Balearic Is
  B.o.oedicnemus
    Europe, SW Asia to N & E Africa
  B.o.theresae
    W Morocco
  B.o.saharae
    N Africa to Israel
  B.o.astutus
    Afghanistan, Pakistan
  B.o.indicus
    India, Sri Lanka to S Indochina
Burhinus senegalensis (Senegal Stone-Curlew)
  B.s.senegalensis
    Senegal to C African Republic & Angola
  B.s.inornatus
    Egypt to N Uganda, Ethiopia
Burhinus vermiculatus (Water Dikkop)
  B.v.buttikoferi
    Liberia to N Zaire
  B.v.vermiculatus
    Kenya to Cape Province
Burhinus capensis (Cape Dikkop)
  B.c.maculosus
    Senegal to Niger, N Nigeria
  B.c.affinis
    Sudan, Ethiopia, Uganda, Somalia
  B.c.ehrenbergi
    Dahlak Is
  B.c.dodsoni
    S Arabia, N Somalia
  B.c.capensis
    Angola to Kenya & Cape Province
  B.c.damarensis
    Namibia
Burhinus bistriatus (Double-striped Stone-Curlew)
  B.b.bistriatus
    S Mexico to W Costa Rica
  B.b.vocifer
    N Colombia to Guyana & Brazil
  B.b.pediacus
    N Colombia

  B.b.dominicensis
    Hispaniola
Burhinus superciliaris (Peruvian Stone-Curlew)
  Ecuador to S Peru
Burhinus magnirostris (Australian Stone-Curlew)
  B.m.rufescens
    NW Australia, Nthn Territory
  B.m.ramsayi
    N Queensland
  B.m.magnirostris
    S Queensland to SW Australia, Tasmania

ESACUS
Esacus recurvirostris (Great Stone Plover)
  India, Burma, Sri Lanka
Esacus magnirostris (Great Australian Stone Plover)
  Malaysia to New Guinea & Australia

55. GLAREOLIDAE (COURSERS, PRATINCOLES)

CURSORIINAE

PLUVIANUS
Pluvianus aegyptius (Egyptian Plover)
  P.a.aegyptius
    Senegal to N Zaire & Egypt
  P.a.angolae
    N Angola, W Zaire

CURSORIUS
Cursorius cursor (Cream-coloured Courser)
  C.c.bogolubovi
    N & E Iran
  C.c.cursor
    N Africa to NW India
  C.c.bannermani
    Canary Is, Morocco
  C.c.exsul
    Cape Verde Is
  C.c.dahlakensis
    Dahlak Is
Cursorius rufus (Burchell's Courser)   55.1
  C.r.somalensis
    N Somalia
  C.r.littoralis
    S Somalia, Kenya
  C.r.meruensis
    C Kenya
  C.r.theresae
    NW Cape Province
  C.r.rufus
    Botswana, Transvaal, Cape Province
Cursorius coromandelicus (Indian Courser)
  India, N Sri Lanka
Cursorius temminckii (Temminck's Courser)
  C.t.temminckii
    Senegal to Ethiopia & Cape Province
  C.t.damarensis
    Namibia

C.t.aridus
N Namibia, S Botswana, W Zimbabwe

RHINOPTILUS
Rhinoptilus africanus (Two-banded Courser)
R.a.raffertyi
C Ethiopia
R.a.hartingi
Somalia
R.a.gracilis
C Kenya, N Tanzania
R.a.illustris
C Tanzania
R.a.bisignatus
Angola
R.a.erlangeri
N Namibia, SW Angola
R.a.traylori
NC Namibia
R.a.africanus
S Namibia, W Cape Province
R.a.granti
Transvaal, C Cape Province
Rhinoptilus cinctus (Heuglin's Courser)
R.c.cinctus
Sudan, Somalia to N Tanzania
R.c.emini
islands in Lake Victoria
R.c.seebohmi
S Angola, Namibia to Zimbabwe
Rhinoptilus chalcopterus (Bronze-winged Courser)
R.c.chalcopterus
Senegal to Sudan & Kenya
R.c.albofasciatus
Angola & Tanzania to Cape Province
Rhinoptilus bitorquatus (Jerdon's Courser)
C India

GLAREOLINAE

STILTIA
Stiltia isabella (Australian Pratincole)
Australia to Borneo, Java, New Guinea

GLAREOLA
Glareola pratincola (Pratincole)
G.p.pratincola
Mediterranean to NW India & N Africa
G.p.boweni
Senegal to Chad & Gabon
G.p.limbata
Sudan, Ethiopia, Somalia, S Arabia
G.p.erlangeri
S Somalia, N Kenya
G.p.fuelleborni
E Zaire, C Kenya to Cape Province
Glareola maldivarus (Eastern Collared Pratincole) 55.2
C & E Asia to Indochina, Malaysia
Glareola nordmanni (Black-winged Pratincole)
SE Europe, C Asia, Africa

Glareola ocularis (Madagascar Pratincole)
E Africa, Madagascar
Glareola nuchalis (White-collared Pratincole)
G.n.liberiae
Sierra Leone to W Cameroun
G.n.nuchalis
Chad to Ethiopia & Mozambique
Glareola cinerea (Cream-coloured Pratincole)
G.c.cinerea
Ghana to C Zaire
G.c.colorata
Upper Niger river
Glareola lactea (Little Pratincole)
India, Sri Lanka to S Indochina

56. CHARADRIIDAE (PLOVERS)

VANELLUS
Vanellus vanellus (Northern Lapwing)
W Europe to China & Japan
Vanellus crassirostris (Long-toed Lapwing)
V.c.crassirostris
Sudan, Uganda
V.c.hybrida
Kenya to Malawi
V.c.leucoptera
Mozambique, N Natal
Vanellus spinosus (Spur-winged Plover)
Middle East, C & E Africa
Vanellus duvaucelii (River Lapwing)
N India to Indochina
Vanellus tectus (Black-headed Plover)
V.t.tectus
Senegal to Ethiopia
V.t.latifrons
S Somalia to E Kenya
Vanellus malabaricus (Yellow-wattled Lapwing)
India, Sri Lanka
Vanellus albiceps (White-crowned Wattled Plover)
Liberia to Sudan & Zimbabwe
Vanellus lugubris (Senegal Plover)
Sierra Leone to Uganda & Natal
Vanellus melanopterus (Black-winged Plover)
V.m.minor
Kenya to Cape Province
V.m.melanopterus
S Arabia, Ethiopia
Vanellus coronatus (Crowned Plover)
V.c.demissus
Somalia
V.c.coronatus
Ethiopia to Angola & Cape Province
V.c.xerophilus
Namibia, N Cape Province, W Transvaal
Vanellus senegallus (Senegal Wattled Plover)
V.s.senegallus
Senegal to Sudan & Uganda
V.s.major
W Ethiopia
V.s.solitaneus
W Kenya to N Namibia

V.s.lateralis
E Zaire, Uganda to Angola & Natal
Vanellus melanocephalus (Spot-breasted Plover)
N Ethiopia
Vanellus superciliosus (Brown-chested Wattled Plover)
Benin to Uganda & Kenya
Vanellus gregarius (Sociable Plover)
C Asia to NE Africa, N India
Vanellus leucurus (White-tailed Plover)
W & C Asia, NE Africa, NW India
Vanellus cayanus (Cayenne Plover)
S Venezuela, the Guianas, Upper Amazonia
Vanellus chilensis (Southern Lapwing)
V.c.cayennensis
Colombia, Venezuela, the Guianas, N Brazil
V.c.lampronotus
S Brazil to C Argentina, Uruguay
V.c.chilensis
C Chile, SW Argentina
V.c.fretensis
S Chile, S Argentina
Vanellus resplendens (Andean Lapwing)
Ecuador to N Chile, NW Argentina
Vanellus cinereus (Grey-headed Lapwing)
China, Japan >> Indochina, Malaysia
Vanellus indicus (Red-wattled Lapwing)
V.i.aigneri
Middle East to Pakistan
V.i.indicus
India, Sri Lanka
V.i.atronuchalis
Burma, Malaysia, Indochina
Vanellus macropterus (Javanese Wattled Lapwing)
Java
Vanellus tricolor (Banded Plover)
S Australia, Tasmania
Vanellus miles (Masked Plover)
V.m.miles
S Moluccas, Kai Is, S New Guinea, N Australia
V.m.novaehollandiae
E Australia

ANITIBYX
Anitibyx armatus (Blacksmith Plover)　56.4
S Angola to Kenya & Natal

PLUVIALIS
Pluvialis apricaria (European Golden Plover)
P.a.apricaria
N Europe & N Asia >> Mediterranean, N India
P.a.oreophilos
N & W British Isles, Denmark, Germany
Pluvialis dominica (American Golden Plover)
N Canada >> C Sth America
Pluvialis fulva (Pacific Golden Plover)　56.2
N Siberia, NE Asia, Alaska, >> SE Asia & Australia

Pluvialis squatarola (Grey Plover)
Circumpolar >> Africa, Australia & S America
Pluvialis obscura (New Zealand Dotterel)
New Zealand

CHARADRIUS
Charadrius hiaticula (Ringed Plover)
C.h.psammodroma
NE Canada, Greenland, Iceland
C.h.hiaticula
British Isles, Sweden to Mediterranean
C.h.tundrae
N Europe, N Asia >> Iran, E Africa
Charadrius semipalmatus (Semi-palmated Plover)
N Canada >> C & S America
Charadrius placidus (Long-billed Ringed Plover)
NE Asia >> China, Burma, Indochina
Charadrius dubius (Little Ringed Plover)
C.d.curonicus
Europe, N Asia >> S Africa, India, China
C.d.jerdoni
India >> Indochina & Lesser Sunda Is
C.d.papuanus
New Ireland, New Guinea
C.d.dubius
S Japan, S China, Philippine Is
Charadrius wilsonia (Wilson's Plover)
C.w.wilsonia
S & SE USA, EC America
C.w.rufinucha
Bahama Is, Gtr Antilles, N Lesser Antilles
C.w.beldingi
Baja California to Peru
C.w.cinnamominus
Colombia to French Guiana, Arubu I, Trinidad
Charadrius vociferous (Killdeer Plover)
C.v.vociferus
W Canada, USA >> W Indies, N Sth America
C.v.ternominatus
Gtr Antilles
C.v.peruvianus
W Peru
Charadrius melodus (Piping Plover)
S Canada, E USA, N Mexico
Charadrius thoracicus (Black-banded Sand Plover)
Madagascar
Charadrius pecuarius (Kittlitz's Sand Plover)
C.p.allenbyi
Nile valley
C.p.tephricolor
N Namibia
C.p.pecuarius
Senegal to Sudan & Cape Province, Madagascar
Charadrius sanctaehelenae (St Helena Sand Plover)
St Helena I

Charadrius tricollaris (Three-banded Plover)
  C.t.forbesi
    Guinea to S Zaire
  C.t.tricollaris
    Sudan to Angola & Cape Province
  C.t.bifrontatus
    Madagascar
Charadrius alexandrinus (Kentish Plover)
  C.a.alexandrinus
    W Europe to C Asia >> Africa, China
  C.a.spatzi
    W African coast
  C.a.dealbatus
    S Japan >> Indochina & S Thailand
  C.a.seebohmi
    Sri Lanka
  C.a.javanicus
    Java
  C.a.hesperius
    Liberia to C African Republic
Charadrius marginatus (White-fronted Sand Plover)
  C.m.pons
    S Somalia
  C.m.tenellus
    E Africa to Natal, Madagascar
  C.m.marginatus
    Angola to Cape Province
Charadrius occidentalis (Snowy Plover)
  C.c.nivosus
    W USA, W Mexico
  C.c.tenuirostris
    C & SE USA, Cuba, Hispaniola, Puerto Rico
  C.c.occidentalis
    Peru to Chile
Charadrius ruficapillus (Red-capped Dotterel)
    S New Guinea, Australia, Tasmania
Charadrius peronii (Malaysian Sand Plover)
    Philippine Is, Sulawesi, Borneo
Charadrius venustus (Chestnut-banded Sand Plover)
  C.v.pallidus
    Angola to S Cape Province
  C.v.venustus
    S Kenya, Tanzania
Charadrius collaris (Collared Plover)
    S Mexico to N Argentina, Trinidad
Charadrius bicinctus (Double-banded Plover)
  C.b.bicinctus
    Australia, Tasmania, New Zealand
  C.b.exilis
    Auckland Is
Charadrius falklandicus (Two-banded Plover)
    S Sth America, Falkland Is
Charadrius alticola (Puna Plover)   56.3
    S Andes
Charadrius mongolus (Lesser Sand Plover)
  C.m.atrifrons
    C Asia >> India, Malaysia, E Africa
  C.m.mongolus
    E Siberia, Japan >> Australia
  C.m.stegmani
    Bering Is

Charadrius leschenaultii (Great Sand Plover)
    E Asia & Red Sea >> S Africa, Australia
Charadrius asiaticus (Caspian Plover)
    SE Russia & Iran >> India, E & S Africa
Charadrius veredus (Eastern Sand Plover)
    N China >> Australia & Sulawesi
Charadrius modestus (Rufous-chested Dotterel)
    S Chile, Argentina, Falkland Is
Charadrius montanus (Mountain Plover)
    W USA >> C Mexico
Charadrius melanops (Black-fronted Plover)
    Australia, Tasmania
Charadrius cinctus (Red-kneed Dotterel)
    Australia
Charadrius rubricollis (Hooded Plover)
    S Australia, Tasmania
Charadrius novaeseelandiae (Long-billed Plover)
    Chatham I

ANARHYNCHUS
Anarhynchus frontalis (Wry-bill)
    New Zealand

PHEGORNIS
Phegornis mitchellii (Mitchell's Plover)
    Peru to N Chile, W Argentina

PELTOHYAS
Peltohyas australis (Australian Courser)
    SW Australia to Victoria & New South Wales

EUDROMIAS
Eudromias morinellus (Dotterel)
    N Europe & N Asia >> Mediterranean & Iran
Eudromias ruficollis (Tawny-throated Dotterel)
  E.r.pallidus
    N Peru
  E.r.ruficollis
    Peru & E Argentina to Tierra del Fuego

PLUVIANELLUS
Pluvianellus socialis (Magellanic Plover)
    Straits of Magellan

57. SCOLOPACIDAE (SANDPIPERS, SNIPE)

TRINGINAE

LIMOSA
Limosa limosa (Black-tailed Godwit)
  L.l.limosa
    Europe, W Asia >> N Africa & India
  L.l.melanuroides
    NE Asia >> China, N Australia
Limosa haemastica (Hudsonian Godwit)
    NW Canada >> S Sth America
Limosa lapponica (Bar-tailed Godwit)

L.l.lapponica
N Europe, N Asia >> tropical Africa, N India
L.l.baueri
NE Asia, NW Canada >> Australia, Pacific islands
Limosa fedoa (Marbled Godwit)
WC Canada >> S USA & Peru

Numenius minutus (Little Curlew)
C & E Siberia >> Moluccas, Australia
Numenius borealis (Eskimo Curlew)
N Canada >> S Sth America e?
Numenius phaeopus (Whimbrel)
N.p.phaeopus
N Europe, N Asia >> Africa, NW India
N.p.alboaxillaris
C Russia >> E Africa, Madagascar
N.p.variegatus
E Siberia >> Australia, Pacific islands
N.p.hudsonicus
N Canada >> N Sth America
Numenius tahitiensis (Bristle-thighed Curlew)
W Alaska >> Hawaii, Society Is
Numenius tenuirostris (Slender-billed Curlew)
SW Siberia to E Europe >> Iran
Numenius arquata (Western Curlew)
N.a.arquata
N Europe & Russia >> Africa & NW India
N.a.orientalis
C Asia >> E Africa, India, Indochina
Numenius madagascariensis (Far Eastern Curlew)
E Siberia >> China & Australia
Numenius americanus (Long-billed Curlew)
N.a.occidentalis
WC Canada to N Mexico
N.a.americanus
WC USA to Guatemala

Bartramia longicauda (Upland Sandpiper)
W & C Canada, NC USA >> C Sth America

Tringa erythropus (Spotted Redshank)
N Europe, N Russia >> Africa & China
Tringa totanus (Common Redshank)
T.t.robusta
Iceland to W Europe >> W Africa
T.t.britannica
British Isles to W Europe
T.t.totanus
N Europe & W Siberia >> Africa & W Asia
T.t.eurhinus
C & E Asia >> India, China, Sulawesi
Tringa stagnatilis (Marsh Sandpiper)
SE Europe to Mongolia >> Africa, Australia
Tringa nebularia (Common Greenshank)
N Palaearctic >> Africa, India to New Zealand

Tringa guttifer (Spotted Greenshank)
E Siberia >> India, Malaysia
Tringa melanoleuca (Greater Yellowlegs)
N Canada >> C & Sth America
Tringa flavipes (Lesser Yellowlegs)
N Canada >> Sth America
Tringa ochropus (Green Sandpiper)
N Palaearctic >> C Africa to Philippine Is
Tringa solitaria (Solitary Sandpiper)
T.s.cinnamomea
NW Canada >> C Sth America
T.s.solitaria
C Canada >> W Indies & N Sth America
Tringa glareola (Wood Sandpiper)
N Palaearctic >> Africa, SE Asia, Australia

Catoptrophorus semipalmatus (Willet)
C.s.inornatus
S Canada, W USA >> Peru
C.s.semipalmatus
E Canada, E USA, Cuba, Puerto Rico

Xenus cinereus (Terek Sandpiper)
NE Europe, W Siberia >>E Africa, India to Australia

Actitis hypoleucos (Common Sandpiper)
Palaearctic >> Africa, NE Asia to Australia
Actitis macularia (Spotted Sandpiper)
N America >> W Indies, C Sth America

Heteroscelus brevipes (Grey-tailed Tattler)
E Siberia >> China, Australia
Heteroscelus incanus (Wandering Tattler)
NW Canada >> W USA, Pacific islands

Prosobonia cancellata (Tuamotu Sandpiper)
Tuamotu Is

ARENARIINAE

Arenaria interpres (Ruddy Turnstone)
A.i.interpres
N Palaearctic >> Africa, SE Asia, Australia S America
A.i.morinella
N Canada >> SE USA, W Indies, E Sth America
Arenaria melanocephala (Black Turnstone)
Alaska to W USA

PHALAROPODINAE

Phalaropus tricolor (Wilson's Phalarope)
SW Canada, W USA >> S Sth America

Phalaropus lobatus (Red-necked Phalarope)
  N America, N Palaearctic >> Southern
  coasts
Phalaropus fulicarius (Grey Phalarope)
  N Holarctic >> coasts of Africa & Chile

SCOLOPACINAE

SCOLOPAX
Scolopax rusticola (Eurasian Woodcock)
  Palaearctic >> India, S China
Scolopax mira (Amami Woodcock)
  Amami-Oshima (Riukiu Is)
Scolopax saturata (Rufous Woodcock)
  S.s.saturata
    Sumatra, Java
  S.s.rosenbergii
    New Guinea
Scolopax celebensis (Sulawesi Woodcock)
  S.c.heinrichi
    N Sulawesi
  S.c.celebensis
    C Sulawesi
Scolopax rochussenii (Obi Woodcock)
  Obi I
Scolopax minor (American Woodcock)
  S Canada, SE USA.Gulf coast

GALLINAGONINAE

COENOCORYPHA
Coenocorypha aucklandica (Sub-Antarctic Snipe)
  C.a.pusilla
    Mangare I
  C.a.iredalei
    Jack Lees I
  C.a.huegeli
    Snares I
  C.a.meinertzhagenae
    Antipodes Is
  C.a.aucklandica
    Auckland Is

GALLINAGO
Gallinago solitaria (Solitary Snipe)
  G.s.solitaria
    C Asia, Himalayas, N Burma
  G.s.japonica
    E Asia, Japan, E China
Gallinago hardwickii (Japanese Snipe)
  Kurile Is, Japan, Australia
Gallinago nemoricola (Wood Snipe)
  Himalayas, Burma, S India
Gallinago stenura (Pintail Snipe)
  NE Asia >> India, S China & TimorI
Gallinago megala (Swinhoe's Snipe)
  EC Asia >> Burma, Borneo, Australia
Gallinago nigripennis (African Snipe)
  G.n.nigripennis
    Ethiopia to Namibia & Cape Province

G.n.angolensis
  Angola, Zambia, Botswana
Gallinago macrodactyla (Madagascar Snipe)
  Madagascar, Mauritius
Gallinago media (Great Snipe)
  N Europe, W Asia >> E Africa
Gallinago gallinago (Common Snipe)
  G.g.faroeensis
    Iceland, Faroe Is
  G.g.gallinago
    N Palaearctic >> E Africa, India, China
  G.g.delicata
    N Canada >> SW USA, C America, W
    Indies
Gallinago paraguaiae (Magellan Snipe)
  G.p.paraguaiae
    Colombia to Uruguay
  G.p.magellanica
    S Sth America, Tierra del Fuego
  G.p.andina
    Peru, N Chile
Gallinago nobilis (Noble Snipe)
  C & E Colombia, N Ecuador
Gallinago undulata (Giant Snipe)
  G.u.undulata
    Guyana, French Guiana, Surinam
  G.u.gigantea
    Brazil, Paraguay, N & E Argentina
Gallinago stricklandii (Cordilleran Snipe)
  S Chile, Falkland Is
Gallinago jamesoni (Andean Snipe)
  N Colombia to Bolivia
Gallinago imperialis (Imperial Snipe)
  C Colombia

LYMNOCRYPTES
Lymnocryptes minima (Jack Snipe)
  N Europe, W Asia >> N Africa, Iran, India

LIMNODROMUS
Limnodromus griseus (Short-billed Dowitcher)
  L.g.caurinus
    Alaska, W USA >> Peru
  L.g.griseus
    NE Canada >> E Caribbean
Limnodromus scolopaceus (Long-billed Dowitcher)
  NW Canada >> S USA to Ecuador, Cuba,
  Jamaica
Limnodromus semipalmatus (Asiatic Dowitcher)
  W Siberia, Mongolia >> China, Japan,
  Indochina

CALIDRIDINAE

APHRIZA
Aphriza virgata (Surfbird)
  SC Alaska, W coast to S Chile

CALIDRIS
Calidris canutus (Red Knot)
C.c.canutus
Spitzbergen, Taimyr Peninsula >> Africa
C.c.rogersi
Siberian islands, E Asia >> Australia
C.c.rufus
Greenland, N Canada >> S Sth America
Calidris tenuirostris (Great Knot)
NE Siberia >> China, India & Australia
Calidris alba (Sanderling)
N Holarctic >> S America, India & Australia
Calidris pusilla (Semipalmated Sandpiper)
N Canada >> S America, W Indies
Calidris mauri (Western Sandpiper)
NW Canada >> W Sth America, Trinidad
Calidris ruficollis (Rufous-necked Stint)
NE Siberia, Alaska >> China, Australia
Calidris minuta (Little Stint)
N Europe >> S Africa, W India
Calidris temminckii (Temminck's Stint)
N Europe, N Asia >> NE Africa to China
Calidris subminuta (Long-toed Stint)
E Siberia >> India, China, Philippine Is
Calidris minutilla (Least Sandpiper)
N Nth America >> S USA & N Sth America
Calidris fuscicollis (White-rumped Sandpiper)
N Canada >> S Sth America
Calidris bairdii (Baird's Sandpiper)
E Siberia & N Canada >> S Sth America
Calidris melanotos (Pectoral Sandpiper)
E Siberia & N Canada >> SC Sth America
Calidris acuminata (Sharp-tailed Sandpiper)
NE Asia >> Australia, Pacific islands
Calidris paramelanotus (Cox's Sandpiper)
57.1
S Australia
Calidris maritima (Purple Sandpiper)
Arctic America & Europe >> NE USA, W
Europe
Calidris ptilocnemis (Rock Sandpiper)
C.p.couesi
NE Siberia, Alaska, W Canada
C.p.ptilocnemis
Bering Sea, SE Alaska
C.p.quarta
Commander Is
C.p.kurilensis
Kurile Is
Calidris alpina (Dunlin)
C.a.arctica
E Greenland
C.a.alpina
N Europe, NW Asia >> NE Africa, SW Asia
C.a.littoralis
S Russia
C.a.schinzii
British Isles, Holland
C.a.centralis
N Siberia, Mongolia >> India
C.a.pacifica
NE Asia, NW Canada >> E China, W USA
& SE USA

Calidris ferruginea (Curlew Sandpiper)
N Asia to Europe >> Africa, India &
Australia

EURYNORHYNCHUS
**Eurynorhynchus pygmeus (Spoon-billed
Sandpiper)**
NE Asia >> S China

LIMICOLA
Limicola falcinellus (Broad-billed Sandpiper)
L.f.falcinellus
N Europe, N Russia >> Middle East, W
India
L.f.sibirica
NE Siberia >> E India & Australia

MICROPALAMA
Micropalama himantopus (Stilt Sandpiper)
N Canada >> C Sth America & W Indies

TRYNGITES
**Tryngites subruficollis (Buff-breasted
Sandpiper)**
N Canada >> SC Sth America

PHILOMACHUS
Philomachus pugnax (Ruff)
N Europe & Asia >> Africa, India Burma

58. THINOCORIDAE (SEED-SNIPE)

ATTAGIS
Attagis gayi (Rufous-bellied Seed-Snipe)
A.g.latreillii
Ecuador
A.g.simonsi
Peru, N Bolivia
A.g.gayi
Chile, Argentina
Attagis malouinus (White-bellied Seed-Snipe)
A.m.cheeputi
C Argentina
A.m.malouinus
Tierra del Fuego

THINOCORUS
**Thinocorus orbignyianus (Grey-breasted
Seed-Snipe)**
T.o.ingae
S Peru, W Bolivia
T.o.orbignyianus
C & S Chile, S Argentina
Thinocorus rumicivorus (Least Seed-Snipe)
T.r.pallidus
SW Ecuador
T.r.cuneicauda
W Peru, N Chile
T.r.bolivianus
SW Bolivia

T.r.rumicivorus
C Chile, C Argentina, Uruguay
T.r.patagonicus
S Argentina

59. CHIONIDIDAE (SHEATHBILLS)

CHIONIS
Chionis alba (Snowy Sheathbill)
S Georgia, S Orkneys, Falkland Is
Chionis minor (Black-faced Sheathbill)
C.m.marionensis
Prince Edward, Marion Is
C.m.crozettensis
Crozet, Possession Is
C.m.minor
Kerguelen I
C.m.nasicornis
Heard I

60. STERCORARIIDAE (SKUAS)

CATHARACTA   60.4
Catharacta skua (Great Skua)
N Atlantic >> C Atlantic
Catharacta chilensis (Chilean Skua)
S Sth America >> Colombia, Uruguay
Catharacta maccormicki (South Polar Skua)
Antarctica >> N Pacific & NW Atlantic
Catharacta antarctica (Antarctic Skua)
C.a.antarctica
Falkland Is, S Argentina
C.a.lonnbergi
Circumpolar >> Tropic of Capricorn
C.a.hamiltoni
Gough I, Tristan da Cunha

STERCORARIUS
Stercorarius pomarinus (Pomarine Skua)
N Holarctic >> Peru, S Africa, India, N
Australia
Stercorarius parasiticus (Arctic Skua)
N Holarctic >> S Sth America, S Africa,
India, Australia
Stercorarius longicaudus (Long-tailed Skua)
N Holarctic >> W Africa, S Sth America,
Mediterranean & Japan

61 LARIDAE (GULLS, TERNS)

LARINAE

GABIANUS
Gabianus pacificus (Pacific Gull)
S coast Australia, Tasmania
Gabianus scoresbii (Magellan Gull)
S coast S America, Falkland Is

PAGOPHILA
Pagophila eburnea (Ivory Gull)   61.3
Circumpolar to N Europe, N Asia, N
America

LARUS
Larus fuliginosus (Dusky Gull)
Galapagos Is
Larus modestus (Grey Gull)
coast of Peru & Chile
Larus heermanni (Heermann's Gull)
W USA, W Mexico
Larus leucophthalmus (White-eyed Gull)
S Red Sea, Somali coast
Larus hemprichii (Sooty Gull)
S Red Sea, Iran & E Africa coast
Larus belcheri (Band-tailed Gull)
L.b.belcheri
coast of Peru
L.b.atlanticus
coast of Argentina
Larus crassirostris (Japanese Gull)
coasts of Japan Sea, China Sea
Larus audouinii (Audouin's Gull)
Mediterranean Is
Larus delawarensis (Ring-billed Gull)
Canada & USA coasts to S Mexico, Cuba
Larus canus (Mew Gull)
L.c.canus
NW Europe to Mediterranean
L.c.brachyrhynchus
Alaska, W Canada, W USA
Larus kamtschatschensis (Eastern Mew Gull)
E Siberia to China, Japan
Larus argentatus (Herring Gull)
L.a.smithsonianus
Canada to W Mexico
L.a.argentatus
NW Europe, Mediterranean
L.a.lusitanius
Portugal
L.a.omissus
White Sea islands
L.a.birulae
Arctic Ocean islands
L.a.heuglini
N Siberia to Persian Gulf
L.a.vegae
NE Siberia to China, Japan
L.a.atlantis
Azores, Madeira, Canary Is
L.a.michahelles
W & C Mediterranean
L.a.cachinnans
S Russia, SC Asia, N Red Sea
L.a.argenteus
UK & North Sea
Larus armenicus (Armenian Gull)   61.1
Mediterranean
Larus thayeri (Thayer's Gull)
Arctic Canada, W USA
Larus fuscus (Lesser Black-backed Gull)

L.f.fuscus
Scandinavia to W & E Africa
L.f.graellsii
British Isles to W Mediterranean & W Africa
Larus californicus (Californian Gull)
L.c.californicus
W USA, W Mexico
L.c.albertaensis
W Canada
Larus occidentalis (Western Gull)
L.o.occidentalis
W USA
L.o.wymani
S California, Baja California
Larus livens (Yellow-footed Gull)   61.3
Gulf of California islands
Larus dominicanus (Southern Black-backed Gull)
S Sth America, S Africa, New Zealand
Larus schistisagus (Slaty-backed Gull)
NE Asia to Alaska & Japan
Larus marinus (Greater Black-backed Gull)
N Atlantic to Cuba, Azores, Mediterranean
Larus glaucescens (Glaucous-winged Gull)
NE Asia & Alaska to W USA & China
Larus hyperboreus (Glaucous Gull)   61.2
L.h.hyperboreus
Europe, W Asia
L.h.pallidissimus
Siberia
L.h.barrovianus
Alaska
L.h.leuceretes
Canada, Greenland
Larus glaucoides (Iceland Gull)
NE Canada, Greenland, N Siberia, Baltic
Larus ichthyaetus (Great Black-headed Gull)
S Russia, Mongolia to Red Sea, India
Larus atricilla (Laughing Gull)
Maine to Brazil & W Central America
Larus brunnicephalus (Indian Black-headed Gull)
C & S Asia
Larus cirrocephalus (Grey-headed Gull)
L.c.cirrocephalus
EC Sth America
L.c.poiocephalus
Ethiopia to Malawi, S Madagascar
Larus serranus (Andean Gull)
coast of Peru & Andean lakes
Larus pipixcan (Franklin's Gull)
S Canada, NC USA to W Sth America
Larus novaehollandiae (Silver Gull)
L.n.forsteri
New Caledonia, N Australia
L.n.novaehollandiae
S Australia, Tasmania
L.n.scopulinus
New Zealand, Chatham I
L.n.hartlaubii
W Cape Province
Larus melanocephalus (Mediterranean Gull)
SE Europe, C & W Asia

Larus relictus (Relict Gull)
C Asia
Larus bulleri (Buller's Gull)
New Zealand
Larus maculipennis (Brown-hooded Gull)
S Sth America, Falkland Is
Larus ridibundus (Black-headed Gull)
Europe, Asia to N Africa, India, Philippine Is
Larus genei (Slender-billed Gull)
Mediterranean, Black Sea, Asia Minor
Larus philadelphia (Bonaparte's Gull)
W Canada, W & E USA
Larus minutus (Little Gull)
N Europe, Siberia to Mediterranean, Black Sea
Larus saundersi (Saunders' Gull)
Mongolia, N China, Japan

RHODOSTETHIA
Rhodostethia rosea (Ross's Gull)
N Siberia, Alaska, Greenland

RISSA
Rissa tridactyla (Black-legged Kittiwake)
R.t.tridactyla
NE Canada, NW Europe to Azores, USA, W Africa
R.t.pollicaris
Bering Sea & islands, Japan, W USA
Rissa brevirostris (Red-legged Kittiwake)
Pribilov, Commander Is

CREAGRUS
Creagrus furcatus (Swallow-tailed Gull)
Galapagos Is

XEMA
Xema sabini (Sabine's Gull)
Arctic regions >> W Africa and W Americas

STERNINAE

CHLIDONIAS
Chlidonias hybrida (Whiskered Tern)
C.h.hybrida
S Europe, SW Asia >> E & W Africa
C.h.swinhoei
S China, Taiwan, Indochina
C.h.indica
Iran to India
C.h.sclateri
Kenya to Cape Province, Madagascar
C.h.javanica
Sri Lanka, Malaysia, Java, Sulawesi
C.h.fluviatilis
Moluccas, New Guinea, Australia
Chlidonias leucoptera (White-winged Black Tern)
SE Europe, C Asia >> S Africa, India, China, Australia
Chlidonias nigra (Black Tern)

C.n.nigra
  Europe W Asia >> SC Africa
C.n.surinamensis
  Canada, N USA >> S America

Phaetusa simplex (Large-billed Tern)
P.s.simplex
  N & E Sth America
P.s.chloropoda
  SC & S Sth America

Gelochelidon nilotica (Gull-billed Tern)
G.n.nilotica
  Europe, C Asia >> N & E Africa & India
G.n.addenda
  S China
G.n.macrotarsa
  Australia
G.n.aranea
  E USA, Cuba, Gulf Coast
G.n.vanrossemi
  S California, W Mexico to Ecuador
G.n.gronvoldi
  Mexiana I, SE Brazil

Hydroprogne caspia (Caspian Tern)
H.c.caspia
  N America, Europe, Africa, C & S Asia
H.c.strenua
  W & S Australia, New Zealand

Sterna aurantia (Indian River Tern)
  Iran, India, Malaysia
Sterna hirundinacea (South American Tern)
  Peru, N Brazil to Tierra del Fuego
Sterna hirundo (Common Tern)
S.h.hirundo
  N America, Europe, W Asia >> S America,
  W Africa
S.h.tibetana
  Turkestan, Tibet >> India & Malaysia
S.h.minussensis
  C Asia, N Mongolia
S.h.longipennis
  NE Asia >> Japan, China, New Guinea
Sterna paradisaea (Arctic Tern)
  Arctic regions >> Chile, S Africa &
  Antarctica
Sterna vittata (Antarctic Tern)
S.v.vittata
  Ascension, St Helena, Gough, Kerguelen Is
S.v.tristanensis
  Tristan da Cunha
S.v.georgiae
  S Georgia, S Orkneys
S.v.gaini
  S Shetlands
S.v.bethunei
  Sub-Antarctic islands of New Zealand

Sterna virgata (Kerguelen Tern)   61.4
  S Indian Ocean
Sterna forsteri (Forster's Tern)
  W Canada, USA, N Central America
Sterna trudeaui (Trudeau's Tern)   61.4
  S Sth America
Sterna dougallii (Roseate Tern)
S.d.dougallii
  E & W Nth Atlantic coasts >> Brazil,
  Azores, S Africa
S.d.korustes
  Sri Lanka, Andaman Is
S.d.arideensis
  Seychelles, Mascarene Is
S.d.bangsi
  Riukiu, Philippine, Kai, Solomon Is
S.d.gracilis
  Moluccas, N & W coasts of Australia
Sterna striata (White-fronted Tern)
S.s.striata
  New Zealand
S.s.incerta
  Tasmania to SE Australia
S.s.aucklandorna
  Auckland, Chatham, Snares Is
Sterna repressa (White-cheeked Tern)
  S Red Sea to Kenya & Persian Gulf
Sterna sumatrana (Black-naped Tern)
S.s.sumatrana
  E Indian & Pacific Ocean islands, N
  Australia
S.s.mathewsi
  W Indian Ocean islands
Sterna melanogaster (Black-bellied Tern)
  India, Burma, S Indochina
Sterna aleutica (Aleutian Tern)
  Bering Sea to Japan
Sterna lunata (Spectacled Tern)
  Moluccas & Fiji to Hawaiian Is
Sterna anaethetus (Bridled Tern)
S.a.anaethetus
  Taiwan to Japan and Australia
S.a.fuligula
  S Red Sea to E Africa & W India
S.a.antarctica
  Seychelles, Mauritius, Maldive Is
S.a.rogersi
  N Western Australia
S.a.novaehollandiae
  Queensland
S.a.nelsoni
  W coast of Mexico & C America
S.a.melanoptera
  W Indies
Sterna fuscata (Sooty Tern)
S.f.fuscata
  W Indies, W African islands
S.f.crissalis
  W Mexican islands, Galapagos Is
S.f.oahuensis
  Hawaii, Bonin Is
S.f.kermadeci
  Kermadec Is

S.f.serrata
  Australia, New Guinea, New Caledonia
S.f.somaliensis
  Mait I, Gulf of Aden
S.f.nubilosa
  Indian Ocean & China Sea islands, Riukiu Is
Sterna nereis (Fairy Tern)
S.n.horni
  Western Australia
S.n.nereis
  S Australia, Victoria, Tasmania
S.n.davisae
  New Zealand
S.n.exsul
  New Caledonia
Sterna albistriata (Black-fronted Tern)
  New Zealand
Sterna superciliaris (Amazon Tern)
  E Sth America
Sterna balaenarum (Damara Tern)
  SW Africa
Sterna lorata (Chilean Tern)
  W Sth America
Sterna albifrons (Little Tern)
S.a.albifrons
  Europe, W Asia >> N Africa, NW India
S.a.guineae
  Ghana to Gabon
S.a.innominata
  Persian Gulf islands
S.a.pusilla
  N India, Burma, Java, Sumatra
S.a.sinensis
  Japan & Indochina >> Philippine Is & New Guinea
S.a.placens
  Australia
Sterna antillarum (Least Tern)   61.3
S.a.antillarum
  E USA >> W Indies & NE Brazil
S.a.mexicana
  Sonora, Sinaloa
S.a.browni
  W American coast from California to Peru
Sterna saundersii (Black-shafted Tern)
  S Red Sea, Somalia to NW India

THALASSEUS
Thalasseus bergii (Greater Crested Tern)
T.b.bergii
  Sthn Africa coast, Madagascar
T.b.thalassinus
  Seychelles, Aldabra, Rodriguez Is
T.b.velox
  NE Africa to Sri Lanka, Red Sea
T.b.cristatus
  Malaysia to Riukiu Is & E Australia
T.b.gwendolenae
  W & NW Australia
Thalasseus maximus (Royal Tern)
T.m.maximus
  California to Peru, Florida to Argentina W Indies

T.m.albidorsalis
  coast of W Africa
Thalasseus bengalensis (Lesser Crested Tern)
T.b.par
  N & E Africa, Madagascar
T.b.bengalensis
  Persian Gulf to Singapore >> Sulawesi
T.b.torresii
  Aru Is, N Australia
Thalasseus bernsteini (Chinese Crested Tern)
  E China, Philippine Is
Thalasseus elegans (Elegant Tern)
  California to Chile
Thalasseus sandvicensis (Sandwich Tern)
T.s.sandvicensis
  W & S Europe >> Africa, NW India
T.s.acuflavidus
  Florida, Gulf Coast >> Brazil, W Indies
T.s.eurygnatha
  E Sth America, Trinidad

LAROSTERNA
Larosterna inca (Inca Tern)
  coast of Peru & Chile

PROCELSTERNA
Procelsterna cerulea (Blue-grey Noddy)
P.c.saxatilis
  W Hawaiian Is, Marcus I
P.c.cerulea
  Christmas, Marquesas Is
P.c.nebouxi
  Phoenix, Ellis, Samoa Is
P.c.teretirostris
  Tuamotu, Society Is
P.c.albivitta
  Kermadec, Friendly, Norfolk Is
P.c.skottsbergii
  Easter I
P.c.imitatrix
  St Ambrose I (Chile)

ANOUS
Anous stolidus (Common Noddy)
A.s.stolidus
  Caribbean & Tropical Atlantic islands
A.s.plumbeigularis
  S Red Sea
A.s.pileatus
  Seychelles to Hawaiian Is & N Australia
A.s.ridgwayi
  islands of W Mexico & W Central America
A.s.galapagensis
  Galapagos Is
Anous tenuirostris (Lesser Noddy)
A.t.tenuirostris
  Seychelles, Madagascar, Mascarene Is
A.t.melanops
  Houtman Abrolhos Is (W Australia)
Anous minutus (White-capped Noddy)
A.m.minutus
  islands from Tuamotu to New Guinea

A.m.worcesteri
Cavilli I (Sulu Sea)
A.m.marcusi
Marcus & Wake Is to Caroline Is
A.m.melanogenys
Hawaiian Is
A.m.diamesus
Clipperton, Cocos Is
A.m.americanus
islands off Belize, C America
A.m.atlanticus
Tropical S Atlantic islands

GYGIS
Gygis alba (White Tern)
G.a.alba
S Atlantic Ocean islands
G.a.monte
Seychelles, Madagascar, Mascarene Is
G.a.royana
Norfolk, Kermadec Is
G.a.candida
SW Pacific Ocean Is
G.a.rothschildi
Laysan I
G.a.leucopes
Henderson I
G.a.microrhyncha
Marquesas Is
G.a.pacifica
S Pacific Ocean islands

62. RYNCHOPIDAE (SKIMMERS)

RYNCHOPS
Rynchops niger (Black Skimmer)
R.n.niger
New Jersey to Gulf Coast & N Brazil
R.n.cinerascens
N & E Sth America
R.n.intercedens
E & S Sth America
Rynchops flavirostris (African Skimmer)
Senegal to Sudan & Transvaal
Rynchops albicollis (Indian Skimmer)
India, Burma, Indochina

63. ALCIDAE (AUKS)

ALLE
Alle alle (Little Auk)
A.a.alle
N Atlantic Ocean >> New Jersey & W
Europe
A.a.polaris
Franz Josef Land, Barents Sea

ALCA
Alca torda (Razorbill)
A.t.pica
NE Canada, NE USA
A.t.islandica
Iceland, Faroe Is, British Isles

A.t.torda
Baltic Sea islands

URIA
Uria lomvia (Brünnich's Guillemot)
U.l.lomvia
Arctic Sea & N Atlantic Ocean
U.l.arra
Bering Sea, N Pacific Ocean
Uria aalge (Common Guillemot)(Murre)
U.a.aalge
Labrador to Orkneys & Norway
U.a.hyperborea
Bear I
U.a.spiloptera
Faroe Is
U.a.albionis
British Isles to Portugal
U.a.intermedia
islands in Baltic Sea
U.a.inornata
Bering Sea, N Pacific Ocean
U.a.californica
California

CEPPHUS
Cepphus grylle (Black Guillemot)
C.g.mandtii
Arctic Sea, Spitzbergen to N Greenland
C.g.arcticus
N Labrador, S Greenland
C.g.grylle
N Atlantic, Baltic & White Sea
Cepphus columba (Pigeon Guillemot)
C.c.columba
Bering Sea, N Pacific Ocean
C.c.snowi
Kurile, N Hokkaido Is
Cepphus carbo (Spectacled Guillemot)
Kurile Is, Okhotsk Sea to N Japan

BRACHYRAMPHUS
Brachyramphus marmoratus (Marbled Murrelet)
B.m.perdix
Kamchatka to Kurile, Hokkaido Is
B.m.marmoratus
Alaska to California
Brachyramphus brevirostris (Kittlitz's Murrelet)
Bering Sea, N Pacific Ocean
Brachyramphus hypoleucus (Xantus' Murrelet)
California & islands
Brachyramphus craveri (Craveri's Murrelet)
Gulf of California, Raza I

SYNTHLIBORAMPHUS
Synthliboramphus antiquus (Ancient Murrelet)
Bering Sea, N Pacific Ocean

Synthliboramphus wumizusume (Crested Murrelet)
coast of Japan

PTYCHORAMPHUS
Ptychoramphus aleuticus (Cassin's Auklet)
Aleutian Is to S California

CYCLORRHYNCHUS
Cyclorrhynchus psittacula (Parakeet Auklet)
Bering Sea, N Pacific Ocean

AETHIA
Aethia cristatella (Crested Auklet)
Bering Sea, N Pacific Ocean
Aethia pusilla (Least Auklet)
Bering Sea, N Pacific Ocean
Aethia pygmaea (Whiskered Auklet)
Kurile, Aleutian Is to N Japan

CERORHINCA
Cerorhinca monocerata (Rhinoceros Auklet)
Aleutian Is & N Pacific coasts

FRATERCULA
Fratercula arctica (Atlantic Puffin)
F.a.naumanni
Greenland to Novaya Zemlya
F.a.arctica
NE Canada to N Norway
F.a.grabae
Faroe Is, British Isles, S Norway
Fratercula corniculata (Horned Puffin)
Bering Sea, N Pacific Ocean

LUNDA
Lunda cirrhata (Tufted Puffin)
Bering Sea, N Pacific Ocean & coasts

COLUMBIFORMES

64. PTEROCLIDIDAE (SANDGROUSE)

SYRRHAPTES
Syrrhaptes tibetanus (Tibetan Sandgrouse)
C Asia, India
Syrrhaptes paradoxus (Pallas' Sandgrouse)
C Asia, N China >> NE China

PTEROCLES
Pterocles alchata (Pintailed Sandgrouse)
P.a.alchata
S Spain, S France
P.a.caudacutus
N Africa, Israel to C Asia & India
Pterocles namaqua (Namaqua Sandgrouse)
Namibia to Transvaal & W Cape
Province
Pterocles exustus (Chestnut-bellied Sandgrouse)
P.e.exustus
Senegal to Ethiopia
P.e.floweri
Egypt
P.e.ellioti
NE Africa to Kenya
P.e.olivascens
Kenya
P.e.erlangeri
SW Saudi Arabia
P.e.hindustan
Iraq, India
Pterocles senegallus (Spotted Sandgrouse)
NE Africa, Middle East, India
Pterocles orientalis (Black-bellied Sandgrouse)
P.o.aragonica
Spain, Canary Is, Morocco
P.o.orientalis
N Africa, Middle East, India
P.o.arenarius
S Russia, N Afghanistan
Pterocles coronatus (Crowned Sandgrouse)
P.c.coronatus
Algeria & Niger to Egypt
P.c.vastitas
Sinai
P.c.saturatus
E Saudi Arabia
P.c.atratus
Iraq to India
P.c.ladas
Sind
Pterocles gutturalis (Yellow-throated Sandgrouse)
P.g.saturatior
N Ethiopia to Tanzania
P.g.gutturalis
Zambia to Mozambique & Transvaal
Pterocles burchelli (Variegated Sandgrouse)
Namibia, Botswana, W Transvaal   64.1
Pterocles personatus (Madagascar Sandgrouse)
W Madagascar
Pterocles decoratus (Black-faced Sandgrouse)
P.d.ellenbecki
S Somalia, N Kenya
P.d.decoratus
S Kenya
P.d.loveridgei
Tanzania
Pterocles lichtensteinii (Lichtenstein's Sandgrouse)
P.l.targius
S Algeria, Niger

P.l.lichtensteinii
Ethiopia, N Sudan, Egypt
P.l.ingramsi
S Saudi Arabia
P.l.sukensis
Kenya
P.l.arabicus
E Saudi Arabia to Afghanistan & Pakistan
**Pterocles bicinctus (Double-banded
Sandgrouse)**   64.1
P.b.ansorgei
S Angola
P.b.bicinctus
C & E Namibia to SW Zambia
P.b.multicolor
Zambia, Malawi, Mozambique to Transvaal
Pterocles indicus (Painted Sandgrouse)
India
**Pterocles quadricinctus (Four-banded
Sandgrouse)**   64.1
Senegal to Sudan, Uganda, NW Kenya

65. COLUMBIDAE (DOVES, PIGEONS)

COLUMBA
Columba livia (Feral Rock Pigeon)
C.l.livia
W Europe, NW Africa
C.l.atlantis
Cape Verde Is, Madeira, Azores
C.l.canariensis
Canary Is
C.l.gymnocyclus
Senegal, Ghana
C.l.targia
Air to Darfur
C.l.lividior
Mali
C.l.butleri
NE Sudan
C.l.daklae
Dakla & Kharga Oases, Libya
C.l.schimperi
Nile Valley
C.l.palestinae
Israel, Sinai, W Arabia
C.l.gaddi
Asia Minor
C.l.neglecta
Turkestan, Pakistan
C.l.intermedia
S India, Sri Lanka
C.l.nigricans
Mongolia, N China
Columba rupestris (Eastern Rock Pigeon)
C.r.turkestanica
C Asia, Himalayas
C.r.rupestris
N China, Manchuria
Columba leuconota (Snow Pigeon)
C.l.leuconota

Himalayas, W China
C.l.gradaria
C China
Columba guinea (Speckled Pigeon)
C.g.guinea
Senegal to Ethiopia, Tanzania
C.g.phaeonota
S Africa
Columba albitorques (White-collared Pigeon)
C & E Ethiopia
Columba oenas (Stock Pigeon)
C.o.oenas
Europe, N Africa, Asia Minor
C.o.hyrcana
N Iran
C.o.yarkandensis
E Turkestan, Tien Shan
Columba eversmanni (Yellow-eyed Pigeon)
Turkestan to NW India
Columba oliviae (Somali Pigeon)
Somalia
Columba palumbus (Wood Pigeon)
C.p.palumbus
Europe, W Russia
C.p.madarensis
Madeira
C.p.azorica
Azores
C.p.excelsa
N Africa
C.p.iranica
Iran
C.p.casiotis
N India
Columba trocaz (Trocaz Pigeon)
Madeira
Columba bollii (Bolle's Pigeon)
Canary Is
Columba unicincta (African Wood Pigeon)
Liberia to Zaire, Uganda
Columba junoniae (Laurel Pigeon)
Palma, Gomera(Canary Is)
Columba sjostedi (Cameroun Olive Pigeon)
SE Nigeria, Cameroun
Columba arquatrix (Olive Pigeon)
Ethiopia & Angola to E Sth Africa
**Columba thomensis (Sao Thome Olive
Pigeon)** Sao Thome I
Columba pollenii (Comoro Olive Pigeon)
Comoro Is
Columba hodgsonii (Speckled Wood Pigeon)
Himalayas, Burma, W China
Columba albinucha (White-naped Pigeon)
E Zaire, W Uganda
Columba pulchricollis (Ashy Wood Pigeon)
Tibet, N Burma, N Thailand
Columba elphinstonii (Nilgiri Wood Pigeon)
SW India
Columba torringtoni (Sri Lanka Wood Pigeon)
Sri Lanka
Columba punicea (Pale-capped Pigeon)
NE India to N Malaysia, Vietnam

Columba argentina (Silver Pigeon)
islands W of Sumatra & N of Borneo
Columba palumboides (Andaman Wood Pigeon)
Andaman, Nicobar Is
Columba janthina (Black Wood Pigeon)
C.j.janthina
S Japanese islands, N Riukiu Is
C.j.stejnegeri
S Riukiu Is
C.j.nitens
Bonin, Volcano Is
Columba vitiensis (White-throated Pigeon)
C.v.halmaheira
Moluccas, New Guinea, Solomon, Sula Is
C.v.leopoldi
New Hebrides
C.v.hypoenochroa
New Caledonia
C.v.griseogularis
Philippine Is, N Bornean islands
C.v.anthracinus
Palawan
C.v.metallica
Lesser Sunda Is
C.v.vitiensis
Fiji Is
C.v.castaneiceps
Samoa
Columba leucomela (White-headed Pigeon)
E Australia
Columba jouyi (Silver-banded Black Pigeon)
Okinawa I e?
Columba pallidiceps (Yellow-legged Pigeon)
Solomon Is, Bismarck Arch
Columba leucocephala (White-crowned Pigeon)
W Indies, S Florida
Columba squamosa (Red-necked Pigeon)
Gtr, Lesser & Dutch Antilles,
Columba speciosa (Scaled Pigeon)
S Mexico to Brazil & Paraguay
Columba picazuro (Picazuro Pigeon)
C.p.marginalis
NE Brazil
C.p.picazuro
E Brazil to NE Argentina
Columba corensis (Bare-eyed Pigeon)
N Colombia, N Venezuela, Dutch Antilles
Columba maculosa (Spotted Pigeon)
C.m.albipennis
S Peru, W Bolivia
C.m.maculosa
N Argentina, Uruguay, Paraguay
Columba fasciata (Band-tailed Pigeon)
C.f.fasciata
W Nth America
C.f.monilis
N Baja California
C.f.vioscae
S Baja California

C.f.letonai
Honduras, El Salvador
C.f.parva
N Nicaragua
C.f.crissalis
Costa Rica, W Panama
C.f.albilinea
N & W Colombia to E Bolivia
C.f.roraimae
Mt Duida, Mt Romaima (Venezuela)
Columba araucana (Chilean Pigeon)
C & S Chile
Columba caribaea (Jamaican Band-tailed Pigeon)
Jamaica
Columba cayennensis (Rufous Pigeon)
C.c.pallidicrissa
S Mexico to Colombia
C.c.cayennensis
Venezuela, the Guianas, N Brazil
C.c.sylvestris
E Peru to N Argentina
C.c.occidentalis
W Colombia
C.c.tamboensis
SW Colombia
Columba flavirostris (Red-billed Pigeon)
C.f.flavirostris
Texas, E & S Mexico to El Salvador
C.f.madrensis
Tres Marias Is
C.f.restricta
W Mexico
C.f.minima
W Costa Rica
Columba oenops (Salvin's Pigeon)
N Peru
Columba inornata (Plain Pigeon)   65.2
Cuba, Hispaniola, Jamaica, Puerto Rico
Columba plumbea (Plumbeous Pigeon)
C.p.bogotensis
Colombia to N Peru
C.p.chapmani
W Ecuador
C.p.pallescens
SE Ecuador to E Brazil
C.p.baeri
C Brazil
C.p.wallacei
Lower Amazon, the Guianas
C.p.plumbea
SE Brazil, Paraguay
Columba subvinacea (Ruddy Pigeon)
C.s.subvinacea
Costa Rica, W Panama
C.s.berlepschi
E Panama to S Ecuador
C.s.ruberrima
NW Colombia
C.s.peninsularis
N Venezuela

C.s.zuliae
W Venezuela
C.s.purpureotincta
E Colombia, E Venezuela, the Guianas
C.s.anolaimae
SC Colombia
C.s.ogilvie-granti
SE Colombia
Columba nigrirostris (Short-billed Pigeon)
SE Mexico to E Panama
Columba goodsoni (Dusky Pigeon)
W Colombia, W Ecuador
Columba delegorguei (Delegorgue's Pigeon)
C.d.sharpei
S Sudan, Kenya, Tanzania
C.d.delegorguei
Natal
Columba iriditorques (Bronze-naped Pigeon)
Sierra Leone to Angola, E Zaire
Columba malherbii (Sao Thome Bronze-naped Pigeon)
Sao Thome, Principé, Annobon Is
Columba mayeri (Pink Pigeon)
Mauritius

STREPTOPELIA
Streptopelia turtur (Turtle Dove)
S.t.turtur
Europe, Asia Minor, Azores Is
S.t.arenicola
N Africa, SW Asia
S.t.hoggara
S Sahara
S.t.isabellina
E Libya, N Egypt
Streptopelia lugens (Dusky Turtle Dove)
S.l.bishaensis
SW Arabia
S.l.lugens
Ethiopia, Somalia
S.l.funebrea
Uganda to Tanzania, Malawi
Streptopelia hypopyrrha (Pink-bellied Turtle Dove)
E Nigeria, Cameroun
Streptopelia orientalis (Eastern Turtle Dove)
S.o.meena
W Himalayas
S.o.erythrocephala
S India
S.o.agricola
Burma, NE India
S.o.orientalis
Siberia, China, Japan
S.o.stimpsoni
Riukiu Is
S.o.orii
Taiwan
Streptopelia bitorquata (Javanese Collared Dove)

S.b.dusumieri
Philippine Is, N Borneo
S.b.bitorquata
Java to Timor I
Streptopelia decaocto (Collared Dove)
S.d.decaocto
Europe to W China
S.d.stoliczkae
Chinese Turkestan
S.d.xanthocyclus
Burma to E China
Streptopelia roseogrisea (African Collared Dove)
S.r.roseogrisea
Sudan, W Ethiopia, Mali, Chad
S.r.arabica
E Ethiopia, Somalia, Arabia
Streptopelia reichenowi (White-winged Collared Dove)
S Somalia, NE Kenya
Streptopelia decipiens (Mourning Collared Dove)
S.d.decipiens
E Chad, Sudan, Ethiopia
S.d.shelleyi
Senegal to N Nigeria
S.d.logonensis
E Cameroun, N Zaire
S.d.elegans
S Ethiopia, N & E Kenya
S.d.ambigua
Angola, S Zaire, W Zambia
S.d.perspicillata
Somalia to Malawi, Mozambique
Streptopelia semitorquata (Red-eyed Dove)
65.3
Ethiopia to Sth Africa
Streptopelia capicola (Ring-necked Dove)
S.c.electa
S Ethiopia
S.c.somalica
S Ethiopia, Somalia, Kenya, Tanzania
S.c.tropica
Uganda, W Tanzania to Mozambique, E Zaire
S.c.ongouati
Namibia, Angola
S.c.damarensis
C & W Sth Africa
S.c.capicola
Transvaal, Natal, Cape Province
Streptopelia vinacea (Vinaceous Dove) 65.3
Senegal to Sudan & N Zaire
Streptopelia tranquebarica (Red-collared Dove)
S.t.humilis
N Tibet to Indochina & N Philippine Is
S.t.murmensis
E Nepal, Sikkim, NE India
S.t.tranquebarica
India

Streptopelia picturata **(Madagascar Turtle Dove)**
S.p.picturata
  Madagascar
S.p.coppingeri
  Glorioso Is
S.p.comorensis
  Anjouan I (Comoro)
S.p.aldabrana
  Aldabra I
S.p.assumptionis
  Assumption I
S.p.saturata
  Amirante I
S.p.rostrata
  Seychelles
S.p.chuni
  Diego Garcia I
Streptopelia chinensis **(Spotted-necked Dove)**
S.c.ceylonensis
  Sri Lanka
S.c.suratensis
  India
S.c.forresti
  NE Burma, NW Yunnan
S.c.chinensis
  E China
S.c.formosa
  Taiwan
S.c.hainana
  Hainan I
S.c.vacillans
  SE Yunnan
S.c.tigrina
  Burma to Palawan, Borneo & Sumatra
Streptopelia senegalensis **(Laughing Dove)**
S.s.phoenicophila
  Morocco, Algeria, Tunisia
S.s.aegyptiaca
  Nile valley, Egypt
S.s.senegalensis
  Senegal to Ethiopia & Cape Province
S.s.sokotrae
  Socotra I
S.s.cambayensis
  Iran, India
S.s.ermanni
  Afghanistan, Turkestan

APLOPELIA
Aplopelia larvata **(Lemon Dove)**   65.3
A.l.bronzina
  Ethiopia
A.l.larvata
  SE Sudan to Cape Province
A.l.jacksoni
  E Zaire, Uganda, W Tanzania, Angola
A.l.inornata
  Cameroun, E Nigeria, Fernando Po, Annobon Is
A.l.principalis
  Principé I

A.l.simplex
  Sao Thome I

MACROPYGIA
Macropygia unchall **(Bar-tailed Cuckoo Dove)**
M.u.tusalia
  Himalayas, W China, N Burma
M.u.minor
  SE China, N Indochina, Hainan I
M.u.unchall
  Malaysia, Sumatra, Java, Lombok, Flores Is
Macropygia amboinensis **(Pink-breasted Cuckoo Dove)**   65.1
M.a.sanghirensis
  Sanghir, Talaut Is
M.a.atrata
  Togian I
M.a.albicapilla
  Sulawesi
M.a.sedecima
  Sula Is
M.a.amboinensis
  Moluccas
M.a.keyensis
  Kai Is
M.a.doreya
  NW New Guinea, W Papuan islands
M.a.maforensis
  Mumfor I
M.a.griseinucha
  Meos Num I
M.a.kerstingi
  N New Guinea, Japen I
M.a.meeki
  Vulcan I
M.a.cinereiceps
  D'Entrecasteaux Arch
M.a.cunctata
  Louisiade Arch
M.a.carteretia
  Bismarck Arch
M.a.huskeri
  New Hanover
Macropygia phasianella **(Large Brown Cuckoo Dove)**
M.p.septentrionalis
  Botel Tobago, Batan Is
M.p.phaea
  Calayan I
M.p.tenuirostris
  Philippine Is, Palawan, Sulu Arch
M.p.borneensis
  N Borneo
M.p.hypopercna
  Simalur I
M.p.modiglianii
  Nias I
M.p.elassa
  Mentawi I
M.p.megala
  Kangean I
M.p.robinsoni
  N Australia

M.p.phasianella
S Queensland, New South Wales
**Macropygia cinnamomea (Enggano Cuckoo
Dove)** 65.1
Enggano I
**Macropygia emiliana (Indonesian Cuckoo
Dove)** 65.1
Sumatra to Flores I
Macropygia magna (Large Cuckoo Dove)
M.m.macassariensis
Saleyer I, S Sulawesi
M.m.longa
Djampea I
M.m.magna
Timor, Alor, Wetar Is
M.m.timorlaoensis
Tanimbar Is
**Macropygia rufipennis (Andaman Cuckoo
Dove)**
M.r.rufipennis
Andaman, Nicobar Is
M.r.tiwarii
Gt Nicobar I
**Macropygia nigrirostris (Lesser Bar-tailed
Cuckoo Dove)**
New Guinea & NE islands
**Macropygia mackinlayi (Mackinlay's Cuckoo
Dove)**
M.m.mackinlayi
Santa Cruz, Banks I, New Hebrides
M.m.arossi
Solomon Is
M.m.krakari
Karkar I
M.m.goodsoni
St Matthias Is, NE New Guinea
Macropygia ruficeps (Little Cuckoo Dove)
M.r.assimilis
S Burma, NW Thailand
M.r.malayana
Malaysia
M.r.engelbachi
N Indochina
M.r.nana
Borneo
M.r.sumatrana
Sumatra
M.r.simalurensis
Simalur I
M.r.ruficeps
Java, Bali I
M.r.orientalis
Sumbawa, Flores, Timor Is

REINWARDTOENA
**Reinwardtoena reinwardtsi (Reinwardt's
Long-tailed Pigeon)**
R.r.reinwardtsi
Moluccas
R.r.griseotincta
New Guinea, W New Guinea islands
R.r.brevis
Biak I

**Reinwardtoena browni (Brown's Long-tailed
Pigeon)**
New Britain, Duke of York I
**Reinwardtoena crassirostris (Crested Long-
tailed Pigeon)**
Solomon Is

TURACOENA
Turacoena manadensis (White-faced Pigeon)
Sulawesi, Peleng, Sula Is
Turacoena modesta (Timor Black Pigeon)
Timor, Wetar Is

TURTUR
**Turtur chalcospilos (Emerald-spotted Wood
Dove)** 65.3
Somalia to Angola & Cape Province
Turtur abyssinicus (Black-billed Wood Dove)
Senegal to N Ethiopia
Turtur afer (Blue-spotted Wood Dove)
Senegal to Ethiopia & Transvaal
Turtur tympanistria (Tambourine Dove) 65.3
Sierra Leone to Tanzania, E Cape Province
Turtur brehmeri (Blue-headed Wood Dove)
T.b.infelix
Sierra Leone to Cameroun Mt
T.b.brehmeri
S Cameroun, Gabon, Zaire

OENA
Oena capensis (Namaqua Dove)
O.c.capensis
Senegal to Arabia & Cape Province
O.c.aliena
Madagascar

CHALCOPHAPS
Chalcophaps indica (Emerald Dove)
C.i.indica
India to the Philippine Is, Moluccas, Gtr
Sunda Is
C.i.robinsoni
Sri Lanka
C.i.maxima
Andaman Is
C.i.natalis
Christmas I
C.i.formosanus
Taiwan
C.i.yamashinae
Riukiu Is
C.i.salimali
Kerala (S India)
C.i.minima
Numfor, Biak, Meos Num Is
C.i.chrysochlora
New Guinea & islands, E Australia
C.i.sandwichensis
Santa Cruz Is, New Hebrides, New
Caledonia
C.i.longirostris
Nthn Territory(Australia), Lesser Sunda Is

C.i.melvillensis
  Melville I
Chalcophaps stephani (Brown-backed Emerald Dove)
  C.s.wallacei
    Sulawesi
  C.s.stephani
    New Guinea & surrounding parts
  C.s.mortoni
    Solomon Is

HENICOPHAPS
Henicophaps albifrons (Black Bronzewing)
  H.a.albifrons
    New Guinea, Waigeu, Misol, Japen Is
  H.a.schlegeli
    Aru Is
Henicophaps foersteri (New Britain Bronzewing)
  New Britain

PHAPS
Phaps chalcoptera (Common Bronzewing)
  P.c.murchisoni
    mid W & SW Australia
  P.c.consobrina
    N Australia
  P.c.chalcoptera
    S Queensland to Tasmania, S Australia
Phaps elegans (Brush Bronzewing)
  P.e.neglecta
    Sthn Australia
  P.e.elegans
    Tasmania
Phaps histrionica (Flock Pigeon)
  P.h.alisteri
    NW Australia
  P.h.histrionica
    W Queensland, W New South Wales

OCYPHAPS
Ocyphaps lophotes (Crested Pigeon)
  O.l.whitlocki
    WC Australia
  O.l.lophotes
    C & EC Australia

PETROPHASSA
Petrophassa plumifera (White-bellied Plumed Pigeon) (Spinifex Pigeon)
  P.p.plumifera
    N Western Australia to NW Queensland
  P.p.mungi
    Derby District N Western Australia
  P.p.proxima
    Upper Fitzroy river, N Western Australia
  P.p.leucogaster
    N Sth Australia, S Nthn Territory
Petrophassa ferruginea (Red-plumed Pigeon)
  NW Australia
Petrophassa scripta (Partridge Bronzewing)
  P.s.peninsulae
    N Queensland

P.s.scripta
  C Queensland, C New South Wales
Petrophassa smithii (Bare-eyed Partridge Bronzewing)
  N & NW Australia
Petrophassa rufipennis (Chestnut-quilled Rock Pigeon)
  N Territory (Australia)
Petrophassa albipennis (White-quilled Rock Pigeon)
  P.a.albipennis
    N Western Australia
  P.a.boothi
    N Northern Territory

GEOPELIA
Geopelia cuneata (Diamond Dove)
  N & C Australia
Geopelia striata (Zebra Dove)
  G.s.striata
    S Burma to Philippine Is, Borneo, Lombok I
  G.s.papua
    S New Guinea
  G.s.tranquilla
    C Australia
  G.s.clelandi
    mid Western Australia
Geopelia maugei (Timor Zebra Dove) 65.1
  G.m.maugei
    Sumbawa to Timor
  G.m.audacis
    Tanimbar, Kai Is
Geopelia placida (Gould's Zebra Dove)
  N Australia
Geopelia humeralis (Bar-shouldered Dove)
  G.h.gregalis
    S New Guinea
  G.h.humeralis
    N & NE Australia

LEUCOSARCIA
Leucosarcia melanoleuca (Wonga Pigeon)
  Queensland to Victoria

ZENAIDA
Zenaida macroura (Mourning Dove)
  Z.m.marginella
    W Nth America & C America
  Z.m.carolinensis
    E Nth America, Bahama Is
  Z.m.macroura
    Cuba, Isle of Pines, Hispaniola
  Z.m.tresmariae
    Tres Marias Is
  Z.m.clarionensis
    Clarion I
Zenaida auriculata (Eared Dove)
  Z.a.caucae
    W Colombia
  Z.a.vulcania
    C Colombia
  Z.a.hypoleuca
    W Ecuador, W Peru

Z.a.auriculata
  Chile, W Argentina
Z.a.chrysauchenia
  Bolivia to Uruguay & S Argentina
Z.a.noronha
  NE Brazil
Z.a.marajoensis
  River Amazon estuary
Z.a.stenura
  Grenada I, Trinidad, NE Sth America
Z.a.penthera
  E Colombia, NW Venezuela
Z.a.antioquiae
  NC Colombia
Z.a.vinaceorufa
  Curacao, Aruba, Bonaire Is
Zenaida aurita (Zenaida Dove)
Z.a.salvadorii
  Yucatan coast & islands
Z.a.zenaida
  Bahama, Gtr Antilles, Virgin Is
Z.a.aurita
  Lesser Antilles
Zenaida galapagoensis (Galapagos Dove)
Z.g.galapagoensis
  Galapagos Is
Z.g.exsul
  Culpepper, Wenman Is
Zenaida asiatica (White-winged Dove)
Z.a.mearnsi
  SW USA, W Mexico, Tres Marias Is
Z.a.asiatica
  S USA, E Mexico, Gtr Antilles Is
Z.a.australis
  W Costa Rica
Z.a.meloda
  SW Ecuador to N Chile

COLUMBINA
Columbina passerina (Common Ground Dove)
C.p.passerina
  SE coast ofUSA
C.p.bahamensis
  Bahama, Bermuda Is
C.p.insularis
  Gtr Antilles, Cayman Is
C.p.jamaicensis
  Jamaica
C.p.umbrina
  Haiti
C.p.portoricensis
  Puerto Rico, Virgin Is
C.p.nigrirostris
  St Croix I, N Lesser Antilles
C.p.trochila
  Martinique
C.p.antillarum
  S Lesser Antilles
C.p.pallescens
  S USA to Guatemala & Belize
C.p.socorroensis
  Socorro I

C.p.neglecta
  Honduras to Costa Rica
C.p.albivitta
  N Colombia, N Venezuela, Dutch Antilles
C.p.parvula
  NC Colombia
C.p.nana
  W Colombia
C.p.quitensis
  C Ecuador
C.p.griseola
  S Venezuela, the Guianas, N Brazil
C.p.tortugensis
  Los Hermanos, La Tortuga Is
Columbina minuta (Plain-breasted Ground Dove)
C.m.interrupta
  SE Mexico, Guatemala, Belize
C.m.elaeodes
  SW Costa Rica, WC Colombia
C.m.minuta
  Venezuela, the Guianas, Peru, Brazil, Paraguay
Columbina buckleyi (Ecuadorean Ground Dove)
  NW Ecuador, NW Peru
Columbina talpacoti (Ruddy Ground Dove)
C.t.eluta
  W Mexico
C.t.rufipennis
  SE Mexico to N Colombia, N Venezuela, Trinidad
C.t.caucae
  Cauca valley, Colombia
C.t.talpacoti
  C & E Sth America from the Guianas to C Argentina
Columbina picui (Picui Dove)
C.p.strepitans
  NE Brazil
C.p.picui
  Bolivia & S Brazil to C Chile & Argentina
Columbina cruziana (Gold-billed Ground Dove)
  N Ecuador to NW Chile
Columbina cyanopis (Blue-eyed Ground Dove)
  C Brazil

CLARAVIS
Claravis pretiosa (Blue Ground Dove)
  SE Mexico to Paraguay & N Argentina
Claravis godefrida (Purple-barred Ground Dove)
  SE Brazil, E Paraguay
Claravis mondetoura (Purple-breasted Ground Dove)
  SE Mexico to Venezuela & E Peru

METRIOPELIA
Metriopelia ceciliae (Bare-faced Ground Dove)

M.c.ceciliae
 W Peru
M.c.obsoleta
 E Peru
M.c.gymnops
 S Peru, Bolivia, N Chile
Metriopelia morenoi (Bare-eyed Ground Dove)
 NW Argentina
Metriopelia melanoptera (Black-winged Ground Dove)
M.m.saturatior
 S Colombia, Ecuador
M.m.melanoptera
 Peru to Chile & W Argentina
Metriopelia aymara (Bronze-winged Ground Dove)
 S Peru to Chile, W Argentina

SCARDAFELLA
Scardafella inca (Inca Dove)
 Arizona to N Costa Rica
Scardafella squammata (Scaly Dove)
S.s.ridgwayi
 coast of Colombia & Venezuela, Trinidad
S.s.squammata
 E & S Brazil

UROPELIA
Uropelia campestris (Long-tailed Ground Dove)
 E Bolivia, C Brazil

LEPTOTILA
Leptotila verreauxi (White-fronted Dove)
L.v.capitalis
 Tres Marias Is
L.v.angelica
 M & C Mexico
L.v.fulviventris
 S Mexico, E Guatemala, Belize
L.v.bangsi
 W Guatemala to W Nicaragua
L.v.nuttingi
 Ometepe I (Lake Nicaragua)
L.v.verreauxi
 SW Nicaragua to N Venezuela, Dutch Antilles
L.v.insularis
 Trinidad
L.v.tobagensis
 Tobago I
L.v.decolor
 W Colombia, W Ecuador, N Peru
L.v.brasiliensis
 the Guianas, N Brazil
L.v.approximans
 E Brazil
L.v.decipiens
 E Peru, E Bolivia, W Brazil

L.v.chalcauchenia
 S Bolivia, Uruguay, N Argentina
Leptotila megalura (White-faced Dove)
L.m.megalura
 N & C Bolivia
L.m.saturata
 S Bolivia, NW Argentina
Leptotila rufaxilla (Grey-fronted Dove)
L.r.pallidipectus
 E Colombia
L.r.dubusi
 E Ecuador to E Venezuela
L.r.rufaxilla
 E Venezuela, French Guiana
L.r.hellmayri
 N Venezuela, Trinidad
L.r.bahiae
 E Brazil
L.r.reichenbachii
 C Brazil to Paraguay & Uruguay
Leptotila plumbeiceps (Grey-headed Dove)
L.p.plumbeiceps
 SE Mexico to W Costa Rica
L.p.notius
 W Panama
L.p.malae
 Mala peninsula, W Panama
L.p.battyi
 Coiba I
Leptotila pallida (Pallid Dove)
 W Colombia, SW Ecuador
Leptotila wellsi (Grenada Dove)
 Grenada I
Leptotila jamaicensis (White-bellied Dove)
L.j.gaumeri
 N Yucatan peninsula & islands
L.j.collaris
 Gd Cayman I
L.j.jamaicensis
 Jamaica
L.j.neoxena
 St Andrews I
Leptotila cassini (Cassin's Dove)
L.c.cerviniventris
 E Guatemala to Panama
L.c.rufinucha
 SW Costa Rica, W Panama
L.c.cassini
 E Panama, N Colombia
Leptotila ochraceiventris (Buff-bellied Dove)
 SW Ecuador
Leptotila conoveri (Tolima Dove)
 C Colombia

GEOTRYGON
Geotrygon lawrencii (Lawrence's Quail Dove)
G.l.carrikeri
 Vera Cruz, Mexico
G.l.lentipes
 NW Costa Rica
G.l.lawrencii
 E Costa Rica, W Panama

Geotrygon costaricensis (Costa Rican Quail Dove)
Costa Rica, W Panama

Geotrygon goldmani (Russet-crowned Quail Dove)
G.g.goldmani
E Darien (E Panama)
G.g.oreas
Quebrada (E Panama)

Geotrygon saphirina (Sapphire Quail Dove)
G.s.purpurata
W Colombia, W Ecuador
G.s.saphirina
E Ecuador
G.s.rothschildi
Marcapata valley, Peru

Geotrygon caniceps (Grey-faced Quail Dove)
G.c.caniceps
Cuba
G.c.leucometopius
Hispaniola

Geotrygon versicolor (Crested Quail Dove)
Jamaica

Geotrygon veraguensis (Olive-backed Quail Dove)
E Costa Rica to NW Ecuador

Geotrygon linearis (Lined Quail Dove)
G.l.albifacies
SE Mexico, NE Guatemala
G.l.rubida
Guerrero, Mexico
G.l.anthonyi
S Mexico, W Guatemala
G.l.silvestris
El Salvador, Honduras, N Nicaragua
G.l.chiriquensis
Costa Rica, W Panama
G.l.infusca
Santa Marta, Colombia
G.l.linearis
E Colombia, W Venezuela
G.l.trinitatis
NE Venezuela, Trinidad

Geotrygon frenata (White-throated Quail Dove)
G.f.bourcieri
Colombia, Ecuador
G.f.subgrisea
SW Ecuador
G.f.frenata
Peru, Bolivia

Geotrygon chrysia (Key West Quail Dove)
Bahama Is, Cuba, Hispaniola

Geotrygon mystacea (Bridled Quail Dove)
Virgin Is, Lesser Antilles

Geotrygon violacea (Violaceous Quail Dove)
G.v.albiventer
Nicaragua to N Colombia
G.v.violacea
Surinam to Paraguay

Geotrygon montana (Ruddy Quail Dove)
G.m.martinica
Lesser Antilles Is
G.m.montana
Mexico to N Argentina, Gtr Antilles, Trinidad

STARNOENAS
Starnoenas cyanocephala (Blue-headed Quail Dove)
Cuba, Isle of Pines

CALOENAS
Caloenas nicobarica (Nicobar Pigeon)
C.n.nicobarica
Nicobar Is to Luzon I, New Guinea, Solomon Is
C.n.pelewensis
Palau Is

GALLICOLUMBA
Gallicolumba luzonica (Luzon Bleeding Heart)
Luzon I, Polillo Is

Gallicolumba criniger (Bartlett's Bleeding Heart)
Mindanao, Leyte, Samar, Basilan Is

Gallicolumba platenae (Mindoro Bleeding Heart)
Mindoro I

Gallicolumba keayi (Negros Bleeding Heart)
Negros I

Gallicolumba menagei (Tawitawi Bleeding Heart)
Tawitawi Is

Gallicolumba rufigula (Golden Heart)
G.r.helviventris
Aru Is
G.r. rufigula
W New Guinea
G.r.septentrionalis
N New Guinea
G.r.alaris
S New Guinea
G.r.orientalis
SE New Guinea

Gallicolumba tristigmata (Sulawesi Quail Dove)
G.t.tristigmata
N Sulawesi
G.t.auripectus
C & SE Sulawesi
G.t.bimaculata
S Sulawesi

Gallicolumba jobiensis (White-breasted Ground Pigeon)
G.j.jobiensis
New Guinea, Bismarck Arch
G.j.chalconota
Vellalavella, Guadalcanal Is

Gallicolumba kubaryi (Truk Is Ground Dove)
E Caroline Is

Gallicolumba erythroptera **(Society Is Ground Dove)**
Society, Tuamotu Is
Gallicolumba xanthonura **(White-throated Dove)**
Mariana, Yap Is
Gallicolumba stairi **(Friendly Quail Dove)**
Fiji, Tonga, Samoan Is
Gallicolumba sanctaecrucis **(Santa Cruz Ground Dove)**
Santa Cruz Is
Gallicolumba salamonis **(Thick-billed Ground Dove)**
San Cristobal, Ramos Is
Gallicolumba rubescens **(Marquesas Ground Dove)**
Marquesas Is
Gallicolumba beccarii **(Grey-breasted Quail Dove)**
G.b.eichhorni
St Matthias Is
G.b.admiralitatis
Admiralty Is
G.b.johannae
Bismarck Arch, Dampier I
G.b.beccarii
New Guinea
G.b.intermedia
W Solomon Is
G.b.solomonensis
Rennell, E Solomon Is
Gallicolumba canifrons **(Palau Ground Dove)**
Palau Is
Gallicolumba hoedtii **(Wetar Ground Dove)**
Wetar, Timor Is

TRUGON
Trugon terrestris **(Thick-billed Ground Pigeon)**
T.t.terrestris
NW New Guinea, Salawati I
T.t.mayri
N New Guinea
T.t.leucopareia
S New Guinea

MICROGOURA
Microgoura meeki **(Solomon Is Ground Pigeon)**
Choiseul I e?

OTIDIPHAPS
Otidiphaps nobilis **(Pheasant Pigeon)**
O.n.nobilis
W New Guinea
O.n.cervicalis
E & SE New Guinea
O.n.insularis
Fergusson I
O.n.aruensis
Aru Is

GOURA
Goura cristata **(Blue Crowned Pigeon)**
G.c.cristata
NW New Guinea
G.c.minor
W Papuan islands
Goura scheepmakeri **(Maroon-breasted Crowned Pigeon)**
G.s.sclaterii
S New Guinea
G.s.wadai
S New Guinea
G.s.scheepmakeri
SE New Guinea
Goura victoria **(Victoria Crowned Pigeon)**
G.v.victoria
Japen, Biak Is
G.v.beccarii
N New Guinea

DIDUNCULUS
Didunculus strigirostris **(Tooth-billed Pigeon)**
Upolu, Savaii Is (Samoa)

PHAPITRERON
Phapitreron leucotis **(Lesser Brown Fruit Dove)**
P.l.leucotis
Catanduanes, Luzon, Mindoro Is
P.l.nigrorum
Tablas, Masbate, Panay, Negros, Cebu Is
P.l.albifrons
Bohoi, Samar, Siquijor Is
P.l.brevirostris
Leyte, Mindanao Is
P.l.occipitalis
Basilan, Sulu Is
Phapitreron amethystina **(Greater Brown Fruit Dove)**
P.a.amethystina
Luzon, Samar, Leyte, Bohol, Mindanao Is
P.a.maculipectus
Negros I
P.a.frontalis
Cebu I
P.a.brunneiceps
Basilan I
P.a.cinereiceps
Tawitawi Is

TRERON
Treron fulvicollis **(Cinnamon-headed Green Pigeon)**
T.f.fulvicollis
Malaysia, Sumatra, S Borneo
T.f.oberholseri
Natuna Is
T.f.melopogenys
Nias I
T.f.baramensis
N Borneo & islands

Treron olax (Little Green Pigeon)
    Malaysia, Sumatra, Borneo, Java
Treron vernans (Pink-necked Green Pigeon)
    T.v.griseicapilla
        Malaysia, S Indochina, W Java, N Borneo
    T.v.parva
        NE Sumatra
    T.v.miza
        Simalur I
    T.v.mesochloa
        Nias, Siberut, Enggano, Pagi Is
    T.v.adina
        Natuna, Anamba Is
    T.v.purpurea
        S Borneo, Java to Sumbawa I, Sulawesi
    T.v.vernans
        Philippine, Palawan Is
Treron bicincta (Orange-breasted Green Pigeon)
    T.b.bicincta
        India to Indochina & Malaysia
    T.b.leggei
        Sri Lanka
    T.b.domvilii
        Hainan I
    T.b.javana
        Java
Treron pompadora (Pompadour Green Pigeon)
    T.p.pompadora
        Sri Lanka
    T.p.affinis
        W India
    T.p.phayrei
        E India to Thailand & S Indochina
    T.p.chloroptera
        Andaman, Nicobar Is
    T.p.axillaris
        Philippine Is
    T.p.everetti
        Sulu Arch
    T.p.pallidior
        Djampea, Kalao Is
    T.p.ada
        Madu, Kalao Tua Is
    T.p.aromatica
        Buru I
Treron curvirostra (Thick-billed Green Pigeon)
    T.c.nipalensis
        W Nepal to Thailand & Indochina
    T.c.curvirostris
        Malaysia, Sumatra
    T.c.harterti
        NE Sumatra
    T.c.hainana
        Hainan I
    T.c.erimacra
        Philippine Is
    T.c.nasica
        Borneo

    T.c.haliploa
        Simalur I
    T.c.pega
        Nias I
    T.c.smicra
        Sipora, Siberut, Batu Is
    T.c.hypothapsina
        Enggano I
Treron griseicauda (Grey-faced Thick-billed Green Pigeon)
    T.g.sanghirensis
        Sanghir Is
    T.g.griseicauda
        Sulawesi
    T.g.goodsoni
        Tukang Besi Is
    T.g.pulverulenta
        S Sumatra, Java, Bali
    T.g.vordermani
        Kangean I
Treron teysmanni (Sumba Is Green Pigeon)
    Sumba I
Treron floris (Flores Green Pigeon)
    Lesser Sunda Is from Lombok to Alor I
Treron psittacea (Timor Green Pigeon)
    Timor, Samau Is
Treron capellei (Large Green Pigeon)
    T.c.magnirostris
        Malaysia, N Sumatra, Borneo
    T.c.capellei
        S Sumatra, Java
Treron phoenicoptera (Yellow-legged Green Pigeon)
    T.p.phoenicoptera
        N India
    T.p.chlorigaster
        S India
    T.p.phillipsi
        Sri Lanka
    T.p.viridifrons
        Burma, W Thailand
    T.p.annamensis
        E Thailand, Indochina
Treron waalia (Yellow-bellied Green Pigeon)
    Senegal to S Arabia
Treron australis (Madagascar Green Pigeon)
    T.a.australis
        E Madagascar
    T.a.xenia
        W Madagascar
    T.a.griveaudi
        Moheli I (Comoro Is)
Treron calva (African Green Pigeon)
    T.c.nudirostris
        Senegal to Guinea
    T.c.sharpei
        Sierra Leone to N Cameroun
    T.c.calva
        Gabon, N Angola, W Zaire
    T.c.poensis
        Fernando Po
    T.c.virescens
        Principé I

T.c.uellensis
  N Zaire, Uganda
T.c.brevicera
  SW Ethiopia, E Kenya
T.c.salvadorii
  Lake Kivu area, NE Zaire
T.c.gibberifrons
  S Zaire, W Kenya
T.c.wakefieldii
  E Kenya, NE Tanzania
T.c.orientalis
  S Tanzania, Mozambique
T.c.schalowi
  Zambia, S Zaire
T.c.ansorgei
  S Angola
T.c.glauca
  E Botswana, N & W Transvaal, S Zimbabwe
T.c.damarensis
  N Namibia
T.c.vylderi
  NE Namibia
T.c.granti
  E Kenya to N Malawi
T.c.delalandii
  Mozambique to Natal
Treron pembaensis (Pemba Is Green Pigeon)
  Pemba I
Treron sanctithomae (Sao Thome Green Pigeon)
  Sao Thome I
Treron apicauda (Pin-tailed Green Pigeon)
T.a.apicauda
  Himalayas, W Burma
T.a.laotinus
  N Indochina
T.a.lowei
  S Vietnam
Treron oxyura (Yellow-bellied Pin-tailed Green Pigeon)
  Sumatra, W Java
Treron seimundi (Yellow-vented Pin-tailed Green Pigeon)
T.s.seimundi
  S Thailand, Malaysia
T.s.modestus
  C Vietnam
Treron sphenura (Wedge-tailed Green Pigeon)
T.s.sphenura
  Kashmir to Burma
T.s.yunnanensis
  SW China, N Vietnam
T.s.annamensis
  C Vietnam
T.s.oblitus
  Hainan I
T.s.robinsoni
  Malaysia
T.s.etorques
  Sumatra
T.s.korthalsi
  Sumatra, Java, Lombok I

Treron sieboldii (Japanese Green Pigeon)
T.s.fopingenis
  Shensi
T.s.sieboldii
  Japan
T.s.sororius
  Taiwan
T.s.murielae
  N & C Vietnam
Treron formosae (Formosan Green Pigeon)
T.f.permagna
  N Riukiu Is
T.f.medioximus
  S Riukiu Is
T.f.formosae
  Taiwan, Botel Tobago I
T.f.australis
  Batan, Calayan, Camiguin Is (Phil Is)

PTILINOPUS
Ptilinopus cincta (Black-backed Fruit Dove)
P.c.albocincta
  Bali to Flores I
P.c.everetti
  Pantar, Alor Is
P.c.cincta
  Timor, Wetar, Roma Is
P.c.lettiensis
  Letti, Moa, Luang, Sermatta Is
P.c.ottonis
  Damar, Babar Is
Ptilinopus alligator (Black-banded Pigeon)
  Nthn Territory (Australia)
Ptilinopus dohertyi (Red-naped Fruit Dove)
  Sumba I
Ptilinopus porphyrea (Pink-necked Fruit Dove)
  Sumatra, Java, Bali
Ptilinopus marchei (Marché's Fruit Dove)
  Luzon, Polillo Is
Ptilinopus merrilli (Merrill's Fruit Dove)
P.m.faustinoi
  Mt Tabuan (N Luzon I)
P.m.merrilli
  E & S Luzon, Polillo Is
Ptilinopus occipitalis (Yellow-breasted Fruit Dove)
  Philippine Is
Ptilinopus fischeri (Fischer's Fruit Dove)
P.f.fischeri
  N Sulawesi
P.f.centralis
  C & SE Sulawesi
P.f.meridionalis
  S Sulawesi
Ptilinopus jambu (Jambu Fruit Dove)
  Malaysia, Sumatra, Borneo
Ptilinopus subgularis (Dark-chinned Fruit Dove)
P.s.epia
  Sulawesi
P.s.subgularis
  Peling, Banggai Is

P.s.mangoliensis
Sula Mangoli I
Ptilinopus leclancheri (Black-chinned Fruit Dove)
P.l.leclancheri
Philippine Is
P.l.gironieri
Palawan I
Ptilinopus bernsteinii (Scarlet-breasted Fruit Dove) 65.1
P.b.micrus
Obi I
P.b.bernsteinii
Halmahera, Ternate Is
Ptilinopus magnificus (Magnificent Fruit Dove)
P.m.puella
NW New Guinea & islands
P.m.interposita
WC & SW New Guinea
P.m.septentrionalis
N & NE New Guinea, Japen, Dampier Is
P.m.poliura
SE New Guinea
P.m.assimilis
N Queensland
P.m.keri
Bellenden Ker, Queensland
P.m.magnificus
S Queensland to Victoria
Ptilinopus perlatus (Pink-spotted Fruit Dove)
P.p.perlatus
NW New Guinea & islands
P.p.plumbeicollis
NE New Guinea
P.p.zonurus
SE New Guinea, Aru, Fergusson Is
Ptilinopus ornatus (Ornate Fruit Dove)
P.o.ornatus
NW New Guinea
P.o.gestroi
C & E New Guinea
P.o.kaporensis
SW New Guinea
Ptilinopus tannensis (Silver-shouldered Fruit Dove)
New Hebrides, Banks Is
Ptilinopus aurantiifrons (Orange-fronted Fruit Dove)
New Guinea & NW islands
Ptilinopus wallacii (Wallace's Fruit Dove)
Babar, Kai, Aru Is, SW New Guinea
Ptilinopus superbus (Superb Fruit Dove)
P.s.temminckii
Sulu Arch, Sulawesi
P.s.superbus
Moluccas to Solomon Is, NE Australia
Ptilinopus perousii (Many-coloured Fruit Dove)
P.p.perousii
Samoan Is
P.p.mariae
Tonga, Fiji Is

Ptilinopus porphyraceus (Purple-capped Fruit Dove)
P.p.fasciatus
Samoan Is
P.p.graeffei
Uvea Is
P.p.ponapensis
Caroline Is
P.p.porphyraceus
Tonga, Fiji Is
Ptilinopus pelewensis (Palau Fruit Dove)
Palau Is
Ptilinopus rarotongensis (Rarotongan Fruit Dove)
P.r.rarotongensis
Rarotonga I
P.r.goodwini
Cook Is
Ptilinopus roseicapilla (Marianas Fruit Dove)
Mariana Is
Ptilinopus regina (Pink-capped Fruit Dove)
P.r.roseipileum
Wetar, Roma, Kissar, Moa, E Timor Is
P.r.xanthogaster
Banda, Kai, Damar, Babar, Tanimbar Is
P.r.flavicollis
Flores, Roti, W Timor Is
P.r.ewingii
Nthn Territory, Melville I
P.r.regina
Cape York to New South Wales
Ptilinopus richardsii (Silver-capped Fruit Dove)
P.r.richardsii
E Solomon Is
P.r.cyanopterus
Rennell I
Ptilinopus purpuratus (Grey-green Fruit Dove)
P.p.chrysogaster
W Society Is
P.p.frater
Moorea I
P.p.purpuratus
Tahiti
P.p.chalcurus
Mahatea I
P.p.coralensis
Tuamotu Is
Ptilinopus greyii (Grey's Fruit Dove)
Santa Cruz Is, New Hebrides, New Caledonia
Ptilinopus huttoni (Rapa I Fruit Dove)
Rapa I
Ptilinopus dupetithouarsii (White-capped Fruit Dove)
P.d.viridior
N Marquesas Is
P.d.dupetithouarsii
S Marquesas Is
Ptilinopus mercierii (Red-moustached Fruit Dove)

P.m.mercierii
  Nukuhiva I
P.m.tristrami
  Hivaoa I
Ptilinopus insularis (Henderson I Fruit Dove)
  Henderson I (Pitcairn Is)
Ptilinopus coronulatus (Lilac-capped Fruit Dove)
P.c.trigeminus
  NW New Guinea, Salawati I
P.c.geminus
  N New Guinea, Japen I
P.c.quadrigeminus
  N New Guinea, Vulcan I
P.c.huonensis
  SE New Guinea
P.c.coronulatus
  S New Guinea, Aru Is
Ptilinopus pulchellus (Crimson-capped Fruit Dove)
P.p.pulchellus
  New Guinea & W islands
P.p.decorus
  N New Guinea
Ptilinopus monacha (Blue-capped Fruit Dove)
  N Moluccas
Ptilinopus rivoli (White-bibbed Fruit Dove)
P.r.buruanus
  Buru I
P.r.prasinorrhous
  Moluccas, Kai Is, W Papuan islands
P.r.rivoli
  Bismarck Arch
P.r.strophium
  Louisiade Arch, Egum Atoll
P.r.miquelii
  Japen, Meos Num Is
P.r.bellus
  New Guinea, Admiralty Is
Ptilinopus solomonensis (Yellow-bibbed Fruit Dove)
P.s.johannis
  St Matthias Is, New Hanover
P.s.meyeri
  New Britain, Rook I
P.s.neumanni
  Nissan I
P.s.bistictus
  Bougainville I
P.s.vulcanorum
  C Solomon Is
P.s.ocularis
  Guadalcanal I
P.s.ambiguus
  Malaita I
P.s.solomonensis
  San Cristobal, Ugi Is
P.s.speciosus
  Numfor, Biak Is
Ptilinopus viridis (Red-bibbed Fruit Dove)
P.v.viridis
  S Moluccas

P.v.vicinus
  Trobriand Is, D'Entrecasteaux Arch
P.v.lewisii
  W Solomon Is
P.v.geelvinkiana
  islands of Geelvink Bay
P.v.pseudogeelvinkiana
  Meos Num I
P.v.pectoralis
  W Papuan islands, NW New Guinea
P.v.salvadorii
  N New Guinea
Ptilinopus eugeniae (White-headed Fruit Dove)
  San Cristobal, Ugi Is
Ptilinopus iozonus (Orange-bellied Fruit Dove)
P.i.humeralis
  W Papuan islands, NW New Guinea
P.i.jobiensis
  Japen, Vulcan Is, N New Guinea
P.i.pseudohumeralis
  C New Guinea
P.i.finschii
  SE New Guinea
P.i.iozonus
  Aru Is
Ptilinopus insolitus (Knob-billed Fruit Dove)
P.i.insolitus
  New Ireland, New Britain, Lihir Is
P.i.inferior
  St Matthias Is
Ptilinopus hyogaster (Grey-headed Fruit Dove) 65.1
  Halmahera, Batjan, Morotai, Ternate Is
Ptilinopus granulifrons (Carunculated Fruit Dove)
  Obi Major I
Ptilinopus melanospila (Black-naped Fruit Dove)
P.m.bangueyensis
  Philippine Is, N Bornean islands
P.m.xanthorrhoa
  Sanghir, Talaut Is
P.m.melanospila
  Togian I, Sulawesi
P.m.chrysorrhoa
  Sula, Seram, Peleng, Banggai Is
P.m.massoptera
  Pulo Mata Siri I
P.m.melanauchen
  Java, Bali to Alor, Butong, Muna Is
Ptilinopus naina (Dwarf Fruit Dove)
P.n.minimus
  W Papuan islands
P.n.naina
  S New Guinea
Ptilinopus arcanus (Ripley's Fruit Dove)
  NC Negros I
Ptilinopus victor (Orange Dove)
  Fiji Is
Ptilinopus luteovirens (Golden Dove)
  Fiji Is

Ptilinopus layardi (Yellow-headed Dove)
Kandavu I (Fiji Is)

DREPANOPTILA
Drepanoptila holosericea (Cloven-feathered Dove)
New Caledonia

ALECTROENAS
Alectroenas madagascariensis (Madagascar Blue Pigeon)
Madagascar
Alectroenas sganzini (Comoro Blue Pigeon)
A.s.minor
Aldabra I
A.s.sganzini
Comoro Is
Alectroenas pulcherrima (Seychelles Blue Pigeon)
Seychelles

DUCULA
Ducula poliocephala (Philippine Zone-tailed Pigeon)
Philippine Is
Ducula forsteni (Green & White Zone-tailed Pigeon)
Sulawesi
Ducula mindorensis (Mindoro Zone-tailed Pigeon)
Mindoro I
Ducula radiata (Grey-headed Zone-tailed Pigeon)
Sulawesi
Ducula carola (Grey-necked Fruit Pigeon)
D.c.carola
Luzon, Mindoro Is
D.c.nigrorum
Negros I
D.c.mindanensis
Mindanao I
Ducula aenea (Green Imperial Pigeon)
D.a.pusilla
S India, Sri Lanka
D.a.sylvatica
N India, Thailand, Indochina
D.a.nicobarica
Nicobar Is
D.a.aenea
Malaysia, Borneo, Sumatra to Bali
D.a.mista
Simalur I
D.a.babiensis
Pulo Babi I
D.a.consobrina
Nias I
D.a.vicina
Mentawi Is
D.a.palawanensis
S Philippine Is
D.a.chalybura
N Philippine Is

D.a.paulina
Sulawesi
D.a.pallidinucha
SE Sulawesi
D.a.polia
Lombok I to Alor I
D.a.intermedia
Talaut I
D.a.sulana
Sula, Badggai, Peleng Is
D.a.aneothorax
Enggano I
Ducula perspicillata (White-eyed Imperial Pigeon)
D.p.perspicillata
N Moluccas
D.p.neglecta
S Moluccas
Ducula concinna (Blue-tailed Imperial Pigeon)
D.c.concinna
Talaut, Tukangbesi, Kai, Tanimbar Is
D.c.aru
Aru Is
Ducula pacifica (Pacific Pigeon)
D.p.tarrali
N New Guinea islands, New Hebrides
D.p.pacifica
Ellis, Tonga Is
D.p.intensitincta
Fiji Is
D.p.microcera
Samoa Is
D.p.sejuncta
Bismarck Arch
Ducula oceanica (Micronesian Pigeon)
D.o.monacha
Yap, Palau Is
D.o.teraokai
Truk I
D.o.townsendi
Ponapé I
D.o.oceanica
Kusaie, Marshall Is
D.o.ratakensis
Arno, Wotje Is (Marshall Is)
Ducula aurorae (Society Is Pigeon)
Society Is
Ducula galeata (Marquesas Pigeon)
Nukuhiva I
Ducula rubricera (Red-knobbed Pigeon)
D.r.rubricera
Bismarck Arch, Lihir Is
D.r.rufigula
Solomon Is
Ducula myristicivora (Black-knobbed Pigeon)
D.m.myristicivora
W Papuan islands
D.m.geelvinkiana
Meos Num, Numfor, Biak Is
Ducula rufigaster (Rufous-bellied Fruit Pigeon)
D.r.rufigaster
W Papuan islands, W New Guinea

D.r.pallida
 S New Guinea
D.r.uropygialis
 N New Guinea
Ducula basilica (Moluccan Rufous-bellied Fruit Pigeon)
D.b.basilica
 N Moluccas
D.b.obiensis
 Obi I
Ducula finschii (Finsch's Rufous-bellied Fruit Pigeon)
 Bismarck Arch
Ducula chalconota (Mountain Rufous-bellied Fruit Pigeon)
D.c.chalconota
 NW New Guinea
D.c.smaragdina
 New Guinea
Ducula pistrinaria (Island Imperial Pigeon)
D.p.rhodinolaema
 Admiralty Is, New Hanover
D.p.vanwyckii
 Bismarck Arch
D.p.postrema
 islands off SE New Guinea
D.p.pistrinaria
 Solomon, Lihir Is
Ducula rosacea (Pink-headed Imperial Pigeon)
 Lesser Sunda Is
Ducula whartoni (Christmas I Imperial Pigeon)
 Christmas I
Ducula pickeringii (Grey Imperial Pigeon)
 N Bornean islands, Sulu Arch, Talaut Is
Ducula latrans (Peale's Pigeon)
 Fiji Is
Ducula brenchleyi (Chestnut-bellied Pigeon)
 Solomon Is
Ducula bakeri (Baker's Pigeon)
 New Hebrides
Ducula goliath (New Caledonian Pigeon)
 New Caledonia
Ducula pinon (Pinon Imperial Pigeon)
D.p.pinon
 W Papuan islands, SW New Guinea
D.p.rubiensis
 C & S New Guinea
D.p.jobiensis
 N New Guinea, Japen, Dampier Is
D.p.salvadorii
 D'Entrecasteaux & Louisiade Arch
Ducula melanochroa (Black Imperial Pigeon)
Bismarck Arch
Ducula mullerii (Black-collared Fruit Pigeon)
D.m.aurantia
 N New Guinea
D.m.mullerii
 S New Guinea, Aru Is
Ducula zoeae (Banded Imperial Pigeon)
 New Guinea & SW & SE islands
Ducula badia (Mountain Imperial Pigeon)

D.b.insignis
 Himalayas
D.b.cuprea
 SW India
D.b.griseicapilla
 Burma, Thailand, Indochina
D.b.obscurata
 SE Thailand
D.b.badia
 Malaysia, Sumatra, Borneo
D.b.capistrata
 W Java
Ducula lacernulata (Dark-backed Imperial Pigeon)
D.l.lacernulata
 W & C Java
D.l.williami
 E Java, Bali
D.l.sasakensis
 Lombok, Flores Is
Ducula cineracea (Timor Imperial Pigeon)
 Timor, Wetar Is
Ducula bicolor (Pied Imperial Pigeon)
 Andaman Is to Philippine Is & Lesser Sunda Is
Ducula luctuosa (Sulawesi Pied Imperial Pigeon)   65.1
 Sulawesi, Sula Is
Ducula spilorrhoa (Australian Pied Imperial Pigeon)
D.s.subflavescens
 Bismarck Arch, Admiralty Is
D.s.spilorrhoa
 Aru Is, W New Guinea & islands
D.s.tarara
 S New Guinea
D.s.melvillensis
 SE New Guinea, N & NE Australia, Lord Howe I

LOPHOLAIMUS
Lopholaimus antarcticus (Top-knot Pigeon)
 N Queensland to Victoria

HEMIPHAGA
Hemiphaga novaeseelandiae (New Zealand Pigeon)
H.n.novaeseelandiae
 New Zealand
H.n.chathamensis
 Chatham I

CRYPTOPHAPS
Cryptophaps poecilorrhoa (Sulawesi Dusky Pigeon)
 N & SE Sulawesi

GYMNOPHAPS
Gymnophaps albertisii (Bare-eyed Mountain Pigeon)
G.a.exsul
 Batjan I (Moluccas)

G.a.albertisii
New Guinea, Bismarck Arch
Gymnophaps mada (Long-tailed Mountain Pigeon)
G.m.mada
Buru I
G.m.stalkeri
Seram I
Gymnophaps solomonensis (Pale Mountain Pigeon)
Solomon Is

PSITTACIFORMES

66. LORIDAE (LORIES)

CHALCOPSITTA
Chalcopsitta atra (Black Lory)
C.a.bernsteini
Misol I
C.a.atra
Batanta, Salawati Is, NW New Guinea
C.a.insignis
Amberpon, NW New Guinea
C.a.spectabilis
NW New Guinea
Chalcopsitta duivenbodei (Duyvenbode's Lory)
C.d.duivenbodei
coast of NW New Guinea
C.d.syringanuchalis
coast of NE New Guinea
Chalcopsitta sintillata (Yellow-streaked Lory)
C.s.rubrifrons
Aru Is
C.s.sintillata
S New Guinea
C.s.chloroptera
SE New Guinea
Chalcopsitta cardinalis (Cardinal Lory)
islands NW of New Ireland, Solomon Is

EOS
Eos cyanogenia (Black-winged Lory)
islands in Geelvink Bay
Eos squamata (Violet-necked Lory)
E.s.obiensis
Obi I
E.s.atrocaerulea
Maju I
E.s.riciniata
N Moluccas
E.s.squamata
W Papuan islands
Eos reticulata (Blue-streaked Lory)
Tanimbar, Kai, Damar Is
Eos histrio (Red & Blue Lory)
E.h.histrio
Gt Sangi, Siao Is
E.h.talautensis
Talaut Is

E.h.challengeri
Nenusa I
Eos bornea (Red Lory)
E.b.bornea
Ambon, Saparua, Seram, Kai Is
E.b.cyanonothus
Buru I
Eos semilarvata (Blue-eared Lory)
C Seram I

PSEUDEOS
Pseudeos fuscata (Dusky Lory)
Salawati, Japen Is, New Guinea

TRICHOGLOSSUS
Trichoglossus ornatus (Ornate Lory)
Sulawesi
Trichoglossus haematodus (Rainbow Lory)
T.h.mitchellii
Bali, Lombok Is
T.h.forsteni
Sumbawa I
T.h.djampeanus
Djampea I
T.h.stresemanni
Kalaotua I
T.h.fortis
Sumba I
T.h.weberi
Flores I
T.h.capistratus
Timor I
T.h.flavotectus
Wetar, Roma Is
T.h.haematodus
S Moluccas, W New Guinea
T.h.rosenbergii
Biak I
T.h.intermedius
N New Guinea
T.h.micropteryx
E New Guinea
T.h.caeruleiceps
S New Guinea
T.h.nigrogularis
E Kai, Aru Is
T.h.brooki
Trangan I (Aru Is)
T.h.massena
New Hebrides, Bismarck Arch, Solomon Is
T.h.flavicans
New Hanover, Admiralty Is
T.h.nesophilus
Ninigo I
T.h.deplanchii
New Caledonia, Loyalty Is
T.h.moluccanus
E Australia, Tasmania
T.h.rubritorquis
N Australia
Trichoglossus rubiginosus (Ponapé Lory)
Ponapé I

Trichoglossus johnstoniae (Johnstone's Lorikeet)
  T.j.johnstoniae
    C & SE Mindanao
  T.j.pistra
    W Mindanao
Trichoglossus flavoviridis (Yellow & Green Lorikeet)
  T.f.meyeri
    Sulawesi
  T.f.flavoviridis
    Sula Is
Trichoglossus chlorolepidotus (Scaly-breasted Lorikeet)
  NE Australia
Trichoglossus euteles (Perfect Lorikeet)
  Timor, Lomblen Is to Babar I
Trichoglossus versicolor (Varied Lorikeet)
  N Australia
Trichoglossus iris (Iris Lorikeet)
  T.i.iris
    Timor I
  T.i.wetterensis
    Wetar I
Trichoglossus goldiei (Goldie's Lorikeet)
  C New Guinea

LORIUS
Lorius hypoinochrous (Purple-bellied Lory)
  L.h.devittatus
    Bismarck Arch, SE New Guinea & islands
  L.h.hypoinochrous
    Misima, Tagula Is
  L.h.rosselianus
    Rossel I
Lorius lory (Black-capped Lory)
  L.l.lory
    NW New Guinea & islands
  L.l.erythrothorax
    C New Guinea
  L.l.somu
    S New Guinea
  L.l.salvadorii
    NE New Guinea
  L.l.viridicrissalis
    N New Guinea
  L.l.jobiensis
    Japen, Meos Num Is
  L.l.cyanauchen
    Biak I
Lorius albidinuchus (White-naped Lory)
  New Ireland
Lorius amabilis (Stresemann's Lory)
  New Britain
Lorius chlorocercus (Yellow-bibbed Lory)
  E Solomon Is
Lorius domicellus (Purple-naped Lory)
  Seram, Ambon Is
Lorius tibialis (Blue-thighed Lory)
  unknown
Lorius garrulus (Chattering Lory)
  L.g.garrulus
    Halmahera, Weda Is

  L.g.flavopalliatus
    Batjan, Obi Is
  L.g.morotaianus
    Morotai I

PHIGYS
Phigys solitarius (Collared Lory)
  Fiji Is

VINI
Vini australis (Blue-crowned Lory)
  Samoa, Tonga Is, Lau Arch
Vini kuhlii (Kühl's Lory)
  Rimitara, Tubuai Is
Vini stepheni (Stephen's Lory)
  Henderson I
Vini peruviana (Tahitian Lory)
  Cook, Society Is
Vini ultramarina (Ultramarine Lory)
  Marquesas Is

GLOSSOPSITTA
Glossopsitta concinna (Musk Lorikeet)
  E & SE Australia, Tasmania
Glossopsitta pusilla (Little Lorikeet)
  E & SE Australia, Tasmania
Glossopsitta porphyrocephala (Purple-crowned Lorikeet)
  SW & SE Australia

CHARMOSYNA
Charmosyna palmarum (Palm Lorikeet)
  New Hebrides, Banks Is
Charmosyna rubrigularis (Red-chinned Lorikeet)
  C.r.rubrigularis
    New Britain, New Ireland
  C.r.krakari
    Karkar I
Charmosyna meeki (Meek's Lorikeet)
  Solomon Is
Charmosyna toxopei (Blue-fronted Lorikeet)
  Buru I
Charmosyna multistriata (Striated Lorikeet)
  WC New Guinea
Charmosyna wilhelminae (Wilhelmina's Lorikeet)
  C New Guinea
Charmosyna rubronotata (Red-spotted Lorikeet)
  C.r.rubronotata
    Salawati I, NW New Guinea
  C.r.kordoana
    Biak I
Charmosyna placentis (Red-flanked Lorikeet)
  C.p.intensior
    N Moluccas
  C.p.placentis
    S Moluccas, Kai, Aru Is, S New Guinea

C.p.ornata
  NW New Guinea & islands
C.p.subplacens
  E New Guinea
C.p.pallidior
  Bismarck Arch, W Solomon Is
Charmosyna diadema (New Caledonian Lorikeet)
  New Caledonia e?
Charmosyna amabilis (Red-throated Lorikeet)
  Viti Levu, Ovalau, Taviuni Is
Charmosyna margarethae (Duchess Lorikeet)
  Solomon Is
Charmosyna pulchella (Fairy Lorikeet)
C.p.pulchella
  NW, C & SE New Guinea
C.p.rothschildi
  NC New Guinea
C.p.bella
  SE New Guinea
Charmosyna josefinae (Josephine's Lory)
C.j.josefinae
  NW New Guinea
C.j.sepikiana
  Sepik Mtn area, New Guinea
C.j.cyclopum
  Cyclops Mtns, New Guinea
Charmosyna papou (Papuan Lory)
C.p.papou
  NW New Guinea
C.p.stellae
  SE New Guinea
C.p.goliathina
  C New Guinea
C.p.wahnesi
  Huon peninsula, New Guinea

OREOPSITTACUS
Oreopsittacus arfaki (Whiskered Lorikeet)
O.a.arfaki
  NW New Guinea
O.a.major
  C New Guinea
O.a.grandis
  SE New Guinea

NEOPSITTACUS
Neopsittacus musschenbroekii (Musschenbroek's Lorikeet)
N.m.musschenbroekii
  NW New Guinea
N.m.medius
  W New Guinea
N.m.major
  SE New Guinea
Neopsittacus pullicauda (Emerald Lorikeet)
N.p.alpinus
  W New Guinea
N.p.socialis
  EC New Guinea
N.p.pullicauda
  SE New Guinea

67 CACATUIDAE (COCKATOOS)

CACATUINAE

PROBOSCIGER
Probosciger aterrimus (Palm Cockatoo)
P.a.goliath
  W Papuan islands, NW to SE New Guinea
P.a.stenolophus
  Japen I, N New Guinea
P.a.aterrimus
  Aru, Misol Is, S New Guinea, Cape York, Queensland

CALYPTORHYNCHUS
Calyptorhynchus funereus (Black Cockatoo)
67.1
C.f.baudinii
  SW Australia
C.f.tenuirostris
  SW Australia
C.f.funereus
  E Australia
C.f.xanthonotus
  SE Australia, Tasmania
Calyptorhynchus magnificus (Red-tailed Cockatoo)
C.m.naso
  SW Australia
C.m.samueli
  SW Queensland, W New South Wales
C.m.macrorhynchus
  N Australia
C.m.magnificus
  Queensland to S New South Wales, W Victoria
Calyptorhynchus lathami (Glossy Cockatoo)
  C Queensland to E Victoria, Kangaroo I

CALLOCEPHALON
Callocephalon fimbriatum (Gang-gang Cockatoo)
  SE Australia, N Tasmania

EOLOPHUS
Eolophus roseicapillus (Galah)
E.r.kuhli
  NW Australia
E.r.assimilis
  Western Australia
E.r.roseicapillus
  NC & E Australia

CACATUA
Cacatua leadbeateri (Major Mitchell's Cockatoo)
C.l.mollis
  mid-Western Australia
C.l.leadbeateri
  interior of Australia
Cacatua sulphurea (Lesser Sulphur-crested Cockatoo)
C.s.sulphurea

PSITTACIFORMES

Sulawesi, Lesser Sunda Is
C.s.abbotti
Solombo Besar I
C.s.parvula
Timor, Lombok, Sumbawa Is
C.s.citrinocristata
Sumba I
Cacatua galerita (Sulphur-crested Cockatoo)
C.g.eleonora
Aru Is
C.g.triton
New Guinea & N & E islands
C.g.fitzroyi
N Australia to W Queensland
C.g.galerita
E & SE Australia
Cacatua ophthalmica (Blue-eyed Cockatoo)
New Britain, New Ireland
Cacatua moluccensis (Salmon-crested Cockatoo)
S Moluccas
Cacatua alba (White Cockatoo)
N & C Moluccas
Cacatua haematuropygia (Red-vented Cockatoo)
Philippine, Palawan Is
Cacatua goffini (Goffin's Cockatoo)
Tanimbar, Kai Is
Cacatua sanguinea (Little Corella)
C.s.normantoni
NW Queensland,
C.s.sanguinea
W & NW Australia,
C.s.gymnopis
EC Australia
C.s.westralensis
WC Australia
C.s.transfreta
S New Guinea
Cacatua tenuirostris (Long-billed Corella)
SE Australia
Cacatua pastinator (Eastern Long-billed Corella)   67.2
SW Australia
Cacatua ducorps (Ducorp's Cockatoo)
E Solomon Is

NYMPHICINAE

NYMPHICUS
Nymphicus hollandicus (Cockatiel)
Australia (Mainly interior)

68. PSITTACIDAE (PARROTS)

NESTORINAE

NESTOR
Nestor notabilis (Kea)
C South I, New Zealand
Nestor meridionalis (Kaka)
New Zealand

MICROPSITTINAE

MICROPSITTA
Micropsitta pusio (Buff-faced Pygmy Parrot)
M.p.beccarii
N New Guinea
M.p.pusio
SE New Guinea, Bismarck Arch
M.p.harteri
Fergusson I
M.p.stresemanni
Misima, Tagula Is
Micropsitta keiensis (Yellow-capped Pygmy Parrot)
M.k.keiensis
Kai, Aru Is
M.k.chloroxantha
W Papuan islands, NW New Guinea
M.k.viridipectus
S New Guinea
Micropsitta geelvinkiana (Geelvink Pygmy Parrot)
M.g.geelvinkiana
Numfor I
M.g.misoriensis
Biak I
Micropsitta meeki (Meek's Pygmy Parrot)
M.m.meeki
Admiralty Is
M.m.proxima
St Matthias, Squally Is
Micropsitta finschii (Finsch's Pygmy Parrot)
M.f.viridifrons
New Hanover, New Ireland, Lihir Is
M.f.finschii
Ugi, San Cristobal, Rennell Is
M.f.aolae
Guadalcanal, Malaita, Russell Is
M.f.tristrami
Vellalavella, Kulambangra, Rendova Is
M.f.nanina
Bougainville, Choiseul, Ysabel Is
Micropsitta bruijnii (Red-breasted Pygmy Parrot)
M.b.pileata
Buru, Seram Is
M.b.bruijnii
New Guinea
M.b.necopinata
New Britain, New Ireland
M.b.rosea
Bougainville, Guadalcanal, Kulambangra Is

PSITTACINAE

OPOPSITTA
Opopsitta gulielmitertii (Orange-breasted Fig Parrot)
O.g.gulielmitertii
Salawati I, NW New Guinea
O.g.nigrifrons
N New Guinea
O.g.ramuensis
Ramu R, N New Guinea
O.g.amabilis
NE New Guinea

O.g.suavissima
SE New Guinea
O.g.fuscifrons
S New Guinea
O.g.melanogenia
Aru Is
Opopsitta diophthalma (Double-eyed Fig Parrot)
O.d.diophthalma
W New Guinea & islands
O.d.coccineifrons
E & NE New Guinea
O.d.aruensis
Aru Is, S New Guinea
O.d.virago
Goodenough, Fergusson Is
O.d.inseparabilis
Tagula I
O.d.marshalli
Cape York Peninsula
O.d.macleayana
coast of N Queensland
O.d.coxeni
coast of N New South Wales

PSITTACULIROSTRIS
Psittaculirostris desmarestii (Desmarest's Fig Parrot)
P.d.blythii
Misol I
P.d.occidentalis
Salawati, Batanta Is, NW New Guinea
P.d.desmarestii
NW New Guinea
P.d.intermedia
Onin Peninsula, NW New Guinea
P.d.godmani
S New Guinea
P.d.cervicalis
SE New Guinea
Psittaculirostris edwardsii (Edwards' Fig Parrot)
NE New Guinea
Psittaculirostris salvadorii (Salvadori's Fig Parrot)
NW New Guinea

BOLBOPSITTACUS
Bolbopsittacus lunulatus (Guaiabero)
B.l.lunulatus
Luzon I
B.l.intermedius
Leyte I
B.l.callainipictus
Samar I
B.l.mindanensis
Mindanao, Panaon Is

PSITTINUS
Psittinus cyanurus (Blue-rumped Parrot)
P.c.cyanurus
SW Thailand, Malaysia, Sumatra, Borneo

P.c.pontius
Siberut, Sipora, Mentawei Is
P.c.abbotti
Simalur, Siumat Is

PSITTACELLA
Psittacella brehmii (Brehm's Parrot)
P.b.brehmii
NW New Guinea
P.b.intermixta
WC New Guinea
P.b.harterti
E New Guinea
P.b.pallida
S & SE New Guinea
Psittacella picta (Painted Parrot)
P.p.picta
SE New Guinea
P.P.excelsa
C New Guinea
P.p.lorentzi
WC New Guinea
Psittacella modesta (Modest Parrot)
P.m.modesta
NW New Guinea
P.m.collaris
WC New Guinea
P.m.subcollaris
C New Guinea
Psittacella maderaszi (Maderasz's Parrot)
P.m.major
WC New Guinea
P.m.hallstromi
C New Guinea
P.m.huonensis
E New Guinea
P.m.maderaszi
SE New Guinea

GEOFFROYUS
Geoffroyus geoffroyi (Red-cheeked Parrot)
G.g.floresianus
Lombok, Sumbawa, Flores, Sumba Is
G.g.geoffroyi
Timor, Wetar Is
G.g.cyanicollis
N Moluccas
G.g.obiensis
C Moluccas
G.g.rhodops
S Moluccas
G.g.keyensis
Kai Is
G.g.timorlaoensis
Tanimbar Is
G.g.aruensis
Aru Is, S New Guinea, NE Queensland
G.g.orientalis
NE New Guinea
G.g.sudestiensis
Misima, Tagula Is

G.g.cyanicarpus
Rossel I
G.g.minor
N New Guinea
G.g.jobiensis
Japen, Meos Num Is
G.g.mysoriensis
Biak, Numfor Is
G.g.pucherani
W Papuan islands, NW New Guinea
Geoffroyus simplex (Blue-collared Parrot)
G.s.simplex
NW New Guinea
G.s.buergersi
C & SE New Guinea
Geoffroyus heteroclitus (Singing Parrot)
G.h.heteroclitus
Lihir Is, New Ireland, New Britain, Solomon Is
G.h.hyacinthinus
Rennell I

PRIONITURUS
Prioniturus luconensis (Green Racket-tailed Parrot)
Luzon, Marinduque Is
Prioniturus discurus (Blue-crowned Racket-tailed Parrot)
P.d.discurus
Mindanao, Basilan, Luzon Is
P.d.whiteheadi
Negros, Bohol, Samar, Leyte, Masbate, Cebu Is
P.d.nesophilus
Catanduanes, Sibuyan, Tablas Is
P.d.mindorensis
Mindoro I
Prioniturus platenae (Palawan Racket-tailed Parrot) 68.1
Palawan, Balabac Is
Prioniturus verticalis (Sulu Racket-tailed Parrot) 68.1
Sulu Arch
Prioniturus waterstradti (Mindanao Racket-tailed Parrot) 68.1
Mindanao
Prioniturus montanus (Mountain Racket-tailed Parrot) 68.1
P.m.montanus
Luzon I
P.m.malindangensis
Mt Malindang, Mindanao I
Prioniturus flavicans (Red-spotted Racket-tailed Parrot)
N Sulawesi
Prioniturus platurus (Golden-mantled Racket-tailed Parrot)
P.p.platurus
Sulawesi, Togian, Peleng, Banggai Is
P.p.talautensis
Talaut Is
P.p.sinerubris
Taliabu I

Prioniturus mada (Buru Racket-tailed Parrot)
Buru I

TANYGNATHUS
Tanygnathus megalorhynchos (Great-billed Parrot)
T.m.megalorhynchos
Talaut, Sanghir Is, N & C Moluccas, Flores, Djampea Is
T.m.affinis
S Moluccas
T.m.subaffinis
Tanibar Is
T.m.hellmayri
W Timor, Semao Is
T.m.sumbensis
Sumba I
Tanygnathus lucionensis (Blue-naped Parrot)
T.l.lucionensis
Luzon, Mindoro Is
T.l.hybridus
Polillo Is
T.l.talautensis
C & S Philippine, Palawan Is, Sulu Arch
Tanygnathus sumatranus (Müller's Parrot)
T.s.duponti
Luzon I
T.s.freeri
Polillo Is
T.s.everetti
Panay, Samar, Leyte, Negros, Mindanao Is
T.s.burbidgii
Sulu Arch
T.s.sumatranus 68.1
Sulawesi, Muna, Buton, Sangihe, Banggai Is
Tanygnathus heterurus (Rufous-tailed Parrot) 68.1
?Sulawesi
Tanygnathus gramineus (Black-lored Parrot)
Buru I

ECLECTUS
Eclectus roratus (Eclectus Parrot)
E.r.vosmaeri
N & C Moluccas
E.r.roratus
S Moluccas
E.r.westermani
unknown
E.r.cornelia
Sumba I
E.r.riedeli
Tanimbar Is
E.r.polychloros
Kai Is, New Guinea & islands
E.r.biaki
Biak I
E.r.aruensis
Aru Is
E.r.macgillivrayi
NE Queensland

E.r.solomonensis
Admiralty Is, Bismarck Arch, Solomon Is

PSITTRICHAS
Psittrichas fulgidus (Pesquet's Parrot)
Mts of New Guinea

PROSOPEIA
Prosopeia tabuensis (Red Shining Parrot)
P.t.taviunensis
Taviuni, Ngamea Is
P.t.tabuensis
Vanua Levu, Kio, Koro, Ngau, Eua Is
**Prosopeia splendens (Kandavu Shining
Parrot)**   68.3
Kandavu, Viti Levu Is
Prosopeia personata (Masked Shining Parrot)
Viti Levu I

ALISTERUS
Alisterus scapularis (Australian King Parrot)
A.s.minor
NE Queensland
A.s.scapularis
Eastern Australia
**Alisterus chloropterus (Green-winged King
Parrot)**
A.c.moszkowskii
N New Guinea
A.c.callopterus
C New Guinea
A.c.chloropterus
E New Guinea
Alisterus amboinensis (Amboina King Parrot)
A.a.amboinensis
Ambon, Seram Is
A.a.sulaensis
Sula Is
A.a.versicolor
Peleng I
A.a.buruensis
Buru I
A.a.hypophonius
Halmahera I
A.a.dorsalis
NW New Guinea & islands

APROSMICTUS
**Aprosmictus erythropterus (Red-winged
Parrot)**
A.e.papua
S New Guinea
A.e.coccineopterus
N Australia
A.e.erythropterus
interior Eastern Australia
**Aprosmictus jonquillaceus (Timor Red-
winged Parrot)**
A.j.jonquillaceus
Timor I
A.j.wetterensis
Wetar I

POLYTELIS
Polytelis swainsonii (Superb Parrot)
interior New South Wales, N Victoria
Polytelis anthopeplus (Regent Parrot)
P.a.anthopeplus
NW Victoria
P.a.westralis
SW Australia
Polytelis alexandrae (Princess Parrot)
interior C & Western Australia

PURPUREICEPHALUS
**Purpureicephalus spurius (Red-capped
Parrot)**
SW Australia

BARNARDIUS
Barnardius barnardi (Mallee Ringneck Parrot)
B.b.macgillivrayi
NW Queensland, E Nthn Territory
B.b.whitei
Flinders Range, South Australia
B.b.barnardi
interior of SE Australia
Barnardius zonarius (Port Lincoln Parrot)
B.z.occidentalis
NW Western Australia
B.z.semitorquatus
SW Western Australia
B.z.dundasi
SW Australia
B.z.myrtae
C Australia
B.z.zonarius
S Australia

PLATYCERCUS
Platycercus caledonicus (Green Rosella)
Tasmania, Bass Strait
Platycercus elegans (Crimson Rosella)
P.e.nigrescens
NE Queensland
P.e.elegans
SE Queensland to SE South Australia
P.e.melanoptera
Kangaroo I
P.e.fleurieuensis
Fleurieu Peninsula, S Australia
Platycercus flaveolus (Yellow Rosella)
interior of SE Australia
Platycercus adelaidae (Adelaide Rosella)
S South Australia
Platycercus eximius (Eastern Rosella)
P.e.cecilae
SE Queensland, NE New South Wales
P.e.eximius
SE Australia
P.e.diemenensis
Tasmania
Platycercus adscitus (Pale-headed Rosella)
P.a.adscitus
N Queensland

P.a.mackaiensis
NE Queensland
P.a.amathusiae
NE Queensland
P.a.palliceps
C Queensland to N New South Wales
Platycercus venustus (Northern Rosella)
NW & N Australia
Platycercus icterotis (Western Rosella)
P.i.icterotis
coast of SW Australia
P.i.xanthogenys
interior of SW Australia

PSEPHOTUS
Psephotus haematonotus (Red-rumped Parrot)
P.h.caeruleus
Innamincka, S Australia
P.h.haematonotus
interior of SE Australia
Psephotus varius (Mulga Parrot)
P.v.varius
interior of S Australia
P.v.orientalis
SW New South Wales, W Victoria
Psephotus haematogaster (Blue Bonnet)
P.h.narethae
SE Western Australia
P.h.pallescens
Lake Eyre Basin
P.h.haematorrhous
S Queensland, N New South Wales
P.h.haematogaster
W & S New South Wales, NW Victoria, SE South Australia
Psephotus chrysopterygius (Golden-shouldered Parrot)
P.c.chrysopterygius
S Cape York Peninsula
P.c.dissimilis
NE Northern Territory
Psephotus pulcherrimus (Paradise Parrot)
C & S Queensland, N New South Wales

CYANORAMPHUS
Cyanoramphus unicolor (Antipodes Green Parakeet)
Antipodes Is
Cyanoramphus novaezelandiae (Red-fronted Parakeet)
C.n.novaezelandiae
New Zealand, Auckland I
C.n.cyanurus
Kermadec Is
C.n.chathamensis
Chatham I
C.n.hochstetteri
Antipodes Is
C.n.cookii
Norfolk I
C.n.saisseti
New Caledonian

Cyanoramphus auriceps (Yellow-fronted Parakeet)
C.a.auriceps
New Zealand, Stewart, Auckland Is
C.a.forbesi
Chatham I
Cyanoramphus cornutus (Horned Parakeet)
C.c.cornutus
New Caledonia
C.c.uvaeensis
Ouvea I

NEOPHEMA
Neophema bourkii (Bourke's Parrot)
interior of C & S Australia
Neophema chrysostoma (Blue-winged Parrot)
SE Australia, Tasmania
Neophema elegans (Elegant Parrot)
SW & SE Australia
Neophema petrophila (Rock Parrot)
N.p.petrophila
coast of W Australia
N.p.zietzi
coast of S Australia
Neophema chrysogaster (Orange-bellied Parrot)
Tasmania, coast of W Victoria
Neophema pulchella (Turquoise Parrot)
SE Queensland to N Victoria
Neophema splendida (Scarlet-chested Parrot)
interior of S Australia

LATHAMUS
Lathamus discolor (Swift Parrot)
SE Australia, Tasmania

MELOPSITTACUS
Melopsittacus undulatus (Budgerigar)
Australia

PEZOPORUS
Pezoporus wallicus (Ground Parrot)
P.w.flaviventris
coast of SW Australia
P.w.wallicus
SE Australia, W Tasmania

GEOPSITTACUS
Geopsittacus occidentalis (Night Parrot)
interior of Australia

CORACOPSIS
Coracopsis vasa (Vasa Parrot)
C.v.drouhardi
W Madagascar
C.v.vasa
E Madagascar
C.v.comorensis
Great Comoro, Moheli, Anjouan Is

Coracopsis nigra (Black Parrot)
C.n.libs
  W Madagascar
C.n.nigra
  E Madagascar
C.n.sibilans
  Gt Comoro, Anjouan Is
C.n.barklyi
  Praslin I

PSITTACUS
Psittacus erithacus (Grey Parrot)
P.e.timneh
  S Guinea to Ivory Coast
P.e.erithacus
  SE Ivory Coast to W Kenya & N Angola
P.e.princeps
  Principé, Fernando Po Is

POICEPHALUS
Poicephalus robustus (Brown-necked Parrot)
P.r.fuscicollis
  Gambia to N Ghana & Togo
P.r.suahelicus
  Angola to S Zaire, Tanzania & Mozambique
P.r.robustus
  E Cape Province to N Natal
Poicephalus gulielmi (Red-fronted Parrot)
P.g.fantiensis
  Liberia to Cameroun
P.g.gulielmi
  S Cameroun to N Angola
P.g.permistus
  C Kenya
P.g.massaicus
  S Kenya, N Tanzania
Poicephalus cryptoxanthus (Brown-headed Parrot)
P.c.tanganyikae
  C Tanzania
P.c.zanzibaricus
  Zanzibar, Pemba Is
P.c.cryptoxanthus
  Natal to SE Kenya
Poicephalus crassus (Niam-Niam Parrot)
  E Cameroun to SW Sudan
Poicephalus senegalus (Senegal Parrot)
P.s.senegalus
  Senegal to Guinea, S Mali
P.s.versteri
  Ivory Coast, Ghana to Nigeria
P.s.mesotypus
  E & NE Nigeria, SW Chad, N Cameroun
Poicephalus rufiventris (Red-bellied Parrot)
P.r.rufiventris
  C Ethiopia to NE Tanzania
P.r.pallidus
  Somalia, E Ethiopia
Poicephalus meyeri (Brown Parrot)
P.m.meyeri
  S Chad, NE Cameroun to W Ethiopia
P.m.saturatus
  Uganda, Kenya, W Tanzania

P.m.matschiei
  SE Kenya to Zambia, Malawi
P.m.transvaalensis
  N Mozambique, Transvaal
P.m.reichenowi
  N & C Angola, SW Zaire
P.m.damarensis
  S Angola, SW Africa
Poicephalus rueppellii (Ruppell's Parrot)
  S Angola, N Namibia
Poicephalus flavifrons (Yellow-faced Parrot)
P.f.flavifrons
  N & C Ethiopia
P.f.aurantiiceps
  SW Ethiopia

AGAPORNIS
Agapornis cana (Grey-headed Lovebird)
A.c.cana
  coast of Madagascar
A.c.ablectanea
  SW Madagascar
Agapornis pullaria (Red-faced Lovebird)
A.p.guineensis
  Guinea to N Zaire
A.p.pullaria
  S Ethiopia & S Sudan to NW Tanzania
A.p.ugandae
  SW Ethiopia to NW Tanzania
Agapornis taranta (Black-winged Lovebird)
  Ethiopia
Agapornis swinderniana (Black-collared Lovebird)
A.s.swinderniana
  Liberia
A.s.zenkeri
  Cameroun, Gabon to C Zaire
A.s.emini
  E Zaire, W Uganda
Agapornis roseicollis (Peach-faced Lovebird)
A.r.roseicollis
  Namibia, NW Cape Province
A.r.catumbella
  S Angola
Agapornis fischeri (Fischer's Lovebird)
  S Kenya, N Tanzania
Agapornis personata (Masked Lovebird)
  NE Tanzania
Agapornis lilianae (Nyasa Lovebird)
  NW Mozambique to E Zambia
Agapornis nigrigenis (Black-cheeked Lovebird)
  SW Zambia

LORICULUS
Loriculus vernalis (Vernal Hanging Parrot)
  SW India S Vietnam
Loriculus beryllinus (Ceylon Hanging Parrot)
  Sri Lanka
Loriculus philippensis (Philippine Hanging Parrot)
L.p.philippensis
  Luzon, Marinduque Is

L.p.mindorensis
  Mindoro I
L.p.bournsi
  Tablas, Romblon, Sibuyan Is
L.p.panayensis
  Ticao, Masbate, Panay Is
L.p.regulus
  Guimaras, Negros Is
L.p.chrysonotus
  Cebu I
L.p.worcesteri
  Samar, Leyte, Bohol Is
L.p.siquijorensis
  Siquijor I
L.p.apicalis
  Mindanao I
L.p.dohertyi
  Basilan I
L.p.bonapartei
  Jolo, Bongao, Tawitawi Is
Loriculus galgulus (Blue-crowned Hanging Parrot)
  Malaysia, Borneo, Sumatra
Loriculus stigmatus (Sulawesi Hanging Parrot)   68.1
  Sulawesi, Togian, Muna, Butong Is
Loriculus catamene (Sangihe Hanging Parrot)
                                                                      68.1
  Sangihe Is
Loriculus amabilis (Moluccan Hanging Parrot)
L.a.amabilis
  Halmahera, Batjan Is
L.a.sclateri
  Sula Is
L.a.ruber
  Peling, Banggai Is
Loriculus exilis (Green Hanging Parrot)
  N & SE Sulawesi
Loriculus flosculus (Wallace's Hanging Parrot)
  Flores I
Loriculus pusillus (Yellow-throated Hanging Parrot)
  Java, Bali
Loriculus aurantiifrons (Orange-fronted Hanging Parrot)
L.a.aurantiifrons
  Misol I
L.a.batavorum
  Waigeu I, W Papuan islands, NW New Guinea
L.a.meeki
  New Guinea, Fergusson, Goodenough Is
L.a.tener
  Bismarck Arch

PSITTACULA
Psittacula eupatria (Alexandrine Parakeet)
P.e.eupatria
  Sri Lanka, S India
P.e.nipalensis
  E Afghanistan to Assam

P.e.magnirostris
  Andaman Is
P.e.avensis
  E Assam, Burma
P.e.siamensis
  N & W Thailand, Indochina
Psittacula krameri (Rose-ringed Parakeet)
P.k.krameri
  Senegal to S Sudan
P.k.parvirostris
  Sudan to NW Somalia
P.k.borealis
  W Pakistan, N India to C Burma
P.k.manillensis
  S India, Sri Lanka
Psittacula echo (Mauritius Parakeet)
  Mauritius
Psittacula himalayana (Slaty-headed Parakeet)
P.h.himalayana
  E Afghanistan to N Assam
P.h.finschii
  S Assam to SW China, N Indochina
Psittacula cyanocephala (Plum-headed Parakeet)
  India, Sri Lanka
Psittacula roseata (Blossom-headed Parakeet)
P.r.roseata
  N Assam, N Burma
P.r.juneae
  S Assam, S Burma to Indochina
Psittacula intermedia (Intermediate Parrot)
  N India
Psittacula columboides (Malabar Parakeet)
  SW India
Psittacula calthorpae (Emerald-collared Parakeet)
  Sri Lanka
Psittacula derbiana (Lord Derby's Parakeet)
  NE Assam, SE Tibet
Psittacula alexandri (Moustached Parakeet)
P.a.alexandri
  Java, Bali
P.a.fasciata
  N India to S China, Indochina
P.a.abbotti
  Andaman Is
P.a.cala
  Simalur I
P.a.major
  Lasia, Babi Is
P.a.perionca
  Nias I
P.a.dammermani
  Karimon Java I
P.a.kangeanensis
  Kangean I
Psittacula caniceps (Blyth's Parakeet)
  Nicobar Is
Psittacula longicauda (Long-tailed Parakeet)
P.l.tytleri
  Andaman Is

P.l.nicobarica
  Nicobar Is
P.l.longicauda
  Malaysia, Borneo, Sumatra
P.l.defontainei
  Natuna Is, Riau Arch
P.l.modesta
  Enggano I

ANODORHYNCHUS
Anodorhynchus hyacinthinus (Hyacinth Macaw)
  SC Brazil
Anodorhynchus glaucus (Glaucous Macaw)
  Paraguay, NE Argentina e?
Anodorhynchus leari (Indigo Macaw)
  NE Brazil

CYANOPSITTA
Cyanopsitta spixii (Spix's Macaw)
  EC Brazil

ARA
Ara ararauna (Blue & Yellow Macaw)
  E Panama to Paraguay & S Brazil
Ara caninde (Wagler's Macaw)
  Bolivia, Paraguay, N Argentina
Ara militaris (Military Macaw)
A.m.mexicana
  N & C Mexico
A.m.militaris
  W Colombia, NE Ecuador, N Peru
A.m.boliviana
  Bolivia, NW Argentina
Ara ambigua (Buffon's Macaw)
  Nicaragua to W Colombia
Ara macao (Scarlet Macaw)
  SC Mexico to Bolivia & C Brazil
Ara chloroptera (Green-winged Macaw)
  E Panama to N Argentina
Ara rubrogenys (Red-fronted Macaw)
  Bolivia
Ara auricollis (Yellow-collared Macaw)
  S Brazil, NW Argentina
Ara severa (Chestnut-fronted Macaw)
A.s.castaneifrons
  E Panama to N Bolivia, C Brazil
A.s.severa
  E Venezuela, the Guianas, NW Brazil
Ara manilata (Red-bellied Macaw)
  S Colombia to N & C Brazil
Ara maracana (Illiger's Macaw)
  E Brazil, NE Argentina
Ara couloni (Blue-headed Macaw)
  E Peru
Ara nobilis (Red-shouldered Macaw)
A.n.nobilis
  the Guianas, E Venezuela, NE Brazil
A.n.cumanensis
  C Brazil

A.n.longipennis
  S Brazil

ARATINGA
Aratinga acuticaudata (Blue-crowned Conure)
A.a.haemorrhous
  E Colombia, N Venezuela to C & SW Brazil
A.a.neoxena
  Margarita I
A.a.acuticaudata
  E Bolivia to N Argentina & Uruguay
A.a.neumanni
  C Bolivia
Aratinga guarouba (Golden Conure)
  NE Brazil
Aratinga holochlora (Green Conure)
A.h.brewsteri
  NW Mexico
A.h.holochlora
  E & S Mexico
A.h.rubritorquis
  E Guatemala, N Nicaragua
A.h.brevipes
  Socorro I
Aratinga strenua (Pacific Parakeet)   68.2
  W Mexico to NW Nicaragua
Aratinga finschi (Finsch's Conure)
  S Nicaragua to W Panama
Aratinga wagleri (Red-fronted Conure)
A.w.wagleri
  NW Venezuela, Colombia
A.w.transilis
  N Venezuela, E Colombia
A.w.frontata
  W Ecuador, W Peru
A.w.minor
  C & S Peru
Aratinga mitrata (Mitred Conure)
A.m.mitrata
  C Peru to NW Argentina
A.m.alticola
  C Peru
Aratinga erythrogenys (Red-masked Conure)
  W Ecuador, NW Peru
Aratinga leucophthalmus (White-eyed Conure)
A.l.leucophthalmus
  the Guianas to Paraguay, Uruguay
A.l.callogenys
  E Ecuador, NE Peru, NW Brazil
A.l.propinquus
  SE Brazil, NE Argentina
A.l.nicefori
  S Colombia
Aratinga chloroptera (Hispaniolan Conure)
  Hispaniola
Aratinga euops (Cuban Conure)
  Cuba
Aratinga auricapilla (Golden-capped Conure)
A.a.auricapilla
  NE Brazil
A.a.aurifrons
  SE Brazil

Aratinga jandaya **(Jandaya Conure)**
  NE Brazil
Aratinga solstitialis **(Sun Conure)**
  the Guianas, NE Brazil
Aratinga weddellii **(Dusky-headed Conure)**
  W Amazonia
Aratinga nana **(Olive-throated Conure)**
  A.n.nana
    Jamaica
  A.n.astec
    E Mexico to SE Costa Rica
  A.n.vicinalis
    NE Mexico
Aratinga canicularis **(Orange-fronted Conure)**
  A.c.eburnirostrum
    SW Mexico
  A.c.clarae
    WC & SW Mexico
  A.c.canicularis
    SW Mexico to W Costa Rica
Aratinga pertinax **(Brown-throated Conure)**
  A.p.ocularis
    W Panama
  A.p.pertinax
    Curacao I
  A.p.xanthogenia
    Bonaire I
  A.p.arubensis
    Aruba I
  A.p.aeruginosa
    N Colombia, NW Venezuela
  A.p.griseipecta
    NE Colombia
  A.p.lehmanni
    E Colombia
  A.p.tortugensis
    Tortuga I
  A.p.margaritensis
    Margarita, Los Frailes Is
  A.p.venezuelae
    Venezuela
  A.p.chrysophrys
    SE Venezuela, S Guyana, NE Brazil
  A.p.surinama
    NE Venezuela, French Guiana, Surinam
  A.p.chrysogenys
    NW Brazil
  A.p.paraensis
    NC Brazil
Aratinga cactorum **(Cactus Conure)**
  A.c.caixana
    NE Brazil
  A.c.cactorum
    NE Brazil (S of A.c.caixana)
Aratinga aurea **(Peach-fronted Conure)**
  A.a.aurea
    C & S Brazil, E Bolivia
  A.a.major
    S Bolivia, NW Argentina

NANDAYUS
Nandayus nenday **(Nanday Conure)**
  SE Bolivia, Paraguay, N Argentina

LEPTOSITTACA
Leptosittaca branickii **(Golden-plumed**
Conure)
  Colombia, Ecuador, Peru

OGNORHYNCHUS
Ognorhynchus icterotis **(Yellow-eared**
Conure)
  S Colombia, N Ecuador

RHYNCHOPSITTA
Rhynchopsitta pachyrhyncha **(Thick-billed**
Parrot)
  NW & C Mexico
Rhynchopsitta terrisi **(Maroon-fronted Parrot)**
  68.2
  Nuevo Leon, Mexico

CYANOLISEUS
Cyanoliseus patagonus **(Patagonian Conure)**
  C.p.byroni
    C Chile
  C.p.andinus
    NW Argentina
  C.p.patagonus
    C & S Argentina

PYRRHURA
Pyrrhura cruentata **(Blue-throated Conure)**
  E Brazil
Pyrrhura devillei **(Blaze-winged Conure)**
  E Bolivia, SW Brazil
Pyrrhura frontalis **(Maroon-bellied Conure)**
  P.f.frontalis
    SE Brazil
  P.f.kriegi
    S & SE Brazil
  P.f.chiripepe
    Paraguay, Uruguay, N Argentina
Pyrrhura perlata **(Pearly Conure)**
  P.p.lepida
    NE Brazil
  P.p.coerulescens
    NE Brazil
  P.p.anerythra
    C Brazil
  P.p.perlata
    unknown
Pyrrhura rhodogaster **(Crimson-bellied**
Conure)
  C Brazil
Pyrrhura molinae **(Green-cheeked Conure)**
  P.m.molinae
    E Bolivia
  P.m.phoenicura
    NE Bolivia, SW Brazil
  P.m.sordida
    S Brazil
  P.m.restricta
    Chiquitos, Bolivia

P.m.australis
S Bolivia, NW Argentina
P.m.hypoxantha
SW Brazil
Pyrrhura leucotis (White-eared Conure)
P.l.emma
N Venezuela
P.l.auricularis
NE Venezuela
P.l.pfrimeri
NE Brazil
P.l.griseipectus
NE Brazil
P.l.leucotis
E & SE Brazil
Pyrrhura picta (Painted Conure)
P.p.eisenmanni
Azuero (Panama)
P.p.subandina
NW Colombia
P.p.caeruleiceps
N Colombia
P.p.roseifrons
NW Brazil
P.p.picta
E Venezuela, the Guianas, N Brazil
P.p.amazonum
N Brazil
P.p.pantchenkoi
NW Venezuela
P.p.microtera
NC Brazil
P.p.lucianii
E Peru, Bolivia, W Brazil
Pyrrhura viridicata (Santa Marta Conure)
N Colombia
Pyrrhura egregia (Fiery-shouldered Conure)
P.e.egregia
W Guyana, SE Venezuela
P.e.obscura
Roraima, N Brazil
Pyrrhura melanura (Maroon-tailed Conure)
P.m.pacifica
SW Colombia
P.m.melanura
NE Peru, NW Brazil, S Venezuela
P.m.souancei
S Colombia, E Ecuador, N Peru
P.m.berlepschi
E Peru
P.m.chapmani
S Colombia
Pyrrhura orcesi (El Oro Parakeet)
SW Ecuador
Pyrrhura rupicola (Black-capped Conure)
P.r.rupicola
C Peru
P.r.sandiae
SE Peru, W Brazil, N Bolivia
Pyrrhura albipectus (White-necked Conure)
SE Ecuador
Pyrrhura calliptera (Brown-breasted Conure)
C Colombia

Pyrrhura hoematotis (Red-eared Conure)
P.h.immarginata
N Venezuela
P.h.hoematotis
NC Venezuela
Pyrrhura rhodocephala (Rose-crowned Conure)
W Venezuela
Pyrrhura hoffmanni (Hoffmann's Conure)
P.h.hoffmanni
S Costa Rica
P.h.gaudens
W Panama

ENICOGNATHUS
Enicognathus ferrugineus (Austral Conure)
E.f.minor
S Chile, SW Argentina
E.f.ferrugineus
S Chile, S Argentina, Tierra del Fuego
Enicognathus leptorhynchus (Slender-billed Conure)
C Chile

MYIOPSITTA
Myiopsitta monachus (Monk Parakeet)
M.m.luchsi
C Bolivia
M.m.cotorra
SE Bolivia, S Brazil, N Argentina
M.m.calita
W Argentina
M.m.monachus
SE Brazil, Uruguay, NE Argentina

BOLBORHYNCHUS
Bolborhynchus aymara (Sierra Parakeet)
E Andes from C Bolivia to NW Argentina
Bolborhynchus aurifrons (Mountain Parakeet)
B.a.robertsi
NW Peru
B.a.aurifrons
coast & W Andes of C Peru
B.a.margaritae
S Peru to N Chile & NW Argentina
B.a.rubrirostrus
WC Argentina, C Chile
Bolborhynchus lineola (Barred Parakeet)
B.l.lineola
S Mexico to W Panama
B.l.tigrinus
NW Venezuela to SW Colombia, C Peru
Bolborhynchus orbygnesius (Andean Parakeet)
Peru & N Bolivia
Bolborhynchus ferrugineifrons (Rufous-fronted Parakeet)
WC Colombia

FORPUS
Forpus cyanopygius (Mexican Parrotlet)
F.c.insularis
Tres Marias Is

F.c.pallidus
SE Sonora, NW Mexico
F.c.cyanopygius
NW Mexico
Forpus passerinus (Green-rumped Parrotlet)
F.p.cyanophanes
N Colombia
F.p.viridissimus
Trinidad, N Venezuela
F.p.passerinus
the Guianas
F.p.cyanochlorus
N Brazil
F.p.deliciosus
NC & E Brazil
Forpus xanthopterygius (Blue-winged Parrotlet)
F.x.spengeli
NW Colombia
F.x.crassirostris
SE Colombia, NE Peru, NW Brazil
F.x.olallae
NW Brazil
F.x.flavissimus
NE Brazil
F.x.flavescens
SE & E Peru, E Bolivia
F.x.xanthopterygius
C & EC Brazil to Paraguay & NE Argentina
Forpus conspicillatus (Spectacled Parrotlet)
F.c.conspicillatus
E Panama, N & C Colombia
F.c.metae
C Colombia to W Venezuela
F.c.caucae
SW Colombia
Forpus sclateri (Dusky-billed Parrotlet)
F.s.eidos
S Venezuela & the Guianas, N Brazil, E Colombia
F.s.sclateri
N & W Brazil to S Colombia, E Peru & N Bolivia
Forpus coelestis (Pacific Parrotlet)
W Ecuador, NW Peru
Forpus xanthops (Yellow-faced Parrotlet)
NE Peru

BROTOGERIS
Brotogeris tirica (Plain Parakeet)
E & SE Brazil
Brotogeris versicolurus (Canary-winged Parakeet)
B.v.versicolurus
E Ecuador to S French Guiana & N Brazil
B.v.chiriri
N Bolivia, N Argentina to E & S Brazil
B.v.behni
C & S Bolivia, NW Argentina
Brotogeris pyrrhopterus (Grey-cheeked Parakeet)
W Ecuador, NW Peru

Brotogeris jugularis (Orange-chinned Parakeet)
B.j.jugularis
SW Mexico to N Colombia & NW Venezuela
B.j.exsul
SE Colombia, W Venezuela
Brotogeris cyanoptera (Cobalt-winged Parakeet)
B.c.cyanoptera
W Upper Amazonia
B.c.gustavi
N Peru
B.c.beniensis
N Bolivia
Brotogeris chrysopterus (Golden-winged Parakeet)
B.c.chrysopterus
E Venezuela, N Brazil, the Guianas
B.c.tuipara
N & NE Brazil
B.c.chrysosema
Madeira River, N Brazil
B.c.solimoensis
Upper Amazon River, N Brazil
B.c.tenuifrons
NW Brazil
Brotogeris sanctithomae (Tui Parakeet)
B.s.sanctithomae
C & W Brazil, SE Colombia, NE Peru
B.s.takatsukasae
NE Brazil

NANNOPSITTACA
Nannopsittaca panychlora (Tepui Parrotlet)
E Venezuela, W Guyana

TOUIT
Touit batavica (Lilac-tailed Parrotlet)
N Venezuela, the Guianas, Trinidad
Touit huetii (Scarlet-shouldered Parrotlet)
C Colombia, N Venezuela, NE Brazil, E Ecuador
Touit dilectissima (Red-winged Parrotlet)
T.d.costaricensis
SE Costa Rica, W Panama
T.d.dilectissima
E Panama, N & W Colombia, NW Ecuador
Touit purpurata (Sapphire-rumped Parrotlet)
T.p.purpurata
S Venezuela, the Guianas, NE Brazil
T.p.viridiceps
SE Colombia, NW & NC Brazil
Touit melanonota (Black-eared Parrotlet)
SE Brazil
Touit surda (Golden-tailed Parrotlet)
T.s.ruficauda
Recife, E Brazil
T.s.surda
SE Brazil
Touit stictoptera (Spot-winged Parrotlet)
SW Colombia, W Ecuador

PIONITES
Pionites melanocephala (Black-headed Caique)
P.m.pallida
S Colombia to NE Peru
P.m.melanocephala
E & S Venezuela, the Guianas, N Brazil
Pionites leucogaster (White-bellied Caique)
P.l.leucogaster
N Brazil
P.l.xanthurus
NW Brazil
P.l.xanthomeria
W Brazil, E Ecuador

PIONOPSITTA
Pionopsitta pileata (Red-capped Parrot)
SE Brazil, E Paraguay, NE Argentina
Pionopsitta haematotis (Brown-hooded Parrot)
P.h.haematotis
S Mexico to W Panama
P.h.coccinicollaris
E Panama, NW Colombia
Pionopsitta pulchra (Rose-faced Parrot)
W Colombia, W Ecuador
Pionopsitta barrabandi (Orange-cheeked Parrot)
P.b.barrabandi
N Upper Amazonia
P.b.aurantiigena
W Upper Amazonia
Pionopsitta pyrilia (Saffron-headed Parrot)
E Panama, N Colombia
Pionopsitta caica (Caica Parrot)
E Venezuela, the Guianas, NE Brazil

GYPOPSITTA
Gypopsitta vulturina (Vulturine Parrot)
Guyana, NE Brazil

HAPALOPSITTACA
Hapalopsittaca melanotis (Black-winged Parrot)
H.m.peruviana
C Peru
H.m.melanotis
WC Bolivia
Hapalopsittaca amazonina (Rusty-faced Parrot)
H.a.amazonina
C Colombia to NW Venezuela
H.a.theresae
NW Venezuela
H.a.fuertesi
C Colombia
H.a.pyrrhops
W Ecuador

GRAYDIDASCALUS
Graydidascalus brachyurus (Short-tailed Parrot)
E Ecuador to NE Brazil

PIONUS
Pionus menstruus (Blue-headed Parrot)
P.m.rubrigularis
S Costa Rica to W Ecuador
P.m.menstruus
the Guianas, Upper Amazonia, E Peru
P.m.reichenowi
NE Brazil
Pionus sordidus (Red-billed Parrot)
P.s.ponsi
N Colombia, NW Venezuela
P.s.sordidus
NW Venezuela
P.s.saturatus
N Colombia
P.s.antelius
NE Venezuela
P.s.corallinus
C Colombia to N Bolivia
P.s.mindoensis
W Ecuador
Pionus maximiliani (Scaly-headed Parrot)
P.m.maximiliani
NE Brazil
P.m.melanoblepharus
C Brazil, E Paraguay, NE Argentina
P.m.siy
S Brazil, E Bolivia, Paraguay
P.m.lacerus
NW Argentina
Pionus tumultuosus (Plum-crowned Parrot)
Andes of E Peru & Bolivia
Pionus seniloides (White-headed Parrot)
NW Venezuela to SW Ecuador
Pionus senilis (White-capped Parrot)
E Mexico to W Panama
Pionus chalcopterus (Bronze-winged Parrot)
P.c.chalcopterus
NW Venezuela, NE & C Colombia
P.c.cyanescens
SW Colombia, W Ecuador, NW Peru
Pionus fuscus (Dusky Parrot)
NE Colombia, S Venezuela, the Guianas, NE Brazil

AMAZONA
Amazona collaria (Yellow-billed Amazon)
Jamaica
Amazona leucocephala (Cuban Amazon)
A.l.palmarum
W Cuba, Isle of Pines
A.l.leucocephala
C & E Cuba
A.l.bahamensis
Bahama Is
A.l.caymanensis
Gd Cayman I
A.l.hesterna
Little Cayman, Cayman Brac Is
Amazona ventralis (Hispaniolan Amazon)
Hispaniola
Amazona albifrons (White-fronted Amazon)

A.a.saltuensis
NW Mexico
A.a.albifrons
WC Mexico to SW Guatemala
A.a.nana
SE Mexico to NW Costa Rica
Amazona xantholora (Yellow-lored Amazon)
Yucatan, SE Mexico & Belize
Amazona agilis (Black-billed Amazon)
Jamaica
Amazona vittata (Puerto Rican Amazon)
Puerto Rico
Amazona tucumana (Tucuman Amazon)
SE Bolivia, N Argentina
Amazona pretrei (Red-spectacled Amazon)
SE Brazil, N Uruguay, NE Argentina
Amazona viridigenalis (Green-cheeked Amazon)
NE Mexico
Amazona finschi (Lilac-crowned Amazon)
A.f.woodi
NW Mexico
A.f.finschi
WC & SW Mexico
Amazona autumnalis (Red-lored Amazon)
A.a.autumnalis
E Mexico to N Nicaragua
A.a.salvini
SE Nicaragua to W Colombia
A.a.lilacina
W Ecuador
A.a.diadema
NW Brazil
Amazona brasiliensis (Red-tailed Amazon)
SE Brazil
Amazona dufresniana (Blue-cheeked Amazon)
A.d.dufresniana
SE Venezuela, the Guianas
A.d.rhodocorytha
E Brazil
Amazona festiva (Festive Amazon)
A.f.bodini
C Venezuela, NW Guyana
A.f.festiva
E Ecuador, NE Peru to C Brazil
Amazona xanthops (Yellow-faced Amazon)
E & C Brazil
Amazona barbadensis (Yellow-shouldered Amazon)
A.b.barbadensis
coast of Venezuela, Aruba I
A.b.rothschildi
Bonaire, Margarita Is
Amazona aestiva (Blue-fronted Amazon)
A.a.aestiva
E Brazil
A.a.xanthopteryx
N & E Bolivia, Paraguay, N Argentina
Amazona auropalliata (Yellow-naped Amazon)   68.2
A.a.auropalliata
S Mexico to NW Costa Rica

A.a.parvipes
E Honduras, NE Nicaragua
A.a.caribaea
Bay Is (N Honduras)
Amazona ochrocephala (Yellow-crowned Amazon)   68.2
A.o.hondurensis
Honduras
A.o.panamensis
E Panama, N Colombia, Pearl Is
A.o.nattereri
S Colombia to E Peru, W Brazil
A.o.xantholaema
Marajo I (N Brazil)
A.o.ochrocephala
W Colombia to Surinam, N Brazil, Trinidad
Amazona oratrix (Yellow-headed Amazon)   68.2
A.o.oratrix
SW & S Mexico
A.o.tresmariae
Tres Marias Is
A.o.belizensis
Belize
Amazona amazonica (Orange-winged Amazon)
A.a.amazonica
Colombia to N Bolivia & C & E Brazil
A.a.tobagensis
Trinidad, Tobago I
Amazona mercenaria (Scaly-naped Amazon)
A.m.canipalliata
NW Venezuela, Colombia, C Ecuador
A.m.mercenaria
N Peru to N Bolivia
Amazona farinosa (Mealy Amazon)
A.f.guatemalae
S Mexico to Honduras
A.f.virenticeps
Nicaragua to W Panama
A.f.inornata
E Panama to NW Ecuador & W Venezuela
A.f.chapmani
SE Colombia to NE Bolivia
A.f.farinosa
S Venezuela, the Guianas, C & E Brazil, N Bolivia
Amazona vinacea (Vinaceous Amazon)
SE Brazil, NE Argentina
Amazona versicolor (St Lucia Amazon)
St Lucia I
Amazona arausiaca (Red-necked Amazon)
Dominica I
Amazona guildingii (St Vincent Amazon)
St Vincent I
Amazona imperialis (Imperial Amazon)
Dominica I

DEROPTYUS
Deroptyus accipitrinus (Hawk-headed Parrot)
D.a.accipitrinus
the Guianas, S Venezuela, N Brazil, NE Peru

D.a.fuscifrons
C & NE Brazil

TRICLARIA
Triclaria malachitacea (Purple-bellied Parrot)
SE Brazil

STRIGOPINAE

STRIGOPS
Strigops habroptilus (Kakapo)
S South I, New Zealand

CUCULIFORMES

69. MUSOPHAGIDAE (TURACOS)

CORYTHAEOLA
Corythaeola cristata (Great Blue Turaco)
Portuguese Guinea to N Angola & W Kenya

CRINIFER
Crinifer piscator (Grey Plantain-eater)
Senegal to N Zaire
Crinifer zonurus (Eastern Grey Plantain-eater)
N Zaire, Ethiopia, NW Tanzania

CORYTHAIXOIDES
Corythaixoides concolor (Go-away Bird)
C.c.pallidiceps
Cabinda to C Namibia
C.c.concolor
S Zaire & Tanzania to S Africa
C.c.chobiensis
S Zambia, N Botswana
Corythaixoides personata (Bare-faced Go-away Bird)
C.p.personata
Ethiopia
C.p.leopoldi
E Zaire, Kenya to Zambia, Malawi

CRINIFEROIDES   69.1
Criniferoides leucogaster (White-bellied Go-away Bird)
Ethiopia to E Tanzania

MUSOPHAGA
Musophaga porphyreolopha (Violet-crested Turaco)   69.1
M.p.chlorochlamys
S Kenya to Mozambique
M.p.porphyreolopha
Zimbabwe, E Transvaal, Natal
Musophaga johnstoni (Ruwenzori Turaco)
                                              69.1
M.j.johnstoni
Ruwenzori Mts
M.j.kivuensis
Kivu Area, E Zaire
Musophaga violacea (Violet Turaco)
Gambia to Nigeria

Musophaga rossae (Lady Ross's Turaco)
N Cameroun to Sudan, N Angola, Zambia

TAURACO
Tauraco persa (Green Turaco)   69.1
T.p.buffoni
Gambia to Sierra Leone
T.p.persa
Ivory Coast to N Angola
T.p.schalowi
SW Kenya, C Angola, S Zaire to Malawi
T.p.chalcolophus
NC Tanzania
T.p.livingstonii
E Tanzania to N Natal & SE Zaire
T.p.phoebus
E Transvaal
T.p.corythaix
Natal, SE Cape Province
Tauraco schuetti (Black-billed Turaco)   69.1
T.s.schuetti
N Zaire
T.s.emini
NE Zaire, Uganda, SW Sudan
Tauraco fischeri (Fischer's Turaco)   69.1
T.f.fischeri
Juba River & SE Kenya
T.f.zanzibaricus
Zanzibar
Tauraco erythrolophus (Red-crested Turaco)
SW Zaire, Angola
Tauraco bannermani (Bannerman's Turaco)
N Cameroun
Tauraco macrorhynchus (Crested Turaco)
T.m.macrorhynchus
Sierra Leone to Ivory Coast
T.m.verreauxii
S Nigeria to W Zaire
Tauraco hartlaubi (Hartlaub's Turaco)
Kenya, N Tanzania
Tauraco leucotis (White-cheeked Turaco)
T.l.leucotis
Ethiopia
T.l.donaldsoni
E Ethiopia, W Somalia
Tauraco ruspolii (Prince Ruspoli's Turaco)
S Ethiopa
Tauraco leucolophus (White-crested Turaco)
Cent African Republic to S Sudan

70. OPISTHOCOMIDAE (HOATZIN)   70.1

OPISTHOCOMUS
Opisthocomus hoazin (Hoatzin)
N Amazonian forest

71. CUCULIDAE (CUCKOOS)

CUCULINAE

CLAMATOR
Clamator glandarius (Great Spotted Cuckoo)
Spain to Iran, NE & S Africa

Clamator coromandus (Red-winged Crested Cuckoo)
Himalayas to S China, Java, Borneo

OXYLOPHUS 71.6
Oxylophus jacobinus (Black & White Cuckoo)
O.j.pica
Senegal to Red Sea
O.j.serratus
Guinea to S Africa, N India, Burma
O.j.jacobinus
S India, Sri Lanka
Oxylophus levaillantii (Levaillant's Cuckoo)
Senegal to Somalia & S Africa

PACHYCOCCYX
Pachycoccyx audeberti (Thick-billed Cuckoo)
71.6
P.a.brazzae
Guinea to S Sudan, S Zaire
P.a.audeberti
Madagascar, Guinea to Cape Province

CUCULUS
Cuculus crassirostris (Sulawesi Hawk Cuckoo)
N & C Sulawesi
Cuculus sparverioides (Large Hawk Cuckoo)
C.s.sparverioides
Himalayas, SE Asia to Philippine Is & Sulawesi
C.s.bocki
Malaysia, Sumatra, Borneo
Cuculus varius (Common Hawk Cuckoo)
C.v.varius
India(except NW)
C.v.ciceliae
Sri Lanka
Cuculus vagans (Small Hawk Cuckoo)
Malaysia, Thailand, Java, Borneo
Cuculus fugax (Fugitive Hawk Cuckoo)
C.f.hyperythrus
NE ASia, China, Indochina
C.f.nisicolor
E Himalayas to Malaysia
C.f.pectoralis
Luzon, Cebu, Mindoro Is
C.f.fugax
Malaysia, Sumatra, Java, Borneo
Cuculus solitarius (Red-chested Cuckoo)
C.s.magnirostris
Cameroun
C.s.solitarius
Ethiopia to Cape Province 71.6
Cuculus clamosus (Black Cuckoo)
C.c.clamosus
Gambia to Ethiopia & Sth Africa
C.c.gabonensis
Nigeria to N Zaire, Uganda
Cuculus micropterus (Short-winged Cuckoo)
C.m.micropterus
India to E Asia & E Asian islands

C.m.ognevi
NE Asia
C.m.concretus
Sumatra, Java, Borneo
Cuculus canorus (Eurasian Cuckoo)
C.c.canorus
Europe & W Siberia >> E & S Africa
C.c.bangsi
Iberia, N Africa
C.c.kleinschmidti
Corsica, Sardina
C.c.johanseni
C Asia
C.c.telephonus
NE Asia & Japan >> India & New Guinea
C.c.fallax
C & S China
C.c.bakeri
NW China, Burma, Indochina
C.c.subtelephonus
Transcaspia to W Chinese Turkestan
Cuculus gularis (African Cuckoo) 71.6
Gambia to Sudan & N Sth Africa
Cuculus saturatus (Oriental Cuckoo)
C.s.horsfieldi
C & E Asia >> SE Asia
C.s.lepidus
Malaysia & Sumatra to Timor I
C.s.insulindae
Borneo
C.s.saturatus
S Himalayas to S China
Cuculus poliocephalus (Small Cuckoo)
Himalayas to India, C China, Japan
Cuculus rochii (Madagascar Cuckoo)
Madagascar
Cuculus pallidus (Pallid Cuckoo)
C.p.occidentalis
W Australia, Nthn Territory
C.p.pallidus
E & S Australia, Tasmania

CERCOCOCCYX
Cercococcyx mechowi (Dusky Long-tailed Cuckoo)
Sierra Leone to N Uganda & N Angola
Cercococcyx olivinus (Olive Long-tailed Cuckoo)
Ghana, Cameroun to N Angola
Cercococcyx montanus (Mountain Long-tailed Cuckoo)
C.m.montanus
Ruwenzori Mts
C.m.patulus
N Tanzania

PENTHOCERYX
Penthoceryx sonneratii (Banded Bay Cuckoo)
P.s.sonneratii
India, Burma, Thailand, S Indochina
P.s.waiti
Sri Lanka

P.s.malayanus
N & C Malaysia
P.s.fasciolatus
S Malaysia, Sumatra, Borneo, Philippine is
P.s.musicus
Java

CACOMANTIS
Cacomantis merulinus (Plaintive Cuckoo)
C.m.passerinus
W Himalayas, India, Sri Lanka
C.m.querulus
E Himalayas to S China, Indochina
C.m.threnodes
Malaysia, Sumatra, Borneo
C.m.subpallidus
Nias I
C.m.lanceolatus
Java
C.m.merulinus
Philippine Is, Sulawesi
Cacomantis variolosus (Brush Cuckoo)   71.3
C.v.whitei
Timor I
C.v.infaustus
N & C New Guinea, N Moluccas
C.v.chivae
Biak I
C.v.obscuratus
Numfor I
C.v.fortior
Goodenough, Fergusson Is
C.v.oreophilus
S New Guinea
C.v.blandus
Admiralty Is
C.v.websteri
New Hanover
C.v.macrocercus
New Britain, New Ireland
C.v.addendus
Kulambangra, Malaita, Rubiana Is
C.v.variolosus
N & E Australia >> Moluccas, New
Guinea
Cacomantis sepulcralis (Indonesian Cuckoo)
71.3
C.s.sepulcralis
Malaysia, Borneo, Philippine Is,
Sumatra to Flores I
C.s.everetti
Basilan I, Sula Arch
C.s.virescens
Sulawesi, Tukangbesi I
C.s.aeruginosus
Buru, Ambon, Seram Is
Cacomantis castaneiventris (Chestnut-breasted Cuckoo)
C.c.arfakianus
W Papuan Is, NW New Guinea
C.c.weiskei
C & E New Guinea

C.c.castaneiventris
Cape York Peninsula
Cacomantis heinrichi (Heinrich's Brush Cuckoo)
Halmahera, Batjan Is
Cacomantis pyrrhophanus (Fan-tailed Cuckoo)
C.p.prionurus
E & S Australia, Tasmania
C.p.excitus
New Guinea
C.p.meeki
Solomon Is
C.p.schistaceigularis
New Hebrides
C.p.pyrrhophanus
New Caledonia, Loyalty Is
C.p.simus
Fiji Is

RHAMPHOMANTIS
Rhamphomantis megarhynchus (Little Long-billed Cuckoo)
R.m.sanfordi
Waigeu I
R.m.megarhynchus
NW & N New Guinea, Aru Is

MISOCALIUS
Misocalius osculans (Black-eared Cuckoo)
Interior of Australia >> Lesser Sunda Is,
New Guinea

CHRYSOCOCCYX
Chrysococcyx cupreus (African Emerald Cuckoo)   71.6
C.c.cupreus
Gambia to Cape Province
C.c.insularum
Sao Thome, Principé Is
Chrysococcyx flavigularis (Yellow-throated Green Cuckoo)
Sierra Leone to N & C Zaire
Chrysococcyx klaas (Klaas' Cuckoo)   71.6
Senegal to Ethiopia & Cape Province, S
Arabia
Chrysococcyx caprius (Didric Cuckoo)
Senegal to Ethiopia & Cape Province

CHALCITES
Chalcites maculatus (Asian Emerald Cuckoo)
Himalayas to China, SE Asia
Chalcites xanthorhynchus (Violet Cuckoo)
C.x.xanthorhynchus
NE India to SE Asia, Borneo, & Java
C.x.limborgi
S Burma
C.x.bangueyensis
Banguey I
C.x.amethystinus
Philippine Is

***Chalcites basalis* (Horsfield's Bronze Cuckoo)**
S Australia >> Gtr Sunda Is
***Chalcites lucidus* (Golden-Bronze Cuckoo)**
C.l.plagosus
S Australia to Lesser Sunda Is & New Guinea
C.l.lucidus
New Zealand to Solomon Is
C.l.layardi
New Caledonia, Loyalty Is
C.l.aeneus
New Hebrides, Banks Is
C.l.harterti
Rennell, Bellona Is
***Chalcites poecilurus* (New Guinea Bronze Cuckoo)**   71.2
New Guinea & islands
***Chalcites russatus* (Gould's Bronze Cuckoo)**
    71.2.3
C.r.aheneus
Borneo, S Philippine Is
C.r.misoriensis
Biak I, New Guinea lowlands
C.r.jungei
C & S Sulawesi
C.r.russatus
Cape York Peninsula
***Chalcites rufomerus* (Green-cheeked Bronze Cuckoo)**   71.2
Lesser Sunda Is
***Chalcites minutillus* (Little Bronze Cuckoo)**
    71.2
C.m.albifrons
N Sumatra, Java
C.m.peninsularis
Malaysia, S Indonesia
C.m.cleis
N & E Borneo
C.m.nieuwenhuysi
Halmahera
C.m.minutillus
Melville I, N Australia
***Chalcites crassirostris* (Pied Bronze Cuckoo)**
Moluccas, Kai Is, New Guinea
***Chalcites ruficollis* (Reddish-throated Bronze Cuckoo)**
NW New Guinea Mts
***Chalcites meyeri* (Meyer's Bronze Cuckoo)**
NW New Guinea

CALIECHTHRUS
***Caliechthrus leucolophus* (White-crowned Koel)**
Salawati I, New Guinea

SURNICULUS
***Surniculus lugubris* (Drongo-Cuckoo)**
S.l.dicruroides
N & C India to S China, Indochina
S.l.stewarti
SW India, Sri Lanka

S.l.barussarum
Malaysia, Sumatra, Borneo
S.l.minimus
Palawan, Balabac Is
S.l.lugubris
Java, Bali
S.l.velutinus
Philippine Is
S.l.musschenbroeki
Sulawesi, Batjan, Halmahera

MICRODYNAMIS
***Microdynamis parva* (Black-capped Cuckoo)**
M.p.parva
SW & E New Guinea
M.p.grisecens
N New Guinea

EUDYNAMYS
***Eudynamys scolopacea* (Koel)**
E.s.scolopacea
India, Sri Lanka, Nicobar Is
E.s.chinensis
W & S China, Indochina
E.s.harterti
Hainan I
E.s.simalurensis
Simalur, Babi Is
E.s.malayana
Assam to Thailand, Malaysia, Sumatra to Flores I
E.s.paraguena
Palawan, Busuanga Is
E.s.dolosa
Andaman Is
E.s.mindanensis
Philippine, Sangir, Talaut Is
E.s.frater
Calayan, Fuga Is
E.s.melanorhyncha
Sulawesi, Lesser Sunda, Timor Is
E.s.corvina
N Moluccas
E.s.orientalis
S Moluccas
E.s.salvadorii
Bismarck Arch
E.s.alberti
Solomon Is
E.s.rufiventer
W Papuan islands, & N & C New Guinea
E.s.minima
S New Guinea
***Eudynamys cyanocephela* (Australian Koel)**
E.c.subcyanocephala
NW Australia, W Queensland
E.c.cyanocephala
N Queensland, N New South Wales

URODYNAMIS
***Urodynamis taitensis* (Long-tailed Koel)**
New Zealand & SW Pacific islands

SCYTHROPS
Scythrops novaehollandiae (Channel-billed Cuckoo)
Flores, Moluccas to Australia, New Guinea

PHAENICOPHAEINAE

COCCYZUS
Coccyzus pumilus (Dwarf Cuckoo)
W Venezuela, E Colombia
Coccyzus cinereus (Ash-coloured Cuckoo)
Paraguay, S Brazil to C Argentina
Coccyzus erythrophthalmus (Black-billed Cuckoo)
S Canada >> NW Sth America
Coccyzus americanus (Yellow-billed Cuckoo)
C.a.americanus
C & E USA >> N Sth America
C.a.occidentalis
SW Canada >> W Mexico
Coccyzus euleri (Pearly-breasted Cuckoo)
NE Sth America
Coccyzus minor (Mangrove Cuckoo)
C.m.palloris
coast of W Mexico to E Panama, Tres Marias Is
C.m.continentalis
coast of E Mexico to Panama
C.m.cozumelae
Cozumel I
C.m.maynardi
S Florida, Bahama Is
C.m.caymanensis
Cayman Is
C.m.nesiotes
Jamaica
C.m.teres
Gtr Antilles
C.m.rileyi
Barbuda, Antigua Is
C.m.dominicae
Montserrat, Guadeloupe, Dominica Is
C.m.vincentis
Martinique, St Lucia, St Vincent Is
C.m.grenadensis
Grenada, Bequia Is
C.m.abbotti
Old Providence, St Andrews Is
C.m.minor
N Sth America
Coccyzus ferrugineus (Cocos Cuckoo) 71.7
Cocos I (Costa Rica)
Coccyzus melacoryphus (Dark-billed Cuckoo)
S America, Galapagos Is
Coccyzus lansbergi (Grey-capped Cuckoo)
Colombia, Venezuela, W Ecuador

HYETORNIS 71.7
Hyetornis rufigularis (Rufous-breasted Cuckoo)
Hispaniola
Hyetornis pluvialis (Chestnut-bellied Cuckoo)
Jamaica

PIAYA
Piaya cayana (Squirrel Cuckoo)
P.c.extima
NW Mexico
P.c.mexicana
W Mexico
P.c.stirtoni
W coast of Guatemala to NW Costa Rica
P.c.thermophila
E Mexico to Panama
P.c.mesura
E Colombia, E Ecuador
P.c.nigricrissa
W Colombia, W Ecuador, NW & EC Peru
P.c.mehleri
NE Colombia, N Venezuela
P.c.circe
W Venezuela
P.c.insulana
Trinidad
P.c.cayana
E & S Venezuela, the Guianas, N Brazil
P.c.boliviana
EC Peru to N Bolivia
P.c.obscura
C Brazil
P.c.hellmayri
C & E Brazil
P.c.pallescens
E Brazil
P.c.cearae
Ceara, Brazil
P.c.cabanisi
SC Brazil
P.c.macroura
SE Brazil, NE Argentina, Paraguay, Uruguay
P.c.mogenseni
S Bolivia, NW Argentina
Piaya melanogaster (Black-bellied Cuckoo)
P.m.melanogaster
N & C Amazonia
P.m.ochracea
SE Colombia, S Peru
Piaya minuta (Little Cuckoo)
P.m.panamensis
E Panama
P.m.gracilis
W Colombia, W Ecuador
P.m.minuta
N & W Amazonia
P.m.chaparensis
N Bolivia

SAUROTHERA
Saurothera merlini (Great Lizard Cuckoo)
S.m.bahamensis
New Providence, Eleuthera Is
S.m.andria
Andros I
S.m.merlini
Cuba

S.m.decolor
Isle of Pines
Saurothera vetula (Jamaican Lizard Cuckoo)
71.7
Jamaica
Saurothera longirostris (Hispaniolan Lizard Cuckoo) 71.7
S.l.petersi
Gonave I
S.l.longirostris
Hispaniola, Tortuga I
S.l.saonae
Saona I
Saurothera vieilloti (Puerto Rican Lizard Cuckoo) 71.7
Puerto Rico

CEUTHMOCHARES
Ceuthmochares aereus (Yellow-bill)
C.a.flavirostris
Senegal to W Nigeria
C.a.aereus
Nigeria to N Angola & W Tanzania
C.a.australis
Kenya to Malawi & Natal

RHOPODYTES
Rhopodytes diardi (Lesser Green-billed Malcoha)
R.d.diardi
S Malaysia, Sumatra
R.d.borneensis
Borneo
Rhopodytes sumatranus (Rufous-bellied Malcoha)
R.s.sumatranus
S Burma, Malaysia, Sumatra
R.s.minor
Borneo
Rhopodytes tristis (Greater Green-billed Malcoha)
R.t.tristis
W Himalayas to N Burma
R.t.saliens
Burma, N Indochina, S China
R.t.longicaudatus
S Burma, Malaysia, S Indochina
R.t.hainanus
Hainan I
R.t.elongatus
Sumatra
R.t.kangeangensis
Kangean I
Rhopodytes viridirostris (Small Green-billed Malcoha)
S India, Sri Lanka

TACCOCUA
Taccocua leschenaultii (Sirkeer Cuckoo)
T.l.sirkee
NW India
T.l.infuscata
W Himalayas

T.l.affinis
NE India
T.l.leschenaultii
S India

RHINORTHA
Rhinortha chlorophaea (Raffles' Malcoha)
R.c.chlorophaea
S Burma, Malaysia, Sumatra
R.c.fuscigularis
N Borneo & islands
R.c.mayri
S Borneo

ZANCLOSTOMUS
Zanclostomus javanicus (Red-billed Malcoha)
Z.j.pallidus
S Burma, Malaysia, Sumatra, Borneo
Z.j.factus
Tanahmasa I
Z.j.javanicus
Java
Z.j.natunensis
Natuna Is

RHAMPHOCOCCYX
Rhamphococcyx calyorhynchus (Fiery-billed Malcoha) 71.3
R.c.calyorhynchus
N Sulawesi, Togian I
R.c.meridionalis
C & S Sulawesi
R.c.rufiloris
Buton I
Rhamphococcyx curvirostris (Chestnut-breasted Malcoha)
R.c.erythrognathus
S Burma, Malaysia, Sumatra
R.c.oeneicaudus
islands off SW Sumatra
R.c.curvirostris
W & C Java
R.c.deningeri
E Java, Bali
R.c.borneensis
Borneo, Natuna Is
R.c.harringtoni
Palawan, Balabac Is

PHAENICOPHAEUS
Phaenicophaeus pyrrhocephalus (Red-faced Malcoha)
S India, Sri Lanka

DASYLOPHUS
Dasylophus superciliosus (Rough-crested Cuckoo)
N Philippine Is

LEPIDOGRAMMUS
Lepidogrammus cumingi (Scale-feathered Cuckoo)
 Luzon, Marinduque Is

CROTOPHAGINAE

CROTOPHAGA
Crotophaga major (Greater Ani)
 E Panama to N Argentina, Trinidad
Crotophaga ani (Smooth-billed Ani)
 Bahama Is, Antilles, N Sth America
Crotophaga sulcirostris (Groove-billed Ani)
 C.s.pallidula
  S Baja California
 C.s.sulcirostris
  Mexico to C Sth America, Curacao I, Trinidad

GUIRA
Guira guira (Guira Cuckoo)
 S & E Brazil, N Argentina, Uruguay

NEOMORPHINAE

TAPERA
Tapera naevia (Striped Cuckoo)
 T.n.excellens
  SE Mexico to Panama
 T.n.naevia
  N Sth America, Trinidad
 T.n.chochi
  S Brazil, N Argentina

MOROCOCCYX
Morococcyx erythropygus (Lesser Ground Cuckoo)
 M.e.dilutus
  W Mexico
 M.e.simulans
  Guerrero, Mexico
 M.e.mexicanus
  SW & S Mexico
 M.e.erythropygus
  S Mexico to N Costa Rica
 M.e.macrourus
  Guatemala

DROMOCOCCYX
Dromococcyx phasianellus (Pheasant Cuckoo)
 D.p.rufigularis
  SE Mexico to Colombia
 D.p.phasianellus
  C & S Brazil, Paraguay, Bolivia
Dromococcyx pavoninus (Pavonine Cuckoo)
 D.p.perijanus
  NW Venezuela

D.p.pavoninus
 N Sth America, intermittently

GEOCOCCYX
Geococcyx californianus (Road-runner)
 G.c.californianus
  S USA to CS Mexico
 G.c.conklingi
  S California
Geococcyx velox (Lesser Road-runner)
 G.v.melanchima
  W Mexico
 G.v.velox
  EC Mexico
 G.v.affinis
  El Salvador, W Guatemala
 G.v.pallidus
  Yucatan, E Guatemala
 G.v.longisignum
  Honduras, N Nicaragua

NEOMORPHUS
Neomorphus geoffroyi (Rufous-vented Ground Cuckoo)
 N.g.salvini
  Nicaragua to W Colombia
 N.g.aequatorialis
  E Ecuador
 N.g.australis
  S Peru, NW Bolivia
 N.g.geoffroyi
  C & S Brazil
 N.g.dulcis
  E Brazil
Neomorphus squamiger (Scaled Ground Cuckoo)
 N.s.squamiger
  EC Brazil
 N.s.iungens
  C Brazil
Neomorphus radiolosus (Banded Ground Cuckoo)
 NW Ecuador
Neomorphus rufipennis (Rufous-winged Ground Cuckoo)
 N.r.rufipennis
  NE Venezuela
 N.r.nigrogularis
  S Venezuela, Guyana, N Brazil
Neomorphus pucheranii (Red-billed Ground Cuckoo)
 N.p.pucheranii
  W Brazil, E Ecuador, NE Peru
 N.p.lepidophanes
  E Peru, SW Brazil

CARPOCOCCYX
Carpococcyx radiceus (Ground Cuckoo)
 C.r.radiceus
  Borneo

C.r.viridis
Sumatra
Carpococcyx renauldi (Coral-billed Ground Cuckoo)
SE Thailand, Indochina

COUINAE

COUA
Coua gigas (Giant Madagascar Coucal)
W & S Madagascar
Coua coquereli (Coquerel's Madagascar Coucal)
W Madagascar
Coua serriana (Rufous-breasted Madagascar Coucal)
NE Madagascar
Coua reynaudii (Red-footed Madagascar Coucal)
NW & E Madagascar
Coua cursor (Running Coucal)
SW Madagascar
Coua ruficeps (Red-capped Madagascar Coucal)
C.r.ruficeps
NW Madagascar
C.r.olivaceiceps
SW Madagascar
Coua cristata (Crested Madagascar Coucal)
C.c.dumonti
W Madagascar
C.c.cristata
N & E Madagascar
C.c.pyropyga
SW Madasgar
C.c.maxima
SE Madagascar
Coua verreauxi (Southern Crested Madagascar Coucal)
SW Madagascar
Coua caerulea (Blue Madagascar Coucal)
NW & E Madagascar

CENTROPODINAE

CENTROPUS
Centropus milo (Buff-headed Coucal)
C.m.albidiventris
Vellalavella, Kulambangra, Gizo, Rendova Is
C.m.milo
Florida, Guadalcanal Is
Centropus goliath (Large Coucal)
N Moluccas
Centropus violaceus (Violet Coucal)
New Ireland, New Britain
Centropus menbeki (Greater Coucal)
C.m.menbeki
W Papuan islands, New Guinea
C.m.jobiensis
Japen I

C.m.aruensis
Aru Is
Centropus ateralbus (New Britain Coucal)
New Britain, New Ireland
Centropus chalybeus (Biak Island Coucal)
Biak, Numfor Is
Centropus phasianinus (Pheasant Coucal)
C.p.mui
Timor I
C.p.propinquus
N New Guinea
C.p.nigricans
SE New Guinea
C.p.obscuratus
Goodenough, Fergusson Is, E New Guinea
C.p.thierfelderi
S New Guinea
C.p.phasianinus
NE Australia
C.p.macrourus
N & mid Western Australia
Centropus spilopterus (Kai Coucal)
Kai Is
Centropus bernsteini (Bernstein's Coucal)
C.b.manam
Vulcan I
C.b.bernsteini
W New Guinea
Centropus chlororhynchus (Ceylon Coucal)
SW Sri Lanka
Centropus rectunguis (Short-toed Coucal)
Malaysia, Sumatra, Borneo
Centropus steerii (Steere's Coucal)
Mindoro I
Centropus sinensis (Common Crow-Pheasant)
C.s.parroti
C & S India, Sri Lanka
C.s.sinensis
N India to S China
C.s.intermedius
Burma, S Thailand, Indochina, Hainan I
C.s.eurycercus
Malaysia, Sumatra, Borneo, Palawan I
C.s.bubutus
Java, Bali
C.s.anonymous
C Philippine Is
C.s.kangeanensis
Kangean I
C.s.andamanensis
Cocos, Andaman Is
Centropus nigrorufus (Sunda Coucal)
Sumatra, Java
Centropus viridis (Philippine Coucal)
C.v.viridis
Philippine Is
C.v.major
Fuga I
C.v.carpenteri
Batan I
C.v.mindorensis
Mindoro, Semirara Is

Centropus toulou (Black Coucal)
  C.t.toulou
    Madagascar
  C.t.insularis
    Aldabra I
  C.t.assumptionis
    Assumption I e?
Centropus bengalensis (Lesser Coucal)
  C.b.bengalensis
    India, Burma to Indochina, SE China
  C.b.chamnongi
    C Thailand, S Vietnam
  C.b.javanensis
    Malaysia, Sumatra, Java, Borneo,
    Philippine Is
  C.b.medius
    Moluccas
Centropus grillii (Black-chested Coucal)   71.6
  Guinea to Natal
Centropus epomidis (Rufous-bellied Coucal)
  Ghana, S Nigeria
Centropus leucogaster (Black-throated Coucal)
  C.l.leucogaster
    Sierra Leone to Nigeria
  C.l.efulenensis
    W Cameroun, Gabon
Centropus neumanni (Smaller Black-throated Coucal)   71.4
  N Zaire
Centropus anselli (Gabon Coucal)
  S Cameroun to Angola, S Zaire
Centropus monachus (Blue-headed Coucal)   71.6
  C.m.fischeri
    Ghana to Kenya & N Angola
  C.m.verheyeni
    Shaba Prov.Zaire
  C.m.monachus
    Ethiopia, Kenya
Centropus cupreicaudus (Coppery-tailed Coucal)
  C.c.songweensis
    S Tanzania, N Malawi
  C.c.cupreicaudus
    S Angola to S Tanzania, Namibia
Centropus senegalensis (Senegal Coucal)
  C.s.aegyptius
    Egypt
  C.s.senegalensis
    Senegal to Sudan, Angola, Tanzania
  C.s.flecki
    Botswana, Zimbabwe, Transvaal
Centropus superciliosus (White-browed Coucal)
  C.s.loandae
    Angola to Uganda & Malawi
  C.s.superciliosus
    Sudan to Somalia & Tanzania
  C.s.fasciipygialis
    E Zimbabwe to S Tanzania
  C.s.sokotrae
    Socotra I

  C.s.burchellii
    S Tanzania to Cape Province
Centropus melanops (Black-faced Coucal)
  C.m.melanops
    Leyte, Bohol, Mindanao, Basilan Is
  C.m.banken
    Samar I
Centropus celebensis (Bay Coucal)
  C.c.celebensis
    N Sulawesi, Togian I
  C.c.rufescens
    C & S Sulawesi, Muna I
Centropus unirufus (Rufous Coucal)
  C.u.unirufus
    Luzon I
  C.u.polillensis
    Polillo Is

STRIGIFORMES

72. TYTONIDAE (BARN OWLS)   72.3

TYTONINAE

TYTO
Tyto soumagnei (Madagascar Masked Owl)
  Madagascar
Tyto alba (Barn Owl)
  T.a.schmitzi
    Madeira
  T.a.gracilirostris
    E Canary Is
  T.a.alba
    W Europe
  T.a.ernesti
    Corsica, Sardinia
  T.a.guttata
    C Europe
  T.a.detorta
    Cape Verde Is
  T.a.affinis
    Gambia to Sudan & Cape Province
  T.a.poensis
    Fernando Po I
  T.a.thomensis
    Sao Thome I
  T.s.erlangeri
    Arabia to Syria & Iraq
  T.a.hypermetra
    Comoro Is, Madagascar
  T.a.stertens
    India, N Burma, Sri Lanka
  T.a.javanica
    Burma to Indochina, Java to Timor I
  T.a.deroepstorffi
    S Andaman Is
  T.a.sumbaensis
    Sumba I
  T.a.bellonae
    Bellona I
  T.a.meeki
    SE New Guinea, Vulcan, Dampier Is

T.a.delicatula
  Australia, Solomon Is
T.a.crassirostris
  Boang I
T.a.interposita
  Santa Cruz, Banks Is, New Hebrides
T.a.lulu
  New Caledonia, Fiji, Tonga, Samoa Is
T.a.pratincola
  C & NE USA to E Nicaragua
T.a.guatemalae
  W Guatemala to Panama
T.a.bondi
  Honduras
T.a.niveicauda
  I de Pinhos
T.a.lucayana
  Bahama Is
T.a.furcata
  Cuba, Cayman Is, Jamaica
T.a.bargei
  Curacao I
T.a.subandeana
  Colombia Ecuador
T.a.contempta
  W Colombia to Venezuela & Peru
T.a.hellmayri
  the Guianas, N Brazil
T.a.tuidara
  C Brazil to Chile, Argentina
T.a.nigrescens
  Dominica I
T.a.insularis
  S Lesser Antilles
T.a.punctatissima
  Galapagos Is
Tyto glaucops (Ashy-faced Barn Owl) 72.2
  Tortuga I, Hispaniola
Tyto rosenbergii (Sulawesi Masked Owl) 72.1
T.r.pelingensis
  Peleng I
T.r.rosenbergii
  Sulawesi, Sangihe Is
Tyto nigrobrunnea (Sula Masked Owl) 72.1
  Sula Is
Tyto inexspectata (Minnahassa Masked Owl)
  72.1
  N Sulawesi
Tyto sororcula (Lesser Masked Owl) 72.1
T.s.sororcula
  Tanimbar Is
T.s.cayelii
  Buru I
Tyto novaehollandiae (Australian Masked Owl)
T.n.calabyi
  S New Guinea
T.n.manusi
  Manus I
T.n.kimberli
  S New Guinea, N Australia
T.n.novaehollandiae
  SE Australia

T.n.perplexa
  SW Australia
T.n.castanops
  Tasmania
Tyto aurantia (New Britain Barn Owl)
  New Britain
Tyto tenebricosa (Greater Sooty Owl)
T.t.arfaki
  New Guinea, Japen I
T.t.tenebricosa
  E & S Australia
Tyto multipunctata (Lesser Sooty Owl) 72.3
  C Queensland
Tyto capensis (Grass Owl) 72.3
T.c.cameroonensis
  Cameroun
T.c.liberatus
  Kenya
T.c.damarensis
  S Angola, N Namibia
T.c.capensis
  SE Zaire to Cape Province
T.c.longimembris
  India to Indochina & Lesser Sunda Is
T.c.melli
  Kwangsi, Kwangtung
T.c.chinensis
  Fukien
T.c.amauronota
  Philippine Is
T.c.walleri
  N & E Australia, Sulawesi, ?Fiji
T.c.papuensis
  SE New Guinea

PHODILINAE

PHODILUS
Phodilus badius (Bay Owl)
P.b.saturatus
  Nepal to Indochina
P.b.parvus
  Billiton I
P.b.badius
  C Burma to Malaysia, Sumatra, Java, Borneo
P.b.assimilis
  Sri Lanka
P.b.arixuthus
  Bunguran I
Phodilus prigoginei (Tanzanian Bay Owl)
  E Zaire, NW Tanzania

73. STRIGIDAE (OWLS) 73.10.11

OTUS 73.1.2.10
Otus sagittatus (White-fronted Scops Owl)
  S Burma, Thailand, Malaysia
Otus rufescens (Rufous Scops Owl)
O.r.malayensis
  Malaysia
O.r.rufescens
  Sumatra, Java, Borneo

O.r.burbidgei
Jolo I
Otus icterorhynchus (Cinnamon Scops Owl)
O.i.icterorhynchus
Ghana
O.i.holerythrus
S Cameroun to N Zaire
Otus ireneae (Sokoke Scops Owl)
SE Kenya
Otus spilocephalus (Spotted Scops Owl)
O.s.huttoni
W Himalayas
O.s.spilocephalus
E Himalayas to Burma
O.s.latouchei
SE China, N Indochina
O.s.hambroecki
Taiwan
O.s.siamensis
Thailand, S Indochina
O.s.vulpes
Malaysia
O.s.vandewateri
Sumatra
O.s.luciae
Borneo
Otus balli (Andaman Scops Owl)
Andaman Is
Otus longicornis (Luzon Scops Owl)
Luzon I
Otus mindorensis (Mindoro Scops Owl) 73.10
Mindoro I
Otus mirus (Mindanao Scops Owl) 73.10
Mindanao I
Otus manadensis (Sulawesi Scops Owl) 73.6
Sulawesi
Otus alfredi (Flores Scops Owl)
Flores I
Otus angelinae (Javan Scops Owl) 73.10
Java
Otus umbra (Simalur Scops Owl)
O.u.umbra
Simalur I
O.u.enganensis
Enggano I
Otus hartlaubi (Sao Thome Scops Owl)
Sao Thome I
Otus brucei (Pallid Scops Owl) 73.8
O.b.exiguus
Israel, C Iraq to W Pakistan
O.b.socotranus
Socotra I
O.b.brucei
NW India
Otus flammeolus (Flammulated Owl)
O.f.meridionalis
S Mexico, Guatemala
O.f.flammeolus
SW Canada to Mexico
O.f.frontalis
Colorado >> C Mexico
O.f.idahoensis
WC USA

O.f.borealis
C Brit.Columbia to NE California >> W Mexico
O.f.rarus
Guatemala
Otus senegalensis (African Scops Owl)
O.s.senegalensis
Senegal to Sudan
O.s.pygmea
S Sudan, NW Ethiopia
O.s.caecus
C & S Ethiopia, Somalia, N Kenya
O.s.socotranus
Socotra I
O.s.pamelae
Saudi Arabia
O.s.ugandae
W Uganda, N & NE Zaire
O.s.feae
Annobon I
O.s.graueri
E Kenya, Tanzania
O.s.hendersoni
Angola, SW Zaire
O.s.pusillus
Mozambique, E Malawi, E Zimbabwe
O.s.intermedius
Namibia to S Mozambique & N Natal
O.s.latipennis
Cape Province
Otus scops (Eurasian Scops Owl)
O.s.scops
W Europe to Russia & C Africa
O.s.cycladum
Cyclades Is, Crete
O.s.cyprius
Cyprus
O.s.turanicus
Transcaspia, N Iran
O.s.pulchellus
Caucasus, Russia to C Asia, NW India
O.s.stictonotus
Manchuria, China, Taiwan
O.s.japonicus
N Japan
O.s.modestus
Andaman Is
O.s.malayanus
S China >> Malaysia
Otus sunia (Oriental Scops Owl)
O.s.sunia
Himalayas, N India
O.s.rufipennis
C & S India
O.s.leggei
Sri Lanka
O.s.distans
Thailand
Otus elegans (Riukiu Scops Owl)
O.e.interpositus
Borodino Is
O.e.elegans
Ryukyu Is

O.e.botelensis
Botel Tobago I
O.e.calayensis
Calayan I
**Otus mantananensis (South Philippines
Scops Owl)**
O.m.mantananensis
Mantanani I
O.m.cuyensis
Cuyo I
O.m.romblonis
Romblon I
O.m.sibutuensis
Sibutu I
O.m.steerei
Tumindao I
Otus magicus (Moluccan Scops Owl)   73.6
O.m.mendeni
Peling I
O.m.siaoensis
Siao I
O.m.sulaensis
Sula Mangoli I
O.m.kalidupae
Kalidupa I
O.m.obira
Obi Is
O.m.leucospilus
Halmahera, Ternate, Morotai, Batjan Is
O.m.magicus
Seram, Ambon, Buru Is
O.m.albiventris
Lombok, Sumbawa, Flores, Lomblen Is
O.m.tempestatis
Wetar I
Otus rutilus (Madagascar Scops Owl)
O.r.capnodes
Anjouan I
O.r.rutilus
Madagascar
O.r.mayottensis
Mayotte I
O.r.pauliani
Comoro Is
Otus pembaensis (Pemba Scops Owl)   73.8
Pemba I
Otus brookii (Rajah Scops Owl)
O.b.brookii
Borneo, Java
O.b.solokensis
Sumatra
Otus bakkamoena (Indian Scops Owl)
O.b.deserticolor
SE Saudi Arabia, S Iran, Pakistan
O.b.plumipes
NW Himalayas
O.b.lettia
E Himalayas to Burma, N Thailand
O.b.manipurensis
Manipur, Assam
O.b.gangeticus

NW & NC India
O.b.marathae
C India
O.b.bakkamoena
S India, Sri Lanka
Otus lempiji (Collared Scops Owl)   73.10
O.l.ussuriensis
S Manchuria to Korea, Sakhalin
O.l.semitorques
Kurile Is, Japan, Quelpart Is
O.l.pryeri
Hachijo, Okinawa Is
O.l.aurorae
N China
O.l.erythrocampe
S China, N Vietnam
O.l.glabripes
Taiwan
O.l.umbratilis
Hainan I
O.l.condorensis
Pulo Condor I
O.l.kangeana
Kangean I
O.l.cnephaeus
Malaysia
O.l.hypnodes
Singapore, Sumatra
O.l.lempiji
Java, Bali, Borneo
O.l.lemurum
Sarawak
Otus megalotis (Philippine Scops Owl)
O.m.megalotis
Luzon
O.m.nigrorum
Negros I
O.m.everetti
Mindanao, Basilan, Bohol Is
Otus fuliginosus (Palawan Scops Owl)   73.10
Palawan I
Otus mentawi (Sipora Scops Owl)
Siberut, Sipora, Pagi Is
Otus silvicola (Wallace's Scops Owl)
Flores, Sumbawa
Otus beccarii (Biak Island Scops Owl)
Biak I
Otus insularis (Bare-legged Scops Owl)
Mahe I
Otus leucotis (White-faced Scops Owl)
O.l.leucotis
Senegal to Ethiopia & Kenya
O.l.margarethae
Sudan
O.l.granti
S Zaire & Tanzania to Cape Province
Otus asio (Eastern Screech Owl)
O.a.swenki
SC Canada to Oklahoma
O.a.maxwelliae
Montana to Colorado

O.a.naevius
  NE & C USA
O.a.asio
  E Virginia to Kansas
O.a.hasbroucki
  C Oklahoma to N Texas
O.a.mccallii
  S Texas to NC Mexico
O.a.floridanus
  S Louisiana, Florida
O.a.semplei
  C Mexico
***Otus kennicotti* (Western Screech Owl)**
O.k.kennicotti
  S Alaska to Washington
O.k.saturatus
  Vancouver I
O.k.macfarlanei
  S Brit.Columbia to W Montana
O.k.brewsteri
  Washington, N California
O.k.myochophilus
  NC Utah to N Arizona
O.k.aikeni
  C Colorado to N Mexico
O.k.bendirei
  W California
O.k.inyoensis
  E California, Nevada, N Utah
O.k.cineraceus
  C Arizona to WC Texas
O.k.quercinus
  S California
O.k.clazus
  S California
O.k.gilmani
  SE California, SW Arizona, NE Baja
  California
O.k.cardonensis
  W Baja California
O.k.xantusi
  S Baja California
O.k.vinaceus
  NE Sinaloa
O.k.sinaloensis
  SE Sonora, NW Sinaloa
O.k.sortilegus
  Jalisco
O.k.suttoni
  Hidalgo
***Otus seductus* (Balsas Screech Owl)**
O.s.seductus
  Michoacan (Mexico)
O.s.colimensis
  Colima
***Otus cooperii* (Pacific Screech Owl)**
O.c.chiapensis
  S Mexico
O.c.cooperii
  El Salvador, NW Costa Rica
***Otus trichopsis* (Whiskered Screech Owl)**

O.t.aspersus
  SE Arizona, NW Mexico
O.t.ridgwayi
  Michoacan
O.t.trichopsis
  W & S Mexico
O.t.guerrerensis
  SW Mexico
O.t.mesamericanus
  Guatemala, El Salvador
O.t.pumilus
  Honduras
O.t.inexpectus
  W Costa Rica, W Panama
***Otus choliba* (Tropical Screech Owl)**
O.c.luctisonus
  Costa Rica to NW Colombia
O.c.margaritae
  N Colombia, N Venezuela, Margarita I
O.c.montanus
  Venezuela, NE Colombia
O.c.kelsoi
  Tobago, Trididad, NE Venezuela
O.c.alticola
  C Colombia
O.c.duidae
  SE Venezuela
O.c.caucae
  S Colombia, N Ecuador
O.c.guyanensis
  S Venezuela, the Guianas
O.c.crucigerus
  Upper Amazonia, C Brazil
O.c.caatingensis
  NE to C Brazil
O.c.portoricensis
  the Guianas
O.c.decussatus
  SC & E Brazil
O.c.choliba
  S Brazil, Paraguay, Uruguay N Argentina
O.c.chapadensis
  S Brazil
O.c.wetmorei
  SE Bolivia, W Paraguay, NW Argentina
O.c.alilucoco
  S Bolivia to C Argentina
O.c.uruguaii
  C & NE Argentina
***Otus sanctaecatarinae* (Long-tufted Screech Owl)**   73.10
  SE Brazil
***Otus roboratus* (Peruvian Screech Owl)**   73.10
  NW Peru
***Otus koepckeae* (Maria Koepcke's Screech Owl)**   73.10
  Peru, Bolivia
***Otus clarkii* (Bare-shanked Screech Owl)**
  Costa Rica, Panama   73.10
***Otus barbarus* (Bearded Screech Owl)**
  N Guatemala

Otus marshalli **(Cloud-forest Screech Owl)**
SE Peru
Otus ingens **(Rufescent Screech Owl)**
O.i.ingens
Ecuador
O.i.venezuelanus
W Venezuela
O.i.minimus
Bolivia
Otus colombianus **(Colombian Screech Owl)**
73.10
W Colombia, N Ecuador, N Peru
Otus peterseni **(Cinnamon Screech Owl)**
N Peru
Otus watsonii **(Tawny-bellied Screech Owl)**
73.2.10
O.w.watsonii
N Amazonia
O.w.morelia
E Colombia, E Ecuador
O.w.ater
C Brazil
O.w.usta
C Brazil to N Argentina
O.w.inambarii
E Peru, E Bolivia
O.w.fulvescens
C Brazil to N Bolivia
Otus atricapillus **(Black-capped Screech Owl)**
73.2
O.a.atricapillus
C & S Brazil
O.a.argentinus
S Brazil
Otus guatemalae **(Vermiculated Screech Owl)**
O.g.tomlini
NW Mexico
O.g.hastatus
W Mexico
O.g.cassini
N Vera Cruz
O.g.fuscus
C Vera Cruz
O.g.pettingilli
C Mexico
O.g.thompsoni
Yucatan, Campeche
O.g.guatemalae
SE Vera Cruz, Guatemala, Honduras
O.g.peteni
N Guatemala, S Mexico
O.g.dacrysistactus
N Nicaragua
O.g.vermiculatus
Costa Rica, Panama
O.g.centralis
E Panama, NW Colombia
O.g.napensis
E Ecuador
O.g.pallidus
N Venezuela
O.g.roraimae
SE Venezuela, S Guyana

O.g.pacificus
NW Peru
O.g.rufus
W Ecuador, N Peru
O.g.helleri
Peru
Otus nudipes **(Puerto Rico Screech Owl)**
O.n.nudipes
Puerto Rico
O.n.newtoni
St Thomas, St John, St Croix Is
O.n.krugii
Lesser Antilles Is
Otus lawrencii **(Cuban Screech Owl)**
O.l.lawrencii
C & E Cuba
O.l.exsul
W Cuba, Isle of Pines
Otus podarginus **(Palau Scops Owl)**   73.10
Palau Is
Otus albogularis **(White-throated Screech Owl)**
O.a.albogularis
Colombia, N Ecuador
O.a.obscurus
NW Venezuela
O.a.meridensis
W Venezuela

MIMIZUKU
Mimizuku gurneyi **(Giant Scops Owl)**
Marinduque, Mindanao Is

JUBULA
Jubula lettii **(Maned Owl)**
Liberia to N Zaire

LOPHOSTRIX
Lophostrix cristata **(Crested Owl)**
L.c.stricklandi
S Mexico to W Guatemala
L.c.wedeli
E Panama
L.c.cristata
the Guianas, N & W Brazil

PULSATRIX
Pulsatrix perspicillata **(Spectacled Owl)**
P.p.saturata
S Mexico to W Panama
P.p.chapmani
E Costa Rica to W Ecuador
P.p.perspicillata
N Sth America
P.p.trinitatis
Trinidad
P.p.pulsatrix
E Brazil, Paraguay
P.p.boliviana
S Bolivia, N Argentina

Pulsatrix koeniswaldiana **(Tawny-browed Owl)**
S Brazil, NE Argentina
Pulsatrix melanota **(Band-bellied Owl)**
  P.m.melanota
    E Ecuador, E Peru
  P.m.philoscia
    Bolivia

BUBO
Bubo virginianus **(Great Horned Owl)**
  B.v.algistus
    W Alaska
  B.v.lagophonus
    C Alaska to NE Oregon & Idaho
  B.v.saturatus
    SW Alaska to California
  B.v.pacificus
    S Oregon, California
  B.v.wapacuthu
    W & C Canada
  B.v.occidentalis
    WC Canada to WC USA
  B.v.pallescens
    SW USA to NC Mexico
  B.v.heterocnemis
    E Canada
  B.v.virginianus
    SE Canada, EC USA
  B.v.elachistus
    S Baja California
  B.v.mayensis
    C Mexico to W Panama
  B.v.elutus
    E Colombia
  B.v.colombianus
    C Colombia
  B.v.nigrescens
    W Ecuador
  B.v.scotinus
    E Venezuela
  B.v.deserti
    E Brazil
  B.v.nacurutu
    Peru & NW Brazil to Tierra del Fuego
Bubo bubo **(Northern Eagle Owl)**
  B.b.bubo
    Scandinavia, W Europe to W Russia
  B.b.hispanus
    Iberian Peninsula
  B.b.interpositus
    SW Russia to Syria
  B.b.ruthenus
    SE Russia
  B.b.sibiricus
    WC & C Asia
  B.b.yenisseensis
    C & EC Siberia
  B.b.dauricus
    N Mongolia
  B.b.jakutensis
    NE Siberia

  B.b.ussuriensis
    Lower Amur, Ussuriland
  B.b.inexpectatus
    Manchuria, N China
  B.b.tenuipes
    Korea, S Kurile Is, Hokkaido I
  B.b.borissowi
    Sakhalin I
  B.b.turcomanus
    Turkestan
  B.b.zaissanensis
    SC Asia
  B.b.nikolskii
    Iran, Iraq
  B.b.tibetanus
    C Tibet to NW China
  B.b.kiautschensis
    C & E China
  B.b.jarlandi
    SE Yunnan
  B.b.swinhoei
    SE China
  B.b.hemachalana
    W Tien Shan, W Himalayas
Bubo ascalaphus **(Desert Eagle Owl)**   73.11
    N Africa, Near East
Bubo bengalensis **(Indian Eagle Owl)**   73.11
    N & C India
Bubo capensis **(Cape Eagle Owl)**
  B.c.dillonii
    Ethiopia
  B.c.mackinderi
    Kenya, Tanzania
  B.c.capensis
    Natal, Cape Province
Bubo africanus **(Spotted Eagle Owl)**
  B.a.cinerascens
    French Guinea to Somalia
  B.a.africanus
    Uganda & Kenya to Angola
  B.a.tanai
    SE Kenya
  B.a.milesi
    S Saudi Arabia
Bubo poensis **(Fraser's Eagle Owl)**
  B.p.poensis
    Ghana to N Zaire
  B.p.vosseleri
    N Tanzania
Bubo nipalensis **(Forest Eagle Owl)**
  B.n.nipalensis
    Himalayas to C Burma, India
  B.n.blighi
    Sri Lanka
Bubo sumatrana **(Malay Eagle Owl)**
  B.s.sumatrana
    S Burma, Malaysia, Sumatra
  B.s.strepitans
    Java, Bali, Borneo
Bubo shelleyi **(Shelley's Eagle Owl)**
    Liberia to S Cameroun
Bubo lacteus **(Verreaux's Eagle Owl)**
    Senegal to Ethiopia to Cape Province

Bubo coromandus (Dusky Eagle Owl)
  B.c.coromandus
    N & C India
  B.c.klossii
    S Burma, Malaysia
Bubo leucostictus (Akun Eagle Owl)
  Sierra Leone to Zaire
Bubo philippensis (Philippine Eagle Owl) 73.11
  B.p.philippensis
    Luzon, Cebu Is
  B.p.mindanensis
    Mindanao I
Bubo blakistoni (Blakiston's Fish Owl) 73.11
  B.b.piscivorus
    W Manchuria
  B.b.doerriesi
    NE Asia
  B.b.karafutonis
    Sakhalin
  B.b.blakistoni
    Hokkaido
Bubo zeylonensis (Brown Fish Owl) 73.11
  B.z.semenowi
    Israel to NW India
  B.z.leschenault
    India, Burma, Thailand
  B.z.zeylonensis
    Sri Lanka
  B.z.orientalis
    NE Burma to SE China, Indochina
Bubo flavipes (Tawny Fish Owl) 73.11
Himalayas to W China, Indochina
Bubo ketupu (Malay Fish Owl) 73.11
  B.k.ketupu
    Malaysia, Sumatra, Java, Borneo
  B.k.aagaardi
    S Assam to S Thailand & Vietnam
  B.k.pageli
    NE Borneo
  B.k.minor
    Nias I

SCOTOPELIA
Scotopelia peli (Pel's Fishing Owl)
  Senegal to Ethiopia & Cape Province
Scotopelia ussheri (Rufous Fishing Owl)
  Sierra Leone to Ghana
Scotopelia bouvieri (Vermiculated Fishing Owl)
  S Cameroun, Congo, N Angola

NYCTEA
Nyctea scandiaca (Snowy Owl)
  N Asia, N Canada, Holarctic region

CICCABA
Ciccaba virgata (Mottled Owl)
  C.v.tamaulipensis
    S Tamaulipas
  C.v.squamulata
    W Mexico

C.v.centralis
  S Mexico to W Panama
C.v.virgata
  E Panama to Venezuela & Ecuador, Trinidad
C.v.macconnelli
  the Guianas
C.v.superciliaris
  NC & NE Brazil
C.v.minuscula
  W Colombia
C.v.borelliana
  S Brazil, Paraguay, NE Argentina
Ciccaba nigrolineata (Black & White Owl)
  S Mexico to W Ecuador
Ciccaba huhula (Black-banded Owl)
  the Guianas to C & S Brazil
Ciccaba albitarsus (Rufous-banded Owl)
  C.a.albitarsus
    Colombia, Ecuador, Venezuela
  C.a.tertia
    Bolivia

STRIX
Strix woodfordii (African Wood Owl) 73.8
  S.w.umbrina
    Ethiopia
  S.w.nigricantior
    Kenya, Tanzania
  S.w.nuchalis
    Sierra Leone to Sudan & N Angola
  S.w.woodfordi
    Zambia & Malawi to Cape Province
Strix seloputo (Spotted Wood Owl)
  S.s.seloputo
    S Burma to S Indochina, Malaysia, Java
  S.s.baweana
    Bawean I
  S.s.wiepkeni
    Palawan I
Strix ocellata (Mottled Wood Owl)
  S.o.ocellata
    Himalayas, N India
  S.o.grandis
    W India
  S.o.grisescens
    NC India
Strix leptogrammica (Brown Wood Owl)
  S.l.newarensis
    Himalayas, N Burma, N Thailand
  S.l.indranee
    S India
  S.l.connectens
    C India
  S.l.ochrogenys
    Sri Lanka
  S.l.maingayi
    S Burma, S Thailand, Malaysia
  S.l.ticehursti
    SE China, N Indochina
  S.l.laotiana
    S Indochina

S.l.caligata
  Taiwan, Hainan I
S.l.myrtha
  Sumatra
S.l.nyctiphasma
  Banjak I
S.l.niasensis
  Nias I
S.l.chaseni
  Billiton I
S.l.bartelsi
  W & C Java
S.l.vaga
  N Borneo
S.l.leptogrammica
  S & C Borneo
Strix aluco (Eurasian Tawny Owl)
S.a.sylvatica
  Britain, W Europe
S.a.mauritanica
  N Africa, Syria, Israel
S.a.aluco
  Scandinavia, C & E Europe
S.a.volhyniae
  SW Russia
S.a.siberiae
  E Russia, W Siberia
S.a.willkonskii
  Caucasus
S.a.obscurata
  S Russia, N Iran
S.a.sanctinicolae
  Iraq, W & SW Iran
S.a.haermsi
  Russian Turkestan
S.a.biddulphi
  Pakistan, NW India
S.a.nivicola
  Himalayas, S & W China
S.a.yamadae
  S Taiwan
S.a.ma
  NE China, Korea
Strix butleri (Hume's Owl)
  SW Asia
Strix varia (Barred Owl)
S.v.varia
  S Canada, EC USA
S.v.georgica
  S & SE USA
S.v.helveola
  SC Texas
S.v.sartorii
  N & C Mexico
Strix fulvescens (Fulvous Owl)   73.9
  S Mexico, W Guatemala, Honduras
Strix occidentalis (Spotted Owl)
S.o.caurina
  S Brit.Columbia to N California
S.o.occidentalis
  S California
S.o.lucida
  SW USA to C Mexico

Strix uralensis (Ural Owl)
S.u.liturata
  N Scandinavia to C Russia
S.u.uralensis
  E Russia to W Siberia
S.u.yenisseensis
  C Siberia
S.u.daurica
  Lake Baikal to W Amurland
S.u.nikolskii
  Sea of Okhotsk too E Amurland
S.u.tatibanai
  Sakhalin I
S.u.coreensis
  SE Manchuria, Korea, Hokkaido I
S.u.hondoensis
  N Honshu I
S.u.momiyamae
  C Honshu I
S.u.fuscescens
  S Honshu, Kyushu Is
S.u.davidi
  W Szechwan   73.11
Strix nebulosa (Great Grey Owl)
S.n.nebulosa
  N Nth America
S.n.lapponica
  N Europe, N Asia, Sakhalin I
S.n.elisabethae
  N Mongolia
Strix hylophila (Rusty Barred Owl)
  Brazil, Paraguay, N Argentina
Strix rufipes (Rufous-legged Owl)
S.r.chacoensis
  Paraguay, N Argentina
S.r.sanborni
  Chiloe I
S.r.rufipes
  S Chile, S Argentina

SURNIA
Surnia ulula (Hawk Owl)
S.u.ulula
  N Europe, N Asia
S.u.tianschanica
  Tien Shan
S.u.caparoch
  W & C Canada, N USA

GLAUCIDIUM
Glaucidium brodiei (Collared Owlet)
G.b.brodiei
  Himalayas to N Indochina & Malaysia
G.b.pardalotum
  Taiwan
G.b.peritum
  Sumatra
G.b.borneense
  Borneo
Glaucidium passerinum (Eurasian Pygmy Owl)

G.p.passerinum
 N Europe, W Asia
G.p.orientale
 E Siberia, Manchuria
Glaucidium perlatum (Pearl-spotted Owlet)
G.p.perlatum
 Senegal to Cameroun
G.p.kilimense
 E & NE Africa
G.p.licua
 Sthn Africa
Glaucidium gnoma (Northern Pygmy Owl)
G.g.grinnelli
 SE Alaska to N California
G.g.swarthi
 Vancouver I
G.g.californicum
 C Brit.Columbia to S California
G.g.pinicola
 WC USA
G.g.hoskinsii
 Baja California
G.g.gnoma
 N & C Mexico
G.g.cobanense
 Guatemala
Glaucidium minutissimum (Least Pygmy Owl)
G.m.oberholseri
 C & S Sinaloa
G.m.palmarum
 W Mexico
G.m.griseiceps
 E Guatemala, Belize, E Honduras
G.m.rarum
 Costa Rica, Panama
G.m.minutissimum
 Guyana, Surinam, Brazil
G.m.griscomi
 SW Morelos, NE Guerrero
G.m.occultum
 E Oaxaca, Chiapas
G.m.sanchezi
 S San Luis Potosi
Glaucidium brasilianum (Ferruginous Pygmy Owl)
G.b.cactorum
 S Arizona, W Mexico
G.b.ridgwayi
 S Texas to C Panama
G.b.medianum
 N Colombia
G.b.phaloenoides
 N Venezuela, Trinidad
G.b.olivaceum
 Mt Augun-tepui (Venezuela)
G.b.margaritae
 Margarita I
G.b.duidae
 Mt Duida (Venezuela)
G.b.ucayalae
 SE Colombia to Peru

G.b.brasilianum
 W & S Amazonia, NE Argentina
G.b.pallens
 E Bolivia, W Paraguay, NW Argentina
G.b.tucumanum
 W Argentina
Glaucidium nanum (Austral Pygmy Owl)   73.7
 S Chile, S Argentina
Glaucidium jardinii (Andean Pygmy Owl)
G.j.jardinii
 Colombia, Ecuador, Peru, Venezuela
G.j.costaricanum
 Costa Rica, Panama
Glaucidium siju (Cuban Pygmy Owl)
G.s.siju
 Cuba
G.s.vittatum
 Isle of Pines
Glaucidium tephronotum (Red-chested Owlet)
G.t.tephronotum
 Ghana
G.t.pycrafti
 S Cameroun
G.t.medje
 N Zaire
G.t.lukolelae
 C Zaire
G.t.kivuense
 E Zaire
G.t.elgonense
 Mt Elgon (Kenya)
Glaucidium sjostedti (Chestnut-backed Owlet)
 Cameroun to C Zaire
Glaucidium radiatum (Jungle Owlet)
G.r.radiatum
 India, Sri Lanka
G.r.malabaricum
 SW India
Glaucidium cuculoides (Cuckoo Owlet)
G.c.castanonotum
 Sri Lanka
G.c.cuculoides
 W Himalayas
G.c.rufescens
 NE India, N Burma
G.c.bruegeli
 S Burma, S Thailand
G.c.austerum
 NE Assam
G.c.delacouri
 N Indochina
G.c.deignani
 SE Thailand, S Indochina
G.c.whitelyi
 W, C & SE China, NE Vietnam
G.c.persimile
 Hainan I
Glaucidium castanopterum (Chestnut-winged Owlet)   73.11
 Java, Bali
Glaucidium capense (African Barred Owlet)

G.c.etchecopari
  Liberia, Ivory Coast
G.c.ngamiense
  S Zaire, NE Angola, W Tanzania, W
  Mozambique
G.c.capense
  Angola, Sthn Africa
Glaucidium castaneum (Ituri Owlet)   73.3
  NE Zaire, SW Uganda
Glaucidium albertinum (Prigogine's Owlet)
                                            73.3
  NE Zaire
Glaucidium scheffleri (Eastern Barred Owlet)
                                            73.4
  G.s.scheffleri
    SE Kenya, NE Tanzania
  G.s.clanceyi
    S Tanzania

XENOGLAUX
Xenoglaux loweryi (Long-whiskered Owlet)
  N Peru

MICRATHENE
Micrathene whitneyi (Elf Owl)
  M.w.whitneyi
    SW USA, NW Mexico
  M.w.idonea
    Texas, C Mexico
  M.w.sanfordi
    Baja California
  M.w.graysoni
    Socorro I

ATHENE
Athene noctua (Little Owl)
  A.n.vidalii
    W Europe
  A.n.noctua
    C Europe
  A.n.sarda
    Sardinia
  A.n.indigena
    N Iran, Greece, S Russia
  A.n.glaux
    N Africa
  A.n.saharae
    S Morocco to N Saudi Arabia
  A.n.solitudinis
    C Sahara
  A.n.lilith
    Syria, Israel
  A.n.bactriana
    Transcaspia to Pakistan
  A.n.orientalis
    NE Russian & Chinese Turkestan
  A.n.ludlowi
    Tibet
  A.n.impasta
    Kokonor, W Kansu
  A.n.plumipes
    EC Asia

A.n.spilogastra
  E Sudan, NE Ethiopia
A.n.somaliensis
  E Ethiopia, Somalia
Athene brama (Spotted Owlet)
  A.b.albida
    Iran
  A.b.indica
    N & C India
  A.b.brama
    S India
  A.b.pulchra
    Burma to SW Indochina
Athene blewitti (Forest Owlet)
  C India
Athene cunicularia (Burrowing Owl)   73.11
  A.c.hypugaea
    SW Canada to W Mexico
  A.c.rostrata
    Clarion I
  A.c.floridana
    C & S Florida, Bahama Is
  A.c.troglodytes
    Hispaniola, Gonave I
  A.c.arubensis
    Aruba I
  A.c.brachyptera
    Margarita I, N Venezuela
  A.c.minor
    S Guyana, S Surinam, NE Brazil
  A.c.carrikeri
    E Colombia
  A.c.tolimae
    W Colombia
  A.c.pichinchae
    W Ecuador
  A.c.punensis
    SW Ecuador, NW Peru
  A.c.intermedia
    W Peru
  A.c.apurensis
    NC Venezuela
  A.c.juninensis
    C Peru, W Bolivia
  A.c.boliviana
    Bolivia
  A.c.nanodes
    SW Peru
  A.c.grallaria
    E & C Brazil
  A.c.cunicularia
    S Bolivia & S Brazil to Tierra del Fuego
  A.c.partridgei
    Corrientes, Argentina

AEGOLIUS
Aegolius funereus (Tengmalm's Owl)
  A.f.funereus
    N & C Europe, W Siberia
  A.f.caucasicus
    N Caucasus

A.f.sibiricus
NC & NE Asia
A.f.pallens
Tien Shan, Tarbagatai
A.f.jakutorum
C Siberia
A.f.beickianus
N Kansu
A.f.magnus
NE Siberia
A.f.richardsoni
N Canada to N USA
Aegolius acadicus (Saw-whet Owl)
A.a.acadicus
Canada, W USA, N Mexico
A.a.brooksi
Queen Charlotte Is
Aegolius ridgwayi (Unspotted Saw-whet Owl)
A.r.tacanensis
S Mexico
A.r.rostratus
Guatemala
A.r.ridgwayi
Costa Rica
Aegolius harrisii (Buff-fronted Owl)
A.h.harrisii
Colombia, Ecuador, Venezuela
A.h.iheringi
SE Brazil, Paraguay, N Argentina

UROGLAUX
Uroglaux dimorpha (Papuan Hawk Owl)
New Guinea, Japen I

NINOX
Ninox rufa (Rufous Hawk Owl)
N.r.humeralis
New Guinea, Waigeu I
N.r.aruensis
Aru Is
N.r.rufa
N Australia
N.r.queenslandica
E Queensland
Ninox strenua (Powerful Owl)
New South Wales, Victoria
Ninox connivens (Barking Owl)
N.c.rufostrigata
N Moluccas
N.c.assimilis
E New Guinea, Vulcan, Dampier Is
N.c.occidentalis
NW Australia, Nthn Territory
N.c.peninsularis
Cape York Peninsula
N.c.enigma
C Nth Queensland
N.c.addenda
SW Australia
N.c.connivens
S & E Australia

Ninox novaeseelandiae (Boobook Owl)
N.n.plesseni
Alor I
N.n.fusca
Timor I
N.n.cinnamomina
Babar I
N.n.moae
Moa, Leti, Roma Is
N.n.remigialis
Kai Is
N.n.pusilla
S New Guinea
N.n.ocellata
N Australia
N.n.marmorata
S & SW Australia
N.n.lurida
NE Queensland
N.n.boobook
E Australia
N.n.leucopsis
Tasmania
N.n.albaria
Lord Howe I
N.n.undulata
Norfolk I
N.n.venatica
North I, New Zealand
N.n.novaeseelandiae
South I, New Zealand
Ninox rudolfi (Sumba Boobook Owl)   73.6
Sumba I
Ninox scutulata (Brown Hawk Owl)
N.s.ussuriensis
NE Asia
N.s.scutulata
Japan, E China to Lesser Sunda Is
N.s.burmanica
S Assam to Malaysia & Indochina
N.s.lugubris
N & C India
N.s.hirsuta
S India, Sri Lanka
N.s.obscura
Andaman, Nicobar Is
N.s.malaccensis
S Malaysia, Sumatra, Bangka I
N.s.japonica
Sulawesi, Lesser Sunda Is
N.s.javanensis
W Java
N.s.borneensis
Borneo, N Natuna Is
N.s.randi
Philippine Is
Ninox affinis (Andaman Hawk Owl)
N.a.affinis
Andaman Is
N.a.isolata
Nicobar Is
N.a.rexpimenti
Gt Nicobar I

Ninox superciliaris **(Madagascar Hawk Owl)**
W Madagascar
Ninox philippensis **(Philippine Hawk Owl)**
N.p.philippensis
Luzon, Marinduque, Leyte Is
N.p.proxima
Ticao, Masbate Is
N.p.centralis
Panay, Guimaras, Negros, Siquijor Is
Ninox spilonota **(Spotted Hawk Owl)**
Mindoro, Tablas, Sibuyan Is
Ninox spilocephala **(Tweedale's Hawk Owl)**
N.s.mindorensis
Mindoro I
N.s.spilocephala
Mindanao, Basilan Is
N.s.reyi
Jolo, Bongao Is
N.s.everetti
Siasi I
Ninox ochracea **(Ochre-bellied Hawk Owl)**
73.6
Sulawesi
Ninox squamipila **(Moluccan Hawk Owl)**
N.s.hypogramma
N Moluccas
N.s.hantu
Buru I
N.s.squamipila
Seram I
N.s.forbesi
Tanimbar Is
N.s.natalis
Christmas I
Ninox theomacha **(Jungle Hawk Owl)**
N.t.hoedtii
Waigeu, Misol Is
N.t.goldii
d'Entrecasteaux Arch
N.t.theomacha
New Guinea
N.t.rosseliana
Louisiade Arch
Ninox meeki **(Admiralty Islands Hawk Owl)**
Admiralty Is
Ninox punctulata **(Speckled Hawk Owl)**
Sulawesi
Ninox variegata **(Bismarck Hawk Owl)**   73.11
N.v.superior
New Hanover
N.v.variegata
New Ireland
Ninox odiosa **(New Britain Hawk Owl)**
New Britain
Ninox jacquinoti **(Solomon Islands Hawk Owl)**
N.j.eichhorni
Bougainville, Choiseul Is
N.j.mono
Mono I
N.j.jacquinoti
Ysabel, St George Is
N.j.floridae
Florida I

N.j.granti
Guadalcanal I
N.j.malaitae
Malaita I
N.j.roseoaxillaris
San Cristobal I

SCELOGLAUX
Sceloglaux albifacies **(White-faced Owl)**
South I, New Zealand

ASIO
Asio clamator **(Striped Owl)**   73.11
A.c.clamator
SE Mexico to C Sth America
A.c.oberi
Tobago I
A.c.midas
S Brazil, Paraguay, Uruguay, N Argentina
Asio otus **(Long-eared Owl)**
A.o.otus
Europe, Asia, NW Africa
A.o.canariensis
Canary Is
A.o.tuftsi
Canada
A.o.wilsonianus
S Canada, W & C USA
Asio stygius **(Stygian Owl)**
A.s.lambi
NW Mexico
A.s.robustus
E Mexico, Guatemala, Nicaragua
A.s.siguapa
Cuba, Isle of Pines
A.s.noctipetens
Hispaniola, Gonave I
A.s.stygius
C & S Brazil
A.s.barberoi
Paraguay, N Argentina
Asio abyssinicus **(Abyssinian Long-eared Owl)**
A.a.abyssinicus
Ethiopia
A.a.graueri
E Zaire, Mt Kenya
Asio madagascariensis **(Madagascar Long-eared Owl)**
Madagascar
Asio flammeus **(Short-eared Owl)**
A.f.flammeus
Europe, N Asia, N Africa, Nth America
A.f.bogotensis
Colombia, Ecuador
A.f.pallidicaudus
Venezuela
A.f.suinda
S Peru, S Brazil to Tierra del Fuego
A.f.sanfordi
Falkland Is
A.f.sandwichensis
Hawaiian Is

A.f.ponapensis
  Ponapé I
A.f.domingensis
  Hispaniola
A.f.portoricensis
  Puerto Rico
A.f.galapagoensis
  Galapagos Is
Asio capensis (African Marsh Owl)
A.c.tingitanus
  NW Africa, Senegal to Cameroun
A.c.capensis
  Ethiopia to Angola & Cape Province
A.c.hova
  Madagascar

PSEUDOSCOPS
Pseudoscops grammicus (Jamaican Owl)
  Jamaica

NESASIO
Nesasio solomonensis (Fearful Owl)
  Bougainville, Choiseul, Ysabel Is

CAPRIMULGIFORMES

74. STEATORNITHIDAE (OILBIRD)

STEATORNIS
Steatornis caripensis (Oilbird)
  Peru, Ecuador to the Guianas, Trinidad

75. PODARGIDAE (FROGMOUTHS)

PODARGUS
Podargus strigoides (Tawny Frogmouth)
P.s.phalaenoides
  NW Australia, Nthn Territory, Melville I
P.s.lilae
  Groote Eylandt I
P.s.gouldi
  W Cape York Peninsula
P.s.cornwalli
  E Queensland
P.s.brachypterus
  NW Victoria, C Australia
P.s.strigoides
  SE Queensland, New South Wales
P.s.victoriae
  S New South Wales, E South Australia
P.s.cuvieri
  Tasmania
Podargus papuensis (Papuan Frogmouth)
  New Guinea & islands, Cape York
  Peninsula
Podargus ocellatus (Marbled Frogmouth)
P.o.ocellatus
  New Guinea & islands
P.o.marmoratus
  Cape York Peninsula
P.o.intermedius
  Triobriand, Fergusson, Goodenough Is

P.o.meeki
  Tagula I
P.o.inexpectatus
  Solomon Is

BATRACHOSTOMUS
Batrachostomus auritus (Large Frogmouth)
  Malaysia, Sumatra, Borneo
Batrachostomus harterti (Dulit Frogmouth)
  C Borneo
Batrachostomus septimus (Philippine Frogmouth)
B.s.microrhynchus
  N Luzon I
B.s.menagei
  Panay, Negros Is
B.s.septimus
  Mindanao, Basilan Is
Batrachostomus stellatus (Gould's Frogmouth)
  Malaysia, Sumatra, Borneo
Batrachostomus moniliger (Ceylon Frogmouth)
  SW India, Sri Lanka
Batrachostomus hodgsoni (Hodgson's Frogmouth)
B.h.hodgsoni
  Sikkim to Assam & N Burma
B.h.indochinae
  C Burma to Indochina
Batrachostomus poliolophus (Pale-headed Frogmouth)
  Sumatra
Batrachostomus mixtus (Sharpe's Frogmouth)
  Borneo
Batrachostomus javensis (Javan Frogmouth)
B.j.continentalis
  S Burma, SE Thailand
B.j.javensis
  W & C Java
B.j.chaseni
  Palawan, Banguey Is
B.j.affinis
  Malaysia, Sumatra, Borneo
Batrachostomus cornutus (Bornean Frogmouth)
B.c.cornutus
  Sumatra, Borneo, Bangka, Billiton Is
B.c.longicaudatus
  Kangean I

76. NYCTIBIIDAE (POTOOS)

NYCTIBIUS
Nyctibius grandis (Great Potoo)
  Panama to Peru & S Brazil
Nyctibius aethereus (Long-tailed Potoo)
N.a.chocoensis
  W Colombia
N.a.longicaudatus
  E Ecuador, E Peru to Guyana

N.a.aethereus
SE Brazil, Paraguay
Nyctibius griseus (Common Potoo)
N.g.mexicanus
S Mexico to Honduras
N.g.costaricensis
Nicaragua to W Panama
N.g.panamensis
C Panama to Peru
N.g.cornutus
C & S Brazil, Paraguay, N Argentina
N.g.griseus
N Brazil, the Guianas, Trinidad
N.g.jamaicensis
Jamaica
N.g.abbotti
Hispaniola, Gonave I
Nyctibius leucopterus (White-winged Potoo)
N.l.maculosus
E Colombia, E Ecuador
N.l.leucopterus
E Brazil
Nyctibius bracteatus (Rufous Potoo)
Guyana, S Colombia, E Ecuador, E Peru

77. AEGOTHELIDAE (OWLET-NIGHTJARS)

AEGOTHELES
Aegotheles crinifrons (Halmahera Owlet-Nightjar)
Halmahera, Batjan Is
Aegotheles insignis (Large Owlet-Nightjar)
A.i.insignis
NW & N New Guinea
A.i.tatei
S New Guinea
A.i.pulcher
SE New Guinea
Aegotheles cristatus (Owlet-Nightjar)
A.c.major
S New Guinea
A.c.leucogaster
N Australia
A.c.cristatus
C & S Australia
A.c.tasmanicus
Tasmania
Aegotheles savesi (New Caledonian Owlet-Nightjar)
New Caledonia
Aegotheles bennettii (Barred Owlet-Nightjar)
A.b.affinis
NW New Guinea
A.b.wiedenfeldi
N New Guinea
A.b.terborghi
EC New Guinea
A.b.bennettii
SE New Guinea
A.b.plumiferus
Fergusson, Goodenough Is
Aegotheles wallacii (Wallace's Owlet-Nightjar)

A.w.wallacii
W New Guinea, Aru Is
A.w.gigas
WC New Guinea
A.w.manni
SW New Guinea
Aegotheles albertisi (Mountain Owlet-Nightjar)
A.a.albertisi
NW New Guinea
A.a.wondiwoi
C New Guinea
A.a.salvadorii
C & S New Guinea
Aegotheles archboldi (Eastern Mountain Owlet-Nightjar)
EC New Guinea

78. CAPRIMULGIDAE (NIGHTJARS)

CHORDEILINAE

LUROCALIS
Lurocalis semitorquatus (Semi-collared Nighthawk)
L.s.stonei
Nicaragua
L.s.noctivagus
Panama
L.s.semitorquatus
N Colombia to the Guianas, N Brazil
L.s.schaeferi
NC Venezuela
L.s.nattereri
C & S Brazil
L.s.rufiventris
W Venezuela, E Colombia to Peru

CHORDEILES
Chordeiles pusillus (Least Nighthawk) 78.5
C.p.septentrionalis
E Venezuela, Guyana
C.p.pusillus
CE & S Brazil
C.p.saturatus
W Brazil
C.p.novaesi
Maranhao (E Brazil)
C.p.xerophilus
Paraiba (E Brazil)
C.p.esmeraldae
S Venezuela, NW Brazil
Chordeiles rupestris (Sand-coloured Nighthawk)
C.r.xyostictus
C Colombia
C.r.rupestris
Upper Amazonia
Chordeiles acutipennis (Lesser Nighthawk) 78.2
C.a.texensis
SW USA >> C America

C.a.littoralis
SC Mexico to Costa Rica
C.a.micromeris
N Yucatan
C.a.acutipennis
N Sth America
C.a.aequatorialis
W Ecuador
C.a.exilis
W Peru
C.a.crissalis
C Colombia
Chordeiles minor (Common Nighthawk)
C.m.minor
Canada, C & E USA >> S Sth America
C.m.hesperis
SW Canada, W USA >> C Sth America
C.m.sennetti
NW USA >> C Sth America
C.m.howelli
WC USA >> C Sth America
C.m.henryi
SW USA >> C Sth America
C.m.aserriensis
SE Texas >> C Sth America
C.m.chapmani
SE USA >> C Sth America
C.m.panamensis
Panama
Chordeiles gundlachii (Antillean Nighthawk)
78.7
C.g.vicinus
Bahama Is
C.g.gundlachii
Cuba, Jamaica, Puerto Rico

NYCTIPROGNE
**Nyctiprogne leucopyga (Band-tailed
Nighthawk)**
N.l.exigua
E Colombia, Venzuela
N.l.pallida
WC Venezuela
N.l.majuscula
C Brazil
N.l.leucopyga
E Venezuela, the Guianas, E & S Brazil
N.l.latifascia
C Venezuela

PODAGER
Podager nacunda (Nacunda Nighthawk)
P.n.minor
N & NE Sth America
P.n.nacunda
E Peru & C Brazil to Patagonia

CAPRIMULGINAE

EUROSTOPODUS
Eurostopodus argus (Spotted Nightjar)   78.4

Babar, Romang Is
**Eurostopodus mystacalis (White-throated
Nightjar)**
E.m.mystacalis
E Australia, New Guinea
E.m.nigripennis
Solomon Is
E.m.exul
New Caledonia
E.m.harterti
NW Australia
E.m.gilberti
Groote Eylandt
E.m.guttatus
E Australia, Aru Is
Eurostopodus diabolicus (Satanic Nightjar)
78.4
NE Sulawesi
Eurostopodus papuensis (Papuan Nightjar)
E.p.papuensis
Salawati I, W New Guinea
E.p.astrolabae
E New Guinea
**Eurostopodus archboldi (Archbold's
Nightjar)**
New Guinea
**Eurostopodus temminckii (Malaysian Eared
Nightjar)**
Malaysia, Sumatra, Borneo
**Eurostopodus macrotis (Great Eared
Nightjar)**
E.m.cerviniceps
Assam to N Malaysia, W China,
Indochina
E.m.bourdilloni
S India
E.m.macrotis
Luzon, Mindoro, Mindanao Is
E.m.jacobsoni
Simalur I
E.m.macropterus
Sulawesi

VELES
Veles binotatus (Brown Nightjar)
Ghana to E Cameroun

NYCTIDROMUS
Nyctidromus albicollis (Common Pauraque)
N.a.insularis
Tres Marias Is
N.a.merrilli
S Texas, E Mexico
N.a.yucatanensis
NW Mexico to Guatemala
N.a.albicollis
W Guatemala to Peru & E Brazil
N.a.gilvus
N Colombia
N.a.derbyanus
C & S Brazil, Paraguay

PHALAENOPTILUS
Phalaenoptilus nuttallii (Common Poorwill)
  P.n.nuttallii
    W & WC USA >> C Mexico
  P.n.californicus
    W California
  P.n.hueyi
    SE California, SW Arizona
  P.n.dickeyi
    S Baja California
  P.n.centralis
    C Mexico

SIPHONORHIS
Siphonorhis brewsteri (Least Pauraque)
  Hispaniola, Gonave I

OTOPHANES
Otophanes mcleodii (Eared Poorwill)
  O.m.mcleodii
    Chihuahua, Jalisco
  O.m.rayi
    Guerrero
Otophanes yucatanicus (Yucatan Poorwill)
  SE Mexico, N Guatemala

NYCTIPHRYNUS
Nyctiphrynus ocellatus (Ocellated Poorwill)
  N.o.lautus
    NE Nicaragua
  N.o.rosenbergi
    W Colombia, NW Ecuador
  N.o.ocellatus
    E Ecuador, C Brazil to NE Argentina

CAPRIMULGUS
Caprimulgus carolinensis (Chuck Will's Widow)
  EC & S USA >> C America
Caprimulgus rufus (Rufus Nightjar)
  C.r.minimus
    Panama to Venezuela
  C.r.rufus
    the Guianas, NE Brazil
  C.r.noctivigulus
    C Colombia
  C.r.rutilus
    S Brazil, Paraguay
Caprimulgus otiosus (St Lucian Nightjar) 78.7
  St Lucia
Caprimulgus cubanensis (Greater Antillean Nightjar)
  C.c.cubanensis
    Cuba, Isle of Pines
  C.c.ekmani
    Hispaniola
Caprimulgus sericocaudatus (Silky-tailed Nightjar)
  C.s.sericocaudatus
    Peru
  C.s.mengeli
    Upper Amazonia

Caprimulgus salvini (Tawny-collared Nightjar)
  E Mexico
Caprimulgus badius (Yucatan Tawny-collared Nightjar)
  Yucatan, Belize
Caprimulgus ridgwayi (Ridgway's Whippoorwill)
  C.r.ridgwayi
    W Mexico
  C.r.troglodytes
    Guatemala, Honduras
Caprimulgus vociferus (Whippoorwill)
  C.v.vociferus
    S Canada, E USA >> Honduras
  C.v.arizonae
    SW USA, N Mexico
  C.v.setosus
    E Mexico
  C.v.oaxacae
    SC Mexico
  C.v.chiapensis
    S Mexico, Guatemala
  C.v.vermiculatus
    Honduras, El Salvador
Caprimulgus noctitherus (Puerto Rican Nightjar) 78.7
  Puerto Rico
Caprimulgus saturatus (Dusky Nightjar)
  Costa Rica, W Panama
Caprimulgus longirostris (Band-winged Nightjar)
  C.l.ruficervix
    Colombia, Venezuela, Ecuador
  C.l.roraimae
    Mt Duida, Mt Roraima (Venezuela)
  C.l.decussatus
    W Peru
  C.l.atripunctatus
    Peru, Bolivia, N Chile
  C.l.bifasciatus
    C Chile
  C.l.longirostris
    Argentina
Caprimulgus cayennensis (White-tailed Nightjar)
  C.c.albicauda
    Costa Rica to N Colombia
  C.c.apertus
    W Colombia
  C.c.insularis
    Curacao, Bonaire, Margarita Is N Venezuela
  C.c.leopetes
    Trinidad, Tobago I
  C.c.cayennensis
    E Colombia, S Venezuela, the Guianas, N Brazil
Caprimulgus candicans (White-winged Nightjar)
  C Brazil, Paraguay
Caprimulgus maculicaudus (Spot-tailed Nightjar)

N & W Amazonia

Caprimulgus parvulus (Little Nightjar)
C.p.anthonyi
W Ecuador
C.p.heterurus
N Colombia
C.p.parvulus
E Peru to E Brazil & C Argentina

Caprimulgus maculosus (Cayenne Nightjar)
French Guiana

Caprimulgus nigrescens (Blackish Nightjar)
W & N Amazonia

Caprimulgus whitelyi (Roraiman Nightjar)
Mt Roraima (Venezuela)

Caprimulgus hirundinaceus (Pygmy Nightjar)
C.h.cearae
E Brazil
C.h.hirundinaceus
E Brazil

Caprimulgus ruficollis (Red-necked Nightjar)
C.r.ruficollis
Portugal, S Spain, Morocco
C.r.desertorum
Algeria, Tunisia >> S Sahara

Caprimulgus indicus (Jungle Nightjar)
C.i.hazarae
Himalayas, Burma, Malaysia
C.i.indicus
India
C.i.kelaarti
Sri Lanka
C.i.jotaka
NE Asia, N China, Japan >> Java, Borneo
C.i.phalaena
Palau Is

Caprimulgus europaeus (European Nightjar)
C.e.europaeus
N Europe, Russia >> C & S Africa
C.e.meridionalis
S Europe, N Africa, Caucasus >> W Africa
C.e.sarudnyi
W Siberia, C Asia
C.e.unwini
SW Asia, Iran, Afghan >> E Africa, NW India
C.e.plumipes
E Turkestan >> SW Africa

Caprimulgus mahrattensis (Sykes' Nightjar)
Afghanistan to NW India

Caprimulgus centralasicus (Vaurie's Nightjar)
W China

Caprimulgus nubicus (Nubian Nightjar)
C.n.tamaricis
Dead Sea to Aden
C.n.nubicus
N Sudan
C.n.torridus
Somalia to N Tanzania
C.n.taruensis
S Somalia, C & S Kenya
C.n.jonesi
Socotra I

Caprimulgus aegyptius (Egyptian Nightjar)
78.6
C.a.aegyptius
Sinai >> W Africa
C.a.arenicolor
Syria to Afghanistan >> NE Africa
C.a.saharae
N Sahara >> Senegal to Nigeria

Caprimulgus eximius (Golden Nightjar)
C.e.simplicior
N Niger, N Chad
C.e.eximius
W & N Sudan

Caprimulgus atripennis (Indian Long-tailed Nightjar)
C.a.albonotatus
N & NE India
C.a.atripennis
S India

Caprimulgus macrurus (Long-tailed Nightjar)
78.1.3
C.m.aequabilis
Sri Lanka
C.m.ambiguus
Burma, Thailand, S Indochina
C.m.bimaculatus
Malaysia, Sumatra
C.m.andamanicus
Andaman Is
C.m.macrurus
Java, Bali
C.m.hainanus
Hainan I
C.m.manillensis
Philippine Is
C.m.delacouri
Mindanao I
C.m.jungei
Sula Is
C.m.celebensis
Sulawesi
C.m.schlegelii
Lesser Sunda Is, Timor, Moluccas
C.m.yorki
New Britain, Aru Is, New Guinea N Australia
C.m.meeki
Tagula I

Caprimulgus pectoralis (African Dusky Nightjar)
78.6
Angola & Tanzania to Cape Province

Caprimulgus nigriscapularis (Black-shouldered Nightjar)
Guinea to E Sudan, Zaire & Uganda

Caprimulgus rufigena (Rufous-cheeked Nightjar)
C.r.damarensis
Namibia, S Angola
C.r.rufigena
Sthn Africa >> W Africa

Caprimulgus fraenatus (Sombre Nightjar)
Ethiopia to S Kenya
78.6

Caprimulgus donaldsoni (Donaldson Smith's Nightjar)

W Somalia & Kenya
Caprimulgus poliocephalus (Abyssinian Nightjar)
C.p.poliocephalus
Ethiopia to N Tanzania
C.p.guttifer
C Tanzania
C.p.koesteri
W Angola
Caprimulgus ruwenzorii (Ruwenzori Nightjar)
78.6
E Zaire, Rwanda
Caprimulgus asiaticus (Indian Nightjar)
C.a.aldabrensis
Aldabra
C.a.madagascariensis
Madagascar
C.a.asiaticus
India to S Indochina
C.a.eidos
Sri Lanka
C.a.siamensis
N Thailand
Caprimulgus natalensis (African White-tailed Nightjar)
C.n.accrae
Liberia to W Cameroun
C.n.chadensis
Chad to Sudan, N Zaire
C.n.gabonensis
Gabon to C Zaire
C.n.carpi
Caprivi Strip, Namibia
C.n.fulviventris
Angola
C.n.mpusa
Zambia
C.n.natalensis
Natal
Caprimulgus inornatus (Plain Nightjar) 78.6
Niger & Nigeria to Tanzania & Yemen
Caprimulgus stellatus (Star-spotted Nightjar)
78.6
Ethiopia, Somalia & Kenya
Caprimulgus ludovicianus (Ludovic's Nightjar)
SW Ethiopia
Caprimulgus monticolus (Franklin's Nightjar)
C.m.monticolus
India
C.m.burmanicus
E Himalayas to Thailand
C.m.amoyensis
SE China
C.m.stictomus
Indochina, Taiwan
Caprimulgus affinis (Allied Nightjar)
C.a.affinis
Sumatra, Borneo, Java
C.a.kasuidori
Savu, Sumba Is
C.a.griseatus
Luzon, Mindoro, Negros, Cebu Is

C.a.mindanensis
Mindanao I
C.a.propinquus
C & S Sulawesi
C.a.undulatus
Lesser Sunda Is
C.a.timorensis
Timor I
Caprimulgus tristigma (Freckled Nightjar)
C.t.sharpei
Senegal to S Sudan
C.t.pallidogriseus
Nigeria
C.t.tristigma
E Sudan & Ethiopia to S Kenya
C.t.lentiginosus
Angola to Tanzania & Transvaal
Caprimulgus concretus (Bonaparte's Nightjar)
Sumatra, Borneo, Billiton I
Caprimulgus pulchellus (Salvadori's Nightjar)
C.p.pulchellus
Sumatra
C.p.bartelsi
Java
Caprimulgus enarratus (Collared Nightjar)
NW & E Madagascar
Caprimulgus batesi (Bates' Nightjar)
S Cameroun to C Zaire
Caprimulgus prigoginei (Prigogine's Nightjar)
78.8
Lake Kivu dist., (E Zaire)

SCOTORNIS
Scotornis fossii (Gabon Nightjar) 78.6
S.f.fossii
Gabon
S.f.welwitschii
Zaire & Kenya to Natal
S.f.griseoplurus
Botswana
Scotornis climacurus (Long-tailed Nightjar)
S.c.climacurus
Senegal to Sudan
S.c.nigricans
W Sudan
S.c.sclateri
Sierra Leone to C Zaire
Scotornis clarus (Slender-tailed Nightjar) 78.6
Ethiopia to C Tanzania

MACRODIPTERYX
Macrodipteryx longipennis (Standard-winged Nightjar)
Senegal to Ethiopia
Macrodipteryx vexillarius (Pennant-winged Nightjar)
78.6
Angola to Transvaal >> Nigeria & Uganda

HYDROPSALIS
Hydropsalis climacocerca (Ladder-tailed Nightjar)

H.c.schomburgki
  E Venezuela, Guyana, Surinam
H.c.climacocerca
  Upper Amazonia
H.c.pallidior
  C Brazil
H.c.intercedens
  C Brazil
H.c.canescens
  C Brazil
Hydropsalis brasiliana (Scissor-tailed Nightjar)
H.b.brasiliana
  C & E Brazil
H.b.furcifera
  E Bolivia, S Brazil, Uruguay

UROPSALIS
Uropsalis segmentata (Swallow-tailed Nightjar)
U.s.segmentata
  Colombia, Ecuador, Peru, Bolivia
U.s.kalinowskii
  C Peru
Uropsalis lyra (Lyre-tailed Nightjar)
U.l.lyra
  Colombia, Ecuador, Venezuela
U.l.peruana
  Peru

MACROPSALIS
Macropsalis creagra (Long-trained Nightjar)
  SE Brazil

ELEOTHREPTUS
Eleothreptus anomalus (Sickle-winged Nightjar)
  Paraguay, N Argentina, SE Brazil

APODIFORMES

79. APODIDAE (SWIFTS)

CYPSELOIDINAE

CYPSELOIDES
Cypseloides fumigatus (Sooty Swift)
  E Panama to S Brazil
Cypseloides cherriei (Spot-fronted Swift)
  Costa Rica
Cypseloides cryptus (White-chinned Swift)
  Costa Rica, Guyana, Peru
Cypseloides lemosi (White-chested Swift)
  Cauca (Colombia)
Cypseloides major (Great Swift)
  S Bolivia, NW Argentina
Cypseloides phelpsi (Tepui Swift)
  S Venezuela
Cypseloides rutilus (Chestnut-collared Swift)
C.r.griseifrons
  W Mexico
C.r.brunnitorques
  SE Mexico to Peru

C.r.rutilus
  the Guianas, Trinidad

NEPHOECETES
Nephoecetes niger (Black Swift)
N.n.borealis
  SE Alaska to SW USA >> Mexico
N.n.costaricensis
  Honduras to Costa Rica
N.n.niger
  W Indies, Trinidad

AERORNIS
Aerornis senex (Great Dusky Swift)
  S Brazil, Paraguay, NE Argentina

STREPTOPROCNE
Streptoprocne zonaris (White-collared Swift)
S.z.mexicana
  S Mexico, Belize to El Salvador
S.z.pallidifrons
  Greater Antilles
S.z.albicincta
  Honduras to NW & C Sth America
S.z.altissima
  Colombia, Ecuador
S.z.zonaris
  S Brazil, Bolivia, W Argentina
Streptoprocne biscutatus (Biscutate Swift)
  E Brazil
Streptoprocne semicollaris (White-naped Swift)
  C Mexico

APODINAE

HYDROCHOUS
Hydrochous gigas (Waterfall Swift)   79.1
  Malaysia, Sumatra, Java

AERODRAMUS
Aerodramus infuscatus (Moluccan Swiftlet)
                                    79.2.3.5
A.i.sororum
  C, S & SE Sulawesi
A.i.infuscatus
  N Moluccas
A.i.ceramensis
  S Moluccas
Aerodramus spodiopygius (White-rumped Swiftlet)
A.s.eichhorni
  Bismarck Arch
A.s.reichenowi
  Guadalcanal I
A.s.terraereginae
  N Queensland
A.s.chillagoensis
  Queensland
A.s.leucopygius
  Loyalty Is, New Hebrides, New Guinea
A.s.assimilis
  Fiji Is
A.s.townsendi
  Tonga

A.s.spodiopygius
Samoa
Aerodramus hirundinaceus (Mountain Swiftlet)
A.h.baru
Japen I
A.h.hirundinaceus
New Guinea, Dampier I
A.h.excelsus
Snowy Mts (New Guinea)
Aerodramus fuciphagus (Edible-nest Swiftlet)
79.3
Andaman Is & Gtr Sunda Is to Timor I
Aerodramus francica (Grey-rumped Swiftlet)
Mauritius, Reunion
Aerodramus germani (Oustalet's Swiftlet)
A.g.germani
Malaysia, Indochina, N Borneo
A.g.amechanus
Anamba Is
Aerodramus amelis (Philippine Swiftlet)
A.a.amelis
Luzon, Cebu, Mindanao Is
A.a.perplexus
E Borneo islands
Aerodramus vanikorensis (Uniform Swiftlet)
79.3

A.v.aenigma
NC, SE, SC Sulawesi, Muna I
A.v.heinrichi
S Sulawesi
A.v.moluccarum
S Moluccas, Kai Is
A.v.coultasi
Admiralty Is
A.v.lihirensis
Lihir, St Matthias Is
A.v.waigeuensis
Waigeu, Morotai, Halmahera Is
A.v.steini
Numfor I
A.v.granti
S & E New Guinea
A.v.tagulae
Louisiade Arch, Tagula I
A.v.yorki
Cape York peninsula
A.v.vanikorensis
Solomon Is, Santa Cruz, New Hebrides, New Caledonia
Aerodramus salangana (Mossy Swiftlet)
A.s.salangana
India, China, Indochina
A.s.natunae
N Borneo, Natuna Is
Aerodramus inquietus (Caroline Swiftlet)
A.i.rukensis
Caroline Is, Yap, Truk Is
A.i.ponapensis
Ponapé
A.i.inquietus
Kusaie
Aerodramus bartschi (Mariana Swiftlet)

A.b.bartschi
Guan
A.b.pelewensis
Palau
Aerodramus mearnsi (Brown-rumped Swiftlet)
Luzon, Mindoro, Panay, Cebu, Mindanao Is
Aerodramus orientalis (Guadalcanal Swiftlet)
Guadalcanal
Aerodramus papuensis (Idenburg River Swiftlet)
C New Guinea
Aerodramus nuditarsus (Schrader Mountain Swiftlet)
NC New Guinea
Aerodramus whiteheadi (Whitehead's Swiftlet)
A.w.tsubame
Palawan
A.w.whiteheadi
Philippine Is, New Guinea
A.w.origenis
Mindanao I
A.w.apoensis
Mt Apo
Aerodramus brevirostris (Himalayan Swiftlet)
A.b.innominatus
C & W China to N Vietnam, Malaysia
A.b.rogersi
NW Thailand, N Laos
A.b.brevirostris
Himalayas, N Burma, SE Tibet
Aerodramus maximus (Lowe's Swiftlet)
A.m.maximus
Malaysia, Anamba Is
A.m.lowi
Sumatra, N & W Borneo
A.m.tichelmani
SE Borneo
A.m.palawanensis
Palawan
A.m.vulcanorum
Java
Aerodramus elaphra (Seychelles Cave Swiftlet)
Seychelles
Aerodramus unicolor (Indian Edible-nest Swiftlet) SW India, Sri Lanka
Aerodramus leucophaeus (Tahitian Swiftlet)
Society Is
Aerodramus ocista (Marquesan Swiftlet)
Marquesas Is
Aerodramus sawtelli (Cook Islands Swiftlet)
Cook Is

COLLOCALIA
Collocalia esculenta (White-bellied Swiftlet)
79.3.5

C.e.affinis
Andaman, Nicobar Is
C.e.elachyptera
S Thailand, Mergui Arch

C.e.cyanoptila
Malaysia, E Sumatra, Billiton I, Borneo
C.e.oberholseri
W Sumatra, Nias, Mentawi Is
C.e.natalis
Christmas I
C.e.isonota
Luzon, Mindoro, Mindanao Is
C.e.bagobo
Mt Apo (Mindanao I)
C.e.sumbawae
Sumbawa, Flores, Sumba Is
C.e.minuta
Tanahdjampea, Kalao Is
C.e.perneglecta
Alor to Damar Is
C.e.neglecta
Timor I
C.e.spilura
N Moluccas
C.e.esculenta
S Moluccas, S Sulawesi, New Guinea
C.e.manadensis
N Sulawesi, Talaut I
C.e.erwini
S New Guinea
C.e.stresemanni
Admiralty Is, Bismarck Arch
C.e.becki
N & C Solomon Is
C.e.makirensis
San Cristobal I
C.e.desiderata
Rennell I
C.e.uropygialis
New Caledonia, New Hebrides
Collocalia marginata (Philippine Swiftlet)
C.m.marginata
Luzon, Mindoro, Masbate, Cebu, Bohol, Palawan Is
C.m.septentrionalis
Babuyan, Calayan, Camiguin (North) Is
Collocalia troglodytes (Pygmy Swiftlet)
Philippine, Palawan Is
Collocalia linchi (Linchi Swiftlet)   79.4
C.l.ripleyi
Sumatra, S Malaysia
C.l.linchi
Java, Bawean Is
C.l.dedii
Bali, Lombok I
C.l.dodgei
Mt Kinabalu

Schoutedenapus myioptilus (Scarce Swift)
S.m.poensis
Fernando Po I
S.m.myioptilus
Kenya to Malawi
S.m.chapini
E Zaire

Schoutedenapus schoutedeni (Schouteden's Swift)
E Zaire

Mearnsia picina (Philippine Spinetailed Swift)
Leyte, Cebu, Mindanao Is
Mearnsia novaeguineae (New Guinea Spinetailed Swift)
M.n.buergersi
New Guinea
M.n.novaeguineae
S New Guinea

Zoonavena grandidieri (Madagascar Spinetailed Swift)
Madagascar
Zoonavena thomensis (Sao Thome Spinetailed Swift)
Sao Thome
Zoonavena sylvatica (Indian White-rumped Spinetailed Swift)
India, Burma

Telacanthura ussheri (Mottle-throated Spinetailed Swift)
T.u.ussheri
Senegal to N Nigeria
T.u.sharpei
S Cameroun to E Zaire
T.u.stictilaema
SW Kenya to S Malawi
T.u.marwitzi
C Tanzania
T.u.benguellensis
Angola
Telacanthura melanopygia (Ituri Mottle-throated Spinetailed Swift)
N Zaire

Raphidura leucopygialis (White-rumped Spinetailed Swift)
S Burma to Sumatra, Java Borneo
Raphidura sabini (Sabine's Spinetailed Swift)
Sierra Leone to NE Zaire

Neafrapus cassini (Cassin's Spinetailed Swift)
S Cameroun to N Zaire
Neafrapus boehmi (Boehm's Spinetailed Swift)
W Angola to Mozambique & Tanzania

Hirundapus caudacuta (White-throated Spinetailed Swift)
H.c.caudacuta
NE Asia, Japan >> E China, Australia

H.c.nudipes
  Himalayas >> Java
H.c.bourreti
  Indochina
H.c.formosanus
  Taiwan
Hirundapus cochinchinensis (White-vented Spinetailed Swift)
  H.c.rupchandi
    Nepal
  H.c.cochinchinensis
    E Himalayas to Indochina, Malaysia, Java, Sumatra
Hirundapus gigantea (Brown Spinetailed Swift)
  H.g.indicus
    E India to Indochina, Andaman Is
  H.g.gigantea
    Malaysia, Sumatra, Java, Borneo, Palawan I
  H.g.dubius
    Luzon, Mindoro, Negros, Mindanao Is
  H.g.ernsti
    W Java
Hirundapus celebensis (Sulawesi Spinetailed Swift)
    N Sulawesi

CHAETURA
Chaetura spinicauda (Band-rumped Swift)
  C.s.fumosa
    W Costa Rica, Panama, N Colombia
  C.s.aetherodroma
    Panama
  C.s.latirostris
    Amacuro (Venezuela)
  C.s.spinicauda
    E Venezuela, the Guianas, N Brazil
  C.s.aethalea
    C Brazil
Chaetura martinica (Lesser Antillean Swift)
    Lesser Antilles
Chaetura cinereiventris (Grey-rumped Swift)
  C.c.phaeopygos
    E Nicaragua to Panama
  C.c.lawrencei
    Grenada, Trinidad, Tobago Is
  C.c.schistacea
    E Colombia, W Venezuela
  C.c.guianensis
    Guyana, E Venezuela
  C.c.occidentalis
    W Colombia, W Ecuador
  C.c.sclateri
    Upper Amazonia
  C.c.cinereiventris
    E Brazil
Chaetura egregia (Pale-rumped Swift)   79.6
    SE Peru, Bolivia, W Brazil
Chaetura pelagica (Chimney Swift)
    S Canada to S USA >> C Sth America
Chaetura vauxi (Vaux's Swift)

C.v.vauxi
  SW Canada to SW USA >> C America
C.v.tamaulipensis
  E Mexico
C.v.richmondi
  S Mexico to Costa Rica
C.v.ochropygia
  E Panama
C.v.gaumeri
  Yucatan peninsula, Cozumel I
C.v.aphanes
  N Venezuela
Chaetura chapmani (Chapman's Swift)
  C.c.chapmani
    French Guiana, Trinidad
  C.c.viridipennis
    C Brazil
Chaetura andrei (Ashy-tailed Swift)
  C.a.andrei
    C Venezuela
  C.a.meridionalis
    C Sth America >> Colombia
Chaetura brachyura (Short-tailed Swift)
  C.b.praevelox
    Grenada, St Vincent Is
  C.b.brachyura
    Upper Amazonia, the Guianas, S Brazil
  C.b.ocypetes
    Peru
  C.b.cinereocauda
    E Brazil

AERONAUTES
Aeronautes saxatilis (White-throated Swift)
  A.s.saxatilis
    SW Canada to SW USA >> Mexico
  A.s.nigrior
    Guatemala, El Salvador
Aeronautes montivagus (White-tipped Swift)
  A.m.montivagus
    N Venezuela, Peru, Bolivia
  A.m.tatei
    Mt Duida (Venezuela)
Aeronautes andecolus (Andean Swift)
  A.a.parvulus
    W Peru, N Chile
  A.a.peruvianus
    SE Peru
  A.a.andecolus
    Bolivia, W Argentina

TACHORNIS
Tachornis phoenicobia (Antillean Palm Swift)
  T.p.iradii
    Cuba, Isle of Pines
  T.p.phoenicobia
    Hispaniola, Jamaica
Tachornis furcata (Pygmy Swift)
  T.f.furcata
    NE Colombia, NW Venezuela
  T.f.nigrodorsalis
    W Venezuela
Tachornis squamata (Fork-tailed Palm Swift)

T.s.semota
E Peru, S Venezuela
T.s.squamata
Trinidad, the Guianas, C & E Brazil

PANYPTILA
Panyptila sanctihieronymi (Great Swallow-tailed Swift)
W Guatemala
Panyptila cayennensis (Lesser Swallow-tailed Swift)
P.c.veraecrucis
E Mexico
P.c.cayennensis
SE Nicaragua to N Sth America

CYPSIURUS
Cypsiurus batasiensis (Asian Palm Swift)
C.b.batasiensis
India, Sri Lanka
C.b.infumatus
Burma to Indochina, Malaysia, Sumatra, Borneo
C.b.bartelsorum
Java, Bali
C.b.pallidior
Philippine Is
Cypsiurus parvus (African Palm Swift)
C.p.parvus
Senegal to N Ethiopia
C.p.brachypterus
Sierra Leone to Angola & S Zaire
C.p.myochrous
E Ethiopia to S Malawi
C.p.gracilis
Madagascar
C.p.hyphaenes
N Namibia, N Botswana
C.p.celer
S Mozambique

TACHYMARPTIS
Tachymarptis melba (Alpine Swift) 79.7
T.m.melba
S Europe to Himalayas >> N Africa
T.m.tuneti
N Africa, Israel to Iran
T.m.archeri
Somalia
T.m.maximus
Mt Ruwenzori (Zaire)
T.m.africanus
E & S Africa
T.m.marjoriae
Namibia
T.m.willsi
Madagascar
T.m.bakeri
S India, Sri Lanka
Tachymarptis aequatorialis (Mottled Swift) 79.7
T.a.aequatorialis
Cameroun & Ethiopia to Angola & Malawi

T.a.furensis
W Sudan
T.a.gelidus
W Zimbabwe
T.a.lowei
Sierra Leone

APUS
Apus alexandri (Alexander's Swift)
Cape Verde Is
Apus barbatus (African Black Swift)
A.b.barbatus
S Malawi to Cape Province
A.b.hollidayi
Zambia
A.b.oreobates
Zimbabwe, Mozambique
A.b.balstoni
Madagascar
A.b.mayottensis
Mayotte I
A.b.sladeniae
Fernando Po I, S Cameroun
A.b.serlei
Cameroun
A.b.roehli
Kivu area (E Zaire)
A.b.glanvillei
Sierra Leone
Apus berliozi (Berlioz' Swift)
A.b.berliozi
Socotra I
A.b.bensoni
N Kenya
Apus bradfieldi (Bradfield's Swift)
A.b.bradfieldi
Namibia
A.b.deserticola
N Sth Africa
Apus niansae (Nyanza Swift)
A.n.niansae
E Ethiopia to Malawi
A.n.somalicus
Somalia
Apus pallidus (Pallid Swift)
A.p.brehmorum
Madeira, Canary Is, SW Europe, C Sahara
A.p.illyricus
Yugoslavia, Cyprus
A.p.pallidus
Egypt, Israel to Iran & Pakistan
Apus apus (Eurasian Swift)
A.a.apus
W Europe to C Asia >> Africa
A.a.pekinensis
Middle East to N China >> India & E & S Africa
Apus unicolor (Plain Swift) 79.7
Madeira, W Canary Is
Apus acuticauda (Dark-backed Swift)
Nepal, Assam
Apus pacificus (Fork-tailed Swift) 79.8

A.p.pacificus
 NE Asia, China, Japan >> SE Asia, Australia
A.p.leuconyx
 Himalayas, N India
A.p.cooki
 C Burma, Malaysia to S China, N Indochina
A.p.kanoi
 Taiwan
Apus affinis (House Swift)
A.a.bannermani
 Sao Thome, Principé, Fernando Po Is
A.a.aerobates
 Gambia to Somalia & Cape Province
A.a.galilejensis
 N Africa, Middle East, Iran
A.a.theresae
 NW Cape Province
A.a.affinis
 India, E Africa
A.a.singalensis
 S India, Sri Lanka
A.a.nipalensis
 Nepal to N Assam
A.a.subfurcatus
 S China, Burma to Philippine Is, Borneo, Java Sumatra
Apus horus (Horus Swift)   79.7
A.h.horus
 Ethiopia to Zimbabwe
A.h.fuscobrunneus
 SW Angola
Apus caffer (White-rumped Swift)   79.7
 N Sudan to Sth Africa
Apus batesi (Bates' Black Swift)
 Cameroun to NE Zaire

80. HEMIPROCNIDAE (TREE SWIFTS)

HEMIPROCNE
Hemiprocne longipennis (Crested Tree Swift)
H.l.coronata
 India to Indochina
H.l.harterti
 S Burma, Malaysia, Sumatra, Borneo
H.l.perlonga
 Simalur I
H.l.ocyptera
 Nias I
H.l.thoa
 Batu, Pagi, Enggano Is
H.l.longipennis
 Java, Bali
H.l.wallacii
 Sula Is, Sulawesi
Hemiprocne mystacea (Whiskered Tree Swift)
H.m.confirmata
 Moluccas, Aru Is
H.m.mystacea
 W Papuan islands & New Guinea
H.m.aeroplanes
 Bismarck Arch

H.m.woodfordiana
 Solomon Is
Hemiprocne comata (Lesser Tree Swift)
H.c.comata
 Malaysia, Sumatra, Borneo
H.c.stresemanni
 Pagi Is
H.c.major
 Philippine Is
H.c.nakamurai
 Mindanao, Basilan Is

81. TROCHILIDAE (HUMMINGBIRDS)

DORYFERA
Doryfera johannae (Blue-fronted Lancebill)
D.j.johannae
 SE Colombia, E Ecuador, NE Peru
D.j.guianensis
 SE Venezuela, S Guyana
Doryfera ludoviciae (Green-fronted Lancebill)
D.l.veraguensis
 Costa Rica, W Panama
D.l.ludoviciae
 C Colombia, W Venezuela, C Peru
D.l.rectirostris
 C Ecuador
D.l.grisea
 NW Bolivia

ANDRODON
Androdon aequatorialis (Tooth-billed Hummingbird)
 E Panama, W Colombia, W Ecuador

RAMPHODON
Ramphodon naevius (Saw-billed Hermit)
 SE Brazil

GLAUCIS
Glaucis dohrnii (Hook-billed Hermit)
 E Brazil e?
Glaucis aenea (Bronzy Hermit)
 Nicaragua to NW Ecuador
Glaucis hirsuta (Rufous-breasted Hermit)
G.h.affinis
 E Panama to W Venezuela & NE Peru
G.h.insularum
 Grenada, Trinidad, Tobago Is
G.h.hirsuta
 N & E Venezuela, the Guianas, N Brazil, Bolivia

THRENETES
Threnetes niger (Sooty Barbthroat)
 French Guiana
Threnetes grzimeki (Black Barbthroat)
 Brazil
Threnetes leucurus (Pale-tailed Barbthroat)
T.l.loehkeni
 N Brazil
T.l.cervinicauda

E Colombia, E Ecuador, NE Peru
T.l.rufigastra
E Peru
T.l.leucurus
the Guianas, S Venezuela, N & C Brazil
T.l.medianus
NE Brazil
Threnetes ruckeri (Band-tailed Barbthroat)
T.r.ventosus
Nicaragua to W Panama
T.r.darienensis
E Panama, N Colombia
T.r.ruckeri
W Colombia, W Ecuador
T.r.venezuelensis
W Venezuela

PHAETHORNIS
Phaethornis yaruqui (White-whiskered Hermit)
P.y.sanctijohannis
W Colombia
P.y.yarqui
W Ecuador
Phaethornis guy (Green Hermit)
P.g.coruscus
Costa Rica, Panama, W Colombia
P.g.apicalis
C Colombia, E Ecuador, E Peru
P.g.guy
NE Venezuela, Trinidad
P.g.emiliae
WC Colombia
Phaethornis syrmatophorus (Tawny-bellied Hermit)
P.s.syrmatophorus
W Colombia, W Ecuador
P.s.columbianus
E Colombia, E Ecuador
P.s.huallagae
NE Peru
Phaethornis superciliosus (Long-tailed Hermit)
P.s.mexicanus
SW Mexico
P.s.veraecrucis
SE Mexico
P.s.longirostris
S Mexico to N Honduras
P.s.cephalus
S Honduras to W Panama
P.s.cassinii
E Panama, NW Colombia
P.s.moorei
E Colombia, E Ecuador, E Peru
P.s.baroni
W Ecuador
P.s.bolivianus
Bolivia
P.s.sussurus
N Colombia

P.s.saturatior
E Venezuela, NW Brazil
P.s.superciliosus
the Guianas, NE Brazil
P.s.muelleri
N Brazil
P.s.insignis
N Brazil
P.s.ochraceiventris
W Brazil
Phaethornis malaris (Great-billed Hermit)
P.m.ucayalii
Rio Ucayali (Peru)
P.m.insolitus
Venezuela
P.m.malaris
French Guiana
Phaethornis margarettae (Klabin Farm Long-tailed Hermit)
Brazil
Phaethornis eurynome (Scale-throated Hermit)
SW Brazil, Paraguay, NE Argentina
Phaethornis nigrirostris (Black-billed Hermit)
Brazil
Phaethornis hispidus (White-bearded Hermit)
Upper Amazonia
Phaethornis anthophilus (Pale-bellied Hermit)
P.a.hyalinus
Pearl Is (Panama)
P.a.anthophilus
N Colombia, W Venezuela
P.a.fuliginosus
S Colombia
Phaethornis koepckeae (Koepcke's Hermit)
Peru
Phaethornis bourcieri (Straight-billed Hermit)
P.b.whitelyi
SE Colombia, S Venezuela, the Guianas
P.b.bourcieri
E Ecuador, NE Peru, W Brazil
Phaethornis philippii (Needle-billed Hermit)
W Brazil
Phaethornis squalidus (Dusky-throated Hermit)
P.s.rupurumii
E Venezuela, Guyana, NW Brazil
P.s.maranhaoensis
EC Brazil
P.s.amazonicus
C Brazil
P.s.squalidus
SE Brazil
Phaethornis augusti (Sooty-capped Hermit)
P.a.augusti
E Colombia, N Venezuela
P.a.vicarius
EC Colombia
P.a.incanescens
SE Venezuela, S Guyana
Phaethornis pretrei (Planalto Hermit)
E Bolivia to SE Brazil

Phaethornis subochraceus (Buff-bellied Hermit)
NE Bolivia
Phaethornis nattereri (Cinnamon-throated Hermit)
C & E Brazil
Phaethornis gounellei (Broad-tipped Hermit)
C & E Brazil
Phaethornis ruber (Reddish Hermit)
P.r.episcopus
E & S Venezuela, Guyana
P.r.ruber
Surinam, French Guiana, N & C Brazil
P.r.nigricinctus
E Ecuador, W Brazil, NE Peru, E Bolivia
P.r.longipennis
EC Peru
Phaethornis stuarti (White-browed Hermit)
Bolivia
Phaethornis griseogularis (Grey-chinned Hermit)
P.g.griseogularis
Colombia, Ecuador, E Peru
P.g.zonura
N Peru
P.g.porcullae
W Peru
Phaethornis longuemareus (Little Hermit)
P.l.adolphi
SE Mexico
P.l.saturatus
Guatemala to C Panama
P.l.subrufescens
E Panama, W Colombia, W Ecuador
P.l.nelsoni
NW Colombia
P.l.striigularis
N & C Colombia
P.l.atrimentalis
E Ecuador, E Peru
P.l.ignobilis
San Esteban to Santa Lucia, Venezuela
P.l.imatacae
Bolivar (Venezuela)
P.l.longuemareus
French Guiana, Surinam, Trinidad
P.l.aethopyga
C Brazil
Phaethornis idaliae (Minute Hermit)
SE Brazil

EUTOXERES
Eutoxeres aquila (White-tipped Sicklebill)
E.a.salvini
E & SW Costa Rica, W Panama
E.a.munda
E Panama, W Colombia
E.a.aquila
E Colombia, E Ecuador
E.a.heterura
SW Colombia, W Ecuador
Eutoxeres condamini (Buff-tailed Sicklebill)

E.c.condamini
SE Colombia, E Ecuador
E.c.gracilis
E Peru

PHAEOCHROA
Phaeochroa cuvierii (Scaly-breasted Hummingbird)
P.c.roberti
E Guatemala to E Nicaragua
P.c.maculicauda
Costa Rica, W Panama
P.c.saturatior
Coiba Is (Panama)
P.c.cuvierii
E Panama
P.c.berlepschi
N Colombia

CAMPYLOPTERUS
Campylopterus curvipennis (Wedge-tailed Sabrewing)
C.c.curvipennis
SE Mexico
C.c.yucatanensis
Yucatan Peninsula
Campylopterus excellens (Long-tailed Sabrewing) 81.2
S Vera Cruz
Campylopterus largipennis (Grey-breasted Sabrewing)
C.l.largipennis
E Venezuela, the Guianas, NW Brazil
C.l.obscurus
NE Brazil
C.l.aequatorialis
Nthn Upper Amazonia
Campylopterus rufus (Rufous Sabrewing)
S Mexico, W Guatemala, El Salvador
Campylopterus hyperythrus (Rufous-breasted Sabrewing)
S Guyana, SE Venezuela
Campylopterus duidae (Buff-breasted Sabrewing)
C.d.duidae
Mt Duida (Venezuela)
C.d.guayquinimae
S Venezuela
Campylopterus hemileucurus (Violet Sabrewing)
C.h.hemileucurus
S Mexico to Nicaragua
C.h.mellitus
Costa Rica, W Panama
Campylopterus ensipennis (White-tailed Sabrewing)
NE Venezuela, Trinidad, Tobago I
Campylopterus falcatus (Lazuline Sabrewing)
E Ecuador, Colombia, W Venezuela
Campylopterus phainopeplus (Santa Marta Sabrewing)
N Colombia

APODIFORMES

Campylopterus villaviscensio (Napo Sabrewing)
E Ecuador

EUPETOMENA
Eupetomena macroura (Swallow-tailed Hummingbird)
E.m.macroura
the Guianas, Brazil, Paraguay
E.m.simoni
NE Brazil
E.m.hirundo
E Peru, NE Bolivia
E.m.boliviana
Beni (Bolivia)

FLORISUGA
Florisuga mellivora (White-necked Jacobin)
F.m.mellivora
C America, N Sth America
F.m.flabellifera
Tobago I

MELANOTROCHILUS
Melanotrochilus fuscus (Black Jacobin)
E Brazil

COLIBRI
Colibri delphinae (Brown Violetear)
Guatemala to Panama, N & W Sth America
Colibri thalassinus (Green Violetear)
C.t.thalassinus
C Mexico to Guatemala
C.t.minor
Honduras
C.t.cabanidis
Costa Rica, W Panama
C.t.cyanotus
Venezuela, Colombia to Peru
C.t.crissalis
Bolivia
Colibri coruscans (Sparkling Violetear)
C.c.coruscans
Venezuela, Colombia to NW Argentina
C.c.germanus
SE Venezuela, S Guyana
C.c.rostratus
S Venezuela
Colibri serrirostris (White-vented Violetear)
E Bolivia, S Brazil, N Argentina

ANTHRACOTHORAX
Anthracothorax viridigula (Green-throated Mango)
NE Venezuela, the Guianas, NE Brazil
Anthracothorax prevostii (Green-breasted Mango)
A.p.prevostii
C Mexico to Guatemala & Belize
A.p.gracilirostris
El Salvador to Costa Rica
A.p.hendersoni
Old Providence I

A.p.pinchoti
St Andrews I
A.p.viridicordatus
NW Venezuela
Anthracothorax nigricollis (Black-throated Mango)
A.n.nigricollis
E Panama, tropical S America
A.n.iridescens
W Colombia, W Ecuador
Anthracothorax veraguensis (Veraguas Mango)
W Panama
Anthracothorax dominicus (Antillean Mango)
A.d.dominicus
Hispaniola
A.d.aurulentus
Puerto Rico, St Thomas I
Anthracothorax viridis (Green Mango)
Puerto Rico
Anthracothorax mango (Jamaican Mango)
Jamaica

AVOCETTULA
Avocettula recurvirostris (Fiery-tailed Awlbill)
Guyana, French Guiana, NE Brazil, E Ecuador

EULAMPIS
Eulampis jugularis (Purple-throated Carib)
Lesser Antilles Is

SERICOTES
Sericotes holosericeus (Green-throated Carib)
S.h.holosericeus
E Puerto Rico, Virgin Is, Lesser Antilles
S.h.chlorolaemus
Grenada I

CHRYSOLAMPIS
Chrysolampis mosquitus (Ruby-Topaz Hummingbird)
N & E Sth America

ORTHORHYNCUS
Orthorhyncus cristatus (Antillean Crested Hummingbird)
O.c.exilis
Virgin Is & Lesser Antilles to St Lucia I
O.c.ornatus
St Vincent I
O.c.cristatus
Barbados I
O.c.emigrans
Union I to Grenada I

KLAIS
Klais guimeti (Violet-headed Hummingbird)
K.g.guimeti
Nicaragua to W Venezuela & E Ecuador
K.g.pallidiventris
E Peru, C Bolivia

ABEILLIA
Abeillia abeillei (Emerald-chinned Hummingbird)
A.a.abeillei
SE Mexico to N Honduras
A.a.aurea
S Honduras, N Nicaragua

STEPHANOXIS
Stephanoxis lalandi (Black-breasted Plovercrest)
S.l.lalandi
SE Brazil
S.l.loddigesii
S Brazil, Paraguay, NE Argentina

LOPHORNIS
Lophornis ornata (Tufted Coquette)
Venezuela, the Guianas, Trinidad
Lophornis gouldii (Dot-eared Coquette)
N & C Brazil
Lophornis magnifica (Frilled Coquette)
C & S Brazil
Lophornis delattrei (Rufous-crested Coquette)
L.d.brachylopha
SW Mexico
L.d.lessoni
W Costa Rica, Panama, C Colombia
L.d.delattrei
NE & C Peru, Bolivia
Lophornis stictolopha (Spangled Coquette)
W Venezuela, E Colombia, E Ecuador
Lophornis chalybea (Festive Coquette)
L.c.verreauxii
C Colombia to C Bolivia
L.c.klagesi
E Venezuela
L.c.chalybea
SE Brazil
Lophornis pavonina (Peacock Coquette)
L.p.punctigula
Venezuela
L.p.pavonina
SE Venezuela, S Guyana
L.p.duidae
Mt Duida (Venezuela)
Lophornis insignibarbis (Bearded Coquette)
Colombia

PAPHOSIA
Paphosia helenae (Black-crested Coquette)
C Mexico to E Costa Rica
Paphosia adorabilis (White-crested Coquette)
SW Costa Rica

POPELAIRIA
Popelairia popelairii (Wire-crested Thorntail)
E Colombia, E Ecuador, NE Peru
Popelairia langsdorffi (Black-bellied Thorntail)

P.l.melanosternon
E Ecuador, E Peru, W Brazil
P.l.langsdorffi
E Brazil
Popelairia letitiae (Coppery Thorntail)
Bolivia
Popelairia conversii (Green Thorntail)
Costa Rica to W Ecuador

DISCOSURA
Discosura longicauda (Racquet-tailed Coquette)
E Venezuela, Guyana, French Guiana, E Brazil

CHLORESTES
Chlorestes notatus (Blue-chinned Sapphire)
C.n.obsoletus
SE Colombia
C.n.notatus
Ecuador & W Brazil to Trinidad & Surinam
C.n.cyanogenys
C & E Brazil

CHLOROSTILBON
Chlorostilbon mellisugus (Blue-tailed Emerald)
C.m.mellisugus
Surinam, French Guiana, NE Brazil
C.m.subfurcatus
E & S Venezuela, Guyana, NW Brazil
C.m.duidae
S Venezuela
C.m.phoeopygus
Upper Amazonia
C.m.peruanus
Peru, E Bolivia
Chlorostilbon vitticeps (Simon's Emerald)
E Ecuador
Chlorostilbon aureoventris (Glittering-bellied Emerald)
C.a.pucherani
E Brazil
C.a.aureoventris
Bolivia, Paraguay, W Argentina
C.a.berlepschi
Uruguay, E Argentina
Chlorostilbon canivetii (Fork-tailed Emerald)
C.c.auriceps
C & W Mexico
C.c.canivetii
SE Mexico, Belize, N Guatemala
C.c.forficatus
Holbox, Cozumel Is
C.c.osberti
C & W Guatemala, El Salvador, Honduras
C.c.salvini
W Nicaragua, W Costa Rica
C.c.caribaeus
Netherlands Antilles, Trinidad, NE Venezuela

C.c.nitens
 N Colombia, NW Venezuela
C.c.nanus
 C Venezuela
Chlorostilbon assimilis (Garden Emerald)
 81.2
 SW Costa Rica, SW Panama, Pearl Is
Chlorostilbon ricordii (Cuban Emerald)
C.r.bracei
 Bahama Is
C.r.ricordii
 Cuba, Isle of Pines
Chlorostilbon swainsonii (Hispaniolan Emerald)
 Hispaniola, Gonave I
Chlorostilbon maugaeus (Puerto Rican Emerald)
 Puerto Rico
Chlorostilbon gibsoni (Red-billed Emerald)
C.g.gibsoni
 C Colombia
C.g.chrysogaster
 N Colombia
C.g.pumilus
 W Colombia, W Ecuador
C.g.melanorhynchus
 SC Colombia, NE Ecuador
Chlorostilbon russatus (Coppery Emerald)
 N Colombia, NW Venezuela
Chlorostilbon stenura (Narrow-tailed Emerald)
C.s.stenura
 Colombia, Venezuela
C.s.ignota
 N Venezuela
Chlorostilbon alice (Green-tailed Emerald)
 N Venezuela
Chlorostilbon poortmani (Short-tailed Emerald)
C.p.poortmani
 Colombia, NW Venezuela
C.p.euchloris
 Colombia

CYNANTHUS
Cynanthus sordidus (Dusky Hummingbird)
 W & S Mexico
Cynanthus latirostris (Broad-billed Hummingbird)
C.l.magicus
 SW USA, NW Mexico
C.l.latirostris
 EC Mexico
C.l.propinquus
 C Mexico
C.l.toroi
 W Mexico
C.l.doubledayi
 W Mexico (S of C.l.toroi)
C.l.lawrencei
 Tres Marias Is
C.l.nitida
 SW Mexico

CYANOPHAIA
Cyanophaia bicolor (Blue-headed Hummingbird)
 Guadelupe, Dominica, Martinique Is

THALURANIA
Thalurania colombica (Crowned Woodnymph) 81.2
T.c.ridgwayi
 W Jalisco (Mexico)
T.c.townsendi
 E Guatemala to SE Honduras
T.c.venusta
 Nicaragua to W Panama
T.c.subtropicalis
 E Panama, W Colombia
T.c.fannyi
 W Colombia
T.c.colombica
 N Colombia, W Venezuela
Thalurania furcata (Fork-tailed Woodnymph)
T.f.viridipectus
 SC Colombia
T.f.verticeps
 S Colombia, N Ecuador
T.f.hypochlora
 W Ecuador
T.f.nigrofasciata
 S Colombia, Ecuador, NW Brazil
T.f.taczanowskii
 NE Peru
T.f.jelskii
 E Peru
T.f.boliviana
 NE Bolivia
T.f.simoni
 W Brazil
T.f.rostrifera
 W Venezuela
T.f.orenocensis
 E Venezuela
T.f.fissilis
 E Venezuela
T.f.refulgens
 NE Venezuela
T.f.furcata
 the Guianas, NE Brazil
T.f.furcatoides
 NE Brazil
T.f.balzani
 W Brazil, E Bolivia
T.f.baeri
 NE C Brazil to SE Bolivia
T.f.eriphile
 E & SE Brazil, Paraguay
Thalurania watertonii (Long-tailed Woodnymph)
 Guyana, E Brazil
Thalurania glaucopis (Violet-capped Woodnymph)
 E & S Brazil, Uruguay, Paraguay
Thalurania lerchi (Lerch's Woodnymph)
 NC Colombia

NEOLESBIA
Neolesbia nehrkorni (Blue-tailed Sylph)
C Colombia

PANTERPE
Panterpe insignis (Fiery-throated Hummingbird)
P.i.insignis
Costa Rica, W Panama
P.i.eisenmanni
NW Costa Rica

DAMOPHILA
Damophila julie (Violet-bellied Hummingbird)
D.j.panamensis
Panama
D.j.julie
N Colombia
D.j.feliciana
E Ecuador

LEPIDOPYGA
Lepidopyga coeruleogularis (Sapphire-throated Hummingbird)
L.c.coeruleogularis
W Panama
L.c.confinis
NE Panama, NW Colombia
L.c.coelina
NE Colombia
Lepidopyga lilliae (Sapphire-bellied Hummingbird)
N Colombia
Lepidopyga goudoti (Shining Green Hummingbird)
L.g.goudoti
NC Colombia
L.g.zuliae
NW Venezuela
L.g.luminosa
N Colombia
L.g.phaeochroa
N Venezuela

HYLOCHARIS
Hylocharis xantusii (Black-fronted Hummingbird)
S Baja California
Hylocharis leucotis (White-eared Hummingbird)
H.l.borealis
SE Arizona, N Mexico
H.l.leucotis
C & S Mexico, Guatemala
H.l.pygmaea
El Salvador, Honduras, Nicaragua
Hylocharis eliciae (Blue-throated Goldentail)
S Mexico to W Panama
Hylocharis sapphirina (Rufous-throated Sapphire)
E Venezuela and the Guianas to N Argentina
Hylocharis cyanus (White-chinned Sapphire)

H.c.viridiventris
N Colombia to the Guianas, N Brazil
H.c.cyanus
E Brazil
H.c.rostrata
E Peru, NE Bolivia, W Brazil
H.c.conversa
Bolivia
Hylocharis pyropygia (Flame-rumped Sapphire)
NE Brazil
Hylocharis chrysura (Gilded Hummingbird)
E Bolivia, S Brazil, Uruguay, N Argentina
Hylocharis grayi (Blue-headed Sapphire)
H.g.grayi
C Colombia to N Ecuador
H.g.humboldtii
W Colombia to NW Ecuador

CHRYSURONIA
Chrysuronia oenone (Golden-tailed Sapphire)
C.o.oenone
N & E Venezuela, C Ecuador
C.o.longirostris
C Colombia
C.o.azurea
Ecuador
C.o.intermedia
Upper Amazon River
C.o.josephinae
E Peru, NE Bolivia

GOLDMANIA
Goldmania violiceps (Violet-capped Hummingbird)
E Panama

GOETHALSIA
Goethalsia bella (Pirre Hummingbird)
E Panama

TROCHILUS
Trochilus polytmus (Streamertail)
T.p.polytmus
Jamaica
T.p.scitulus
NE Jamaica

LEUCOCHLORIS
Leucochloris albicollis (White-throated Hummingbird)
SE Brazil, Paraguay, N Argentina

POLYTMUS
Polytmus guainumbi (White-tailed Goldenthroat)
P.g.doctus
Colombia
P.g.guainumbi
Venezuela, the Guianas, Trinidad
P.g.thaumantias
E & C Brazil, Paraguay, Bolivia

Polytmus millerii (Tepui Goldenthroat)
SE Venezuela
Polytmus theresiae (Green-tailed Goldenthroat)
P.t.theresiae
the Guianas, N Brazil
P.t.leucorrhous
N Peru, NW Brazil

LEUCIPPUS
Leucippus fallax (Buffy Hummingbird)
L.f.cervina
NE Colombia, NW Venezuela
L.f.richmondi
N Venezuela, Margarita I
L.f.fallax
N Venezuela
Leucippus baeri (Tumbes Hummingbird)
W Peru
Leucippus taczanowskii (Spot-throated Hummingbird)
L.t.fractus
N Peru
L.t.taczanowskii
C Peru
Leucippus chlorocercus (Olive-spotted Hummingbird)
E Peru

TAPHROSPILUS
Taphrospilus hypostictus (Many-spotted Hummingbird)
T.h.hypostictus
E Ecuador
T.h.peruvianus
E Peru, N Bolivia

AMAZILIA
Amazilia chionogaster (White-bellied Hummingbird)
A.c.chionogaster
N & C Peru
A.c.hypoleucus
Bolivia, NW Argentina
Amazilia viridicauda (Green & White Hummingbird)
Peru
Amazilia candida (White-bellied Emerald)
A.c.genini
EC Mexico
A.c.candida
SE Mexico to Nicaragua
A.c.pacifica
W Guatemala
Amazilia chionopectus (White-chested Emerald)
A.c.chionopectus
E Venezuela to Surinam, Trinidad
A.c.whitelyi
Guyana
A.c.orienticola
French Guiana
Amazilia versicolor (Versicoloured Emerald)

A.v.millerii
E Colombia, Venezuela, W Brazil
A.v.hollandi
E Venezuela
A.v.nitidifrons
NE Brazil
A.v.versicolor
E Bolivia, Paraguay
Amazilia luciae (Honduras Emerald)
Honduras
Amazilia fimbriata (Glittering-throated Emerald)
A.f.elegantissima
N Venezuela
A.f.obscuricauda
Venezuela
A.f.maculicauda
E Venezuela to Surinam
A.f.fimbriata
French Guiana, E Brazil
A.f.apicalis
E Colombia, W Venezuela
A.f.fluviatilis
S Colombia, E Ecuador
A.f.laeta
NE Peru, W Brazil
A.f.alia
C Brazil
A.f.nigricauda
Bolivia to E Brazil
A.f.tephrocephala
SE Brazil
Amazilia distans (Tachira Emerald)
W Venezuela
Amazilia lactea (Sapphire-spangled Emerald)
A.l.bartletti
E & SE Peru, N Bolivia
A.l.zimmeri
SE Venezuela
A.l.lactea
E Brazil
Amazilia decora (Charming Hummingbird)
81.2
SW Costa Rica, W Panama
Amazilia amabilis (Blue-chested Hummingbird)
A.a.costaricensis
E Nicaragua to C Panama
A.a.amabilis
E Panama, W Colombia, W Ecuador
Amazilia rosenbergi (Purple-chested Hummingbird)
W Colombia, NW Ecuador
Amazilia boucardi (Mangrove Hummingbird)
W Costa Rica
Amazilia franciae (Andean Emerald)
A.f.franciae
C Colombia
A.f.viridiceps
SW Colombia, W Ecuador
A.f.cyanocollis
N Peru

A.f.veneta
Colombia?
Amazilia leucogaster (Plain-bellied Emerald)
A.l.leucogaster
the Guianas, N Brazil
A.l.bahiae
E Brazil
**Amazilia cyanocephala (Red-billed
Azurecrown)**
A.c.cyanocephala
SE Mexico, NW Guatemala
A.c.guatemalensis
Guatemala to N Nicaragua
**Amazilia microrhyncha (Small-billed
Azurecrown)**
Honduras?
**Amazilia cyanifrons (Indigo-capped
Hummingbird)**
A.c.alfaroana
Costa Rica
A.c.cyanifrons
N Colombia
Amazilia beryllina (Berylline Hummingbird)
A.b.viola
N & W Mexico
A.b.beryllina
E & C Mexico
A.b.lichtensteinei
S Mexico
A.b.devillei
S Mexico to Belize, El Salvador
Amazilia cyanura (Blue-tailed Hummingbird)
A.c.guatemalae
S Mexico, Guatemala
A.c.cyanura
El Salvador to W Nicaragua
**Amazilia saucerrottei (Steely-vented
Hummingbird)**
A.s.hoffmanni
W & S Nicaragua, Costa Rica
A.s.saucerrottei
C Colombia
A.s.australis
S Colombia
A.s.warscewiczi
N & E Colombia
A.s.braccata
W Venezuela
**Amazilia tobaci (Copper-rumped
Hummingbird)**
A.t.feliciae
N Venezuela
A.t.monticola
NW Venezuela
A.t.caudata
NE Venezuela
A.t.caurensis
E & SE Venezuela
A.t.aliciae
NE Venezuela, Margarita I
A.t.erythronotos
Trinidad

A.t.tobaci
Tobago I
**Amazilia virdigaster (Green-bellied
Hummingbird)**
A.v.viridigaster
E Colombia, W Venezuela
A.v.duidae
Mt Duida (Venezuela)
A.v.cupreicauda
S Venezuela, S Guyana
**Amazilia edward (Snowy-breasted
Hummingbird)**
A.e.niveoventer
SW Costa Rica, W Panama
A.e.edward
C Panama
A.e.margaritarum
Pearl Is (Panama)
A.e.crosbyi
SE Panama
Amazilia rutila (Cinnamon Hummingbird)
A.r.diluta
NW Mexico
A.r.rutila
W & S Mexico to W Costa Rica
A.r.corallirostris
SW Mexico to El Salvador
A.r.graysoni
Maria Madre I
**Amazilia yucatanensis (Buff-bellied
Hummingbird)**
A.y.chalconota
S Texas, NE Mexico
A.y.cerviniventris
S Mexico
A.y.yucatanensis
SE Mexico, N Guatemala, Belize
Amazilia tzacatl (Rufous-tailed Hummingbird)
A.t.tzacatl
E Mexico to W Venezuela
A.t.jucunda
SW Colombia, E Ecuador
**Amazilia castaneiventris (Chestnut-bellied
Hummingbird)**
NC Colombia
Amazilia amazilia (Amazilia Hummingbird)
A.a.dumerilii
W Ecuador, NW Peru
A.a.alticola
S Ecuador, N Peru
A.a.amazilia
W Peru
A.a.caeruleigularis
W Peru
A.a.leucophoea
E & S Peru
**Amazilia viridifrons (Green-fronted
Hummingbird)**
SW & S Mexico
**Amazilia violiceps (Violet-crowned
Hummingbird)**
A.v.ellioti
W Mexico

A.v.violiceps
SC Mexico

EUPHERUSA
Eupherusa poliocerca (White-tailed Hummingbird)
W Mexico
Eupherusa eximia (Stripe-tailed Hummingbird)
E.e.nelsoni
SE Mexico
E.e.eximia
S Mexico to N Nicaragua
E.e.egregia
Costa Rica, W Panama
Eupherusa cyanophrys (Oaxaca Hummingbird)
Oaxaca (Mexico)
Eupherusa nigriventris (Black-bellied Hummingbird)
E Costa Rica

ELVIRA
Elvira chionura (White-tailed Emerald)
SW Costa Rica
Elvira cupreiceps (Coppery-headed Emerald)
E Costa Rica

MICROCHERA
Microchera albocoronata (Snowcap)
M.a.parvirostris
E Nicaragua, E Costa Rica
M.a.albocoronata
NW Panama

CHALYBURA
Chalybura buffonii (White-vented Plumeleteer)
C.b.micans
E Panama, NW Colombia
C.b.buffonii
NC Colombia, W Venezuela
C.b.aeneicauda
NE Colombia, N Venezuela
C.b.caeruleogaster
E Colombia
C.b.intermedia
SW Ecuador
Chalybura urochrysia (Bronze-tailed Plumeleteer)
C.u.melanorrhoa
E Nicaragua
C.u.isaurae
NW Panama
C.u.incognita
E Panama
C.u.urochrysia
W Colombia, NW Ecuador

APHANTOCHROA
Aphantochroa cirrochloris (Sombre Hummingbird)
C & E Brazil

LAMPORNIS
Lampornis clemenciae (Blue-throated Hummingbird)
L.c.bessophilus
SW USA, NW Mexico
L.c.clemenciae
S USA, N & C Mexico
Lampornis amethystinus (Amethyst-throated Hummingbird)
L.a.amethystinus
E Mexico
L.a.brevirostris
W Mexico
L.a.margaritae
SW Mexico
L.a.salvini
S Mexico, C Guatemala
L.a.nobilis
Honduras
Lampornis viridipallens (Green-throated Mountain Gem)
L.v.ovandensis
S Mexico
L.v.viridipallens
S Mexico, Guatemala
L.v.nubivagus
El Salvador, W Honduras
L.v.connectens
El Salvador
Lampornis sybillae (Green-breasted Mountain Gem) 81.2
C Honduras, N Nicaragua
Lampornis hemileucus (White-bellied Mountain Gem)
NE Costa Rica
Lampornis castaneoventris (White-throated Mountain Gem)
W Panama
Lampornis calolaema (Purple-throated Mountain Gem) 81.2
L.c.pectoralis
W Nicaragua, NW Costa Rica
L.c.calolaema
C Costa Rica, W Panama
Lampornis cinereicauda (Grey-tailed Mountain Gem)
SW Costa Rica

LAMPROLAIMA
Lamprolaima rhami (Garnet-throated Hummingbird)
L.r.rhami
S Mexico
L.r.saturatior
Honduras & N El Salvador

ADELOMYIA
Adelomyia melanogenys (Speckled Hummingbird)
A.m.cervina
W & C Colombia

A.m.connectens
  Colombia
A.m.melanogenys
  E Colombia, W Venezuela
A.m.aeneosticta
  C & N Venezuela
A.m.maculata
  C Ecuador, N Peru
A.m.chlorospila
  SE Peru
A.m.inornata
  Bolivia, NW Argentina

ANTHOCEPHALA
Anthocephala floriceps (Blossomcrown)
  A.f.berlepschi
    C Colombia
  A.f.floriceps
    N Colombia

UROSTICTE
Urosticte benjamini (Whitetip)
  U.b.rostrata
    W Colombia
  U.b.benjamini
    W Ecuador, SW Colombia
  U.b.ruficrissa
    E Ecuador
  U.b.intermedia
    NE Peru

PHLOGOPHILUS
Phlogophilus hemileucurus (Ecuadorean Piedtail)
  E Ecuador
Phlogophilus harterti (Peruvian Piedtail)
  S Peru

CLYTOLAEMA
Clytolaema rubricauda (Brazilian Ruby)
  SE Brazil

POLYPLANCTA
Polyplancta aurescens (Gould's Jewel-front)
  E Ecuador, E Peru, S Venezuela, N Brazil

HELIODOXA
Heliodoxa rubinoides (Fawn-breasted Brilliant)
  H.r.rubinoides
    E Colombia
  H.r.aequatorialis
    W Colombia, W Ecuador
  H.r.cervinigularis
    E Ecuador, E Peru
Heliodoxa leadbeateri (Violet-fronted Brilliant)
  H.l.leadbeateri
    Upper Amazonia
  H.l.sagitta
    SC Colombia
  H.l.parvula
    W Venezuela
Heliodoxa jacula (Green-crowned Brilliant)

H.j.henryi
  Costa Rica, W Panama
H.j.jacula
  E Panama, E Colombia
H.j.jamesoni
  W Ecuador
Heliodoxa xanthogonys (Velvet-browed Brilliant)
  SE Venezuela, S Guyana
Heliodoxa schreibersii (Black-throated Brilliant)
  H.s.schreibersii
    E Ecuador, NE Peru, NW Brazil
  H.s.whitelyana
    E Peru
Heliodoxa gularis (Pink-throated Brilliant)
  E Ecuador, NE Peru
Heliodoxa branickii (Rufous-webbed Brilliant)
  C Peru
Heliodoxa imperatrix (Empress Brilliant)
  W Ecuador

EUGENES
Eugenes fulgens (Rivoli's Hummingbird)
  E.f.fulgens
    SW USA, N & C Mexico
  E.f.viridiceps
    S Mexico to Nicaragua
  E.f.spectabilis
    Costa Rica, W Panama

HYLONYMPHA
Hylonympha macrocerca (Scissor-tailed Hummingbird)
  Venezuela

STERNOCLYTA
Sternoclyta cyanopectus (Violet-chested Hummingbird)
  NW Venezuela

TOPAZA
Topaza pella (Crimson Topaz)
  T.p.pella
    N Brazil, S Venezuela, Guyana, Surinam
  T.p.smaragdula
    French Guiana
  T.p.microrhyncha
    NE Brazil
  T.p.pamprepta
    E Ecuador
Topaza pyra (Fiery Topaz)
  SE Colombia, E Ecuador, W Brazil

OREOTROCHILUS
Oreotrochilus melanogaster (Black-breasted Hillstar)
  C Peru
Oreotrochilus estella (Andean Hillstar)
  O.e.jamesonii
    E Ecuador
  O.e.soderstromi
    Mt Quillotoa (Ecuador)

APODIFORMES

O.e.chimborazo
  Mt Chimborazo (Ecuador)
O.e.stolzmanni
  N Peru
O.e.estella
  S Peru to N Chile & NW Argentina
O.e.boliviana
  C Bolivia
Oreotrochilus leucopleurus (White-sided Hillstar)
  S Bolivia, Chile, W Argentina
Oreotrochilus adela (Wedgetailed Hillstar)
  C Bolivia

UROCHROA
Urochroa bougueri (White-tailed Hillstar)
U.b.bougueri
  SW Colombia, NW Ecuador
U.b.eulcura
  SW Colombia
U.b.leucura
  E Ecuador

PATAGONA
Patagona gigas (Giant Hummingbird)
P.g.peruviana
  Ecuador to N Chile, NW Argentina
P.g.gigas
  C Chile, W Argentina

AGLAEACTIS
Aglaeactis cupripennis (Shining Sunbeam)
A.c.cupripennis
  Colombia, N & C Ecuador
A.c.parvulus
  S Ecuador, N Peru
A.c.ruficauda
  C Peru
A.c.caumatonotus
  SC Peru
Aglaeactis aliciae (Purple-backed Sunbeam)
  N Peru
Aglaeactis castelnaudii (White-tufted Sunbeam)
  C Peru
Aglaeactis pamela (Black-hooded Sunbeam)
  Bolivia

LAFRESNAYA
Lafresnaya lafresnayi (Mountain Velvetbreast)
L.l.tamai
  N Venezuela
L.l.liriope
  N Colombia, W Venezuela
L.l.lafresnayi
  C Colombia
L.l.greenewalti
  W Venezuela
L.l.saul
  W Colombia, Ecuador, N Peru
L.l.rectirostris
  C Peru

PTEROPHANES
Pterophanes cyanopterus (Great Sapphirewing)
P.c.cyanopterus
  Colombia, Ecuador
P.c.caeruleus
  S Colombia
P.c.peruvianus
  Peru, N Bolivia

COELIGENA
Coeligena coeligena (Bronzy Inca)
C.c.ferruginea
  W Colombia
C.c.columbiana
  C Colombia to C Ecuador & W Venezuela
C.c.zuloagae
  NW Venezuela
C.c.boliviana
  C Peru, C Bolivia
C.c.coeligena
  N Venezuela
C.c.zuliana
  W Venezuela, NE Colombia
Coeligena wilsoni (Brown Inca)
  SW Colombia, W Ecuador
Coeligena prunellei (Black Inca)
  E Colombia
Coeligena torquata (Collared Inca)
C.t.torquata
  Colombia, E Ecuador
C.t.fuligidigula
  W Ecuador
C.t.margaretae
  N Peru
C.t.eisenmanni
  C Peru
C.t.insectivora
  NC Peru
C.t.omissa
  S Peru
C.t.conradii
  NE Colombia, NW Venezuela
C.t.inca
  S Peru, N Bolivia
Coeligena phalerata (White-tailed Starfrontlet)
  N Colombia
Coeligena bonapartei (Golden-bellied Starfrontlet)
C.b.consita
  NE Colombia, NW Venezuela
C.b.bonapartei
  E Colombia
C.b.eos
  NW Venezuela
Coeligena orina (Dusky Starfrontlet)
  Colombia
Coeligena helianthea (Blue-throated Starfrontlet)
C.h.tamae
  NE Colombia
C.h.helianthea
  E Colombia

Coeligena lutetiae (Buff-winged Starfrontlet)
C Colombia, Ecuador
Coeligena violifer (Violet-throated
Starfrontlet)
C.v.dichroura
N & C Peru
C.v.osculans
S Peru
C.v.violifer
NW Bolivia
Coeligena iris (Rainbow Starfrontlet)
C.i.iris
S Ecuador
C.i.aurora
S Ecuador, N Peru
C.i.fulgidiceps
N Peru
C.i.flagrans
N Peru
C.i.hypocrita
N Peru
C.i.eva
N Peru
C.i.hesperus
SC Ecuador

ENSIFERA
Ensifera ensifera (Sword-billed Hummingbird)
Venezuela, Colombia to N Bolivia

SEPHANOIDES
**Sephanoides sephaniodes (Green-backed
Firecrown)**
Chile, SW Argentina
**Sephanoides fernandensis (Fernandez
Firecrown)**
S.f.fernandensis
Masatierra I
S.f.leyboldi
Masafuera I

BOISSONNEAUA
**Boissonneaua flavescens (Buff-tailed
Coronet)**
B.f.flavescens
Colombia, W Venezuela
B.f.tinochlora
W Ecuador
**Boissonneaua matthewsii (Chestnut-breasted
Coronet)**
Ecuador, Peru
Boissonneaua jardini (Velvet-Purple Coronet)
W Colombia, W Ecuador

HELIANGELUS
**Heliangelus mavors (Orange-throated
Sunangel)**
NE Colombia W Venezuela
Heliangelus spencei (Merida Sunangel)
W Venezuela
**Heliangelus amethysticollis (Amethyst-
throated Sunangel)**

H.a.clarisse
N Colombia, W Venezuela
H.a.violiceps
NE Colombia
H.a.viridiscutatus
NE Colombia
H.a.laticlavius
S Ecuador to C Peru
H.a.amethysticollis
S Peru, Bolivia
Heliangelus strophianus (Gorgeted Sunangel)
W Ecuador
Heliangelus regalis (Royal Sunangel)
N Peru
Heliangelus exortis (Tourmaline Sunangel)
Colombia, E Ecuador
Heliangelus viola (Purple-throated Sunangel)
W Ecuador, NW Peru
Heliangelus micraster (Little Sunangel)
H.m.micraster
S Ecuador
H.m.cutervensis
N Peru
**Heliangelus squamigularis (Olive-throated
Sunangel)**
Colombia

ERIOCNEMIS
**Eriocnemis nigrivestris (Black-breasted
Puffleg)**
N W Ecuador
**Eriocnemis soderstromi (Söderström's
Puffleg)**
Ecuador
Eriocnemis vestitus (Glowing Puffleg)
E.v.vestitus
E Colombia, W Venezuela
E.v.paramillo
NW Colombia
E.v.smaragdinipectus
SC Colombia, E Ecuador
**Eriocnemis godini (Turquoise-throated
Puffleg)**
Ecuador
**Eriocnemis cupreoventris (Coppery-bellied
Puffleg)**
E Colombia, W Venezuela
Eriocnemis luciani (Sapphire-vented Puffleg)
E.l.luciani
W Ecuador
E.l.catharina
N Peru
E.l.sapphiropygia
C Peru
Eriocnemis isaacsonii (Isaacson's Puffleg)
Colombia
**Eriocnemis mosquera (Golden-breasted
Puffleg)**
Colombia, N Ecuador
**Eriocnemis glaucopoides (Blue-capped
Puffleg)**
Bolivia, N Argentina

Eriocnemis mirabilis (Colourful Puffleg)
  N Peru
Eriocnemis alinae (Emerald-bellied Puffleg)
  E.a.alinae
    Colombia, N Ecuador
  E.a.dybowskii
    N & C Peru
Eriocnemis derbyi (Black-thighed Puffleg)
  E.d.longirostris
    NC Colombia
  E.d.derbyi
    S Colombia, N Ecuador

HAPLOPHAEDIA
Haplophaedia aureliae (Greenish Puffleg)
  H.a.caucensis
    E Panama, W Colombia
  H.a.aureliae
    C & E Colombia
  H.a.russata
    E Ecuador
  H.a.affinis
    N Peru
  H.a.assimilis
    SE Peru
Haplophaedia lugens (Hoary Puffleg)
  WC Colombia, C Ecuador

OCREATUS
Ocreatus underwoodii (Booted Racquet-tail)
  O.u.polystictus
    N Venezuela
  O.u.underwoodii
    Colombia, W Venezuela
  O.u.ambiguus
    S Colombia
  O.u.discifer
    W Venezuela
  O.u.melanantherus
    E Ecuador
  O.u.peruanus
    E Ecuador, NE Peru
  O.u.annae
    C Peru
  O.u.addae
    Bolivia

LESBIA
Lesbia victoriae (Black-tailed Trainbearer)
  L.v.victoriae
    S & E Colombia
  L.v.eucharis
    Colombia
  L.v.aequatorialis
    W Ecuador
  L.v.juliae
    N & C Peru
  L.v.berlepschi
    SW Peru
Lesbia nuna (Green-tailed Trainbearer)
  L.n.gouldii
    Colombia, W Venezuela

  L.n.gracilis
    Ecuador
  L.n.pallidiventris
    N Peru
  L.n.chlorura
    C Peru
  L.n.nuna
    SW Peru
  L.n.boliviana
    N Bolivia

SAPPHO
Sappho sparganura (Red-tailed Comet)
  S.s.sparganura
    N & C Bolivia
  S.s.sapho
    S Bolivia, N & W Argentina

POLYONYMUS
Polyonymus caroli (Bronze-tailed Comet)
  Peru

ZODALIA
Zodalia glyceria (Purple-tailed Comet)
  Colombia, Ecuador

RAMPHOMICRON
Ramphomicron microrhynchum (Purple-backed Thornbill)
  R.m.andicolum
    W Venezuela
  R.m.microrhynchum
    Colombia, Ecuador
  R.m.albiventre
    Peru
  R.m.bolivianum
    N Bolivia
Ramphomicron dorsale (Black-backed Thornbill)
  N Colombia

METALLURA
Metallura phoebe (Black Metaltail)
  Peru, Bolivia, N Chile
Metallura theresiae (Coppery Metaltail)
  N Peru
Metallura purpureicauda (Purple-tailed Thornbill)
  Ecuador
Metallura aeneocauda (Scaled Metaltail)
  M.a.aeneocauda
    S Peru, Bolivia
  M.a.malagae
    Bolivia
Metallura baroni (Violet-throated Metaltail)
  SW Ecuador
Metallura eupogon (Fire-throated Metaltail)
  N & C Peru
Metallura odomae (Neblina Metaltail)
  N Peru
Metallura williami (Viridian Metaltail)
  M.w.williami
    C Colombia

M.w.primolina
NE Ecuador
M.w.atrigularis
S Ecuador
Metallura tyrianthina (Tyrian Metaltail)
M.t.chloropogon
N Venezuela
M.t.oreopola
W Venezuela
M.t.districta
N Colombia
M.t.tyrianthina
Colombia, E & S Ecuador
M.t.quitensis
NW Ecuador
M.t.septentrionalis
N Peru
M.t.peruviana
C Peru
M.t.smaragdinicollis
S Peru, Bolivia
Metallura iracunda (Perija Metaltail)
E Colombia, W Venezuela

CHALCOSTIGMA
Chalcostigma ruficeps (Rufous-capped Thornbill)
C.r.aureofastigata
S Ecuador
C.r.ruficeps
Peru, N Bolivia
Chalcostigma olivaceum (Olivaceous Thornbill)
C.o.olivaceum
N & C Peru, Bolivia
C.o.pallens
WC Peru
Chalcostigma stanleyi (Blue-mantled Thornbill)
C.s.stanleyi
Ecuador
C.s.versigularis
C Peru
C.s.vulcani
SE Peru, Bolivia
Chalcostigma heteropogon (Bronze-tailed Thornbill)
E Colombia, W Venezuela
Chalcostigma herrani (Rainbow-bearded Thornbill)
C.h.tolimae
W Colombia
C.h.herrani
S Colombia, N Ecuador

OXYPOGON
Oxypogon guerinii (Bearded Helmetcrest)
O.g.stubelii
C Colombia
O.g.guerinii
E Colombia
O.g.cyanolaemus
N Colombia

O.g.lindenii
W Venezuela

OPISTHOPRORA
Opisthoprora euryptera (Mountain Avocetbill)
S Colombia, NE Ecuador

TAPHROLESBIA
Taphrolesbia griseiventris (Grey-billed Comet)
Peru

AGLAIOCERCUS
Aglaiocercus kingi (Long-tailed Sylph)
A.k.berlepschi
Venezuela
A.k.margarethae
Venezuela
A.k.kingi
E Colombia
A.k.mocoa
C Colombia to N Peru
A.k.smaragdinus
C Peru, N Bolivia
A.k.emmae
NC Colombia
A.k.caudata
Colombia, W Venezuela
Aglaiocercus coelestis (Violet-tailed Sylph)
A.c.coelestis
W Colombia, NW Ecuador
A.c.pseudocoelestis
SW Colombia
A.c.aethereus
SW Ecuador

OREONYMPHA
Oreonympha nobilis (Bearded Mountaineer)
O.n.albolimbata
WC Peru
O.n.nobilis
S Peru

AUGASTES
Augastes scutatus (Hyacinth Visor-bearer)
Brazil
Augastes lumachellus (Hooded Visor-bearer)
E Brazil

SCHISTES
Schistes geoffroyi (Wedge-billed Hummingbird)
S.g.albogularis
WC Colombia, W Ecuador
S.g.geoffroyi
N Venezuela, E Colombia to E Peru

HELIOTHRYX
Heliothryx barroti (Purple-crowned Fairy)
Guatemala to W Ecuador
Heliothryx aurita (Black-eared Fairy)
H.a.aurita
N Colombia to the Guianas, N Brazil

H.a.major
  W Ecuador
H.a.phainolaema
  NE Brazil
H.a.auriculata
  E Peru, Bolivia, C & S Brazil

HELIACTIN
Heliactin cornuta (Horned Sungem)
  C & E Brazil

LODDIGESIA
Loddigesia mirabilis (Marvellous Spatuletail)
  N Peru

HELIOMASTER
Heliomaster constantii (Plain-capped
  Starthroat)
  H.c.surdus
    NW Mexico
  H.c.pinicola
    W Mexico
  H.c.leocadiae
    W Mexico, W Guatemala
  H.c.constantii
    El Salvador to Costa Rica
Heliomaster longirostris (Long-billed
  Starthroat)
  H.l.pallidiceps
    S Mexico to Nicaragua
  H.l.longirostris
    E Costa Rica to N & NW Sth America
  H.l.stuartae
    Colombia
  H.l.albicrissa
    W Ecuador, NW Peru
Heliomaster squamosus (Stripe-breasted
  Starthroat)
  EC Brazil
Heliomaster furcifer (Blue-tufted Starthroat)
  C Brazil to Bolivia & N Argentina

RHODOPIS
Rhodopis vesper (Oasis Hummingbird)
  R.v.tertia
    N Peru
  R.v.koepckeae
    Peru
  R.v.vesper
    SW Peru, N Chile
  R.v.atacamensis
    N Chile

THAUMASTURA
Thaumastura cora (Peruvian Sheartail)
  W Peru

PHILODICE
Philodice evelynae (Bahama Woodstar)
  P.e.evelynae
    Bahama Is
  P.e.lyrura
    Gt Inagua I

P.e.salita
  Caicos Is
Philodice bryantae (Magenta-throated
  Woodstar)
  Costa Rica, W Panama
Philodice mitchellii (Purple-throated
  Woodstar)
  W Colombia, W Ecuador

DORICHA
Doricha enicura (Slender Sheartail)
  S Mexico, Guatemala, El Salvador
Doricha eliza (Mexican Sheartail)
  SE Mexico, Holbox I

TILMATURA
Tilmatura dupontii (Sparkling-tailed
  Hummingbird)   81.2
  S Mexico to N Nicaragua

MICROSTILBON
Microstilbon burmeisteri (Slender-tailed
  Woodstar)
  C & S Bolivia, N Argentina

CALOTHORAX
Calothorax lucifer (Lucifer Hummingbird)
  SW USA to SC Mexico
Calothorax pulcher (Beautiful Hummingbird)
  S Mexico

ARCHILOCHUS
Archilochus colubris (Ruby-throated
  Hummingbird)
  S Canada, E USA to Panama
Archilochus alexandri (Black-chinned
  Hummingbird)
  SW Canada, W USA to W Mexico

CALLIPHLOX
Calliphlox amethystina (Amethyst Woodstar)
  Ecuador & Bolivia to the Guianas, NE
  Argentina

MELLISUGA
Mellisuga minima (Vervain Hummingbird)
  M.m.minima
    Jamaica
  M.m.vieilloti
    Gonave I, Hispaniola

CALYPTE
Calypte anna (Anna's Hummingbird)
  C California, NW Baja California
Calypte costae (Costa's Hummingbird)
  SW USA, NW Mexico
Calypte helenae (Bee Hummingbird)
  Cuba, Isle of Pines

STELLULA
Stellula calliope (Calliope Hummingbird)
  S.c.calliope
    W USA to SC Mexico

S.c.lowei
Guerrero (Mexico)

ATTHIS
Atthis heloisa (Bumblebee Hummingbird)
A.h.margarethae
MW Mexico
A.h.heloisa
C & S Mexico
Atthis ellioti (Wine-throated Hummingbird)
A.e.ellioti
S Mexico, Guatemala
A.e.selasphoroides
Honduras

MYRTIS
Myrtis fanny (Purple-collared Woodstar)
Ecuador, W Peru

EULIDIA
Eulidia yarrellii (Chilean Woodstar)
N Chile

MYRMIA
Myrmia micrura (Short-tailed Woodstar)
W Ecuador, W Peru

ACESTRURA
Acestrura mulsant (White-bellied Woodstar)
Colombia
Acestrura bombus (Little Woodstar)
Ecuador, N Peru
Acestrura heliodor (Gorgeted Woodstar) 81.1
A.h.heliodor
W Ecuador to Venezuela
A.h.cleavesi
NE Ecuador
Acestrura astreans (Colombian Woodstar)
81.1
N Colombia
Acestrura berlepschi (Esmeralda's Woodstar)
W Ecuador
Acestrura harterti (Hartert's Woodstar)
Ecuador

CHAETOCERCUS
Chaetocercus jourdanii (Rufous-shafted Woodstar)
C.j.andinus
W Venezuela, NE Colombia
C.j.jourdanii
NE Venezuela, Trinidad
C.j.rosae
E Colombia, N & W Venezuela

SELASPHORUS
Selasphorus platycercus (Broad-tailed Hummingbird)
S.p.platycercus
W USA to WC Mexico
S.p.guatemalae
W Guatemala

Selasphorus rufus (Rufous Hummingbird)
SE Alaska to WC Mexico
Selasphorus sasin (Allen's Hummingbird)
S.s.sasin
SW California, NW Mexico
S.s.sedentarius
Santa Barbara I (California)
Selasphorus flammula (Volcano Hummingbird) 81.2
S.f.flammula
Costa Rica
S.f.torridus
Costa Rica, W Panama
S.f.simoni
Costa Rica
Selasphorus ardens (Glow-throated Hummingbird)
W Panama
Selasphorus scintilla (Scintillant Hummingbird)
Costa Rica, W Panama

COLIIFORMES

82. COLIIDAE (MOUSEBIRDS)

COLIUS
Colius striatus (Speckled Mousebird)
C.s.nigricollis
N Nigeria to W Zaire
C.s.leucophthalmus
S Sudan, NE Zaire
C.s.leucotis
W Sudan, Ethiopia
C.s.hilgerti
E Ethiopia, Somalia
C.s.jebelensis
SW Sudan, NE Zaire
C.s.mossambicus
S Somalia to NE Tanzania
C.s.kikuyensis
W & C Kenya
C.s.cinerascens
N Tanzania
C.s.affinis
S Somalia, E Kenya, E Tanzania
C.s.berlepschi
S Tanzania, Malawi, NW Mozambique
C.s.kiwuensis
Kivu (Rwanda), Uganda, NW Tanzania
C.s.congicus
S Zaire, W Zambia
C.s.simulans
C Mozambique
C.s.integralis
S Mozambique, N Sth Africa
C.s.rhodesiae
E Zimbabwe
C.s.minor
Mozambique to E Cape Province
C.s.striatus
S Cape Province

***Colius castanotus* (Red-backed Mousebird)**
  Angola
***Colius colius* (White-backed Mousebird)**
  W Sth Africa
***Colius leucocephalus* (White-headed
  Mousebird)**
  C.l.turneri
    N & NE Kenya
  C.l.leucocephalus
    SE Kenya to NE Tanzania

UROCOLIUS
***Urocolius indicus* (Red-faced Mousebird)** 82.1
  U.i.mossambicus
    S Zaire, S Angola
  U.i.lacteifrons
    S Angola, N Namibia
  U.i.pallidus
    S Tanzania, N Malawi
  U.i.transvaalensis
    Zimbabwe, Transvaal, Natal
  U.i.indicus
    Cape Province
***Urocolius macrourus* (Blue-naped Mousebird)**
  U.m.macrourus
    Senegal to Somalia
  U.m.laeneni
    Niger to Air Mountains
  U.m.abyssinicus
    C & S Ethiopia
  U.m.pulcher
    W Uganda, S Ethiopia to N Tanzania
  U.m.massaicus
    E Tanzania
  U.m.griseogularis
    Sudan to E Zaire

TROGONIFORMES

83. TROGONIDAE (TROGONS)

PHAROMACHRUS
***Pharomachrus mocinno* (Resplendent
  Quetzal)**
  P.m.mocinno
    S Mexico to N Nicaragua
  P.m.costaricensis
    Costa Rica to W Panama
***Pharomachrus antisianus* (Crested Quetzal)**
  W Venezuela, Colombia to Brazil
***Pharomachrus fulgidus* (White-tipped
  Quetzal)**
  P.f.festatus
    N Colombia
  P.f.fulgidus
    NE & NC Venezuela
***Pharomachrus auriceps* (Golden-headed
  Quetzal)**
  Venezuela, Colombia to N Bolivia
***Pharomachrus pavoninus* (Pavonine Quetzal)**
  P.p.hargitti
    E Colombia, W Venezuela

P.p.heliactin
  W Ecuador
P.p.pavoninus
  Upper Amazonia
P.p.viridiceps
  NE Brazil

EUPTILOTIS
***Euptilotis neoxenus* (Eared Trogon)**
  C Mexico

PRIOTELUS
***Priotelus temnurus* (Cuban Trogon)**
  P.t.temnurus
    Cuba
  P.t.vescus
    Isle of Pines

TEMNOTROGON
***Temnotrogon roseigaster* (Hispaniolan
  Trogon)**
  Hispaniola

TROGON
***Trogon massena* (Slaty-tailed Trogon)**
  T.m.massena
    S Mexico to Nicaragua
  T.m.hoffmanni
    Costa Rica, Panama
  T.m.australis
    W Colombia
***Trogon clathratus* (Lattice-tailed Trogon)**
  E Costa Rica, W Panama
***Trogon melanurus* (Black-tailed Trogon)**
  T.m.macroura
    E Panama, N Colombia
  T.m.mesurus
    W Ecuador, NW Peru
  T.m.eumorphus
    Peru
  T.m.occidentalis
    Brazil
  T.m.melanurus
    Colombia to Bolivia & N Brazil
***Trogon comptus* (Blue-tailed Trogon)**
  Colombia
***Trogon bairdii* (Baird's Trogon)** 83.2
  SW Costa Rica, W Panama
***Trogon viridis* (White-tailed Trogon)**
  T.v.chionurus
    E Panama, W Colombia, W Ecuador
  T.v.viridis
    Colombia to Peru, Brazil & Trinidad
  T.v.melanopterus
    SE Brazil
***Trogon citreolus* (Citreoline Trogon)**
  T.c.citreolus
    W Mexico
  T.c.sumichrasti
    S Mexico
***Trogon melanocephalus* (Black-headed
  Trogon)** 83.2

T.m.melanocephalus
  E Mexico to NE Costa Rica
T.m.illaetabilis
  W Costa Rica
Trogon mexicanus (Mexican Trogon)
T.m.clarus
  NW Mexico
T.m.mexicanus
  C Mexico to W Guatemala
T.m.lutescens
  Honduras
Trogon elegans (Elegant Trogon)   83.2
T.e.canescens
  S Arizona, NW Mexico
T.e.ambiguus
  S Texas, E & C Mexico
T.e.elegans
  Guatemala
T.e.lubricus
  Nicaragua, Costa Rica
T.e.goldmani
  Tres Marias Is
Trogon collaris (Collared Trogon)
T.c.puella
  C Mexico to W Panama
T.c.extimus
  E Panama
T.c.virginalis
  W Colombia, W Ecuador, NW Peru
T.c.subtropicalis
  C Colombia
T.c.exoptatus
  N Venezuela
T.c.collaris
  Colombia to Bolivia, Trinidad, S Brazil
T.c.castaneus
  SE Colombia, NW Brazil
Trogon aurantiiventris (Orange-bellied Trogon)
T.a.underwoodi
  NW Costa Rica
T.a.aurantiiventris
  C Costa Rica to W Panama
T.a.flavidior
  E Panama
Trogon personatus (Masked Trogon)
T.p.sanctaemartae
  N Colombia
T.p.ptaritepui
  Venezuela
T.p.personatus
  W Venezuela, E Colombia, E Peru
T.p.assimilis
  W Ecuador
T.p.temperatus
  C Colombia, Ecuador
T.p.submontanus
  Bolivia
T.p.duidae
  Mt Duida (Venezuela)
T.p.roraimae
  SE Venezuela, S Guyana
Trogon rufus (Black-throated Trogon)

T.r.tenellus
  SE Honduras to NW Colombia
T.r.cupreicauda
  W Colombia, W Ecuador
T.r.rufus
  E Venezuela, the Guianas, N Brazil
T.r.sulphureus
  E Peru, W Brazil
T.r.amazonicus
  NE Brazil
T.r.chrysochloros
  S Brazil, Paraguay, NE Argentina
Trogon surrucura (Surucua Trogon)
T.s.aurantius
  E Brazil
T.s.surrucura
  S Brazil, Paraguay, Uruguay, N Argentina
Trogon curucui (Blue-crowned Trogon)
T.c.bolivianus
  S Colombia to Bolivia, W Brazil
T.c.peruvianus
  SC Colombia
T.c.curucui
  E Brazil
T.c.behni
  E Bolivia, S Brazil, Paraguay, N Argentina
Trogon violaceus (Violaceous Trogon)
T.v.braccatus
  C Mexico to Nicaragua
T.v.concinnus
  Costa Rica to W Ecuador
T.v.caligatus
  N Colombia, W Venezuela
T.v.violaceus
  Venezuela, the Guianas, N Brazil, Trinidad
T.v.ramonianus
  Upper Amazonia
T.v.crissalis
  E Brazil

APALODERMA
Apaloderma narina (Narina's Trogon)
A.n.constantia
  Liberia to Ghana
A.n.brachyurum
  S Cameroun to Uganda, Zaire
A.n.narina
  Sudan & Ethiopia to Cape Province
A.n.littoralis
  E Kenya, E Tanzania, Zanzibar I
Apaloderma aequatoriale (Bare-cheeked Trogon)
  Cameroun to S Zaire
Apaloderma vittatum (Bar-tailed Trogon)   83.1
  Cameroun, Angola to Kenya, Tanzania

HARPACTES
Harpactes reinwardtii (Reinwardt's Blue-tailed Trogon)
H.r.mackloti
  Sumatra
H.r.reinwardtii
  Java

Harpactes fasciatus (Malabar Trogon)
H.f.malabaricus
W & S India
H.f.fasciatus
Sri Lanka
H.f.parvus
Sri Lanka
Harpactes kasumba (Red-naped Trogon)
H.k.kasumba
Malaysia, Sumatra
H.k.impavidus
Borneo
Harpactes diardii (Diard's Trogon)
H.d.sumatranus
Malaysia, Sumatra
H.d.diardii
Borneo, Bangka I
Harpactes ardens (Philippine Trogon)
H.a.ardens
Philippine Is
H.a.luzoniensis
Luzon, Bataan Is
Harpactes whiteheadi (Whitehead's Trogon)
Mt Kinabalu (N Borneo)
Harpactes orrhophaeus (Cinnamon-rumped Trogon)
H.o.orrhophaeus
Malaysia, Sumatra
H.o.vidua
NW Borneo
Harpactes duvaucelii (Scarlet-rumped Trogon)
S Burma, Malaysia, Sumatra
Harpactes oreskios (Orange-breasted Trogon)
H.o.stellae
S Burma, S Indochina
H.o, uniformis
S Burma, Malaysia, Sumatra
H.o.oreskios
Java
H.o.dulitensis
NW Borneo
H.o.nias
Nias I
Harpactes erythrocephalus (Red-headed Trogon)
H.e.erythrocephalus
Himalayas, Burma, NW Thailand
H.e.helenae
W Yunnan, N Burma
H.e.yamakanensis
SE China
H.e.rosa
SC China
H.e.intermedius
N Laos, N Vietnam
H.e.annamensis
NE Thailand, S Indochina
H.e.klossi
W Cambodia
H.e.chaseni
Malaysia

H.e.hainanus
Hainan I
H.e.flagrans
Sumatra
Harpactes wardi (Ward's Trogon)
N Burma, NE Vietnam

CORACIIFORMES

84. ALCEDINIDAE (KINGFISHERS)

CERYLINAE

MEGACERYLE   84.4
Megaceryle lugubris (Greater Pied Kingfisher)
M.l.guttulata
Himalayas, China, Burma, Thailand
M.l.pallida
Hokkaido I
M.l.lugubris
Japan, Korea
Megaceryle maxima (Giant Kingfisher)
M.m.maxima
Senegal to Nigeria, Ethiopia to Cape Province
M.m.gigantea
S Nigeria to Zaire, Tanzania
Megaceryle torquata (Ringed Kingfisher)
M.t.torquata
Mexico to N Argentina, Peru
M.t.stictipennis
Guadeloupe I, Dominica I
M.t.stellata
Chile, Argentina
Megaceryle alcyon (Belted Kingfisher)
M.a.caurina
Alaska to W Mexico
M.a.alcyon
C & E Canada to N Sth America & W Indies

CERYLE
Ceryle rudis (Lesser Pied Kingfisher)
C.r.rudis
Asia Minor, Iran, Africa
C.r.leucomelanura
Pakistan to Burma, Indochina
C.r.travancoreensis
SW India
C.r.insignis
SE China, Hainan I

CHLOROCERYLE
Chloroceryle amazona (Amazon Kingfisher)
C.a.mexicana
S Mexico to E Panama
C.a.amazona
N Sth America
Chloroceryle americana (Green Kingfisher)
C.a.hachisukai
W Texas, NW Mexico
C.a.septentrionalis
SE Texas, E Mexico, Guatemala

C.a.isthmica
Honduras to N Colombia, Pearl Is(Panama)
C.a.americana
N Sth America (East of Andes)
C.a.hellmayri
W Colombia
C.a.ecuadorensis
W Ecuador
C.a.cabanisii
W Peru
C.a.croteta
Trinidad, Tobago I
C.a.mathewsii
C & SC Sth America
Chloroceryle inda (Green & Rufous Kingfisher)
C.i.inda
Panama & tropical Sth America
C.i.chocoensis
W Colombia, W Ecuador
Chloroceryle aenea (Pygmy Kingfisher)
C.a.stictoptera
S Mexico to Nicaragua
C.a.aenea
Costa Rica, Panama, N Sth America, Trinidad

ALCEDININAE

ALCEDO
Alcedo hercules (Blyth's Kingfisher)
E Himalayas to N Vietnam
Alcedo atthis (River Kingfisher)
A.a.ispida
Europe
A.a.atthis
Mediterranean, Syria, Arabia
A.a.bengalensis
N India to Philippine Is & Gtr Sunda Is
A.a.taprobana
S India, Sri Lanka
A.a.japonica
Sakhalin I, Japan, Taiwan
A.a.floresiana
Bali to Timor I
A.a.hispidoides
Sulawesi to Bismarck Arch & New Guinea
A.a.salomonensis
Solomon Is
Alcedo semitorquata (Half-collared Kingfisher)
A.s.semitorquata
Ethiopia to Angola & Cape Province
A.s.tephria
SC Africa
A.s.heuglini
Ethiopia
Alcedo meninting (Blue-eared Kingfisher)
A.m.coltarti
Sikkim to S Burma & Laos
A.m.laubmanni
E India
A.m.phillipsi
SW India, Sri Lanka

A.m.scintillans
S Burma, S Thailand
A.m.rufigaster
Andaman Is
A.m.verreauxii
Malaysia, Sumatra, Borneo
A.m.proxima
Pagi Is
A.m.subviridis
Banjak, Nias Is
A.m.callima
Batu Is
A.m.meninting
Java to Lombok & Sula Is
Alcedo quadribrachys (Shining-blue Kingfisher)
A.q.quadribrachys
Gambia to N Nigeria
A.q.guentheri
S Nigeria to Uganda, S Zaire & Angola
Alcedo euryzona (Blue-banded Kingfisher)
A.e.peninsulae
Malaysia, Sumatra
A.e.euryzona
Java, Borneo
Alcedo coerulescens (Small Blue Kingfisher)
Java to Sumbawa I
Alcedo azureus (Azure Kingfisher)   84.1
A.a.affinis
N Moluccas
A.a.wallaceanus
Aru Is
A.a.lessonii
W Papuan Is, New Guinea
A.a.ochrogaster
N New Guinea to Geelvink Bay Is
A.a.ruficollaris
Tanimbar Is, N Australia
A.a.mixtus
NE Queensland
A.a.azureus
E & S Australia, Tasmania
Alcedo pusilla (Mangrove Kingfisher)   84.1
A.p.pusilla
W Papuan Is, New Guinea, Kai, Halmahera, Obi Is
A.p.laetior
N New Guinea
A.p.ramsayi
N Territory
A.p.halli
N Queensland
A.p.masauji
Bismarck Arch
A.p.bouganvillei
Bougainville, Ysabel, Choiseul Is
A.p.richardsi
Vellalavella, Kulambangra Is
A.p.aolae
Guadalcanal
Alcedo cristata (Malachite Kingfisher)   84.3
A.c.galerita
Senegal to Ghana

A.c.cristata
  Nigeria to Uganda & Cape Province
A.c.thomensis
  Sao Thome
A.c.stuartkeithi
  Sudan, Ethiopia
**Alcedo vintsioides (Madagascar Malachite
Kingfisher)**   84.3
  A.v.vintsioides
    Madagascar
  A.v.johannae
    Comoro Is
Alcedo leucogaster (White-bellied Kingfisher)
                                             84.3
  A.l.leucogaster
    Fernando Po I
  A.l.bowdleri
    Guinea to Ghana
  A.l.batesi
    Nigeria to Angola
  A.l.nais
    Principé I
  A.l.leopoldi
    WC Zaire

CEYX
Ceyx lecontei (African Dwarf Kingfisher) 84.4
  C.l.lecontei
    Sierra Leone to Angola & E Zaire
  C.l.ugandae
    W Uganda
Ceyx picta (African Pygmy Kingfisher)   84.4
  C.p.picta
    Senegal to Angola, Kenya & Ethiopia
  C.p.natalensis
    S Zaire & Tanzania to Natal
**Ceyx madagascariensis (Madagascar Pygmy
Kingfisher)**   84.4
  C.m.madagascariensis
    Madagascar
  C.m.diluta
    Sakaraha, Madagascar
Ceyx cyanopectus (Dwarf River Kingfisher)
  C.c.cyanopectus
    Luzon, Mindoro, Masbate, Ticao Is
  C.c.nigrirostris
    Panay, Negros, Cebu Is
Ceyx argentatus (Silvery Kingfisher)
  C.a.argentatus
    Panay, Negros, Cebu, Mindanao Is
  C.a.flumenicolus
    Samar, Leyte Is
Ceyx goodfellowi (Goodfellow's Kingfisher)
  C.g.goodfellowi
    Mindanao I
  C.g.virgicapitus
    Tawitawi Is
Ceyx lepidus (Dwarf Kingfisher)
  C.l.margarethae
    S Philippine Is
  C.l.wallacii
    Sula Is

C.l.uropygialis
  N Moluccas
C.l.lepidus
  Moluccas
C.l.cajeli
  Buru I
C.l.solitarius
  New Guinea & islands, Aru Is
C.l.dispar
  Admiralty Is
C.l.mulcatus
  New Hanover, New Ireland
C.l.sacerdotis
  New Britain, Rook I
C.l.pallidus
  Bougainville, Buka Is
C.l.collectoris
  Choiseul, Vellalavella, New Georgia,
  Rendova Is
C.l.meeki
  Choiseul, Ysabel Is
C.l.malaitae
  Malaita I
C.l.nigromaxilla
  Guadalcanal I
C.l.gentianus
  San Cristobal I
Ceyx websteri (Bismarck Pygmy Kingfisher)
  Bismarck Arch
Ceyx erithacus (Three-toed Kingfisher)   84.2
  C.e.erithacus
    India to SE China, Indochina, Sumatra
  C.e.macrocarus
    Andaman, Nicobar Is
  C.e.motleyi
    N Bornean Is, Borneo
  C.e.captus
    Nias I
  C.e.vargasi
    Mindoro I
  C.e.rufidorsus
    Malaysia to W Philippine I, Borneo, Lesser
    Sunda Is
  C.e.jungei
    Batu, Simalur Is
**Ceyx melanurus (Philippine Forest
Kingfisher)**
  C.m.melanurus
    Luzon I
  C.m.samarensis
    Samar, Leyte Is
  C.m.mindanensis
    Mindanao, Basilan Is
Ceyx fallax (Sulawesi Pygmy Kingfisher)   84.2
  C.f.sangirensis
    Sanghir Is
  C.f.fallax
    Sulawesi

LACEDO
Lacedo pulchella (Banded Kingfisher)
  L.p.amabilis
    S Burma, Thailand, S Vietnam

L.p.deignani
  S Thailand
L.p.pulchella
  Malaysia, Sumatra, Java
L.p.melanops
  Borneo, Bangka I

DACELO
Dacelo novaeguineae (Laughing Kookaburra)
D.n.minor
  N Queensland
D.n.novaeguineae
  Australia, Tasmania
Dacelo leachii (Blue-winged Kookaburra)
D.l.superflua
  S New Guinea
D.l.intermedia
  SE New Guinea
D.l.cliftoni
  NW Australia
D.l.kempi
  N Queensland
D.l.cervina
  Melville I, N Nthn Territory
D.l.leachii
  C Nthn Territory, S Queensland
Dacelo tyro (Aru Giant Kingfisher)
D.t.archboldi
  S New Guinea
D.t.tyro
  Aru Is
Dacelo gaudichaud (Rufous-bellied Giant Kingfisher)
  New Guinea, New Guinea Is, Aru Is

CLYTOCEYX
Clytoceyx rex (Shovel-billed Kingfisher)
C.r.rex
  E New Guinea
C.r.imperator
  C New Guinea

MELIDORA
Melidora macrorrhina (Hook-billed Kingfisher)
M.m.waigiuensis
  Waigeu I
M.m.macrorrhina
  Misool, Batanta Is, New Guinea
M.m.jobiensis
  Japen I, N New Guinea

CITTURA
Cittura cyanotis (Sulawesi Blue-eared Kingfisher)
C.c.sanghirensis
  Sanghir Is
C.c.cyanotis
  Sulawesi

HALCYON
Halcyon amauroptera (Brown-winged Kingfisher)
  NE India to Malaysia
Halcyon capensis (Stork-billed Kingfisher)   84.1
H.c.capensis
  Nepal, India, Sri Lanka
H.c.burmanica
  Burma to Indochina, N Malaya
H.c.intermedia
  Nicobar Is
H.c.malaccensis
  S Malaysia
H.c.cyanopteryx
  Sumatra, Bangka, Billiton Is
H.c.simalurensis
  Simalur I
H.c.sodalis
  Banjak I
H.c.nesoeca
  Nias, Batu Is
H.c.isoptera
  Pagi, Siberut, Sipora Is
H.c.fraseri
  Java
H.c.floresiana
  Bali to Flores I
H.c.javana
  Borneo
H.c.gouldi
  N & W Philippine Is
H.c.smithi
  SE Luzon, Masbate, Panay, Negros Is
H.c.gigantea
  S Philippine Is
Halcyon melanorhyncha (Great-billed Kingfisher)   84.1
H.m.melanorhyncha
  N, NC & SE Sulawesi
H.m.dicrorhyncha
  Peleng, Banggai Is
H.m.eutreptorhyncha
  Sula Is
Halcyon coromanda (Ruddy Kingfisher)
H.c.major
  Korea, Japan to S China & Sulawesi
H.c.coromanda
  Nepal to Malaysia & Indochina
H.c.mizorhina
  Andaman & Nicobar Is
H.c.minor
  S Malaysia, Borneo, Sumatra, Java
H.c.bangsi
  Riukiu Is, Taiwan, N & W Philippine Is
H.c.ochrothorectis
  S Philippine is
H.c.pelingensis
  Peling Is, N Sulawesi
H.c.rufa
  S Sulawesi, Sangihe Is
H.c.sulana
  Sula Is

Halcyon badia (**Chocolate-backed Kingfisher**)
Liberia to Gabon & E Zaire, W Uganda

Halcyon smyrnensis (White-throated Kingfisher)
H.s.smyrnensis
Asia Minor to S Yemen & India
H.s.fusca
W India, Sri Lanka
H.s.saturatior
Andaman Is
H.s.perpulchra
Burma, Malaysia to Indochina
H.s.fokiensis
S & E China, Taiwan
H.s.gularis
Philippine Is

Halcyon pileata (Black-capped Kingfisher)
India to China, Philippine Is, Sulawesi

Halcyon cyanoventris (Java Kingfisher)
Java, Bali

Halcyon leucocephala (Grey-headed Kingfisher)
H.l.acteon
Cape Verde Is
H.l.leucocephala
Senegal to N Zaire & Ethiopia
H.l.semicaerulea
Yemen & S Yemen
H.l.centralis
Kenya, N Tanzania
H.l.hyacinthina
E Kenya to Mozambique
H.l.pallidiventris
S Zaire to Namibia & Zimbabwe

Halcyon senegalensis (Woodland Kingfisher)
H.s.fuscopilea
Sierra Leone to Angola & S Zaire
H.s.senegalensis
Senegal to Ethiopia & Kenya
H.s.cyanoleuca
S Angola to Malawi & Transvaal

Halcyon senegaloides (African Mangrove Kingfisher)
H.s.ranivorus
coast of E Africa
H.s.senegaloides
coast of SE Africa

Halcyon malimbica (Blue-breasted Kingfisher)
H.m.fortis
Senegal
H.m.torquata
Gambia, Guinea
H.m.forbesi
Sierra Leone to S Cameroun
H.m.dryas
Principé I
H.m.malimbica
S Cameroun to Angola & Zaire
H.m.prenticei
Sudan, E Zaire, Uganda

Halcyon albiventris (Brown-hooded Kingfisher)

H.a.erlangeri
S Somalia, NE Kenya
H.a.orientalis
S Gabon & Angola to N Kenya & Mozambique
H.a.albiventris
Zimbabwe to E Cape Province
H.a.vociferans
NE Sth Africa
H.a.hylophila
Angola
H.a.prentissgrayi
Angola to SE Kenya

Halcyon chelicuti (Striped Kingfisher)
H.c.eremogiton
S Niger to E Sudan
H.c.chelicuti
Senegal to Ethiopia & Zambia
H.c.damarensis
S Angola to Transvaal & Mozambique

Halcyon nigrocyanea (Blue-black Kingfisher)
H.n.nigrocyanea
W New Guinea
H.n.quadricolour
Japen I, N New Guinea
H.n.stictolaema
S New Guinea

Halcyon winchelli (Winchell's Kingfisher)
H.w.nigrorum
Negros I
H.w.winchelli
S Philippine Is

Halcyon diops (Moluccan Kingfisher)
N Moluccas

Halcyon lazuli (South Moluccan Kingfisher)
S Moluccas

Halcyon macleayii (Forest Kingfisher)
H.m.elizabeth
E New Guinea
H.m.insularis
Aru Is
H.m.macleayii
N & E Australia to New Guinea
H.m.incincta
E Queensland >> E New Guinea, Kai Is

Halcyon albonotata (White-backed Kingfisher)
New Britain

Halcyon leucopygia (Ultramarine Kingfisher)
Solomon Islands

Halcyon farquhari (Chestnut-bellied Kingfisher)
C New Hebrides

Halcyon pyrrhopygia (Red-backed Kingfisher)
dry areas of Australia

Halcyon torotoro (Lesser Yellow-billed Kingfisher)
H.t.torotoro
W Papuan islands & W New Guinea
H.t.tentelare
Aru Is
H.t.pseustes
S New Guinea

H.t.brevirostris
S New Guinea
H.t.meeki
SE New Guinea
H.t.flavirostris
N Queensland
H.t.ochracea
D'Entrecasteaux Arch
Halcyon megarhyncha (Mountain Yellow-billed Kingfisher)
H.m.wellsi
C New Guinea
H.m.sellamontis
E New Guinea
H.m.megarhyncha
SE New Guinea
Halcyon australasia (Timor Kingfisher)
H.a.australasia
Lombok, Sumba, Timor, Wetar, Roma Is
H.a.dammeriana
Damar, Babar, Leti, Moa Is
H.a.odites
Timorlaut I
Halcyon sancta (Sacred Kingfisher)
H.s.sancta
Australia to Sumatra, Borneo, Philippine Is
H.s.vagans
New Zealand, Kermadec Is
H.s.norfolkiensis
Norfolk I
H.s.adamsi
Lord Howe I
H.s.canacorum
New Caledonia
H.s.macmillani
Loyalty I
Halcyon cinnamomina (Micronesian Kingfisher)
H.c.pelewensis
Palau Is
H.c.reichenbachii
Ponapé I
H.c.cinnamomina
Guam I
Halcyon funebris (Sombre Kingfisher)
Halmahera, Ternate Is
Halcyon chloris (White-collared Kingfisher)
H.c.abyssinica
Red Sea
H.c.kalbaensis
E Arabia
H.c.vidali
W India
H.c.davisoni
Andaman Is
H.c.occipitalis
Nicobar Is
H.c.humii
Burma, Malaysia, NE Sumatra
H.c.armstrongi
S Thailand, S Vietnam
H.c.chloroptera
W Sumatran islands

H.c.azela
Enggano I
H.c.palmeri
Java, Bali
H.c.laubmanniana
Borneo & Bornean islands, S Sumatra
H.c.collaris
Philippine Is
H.c.chloris
Sulawesi to NW New Guinea, Lesser Sunda Is
H.c.sordida
S New Guinea, Aru Is, N Australia
H.c.colona
SE New Guinea, Louisiade Arch
H.c.teraokai
Palau Is
H.c.owstoni
Ascuncion, Pagan, Alamagan Is (Mariana Is)
H.c.albicilla
Saipan, Tinan Is (Mariana Is)
H.c.orii
Rota I
H.c.matthiae
St Matthias Is
H.c.stresemanni
French, Rook Is
H.c.nusae
Bismarck Arch
H.c.novaehiberniae
SW New Ireland
H.c.bennetti
Nissan I
H.c.tristrami
New Britain
H.c.alberti
N & C Solomon Is
H.c.mala
Malaita I
H.c.pavuvu
Pavuvu I
H.c.solomonis
Ugi, San Cristobal, St Anna Is
H.c.amoena
Rennell I
H.c.brachyura
Reef I
H.c.vicina
Duff I
H.c.ornata
Santa Cruz, Tinakula Is
H.c.utupuae
Utupua I
H.c.melanodera
Vanikoro I
H.c.torresiana
Torres I
H.c.santoensis
Espiritu Santo, Banks Is
H.c.erromangae
Erromanga I

H.c.tannensis
Tanna I
H.c.juliae
Aneitum, Efate Is
H.c.vitiensis
Viti Levu, Vanua Levu, Taviuni Is
H.c.eximia
Kandavu, Ono, Vanua Kula Is
H.c.marina
Lau Arch
H.c.sacra
Tonga I
H.c.regina
Futuna I
H.c.pealei
Tutuila, Samoa Is
H.c.manuae
Ofu, Olosinga, Tau Is
Halcyon enigma (Obscure Kingfisher) 84.2
Talaut Is
Halcyon saurophaga (White-headed Kingfisher)
H.s.saurophaga
N Moluccas to Solomon Is
H.s.anachoreta
Hermit, Ninigo Is
H.s.admiralitatis
Admiralty Is
Halcyon recurvirostris (Flat-billed Kingfisher)
Samoa
Halcyon venerata (Tahitian Kingfisher)
H.v.venerata
Tahiti I
H.v.youngi
Moorea I
Halcyon tuta (Borabora Kingfisher)
H.t.tuta
Borabora I
H.t.atiu
Atiu I (Cook Is)
H.t.mauke
Mauke I (Cook Is)
Halcyon ruficollaris (Mangaia Kingfisher)
Mangaia I (Cook Is
Halcyon gambieri (Tuamotu Kingfisher)
H.g.gambieri
Mangareva I
H.g.gertrudae
Niau I
Halcyon godeffroyi (Marquesas Kingfisher)
Marquesas Is

CARIDONAX 84.2
Caridonax fulgida (Glittering Kingfisher)
Lombok, Sumbawa, Flores Is

ACTENOIDES 84.1
Actenoides bougainvillei (Moustached Kingfisher) 84.1
A.b.bougainvillei
Bougainville I
A.b.excelsa
Guadalcanal I

Actenoides concreta (Chestnut-collared Kingfisher)
A.c.concreta
S Burma, Malaysia, Sumatra, Bangka I
A.c.peristephes
S Thailand
A.c.borneana
Borneo
Actenoides lindsayi (Spotted Wood Kingfisher)
A.l.lindsayi
Luzon I
A.l.moseleyi
Negros I
A.l.hombroni
Mindanao I
A.l.burtoni
Mindanao I
Actenoides monacha (Blue-headed Wood Kingfisher) 84.1
A.m.monacha
N Sulawesi
A.m.intermedia
NC Sulawesi
A.m.capucina
E, SE & S Sulawesi
Actenoides princeps (Bar-headed Wood Kingfisher) 84.1
A.p.princeps
NE Sulawesi
A.p.erythrorohamphus
NW & C Sulawesi
A.p.regalis
SE Sulawesi

TANYSIPTERA
Tanysiptera hydrocharis (Aru Paradise Kingfisher)
S New Guinea, Aru Is
Tanysiptera galatea (Common Paradise Kingfisher)
T.g.emiliae
Rau I, Moluccas
T.g.doris
Morotai I
T.g.browningi
Halmahera
T.g.margarethae
Batjan I
T.g.sabrina
Kayoa I, Moluccas
T.g.obiensis
Obi, Oblilatu I
T.g.acis
Buru I
T.g.boanensis
Boano I
T.g.brunhildae
Doi I
T.g.nais
S Moluccas
T.g.galatea
NW New Guinea & W Papuan islands

T.g.meyeri
N New Guinea
T.g.minor
S & SE New Guinea
T.g.vulcani
Vulcan I
T.g.rosseliana
Rossel I
Tanysiptera riedelii (Biak Paradise Kingfisher)
Biak I
Tanysiptera carolinae (Numfor Paradise Kingfisher)
Numfor I
Tanysiptera ellioti (Kofiau Paradise Kingfisher)
Kofiau I
Tanysiptera nympha (Pink-breasted Paradise Kingfisher)
New Guinea
Tanysiptera danae (Brown-backed Paradise Kingfisher)
SE New Guinea
Tanysiptera sylvia (White-tailed Kingfisher)
T.s.leucura
Rook I
T.s.nigriceps
New Britain, Duke of York I
T.s.salvadoriana
SE New Guinea
T.s.sylvia
N Queensland to S New Guinea

85. TODIDAE (TODIES)

TODUS
Todus multicolor (Cuban Tody)
Cuba, Isle of Pines
Todus angustirostris (Narrow-billed Tody)
Hispaniola
Todus todus (Jamaican Tody)
Jamaica
Todus mexicanus (Puerto Rican Tody)
Puerto Rico
Todus subulatus (Broad-billed Tody)
Hispaniola, Gonave I

86. MOMOTIDAE (MOTMOTS)

HYLOMANES
Hylomanes momotula (Tody-Motmot)
H.m.chiapensis
S Mexico (Pacific)
H.m.momotula
S Mexico to Honduras (Caribbean)
H.m.obscurus
NW Costa Rica to NW Colombia

ASPATHA
Aspatha gularis (Blue-throated Motmot)
S Mexico to Honduras, El Salvador

ELECTRON
Electron platyrhynchum (Broad-billed Motmot)
E.p.minor
E Honduras to C Colombia
E.p.platyrhynchum
W Colombia, W Ecuador
E.p.pyrrholaemum
E Colombia, E Ecuador, Peru, N Bolivia
E.p.orienticola
W Brazil
E.p.chlorophrys
SW & C Brazil
E.p.colombianum
Colombia
Electron carinatum (Keel-billed Motmot)
S Mexico to NW Costa Rica

EUMOMOTA
Eumomota superciliosa (Turquoise-browed Motmot)
E.s.bipartita
S Mexico, W Guatemala
E.s.superciliosa
SE Mexico
E.s.vanrossemi
C Guatemala
E.s.sylvestris
E Guatemala
E.s.apiaster
El Salvador, W Honduras, NW Nicaragua
E.s.euroaustris
N Honduras
E.s.australis
NW Costa Rica

BARYPHTHENGUS
Baryphthengus ruficapillus (Rufous-capped Motmot)
B.r.semirufus
Panama to W Ecuador
B.r.costaricensis
E Nicaragua, E Costa Rica
B.r.ruficapillus
S & E Brazil, Paraguay, NE Argentina
Baryphthengus martii (Rufous Motmot)
Upper Amazonia

MOMOTUS
Momotus mexicanus (Russet-crowned Motmot)
M.m.vanrossemi
NW Mexico
M.m.mexicanus
NC & C Mexico
M.m.saturatus
SW Mexico
M.m.castaneiceps
C Guatemala
Momotus momota (Blue-crowned Motmot)
M.m.coeruliceps
NE & C Mexico

M.m.goldmani
SE Mexico, N Guatemala
M.m.exiguus
S Mexico
M.m.lessonii
S Mexico to W Panama
M.m.conexus
S Panama, NW Colombia
M.m.reconditus
E Panama, N Colombia
M.m.spatha
Colombia
M.m.olivaresi
Colombia
M.m.subrufescens
N Colombia, N Venezuela
M.m.osgoodi
W Venezuela
M.m.bahamensis
Trinidad, Tobago I
M.m.aequatorialis
WC Colombia, E Ecuador
M.m.chlorolaemus
E Peru
M.m.microstephanus
SE Colombia, E Ecuador, NW Brazil
M.m.momota
E Venezuela, the Guianas, N Brazil
M.m.argenticinctus
W Ecuador, NW Peru
M.m.ignobilis
E Peru, W Brazil
M.m.nattereri
NE Bolivia
M.m.simplex
W Brazil
M.m.cametensis
NC Brazil
M.m.parensis
NE Brazil
M.m.pilcomajensis
S Bolivia, S Brazil, NW Argentina

87 MEROPIDAE (BEE EATERS)

NYCTYORNIS
Nyctyornis amicta (Red-bearded Bee Eater)
Malaysia, Sumatra, Borneo
Nyctyornis athertoni (Blue-bearded Bee Eater)
N.a.athertoni
SW India & S Himalayas to Indochina
N.a.brevicaudata
Hainan I

MEROPOGON
Meropogon forsteni (Sulawesi Bearded Bee Eater)
Sulawesi

MEROPS
Merops gularis (Black Bee Eater)
M.g.gularis
Sierra Leone to Niger river

M.g.australis
Cameroun to Uganda & Angola
Merops muelleri (Blue-headed Bee Eater)
M.m.mentalis
Sierra Leone to Cameroun
M.m.muelleri
S Cameroun to E Zaire, W Kenya
Merops bulocki (Red-throated Bee Eater)
M.b.bulocki
Senegal to Central African Republic
M.b.frenatus
E Sudan to N Zaire & N Uganda
Merops bullockoides (White-fronted Bee Eater)
M.b.bullockoides
Gabon to Angola, Kenya & Natal
M.b.randorum
S Tanzania
Merops pusillus (Little Bee Eater)
M.p.pusillus
Senegal to N Zaire
M.p.ocularis
Sudan, W Ethiopia
M.p.cyanostictus
Ethiopia & Somalia to Tanzania
M.p.meridionalis
S Zaire & Tanzania to Namibia & Natal
M.p.argutus
SW Angola, Botswana, SW Zambia
Merops variegatus (Blue-breasted Bee Eater)
M.v.loringi
Cameroun to Uganda
M.v.variegatus
Gabon to Zaire & Angola
M.v.bangweoloensis
S Zaire & Zambia
M.v.lafresnayii
Ethiopia
Merops oreobates (Cinnamon-chested Bee Eater)
S Sudan, E Zaire, N Tanzania
Merops hirundineus (Swallow-tailed Bee Eater)
M.h.chrysolaimus
Senegal to Ghana & Central African Republic
M.h.heuglini
N Zaire, Uganda to Tanzania
M.h.hirundineus
SE Kenya to Angola & Natal
Merops breweri (Black-headed Bee Eater)
Cameroun, Gabon, W Zaire
Merops revoilii (Somali Bee Eater)
SE Ethiopia, Somalia, NE Kenya
Merops albicollis (White-throated Bee Eater)
Senegal to Ethiopia & Tanzania
Merops orientalis (Little Green Bee Eater)
M.o.viridissimus
Senegal to Air & N Ethiopia
M.o.cleopatra
Nile valley, Egypt
M.o.cyanophrys
W Saudi Arabia

M.o.muscatensis
Oman
M.o.najdanus
C Saudi Arabia
M.o.beludschicus
SE Iran to NW India
M.o.orientalis
India
M.o.ceylonicus
Sri Lanka
M.o.birmanus
Assam, Burma, W China, Indochina
Merops boehmi (Boehm's Bee Eater)
Tanzania, E Zambia, Malawi
Merops viridis (Chestnut-headed Bee Eater)
M.v.viridis
SE China to Sumatra, Java, Borneo
M.v.americanus
Philippine Is
Merops superciliosus (Blue-cheeked Bee Eater)   87.1.2
M.s.persicus
Israel to C Asia >> NW India & Africa
M.s.chrysocercus
N Africa to W Africa
M.s.superciliosus
Madagascar to S, C & E Africa
M.s.philippinus
India to SE China, Borneo & Lesser Sunda Is
M.s.celebensis
S Sulawesi
M.s.salvadorii
E New Guinea, New Britain
Merops ornatus (Australian Bee Eater)
Australia to Lesser Sunda Is
Merops apiaster (European Bee Eater)
S Europe & C Asia >> N India & Africa
Merops leschenaulti (Bay-headed Bee Eater)
M.l.leschenaulti
W India to Malaysia & Indochina
M.l.quinticolor
Java, Bali
M.l.andamanensis
S Andaman Is
Merops malimbicus (Rosy Bee Eater)
Ghana to N & W Zaire
Merops nubicus (Carmine Bee Eater)
M.n.nubicus
Senegal to Zaire & Ethiopia
M.n.nubicoides
Angola to Natal >> S Zaire & Tanzania

88.   CORACIIDAE (ROLLER)

CORACIAS
Coracias garrulus (European Roller)
C.g.garrulus
S Europe & C Asia >> Africa & India
C.g.semenowi
Transcaspia to NW India
Coracias abyssinica (Abyssinian Roller)
Senegal to S Yemen & Kenya

Coracias caudata (Lilac-breasted Roller)
C.c.lorti
S Ethiopia, Somalia, N Kenya
C.c.caudata
Uganda to Angola & N & E Sth Africa
Coracias spatulata (Racquet-tailed Roller)
Angola to Tanzania & Mozambique
Coracias naevia (Rufous-crowned Roller)
C.n.naevia
Senegal to Ethiopia & N Tanzania
C.n.mosambica
Angola & S Zaire to E Cape Province
Coracias benghalensis (Indian Roller)
C.b.benghalensis
E Saudi Arabia to NE India
C.b.indica
S India, Sri Lanka
C.b.affinis
Bhutan to Malaysia & Indochina
Coracias temminckii (Sulawesi Roller)   88.1
Sulawesi
Coracias cyanogaster (Blue-bellied Roller)
Senegal to N Zaire & Sudan

EURYSTOMUS
Eurystomus glaucurus (African Broad-billed Roller)
E.g.afer
Senegal to NE Zaire & E Ethiopia
E.g.aethiopicus
Sudan, W Ethiopia
E.g.suahelicus
Somalia to C Zaire & NE Zambia, Angola, Natal
E.g.glaucurus
Madagascar to C Africa
Eurystomus gularis (Blue-throated Roller)
E.g.gularis
Senegal to Mali & Togo
E.g.neclectus
Cameroun to Angola & E Zaire
Eurystomus orientalis (Eastern Broad-billed Roller)(Dollar Bird)
E.o.abundus
N Himalayas to Korea, Indochina
E.o.deignani
N Thailand to Sumatra, Java, Borneo
E.o.orientalis
S Himalayas to Sumatra & N Sulawesi
E.o.gigas
S Andaman Is
E.o.oberholseri
Simalur I
E.o.latouchei
NE China
E.o, waigiouensis
New Guinea & islands
E.o.pacificus
Australia, New Guinea, Lesser Sunda Is, S Sulawesi
E.o.crassirostris
Bismarck Arch

E.o.solomonensis
  Feni, Solomom Is
Eurystomus azureus (Azure Roller)   88.1
  N Moluccas

89. BRACHYPTERACIIDAE (GROUND ROLLERS)

BRACHYPTERACIAS
Brachypteracias leptosomus (Short-legged Ground Roller)
  NE Madagascar
Brachypteracias squamigera (Scaly Ground Roller)
  NE Madagascar

ATELORNIS
Atelornis pittoides (Pitta-like Ground Roller)
  Madagascar
Atelornis crossleyi (Crossley's Ground Roller)
  NE Madagascar

URATELORNIS
Uratelornis chimaera (Long-tailed Ground Roller)
  SW Madagascar

90 LEPTOSOMATIDAE (COUROLS)

LEPTOSOMUS
Leptosomus discolor (Courol)
  L.d.gracilis
    Gd Comoro I
  L.d.intermedius
    Anjouan I
  L.d.discolor
    Mayotte I, Madagascar

91. UPUPIDAE (HOOPOES)

UPUPA
Upupa epops (Hoopoe)
  U.e.epops
    Europe & W Asia to W & C Africa & India
  U.e.major
    Egypt
  U.e.senegalensis
    Senegal to Somalia
  U.e.waibeli
    Cameroun to Kenya
  U.e.orientalis
    NW India
  U.e.ceylonensis
    C & S India, Sri Lanka
  U.e.saturata
    E Siberia to N & E China
  U.e.longirostris
    Assam to Indochina, Malaysia, Sumatra
  U.e.marginata
    Madagascar
  U.e.africana
    S Zaire to Uganda, Kenya & Cape Province

92. PHOENICULIDAE (WOOD HOOPOES)

PHOENICULUS
Phoeniculus purpureus (Green Wood Hoopoe)
  P.p.senegalensis
    Senegal to Ghana
  P.p.guineensis
    Mali to Ivory Coast & Central African Republic
  P.p.niloticus
    N Sudan to E Zaire
  P.p.marwitzi
    E Uganda, E Zaire to Natal, Zanzibar
  P.p.angolensis
    Angola
  P.p.purpureus
    C & SW Sth Africa
Phoeniculus somaliensis (Black-billed Wood Hoopoe)   92.1
  P.s.abyssinicus
    N Ethiopia
  P.s.somaliensis
    Somalia
  P.s.neglectus
    S Ethiopia
Phoeniculus damarensis (Violet Wood Hoopoe)
  P.d.damarensis
    N & C Namibia
  P.d.granti
    S Ethiopia, W Kenya
Phoeniculus bollei (White-headed Wood Hoopoe)
  P.b.bollei
    Ghana to Cameroun
  P.b.jacksoni
    E Zaire, W Kenya
  P.b.okuensis
    S Cameroun
Phoeniculus castaneiceps (Forest Wood Hoopoe)
  P.c.castaneiceps
    Ghana, SW Nigeria
  P.c.brunneiceps
    S Cameroum, Zaire, Uganda
Phoeniculus aterrimus (Black Wood Hoopoe)
  P.a.aterrimus
    Senegal to N Zaire, S Sahara
  P.a.emini
    S Sudan, NE Zaire
  P.a.notatus
    N Ethiopia
  P.a.anchietae
    SW Zaire, N Angola

RHINOPOMASTUS
Rhinopomastus minor (Abyssinian Scimitar-bill)
  R.m.minor
    Ethiopia to C Somalia, E Kenya
  R.m.cabanisi
    Uganda, Kenya, C Tanzania

Rhinopomastus cyanomelas (Scimitar-bill)
  R.c.schalowi
    N Kenya to E Zaire, NE Transvaal &
    Mozambique
  R.c.cyanomelas
    Angola to Namibia, Botswana & Natal

93 BUCEROTIDAE (HORNBILLS)

TOCKUS
Tockus birostris (Indian Grey Hornbill)
    N & C India
Tockus fasciatus (Pied Hornbill)
  T.f.semifasciatus
    Senegal to Ghana
  T.f.fasciatus
    Cameroun to SW Sudan, Zaire, Angola
Tockus alboterminatus (Crowned Hornbill)
                                                93.1
    Ethiopia to Sth Africa
Tockus bradfieldi (Bradfield's Hornbill)   93.1
    S Angola, N Namibia, Botswana, W
    Zimbabwe
Tockus pallidirostris (Pale-billed Hornbill)
  T.p.pallidirostris
    Angola to SW Tanzania
  T.p.neumanni
    S Kenya, E Tanzania, Malawi
Tockus nasutus (African Grey Hornbill)
  T.n.nasutus
    Senegal to Ethiopia & N Kenya
  T.n.forskalii
    NE Ethiopia, W & S Saudi Arabia
  T.n.epirhinus
    S Kenya to Angola, Botswana & Natal
  T.n.dorsalis
    S Angola, Namibia
Tockus hemprichii (Hemprich's Hornbill) 93.1
    Ethiopia, Somalia, NW Kenya
Tockus monteiri (Monteiro's Hornbill)
    S Angola, N Namibia
Tockus griseus (Malabar Grey Hornbill)
  T.g.griseus
    W India
  T.g.gingalensis
    Sri Lanka
Tockus hartlaubi (Black Dwarf Hornbill)
  T.h.hartlaubi
    Guinea to S Cameroun
  T.h.granti
    WC to NE Zaire
Tockus camurus (Red-billed Dwarf Hornbill)
                                                93.1
    Liberia to NE Zaire
Tockus erythrorhynchus (Red-billed Hornbill)
  T.e.erythrorhynchus
    Senegal to Somalia & Tanzania
  T.e.rufirostris
    Angola to Malawi & Transvaal
  T.e.damarensis
    N Namibia
Tockus flavirostris (Yellow-billed Hornbill)
    Ethiopia & Somalia to S Kenya

Tockus leucomelas (Southern Yellow-billed
    Hornbill)   93.1
  T.l.elegans
    Angola
  T.l.leucomelas
    Namibia to Mozambique & Natal
Tockus deckeni (Von der Decken's Hornbill)
    C Ethiopia to C Tanzania
Tockus jacksoni (Jackson's Hornbill)
    C Ethiopia to C Tanzania

BERENICORNIS
Berenicornis comatus (Long-crested
    Hornbill)
    S Vietnam, Malaysia, Sumatra, Borneo
Berenicornis albocristatus (African White-
    crested Hornbill)
  B.a.albocristatus
    Sierra Leone to Ivory Coast
  B.a.macrourus
    Ghana & Togo
  B.a.cassini
    W Nigeria to Gabon & Uganda

PTILOLAEMUS
Ptilolaemus tickelli (Tickell's Hornbill)
  P.t.austeni
    S Assam
  P.t.tickelli
    S Burma
  P.t.indochinensis
    Indochina

ANORRHINUS
Anorrhinus galeritus (Bushy-crested Hornbill)
  A.g.carinatus
    S Burma, S Thailand, Malaysia
  A.g.galeritus
    Sumatra, N Borneo
  A.g.minor
    S Borneo

PENELOPIDES
Penelopides panini (Rufous-tailed Hornbill)
  P.p.manillae
    Luzon, Marinduque Is
  P.p.subnigra
    Polillo Is
  P.p.mindorensis
    Mindoro I
  P.p.ticaensis
    Ticao I
  P.p.panini
    Masbate, Panay, Negros Is
  P.p.samarensis
    Samar, Leyte, Bohol Is
  P.p.affinis
    Dinagat, Mindanao Is
  P.p.basilanica
    Basilan I
Penelopides exarhatus (Temminck's Hornbill)
  P.e.exarhatus
    N Sulawesi

P.e.sanfordi
C, SE, & S Sulawesi

ACEROS
Aceros nipalensis (Rufous-necked Hornbill)
Nepal to N Indochina
Aceros corrugatus (Wrinkled Hornbill)
A.c.corrugatus
Malaysia, Borneo
A.c.megistus
Sumatra, Batu Is
**Aceros leucocephalus (White-headed
Hornbill)**
A.l.waldeni
Panay, Guimaras, Negros Is
A.l.leucocephalus
Mindanao, Camiguin (South) Is
Aceros cassidix (Sulawesi Hornbill)
Sulawesi
Aceros undulatus (Wreathed Hornbill)
A.u.ticehursti
S Assam to Indochina & N Malaysia
A.u.undulatus
S Malaysia, Sumatra, Java, Bali, Borneo
Aceros plicatus (Blyth's Hornbill)
A.p.subruficollis
S Burma, S Thailand, Malaysia, Sumatra,
Borneo
A.p.plicatus
S Moluccas
A.p.ruficollis
N Moluccas, W & N New Guinea & islands
A.p.jungei
E New Guinea, Fergusson, Goodenough Is
A.p.dampieri
Bismarck Arch
A.p.harterti
Buka, Bougainville, Shortland Is
A.p.mendanae
Choiseul, Ysabel, Guadalcanal, Malaita Is
Aceros everetti (Everett's Hornbill)
Sumba I
Aceros narcondami (Narcondam Hornbill)
Narcondam I

ANTHRACOCEROS
Anthracoceros malayanus (Black Hornbill)
S Malaysia, Sumatra, Borneo
**Anthracoceros malabaricus (Indian Pied
Hornbill)**
A.m.malabaricus
S Himalayas, N Burma
A.m.leucogaster
Burma to Malaysia, Indochina, SE China
**Anthracoceros coronatus (Malabar Pied
Hornbill)**
A.c.coronatus
S India, Sri Lanka
A.c.convexus
Malaysia, Sumatra, Java, Borneo
Anthracoceros montani (Sulu Hornbill)
Jolo, Tawitawi Is

Anthracoceros marchei (Palawan Hornbill)
Calamian, Palawan, Balabac Is

CERATOGYMNA
Ceratogymna fistulator (Piping Hornbill)   93.1
C.f.fistulator
Senegal to W Nigeria
C.f.sharpii
SE Nigeria to C Zaire & N Angola
C.f.duboisi
Cameroun to W Uganda
Ceratogymna bucinator (Trumpeter Hornbill)
93.1
Angola to N Kenya & E Cape Province
**Ceratogymna cylindricus (Brown-cheeked
Hornbill)**   93.1
C.c.cylindricus
Sierra Leone to Benin
C.c.albotibialis
Nigeria & Cameroun to Uganda
**Ceratogymna subcylindricus (Black & White
Casqued Hornbill)**   93.1
C.s.subcylindricus
Ghana to S Nigeria
C.s.subquadratus
Kenya & E Zaire to Cameroun >> Angola
**Ceratogymna brevis (Silvery-cheeked
Hornbill)**   93.1
Ethiopia, S Kenya to Malawi & Zimbabwe
Ceratogymna atrata (Black-casqued Hornbill)
Liberia to Sudan & Angola
Ceratogymna elata (Yellow-casqued Hornbill)
Guinea to W Cameroun

BUCEROS
Buceros rhinoceros (Rhinoceros Hornbill)
B.r.rhinoceros
Malaysia
B.r.sumatranus
Sumatra, Billiton I
B.r.silvestris
Java
B.r.borneoensis
Borneo
Buceros bicornis (Great Indian Hornbill)
B.b.bicornis
W India, Himalayas to Indochina, Malaysia
B.b.cavatus
SW India
Buceros hydrocorax (Rufous Hornbill)
B.h.hydrocorax
Luzon, Marinduque Is
B.h.semigaleatus
Samar, Leyte, Bohol Is
B.h.mindanensis
Mindanao I
B.h.basilanicus
Basilan

RHINOPLAX
Rhinoplax vigil (Helmeted Hornbill)
Malaysia, Sumatra, Borneo

BUCORVUS
Bucorvus abyssinicus (Abyssinian Ground Hornbill)
Gambia to C Kenya
Bucorvus leadbeateri (Southern Ground Hornbill)
N Angola to Kenya & Cape Province

PICIFORMES

94. GALBULIDAE (JACAMARS)

GALBALCYRHYNCHUS
Galbalcyrhynchus leucotis (Chestnut Jacamar)
N Upper Amazonia
Galbalcyrhynchus purusianus (Purus Jacamar)
E & S Peru, W Brazil, N Bolivia

BRACHYGALBA
Brachygalba albogularis (White-throated Jacamar)
E Peru
Brachygalba lugubris (Brown Jacamar)
B.l.fulviventris
E Colombia
B.l.caquetae
SE Colombia to E Peru
B.l.lugubris
E & S Venezuela, the Guianas, N Brazil
B.l.obscuriceps
S Venezuela, NW Brazil
B.l.naumbergi
NE Brazil
B.l.melanosterna
E Bolivia, C & SW Brazil
B.l.phaeonota
C Brazil
Brachygalba goeringi (Pale-headed Jacamar)
E Colombia, N Venezuela
Brachygalba salmoni (Dusky-backed Jacamar)
E Panama, NW Colombia

JACAMARALCYON
Jacamaralcyon tridactyla (Three-toed Jacamar)
SE Brazil

GALBULA
Galbula albirostris (Yellow-billed Jacamar)
G.a.chalcocephala
S Colombia, Ecuador, NW Peru, W Brazil
G.a.albirostris
E Venezuela, the Guianas, N Brazil
Galbula cyanicollis (Blue-necked Jacamar)
C Brazil
Galbula galbula (Green-tailed Jacamar)
E & S Venezuela, the Guianas, N & C Brazil
Galbula ruficauda (Rufous-tailed Jacamar)
G.r.melanogenia
S Mexico to W Ecuador

G.r.ruficauda
C Colombia, Venezuela, the Guianas, N Brazil
G.r.pallens
N Colombia
G.r.brevirostris
NW Venezuela
G.r.rufoviridis
S Brazil, N Bolivia, Paraguay, N Argentina
G.r.heterogyna
E Bolivia, SW Brazil
Galbula tombacea (White-chinned Jacamar)
G.t.tombacea
S Colombia, Ecuador, Peru, E Brazil
G.t.mentalis
C & WC Brazil
Galbula cyanescens (Bluish-fronted Jacamar)
W Brazil, E Peru
Galbula pastazae (Coppery-chested Jacamar)
W Brazil, E Ecuador
Galbula leucogastra (Bronzy Jacamar)
G.l.chalcothorax
W Brazil, E Ecuador, E Peru
G.l.leucogastra
W Venezuela, the Guianas, W Brazil
G.l.viridissima
C Brazil
Galbula dea (Paradise Jacamar)
G.d.dea
Venezuela, the Guianas, N Brazil
G.d.amazonum
N Bolivia, SW Brazil
G.d.brunneiceps
E Colombia, E Peru, W Brazil
G.d.phainopepla
W C Brazil

JACAMEROPS
Jacamerops aurea (Great Jacamar)
J.a.penardi
Costa Rica to W Colombia
J.a.aurea
E Colombia, the Guianas, Venezuela
J.a.ridgwayi
NE & C Brazil
J.a.isidori
E Ecuador, E Peru, N Bolivia, W Brazil

95. BUCCONIDAE (PUFFBIRDS)

NOTHARCHUS
Notharchus macrorhynchos (White-necked Puffbird)
N.m.cryptoleucus
El Salvador, NW Nicaragua
N.m.hyperrynchus
S Mexico to NW Sth America
N.m.macrorhynchos
the Guianas, N Brazil
N.m.paraensis
E Brazil

N.m.swainsoni
SE Brazil, E Paraguay, NE Argentina
Notharchus pectoralis (Black-breasted Puffbird)
E Panama to NW Ecuador
Notharchus ordii (Brown-banded Puffbird)
S Venezuela, NW Brazil
Notharchus tectus (Pied Puffbird)
N.t.subtectus
E Panama to C Colombia & SW Ecuador
N.t.picatus
E Ecuador, E Peru
N.t.tectus
S Venezuela, the Guianas, N Brazil

BUCCO
Bucco macrodactylus (Chestnut-capped Puffbird)
B.m.macrodactylus
N & W Amazonia
B.m.caurensis
S Venezuela
Bucco tamatia (Spotted Puffbird)
B.t.pulmentum
S Colombia, E Ecuador, E Peru, W Brazil
B.t.tamatia
E Colombia, Venezuela, the Guianas, N Brazil
B.t.inexpectatus
NC Brazil
B.t.punctuliger
C Brazil
B.t.hypneleus
EC Brazil
B.t.interior
SW Brazil
Bucco noanamae (Sooty-capped Puffbird)
W Colombia
Bucco capensis (Collared Puffbird)
B.c.dugandi
SE Colombia, Ecuador, C Peru
B.c.capensis
the Guianas, Brazil, E Peru

NYSTALUS
Nystalus radiatus (Barred Puffbird)
W Panama to W Ecuador
Nystalus chacuru (White-eared Puffbird)
N.c.uncirostris
E Peru, E Bolivia
N.c.chacuru
S Brazil, Paraguay, NE Argentina
Nystalus striolatus (Striolated Puffbird)
N.s.striolatus
W Amazonia
N.s.torridus
S Brazil
Nystalus maculatus (Spot-backed Puffbird)
N.m.maculatus
E Brazil
N.m.parvirostris
C Brazil

N.m.pallidigula
SW Brazil
N.m.striatipectus
E & S Bolivia, N Argentina

HYPNELUS
Hypnelus ruficollis (Russet-throated Puffbird)
H.r.ruficollis
N Colombia, W Venezuela
H.r.decolor
NE Colombia, NW Venezuela
H.r.striaticollis
NW Venezuela
H.r.coloratus
W Venezuela
H.r.bicinctus
N Venezuela
H.r.stoicus
Margarita I

MALACOPTILA
Malacoptila striata (Crescent-chested Puffbird)
M.s.minor
E Brazil
M.s.striata
SE Brazil
Malacoptila fusca (White-chested Puffbird)
M.f.fusca
N & NW Amazonia
M.f.venezuelae
S Venezuela
Malacoptila semicincta (Semicollared Puffbird)
S Peru, N Bolivia, W Brazil
Malacoptila fulvogularis (Black-streaked Puffbird)
M.f.substriata
Colombia
M.f.huilae
N Colombia
M.f.fulvogularis
E Ecuador to Bolivia
Malacoptila rufa (Rufous-necked Puffbird)
M.r.rufa
E Ecuador, E Peru, W Brazil
M.r.brunnescens
C Brazil
Malacoptila panamensis (White-whiskered Puffbird)
M.p.inornata
S Mexico to N Nicaragua
M.p.fuliginosa
SE Nicaragua to W Panama
M.p.panamensis
SW Costa Rica to W Colombia
M.p.poliopis
SW Colombia, W Ecuador
M.p.magdalenae
N Colombia
Malacoptila mystacalis (Moustached Puffbird)
Colombia, NW Venezuela

MICROMONACHA
Micromonacha lanceolata (Lanceolated Monklet)
M.l.austinsmithi
E Costa Rica, W Panama
M.l.lanceolata
W Colombia, E Ecuador, E Peru, W Brazil

NONNULA
Nonnula rubecula (Rusty-breasted Nunlet)
N.r.duidae
S Venezuela
N.r.interfluvialis
S Venezuela, N Brazil
N.r.cineracea
W Brazil, NE Peru
N.r.simplex
NE Brazil
N.r.rubecula
SE Brazil, Paraguay, NE Argentina
Nonnula sclateri (Fulvous-chinned Nunlet)
W Brazil
Nonnula brunnea (Brown Nunlet)
S Colombia, E Ecuador, E Peru
Nonnula ruficapilla (Grey-cheeked Nunlet)
N.r.frontalis
E Panama, N Colombia
N.r.pallescens
NE Colombia
N.r.rufipectus
NE Peru
N.r.ruficapilla
E Peru, W Brazil
N.r.nattereri
SW Brazil
Nonnula amaurocephala (Chestnut-headed Nunlet)
W Brazil

HAPALOPTILA
Hapaloptila castanea (White-faced Nunbird)
W Colombia, W Ecuador

MONASA
Monasa atra (Black Nunbird)
S Venezuela, the Guianas, N Brazil
Monasa nigrifrons (Black-fronted Nunbird)
M.n.nigrifrons
W & NW Amazonia
M.n.canescens
E Bolivia
Monasa morphoeus (White-fronted Nunbird)
M.m.grandior
Nicaragua to NW Panama
M.m.fidelis
E Panama
M.m.pallescens
E Panama, NW Colombia
M.m.sclateri
N Colombia
M.m.peruana
SE Colombia, E Peru, NW Brazil

M.m.morphoeus
S & E Brazil
M.m.boliviana
NE Bolivia
Monasa flavirostris (Yellow-billed Nunbird)
E Colombia, E Ecuador, E Peru, W Brazil

CHELIDOPTERA
Chelidoptera tenebrosa (Swallow-wing Puffbird)
C.t.tenebrosa
NE & NC Sth America
C.t.brasiliensis
E & SE Brazil
C.t.pallida
W Venezuela

96. CAPITONIDAE (BARBETS)

CAPITO
Capito aurovirens (Scarlet-crowned Barbet)
Upper Amazonia
Capito maculicoronatus (Spot-crowned Barbet)
C.m.maculicoronatus
W Panama
C.m.pirrensis
E Panama (Pacific), NW Colombia
C.m.melas
E Panama (Caribbean)
C.m.rubrilateralis
W Colombia
Capito squamatus (Orange-fronted Barbet)
SW Colombia, W Ecuador
Capito hypoleucus (White-mantled Barbet)
96.1
C.h.hypoleucus
N Colombia
C.h.carrikeri
NC Colombia
C.h.extinctus
C Colombia e?
Capito dayi (Black-girdled Barbet)
W Brazil
Capito quinticolor (Five-coloured Barbet)
W Colombia
Capito niger (Black-spotted Barbet)
C.n.niger
the Guianas, NE Brazil
C.n.punctatus
E Colombia to E Peru
C.n.intermedius
WC Venezuela
C.n.aurantiicinctus
S Venezuela
C.n.auratus
NW Peru
C.n.orosae
NW Peru, NW Brazil
C.n.amazonicus
NW Brazil
C.n.transilens
NW Brazil

C.n.nitidior
NW Brazil
C.n.hypochondriacus
NC Brazil
C.n.novaolindae
NW Brazil
C.n.arimae
NW Brazil
C.n.insperatus
SE Peru, N Bolivia, W Brazil
C.n.bolivianus
W Bolivia
C.n.brunneipectus
NC Brazil

EUBUCCO
Eubucco richardsoni (Lemon-throated Barbet)
E.r.richardsoni
SE Colombia, E Ecuador, E Peru
E.r.nigriceps
NW Peru
E.r.aurantiicollis
E Peru, W Brazil
E.r.coccineus
C Peru
Eubucco bourcierii (Red-headed Barbet)
E.b.salvini
Costa Rica, W Panama
E.b.anomalus
E Panama
E.b.occidentalis
W Colombia
E.b.bourcierii
C & E Colombia
E.b.aequatorialis
W Ecuador
E.b.orientalis
E Ecuador
Eubucco tucinkae (Scarlet-hooded Barbet)
SE Peru
Eubucco versicolor (Versicoloured Barbet)
E.v.steerii
N Peru
E.v.glaucogularis
C Peru
E.v.versicolor
S Peru, NW Bolivia

SEMNORNIS
Semnornis frantzii (Prong-billed Barbet)
Costa Rica, W Panama
Semnornis ramphastinus (Toucan Barbet)
S.r.caucae
W Colombia
S.r.ramphastinus
C Ecuador

PSILOPOGON
Psilopogon pyrolophus (Fire-tufted Barbet)
Malaysia, Sumatra

MEGALAIMA
Megalaima virens (Great Barbet)
M.v.marshallorum
NW Himalayas
M.v.magnifica
Assam, N Burma
M.v.clamator
C Burma
M.v.virens
E & S China to S Burma, Thailand N Indochina
Megalaima lagrandieri (Red-vented Barbet)
M.l.rothschildi
N Laos, N Vietnam
M.l.lagrandieri
S Laos, S Vietnam
Megalaima zeylanica (Oriental Green Barbet)
M.z.kangrae
W Himalayas
M.z.inornata
W India
M.z.caniceps
C & E India
M.z.zeylanica
S India, Sri Lanka
Megalaima lineata (Lineated Barbet)
M.l.hodgsoni
W Himalayas to Malaysia & Indochina
M.l.lineata
Java, Bali
Megalaima viridis (Small Green Barbet)
S India
Megalaima faiostricta (Green-eared Barbet)
M.f.praetermissa
S China, N Vietnam
M.f.faiostricta
Thailand, Laos, S Vietnam
Megalaima corvina (Brown-throated Barbet)
Java
Megalaima chrysopogon (Gold-whiskered Barbet)
M.c.laeta
S Thailand, Malaysia
M.c.chrysopogon
Sumatra
M.c.chrysopsis
Borneo
Megalaima rafflesii (Many-coloured Barbet)
M.r.malayensis
S Burma, Malaysia
M.r.rafflesii
Sumatra, Bangka I
M.r.billitonis
Billiton, Mendanau Is
M.r.borneensis
Borneo
Megalaima mystacophanos (Gaudy Barbet)
M.m.mystacophanos
S Burma, Malaysia, Sumatra
M.m.humii
Borneo
M.m.ampala
Batu Is

Megalaima javensis (Black-banded Barbet)
   Java
Megalaima flavifrons (Yellow-fronted Barbet)
   Sri Lanka
Megalaima franklinii (Golden-throated Barbet)
   M.f.franklinii
   E Himalayas to S China & N Vietnam
   M.f.ramsayi
   S Burma, N & W Thailand
   M.f.auricularis
   S Laos, S Vietnam
   M.f.trangensis
   S Thailand
   M.f.minor
   Malaysia
Megalaima oorti (Müller's Barbet)
   M.o.oorti
   Malaysia, Sumatra
   M.o.annamensis
   S Indochina
   M.o.nuchalis
   Taiwan
   M.o.faber
   Hainan I
   M.o.sini
   SE China
Megalaima asiatica (Blue-throated Barbet)
   M.a.asiatica
   N India, Assam, N & C Burma
   M.a.rubescens
   S Assam, NW Burma
   M.a.davisoni
   S Burma to S China, N Indochina
   M.a.chersonesus
   S Thailand
   M.a.monticola
   N Borneo
Megalaima incognita (Hume's Blue-throated Barbet)
   M.i.incognita
   S Burma
   M.i.elbeli
   S Thailand
   M.i.euroa
   S Thailand, Indochina
Megalaima henricii (Yellow-crowned Barbet)
   M.h.henricii
   S Thailand, Malaysia, Sumatra
   M.h.brachyrhyncha
   Borneo
Megalaima armillaris (Blue-crowned Barbet)
   M.a.armillaris
   W & C Java
   M.a.baliensis
   E Java, Bali
Megalaima pulcherrima (Golden-naped Barbet)
   NW Borneo
Megalaima australis (Blue-eared Barbet)
   M.a.cyanotis
   Himalayas to Indochina
   M.a.stuarti
   S Burma, S Thailand

   M.a.duvaucelii
   Malaysia, Sumatra, Borneo
   M.a.gigantorhinus
   Nias I
   M.a.tanamassae
   Batu Is
   M.a.australis
   Java
   M.a.hebereri
   Bali
Megalaima eximia (Black-throated Barbet)
   M.e.eximia
   N Borneo
   M.e.cyanea
   NE Borneo
Megalaima rubricapilla (Crimson-throated Barbet)
   M.r.malabarica
   SW India
   M.r.rubricapilla
   Sri Lanka
Megalaima haemacephala (Crimson-breasted Barbet)
   M.h.indica
   NW India to Malaysia & Indonesia
   M.h.delica
   Sumatra
   M.h.rosea
   Java, Bali
   M.h.haemacephala
   Luzon, Mindoro, Samar, Leyte, Mindanao Is
   M.h.intermedia
   Tablas, Romblon, Masbate, Negros, Cebu Is

CALORHAMPHUS
Calorhamphus fuliginosus (Brown Barbet)
   C.f.hayii
   S Burma, Malaysia, Sumatra
   C.f.fuliginosus
   Borneo (except North)
   C.f.tertius
   N Borneo

GYMNOBUCCO
Gymnobucco calvus (Naked-faced Barbet)
   G.c.calvus
   Sierra Leone to S Nigeria, Gabon
   G.c.congicus
   SW Zaire, N Angola
   G.c.vernayi
   S Angola
Gymnobucco peli (Bristle-nosed Barbet)
   Ghana to Gabon
Gymnobucco sladeni (Sladen's Barbet)
   N & E Zaire
Gymnobucco bonapartei (Grey-throated Barbet)
   G.b.bonapartei
   W Cameroun to E Zaire

G.b.cinereiceps
S Sudan to NW Tanzania

SMILORHIS
Smilorhis leucotis (White-eared Barbet)
S.l.kilimensis
C & S Kenya, N Tanzania
S.l.leucogrammicus
Tanzania
S.l.leucotis
Malawi to Natal

CRYPTOLYBIA
Cryptolybia olivacea (Green Barbet)
C.o.olivacea
E Kenya, Tanzania
C.o.howelli
Morogoro, E Tanzania
C.o.woodwardi
N Natal
C.o.rungweensis
SW Tanzania, N Malawi
C.o.belcheri
Malawi, N Mozambique

STACTOLAEMA
Stactolaema anchietae (Anchieta's Barbet)
S.a.rex
Angola
S.a.anchietae
Angola, S Zaire
S.a.katangae
S Zaire, N Zambia
Stactolaema whytii (Whyte's Barbet)
S.w.stresemanni
S Tanzania
S.w.whytii
S Tanzania, S Malawi
S.w.terminatum
Iringa (E Tanzania)
S.w.sowerbyi
S Malawi, Zimbabwe
S.w.buttoni
N Zambia
S.w.angoniensis
W Malawi, NE Zambia

BUCCANODON   96.3
Buccanodon duchaillui (Yellow-spotted Barbet)
Sierra Leone & Gabon to Uganda

POGONIULUS
Pogoniulus scolopaceus (Speckled Tinkerbird)
P.s.scolopaceus
Sierra Leone & E Nigeria
P.s.stellatus
Fernando Po I
P.s.flavisquamatus
Cameroun to W Kenya to N Angola

Pogoniulus leucomystax (Moustached Green Tinkerbird)
W Kenya to W Malawi
Pogoniulus simplex (Green Tinkerbird)
S Kenya, Tanzania to S Malawi
Pogoniulus coryphaeus (Western Green Tinkerbird)
P.c.coryphaeus
N & W Cameroun
P.c.hildamariae
E Zaire, W Uganda
P.c.angolensis
W Angola
Pogoniulus pusillus (Red-fronted Tinkerbird)
P.p.uropygialis
Ethiopia, N Somalia
P.p.affinis
Kenya, S Somalia, N Tanzania
P.p.pusillus
SE Sth Africa
Pogoniulus chrysoconus (Yellow-fronted Tinkerbird)   96.3
P.c.chrysoconus
Senegal to Sudan, Uganda
P.c.xanthostictus
Ethiopia
P.c.extoni
S Angola, Namibia to SE Malawi & N Transvaal
Pogoniulus bilineatus (Yellow-rumped Tinkerbird)   96.3
P.b.leucolaima
Gambia to Uganda & N Angola
P.b.poensis
Fernando Po I
P.b.mfumbiri
S Uganda, S Zaire, SW Tanzania, NE Zambia
P.b.jacksoni
C & W Kenya
P.b.fischeri
E Kenya, Tanzania, Zanzibar I
P.b.bilineatus
Malawi, Mozambique, E Sth Africa
Pogoniulus makawai (White-chested Tinkerbird)
NW Zambia
Pogoniulus subsulphureus (Yellow-throated Tinkerbird)
P.s.chrysopygus
Guinea to Ghana
P.s.flavimentum
S Nigeria to Uganda & SW Zaire
P.s.subsulphureus
Fernando Po I
Pogoniulus atroflavus (Red-rumped Tinkerbird)
Guinea to NE Zaire

TRICHOLAEMA
Tricholaema lacrymosa (Spotted-flanked Barbet)

T.l.lacrymosa
N Uganda & S Ethiopia to C Tanzania
T.l.radcliffei
E Zaire, Rwanda, W Tanzania, SW Kenya
Tricholaema leucomelaina (Pied Barbet)
T.l.centralis
Transvaal, N Botswana
T.l.leucomelaina
Orange Free State, N Botswana
T.l.affinis
E Zimbabwe to Natal
Tricholaema diademata (Red-fronted Barbet)
T.d.diademata
S Sudan, Ethiopia, N Somalia, Kenya
T.d.massaica
S Kenya, N Tanzania
Tricholaema frontata (Miombo Pied Barbet)
96.3
Angola to Malawi
Tricholaema melanocephala (African Black-throated Barbet)
T.m.melanocephala
N Ethiopia, NW Somalia
T.m.stigmatothorax
S Ethiopia, S Somalia to C Tanzania
T.m.blandi
C & E Somalia, SE Ethiopia
T.m.flavibuccalis
N Tanzania 96.3
Tricholaema hirsuta (Hairy-breasted Barbet)
T.h.hirsusta
Sierra Leone to Togo
T.h.flavipunctata
S Nigeria, Cameroun, Gabon, Cabinda
T.h.angolensis
S Zaire, N Angola
T.h.ansorgii
Cameroun to E Zaire, Uganda

LYBIUS
Lybius undatus (Banded Barbet)
L.u.thiogaster
N Ethiopia
L.u.undatus
E & C Ethiopia
L.u.leucogenys
W & SW Ethiopia
L.u.salvadori
SE Ethiopia
Lybius vieilloti (Vieillot's Barbet)
L.v.buchanani
S Sahara
L.v.rubescens
Senegal to N Cameroun
L.v.vieilloti
Ethiopia, S Sudan, N Zaire
Lybius torquatus (Black-collared Barbet)
L.t.zombae
S Malawi, S Tanzania
L.t.pumilio
E Zaire, Burundi
L.t.irroratus
E Kenya, NE Tanzania

L.t.congicus
SE Zaire, N Angola, N Zambia, SW Tanzania
L.t.vivacens
Mozambique
L.t.bocagei
SW Angola
L.t.torquatus
S Angola, Zimbabwe, Sth Africa
Lybius guifsobalito (Black-billed Barbet)
Ethiopia, Sudan, Uganda, NE Zaire
Lybius rubrifacies (Red-faced Barbet)
S Uganda, NW Tanzania
Lybius chaplini (Chaplin's Barbet)
S Zambia
Lybius leucocephalus (White-headed Barbet)
L.l.leucocephalus
S Sudan, Uganda, S Zaire
L.l.adamauae
N Nigeria to N Zaire
L.l.albicauda
S Kenya, S Tanzania
L.l.senex
C Kenya
L.l.lynesi
C & SC Tanzania
L.l.leucogaster
SW Angola
Lybius minor (Black-backed Barbet)
L.m.minor
SW Zaire, NW Angola 96.2
L.m.macclounii
C & SE Zaire, Zambia, N Malawi
Lybius melanopterus (Brown-breasted Barbet)
S Somalia to Malawi, Mozambique
Lybius bidentatus (Double-toothed Barbet)
L.b.bidentatus
Guinea to N Angola
L.b.aequatorialis
N Zaire, W Kenya, NW Tanzania, E Ethiopia
Lybius dubius (Bearded Barbet)
Senegal to S Chad
Lybius rolleti (Black-breasted Barbet)
Central African Republic to W & SW Sudan

TRACHYPHONUS
Trachyphonus purpuratus (Yellow-billed Barbet)
T.p.goffinii
Sierra Leone to Ghana
T.p.togoensis
Togo, SW Nigeria
T.p.purpuratus
Cameroun to N Zaire & N Angola
T.p.elgonensis
NE Zaire & Uganda
Trachyphonus vaillantii (Levaillant's Barbet)
T.v.suahelicus
N Angola to Tanzania
T.v.vaillantii
Sthn Africa

Trachyphonus erythrocephalus (Red & Yellow Barbet)
T.e.shelleyi
E Ethiopia, N Somalia
T.e.versicolor
N Uganda, W Kenya, S Ethiopia
T.e.erythrocephalus
S Kenya, N Tanzania
Trachyphonus darnaudii (d'Arnaud's Barbet)
T.d.darnaudii
E Sudan to W Kenya
T.d.boehmi
S Somalia, E Kenya, , NE Tanzania
T.d.emini
C & SW Tanzania
Trachyphonus usambiro (Usambiro Barbet)
C Kenya to NW Tanzania
Trachyphonus margaritatus (Yellow-breasted Barbet)
T.m.margaritatus
Niger, N.Nigeria to N Sudan, W Ethiopia
T.m.somalicus
C Ethiopia, N Somalia

97 INDICATORIDAE (HONEYGUIDES)

PRODOTISCUS
Prodotiscus insignis (Cassin's Honeybird)
P.i.flavodorsalis
Sierra Leone to Togo
P.i.insignis
S Cameroun to W Zaire, Uganda
Prodotiscus zambesiae (Green-backed Honeybird)   97.3
P.z.ellenbecki
S Ethiopia to C Tanzania
P.z.zambesiae
SE Zaire, N Angola to Zimbabwe & Mozambique
Prodotiscus regulus (Wahlberg's Honeybird)
P.r.camerunensis
Cameroun
P.r.regulus
Ethiopia to N Angola & Natal

MELIGNOMON
Melignomon zenkeri (Zenker's Honeyguide)
S Cameroun to Uganda
Melignomon eisentrauti (Yellow-footed Honeyguide)   97.1
Liberia to Mt Cameroun

INDICATOR
Indicator maculatus (Spotted Honeyguide)
I.m.maculatus
Gambia to Ghana
I.m.stictithorax
Cameroun to NE Zaire
Indicator variegatus (Scaly-throated Honeyguide)   97.2.3
Somalia to Cape Province

Indicator indicator (Black-throated Honeyguide)
Senegal to Ethiopia & Cape Province
Indicator minor (Lesser Honeyguide)
I.m.senegalensis
Senegal to W Sudan
I.m.riggenbachi
W Cameroun to N Zaire
I.m.diadematus
E Sudan, Ethiopia, Somalia
I.m.teitensis
Uganda, Kenya to Malawi, Zambia
I.m.minor
N Angola to Natal & Cape Province
I.m.damarensis
Namibia
Indicator conirostris (Thick-billed Honeyguide)   97.3
I.c.ussheri
Ghana
I.c.conirostris
Nigeria to W Uganda
Indicator exilis (Least Honeyguide)
I.e.poensis
Fernando Po I
I.e.exilis
S Nigeria to E Zaire, Angola, Zambia
I.e.pachyrhynchus
Sudan, Uganda to W Tanzania
Indicator willcocksi (Willcocks' Honeyguide)
I.w.ansorgei
Guinea
I.w.willcocksi
Ghana
I.w.hutsoni
Nigeria, N Cameroun
Indicator meliphilus (Eastern Least Honeyguide)
Kenya to Malawi, Zimbabwe
Indicator pumilio (Pygmy Honeyguide)
E Zaire, Rwanda
Indicator xanthonotus (Indian Honeyguide)
I.x.xanthonotus
W Himalayas to E Assam
I.x.fulvus
Naga Hills (E Assam)
Indicator archipelagus (Malay Honeyguide)
Malaysia, Sumatra, Borneo

MELICHNEUTES
Melichneutes robustus (Lyre-tailed Honeyguide)
Cameroun to C Zaire

98. RAMPHASTIDAE (TOUCANS)

AULACORHYNCHUS
Aulacorhynchus prasinus (Emerald Toucanet)
A.p.wagleri
SW Mexico
A.p.prasinus
SE Mexico

A.p.stenorhabdus
  S Mexico, W Guatemala, El Salvador
A.p.virescens
  N Guatemala, Belize to N Nicaragua
A.p.volcanius
  E El Salvador
A.p.maxillaris
  Costa Rica, W Panama
A.p.caeruleogularis
  EC Panama
A.p.cognatus
  E Panama
A.p.griseigularis
  NW Colombia
A.p.phaeolaemus
  W Colombia
A.p.lautus
  N Colombia
A.p.albivitta
  E Colombia, E Ecuador, W Venezuela
A.p.cyanolaemus
  SE Ecuador, N Peru
A.p.dimidiatus
  N Peru
A.p.atrogularis
  E Peru
Aulacorhynchus sulcatus (Groove-billed Toucanet)
A.s.sulcatus
  N Venezuela
A.s.erythrognathus
  NE Venezuela
A.s.calorhynchus
  N Colombia, NW Venezuela
Aulacorhynchus derbianus (Chestnut-tipped Toucanet)
A.d.derbianus
  E Ecuador, E Peru, NE Bolivia
A.d.nigrirostris
  C Peru
A.d.duidae
  S Venezuela
A.d.whitelianus
  S Venezuela, S Guyana
A.d.osgoodi
  S Guyana
Aulacorhynchus haematopygus (Crimson-rumped Toucanet)
A.h.sexnotatus
  SW Colombia, W Ecuador
A.h.haematopygus
  N Colombia, W Venezuela
Aulacorhynchus huallagae (Yellow-browed Toucanet)
  NC Peru
Aulacorhynchus coeruleicinctis (Blue-banded Toucanet)
  S Peru, N Bolivia

PTEROGLOSSUS
Pteroglossus viridis (Green Araçari)
P.v.humboldti
  SE Colombia to N Bolivia, W Brazil

P.v.didymus
  ?
P.v.viridis
  E Venezuela, the Guianas, N Brazil
Pteroglossus inscriptus (Lettered Araçari)
  C & S Brazil
Pteroglossus bitorquatus (Red-necked Araçari)
P.b.sturmii
  WC Brazil
P.b.reichenowi
  NC Brazil
P.b.bitorquatus
  NE Brazil
Pteroglossus flavirostris (Ivory-billed Araçari)
P.f.flavirostris
  N Upper Amazonia
P.f.azara
  NW Brazil
P.f.mariae
  NE Peru, W Brazil, N Bolivia
Pteroglossus aracari (Black-necked Araçari)
P.a.roraimae
  S & E Venezuela, Guyana, Surinam
P.a.atricollis
  French Guiana, N Brazil
P.a.aracari
  C & E Brazil
P.a.vergens
  S & SE Brazil
Pteroglossus castanotis (Chestnut-eared Araçari)
P.c.castanotis
  E Colombia, E Ecuador, E Peru, NW Brazil
P.c.australis
  E Bolivia, W Brazil, NE Argentine
Pteroglossus pluricinctus (Many-banded Araçari)
  E Peru & E Colombia to E Venezuela
Pteroglossus torquatus (Collared Araçari)
P.t.torquatus
  S Mexico to W Panama
P.t.erythrozonus
  SE Mexico, N Guatemala, Belize
P.t.nuchalis
  N Colombia, N Venezuela
P.t.pectoralis
  NW Venezuela
P.t.sanguineus
  W Colombia, NW Ecuador
P.t.erythropygius
  W Ecuador
Pteroglossus frantzii (Fiery-billed Araçari)
  98.1
  Costa Rica, W Panama
Pteroglossus beauharnaesii (Curl-crested Araçari)
  E Peru, N Bolivia, W Brazil

SELENIDERA
Selenidera maculirostris (Spot-billed Toucanet)

S.m.hellmayri
  NC Brazil
S.m.maculirostris
  SE Brazil
Selenidera gouldii (Gould's Toucanet)
  NE Brazil
Selenidera reinwardtii (Golden-collared Toucanet)
S.r.reinwardtii
  S Colombia, E Ecuador, NE Peru
S.r.langsdorffi
  E Peru, W Brazil
Selenidera nattereri (Tawny-tufted Toucanet)
  E Venezuela, NW Brazil
Selenidera culik (Guianan Toucanet)
  the Guianas, N Brazil
Selenidera spectabilis (Yellow-eared Toucanet)
  Honduras to Panama, NC & NW Colombia

BAILLONIUS
Baillonius bailloni (Saffron Toucanet)
  SE Brazil

ANDIGENA
Andigena hypoglauca (Grey-breasted Mountain Toucan)
A.h.hypoglauca
  C Colombia
A.h.lateralis
  E Ecuador, E Peru
Andigena laminirostris (Plate-billed Mountain Toucan)
  SW Colombia, W Ecuador
Andigena cucullata (Hooded Mountain Toucan)
  S Peru, W Bolivia
Andigena nigrirostris (Black-billed Mountain Toucan)
A.n.occidentalis
  W Colombia
A.n.spilorhynchus
  S Colombia, NE Ecuador
A.n.nigrirostris
  E Colombia

RAMPHASTOS
Ramphastos dicolorus (Red-breasted Toucan)
  SE Brazil, Paraguay, NE Argentina
Ramphastos vitellinus (Channel-billed Toucan)
R.v.culminatus
  Upper Amazonia
R.v.citreolaemus
  C Colombia
R.v.vitellinus
  Trinidad, Venezuela, the Guianas, N Brazil
R.v.ariel
  C & S Brazil
R.v.pintoi
  SE Brazil

R.v.theresae
  NE Brazil
Ramphastos brevis (Choco Toucan)
  E Panama, W Colombia, W Ecuador
Ramphastos sulfuratus (Keel-billed Toucan)
R.s.sulfuratus
  S Mexico, N Guatemala, Belize
R.s.brevicarinatus
  SE Guatemala to N Colombia, NW Venezuela
Ramphastos toco (Toco Toucan)
R.t.toco
  the Guianas, N & E Brazil
R.t.albogularis
  E & S Brazil, Paraguay, Bolivia, N Argentina
Ramphastos tucanus (Red-billed Toucan)
R.t.tucanus
  SE Venezuela, the Guianas, N Brazil
R.t.cuvieri
  Upper Amazonia
R.t.oblitus
  NC Brazil
R.t.inca
  E Bolivia
Ramphastos swainsonii (Chestnut-mandibled Toucan) 98.1
  SE Honduras to W Ecuador
Ramphastos ambiguus (Black-mandibled Toucan)
R.a.ambiguus
  N Upper Amazonia
R.a.abbreviatus
  W Venezuela, NE Colombia

99.   PICIDAE (WOODPECKERS)   99.1

JYNX
Jynx torquilla (Northern Wryneck)
J.t.torquilla
  W Europe to SE Asia & Japan
J.t.tschusii
  Italy, Sardinia, Corsica
J.t.mauretanica
  N Algeria
J.t.himalayana
  NW India
Jynx ruficollis (Rufous-necked Wryneck)
J.r.ruficollis
  Gabon to Kenya & Sth Africa
J.r.pulchricollis
  W Zaire to SW Sudan, W Uganda
J.r.aequatorialis
  Ethiopia
J.r.pectoralis
  N Angola, Cabinda

PICUMNUS
Picumnus innominatus (Speckled Piculet)
P.i.innominatus
  W Himalayas to Assam
P.i.malayorum
  India to Indochina, Sumatra, Borneo

P.i.chinensis
  W & S China
Picumnus aurifrons (Bar-breasted Piculet)
P.a.aurifrons
  C Brazil
P.a.transfasciatus
  EC Brazil
P.a.borbae
  C Brazil
P.a.wallacii
  WC Brazil
P.a.purusianus
  W Brazil
P.a.flavifrons
  E Peru, W Brazil
P.a.juruanus
  E Peru, W Brazil
Picumnus lafresnayi (Lafresnaye's Piculet)
P.l.lafresnayi
  S Colombia, E Ecuador, S Peru
P.l.punctifrons
  E Peru
P.l.taczanowskii
  NE Peru
P.l.pusillus
  W Brazil
P.l.pumilus
  N Brazil, S Colombia, S Venezuela
Picumnus exilis (Golden-spangled Piculet)
P.e.salvini
  E Venezuela
P.e.clarus
  EC Venezuela
P.e.undulatus
  SE Venezuela, S Guyana, N Brazil
P.e.buffoni
  Surinam, Cayenne, NE Brazil
P.e.pernambucensis
  E Brazil
P.e.alegriae
  NE Brazil
P.e.exilis
  E Brazil
Picumnus sclateri (Ecuadorean Piculet)
P.s.parvistriatus
  W Ecuador
P.s.sclateri
  SW Ecuador, NW Peru
P.s.porcullae
  N Peru
Picumnus squamulatus (Scaled Piculet)
P.s.squamulatus
  N & E Colombia
P.s.roehli
  NE Colombia, N Venezuela
P.s.obsoletus
  NE Venezuela
P.s.apurensis
  Venezuela
P.s.lovejoyi
  Venezuela
Picumnus spilogaster (White-bellied Piculet)

P.s.orinocensis
  E Venezuela
P.s.spilogaster
  NE Brazil, the Guianas
P.s.pallidus
  NE Brazil
Picumnus minutissimus (Guianan Piculet)
  the Guianas
Picumnus pygmaeus (Spotted Piculet)
  NE Brazil
Picumnus steindachneri (Speckle-chested Piculet)
  NE Peru
Picumnus varzeae (Varzea Piculet)
  C Amazon islands
Picumnus cirratus (White-barred Piculet)
P.c.macconnelli
  NE Brazil
P.c.confusus
  Guyana, Cayenne
P.c.jelskii
  E Peru
P.c.cirratus
  S Brazil, E Paraguay
P.c.temminckii
  E Paraguay, SE Brazil, NE Argentina
P.c.pilcomayensis
  SW Brazil, Paraguay, N Argentina
P.c.tucumanus
  N Argentina
P.c.thamnophiloides
  S Bolivia, NW Argentina
P.c.dorbygnianus
  N & C Bolivia
Picumnus albosquamatus (White-wedged Piculet)
P.a.albosquamatus
  Bolivia, W Brazil
P.a.guttifer
  C Brazil
Picumnus fuscus (Rusty-necked Piculet)
  S Brazil
Picumnus rufiventris (Rufous-breasted Piculet)
P.r.rufiventris
  SW Colombia, E Ecuador
P.r.grandis
  E Peru
P.r.brunneifrons
  NW Bolivia
Picumnus fulvescens (Tawny Piculet)
  NE Brazil
Picumnus limae (Ochraceous Piculet)
  E Brazil
Picumnus nebulosus (Mottled Piculet)
  S Brazil
Picumnus castelnau (Plain-breasted Piculet)
  E Ecuador, NE Peru
Picumnus subtilis (Fine-barred Piculet)
  Cuzco (SE Peru)
Picumnus olivaceus (Olivaceous Piculet)
P.o.dimotus
  E Guatemala to Nicaragua

P.o.flavotinctus
Costa Rica to E Panama
P.o.olivaceus
N Colombia
P.o.eisenmanni
NE Colombia, NW Venezuela
P.o.tachirensis
C Colombia, NE Venezuela
P.o.harterti
SW Colombia, W Ecuador
Picumnus granadensis (Greyish Piculet)
P.g.antioquensis
N Colombia
P.g.granadensis
NC Colombia
Picumnus cinnamomeus (Chestnut Piculet)
P.c.cinnamomeus
N Colombia
P.c.perijanus
NW Venezuela
P.c.persaturatus
C Colombia
P.c.venezuelensis
WC Venezuela

SASIA
Sasia africana (African Piculet)
Nigeria to C Zaire
Sasia abnormis (Rufous Piculet)
S.a.abnormis
S Burma, Malaysia, Sumatra to Borneo
S.a.magnirostris
Nias I
Sasia ochracea (White-browed Piculet)
S.o.ochracea
Himalayas to S Vietnam
S.o.reichenowi
S Burma, Thailand
S.o.kinneari
S China, N Vietnam

NESOCTITES
Nesoctites micromegas (Antillean Piculet)
N.m.micromegas
Hispaniola
N.m.abbotti
Gonave I

MELANERPES
Melanerpes candidus (White Woodpecker)
N & E Brazil to C Argentina
Melanerpes lewis (Lewis' Woodpecker)
W Canada to NW Mexico
Melanerpes herminieri (Guadeloupe Woodpecker)
Guadeloupe I
Melanerpes portoricensis (Puerto Rican Woodpecker)
Puerto Rico
Melanerpes erythrocephalus (Red-headed Woodpecker)
M.e.erythrocephalus
SC Canada, E USA

M.e.caurinus
WC USA
Melanerpes formicivorus (Acorn Woodpecker)
M.f.bairdi
Oregon to N Baja California
M.f.angustifrons
S Baja California
M.f.formicivorus
SW USA to C Mexico
M.f.albeolus
SE Mexico to Belize
M.f.lineatus
S Mexico to N Nicaragua
M.f.striatipectus
Nicaragua to W Panama
M.f.flavigula
Colombia
Melanerpes cruentatus (Red-fronted Woodpecker)
the Guianas to NE Bolivia
Melanerpes flavifrons (Yellow-fronted Woodpecker)
S Brazil, Paraguay, N Argentina
Melanerpes chrysauchen (Golden-naped Woodpecker)
M.c.chrysauchen
SW Costa Rica, W Panama
M.c.pulcher
N Colombia
Melanerpes pucherani (Black-cheeked Woodpecker)
S Mexico to W Ecuador
Melanerpes cactorum (White-fronted Woodpecker)
S Peru to C Argentina
Melanerpes striatus (Hispaniolan Woodpecker)
Hispaniola
Melanerpes radiolatus (Jamaican Woodpecker)
Jamaica
Melanerpes chrysogenys (Golden-cheeked Woodpecker)
M.c.chrysogenys
NW Mexico
M.c.flavinuchus
SW Mexico
Melanerpes hypopolius (Grey-breasted Woodpecker)
Mexico
Melanerpes pygmaeus (Yucatan Woodpecker) 99.2
M.p.tysoni
Bonacca I
M.p.rubricomus
Yucatan
M.p.pygmaeus
Cozumel I
Melanerpes rubricapillus (Red-crowned Woodpecker)
M.r.rubricapillus
SW Costa Rica to Surinam & Tobago

M.r.subfusculus
  Coiba I
M.r.seductus
  San Miguel I
Melanerpes hoffmanni (Hoffmann's Woodpecker)
  Nicaragua, W Costa Rica
Melanerpes uropygialis (Gila Woodpecker)
M.u.uropygialis
  SW Nth America
M.u.cardonensis
  N Baja California
M.u.brewsteri
  S Baja California
M.u.fuscescens
  S Sonora
Melanerpes aurifrons (Golden-fronted Woodpecker)
M.a.aurifrons
  Texas to SC Mexico
M.a.polygrammus
  SW Mexico
M.a.grateloupensis
  E Mexico
M.a.dubius
  SE Mexico, Belize
M.a.leei
  Mecos, Cozumel I
M.a.santacruzi
  S Chiapas to N Nicaragua
M.a.pauper
  E Honduras
M.a.insulanus
  Utilla I
M.a.canescens
  Ruatan, Barburat Is
Melanerpes carolinus (Red-bellied Woodpecker)
  C, E & S USA
Melanerpes superciliaris (Great Red-bellied Woodpecker)
M.s.bahamensis
  Gt Bahama, Watling's Is
M.s.blakei
  Abaco I
M.s.superciliaris
  Cuba
M.s.murceus
  Isle of Pines
M.s.caymanensis
  Grand Cayman I

SPHYRAPICUS
Sphyrapicus varius (Yellow-bellied Sapsucker)
  S Canada, E USA to C America
Sphyrapicus nuchalis (Red-naped Sapsucker)
  W Canada, W USA, W Mexico
Sphyrapicus ruber (Red-breasted Sapsucker)
S.r.ruber
  SE Alaska to W Oregon
S.r.daggetti
  S Oregon to SC California

Sphyrapicus thyroideus (Williamson's Sapsucker)
S.t.thyroideus
  S Brit.Columbia to S California
S.t.nataliae
  SE Brit.Columbia, WC USA, .C Mexico

XIPHIDIOPICUS
Xiphidiopicus percussus (Cuban Green Woodpecker)
X.p.percussus
  Cuba
X.p.insulaepinorum
  Isle of Pines

CAMPETHERA
Campethera punctuligera (Fine-spotted Woodpecker)
C.p.punctuligera
  Senegal to Niger & Central African Republic
C.p.balia
  S Chad, S Sudan, N Zaire
Campethera bennettii (Bennett's Woodpecker)
C.b.scriptoricauda
  N Angola to N Mozambique
C.b.bennettii
  Zimbabwe to Malawi & Natal
C.b.capricorni
  S Angola, N Namibia
Campethera nubica (Nubian Woodpecker)
C.n.nubica
  Sudan to Kenya & S Tanzania
C.n.pallida
  Somalia
Campethera abingoni (Golden-tailed Woodpecker)
C.a.chrysura
  Senegal to Sudan & Malawi
C.a.kavirondensis
  SW Kenya to C Tanzania
C.a.mombassica
  coast from S Somalia to N Tanzania
C.a.suahelica
  S Tanzania to Zimbabwe & Mozambique
C.a.abingoni
  Botswana to S Mozambique & Natal
C.a.anderssoni
  Namibia
C.a.constricta
  Natal
Campethera notata (Knysna Woodpecker)
  S Natal, S & E Cape Province
Campethera cailliautii (Green-backed Woodpecker)
C.c.permista
  Cameroun to Angola & Ethiopia
C.c.nyansae
  Angola, E Zaire, Uganda
C.c.cailliautii
  E Kenya, NE Tanzania
C.c.loveridgi
  EC Tanzania to Mozambique

Campethera maculosa (Little Green Woodpecker)
Guinea to Ghana
Campethera tullbergi (Tullberg's Woodpecker)
C.t.tullbergi
Cameroun
C.t.taeniolaema
W Kenya, W Uganda to SE Zaire
C.t.hausbergi
E & C Kenya
Campethera nivosa (Buff-spotted Woodpecker)
C.n.nivosa
Guinea to Zaire
C.n.poensis
Fernando Po I
C.n.herberti
N Zaire to Kenya
Campethera caroli (Brown-eared Woodpecker)
C.c.arizelus
Sierra Leone, Liberia
C.c.caroli
Ghana to Sudan, W Kenya, Angola

GEOCOLAPTES
Geocolaptes olivaceus (Ground Woodpecker)
99.3
G.o.olivaceus
W Cape Province to S Transvaal, Natal
G.o.prometheus
E Cape Province, S Natal

DENDROPICOS
Dendropicos elachus (Little Grey Woodpecker)
Senegal to W Sudan
Dendropicos poecilolaemus (Speckle-breasted Woodpecker)
N Cameroun to Uganda, Kenya
Dendropicos abyssinicus (Gold-mantled Woodpecker)
Ethiopia
Dendropicos fuscescens (Cardinal Woodpecker)
D.f.lafresnayei
Senegal to Nigeria
D.f.sharpei
E Nigeria to Sudan & N Angola
D.f.lepidus
E Zaire to Ethiopia, Kenya, Tanzania
D.f.massaicus
S Ethiopia to W Tanzania
D.f.hemprichii
Ethiopia, Somalia, N Kenya
D.f.centralis
Angola to W Tanzania, Zambia, Namibia
D.f.hartlaubii
SE Kenya to Mozambique coast
D.f.intermedius
E Transvaal, Natal

D.f.fuscescens
Cape Province to Zimbabwe
Dendropicos gabonensis (Gabon Woodpecker)
D.g.lugubris
Guinea to Ghana
D.g.reichenowi
NW Cameroun
D.g.gabonensis
Cameroun and Gabon, NE & C Zaire
Dendropicos stierlingi (Stierling's Woodpecker)
S Tanzania, S Mozambique
Dendropicos namaquus (Bearded Woodpecker)
D.n.schoensis
W Sudan, S Ethiopia, N Kenya
D.n.namaquus
SW Uganda & N Kenya to Malawi & W Cape Province
D.n.coalescens
E Cape Province to S Mozambique
Dendropicos xantholophus (Yellow-crested Woodpecker)
Cameroun to S Angola & W Kenya
Dendropicos pyrrhogaster (Fire-bellied Woodpecker)
Sierra Leone to SE Nigeria
Dendropicos elliotii (Elliot's Woodpecker)
D.e.johnstoni
E Nigeria, Cameroun Mt, Fernando Po I
D.e.elliotii
Cameroun to E Uganda & Angola
D.e.kupeensis
Kupe Mt (Cameroun)
D.e.gabela
NW Angola
Dendropicos goertae (Grey Woodpecker)
D.g.koenigi
Mali to N Sudan
D.g.goertae
Senegal to Sudan & Angola
D.g.abessinicus
N Ethiopia
D.g.spodocephalus
S Ethiopia
D.g.rhodeogaster
Kenya, N Tanzania
Dendropicos griseocephalus (Olive Woodpecker)
D.g.ruwenzori
E Zaire to W Angola, N Malawi
D.g.kilimensis
N Tanzania
D.g.griseocephalus
N Transvaal to Natal & Cape Province

PICOIDES
Picoides temminckii (Temminck's Pygmy Woodpecker)
Sulawesi, Togian Is
Picoides maculatus (Philippine Pygmy Woodpecker)

P.m.validirostris
Luzon, Catanduanes, Mindoro Is
P.m.maculatus
Panay, Cebu, Negros, Sibuyan Is
P.m.fulvifasciatus
Samar, Leyte, Bohol, Basilan, Mindanao Is
P.m.ramsayi
Jolo, Bongao, Siasi Is
Picoides moluccensis (Brown-capped Woodpecker)
P.m.nanus
N & C India
P.m.cinereigula
S India
P.m.gymnophthalmus
Sri Lanka
P.m.moluccensis
Malaysia, Sumatra, Java, Borneo
P.m.grandis
Lombok to Lomblen, Alor, Flores Is
Picoides obsoletus (Brown-backed Woodpecker)
P.o.obsoletus
Gambia to NE Zaire
P.o.heuglini
N Sudan to NE Ethiopia
P.o.ingens
S Ethiopia, Uganda, W Kenya
P.o.crateri
N Tanzania
Picoides kizuki (Japanese Spotted Woodpecker)
P.k.ijimae
NE Asia, N Japan
P.k.seebohmi
Korea, C Japan
P.k.amamii
N Riukiu Is
P.k.kizuki
N China, S Japan, Okinawa
Picoides canicapillus (Grey-capped Woodpecker)
P.c.doerriesi
NE Asia
P.c.scintilliceps
N China
P.c.kaleensis
S China, N Burma, N Vietnam, Taiwan
P.c.swinhoei
Hainan I
P.c.semicoronatus
N India
P.c.mitchelli
N Pakistan to W Nepal
P.c.canicapillus
E Assam, C & S Burma, Thailand, Laos
P.c.auritus
S Thailand, Malaysia
P.c.volzi
NW Sumatra
P.c.aurantiiventris
N & E Borneo
Picoides minor (Lesser Spotted Woodpecker)

P.m.minor
Scandinavia to Ural Mts
P.m.amurensis
NE Asia, N Japan
P.m.kamtschatkensis
Siberia
P.m.hortorum
NC Europe
P.m.buturlini
Spain to Romania, Algeria, Greece
P.m.comminutus
England, Wales
P.m.danfordi
Asia Minor
P.m.colchicus
Caucasus
P.m.quadrifasciatus
S Caucasus
P.m.morgani
SW Iran
Picoides macei (Streak-bellied Woodpecker)
P.m.westermanni
W Himalayas
P.m.macei
E Himalayas, N Burma
P.m.longipennis
C & S Burma to S Vietnam
P.m.andamanensis
Andaman Is
P.m.analis
Sumatra, Java, Bali
Picoides atratus (Stripe-breasted Woodpecker)
P.a.atratus
Burma, N Thailand, Laos
P.a.vietnamensis
Vietnam
Picoides auriceps (Brown-fronted Woodpecker)
NE Himalayas
Picoides mahrattensis (Yellow-crowned Woodpecker)
P.m.pallescens
Pakistan, NW India
P.m.mahrattensis
India, Sri Lanka, Burma
Picoides dorae (Arabian Woodpecker)
C Arabia
Picoides hyperythrus (Rufous-bellied Woodpecker)
P.h.marshalli
NW Himalayas, W Tibet
P.h.hyperythrus
E Himalayas to N Thailand
P.h.subrufinus
C Manchuria to C China
P.h.annamensis
S Laos, S Vietnam
Picoides cathpharius (Crimson-breasted Woodpecker)
P.c.cathpharius
E Himalayas

P.c.ludlowi
  China, SW Sikang
P.c.pyrrhothorax
  S Assam, W Burma
P.c.tenebrosus
  N Burma to N Vietnam, S China
P.c.pernyii
  SW China
P.c.innixus
  C Hupeh
Picoides darjellensis (Brown-throated Woodpecker)
  Nepal, N Burma, W China, N Vietnam
Picoides medius (Middle-spotted Woodpecker)
P.m.caucasicus
  Spain to N Iran
P.m.medius
  N Europe to Caucasus
P.m.sanctijohannis
  SW Iran
Picoides leucotos (White-backed Woodpecker)
P.l.leucotos
  N Europe to NE Asia, Korea, Sakhalin
P.l.lilfordi
  Pyrenees, Balkans, Asia Minor
P.l.tangi
  W China
P.l.subcirrus
  S Kurile Is, Hokkaido
P.l.stejnegeri
  NC Honshu
P.l.namiyei
  S Japan
P.l.takahashii
  Dagelat Is
P.l.owstoni
  N Riukiu Is
P.l.fohkiensis
  SE China
P.l.insularis
  Taiwan
Picoides himalayensis (Himalayan Woodpecker)
P.h.albescens
  W Himalayas
P.h.himalayensis
  C Himalayas
Picoides assimilis (Sind Woodpecker)
  SE Iran to W Punjab
Picoides syriacus (Syrian Woodpecker)
  E Europe to SE Iran
Picoides leucopterus (White-winged Woodpecker)
  Caspian Sea to Turkestan, N Sinkiang
Picoides major (Great Spotted Woodpecker)
P.m.major
  Scandinavia, N Russia
P.m.pinetorum
  England to Italy, Bulgaria, Caucasus
P.m.harterti
  Corsica, Sardinia

P.m.hispanus
  Portugal, S Spain
P.m.canariensis
  Teneriffe I
P.m.thanneri
  Gran Canaria I
P.m.mauritanus
  Morocco
P.m.numidus
  Algeria, Tunisia
P.m.poelzami
  N Iran
P.m.brevirostris
  Siberia, N Mongolia
P.m.kamtschaticus
  Kamchatka
P.m.japonicus
  Sakhalin, Korea, N Japan
P.m.cabanisi
  Manchuria, N, S & E China, Hainan I
P.m.stresemanni
  W China, N Burma
Picoides mixtus (Checkered Woodpecker)
P.m.cancellatus
  SE Brazil
P.m.mixtus
  E Paraguay, Uruguay, E Argentina
P.m.malleator
  W Paraguay, SE Bolivia, N & W Argentina
P.m.berlepschi
  S & W Argentina
Picoides lignarius (Striped Woodpecker)
  Bolivia, Chile, S & W Argentina
Picoides scalaris (Ladder-backed Woodpecker)
P.s.cactophilus
  SW USA to C & E Mexico
P.s.eremicus
  N Baja California
P.s.lucasanus
  S Baja California
P.s.graysoni
  Tres Marias Is
P.s.sinaloensis
  W Mexico
P.s.scalaris
  S Mexico
P.s.parvus
  N Yucatan, Cozumel I
P.s.leucoptilurus
  Belize
Picoides nuttallii (Nuttall's Woodpecker)
  W California, NW Baja California
Picoides pubescens (Downy Woodpecker)
P.p.glacialis
  SE Alaska
P.p.medianus
  C Alaska to SC & E USA
P.p.gairdnerii
  W Brit.Columbia to NW California
P.p.turati
  Washington, Oregon

P.p.leucurus
Rocky Mts
P.p.pubescens
SE USA
Picoides borealis (Red-cockaded Woodpecker)
SC & SE USA
Picoides stricklandii (Strickland's Woodpecker)
P.s.arizonae
Arizona, NW Mexico
P.s.fraterculus
W Mexico
P.s.stricklandi
C & E Mexico
Picoides villosus (Hairy Woodpecker)
P.v.septentrionalis
SC Alaska, S Canada, NW USA
P.v.terraenovae
Newfoundland
P.v.villosus
SE Canada, NC & E USA
P.v.audubonii
SE USA
P.v.piger
Gd Bahama, Mores, Abaco Is
P.v.maynardi
New Providence, Andros Is
P.v.picoideus
Queen Charlotte Is
P.v.sitkensis
SE Alaska, N Brit.Columbia
P.v.harrisi
S Brit.Columbia to N California
P.v.hyloscopus
W & S California
P.v.orius
SC Washington to W Texas
P.v.icastus
SE Arizona, NW Mexico
P.v.jardinii
E & S Mexico
P.v.sanctorum
S Mexico to N Nicaragua
Picoides albolarvatus (White-headed Woodpecker)
P.a.albolarvatus
W USA
P.a.gravirostris
S California
Picoides tridactylus (Three-toed Woodpecker)
P.t.tridactylus
Scandinavia to NE Asia
P.t.alpinus
Alps to N Japan
P.t.crissoleucus
N Siberia, N Mongolia
P.t.albidior
Kamtchatka
P.t.funebris
W China

P.t.fasciatus
N Alaska, W Canada, NW USA
P.t.dorsalis
N Montana to New Mexico
P.t.bacatus
C & E Canada, NE USA
Picoides arcticus (Black-backed Woodpecker)
Canada, N & W USA

VENILIORNIS
Veniliornis callonotus (Scarlet-backed Woodpecker)
V.c.callonotus
W Ecuador
V.c.major
SW Ecuador, N Peru
Veniliornis dignus (Yellow-vented Woodpecker)
V.d.dignus
W Colombia, SW Venezuela
V.d.baezae
E Ecuador
V.d.valdizani
C Peru
Veniliornis nigriceps (Bar-bellied Woodpecker)
V.n.equifasciatus
WC Colombia, N Ecuador
V.n.pectoralis
C Peru
V.n.nigriceps
W Bolivia
Veniliornis fumigatus (Smoky-brown Woodpecker)
V.f.oleagineus
E Mexico
V.f.sanguinolentus
C & S Mexico to W Panama
V.f.exsul
N Colombia
V.f.reichenbachi
N Venezuela
V.f.tectricialis
NE Venezuela
V.f.fumigatus
Upper Amazonia
V.f.obscuratus
NW Peru
Veniliornis passerinus (Little Woodpecker)
V.p.fidelis
E Colombia, W Venezuela
V.p.modestus
NE Venezuela
V.p.diversus
N Brazil
V.p.agilis
E Ecuador to N Bolivia, W Brazil
V.p.insignis
WC Brazil
V.p.tapajozensis
C Brazil

V.p.passerinus
the Guianas, NE Brazil
V.p.taenionotus
E Brazil
V.p.olivinus
S Bolivia, S Brazil, Paraguay, N Argentina
Veniliornis frontalis (Dot-fronted Woodpecker)
NW Argentina
Veniliornis spilogaster (White-spotted Woodpecker)
S Brazil, Paraguay, NE Argentina
Veniliornis sanguineus (Blood-coloured Woodpecker)
the Guianas
Veniliornis maculifrons (Yellow-eared Woodpecker)
SE Brazil
Veniliornis affinis (Red-stained Woodpecker)
V.a.chocoensis
W Colombia
V.a.orenocensis
E Colombia, E Venezuela, N Brazil
V.a.caquetanus
S Colombia
V.a.hilaris
E Ecuador, E Peru, N Bolivia, W Brazil
V.a.ruficeps
C & E Brazil
V.a.affinis
E Brazil
Veniliornis cassini (Golden-collared Woodpecker)
Venezuela, the Guianas, N Brazil
Veniliornis kirkii (Red-rumped Woodpecker)
V.k.neglectus
SW Costa Rica, Panama
V.k.cecilii
E Panama, W Colombia, W Ecuador
V.k.continentalis
N & W Venezuela
V.k.monticola
Mt Roraima (Venezuela)
V.k.kirkii
Trinidad, Tobago I

PICULUS
Piculus leucolaemus (White-throated Woodpecker)
P.l.simplex
Honduras to W Panama
P.l.callopterus
E Panama
P.l.litae
W Colombia, NW Ecuador
P.l.leucolaemus
E Colombia to Bolivia, W Brazil
Piculus flavigula (Yellow-throated Woodpecker)
P.f.flavigula
N Amazonia
P.f.magnus

SE Colombia, NW Brazil
P.f.erythropis
E & SE Brazil
Piculus chrysochloros (Golden-green Woodpecker)
P.c.aurosus
Panama
P.c.xanthochlorus
NE Colombia, NW Venezuela
P.c.capistratus
C Colombia to Guyana, NW Brazil
P.c.guianensis
Cayenne
P.c.laemostictus
W Brazil
P.c.hypochryseus
W Brazil, N Bolivia
P.c.paraensis
NE Brazil
P.c.polyzonus
SE Brazil
P.c.chrysochloros
C & S Brazil, Bolivia, N Argentina
Piculus aurulentus (White-browed Woodpecker)
SE Brazil, N Argentina
Piculus rubiginosus (Golden-olive Woodpecker)
P.r.aeruginosus
NE Mexico
P.r.yucatanensis
C Mexico to W Panama
P.r.alleni
N Colombia
P.r.buenavistae
E Colombia, E Ecuador
P.r.meridensis
W Venezuela
P.r.rubiginosus
N Venezuela
P.r.deltanus
NE Venezuela
P.r.paraquensis
E Venezuela
P.r.guianae
SE Venezuela, S Guyana
P.r.viridissimus
S Venezuela
P.r.nigriceps
Guyana, Surinam
P.r.trinitatis
Trinidad
P.r.tobagensis
Tobago I
P.r.gularis
C & SW Colombia
P.r.rubripileus
SW Colombia, NW Peru
P.r.coloratus
NC Peru
P.r.chrysogaster
C Peru

P.r.canipileus
  N Bolivia
P.r.tucumanus
  C Bolivia to NW Argentina
Piculus auricularis (Grey-crowned Woodpecker)
  W & NW Mexico
Piculus rivolii (Crimson-mantled Woodpecker)
P.r.quindiuna
  NC Colombia
P.r.rivolii
  W Venezuela, EC Colombia
P.r.meridae
  W Venezuela
P.r.brevirostris
  SW Colombia to C Peru
P.r.atriceps
  SE Peru, Bolivia

COLAPTES
Colaptes atricollis (Black-necked Flicker)
C.a.atricollis
  W Peru
C.a.peruvianus
  E Peru
Colaptes punctigula (Spot-breasted Flicker)
C.p.ujhelyii
  N Colombia
C.p.striatigularis
  E Panama to WC Colombia
C.p.punctipectus
  E Colombia, Venezuela
C.p.zuliae
  NW Venezuela
C.p.punctigula
  Surinam, Cayenne
C.p.guttatus
  Upper Amazonia
Colaptes melanochloros (Green-barred Flicker)
C.m.nattereri
  E Bolivia, S Brazil
C.m.melanochloros
  C Brazil
C.m.melanolaimus
  Bolivia, W Argentina
C.m.nigroviridis
  S Bolivia, Paraguay
C.m.leucofrenatus
  C & S Argentina
Colaptes auratus (Common Flicker)
C.a.cafer
  S Alaska to N California
C.a.collaris
  SE Brit.Columbia to NW Mexico
C.a.nanus
  NE Mexico
C.a.mexicanus
  C Mexico
C.a.mexicanoides
  S Mexico to N Nicaragua

C.a.luteus
  Canada, E, S & C USA
C.a.auratus
  SE USA
C.a.chrysocaulosus
  Cuba
C.a.gundlachi
  Gd Cayman I
C.a.mearnsi
  SE California, NW Baja California
C.a.tenebrosus
  NW Mexico
C.a.brunnescens
  C Baja California
C.a.chrysoides
  S & W Baja California
Colaptes fernandinae (Fernandina's Flicker)
  Cuba
Colaptes pitius (Chilean Flicker)
  C & S Chile, S Argentina
Colaptes rupicola (Andean Flicker)
C.r.cinereicapilla
  N Peru
C.r.puna
  S & C Peru
C.r.rupicola
  Bolivia, N Chile, NW Argentina
Colaptes campestris (Campo Flicker)
  E Bolivia, C & E Brazil
Colaptes campestroides (Field Flicker)
  SC Sth America

CELEUS
Celeus brachyurus (Rufous Woodpecker)
C.b.humei
  NW Himalayas
C.b.phaioceps
  E Himalayas, NE India
C.b.jerdonii
  W India, Sri Lanka
C.b.annamensis
  Laos, Cambodia, S Vietnam
C.b.fokiensis
  SE China, N Vietnam
C.b.holroydi
  Hainan I
C.b.squamigularis
  Malaysia, Sumatra, Bangka, Billiton, Nias Is
C.b.brachyurus
  Java
C.b.badiosus
  Borneo, N Natuna Is
Celeus loricatus (Cinnamon Woodpecker)
C.l.diversus
  SE Costa Rica, W Panama
C.l.mentalis
  E Panama, NW Colombia
C.l.innotatus
  N & NC Colombia
C.l.loricatus
  W Colombia, W Ecuador

Celeus undatus (Waved Woodpecker)
  C.u.amarcurensis
    NE Venezuela
  C.u.undatus
    the Guianas, N Brazil
  C.u.multifasciatus
    NE Brazil
Celeus grammicus (Scale-breasted Woodpecker)
  C.g.verreauxii
    SE Colombia, E Ecuador
  C.g.grammicus
    S Venezuela, NE Peru, W Brazil
  C.g.subcervinus
    NC Brazil
  C.g.latifasciatus
    SE Peru, NE Bolivia
Celeus castaneus (Chestnut-coloured Woodpecker)
    SE Mexico to C Panama
Celeus elegans (Chestnut Woodpecker)
  C.e.hellmayri
    E Venezuela, NE Brazil, Guyana, Surinam
  C.e.deltanus
    NE Venezuela
  C.e.leotaudi
    Trinidad
  C.e.elegans
    NE Brazil, Cayenne
  C.e.citreopygius
    SE Colombia, E Ecuador, Peru
  C.e.jumana
    E Colombia to N Bolivia
Celeus lugubris (Pale-crested Woodpecker)
  C.l.lugubris
    WC Brazil, E Bolivia
  C.l.kerri
    Paraguay, S Brazil
Celeus flavescens (Blond-crested Woodpecker)
  C.f.intercedens
    NE Brazil
  C.f.ochraceus
    E Brazil
  C.f.flavescens
    S Brazil, SE Bolivia, E Paraguay
Celeus flavus (Cream-coloured Woodpecker)
  C.f.flavus
    C Colombia to the Guianas, N Brazil
  C.f.peruvianus
    W Brazil, E Peru
  C.f.tectricialis
    NE Brazil
  C.f.subflavus
    E Brazil
Celeus spectabilis (Rufous-headed Woodpecker)
  C.s.spectabilis
    E Ecuador
  C.s.obrieni
    Piauhy (Brazil)
  C.s.exsul
    C Bolivia

Celeus torquatus (Ringed Woodpecker)
  C.t.torquatus
    E Venezuela, the Guianas, N Brazil
  C.t.occidentalis
    E Peru to C Brazil
  C.t.tinnunculus
    E Brazil

DRYOCOPUS
Dryocopus galeatus (Helmeted Woodpecker)
    S Brazil, Paraguay, NE Argentina
Dryocopus schulzi (Black-bodied Woodpecker)
    N Argentina
Dryocopus lineatus (Lineated Woodpecker)
  D.l.scapularis
    W Mexico
  D.l.similis
    NE Mexico to NW Costa Rica
  D.l.lineatus
    E Costa Rica to C Brazil
  D.l.fuscipennis
    W Ecuador, NW Peru
  D.l.erythrops
    SE & S Brazil
Dryocopus pileatus (Pileated Woodpecker)
  D.p.abieticola
    Canada, W to NE USA
  D.p.pileatus
    C & E USA
Dryocopus javensis (White-bellied Woodpecker)
  D.j.hodgei
    Andaman Is
  D.j.hodgsonii
    W India
  D.j.richardsi
    C & S Korea
  D.j.forresti
    SE Sikang, W Yunnan, N Vietnam
  D.j.feddeni
    Burma, N Thailand, S Indochina
  D.j.javensis
    Gtr. Sunda Is, Malaysia
  D.j.parvus
    Simalur I
  D.j.hargitti
    Palawan
  D.j.confusus
    Luzon I
  D.j.mindorensis
    Mindoro I
  D.j.philippensis
    Masbate, Panay, Negros Is
  D.j.pectoralis
    Leyte, Samar, Bohol Is
  D.j.multilunatus
    Basilan, Mindanao Is
  D.j.suluensis
    Sulu Arch

Dryocopus martius (Black Woodpecker)
  D.m.martius
    NW Europe to NE Asia, Japan
  D.m.khamensis
    W China

CAMPEPHILUS
Campephilus pollens (Powerful Woodpecker)
  C.p.pollens
    Colombia, Ecuador
  C.p.peruvianus
    E Peru
Campephilus haematogaster (Crimson-bellied Woodpecker)
  C.h.splendens
    Panama to NW Ecuador
  C.h.haematogaster
    N Colombia to S Peru
Campephilus rubricollis (Red-necked Woodpecker)
  C.r.rubricollis
    C Colombia to the Guianas, N Brazil
  C.r.trachelopyrus
    E Peru, N Bolivia, W Brazil
  C.r.olallae
    W Bolivia to C & E Brazil
Campephilus robustus (Robust Woodpecker)
    SE Brazil, Paraguay, NE Argentina
Campephilus guatemalensis (Pale-billed Woodpecker)
  C.g.regius
    E & C Mexico
  C.g.nelsoni
    N & W Mexico
  C.g.guatemalensis
    S Mexico to W Panama
Campephilus melanoleucos (Crimson-crested Woodpecker)
  C.m.malherbii
    E Panama to W Venezuela
  C.m.melanoleucos
    C & W Amazonia
  C.m.cearae
    NE Brazil
Campephilus gayaquilensis (Guayaquil Woodpecker)
    W Ecuador, NW Peru
Campephilus leucopogon (Cream-backed Woodpecker)
    Bolivia to Uruguay, NE Argentina
Campephilus magellanicus (Magellanic Woodpecker)
    S Sth America
Campephilus principalis (Ivory-billed Woodpecker)
  C.p.principalis
    N Louisiana
  C.p.bairdii
    Cuba
Campephilus imperialis (Imperial Woodpecker)
    NW & W Mexico

PICUS
Picus miniaceus (Banded Red Woodpecker)
  P.m.perlutus
    S Burma, SW Thailand
  P.m.malaccensis
    Malaysia, Gtr Sunda Is
  P.m.niasensis
    Nias I
  P.m.miniaceus
    W & C Java
Picus puniceus (Crimson-winged Woodpecker)
  P.p.observandus
    S Burma, Malaysia, Sumatra, Borneo
  P.p.soligae
    Nias I
  P.p.puniceus
    Java
Picus chlorolophus (Lesser Yellow-naped Woodpecker)
  P.c.simlae
    NW Himalayas
  P.c.chlorolophus
    E Himalayas to N Laos
  P.c.chlorigaster
    C & S India
  P.c.wellsi
    Sri Lanka
  P.c.citrinocristatus
    SE China, N Indochina
  P.c.longipennis
    Hainan I
  P.c.annamensis
    S Indochina
  P.c.rodgeri
    NW Malaysia
  P.c.vanheysti
    Sumatra
Picus mentalis (Checker-throated Woodpecker)
  P.m.humii
    S Burma, Malaysia, Sumatra, Borneo
  P.m.mentalis
    Java
Picus flavinucha (Greater Yellow-naped Woodpecker)
  P.f.flavinucha
    E Himalayas to N Vietnam
  P.f.styani
    SE China, NE Vietnam, Hainan I
  P.f.pierrei
    S & E Thailand, S Indochina
  P.f.ricketti
    C Fukien
  P.f.mystacalis
    N Sumatra
  P.f.korinchi
    S Sumatra
Picus vittatus (Laced Woodpecker)
  P.v.viridanus
    Burma, S Thailand, N Malaysia
  P.v.vittatus
    S Malaysia, Indochina, Sumatra, Java, Bali

Picus xanthopygaeus (Streak-throated Woodpecker)
  Himalayas, India to Indochina
Picus squamatus (Scaly-bellied Woodpecker)
  P.s.flavirostris
  Caspian Sea to Afghanistan
  P.s.squamatus
  W & C Himalayas
Picus awokera (Wavy-bellied Woodpecker)
  P.a.awokera
  Honshu I
  P.a.horii
  S Japan & islands
Picus viridis (Green Woodpecker)
  P.v.viridis
  Britain to N Iran
  P.v.sharpei
  Iberia
  P.v.innominatus
  W Iran
  P.v.bampurensis
  Baluchistan
  P.v.vaillantii
  Morocco to Tunisia
Picus rabieri (Red-collared Woodpecker)
  Indochina
Picus erythropygius (Black-headed Woodpecker)
  P.e.nigrigenis
  Burma, S Thailand
  P.e.erythropygius
  SE Thailand, S Indochina
Picus canus (Grey-faced Woodpecker)
  P.c.canus
  Norway, E Europe to C Asia, N Mongolia
  P.c.jessoensis
  NE Asia, Hokkaido
  P.c.kogo
  W China
  P.c.guerini
  E China
  P.c.sobrinus
  SE China, N Vietnam
  P.c.tancolo
  Taiwan, Hainan I
  P.c.sordidior
  NW Yunnan
  P.c.sanguiniceps
  NW Himalayas
  P.c.hessei
  E Nepal to Indochina
  P.c.robinsoni
  Malaysia
  P.c.dedemi
  Sumatra

DINOPIUM
Dinopium rafflesii (Olive-backed Woodpecker)
  D.r.rafflesii
  S Burma, S Thailand, Malaysia, Sumatra
  D.r.dulitense
  Borneo

Dinopium shorii (Himalayan Gold-backed Woodpecker)
  D.s.shorii
  Himalayas
  D.s.angusto
  Assam, Burma
Dinopium javanense (Common Gold-backed Woodpecker)
  D.j.malabaricum
  SW India
  D.j.intermedium
  NE India to Indochina
  D.j.javanense
  Malaysia, Borneo, Sumatra, W Java
  D.j.exsul
  E Java, Bali
  D.j.raveni
  NE Borneo
  D.j.everetti
  SW Philippine Is
Dinopium benghalense (Lesser Flame-backed Woodpecker)
  D.b.benghalense
  E & C India
  D.b.dilutum
  Baluchistan, Pakistan
  D.b.puncticolle
  S India
  D.b.tehminae
  W India
  D.b.psarodes
  Sri Lanka

CHRYSOCOLAPTES
Chrysocolaptes lucidus (Greater Flame-backed Woodpecker)
  C.l.socialis
  W India
  C.l.guttacristatus
  NW Himalayas to S Indochina
  C.l.chersonesus
  S India, S Malaysia, Sumatra, Java
  C.l.stricklandi
  Sri Lanka
  C.l.strictus
  S Java, Bali, Kangean I
  C.l.andrewsi
  Borneo
  C.l.erythrocephalus
  Palawan, Balabac Is
  C.l.haematribon
  N Philippine Is
  C.l.xanthocephalus
  Masbate, Panay, Negros Is
  C.l.rufopunctatus
  Samar, Leyte, Bohol Is
  C.l.lucidus
  Mindanao, Basilan Is
  C.l.montanus
  E Mindanao I
Chrysocolaptes festivus (Black-rumped Woodpecker)

C.f.festivus
  C & S India
C.f.tantus
  Sri Lanka

GECINULUS
Gecinulus grantia (Bamboo Woodpecker)
  G.g.grantia
    Nepal to N Burma
  G.g.indochinensis
    Indochina
  G.g.viridis
    C Burma to Malaysia, Thailand
  G.g.viridanus
    S China

SAPHEOPIPO
**Sapheopipo noguchii (Okinawan
  Woodpecker)**
    Okinawa I

BLYTHIPICUS
**Blythipicus rubiginosus (Maroon
  Woodpecker)**
    S Burma, Thailand, Malaysia, Sumatra,
    Borneo
Blythipicus pyrrhotis (Bay Woodpecker)
  B.p.pyrrhotis
    NE Himalayas to Indochina
  B.p.cameroni
    Malaysia
  B.p.sinensis
    SE China
  B.p annamensis
    S Indochina
  B.p.hainanus
    Hainan I

REINWARDTIPICUS
**Reinwardtipicus validus (Orange-backed
  Woodpecker)**
  R.v.xanthopygius
    Malaysia, Sumatra, Borneo
  R.v.validus
    W & C Java

MEIGLYPTES
Meiglyptes tristis (Buff-rumped Woodpecker)
  M.t.grammithorax
    S Burma, S Thailand, Malaysia Sumatra,
    Borneo
  M.t.tristis
    Java
**Meiglyptes jugularis (Black & Buff
  Woodpecker)**
    Burma to S Indochina
Meiglyptes tukki (Buff-necked Woodpecker)
  M.t.tukki
    Malaysia, Sumatra, Borneo
  M.t.pulonis
    Banggai I
  M.t.percnerpes
    S Borneo

M.t.infuscatus
  Nias I
M.t.batu
  Batu I

HEMICIRCUS
**Hemicircus concretus (Grey and Buff
  Woodpecker)**
  H.c.sordidus
    S Burma, S Thailand, Malaysia, Sumatra,
    Borneo
  H.c.concretus
    W & C Java
**Hemicircus canente (Heart-spotted
  Woodpecker)**
    W India to S Indochina

MÜLLERIPICUS
Mülleripicus fulvus (Fulvous Woodpecker)
  M.f.fulvus
    N Sulawesi, Togian Is
  M.f.wallacei
    C & S Sulawesi
Mülleripicus funebris (Sooty Woodpecker)
  M.f.funebris
    N Luzon, Marinduque Is
  M.f.mayri
    S Luzon
  M.f.parkesi
    Polillo Is
  M.f.fuliginosus
    Samar, Leyte, Minanao Is
**Mülleripicus pulverulentus (Great Slaty
  Woodpecker)**
  M.p.harterti
    N India to C & S Indochina
  M.p.pulverulentus
    Malaysia, Gtr Sunda, Palawan Is

PASSERIFORMES

100. EURYLAIMIDAE (BROADBILLS)

EURYLAIMINAE

SMITHORNIS
Smithornis capensis (African Broadbill)
  S.c.delacouri
    Liberia, Ghana, Ivory Coast
  S.c.camerunensis
    S Cameroun, Gabon
  S.c.albigularis
    N Angola, S Zaire
  S.c.medianus
    E Zaire, Uganda, Kenya
  S.c.cryptoleucus
    E Zimbabwe, NE Transvaal, S Malawi, SE
    Tanzania
  S.c.conjunctus
    N Botswana, S Angola, SE Zambia, S
    Mozambique
  S.c.capensis
    Natal

Smithornis rufolateralis (Rufous-sided Broadbill)
  S.r.rufolateralis
    W Africa, Cameroun, Central African Republic
  S.r.budongoensis
    NE Zaire, W Uganda
Smithornis sharpei (Grey-headed Broadbill)
  S.s.sharpei
    Fernando Po I
  S.s.zenkeri
    Cameroun
  S.s.eurylaemus
    E Zaire

PSEUDOCALYPTOMENA
Pseudocalyptomena graueri (African Green Broadbill)
    E Zaire

CORYDON
Corydon sumatranus (Dusky Broadbill)
  C.s.laoensis
    S Burma, Thailand, Laos, N Vietnam
  C.s.morator
    S Thailand
  C.s.pallescens
    Malaysia
  C.s.ardescens
    SE Thailand
  C.s.sumatranus
    Sumatra
  C.s.brunnescens
    N Borneo, N Natuna I
  C.s.orientalis
    S Borneo

CYMBIRHYNCHUS
Cymbirhynchus macrorhynchos (Black & Red Broadbill)
  C.m.affinis
    SE Burma
  C.m.siamensis
    S Burma, S Thailand, Cambodia, S Vietnam
  C.m.malaccensis
    S Malaysia
  C.m.lemniscatus
    Sumatra, Bangka, Billiton Is
  C.m.tenebrosus
    SE Sumatra
  C.m.macrorhynchos
    Borneo

EURYLAIMUS
Eurylaimus javanicus (Banded Broadbill)
  E.j.pallidus
    S Burma, Malaysia, Thailand, Laos, S Vietnam

E.j.harterti
    Sumatra
E.j.javanicus
    Java
E.j.billitonis
    Billiton I
E.j.brookei
    Borneo
Eurylaimus ochromalus (Black & Yellow Broadbill)
  E.o.ochromalus
    S Burma, Malaysia, Sumatra
  E.o.mecistus
    Banjak I
  E.o.kalamantan
    Borneo
Eurylaimus steerii (Wattled Broadbill)
  E.s.samarensis
    Samar, Leyte, Philippine Is
  E.s.steerii
    Mindanao, Basilan Is
  E.s.mayri
    Mindanao I

SERILOPHUS
Serilophus lunatus (Silver-breasted Broadbill)
  S.l.rubropygius
    Nepal, N Borneo
  S.l.atrestus
    E Burma, E Thailand, N Laos, N Vietnam
  S.l.polionotus
    Hainan I
  S.l.impavidus
    S Laos
  S.l.lunatus
    S Burma
  S.l.intrepidus
    NE Thailand
  S.l.stolidus
    S Burma, S Thailand
  S.l.rothschildi
    Malaysia
  S.l.moderatus
    N Sumatra
  S.l.intensus
    S Sumatra

PSARISOMUS
Psarisomus dalhousiae (Long-tailed Broadbill)
  P.d.dalhousiae
    Himalayas, Burma, N & W Thailand, Laos, N Vietnam
  P.d.cyanicauda
    SE Thailand
  P.d.divinus
    Cambodia, S Vietnam
  P.d.psittacinus
    Malaysia, Sumatra
  P.d.borneensis
    mountains of NW Borneo

CALYPTOMENINAE

CALYPTOMENA
Calyptomena viridis (Lesser Green Broadbill)
C.v.continentis
S Burma, S Thailand, Malaysia
C.v.viridis
Nias, N Natuna Is, Sumatra, Borneo
C.v.siberu
Siberut, S Pagi Is
Calyptomena hosii (Magnificent Green Broadbill)
mountains of N Borneo
Calyptomena whiteheadi (Black-throated Green Broadbill)
Mt Kinabalu (N Borneo)

101. DENDROCOLAPTIDAE (WOODCREEPERS)

DENDROCINCLA
Dendrocincla tyrannina (Tyrannine Woodcreeper)
D.t.tyrannina
Colombia
D.t.hellmayri
E Colombia, W Venezuela
Dendrocincla macrorhyncha (Large Tyrannine Woodcreeper)
E Ecuador
Dendrocincla fuliginosa (Plain-brown Woodcreeper)
D.f.ridgwayi
E Central America, W Colombia, W Ecuador
D.f.lafresnayei
N & E Colombia, NW Venezuela
D.f.meruloides
coast of N Venezuela, Trinidad
D.f.deltana
delta of Orinoco
D.f.barinensis
C Venezuela
D.f.phaeochroa
S Colombia, E Ecuador, E Peru
D.f.neglecta
W Brazil
D.f.atrirostris
NE Bolivia, SW Brazil
D.f.fuliginosa
E Venezuela, the Guianas, N Brazil
D.f.rufoolivacea
EC Brazil
D.f.taunayi
NE Brazil
D.f.turdina
E & S Brazil, E Paraguay, NE Argentina
D.f.brumaii
C Brazil
Dendrocincla anabatina (Tawny-winged Woodcreeper)
D.a.anabatina
S Mexico to Costa Rica

D.a.typhla
SE Mexico
Dendrocincla merula (White-chinned Woodcreeper)
D.m.bartletti
S Venezuela, W Brazil, NE Paraguay
D.m.merula
the Guianas, N Brazil
D.m.obidensis
Brazil
D.m.remota
EC Bolivia
D.m.olivascens
Brazil
D.m.castanoptera
C Brazil
D.m.badia
C Brazil
Dendrocincla homochroa (Ruddy Woodcreeper)
D.h.homochroa
S Mexico to Honduras
D.h.acedesta
SW Nicaragua, W Costa Rica
D.h.ruficeps
Panama, W Venezuela
D.h.meridionalis
NE Colombia

DECONYCHURA
Deconychura longicauda (Long-tailed Woodcreeper)
D.l.typica
SW Costa Rica, W Panama
D.l.darienensis
E Panama
D.l.minor
N Colombia
D.l.longicauda
the Guianas, N Brazil
D.l.connectens
NW Amazonia
D.l.pallida
SE Peru, N Bolivia, W Brazil
D.l.zimmeri
Para, Brazil
Deconychura stictolaema (Spot-throated Woodcreeper)
D.s.clarior
French Guiana, NE Brazil
D.s.secunda
E Ecuador, NE Peru, W Brazil, S Venezuela
D.s.stictolaema
C Brazil

SITTASOMUS
Sittasomus griseicapillus (Olivaceous Woodcreeper)
S.g.jaliscensis
SW Mexico
S.g.sylvioides
SE Mexico to Costa Rica

S.g.gracileus
  E Mexico
S.g.levis
  W Panama, N Colombia
S.g.veraguensis
  E Panama
S.g.aequatorialis
  W Ecuador, NW Peru
S.g.perijanus
  NW Venezuela
S.g.tachiranus
  W Venezuela
S.g.griseus
  coast of N Venezuela, Tobago I
S.g.enochrus
  Colombia
S.g.amazonus
  SE Colombia, S Venezuela, E Ecuador, W
  Brazil
S.g.axillaris
  E Venezuela, N Brazil
S.g.viridis
  N & E Bolivia
S.g.viridior
  E Bolivia
S.g.transitivus
  C Brazil
S.g.griseicapillus
  S Bolivia, W Brazil, W Paraguay, NW
  Argentina
S.g.reiseri
  NE Brazil
S.g.olivaceus
  E Brazil
S.g.sylviellus
  SE Brazil, NE Argentina

GLYPHORHYNCHUS
**Glyphorhynchus spirurus (Wedge-billed
Woodcreeper)**
  G.s.pectoralis
    S Mexico to Nicaragua
  G.s.sublestus
    Costa Rica to W Ecuador, W Venezuela
  G.s.subrufescens
    W Colombia
  G.s.integratus
    NE Colombia
  G.s.rufigularis
    E Colombia, S Venezuela, N Ecuador
  G.s.amacurensis
    NE Venezuela
  G.s.coronobscurus
    S Venezuela
  G.s.spirurus
    E Venezuela, the Guianas, N Brazil
  G.s.pallidus
    Panama
  G.s.castelnaudii
    E Ecuador, N Peru
  G.s.albigularis
    SE Peru, N Bolivia

G.s.inornatus
  C Brazil
G.s.cuneatus
  EC Brazil
G.s.paraensis
  Belem (Brazil)

DRYMORNIS
**Drymornis bridgesii (Scimitar-billed
Woodhewer)**
  Paraguay, Uruguay, Argentina

NASICA
**Nasica longirostris (Long-billed
Woodcreeper)**
  W Amazonia to French Guiana

DENDREXETASTES
**Dendrexetastes rufigula (Cinnamon-throated
Woodcreeper)**
  D.r.devillei
    W Amazonia
  D.r.rufigula
    the Guianas, N Brazil
  D.r.moniliger
    W Brazil
  D.r.paraensis
    NE Brazil

HYLEXETASTES
**Hylexetastes perrotii (Red-billed
Woodcreeper)**
  H.p.perrotii
    E Venezuela, Guyana, French Guiana, N
    Brazil
  H.p.uniformis
    N Brazil
**Hylexetastes stresemanni (Bar-bellied
Woodcreeper)**
  H.s.insignis
    NW Brazil
  H.s.stresemanni
    NW Brazil
  H.s.undulatus
    E Peru, W Brazil

XIPHOCOLAPTES
**Xiphocolaptes promeropirhynchus (Strong-
billed Woodcreeper)**
  X.p.omiltemensis
    SW Mexico
  X.p.sclateri
    S Mexico
  X.p.emigrans
    S Mexico to N Nicaragua
  X.p.costaricensis
    Costa Rica
  X.p.panamensis
    S Panama
  X.p.rostratus
    N Colombia

X.p.sanctaemartae
  N Colombia, W Venezuela
X.p.virgatus
  C Colombia
X.p.macarenae
  WC Colombia
X.p.promeropirhynchus
  EC Colombia, W Venezuela
X.p.procerus
  N Venezuela
X.p.tenebrosus
  E Venezuela
X.p.neblinae
  S Venezuela
X.p.ignotus
  W Ecuador
X.p.crassirostris
  SW Ecuador, NW Peru
X.p.compressirostris
  N Peru
X.p.phaeopygus
  E Peru
X.p.solivagus
  E Peru
X.p.lineatocephalus
  SE Peru, N & W Bolivia
X.p.orenocensis
  Venezuela, E Ecuador, E Peru, W Brazil
X.p.berlepschi
  E Peru, W Brazil
X.p.paraensis
  C Brazil
X.p.obsoletus
  N & E Bolivia

Xiphocolaptes albicollis (White-throated Woodcreeper)
X.a.bahiae
  NE Brazil
X.a.albicollis
  SE Brazil, E Paraguay, NE Argentina

Xiphocolaptes villanovae (Vila Nova Woodcreeper)
  E Brazil

Xiphocolaptes falcirostris (Moustached Woodcreeper)
  NE Brazil

Xiphocolaptes franciscanus (Snethlage's Woodcreeper)
  C Brazil

Xiphocolaptes major (Great Rufous Woodcreeper)
X.m.remoratus
  C Brazil
X.m.castaneus
  S Brazil, E & S Bolivia, NW Argentina
X.m.major
  Paraguay, N Argentina

DENDROCOLAPTES
Dendrocolaptes certhia (Barred Woodcreeper)
D.c.sanctithomae
  S Mexico to Nicaragua

D.c.scheffleri
  Oaxaca (Mexico)
D.c.nigrirostris
  Costa Rica, Panama
D.c.hesperius
  SW Costa Rica
D.c.colombianus
  W Colombia, NW Ecuador
D.c.hyleorus
  NE Colombia
D.c.radiolatus
  SE Colombia, E Ecuador, NE Peru, NW Brazil
D.c.punctipectus
  NW Venezuela
D.c.certhia
  S Venezuela, the Guianas, N Brazil
D.c.juruanus
  E Peru, E Bolivia, W Brazil
D.c.polyzonus
  N Bolivia
D.c.ridgwayi
  N Brazil
D.c.medius
  N Brazil
Dendrocolaptes concolor (Concolor Woodcreeper)
  N Brazil
Dendrocolaptes hoffmannsi (Hoffmann's Woodcreeper)
  C Brazil
Dendrocolaptes picumnus (Black-banded Woodcreeper)
D.p.puncticollis
  Guatemala, Honduras
D.p.costaricensis
  Costa Rica, W Panama
D.p.veraguensis
  S Panama
D.p.multistrigatus
  E Colombia, W Venezuela
D.p.seilerni
  N Colombia, N Venezuela
D.p.picumnus
  E Venezuela, the Guianas, N Brazil
D.p.validus
  W Amazonia
D.p.transfasciatus
  C Brazil
D.p.olivaceus
  C Bolivia
D.p.pallescens
  S Brazil, S Bolivia, Paraguay
D.p.extimus
  E Paraguay
D.p.casaresi
  NW Argentina
Dendrocolaptes platyrostris (Planalto Woodcreeper)
D.p.intermedius
  NE Brazil, N Paraguay
D.p.platyrostris
  SE Brazil, Paraguay, N Argentina

XIPHORHYNCHUS

Xiphorhynchus picus (Straight-billed Woodcreeper)

X.p.extimus
S Panama

X.p.dugandi
N Colombia

X.p.picirostris
NW Colombia, W Venezuela

X.p.saturatior
E Colombia, W Venezuela

X.p.borreroi
SW Colombia

X.p.choicus
N Venezuela

X.p.parguanae
NW Venezuela

X.p.longirostris
Margarita I

X.p.altirostris
Trinidad

X.p.phalera
S Venezuela

X.p.deltans
NE Venezuela

X.p.picus
E Colombia to the Guianas, N Brazil

X.p.duidae
S Venezuela, NW Brazil

X.p.peruvianus
E Peru, W Brazil, N Bolivia

X.p.kienerii
W Brazil

X.p.rufescens
C Brazil

X.p.bahiae
NE Brazil

Xiphorhynchus necopinus (Zimmer's Woodcreeper)
C & NE Brazil

Xiphorhynchus obsoletus (Striped Woodcreeper)

X.o.palliatus
W Amazonia, Bolivia

X.o.notatus
E Colombia, SW Venezuela, NW Brazil

X.o.obsoletus
E Venezuela to French Guiana, N Brazil

X.o.caicarae
NE Venezuela

Xiphorhynchus ocellatus (Ocellated Woodcreeper)

X.o.napensis
SE Colombia, E Ecuador, NE Peru

X.o.lineatocapillus
E Venezuela

X.o.ocellatus
E Colombia, S Venezuela, NE Peru, NW Brazil

X.o.perplexus
NE Peru, W Brazil

X.o.chunchotambo
E Peru

X.o.brevirostris
SE Peru, NE Bolivia

Xiphorhynchus spixii (Spix's Woodcreeper)

X.s.buenavistae
W Colombia

X.s.insignis
EC Peru

X.s.juruanus
SE Peru, NE Bolivia, W Brazil

X.s.spixii
S Brazil

Xiphorhynchus elegans (Elegant Woodcreeper)

X.e.ornatus
SE Colombia, E Ecuador, NE Peru, W Brazil

X.e.elegans
S Brazil

Xiphorhynchus pardalotus (Chestnut-rumped Woodcreeper)

X.p.caurensis
SE Venezuela, W Guyana

X.p.pardalotus
the Guianas, N Brazil

Xiphorhynchus guttatus (Buff-throated Woodcreeper)

X.g.confinis
E Guatemala, N Honduras

X.g.costaricensis
SW Honduras to Panama

X.g.marginatus
E Panama

X.g.nanus
E Panama, N Colombia, W Venezuela

X.g.rosenbergi
W Colombia

X.g.demonstratus
E Colombia, NW Venezuela

X.g.susurrans
NE Venezuela, Trinidad, Tobago I

X.g.jardinei
NE Venezuela

X.g.margaritae
Margarita I

X.g.polystictus
E Colombia to the Guianas, N Brazil

X.g.connectens
N Brazil

X.g.guttatoides
W Amazonia

X.g.vicinalis
S Brazil

X.g.dorbignyanus
NE Bolivia, SW Brazil

X.g.guttatus
coast of E Brazil

Xiphorhynchus eytoni (Dusky-billed Woodcreeper)
E & S Brazil

Xiphorhynchus flavigaster (Ivory-billed Woodcreeper)

X.f.tardus
NW Mexico

X.f.mentalis
W Mexico
X.f.flavigaster
SW Mexico
X.f.saltuarius
NE Mexico
X.f.yucatanensis
Yucatan peninsula, Meco I
X.f.ascensor
S Mexico
X.f.eburneirostris
SE Mexico to NW Costa Rica
X.f.ultimus
NW Costa Rica
Xiphorhynchus striatigularis (Stripe-throated Woodcreeper)
NE Mexico
Xiphorhynchus lachrymosus (Black-striped Woodcreeper)
X.l.lachrymosus
E Nicaragua to W Ecuador
X.l.alarum
N Colombia
Xiphorhynchus erythropygius (Spotted Woodcreeper)
X.e.erythropygius
S Mexico
X.e.parvus
S Mexico to N Nicaragua
X.e.punctigula
S Nicaragua to W Panama
X.e.insolitus
E Panama, NW Colombia
X.e.aequatorialis
W Colombia, E Ecuador
Xiphorhynchus triangularis (Olive-backed Woodcreeper)
X.t.triangularis
Colombia, W Venezuela, E Ecuador, N Peru
X.t.hylodromus
N Venezuela
X.t.intermedius
C & S Peru
X.t.bangsi
Bolivia

LEPIDOCOLAPTES
Lepidocolaptes leucogaster (White-striped Woodcreeper)
L.l.umbrosus
NW Mexico
L.l.leucogaster
C & S Mexico
Lepidocolaptes souleyetii (Streak-headed Woodcreeper)
L.s.guerrerensis
W Mexico
L.s.insignis
SE Mexico to N Honduras
L.s.compressus
S Mexico to W Panama
L.s.lineaticeps
E Panama, N Colombia, W Venezuela

L.s.littoralis
N Colombia to Guyana, Trinidad, N Brazil
L.s.uaireni
SE Venezuela
L.s.esmeraldae
SW Colombia, W Ecuador
L.s.souleyetii
SW Ecuador, NW Peru
Lepidocolaptes angustirostris (Narrow-billed Woodcreeper)
L.a.griseiceps
Sipaliwini (Surinam)
L.a.coronatus
N Brazil
L.a.bahiae
NE Brazil
L.a.bivittatus
E Bolivia, C & E Brazil
L.a.hellmayri
WC Bolivia
L.a.certhiolus
C Bolivia to Paraguay, NW Argentina
L.a.dabbenei
SW Paraguay, N Argentina
L.a.angustirostris
E Paraguay, SW Brazil, N Argentina
L.a.praedatus
W Uruguay, E & S Argentina
Lepidocolaptes affinis (Spot-crowned Woodcreeper)
L.a.lignicida
NE Mexico
L.a.affinis
S Mexico to N Nicaragua
L.a.neglectus
Costa Rica, W Panama
L.a.sanctaemartae
N Colombia
L.a.sneiderni
W Colombia
L.a.lacrymiger
E Colombia, W Venezuela
L.a.lafresnayi
N Venezuela
L.a.aequatorialis
SW Colombia, Ecuador
L.a.frigidus
SW Colombia
L.a.warscewiczi
N & C Peru
L.a.carabayae
SE Peru
L.a.bolivianus
Bolivia
Lepidocolaptes squamatus (Scaled Woodcreeper)
L.s.wagleri
NE Brazil
L.s.squamatus
E Brazil
L.s.falcinellus
SE Brazil, Paraguay
Lepidocolaptes fuscus (Lesser Woodcreeper)

L.f.atlanticus
  E Brazil
L.f.brevirostris
  NE Brazil
L.f.tenuirostris
  E Brazil
L.f.fuscus
  SE Brazil, E Paraguay, NE Argentina
Lepidocolaptes albolineatus (Lineated Woodcreeper)
  L.a.albolineatus
  E Venezuela, the Guianas, N Brazil
  L.a.duidae
  S Venezuela, NW Brazil
  L.a.fuscicapillus
  SW Amazonia
  L.a.madeirae
  C Brazil
  L.a.layardi
  C Brazil

CAMPYLORHAMPHUS
Campylorhamphus pucheranii (Greater Scythebill)
  S Colombia, E Ecuador
Campylorhamphus trochilirostris (Red-billed Scythebill)
  C.t.brevipennis
  E Panama, W Colombia
  C.t.venezuelensis
  N Colombia, N Venezuela
  C.t.thoracicus
  SW Colombia, W Ecuador
  C.t.zarumillanus
  NW Peru
  C.t.napensis
  E Ecuador, E Peru
  C.t.notabilis
  W Brazil
  C.t.snethlageae
  C Brazil
  C.t.devius
  N Bolivia
  C.t.lafresnayanus
  E Bolivia, N Paraguay
  C.t.major
  NE Brazil
  C.t.omissus
  NE Brazil
  C.t.trochilirostris
  NE Brazil
  C.t.hellmayri
  N Argentina
Campylorhamphus falcularius (Black-billed Scythebill)
  SE Brazil, Paraguay, NE Argentina
Campylorhamphus pusillus (Brown-billed Scythebill)
  C.p.borealis
  Costa Rica, W Panama
  C.p.olivaceus
  C Panama

C.p.tachirensis
  NE Colombia
C.p.pusillus
  Colombia, W Ecuador
Campylorhamphus procurvoides (Curve-billed Scythebill)
  C.p.sanus
  E Colombia, Venezuela, W Guyana, N Brazil
  C.p.procurvoides
  French Guiana, N Brazil
  C.p.probatus
  C Brazil
  C.p.multostriatus
  C Brazil

102. FURNARIIDAE (OVENBIRDS)

FURNARIINAE

GEOSITTA
Geositta poeciloptera (Campo Miner)
  Brazil
Geositta cunicularia (Common Miner)
  G.c.juninensis
  C Peru
  G.c.titicacae
  S Peru to N Chile, NW Argentina
  G.c.georgei
  Peru
  G.c.frobeni
  S Peru
  G.c.deserticolor
  SW Peru, N Chile
  G.c.fissirostris
  C Chile
  G.c.hellmayri
  W Argentina
  G.c.cunicularia
  S Brazil, Uruguay, Argentina
Geositta maritima (Greyish Miner)
  Peru, N Chile
Geositta peruviana (Coastal Miner)
  G.p.paytae
  N Peru
  G.p.peruviana
  C Peru
  G.p.rostrata
  SC Peru
Geositta punensis (Puna Miner)
  Bolivia, Peru, Chile, N Argentina
Geositta saxicolina (Dark-winged Miner)
  C Peru
Geositta isabellina (Creamy-rumped Miner)
  C Chile, W Argentina
Geositta antarctica (Short-billed Miner)
  S Chile, S Argentina
Geositta rufipennis (Rufous-banded Miner)
  G.r.fasciata
  W Bolivia, C Chile, W Argentina
  G.r.harrisoni
  N Chile
  G.r.hellmayri
  N Chile

G.r.rufipennis
W Argentina
G.r.giaii
Bariloche (Argentina)
G.r.fragae
Argentina
G.r.ottowi
Cordoba (Argentina)
Geositta crassirostris (Thick-billed Miner)
central coasts of Peru
Geositta tenuirostris (Slender-billed Miner)
S Peru, W Bolivia, NW Argentina

UPUCERTHIA
Upucerthia certhioides (Chaco Earthcreeper)
U.c.luscinia
W Argentina
U.c.estebani
S Paraguay, N Argentina
U.c.certhioides
N & C Argentina
Upucerthia harterti (Bolivian Earthcreeper)
102.1
S Bolivia
**Upucerthia ruficauda (Straight-billed
Earthcreeper)**
U.r.montana
S Peru
U.r.ruficauda
W Bolivia, N Chile, NW Argentina
U.r.famatinae
N Argentina
Upucerthia andaecola (Rock Earthcreeper)
W Bolivia, NW Argentina, Chile
**Upucerthia albigula (White-throated
Earthcreeper)**
S Peru, N Chile
Upucerthia serrana (Striated Earthcreeper)
U.s.serrana
N Peru
U.s.huancavelicae
SW Peru
**Upucerthia dumetaria (Scale-throated
Earthcreeper)**
U.d.hypoleuca
SW Bolivia, N Chile, W Argentina
U.d.hallinani
N Chile, NW Argentina
U.d.saturatior
C Chile
U.d.dumetaria
Tierra del Fuego, C & S Argentina
**Upucerthia validirostris (Buff-breasted
Earthcreeper)**
U.v.saturata
W Peru
U.v.pallida
S Peru, Bolivia, N Chile, NW Argentina
U.v.rufescens
N Argentina
U.v.validirostris
W Argentina

**Upucerthia jelskii (Plain-breasted
Earthcreeper)**
C Peru

CINCLODES
Cinclodes fuscus (Bar-winged Cinclodes)
C.f.heterurus
W Venezuela
C.f.oreobates
N Colombia
C.f.paramo
SW Colombia
C.f.albidiventris
Ecuador
C.f.longipennis
N Peru
C.f.rivularis
C & S Peru
C.f.albiventris
S Peru, Bolivia, N Chile, NW Argentina
C.f.riojanus
NE Argentina
C.f.rufus
C Argentina
C.f.yzurietae
S Argentina
C.f.fuscus
S Brazil, Uruguay, S Chile, S Argentina
Cinclodes excelsior (Stout-billed Cinclodes)
102.1
C.e.colombiana
C Colombia
C.e.excelsior
SW Colombia, Ecuador
C.e.aricomae
N Peru 102.3
Cinclodes olrogi (Olrog's Cinclodes)
C Argentina
**Cinclodes comechingonus (Comechingones
Cinclodes)**
C Argentina
Cinclodes pabsti (Long-tailed Cinclodes)
S Brazil
**Cinclodes atacamensis (White-winged
Cinclodes)**
C.a.atacamensis
S Peru, W Bolivia, N Chile, N Argentina
C.a.schocolatinus
C Argentina
Cinclodes palliatus (White-bellied Cinclodes)
C Peru
Cinclodes oustaleti (Grey-flanked Cinclodes)
C.o.oustaleti
S Chile
C.o.hornensis
Cape Horn islands, Tierra del Fuego
C.o.baeckstroemii
Juan Fernandez I
**Cinclodes patagonicus (Dark-bellied
Cinclodes)**
C.p.chilensis
C Chile, W Argentina

C.p.patagonicus
S Chile, S Argentina
Cinclodes nigrofumosus (Seaside Cinclodes)
coast of N Chile
Cinclodes taczanowskii (Taczanowski's Cinclodes)
central coast of Peru
Cinclodes antarcticus (Blackish Cinclodes)
C.a.maculirostris
Cape Horn islands
C.a.antarcticus
Falkland Is

CHILIA
Chilia melanura (Crag Chilia)
C.m.atacamae
N Chile
C.m.melanura
C Chile

FURNARIUS
Furnarius minor (Lesser Hornero)
NE Peru, W Brazil, E Ecuador
Furnarius figulus (White-banded Hornero)
F.f.pileatus
C Brazil
F.f.figulus
E Brazil
Furnarius tricolor (Tricolour Hornero)
E Peru, W Brazil, N Bolivia
Furnarius leucopus (Pale-legged Hornero)
F.l.longirostris
N Colombia, NW Venezuela
F.l.endoecus
C Colombia, W Venezuela
F.l.leucopus
Guyana, N Brazil
F.l.cinnamomeus
SW Ecuador, NW Peru
F.l.assimilis
E & S Brazil, SE Bolivia
Furnarius torridus (Pale-billed Hornero) 102.1
NE Peru, W Brazil
Furnarius rufus (Rufous Hornero)
F.r.albogularis
SE Brazil
F.r.commersoni
Bolivia, W Brazil
F.r.schuhmacheri
S Bolivia
F.r.paraguayae
Paraguay, N Argentina
F.r.rufus
S Brazil, Uruguay, C & E Argentina
Furnarius cristatus (Crested Hornero)
Paraguay, N Argentina, Bolivia

SYNALLAXINAE

SYLVIORTHORHYNCHUS
Sylviorthorhynchus desmursii (Des Murs' Wiretail)
S Chile, W Argentina

APHRASTURA
Aphrastura spinicauda (Thorn-tailed Rayadito)
A.s.spinicauda
Tierra del Fuego, S Chile, W Argentina
A.s.bullocki
Moncha I (Chile)
A.s.fulva
Chiloe I (Chile)
Aphrastura masafuerae (Masafuera Rayadito)
Mas Afuera Is

LEPTASTHENURA
Leptasthenura fuliginiceps (Brown-capped Tit-Spinetail)
L.f.fuliginiceps
W Bolivia
L.f.paranensis
W Argentina
Leptasthenura yanacensis (Tawny Tit-Spinetail)
W Peru, W Bolivia
Leptasthenura platensis (Tufted Tit-Spinetail)
S Brazil, Uruguay, Argentina
Leptasthenura aegithaloides (Plain-mantled Tit-Spinetail)
L.a.grisescens
S Peru, N Chile
L.a.berlepschi
S Peru, Bolivia, N Chile, W Argentina
L.a.aegithaloides
C Chile
L.a.pallida
W & S Argentina
Leptasthenura setaria (Araucaria Tit-Spinetail)
S Brazil
Leptasthenura striata (Streaked Tit-Spinetail)
L.s.superciliaris
C Peru
L.s.albigularis
W Peru
L.s.striata
SW Peru, N Chile
Leptasthenura striolata (Striolated Tit-Spinetail)
SE Brazil
Leptasthenura pileata (Rusty-crowned Tit-Spinetail)
L.p.latistriata
W Peru
L.p.cajabambae
C Peru
L.p.pileata
central coast of Peru
Leptasthenura xenothorax (White-browed Tit-Spinetail) 102.1
S Peru
Leptasthenura andicola (Andean Tit-Spinetail)
L.a.certhia
W Venezuela

L.a.extima
  N Colombia
L.a.exterior
  E Colombia
L.a.andicola
  C Colombia, Ecuador
L.a.peruviana
  C Peru, N Bolivia

Schizoeaca fuliginosa (White-chinned Spinetail)
  S.f.fumigata
    Colombia
  S.f.fuliginosa
    W Venezuela, E Colombia, N Ecuador
  S.f.peruviana
    N Peru
  S.f.avacuchensis
    Avacucho (Peru)
  S.f.vilcabambae
    Cuzco (Peru)
  S.f.plengei
    C Peru
Schizoeaca coryi (Ochre-browed Thistletail)
  102.1
  NW Venezuela
Schizoeaca griseomurina (Mouse-coloured Thistletail) 102.1
  S Ecuador, N Peru
Schizoeaca palpebralis (Eye-ringed Thistletail) 102.1
  S.p.palpebralis
    C Peru
  S.p.helleri 102.1
    SE Peru
  S.p.harterti 102.1
    N Bolivia
Schizoeaca moreirae (Itatiaia Spinetail)
  Rio de Janeiro

Schoeniophylax phryganophila (Chotoy Spinetail)
  S.p.phryganophila
    E Bolovia, S Brazil, N Argentina
  S.p.petersi
    E Brazil

Synallaxis ruficapilla (Rufous-capped Spinetail)
  SE Brazil to N Argentina
Synallaxis superciliosa (Buff-browed Spinetail)
  S.s.samaipatae
    S Bolivia
  S.s.superciliosa
    NW Argentina
Synallaxis frontalis (Sooty-fronted Spinetail)

S.f.poliophrys
  French Guiana
S.f.fuscipennis
  Bolivia, NW Argentina
S.f.frontalis
  Brazil, Paraguay, Uruguay, N Argentina
Synallaxis azarae (Azara's Spinetail)
  S.a.media
    W Colombia, N Ecuador
  S.a.ochracea
    E Ecuador, NW Peru
  S.a.fruticicola
    N Peru
  S.a.infumata
    NC Peru
  S.a.urubambae
    SE Peru
  S.a.carabayae
    SE Peru, N Bolivia
  S.a.azarae
    N Bolivia
Synallaxis elegantior (Elegant Spinetail)
  E Colombia, W Venezuela
Synallaxis albigularis (Dark-breasted Spinetail)
  S.a.rodolphei
    S Colombia
  S.a.albigularis
    S Colombia, E Ecuador, E Peru
Synallaxis albescens (Pale-breasted Spinetail)
  S.a.latitabunda
    SW Costa Rica
  S.a.hypoleuca
    S Panama, NW Colombia
  S.a.insignis
    C Colombia
  S.a.occipitalis
    E Colombia, NW Venezuela
  S.a.littoralis
    coast of N Colombia
  S.a.perpallida
    NE Colombia, NW Venezuela,
  S.a.nesiotis
    N Colombia, N Venezuela, Margarita I
  S.a.trinitatis
    E Venezuela, Trinidad
  S.a.josephinae
    S Venezuela, Guyana, Surinam, N Brazil
  S.a.inaequalis
    French Guiana
  S.a.griseonota
    C Brazil
  S.a.albescens
    E Brazil, Paraguay, NE Argentina
  S.a.australis
    E Bolivia, W Paraguay, NW Argentina
Synallaxis spixi (Chicli Spinetail)
  S Brazil to N Argentina
Synallaxis hypospodia (Cinereous-breasted Spinetail)
  E Peru, N Bolivia, W Brazil

Synallaxis infuscata (Plain Spinetail)
E Brazil
Synallaxis brachyura (Slaty Spinetail)
S.b.nigrofumosa
E Honduras to Panama
S.b.chapmani
SW Costa Rica to W Ecuador
S.b.caucae
C Colombia
S.b.brachyura
E Colombia
S.b.jaraguana
N Brazil
Synallaxis courseni (Apurimac Spinetail)
Apurimac, Peru
Synallaxis moesta (Dusky Spinetail)
S.m.moesta
E Colombia
S.m.obscura
SE Colombia
S.m.brunneicaudalis
E Ecuador, NE Peru
S.m.yavii
Venezuela
S.m.obscurior
French Guiana
S.m.cabanisi
Peru
S.m.fulviventris
Bolivia
Synallaxis macconnelli (McConnell's Spinetail)
Mt Roraima (Venezuela)
Synallaxis subpudica (Silvery-throated Spinetail)
E Colombia, Ecuador
Synallaxis tithys (Blackish-headed Spinetail)
SW Ecuador, NW Peru
Synallaxis cinerascens (Grey-bellied Spinetail)
SE Brazil, Paraguay, NE Argentina
Synallaxis maranonica (Maranon Spinetail)
N Peru
Synallaxis propinqua (White-bellied Spinetail)
N & W Amazonia
Synallaxis hellmayri (Reiser's Spinetail)
NE Brazil
Synallaxis gujanensis (Plain-crowned Spinetail)
S.g.colombiana
E Colombia
S.g.gujanensis
Venezuela, the Guianas, N Brazil
S.g.huallagae
NE Peru
S.g.canipileus
SE Peru
S.g.inornata
W Bolivia, W & C Brazil
S.g.certhiola
N Bolivia
S.g.simoni
C Brazil

Synallaxis albilora (Ochre-breasted Spinetail)
SE Brazil, N Paraguay
Synallaxis rutilans (Ruddy Spinetail)
S.r.caquetensis
SE Colombia, E Ecuador, NE Peru
S.r.confinis
NW Brazil
S.r.dissors
E Colombia to the Guianas, N Brazil
S.r.amazonica
E Peru, N Bolivia, W & C Brazil
S.r.rutilans
C & S Brazil
S.r.omissa
EC Brazil
S.r.tertia
NE Bolivia, SW Brazil
Synallaxis cherriei (Chestnut-throated Spinetail)
S.c.napoensis
E Ecuador, N Peru
S.c.cherriei
S Brazil
Synallaxis unirufa (Rufous Spinetail)
S.u.unirufa
N Colombia, E Ecuador
S.u.munotztebari
NE Colombia
S.u.meridana
E Colombia, W Venezuela
S.u.ochrogaster
Peru
Synallaxis castanea (Black-throated Spinetail)
N Venezuela
Synallaxis fuscorufa (Santa Marta Spinetail)
N Colombia
Synallaxis zimmeri (Russet-bellied Spinetail)
C Peru
Synallaxis erythrothorax (Rufous-breasted Spinetail)
S.e.furtiva
SE Mexico
S.e.erythrothorax
SE Mexico to NW Honduras
S.e.pacifica
S Mexico to El Salvador
Synallaxis cinnamomea (Stripe-breasted Spinetail)
S.c.carri
Trinidad
S.c.terrestris
Tobago I
S.c.cinnamomea
E Colombia, NW Venezuela
S.c.aveledoi
W Venezuela
S.c.bolivari
N Venezuela
S.c.striatipectus
NE Venezuela
S.c.pariae
Paria peninsula (Venezuela)

Synallaxis stictothorax **(Necklaced Spinetail)**
  S.s.stictothorax
    NW Ecuador, Puna Is
  S.s.maculata
    NW Peru
  S.s.chinchipensis
    N Peru
Synallaxis candei **(White-whiskered Spinetail)**
  S.c.candei
    N Colombia, W Venezuela
  S.c.atrigularis
    NC Colombia
  S.c.venezuelensis
    NE Colombia, NW Venezuela
Synallaxis kollari **(Hoary-throated Spinetail)**
  N Brazil
Synallaxis scutatus **(Ochre-cheeked Spinetail)**
  S.s.scutatus
    E & C Brazil
  S.s.whitii
    E Bolivia, S Brazil, NW Argentina
  S.s.teretiala
    Para (Brazil)

HELLMAYREA   102.5
Hellmayrea gularis **(Lafresnaye's White-browed Spinetail)**
  H.g.gularis
    Colombia, N Ecuador, W Venezuela
  H.g.brunneidorsalis
    NE Colombia
  H.g.cinereiventris
    W Venezuela
  H.g.rufiventris
    C Peru

CERTHIAXIS
Certhiaxis erythrops **(Red-faced Spinetail)**
  C.e.rufigenis
    Costa Rica
  C.e.griseigularis
    W Colombia
  C.e.erythrops
    W Ecuador
Certhiaxis demissa **(Tepui Spinetail)**
  S Venezuela
Certhiaxis antisiensis **(Line-cheeked Spinetail)**
  C.a.antisiensis
    S Ecuador
  C.a.palamblae
    N Peru
  C.a.furcata
    N Peru
Certhiaxis pallida **(Pallid Spinetail)**
  SE Brazil
Certhiaxis curtata **(Ash-browed Spinetail)**
  C.c.curtata
    E Colombia
  C.c.cisandina
    C Colombia, E Ecuador, N Peru

  C.c.debilis
    C Peru
Certhiaxis obsoleta **(Olive Spinetail)**
  SE Brazil, E Paraguay, N Argentina
Certhiaxis hellmayri **(Streak-capped Spinetail)**
  N Colombia
Certhiaxis subcristata **(Crested Spinetail)**
  C.s.fuscivertex
    NE Colombia
  C.s.subcristata
    E Colombia, N Venezuela
Certhiaxis pyrrhophia **(Stripe-crowned Spinetail)**
  C.p.rufipennis
    C Bolivia
  C.p.striaticeps
    E Bolivia
  C.p.pyrrhophia
    S Bolivia to Uruguay, N Argentina
Certhiaxis marcapatae **(Marcapata Spinetail)**
  C.m.marcapatae
    Cuzco (SE Peru)
  C.m.weskei
    Peru
Certhiaxis albiceps **(Light-crowned Spinetail)**
  C.a.albiceps
    La Paz (Bolivia)
  C.a.discolor
    Cochabamba (Bolivia)
Certhiaxis semicinerea **(Grey-headed Spinetail)**
  C.s.semicinerea
    Ceara (Brazil), Baia (Brazil)
  C.s.goyana
    Goiaz (N Brazil)
Certhiaxis albicapilla **(Creamy-chested Spinetail)**
  C.a.albicapilla
    C Peru
  C.a.albigula
    SE Peru
Certhiaxis vulpina **(Rusty-backed Spinetail)**
  C.v.apurensis
    SW Venezuela
  C.v.alopecias
    E Colombia, W Venezuela, N Brazil
  C.v.vulpecula
    NE Peru, W Brazil, NE Bolivia
  C.v.vulpina
    W & C Brazil
  C.v.foxi
    W Bolivia
  C.v.reiseri
    NE Brazil
Certhiaxis muelleri **(Scaled Spinetail)**
  N Brazil
Certhiaxis gutturata **(Speckled Spinetail)**
  C.g.peruviana
    S Colombia to N Bolivia
  C.g.hyposticta
    S Venezuela to Surinam, N Brazil
  C.g.gutturata
    S Venezuela to French Guiana, N Brazil

Certhiaxis sulphurifera (Sulphur-throated Spinetail)
E Argentina, Uruguay
Certhiaxis cinnamomea (Yellow-throated Spinetail)
C.c.fuscifrons
N Colombia
C.c.marabina
NW Venezuela
C.c.valenciana
Venezuela
C.c.orenocensis
Orinoco valley (Venezuela)
C.c.cinnamomea
Trinidad, NE Venezuela, the Guianas, NE Brazil
C.c.pallida
N Brazil
C.c.cearensis
E Brazil
C.c.russeola
W Bolivia, S Brazil, Paraguay, NW Argentina
Certhiaxis mustelina (Red & White Spinetail)
NE Peru

THRIPOPHAGA
Thripophaga pyrrholeuca (Lesser Canastero)
T.p.affinis
S Bolivia, W Argentina
T.p.pyrrholeuca
Paraguay, NW Argentina
T.p.sordida
Chile, W Argentina
T.p.flavogularis
E & S Argentina
Thripophaga baeri (Short-billed Canastero)
T.b.chacoensis
NW Paraguay
T.b.baeri
S Brazil, NE & C Argentina
Thripophaga pudibunda (Canyon Canastero)
T.p.neglecta
W Peru
T.p.pudibunda
SC Peru
Thripophaga ottonis (Rusty-fronted Canastero)
SE Peru
Thripophaga heterura (Iquico Canastero)
Bolivia
Thripophaga modesta (Cordillera Canastero)
T.m.monticola
W Peru
T.m.proxima
C & S Peru
T.m.modesta
S Peru to N Chile, W Argentina
T.m.rostrata
N Bolivia
T.m.serrana
C Argentina

T.m.australis
Chile, C Argentina
Thripophaga cactorum (Cactus Canastero) 102.1
T.c.cactorum
Peru
T.c.lachayensis
central coast of Peru
Thripophaga dorbignyi (Creamy-breasted Canastero)
T.d.huancavelicae
C Peru
T.d.usheri
SC Peru
T.d.arequipae
S Peru, N Chile, W Bolivia
T.d.consobrina
NW Bolivia
T.d.dorbignyi
E Bolivia, NW Argentina
Thripophaga berlepschi (Berlepsch's Canastero)
N Bolivia
Thripophaga steinbachi (Chestnut Canastero)
W Argentina
Thripophaga humicola (Dusky-tailed Canastero)
T.h.goodalli
N Chile
T.h.humicola
N & C Chile
T.h.polysticta
S Chile
Thripophaga patagonica (Patagonian Canastero)
S Argentina
Thripophaga humilis (Streak-throated Canastero)
T.h.cajamarcae
W Peru
T.h.humilis
C Peru
T.h.robusta
SE Peru, W Bolivia
Thripophaga usheri (White-tailed Canastero)
Apurimac (Peru) 102.6
Thripophaga anthoides (Austral Canastero)
S Chile, S Argentina
Thripophaga wyatti (Streak-backed Canastero)
T.w.wyatti
NC Colombia
T.w.sanctaemartae
N Colombia
T.w.mucuchiesi
NW Venezuela
T.w.perijanus
W Venezuela
T.w.aequatorialis
C Ecuador
T.w.azuay
S Ecuador

T.w.graminicola
 Peru
***Thripophaga sclateri* (Cordoba Canastero)**
102.2
T.s.punensis
 SE Peru, W Bolivia
T.s.cuchacanchae
 Bolivia
T.s.lilloi
 NW Argentina
T.s.sclateri
 C Argentina
***Thripophaga urubambensis* (Line-fronted Canastero)**
T.u.huallagae
 N Peru
T.u.urubambensis
 SE Peru, N Bolivia
***Thripophaga virgata* (Junin Canastero)**
 C Peru
***Thripophaga maculicauda* (Scribble-tailed Canastero)**
 Peru, Bolivia, NW Argentina
***Thripophaga flammulata* (Many-striped Canastero)**
T.f.multostriata
 E Colombia
T.f.quindiana
 C Colombia
T.f.flammulata
 S Colombia, Ecuador
T.f.pallida
 NE Peru
T.f.taczanowskii
 NC Peru
***Thripophaga cherriei* (Orinoco Softtail)**
 Venezuela
***Thripophaga macroura* (Striated Softtail)**
 E Brazil
***Thripophaga hudsoni* (Hudson's Canastero)**
 Uruguay, NE Argentina
***Thripophaga hypochondriacus* (Great Spinetail)**

PHACELLODOMUS
***Phacellodomus rufifrons* (Rufous-fronted Thornbird)**
P.r.inornatus
 N Venezuela
P.r.castilloi
 S Venezuela
P.r.peruvianus
 N Peru
P.r.specularis
 NE Brazil
P.r.rufifrons
 E Brazil
P.r.fargoi
 S Brazil, Paraguay
P.r.sincipitalis
 Bolivia, W Argentina
***Phacellodomus sibilatrix* (Little Thornbird)**
 Paraguay, Argentina

***Phacellodomus striaticeps* (Streak-fronted Thornbird)**
P.s.griseipectus
 SE Peru
P.s.striaticeps
 Bolivia, N Argentina
***Phacellodomus erythrophthalmus* (Red-eyed Thornbird)**
P.e.erythrophthalmus
 coast of E Brazil
P.e.ferrugineigula
 SE Brazil
***Phacellodomus striaticollis* (Freckle-breasted Thornbird)**
P.s.maculipectus
 E Bolivia, NW Argentina
P.s.striaticollis
 SE Brazil, Uruguay, E Argentina
***Phacellodomus dorsalis* (Chestnut-backed Thornbird)**
 N Peru
***Phacellodomus ruber* (Greater Thornbird)**
 Bolivia, C Brazil to Argentina
***Phacellodomus fusciceps* (Plain Softtail)**
P.f.dimorpha
 Ecuador, E Peru
P.f.obidensis
 N Brazil
P.f.fusciceps
 Bolivia
***Phacellodomus berlepschi* (Russet-mantled Softtail)**
 N Peru
***Phacellodomus dendrocolaptoides* (Canebrake Groundcreeper)**
 SE Brazil, Paraguay, NE Argentina

SPARTONOICA
***Spartonoica maluroides* (Bay-capped Wren Spinetail)**
 S Brazil, Uruguay, Argentina

PHLEOCRYPTES
***Phleocryptes melanops* (Wren-like Rushbird)**
P.m.brunnescens
 S Peru
P.m.juninensis
 C Peru
P.m.schoenobaenus
 S Peru, W Bolivia, NW Argentina
P.m.loaensis
 N Chile
P.m.melanops
 S Brazil to C Chile, C Argentina

LIMNORNIS
***Limnornis curvirostris* (Curve-billed Reedhaunter)**
 S Brazil, E Argentina
***Limnornis rectirostris* (Straight-billed Reedhaunter)**
 SE Uruguay, NE Argentina

PASSERIFORMES

ANUMBIUS
Anumbius annumbi (Firewood Gatherer)
SE Brazil to N Argentina

CORYPHISTERA
Coryphistera alaudina (Lark-like Brushrunner)
C.a.campicola
E Bolivia, W Paraguay
C.a.alaudina
S Bolivia, NW Argentina

EREMOBIUS
Eremobius phoenicurus (Band-tailed Earthcreeper)
W & S Argentina

SIPTORNIS
Siptornis striaticollis (Spectacled Prickletail)
Colombia, E Ecuador

METOPOTHRIX
Metopothrix aurantiacus (Orange-fronted Plushcrown)
W Amazonia

XENERPESTES
Xenerpestes minlosi (Double-banded Greytail)
X.m.minlosi
E Panama, Colombia
X.m.umbraticus
W Colombia
Xenerpestes singularis (Equatorial Greytail)
C Ecuador, N Peru

PHILYDORINAE

MARGARORNIS
Margarornis adustus (Roraima Barbtail)
M.a.obscurodorsalis
SE Venezuela
M.a.duidae
Mt Duida (S Venezuela
M.a.adustus
E Venezuela, W Guyana
Margarornis guttuligera (Rusty-winged Barbtail)
M.g.guttuligera
Colombia, Ecuador, Peru
M.g.venezuelana
NW Venezuela
Margarornis brunnescens (Spotted Barbtail)
M.b.brunneicauda
Costa Rica, W Panama
M.b.distinctus
C Panama
M.b.mnionophilus
Panama
M.b.albescens
E Panama
M.b.coloratus
N Colombia

M.b.brunnescens
Venezuela, Colombia, Ecuador, N Peru
M.b.stictonotus
SE Peru, W Bolivia
M.b.rostratus
N Venezuela
Margarornis tatei (White-throated Barbtail)
M.t.tatei
NE Venezuela
M.t.pariae
NE Venezuela
Margarornis rubiginosus (Ruddy Treerunner)
M.r.rubiginosus
Costa Rica, W Panama
M.r.boultoni
C Panama
Margarornis stellatus (Fulvous-dotted Treerunner)
W Colombia, NW Ecuador
Margarornis bellulus (Beautiful Treerunner)
E Panama
Margarornis squamiger (Pearled Treerunner)
M.s.perlatus
Venezuela, Colombia, Ecuador, N Peru
M.s.peruvianus
C Peru
M.s.squamiger
SE Peru, W Bolivia

LOCHMIAS
Lochmias nematura (Sharp-tailed Streamcreeper)
L.n.nelsoni
E Panama
L.n.sororia
Venezuela, Colombia, E Ecuador, NE Peru
L.n.chimantae
SE Venezuela
L.n.castanonota
E Venezuela
L.n.obscurata
Peru, Bolivia
L.n.nematura
Brazil, Paraguay, Uruguay, NE Argentina

PSEUDOSEISURA
Pseudoseisura cristata (Rufous Cachalote)
P.c.cristata
E Brazil
P.c.unirufa
E Bolivia, S Brazil
Pseudoseisura lophotes (Brown Cachalote)
Paraguay, Uruguay, N Argentina
Pseudoseisura gutturalis (White-throated Cachalote)
W Argentina

PSEUDOCOLAPTES
Pseudocolaptes lawrencii (Buffy Tuftedcheek)
P.l.lawrencii
Costa Rica, W Panama

P.l.panamensis
C Panama
P.l.johnsoni
W Colombia, W Ecuador
Pseudocolaptes boissonneautii (Streaked Tuftedcheek)
P.b.striaticeps
N Venezuela
P.b.meridae
W Venezuela
P.b.boissonneautii
C Colombia, NW Ecuador
P.b.oberholseri
S Colombia
P.b.orientalis
S Ecuador
P.b.intermedianus
NW Peru
P.b.pallidus
NW Peru
P.b.medianus
N Peru
P.b.auritus
C Peru
P.b.carabayae
SE Peru, W Bolivia

BERLEPSCHIA
Berlepschia rikeri (Point-tailed Palmcreeper)
S Venezuela, Guyana, N Brazil

PHILYDOR
Philydor strigilatus (Chestnut-winged Hookbill)
P.s.strigilatus
NW Amazonia
P.s.cognitus
NC Brazil
Philydor subulatus (Striped Woodhaunter)
P.s.nicaraguae
E Nicaragua
P.s.virgatus
Costa Rica, W Panama
P.s.assimilis
E Panama, W Colombia, W Ecuador
P.s.cordobae
NW Colombia
P.s.lemae
SE Venezuela
P.s.subulatus
N & W Amazonia
Philydor guttulatus (Guttulated Foliage-gleaner)
P.g.guttulatus
N Venezuela
P.g.pallidus
NE Venezuela
P.g.mirandae
C Brazil
Philydor subalaris (Lineated Foliage-gleaner)
P.s.lineata
Costa Rica, W Panama

P.s.tacarcunae
E Panama
P.s.subalaris
W Colombia, W Ecuador
P.s.striolatus
N Colombia, W Venezuela
P.s.mentalis
E Ecuador
P.s.colligatus
NW Peru
P.s.ruficrissus
C Peru
Philydor rufosuperciliatus (Buff-browed Foliage-gleaner)
P.r.similis
N Peru
P.r.cabanisi
W Peru, W Bolivia
P.r.oleagineus
SE Bolivia, NW Argentina
P.r.rufosuperciliatus
SE Brazil
P.r.acritus
S Brazil, Paraguay, Uruguay, N Argentina
Philydor striaticollis (Montane Foliage-gleaner)
P.s.striaticollis
Colombia, Venezuela
P.s.anxius
N Colombia
P.s.perijanus
NW Venezuela
P.s.venezuelanus
N Venezuela
P.s.montanus
E Ecuador, C Peru
P.s.yungae
SE Peru, NW Bolivia
Philydor amaurotis (White-browed Foliage-gleaner)
SE Brazil
Philydor variegaticeps (Scaly-throated Foliage-gleaner)
P.v.variegaticeps
S Mexico to W Panama
P.v.temporalis
W Colombia, Ecuador
Philydor ruficaudatus (Rufous-tailed Foliage-gleaner)
P.r.ruficaudatus
N & W Amazonia
P.r.flavipectus
S Venezuela, N Brazil
Philydor erythrocercus (Rufous-rumped Foliage-gleaner)
P.e.fuscipennis
W Panama
P.e.subfulvus
SE Colombia, E Ecuador, N Peru
P.e.ochrogaster
C & SE Peru, Bolivia
P.e.lyra
E Peru, NE Bolivia, S Brazil

P.e.suboles
W Brazil
P.e.erythrocercus
Guyana, French Guiana, N Brazil
Philydor erythropterus (Chestnut-winged Foliage-gleaner)
P.e.erythropterus
W Amazonia
P.e.diluvialis
NC Brazil
Philydor lichtensteini (Lichtenstein's Foliage-gleaner)
SE Brazil, Paraguay, NE Argentina
Philydor erythronotus (Rufous-backed Foliage-gleaner)
E Panama to NW Ecuador
Philydor novaesi (Novaes' Foliage-gleaner) 102.4
NE Brazil
Philydor atricapillus (Black-capped Foliage-gleaner)
SE Brazil, E Paraguay, NE Argentina
Philydor rufus (Buff-fronted Foliage-gleaner)
P.r.panerythrus
Costa Rica to E Colombia
P.r.riveti
W Colombia, NW Ecuador
P.r.colombianus
N Venezuela
P.r.cuchiverus
Venezuela
P.r.bolivianus
E Peru, Bolivia
P.r.chapadensis
S Brazil
P.r.rufus
C & S Brazil, E Paraguay, NE Argentina
Philydor pyrrhodes (Cinnamon-rumped Foliage-gleaner)
N & C South America
Philydor dimidiatus (Russet-mantled Foliage-gleaner)
P.d.dimidiatus
SC Brazil
P.d.baeri
S & SE Brazil, N Paraguay
Philydor fuscus (White-collared Foliage-gleaner)
SE Brazil
Philydor ucayalae (Peruvian Recurvebill)
Peru
Philydor striatus (Bolivian Recurvebill) 102.1
N Bolivia

CICHLOCOLAPTES
Cichlocolaptes leucophrus (Pale-browed Treehunter)
SE Brazil

THRIPADECTES
Thripadectes ignobilis (Uniform Treehunter)
W Colombia, NW Ecuador

Thripadectes rufobrunneus (Streak-breasted Treehunter)
Costa Rica, W Panama
Thripadectes melanorhynchus (Black-billed Treehunter)
T.m.melanorhynchus
E Ecuador, E Peru
T.m.striaticeps
Colombia
Thripadectes holostictus (Striped Treehunter)
T.h.striatidorsus
SW Colombia, W Ecuador
T.h.holostictus
Colombia, E Ecuador, N Peru
T.h.moderatus
E Peru, N Bolivia
Thripadectes virgaticeps (Streak-capped Treehunter)
T.v.sclateri
W Colombia
T.v.magdelenae
N Colombia
T.v.virgaticeps
NE Ecuador
T.v.sumaco
E Ecuador
T.v.klagesi
N Venezuela
T.v.tachirensis
NW Venezuela
Thripadectes scrutator (Buff-throated Treehunter)
C Peru
Thripadectes flammulatus (Flammulated Treehunter)
T.f.flammulatus
Colombia, Ecuador
T.f.bricenoi
Venezuela

AUTOMOLUS
Automolus ruficollis (Rufous-necked Foliage-gleaner)
A.r.celicae
SW Ecuador
A.r.ruficollis
NW Peru
Automolus ochrolaemus (Buff-throated Foliage-gleaner)
A.o.cervinigularis
S Mexico to Nicaragua
A.o.amusos
SE Guatemala to Honduras
A.o.hypophaeus
E Nicaragua to NW Panama
A.o.exsertus
SW Costa Rica, W Panama
A.o.pallidigularis
E Panama, Colombia, NW Ecuador
A.o.turdinus
N & W Amazonia
A.o.ochrolaemus
E Peru, N Bolivia, W Brazil

A.o.auricularis
NE Bolivia, W Brazil
Automolus infuscatus (Olive-backed Foliage-gleaner)
A.i.infuscatus
SE Colombia, E Ecuador, E Peru
A.i.badius
S Venezuela, E Colombia, NW Brazil
A.i.cervicalis
E Venezuela, Guyana, Surinam, N Brazil
A.i.perusianus
W Brazil
A.i.paraensis
C Brazil
Automolus dorsalis (Crested Foliage-gleaner)
SE Colombia, E Ecuador, Peru
Automolus leucophthalmus (White-eyed Foliage-gleaner)
A.l.lammi
E Brazil
A.l.leucophthalmus
EC Brazil
A.l.sulphurascens
S Brazil, NE Paraguay, NE Argentina
Automolus melanopezus (Brown-rumped Foliage-gleaner)
SE Colombia, E Ecuador, W Brazil
Automolus albigularis (White-throated Foliage-gleaner)
A.a.paraquensis
S Venezuela
A.a.duidae
Mt Duida (S Venezuela)
A.a.albigularis
SE Venezuela
A.a.roraimae
Mt Roraima (SE Venezuela)
Automolus rubiginosus (Ruddy Foliage-gleaner)
A.r.guerrerensis
SW Mexico
A.r.rubiginosus
E Mexico
A.r.veraepacis
N Guatemala
A.r.umbrinus
S Mexico to N Nicaragua
A.r.fumosus
W Panama
A.r.saturatus
E Panama, NW Colombia
A.r.sasaimae
N Colombia
A.r.nigricauda
W Colombia, W Ecuador
A.r.rufipectus
N Colombia
A.r.cinnamomeigula
E Colombia
A.r.caquetae
SE Colombia
A.r.venezuelanus
S Venezuela

A.r.obscurus
French Guiana
A.r.brunnescens
E Ecuador, NE Peru
A.r.moderatus
N Peru
A.r.watkinsi
SE Peru, N Bolivia
Automolus rufipileatus (Chestnut-crowned Foliage-gleaner)
A.r.consobrinus
N & W Amazonia
A.r.rufipileatus
C Brazil
Automolus rectirostris (Chestnut-capped Foliage-gleaner)
C Brazil
Automolus erythrocephalus (Henna-hooded Foliage-gleaner)
A.e.erythrocephalus
SW Ecuador, N Peru
A.e.palamblae
W Peru

SCLERURUS
Sclerurus mexicanus (Tawny-throated Leafscraper)
S.m.mexicanus
SE Mexico to Honduras
S.m.pullus
Costa Rica to W Panama
S.m.andinus
E Panama, Colombia, to Guyana
S.m.obscurior
W Colombia, W Ecuador
S.m.peruvianus
W Amazonia
S.m.macconnelli
N Brazil, Guyana, French Guiana
S.m.bahiae
E Brazil
Sclerurus rufigularis (Short-billed Leafscraper)
S.r.fulvigularis
E & S Venezuela to French Guiana, N Brazil
S.r.brunnescens
N Brazil
S.r.furfurosus
C Brazil
S.r.rufigularis
N Bolivia, SW Brazil
Sclerurus albigularis (Grey-throated Leafscraper)
S.a.canigularis
Costa Rica, W Panama
S.a.propinquus
N Colombia
S.a.albigularis
E Colombia, Venezuela, Trinidad, Tobago I
S.a.kunanensis
Venezuela
S.a.zamorae
SE Ecuador, N Peru

S.a.albicollis
N Bolivia
**Sclerurus caudacutus (Black-tailed
Leafscraper)**
S.c.caudacutus
Guyana, French Guiana
S.c.insignis
S Venezuela, N Brazil
S.c.brunneus
W Amazonia
S.c.olivascens
Peru, Bolivia
S.c.pallidus
N Brazil
S.c.umbretta
coast of E Brazil
**Sclerurus scansor (Rufous-breasted
Leafscraper)**
S.s.cearensis
NE Brazil
S.s.scansor
C & E Brazil, Paraguay, NE Argentina
**Sclerurus guatemalensis (Scaly-throated
Leafscraper)**
S.g.guatemalensis
S Mexico to Panama
S.g.salvini
W Colombia, W Ecuador
S.g.ennosiphyllus
C Colombia

XENOPS
**Xenops contaminatus (Sharp-billed
Treehunter)**
SE Brazil, Paraguay, NE Argentina
Xenops milleri (Rufous-tailed Xenops)
N Amazonia
Xenops tenuirostris (Slender-billed Xenops)
X.t.acutirostris
SE Colombia, S Venezuela, E Ecuador, NE
Peru
X.t.hellmayri
French Guiana, Surinam
X.t.tenuirostris
S Venezuela, E Peru, N Bolivia, N & W
Brazil
Xenops minutus (Plain Xenops)
X.m.mexicanus
S Mexico to Honduras
X.m.ridgwayi
Nicaragua to W Panama
X.m.littoralis
E Panama to W Ecuador
X.m.neglectus
N Colombia, N Venezuela
X.m.remoratus
E Colombia, Venezuela, N Brazil
X.m.ruficaudus
E Colombia, Venezuela, the Guianas, N
Brazil
X.m.olivaceus

NE Colombia
X.m.obsoletus
E Ecuador, E Peru, N Bolivia, W Brazil
X.m.genibarbis
N Brazil
X.m.minutus
E & SE Brazil, E Paraguay
Xenops rutilans (Streaked Xenops)
X.r.septentrionalis
Costa Rica, W Panama
X.r.heterurus
E Panama to NE Ecuador, Venezuela
X.r.incomptus
Darien (Panama)
X.r.perijanus
NE Colombia
X.r.phelpsi
N Colombia
X.r.guayae
W Ecuador, NW Peru
X.r.peruvianus
E Ecuador, E Peru
X.r.purusianus
C Brazil
X.r.connectens
E Bolivia, NW Argentina
X.r.chapadensis
SW Brazil, N Bolivia
X.r.rutilans
SE Brazil, Paraguay, NE Argentina

MEGAXENOPS
Megaxenops parnaguae (Great Xenops)
NE Brazil

PYGARRHICHAS
**Pygarrhichas albogularis (White-throated
Treerunner)**
S Chile, SW Argentina, Tierra del Fuego

103. FORMICARIIDAE (ANTBIRDS)

CYMBILAIMUS
Cymbilaimus lineatus (Fasciated Antshrike)
C.l.fasciatus
Nicaragua to NW Ecuador
C.l.intermedius
Upper Amazonia
C.l.lineatus
SE Venezuela, the Guianas, NE Brazil
C.l.sanctaemariae
NE Bolivia

HYPOEDALEUS
**Hypoedaleus guttatus (Spot-backed
Antshrike)**
H.g.leucogaster
E Brazil

H.g.guttatus
S Brazil, Paraguay, NE Argentina

BATARA
Batara cinerea (Giant Antshrike)
B.c.excubitor
E Bolivia
B.c.argentina
S Bolivia, NW Argentina
B.c.cinerea
SE Brazil, NE Argentina

MACKENZIAENA
Mackenziaena leachii (Large-tailed Antshrike)
SE Brazil, NE Argentina
Mackenziaena severa (Tufted Antshrike)
SE Brazil, NE Argentina

FREDERICKENA
Frederickena viridis (Black-throated Antshrike)
E Venezuela, Guyana
Frederickena unduligera (Undulated Antshrike)
F.u.fulva
SE Colombia, E Ecuador
F.u.diversa
E & SE Peru
F.u.unduligera
NW Brazil
F.u.pallida
Brazil

TARABA
Taraba major (Great Antshrike)
T.m.melanocrissa
SE Mexico to W Panama
T.m.obscura
SW Costa Rica to NW Colombia
T.m.transandeana
SW Colombia, W Ecuador, NW Peru
T.m.granadensis
E Colombia, Venezuela
T.m.semifasciata
E Colombia to the Guianas, N & E Brazil
T.m.duidae
SE Venezuela
T.m.melanura
Upper Amazonia
T.m.borbae
C Brazil
T.m.stagura
E & NE Brazil
T.m.major
E Bolivia, S Brazil, Paraguay, N Argentina

SAKESPHORUS
Sakesphorus canadensis (Black-crested Antshrike)

S.c.pulchellus
N Colombia, W Venezuela
S.c.paraguanae
N E Colombia, NW Venezuela
S.c.intermedius
S Venezuela, N Brazil
S.c.fumosus
S Venezuela
S.c.trinitatis
NE Venezuela, Guyana, Trinidad
S.c.canadensis
Surinam, French Guiana
S.c.loretoyacuensis
SE Colombia, NW Brazil
Sakesphorus cristatus (Silvery-cheeked Antshrike)
E Brazil
Sakesphorus bernardi (Collared Antshrike)
S.b.bernardi
W Ecuador
S.b.piurae
SW Ecuador, N Peru
S.b.cajamarcae
W Peru
S.b.shumbae
N Peru
Sakesphorus melanonotus (Black-backed Antshrike)
NE Colombia, W Venezuela
Sakesphorus melanothorax (Band-tailed Antshrike)
French Guiana
Sakesphorus luctuosus (Glossy Antshrike)
S.l.luctuosus
NE Brazil
S.l.araguayae
C Brazil

BIATAS
Biatas nigropectus (White-bearded Antshrike)
SE Brazil

THAMNOPHILUS
Thamnophilus doliatus (Barred Antshrike)
T.d.intermedius
E Mexico to E Costa Rica
T.d.yucatanensis
S Mexico, N Guatemala
T.d.pacificus
W Honduras to W Costa Rica
T.d.nigricristatus
E Panama, N Colombia
T.d.albicans
N Colombia
T.d.zarumae
SW Ecuador, NW Peru
T.d.palamblae
NW Peru
T.d.nigrescens
NE Colombia, NW Venezuela
T.d.tobagensis
Tobago I

T.d.nesiotes
Isla del Rey
T.d.fraterculus
E Colombia, N Venezuela, Trinidad
T.d.doliatus
E Venezuela, the Guianas, N Brazil
T.d.subradiatus
E Peru, W Brazil
T.d.signatus
NE Bolivia, SW Brazil
T.d.difficilis
E Brazil
T.d.capistratus
E Brazil
T.d.radiatus
E Bolivia, S Brazil, N Argentina
T.d.cadwaladeri
S Bolivia

Thamnophilus multistriatus (Bar-crested Antshrike)
T.m.brachyurus
W Colombia
T.m.selvae
W Colombia
T.m.multistriatus
C Colombia
T.m.oecotonophilus
NE Colombia

Thamnophilus palliatus (Lined Antshrike)
T.p.tenuepunctatus
E Colombia
T.p.tenuifasciatus
SE Colombia, E Ecuador
T.p.berlepschi
SE Ecuador, N Peru
T.p.similis
C Peru
T.p.puncticeps
SE Peru, Bolivia, W Brazil
T.p.palliatus
C & E Brazil

Thamnophilus bridgesi (Black-hooded Antshrike)
SW Costa Rica, W Panama

Thamnophilus nigriceps (Black Antshrike)
T.n.nigriceps
E Panama, NW Colombia
T.n.magdalenae
N Colombia

Thamnophilus praecox (Cocha Antshrike)
S Ecuador

Thamnophilus nigrocinereus (Blackish-grey Antshrike)
T.n.cinereoniger
Colombia, Venezuela, NW Brazil
T.n.kulczynskii
French Guiana
T.n.nigrocinereus
NE Brazil
T.n.tschudii
W Brazil
T.n.huberi
N Brazil

Thamnophilus cryptoleucus (Castelnau's Antshrike)
NE Peru, W Brazil

Thamnophilus aethiops (White-shouldered Antshrike)
T.a.aethiops
E Ecuador, NE Peru
T.a.wetmorei
SE Colombia
T.a.polionotus
S & E Venezuela, NW Brazil
T.a.kapouni
E & SE Peru, N Bolivia, W Brazil
T.a.juruanus
W Brazil
T.a.injunctus
NE Brazil
T.a.punctuliger
C Brazil
T.a.atriceps
C Brazil
T.a.incertus
NE Brazil

Thamnophilus unicolor (Uniform Antshrike)
T.u.unicolor
E Ecuador
T.u.grandior
Colombia, E Ecuador, N Peru
T.u.caudatus
N Peru

Thamnophilus schistaceus (Black-capped Antshrike)
T.s.capitalis
Upper Amazonia
T.s.dubius
S Ecuador, N Peru
T.s.schistaceus
SE Peru, N Bolivia, W Brazil
T.s.heterogynus
W Brazil
T.s.inornatus
C Brazil

Thamnophilus murinus (Mouse-coloured Antshrike)
T.m.murinus
E Ecuador, E Colombia to Surinam
T.m.cayennensis
French Guiana, N Brazil
T.m.canipennis
NE Peru, W Brazil

Thamnophilus aroyae (Upland Antshrike)
SE Peru, NW Bolivia

Thamnophilus punctatus (Slaty Antshrike)
T.p.atrinucha
Honduras to Ecuador & Venezuela
T.p.gorgonae
Gorgona I
T.p.subcinereus
N Colombia, NW Venezuela
T.p.interpositus
E Colombia
T.p.punctatus
E Venezuela, the Guianas, N Brazil

T.p.leucogaster
N Peru
T.p.saturatus
C Brazil
T.p.zimmeri
C Brazil
T.p.stictocephalus
C Brazil
T.p.sticturus
Bolivia, S Brazil
T.p.pelzelni
C & E Brazil
T.p.ambiguus
coastal SE Brazil
Thamnophilus amazonicus (Amazonian Antshrike)
T.a.cinereiceps
Venezuela, Colombia
T.a.huallagae
Peru
T.a.amazonicus
S Colombia to N Bolivia
T.a.obscurus
C Brazil
T.a.paraensis
E Venezuela, the Guianas, N Brazil
Thamnophilus insignis (Streak-backed Antshrike)
T.i.insignis
S Venezuela
T.i.nigrofrontalis
S Venezuela
Thamnophilus caerulescens (Variable Antshrike)
T.c.subandinus
N Peru
T.c.melanochrous
C & S Peru
T.c.aspersiventer
N Bolivia
T.c.connectens
E Bolivia
T.c.dinellii
C & S Bolivia, NW Argentina
T.c.paraguayensis
S Brazil, N Paraguay
T.c.gilvigaster
SE Brazil, NE Argentina
T.c.caerulescens
S Brazil, E Paraguay, NE Argentina
T.c.albonotatus
EC Brazil
T.c.ochraceiventer
SC Brazil
T.c.pernambucensis
E Brazil
T.c.cearensis
E Brazil
Thamnophilus torquatus (Rufous-winged Antshrike)
Brazil
Thamnophilus ruficapillus (Rufous-capped Antshrike)

T.r.jaczewskii
N Peru
T.r.marcapatae
SE Peru
T.r.subfasciatus
W Bolivia
T.r.cochabambae
W Bolivia, NW Argentina
T.r.ruficapillus
S & E Brazil, NE Argentina

PYGIPTILA
Pygiptila stellaris (Spot-winged Antshrike)
P.s.maculipennis
SE Colombia to NE Peru
P.s.occipitalis
S Colombia to the Guianas, N Brazil
P.s.purusiana
W Brazil
P.s.stellaris
C Brazil

MEGASTICTUS
Megastictus margaritatus (Pearly Antshrike)
N Upper Amazonia

NEOCTANTES
Neoctantes niger (Black Bushbird)
E Ecuador, N Peru, W Brazil

CLYTOCTANTES
Clytoctantes alixii (Recurvebill Bushbird)
Colombia

XENORNIS
Xenornis setifrons (Speckle-breasted Antshrike)
E Panama, NW Colombia

THAMNISTES
Thamnistes anabatinus (Russet Antshrike)
T.a.anabatinus
SE Mexico to Honduras
T.a.saturatus
Nicaragua to W Panama
T.a.coronatus
C & E Panama
T.a.intermedius
W Colombia, W Ecuador
T.a.gularis
C Colombia
T.a.aequatorialis
SE Colombia, E Ecuador
T.a.rufescens
C & SE Peru, N Bolivia

DYSITHAMNUS
Dysithamnus stictothorax (Spot-breasted Antvireo)
SE Brazil
Dysithamnus mentalis (Plain Antvireo)
D.m.septentrionalis
S Mexico to W Panama

D.m.suffusus
E Panama, NW Colombia
D.m.extremus
C Colombia
D.m.semicinereus
NE Colombia
D.m.viridis
NE Colombia, NW Venezuela
D.m.cumbreanus
N Venezuela
D.m.andrei
E Venezuela, Trinidad
D.m.oberi
Tobago I
D.m.ptaritepui
S Venezuela
D.m.spodionotus
S & E Venezuela
D.m.aequatorialis
W Ecuador, NW Peru
D.m.napensis
E Ecuador
D.m.tambillanus
N Peru
D.m.olivaceus
C Peru
D.m.tavarae
SE Peru, Bolivia
D.m.emiliae
E Brazil
D.m.affinis
C Brazil
D.m.mentalis
S & SE Brazil, E Paraguay, NE Argentina
Dysithamnus striaticeps (Streak-crowned Antvireo)
E Nicaragua, E Costa Rica
Dysithamnus puncticeps (Spot-crowned Antvireo)
D.p.puncticeps
E Costa Rica, Panama, N Colombia
D.p.intensus
S Panama, W Colombia
D.p.flemmingi
SW Colombia, W Ecuador
Dysithamnus xanthopterus (Rufous-backed Antvireo)
SE Brazil
Dysithamnus ardesiacus (Grey-throated Antvireo)
D.a.ardesiacus
SE Colombia, E Ecuador, N & C Peru
D.a.obidensis
S Venezuela, the Guianas, N Brazil

THAMNOMANES
Thamnomanes saturninus (Saturnine Antshrike)
T.s.huallagae
E Peru, W Brazil
T.s.saturninus
WC Brazil

Thamnomanes occidentalis (Western Antshrike)
T.o.occidentalis
S Colombia
T.o.punctitectus
S Ecuador
Thamnomanes plumbeus (Plumbeous Antshrike)
T.p.tucuyensis
NW Venezuela
T.p.leucostictus
E Colombia, E Ecuador
T.p.plumbeus
SE Brazil
Thamnomanes caesius (Cinereous Antshrike)
T.c.glaucus
N & W Amazonia
T.c.intermedius
C Peru
T.c.persimilis
C Brazil
T.c.hoffmannsi
SC Brazil
T.c.caesius
E Brazil
Thamnomanes schistogynus (Bluish-slate Antshrike)
SE Peru, W Brazil

MYRMOTHERULA
Myrmotherula brachyura (Pygmy Antwren)
M.b.ignota
C & E Panama, NW Colombia
M.b.brachyura
N & W Amazonia
Myrmotherula obscura (Short-billed Antwren)
SE Colombia, NE Peru, W Brazil
Myrmotherula sclateri (Sclater's Antwren)
C Brazil
Myrmotherula klagesi (Klages' Antwren)
NE Brazil
Myrmotherula surinamensis (Streaked Antwren)
M.s.pacifica
E Panama to E Ecuador
M.s.surinamensis
S Venezuela, the Guianas, N Brazil
M.s.multostriata
W Amazonia
Myrmotherula ambigua (Yellow-throated Antwren)
E Colombia, S Venezuela, NW Brazil
Myrmotherula cherriei (Cherrie's Antwren)
E Colombia, W Venezuela, N Brazil
Myrmotherula guttata (Rufous-bellied Antwren)
N Amazonia
Myrmotherula longicauda (Stripe-chested Antwren)
M.l.soderstromi
E Ecuador
M.l.pseudoaustralis
E Ecuador, N Peru

M.l.longicauda
E Peru
M.l.australis
SE Peru, N Bolivia
Myrmotherula hauxwelli (Plain-throated Antwren)
M.h.suffusa
SE Colombia, E Ecuador, NE Peru
M.h.hauxwelli
E Peru, W Brazil
M.h.clarior
C Brazil
M.h.hellmayri
NE Brazil
Myrmotherula gularis (Star-throated Antwren)
SE Brazil
Myrmotherula gutturalis (Brown-bellied Antwren)
the Guianas, Venezuela, N Brazil
Myrmotherula fulviventris (Checker-throated Antwren)
M.f.costaricensis
S Honduras to W Panama
M.f.fulviventris
E Panama to W Ecuador
M.f.salmoni
C Colombia
Myrmotherula leucophthalma (White-eyed Antwren)
M.l.dissita
SE Peru, N Bolivia
M.l.leucophthalma
WC Brazil
M.l.phaeonota
NC Brazil
M.l.sordida
NE Brazil
Myrmotherula haematonota (Stipple-throated Antwren)
M.h.pyrrhonota
SE Colombia, S Venezuela, NW Brazil
M.h.spodionota
E Ecuador
M.h.haematonota
NE Peru
M.h.sororia
N & C Peru
M.h.amazonica
W Brazil
Myrmotherula ornata (Ornate Antwren)
M.o.ornata
E Colombia
M.o.saturata
SE Colombia, E Ecuadoe, NE Peru
M.o.atrogularis
N & C Peru
M.o.meridionalis
SE Peru, N Bolivia
M.o.hoffmannsi
C Brazil
Myrmotherula erythrura (Rufous-tailed Antwren)

M.e.erythrura
NW Amazonia
M.e.septentrionalis
E Peru, W Brazil
Myrmotherula erythronotos (Black-hooded Antwren)
SE Brazil
Myrmotherula axillaris (White-flanked Antwren)
M.a.albigula
C America, W Colombia, Ecuador
M.a.melaena
NW Amazonia
M.a.heterozyga
E Peru, W Brazil
M.a.axillaris
Trinidad, Venezuela, the Guianas, N Brazil
M.a.fresnayana
SE Peru, Bolivia
M.a.luctuosa
E Brazil
Myrmotherula schisticolor (Slaty Antwren)
M.s.schisticolor
Mexico to W Ecuador
M.s.sanctaemartae
N Colombia, N Venezuela
M.s.interior
E Colombia, E Ecuador, N Peru
Myrmotherula sunensis (Rio Suno Antwren)
M.s.sunensis
E Ecuador, NE Peru
M.s.yessupi
W Peru
Myrmotherula longipennis (Long-winged Antwren)
M.l.longipennis
N & W Amazonia
M.l.zimmeri
E Ecuador, NE Peru
M.l.garbei
NE Peru, W Brazil
M.l.transitiva
WC Brazil
M.l.ochrogyna
C Brazil
M.l.paraensis
EC Brazil
Myrmotherula minor (Salvadori's Antwren)
NE Peru, W Brazil
Myrmotherula iheringi (Ihering's Antwren)
M.i.heteroptera
SC Brazil
M.i.iheringi
C Brazil
Myrmotherula fluminensis (Rio de Janeiro Antwren) 103.1
Rio de Janeiro, Brazil
Myrmotherula grisea (Ashy Antwren)
W Bolivia
Myrmotherula unicolor (Unicoloured Antwren)
M.u.unicolor
SE Brazil

M.u.snowi
S Brazil
Myrmotherula behni (Plain-winged Antwren)
M.b.behni
E Colombia
M.b.yavii
S Venezuela
M.b.inornata
SE Venezuela, Guyana
M.b.camanii
S Venezuela
Myrmotherula urosticta (Band-tailed Antwren)
SE Brazil
Myrmotherula menetriesii (Grey Antwren)
M.m.pallida
NW & W Amazonia
M.m.cinereiventris
E Venezuela, the Guianas, N Brazil
M.m.menetriesii
E Peru, N Bolivia, W Brazil
M.m.berlepschi
WC Brazil
M.m.omissa
NE Brazil
Myrmotherula assimilis (Leaden Antwren)
NE Peru, W Brazil

DICHROZONA
Dichrozona cincta (Banded Antcatcher)
D.c.cincta
E Colombia, S Venezuela, NW Brazil
D.c.stellata
E Ecuador, W Brazil
D.c.zononota
C Brazil, N Bolivia

MYRMORCHILUS
Myrmorchilus strigilatus (Stripe-backed Antbird)
M.s.strigilatus
E Brazil
M.s.suspicax
S Bolivia, S Brazil, W Paraguay, N Argentina

HERPSILOCHMUS
Herpsilochmus pileatus (Black-capped Antwren)
H.p.pileatus
E Brazil
H.p.atricapillus
C & S Brazil, Bolivia, Paraguay, N Argentina
H.p.motacilloides
C Peru
Herpsilochmus sticturus (Spot-tailed Antwren)
H.s.dugandi
Colombia
H.s.sticturus
S Venezuela, the Guianas, N Brazil
Herpsilochmus stictocephalus (Todd's Antwren)
E Venezuela, the Guianas

Herpsilochmus dorsimaculatus (Spot-backed Antwren)
S Venezuela, NW Brazil
Herpsilochmus roraimae (Roraiman Antwren)
SE Venezuela, SW Guyana
Herpsilochmus pectoralis (Pectoral Antwren)
E Brazil
Herpsilochmus longirostris (Large-billed Antwren)
SC Brazil
Herpsilochmus parkeri (Ash-throated Antwren)   103.5
Peru
Herpsilochmus axillaris (Yellow-breasted Antwren)
H.a.senex
SW Colombia
H.a.aequatorialis
E Ecuador
H.a.puncticeps
N Peru
H.a.axillaris
S Peru
Herpsilochmus rufimarginatus (Rufous-winged Antwren)
H.r.exiguus
E Panama
H.r.frater
N & W Amazonia
H.r.scapularis
E Brazil
H.r.rufimarginatus
SE Brazil, E Paraguay, NE Argentina

MICRORHOPIAS
Microrhopias quixensis (Dot-winged Antwren)
M.q.boucardi
S Mexico to SW Honduras
M.q.virgata
Nicaragua to W Panama
M.q.consobrina
E Panama to W Ecuador
M.q.quixensis
E Ecuador, NE Peru
M.q.intercedens
N Peru
M.q.nigriventris
C Peru
M.q.albicauda
SE Peru
M.q.microsticta
French Guiana
M.q.bicolor
N Bolivia, C Brazil
M.q.emiliae
C Brazil

FORMICIVORA
Formicivora iheringi (Narrow-billed Antwren)
E Brazil
Formicivora grisea (White-fringed Antwren)

F.g.alticincta
Pearl I (Panama)
F.g.hondae
N Colombia
F.g.fumosa
E Colombia, W Venezuela
F.g.intermedia
NE Colombia, NW Venezuela, Margarita I
F.g.tobagensis
Tobago I
F.g.orenocensis
S Venezuela
F.g.rufiventris
E Colombia, S Venezuela
F.g.grisea
the Guianas, NE Brazil
F.g.deluzae
SE Brazil
Formicivora serrana (Serra Antwren)
F.s.serrana
E Brazil
F.s.interposita
SE Brazil
F.s.littoralis
Rio de Janeiro
Formicivora melanogaster (Black-bellied Antwren)
F.m.melanogaster
E Bolivia, W & C Brazil
F.m.bahiae
E Brazil
Formicivora rufa (Rusty-backed Antwren)
F.r.urubambae
E Peru
F.r.chapmani
E Brazil
F.r.rufa
E Bolivia, E Paraguay, C & S Brazil

DRYMOPHILA
Drymophila ferruginea (Ferruginous Antbird)
SE Brazil to NE Argentina
Drymophila rubricollis (Bertoni's Antwren)
103.7
SE Brazil
Drymophila genei (Rufous-tailed Antbird)
SE Brazil
Drymophila ochropyga (Ochre-rumped Antbird)
SE Brazil
Drymophila devillei (Striated Antbird)
D.d.devillei
E Ecuador, E Peru, N Bolivia
D.d.subochracea
C Brazil
Drymophila caudata (Long-tailed Antbird)
D.c.hellmayri
NE Colombia
D.c.klagesi
N Venezuela
D.c.caudata
Colombia to N Bolivia
Drymophila malura (Dusky-tailed Antbird)

SE Brazil, N Argentina
Drymophila squamata (Scaled Antbird)
C & SE Brazil

TERENURA
Terenura maculata (Streak-capped Antwren)
SE Brazil, NE Argentina
Terenura sicki (Orange-bellied Antwren) 103.2
NE Brazil
Terenura callinota (Rufous-rumped Antwren)
T.c.callinota
Panama to N Peru
T.c.peruviana
C Peru
T.c.guianensis
Guyana
Terenura humeralis (Chestnut-shouldered Antwren)
T.h.humeralis
E Ecuador, NE Peru, W Brazil
T.h.transfluvialis
C Brazil
Terenura sharpei (Yellow-rumped Antwren)
SE Peru
Terenura spodioptila (Ash-winged Antwren)
T.s.signata
SE Colombia, NW Brazil
T.s.spodioptila
S Venezuela, Guyana, NW Brazil
T.s.elaopteryx
French Guiana, NE Brazil
T.s.meridionalis
C Brazil

CERCOMACRA
Cercomacra cinerascens (Grey Antbird)
C.c.cinerascens
Upper Amazonia
C.c.immaculata
E Venezuela, the Guianas, N Brazil
C.c.sclateri
SW Amazonia
C.c.iterata
C Brazil
Cercomacra brasiliana (Rio de Janeiro Antbird)
SE Brazil
Cercomacra tyrannina (Dusky Antbird)
C.t.crepera
S Mexico to W Panama
C.t.rufiventris
E Panama to W Ecuador
C.t.tyrannina
E Colombia, S Venezuela, NW Brazil
C.t.vicina
NE Colombia, W Venezuela
C.t.saturatior
E Venezuela, Guyana, Surinam
C.t.laeta
N & C Brazil
C.t.sabinoi
NE Brazil
Cercomacra nigrescens (Blackish Antbird)

225

C.n.nigrescens
Surinam, French Guiana
C.n.aequatorialis
E Ecuador, N Peru
C.n.notata
C Peru
C.n.fuscicauda
E Peru, W Brazil, N Bolivia
C.n.approximans
C Brazil
C.n.ochrogyna
C Brazil
Cercomacra serva (Black Antbird)
C.s.serva
E Ecuador, NE Peru
C.s.hypomelaena
E & SE Peru, N Bolivia, W Brazil
Cercomacra nigricans (Jet Antbird)
C.n.nigricans
E Panama to N Brazil
C.n.atrata
NW Colombia
Cercomacra carbonaria (Rio Branco Antbird)
Rio Branco, Brazil
Cercomacra melanaria (Matto Grosso Antbird)
N Bolivia, C & S Brazil
Cercomacra ferdinandi (Bananal Antbird)
SE Brazil

SIPIA
Sipia berlepschi (Stub-tailed Antbird)
W Colombia, NW Ecuador
Sipia rosenbergi (Esmeralda's Antbird)
W Colombia, NW Ecuador

PYRIGLENA
Pyriglena leuconota (White-backed Fire-eye)
P.l.pacifica
W Ecuador
P.l.castanoptera
E Ecuador, N Peru
P.l.picea
N & C Peru
P.l.similis
C Brazil
P.l.marcapatensis
SE Peru
P.l.hellmayri
W Bolivia
P.l.maura
SE Bolivia, SE Brazil
P.l.interposita
EC Brazil
P.l.leuconota
S Brazil
P.l.pernambucensis
E Brazil
Pyriglena atra (Swainson's Fire-eye)
E Brazil
Pyriglena leucoptera (White-shouldered Fire-eye)
E Paraguay, SE Brazil

RHOPORNIS
Rhopornis ardesiaca (Slender Antbird)
SE Brazil

MYRMOBORUS
Myrmoborus leucophrys (White-browed Antcreeper)
M.l.erythrophrys
E Colombia
M.l.leucophrys
N & W Amazonia
M.l.griseigula
C Brazil, N Bolivia
M.l.angustirostris
N Amazonia
Myrmoborus lugubris (Ash-breasted Antcreeper)
M.l.berlepschi
NE Peru, W Brazil
M.l.stictopterus
C Brazil
M.l.femininus
C Brazil
M.l.lugubris
C Brazil
Myrmoborus myotherinus (Black-faced Antcreeper)
M.m.elegans
E Colombia, S Venezuela, NW Brazil
M.m.napensis
E Ecuador, NE Peru
M.m.myotherinus
S Peru, NE Bolivia, S & W Brazil
M.m.incanus
WC Brazil
M.m.ardesiacus
W Brazil
M.m.proximus
W Brazil
M.m.ochrolaema
C Brazil
M.m.sororius
C & S Brazil
Myrmoborus melanurus (Black-tailed Antcreeper) NE Peru

HYPOCNEMIS
Hypocnemis cantator (Warbling Antbird)
H.c.flavescens
S Venezuela, E Colombia, NW Brazil
H.c.notaea
SE Venezuela, Guyana, N Brazil
H.c.cantator
Surinam, French Guiana, N Brazil
H.c.saturata
SE Colombia, E Ecuador, NE Peru
H.c.peruviana
E Peru, W Brazil
H.c.implicata
C Brazil
H.c.striata

EC Brazil
H.c.affinis
EC Brazil
H.c.subflava
C Peru
H.c.collinsi
SE Peru, N Bolivia
H.c.ochrogyna
NE Bolivia, S Brazil
Hypocnemis hypoxantha (Yellow-browed Antbird)
H.h.hypoxantha
SE Colombia, E Ecuador, NE Peru
H.h.ochraceiventris
E Brazil

HYPOCNEMOIDES
Hypocnemoides melanopogon (Black-chinned Antcreeper)
H.m.occidentalis
E & S Colombia, Venezuela, NW Brazil
H.m.melanopogon
the Guianas, N Brazil
H.m.minor
C Brazil
Hypocnemoides maculicauda (Band-tailed Antcreeper)
H.m.maculicauda
Peru, N Bolivia, N Paraguay, W & S Brazil
H.m.orientalis
C & SE Brazil

MYRMOCHANES
Myrmochanes hemileucus (Black & White Antcatcher)
Peru, N Bolivia, W Brazil

GYMNOCICHLA
Gymnocichla nudiceps (Bare-crowned Antcatcher)
G.n.chiroleuca
E Guatemala to Costa Rica
G.n.erratilis
SW Costa Rica, W Panama
G.n.nudiceps
E Panama, NW Colombia
G.n.sanctamartae
N Colombia

SCLATERIA
Sclateria naevia (Silvered Antcatcher)
S.n.naevia
NE Venezuela, the Guianas, N Brazil, Trinidad
S.n.diaphora
Venezuela
S.n.argentata
Upper Amazonia
S.n.toddi
C Brazil

PERCNOSTOLA
Percnostola rufifrons (Black-headed Antbird)
P.r.rufifrons
the Guianas, NE Brazil
P.r.subcristata
N Brazil
P.r.minor
NW Amazonia
Percnostola macrolopha (White-lined Antbird)
Peru
Percnostola schistacea (Slate-coloured Antbird)
SE Colombia, NE Peru, W Brazil
Percnostola leucostigma (Spot-winged Antbird)
P.l.subplumbea
E Colombia, E Ecuador, NE Peru
P.l.obscura
C Venezuela
P.l.saturata
SE Venezuela
P.l.leucostigma
S Venezuela, the Guianas, N Brazil
P.l.intensa
C Peru
P.l.brunneiceps
SE Peru
P.l.infuscata
S Venezuela, N & W Brazil
P.l.humaythae
C Brazil
P.l.rufifacies
C Brazil
Percnostola caurensis (Caura Antbird)
P.c.caurensis
C Venezuela
P.c.australis
S Venezuela, N Brazil
Percnostola lophotes (Rufous-crested Antbird)
Peru

MYRMECIZA
Myrmeciza longipes (Swainson's Antcatcher)
M.l.panamensis
E Panama, N Colombia
M.l.longipes
E Colombia, N Venezuela, Trinidad
M.l.boucardi
NC Colombia
M.l.griseipectus
SE Colombia, S Venezuela, Guyana, NE Brazil
Myrmeciza exsul (Chestnut-backed Antbird)
M.e.exsul
E Nicaragua to W Panama
M.e.occidentalis
W Costa Rica, S Panama
M.e.cassini
E Panama, N Colombia

M.e.niglarus
NW Colombia
M.e.maculifer
W Colombia, W Ecuador
Myrmeciza ferruginea (Ferruginous-backed Antbird)
M.f.ferruginea
the Guianas, N Brazil
M.f.eluta
C Brazil
Myrmeciza ruficauda (Scalloped Antbird)
M.r.soror
NE Brazil
M.r.ruficauda
SE Brazil
Myrmeciza loricata (White-bibbed Antbird)
SE Brazil
Myrmeciza squamosa (Squamate Antbird)
SE Brazil
Myrmeciza laemosticta (Dull-mantled Antbird)
M.l.laemosticta
E Costa Rica
M.l.palliata
E Panama, NW Colombia
M.l.bolivari
C Colombia
M.l.nigricauda
SW Colombia, NW Ecuador
M.l.venezuelae
W Venezuela
Myrmeciza disjuncta (Yapacana Antbird)
Venezuela
Myrmeciza pelzelni (Grey-bellied Antbird)
E Colombia, Venezuela, N Brazil
Myrmeciza hemimelaena (Chestnut-tailed Antbird)
M.h.hemimelaena
W Amazonia
M.h.pallens
C Brazil
Myrmeciza hyperythra (Plumbeous Antbird)
W Amazonia
Myrmeciza goeldii (Goeldi's Antbird)
W Brazil
Myrmeciza melanoceps (White-shouldered Antbird)
W Amazonia
Myrmeciza fortis (Sooty Antbird)
M.f.fortis
SE Colombia, Peru, W Brazil
M.f.incanescens
C Brazil
Myrmeciza immaculata (Immaculate Antbird)
M.i.zeledoni
E Costa Rica, W Panama
M.i.berlepschi
E Panama to W Ecuador
M.i.immaculata
E Colombia, W Venezuela
M.i.brunnea
NW Venezuela
Myrmeciza griseiceps (Grey-headed Antbird)
SW Ecuador, NW Peru

Myrmeciza atrothorax (Black-throated Antbird)
M.a.metae
E Colombia
M.a.atrothorax
S Venezuela, the Guianas, N Brazil
M.a.tenebrosa
NE Peru, W Brazil
M.a.maynana
N Peru
M.a.obscurata
E Peru, W Brazil
M.a.griseiventris
W Bolivia
M.a.melanura
E Bolivia, SW Brazil
Myrmeciza stictothorax (Spot-breasted Antbird)
C Brazil

PITHYS
Pithys albifrons (White-faced Antcatcher)
P.a.albifrons
S Venezuela, the Guianas, N Brazil
P.a.brevibarba
NW Amazonia
P.a.peruviana
N & E Peru
Pithys castanea (White-masked Antcatcher)
W Ecuador

GYMNOPITHYS
Gymnopithys rufigula (Rufous-throated Antcatcher)
G.r.pallida
S Venezuela
G.r.pallidigula
S Venezuela
G.r.rufigula
E Venezuela, the Guianas, N Brazil
Gymnopithys salvini (White-throated Antcatcher)
G.s.maculata
E Peru, W Brazil
G.s.salvini
Bolivia, SW Brazil
Gymnopithys lunulata (Lunulated Antcatcher)
E Peru
Gymnopithys leucaspis (Bicoloured Antcatcher)
G.l.olivascens
Honduras to W Panama
G.l.bicolor
E Panama, NW Colombia
G.l.daguae
W Colombia
G.l.aequatorialis
SW Colombia, W Ecuador
G.l.ruficeps
C Colombia
G.l.leucaspis
E Colombia

G.l.castanea
E Ecuador, NE Peru
G.l.peruana
N Peru
G.l.lateralis
W Brazil

RHEGMATORHINA
Rhegmatorhina gymnops (Bare-eyed Antcatcher)
N Brazil
Rhegmatorhina berlepschi (Harlequin Antcatcher)
N Brazil
Rhegmatorhina cristata (Chestnut-crested Antcatcher)
NW Brazil
Rhegmatorhina hoffmannsi (White-breasted Antcatcher)
W Brazil
Rhegmatorhina melanosticta (Hairy-crested Antcatcher)
R.m.melanosticta
E Ecuador
R.m.brunneiceps
N Peru
R.m.purusiana
E Peru, W Brazil
R.m.badia
SE Peru, N Bolivia, W Brazil

HYLOPHYLAX
Hylophylax naevioides (Spotted Antbird)
H.n.capnitis
E Nicaragua to W Panama
H.n.naevioides
E Panama to W Ecuador
H.n.subsimilis
WC Colombia
Hylophylax naevia (Spot-backed Antbird)
H.n.theresae
W Amazonia
H.n.peruviana
N Peru
H.n.consobrina
S Venezuela, NW Brazil
H.n.naevia
S Venezuela, the Guianas
H.n.obscura
N Brazil
H.n.ochracea
N Brazil
Hylophylax punctulata (Dot-backed Antbird)
H.p.punctulata
S & E Venezuela, NE Peru, Brazil
H.p.subochracea
C Brazil
Hylophylax poecilonota (Scale-backed Antbird)
H.p.poecilonota
SE Venezuela, the Guianas, N Brazil

H.p.duidae
E Colombia, S Venezuela, NW Brazil
H.p.lepidonota
SE Colombia, E Ecuador, NE Peru
H.p.griseiventris
SE Peru, N Bolivia, SW Brazil
H.p.gutturalis
W Brazil
H.p.nigrigula
C Brazil
H.p.vidua
EC Brazil

PHLEGOPSIS
Phlegopsis nigromaculata (Black-spotted Bare-eye)
P.n.nigromaculata
E Peru, N Bolivia, W & SW Brazil
P.n.bowmani
N Brazil
P.n.confinis
N Brazil
P.n.paraensis
N Brazil
Phlegopsis barringeri (Argus Bare-eye)
Colombia
Phlegopsis erythroptera (Reddish-winged Bare-eye)
P.e.erythroptera
N & W Amazonia
P.e.ustulata
E Peru, W Brazil

SKUTCHIA
Skutchia borbae (Pale-faced Bare-eye)
C Brazil

PHAENOSTICTUS
Phaenostictus mcleannani (Ocellated Antbird)
P.m.saturatus
SE Nicaragua to W Panama
P.m.mcleannani
EC Panama
P.m.chocoanus
E Panama, NW Colombia
P.m.pacificus
SW Colombia, NW Ecuador

FORMICARIUS
Formicarius colma (Rufous-capped Ant-thrush)
F.c.colma
E Colombia, S Venezuela, the Guianas, N Brazil
F.c.nigrifrons
E Ecuador, E Peru, NW Brazil
F.c.amazonicus
C Brazil
F.c.ruficeps
E & SE Brazil
Formicarius analis (Black-faced Ant-thrush)

F.a.moniliger
S Mexico, E Guatamela
F.a.pallidus
SE Mexico
F.a.intermedius
Belize, Honduras
F.a.umbrosus
E Nicaragua to W Panama
F.a.hoffmanni
S Costa Rica, SW Panama
F.a.panamensis
E Panama, NW Colombia
F.a.virescens
N Colombia
F.a.saturatus
NC Colombia, NW Venezuela, Trinidad
F.a.griseoventris
NE Colombia
F.a.connectens
E Colombia
F.a.zamorae
E Ecuador, NE Peru, W Brazil
F.a.olivaceus
N Peru
F.a.crissalis
SE Venezuela, the Guianas, N Brazil
F.a.analis
Peru, W Brazil, N & E Bolivia
Formicarius rufifrons (Rufous-fronted Ant-thrush)
E Peru
Formicarius nigricapillus (Black-headed Ant-thrush)
F.n.nigricapillus
E Costa Rica, W Panama
F.n.destructus
W Colombia, W Ecuador
Formicarius rufipectus (Rufous-breasted Ant-thrush)
F.r.rufipectus
E Costa Rica, Panama
F.r.carrikeri
W Colombia, W Ecuador
F.r.lasallei
NW Venezuela
F.r.thoracicus
E Ecuador, E Peru

CHAMAEZA
Chamaeza campanisona (Short-tailed Ant-thrush)
C.c.colombiana
E Colombia
C.c.punctigula
E Ecuador, N Peru
C.c.olivacea
C Peru
C.c.huachamacarii
C Peru
C.c.berlepschi
SE Peru
C.c.venezuelana
N Venezuela

C.c.yavii
SC Venezuela
C.c.obscura
E Venezuela
C.c.fulvescens
SE Venezuela, Guyana
C.c.boliviana
W Bolivia
C.c.campanisona
SE Brazil, E Paraguay
Chamaeza nobilis (Striated Ant-thrush)
C.n.rubida
SE Colombia, E Ecuador, NE Peru
C.n.nobilis
NE Peru, W Brazil
C.n.fulvipectus
C Brazil
Chamaeza ruficauda (Rufous-tailed Ant-thrush)
C.r.turdina
C Colombia
C.r.chionogaster
N Venezuela
C.r.ruficauda
SE Brazil
Chamaeza mollissima (Barred Ant-thrush)
C.m.mollissima
Colombia, Ecuador
C.m.yungae
N Brazil

MYRMORNIS
Myrmornis torquata (Wing-banded Ant-thrush)
M.t.stictoptera
SE Nicaragua, Panama, NW Colombia
M.t.torquata
N & E Amazonia

PITTASOMA
Pittasoma michleri (Black-crowned Antpitta)
P.m.zeledoni
Costa Rica, W Panama
P.m.michleri
E Panama, NW Colombia
Pittasoma rufopileatum (Rufous-crowned Antpitta)
P.r.rosenbergi
W Colombia
P.r.harterti
SW Colombia
P.r.rufopileatum
NW Ecuador

GRALLARIA
Grallaria squamigera (Undulated Antpitta)
G.s.squamigera
Colombia, Venezuela, Ecuador
C.s.canicauda
Peru, NW Bolivia
Grallaria gigantea (Giant Antpitta)
G.g.lehmanni
C Colombia

G.g.hylodroma
W Ecuador

G.g.gigantea
E Ecuador

Grallaria excelsa (Great Antpitta)

G.e.excelsa
W Venezuela

G.e.phelpsi
N Venezuela

Grallaria varia (Variegated Antpitta)

G.v.cinereiceps
S Venezuela, NW Brazil

G.v.varia
the Guianas, NW Brazil

G.v.distincta
NC Brazil

G.v.intercedens
E Brazil

G.v.imperator
SE Brazil, Paraguay, NE Argentina

Grallaria alleni (Moustached Antpitta)
C Colombia

Grallaria carrikeri (Pale-billed Antpitta)   103.3
N Peru

Grallaria guatimalensis (Scaled Antpitta)

G.g.ochraceiventris
SW Mexico

G.g.guatimalensis
S Mexico to N Nicaragua

G.g.princeps
Costa Rica, W Panama

G.g.chocoensis
mountains of E Panama, NW Colombia

G.g.regulus
Ecuador, Peru

G.g.carmelitae
NE Colombia, NW Venezuela

G.g.aripoensis
N Trinidad

G.g.roraimae
S Venezuela, N Brazil

Grallaria chthonia (Tachira Antpitta)
Venezuela

Grallaria haplonota (Plain-backed Antpitta)

G.h.haplonota
N Venezuela

G.h.pariae
NE Venezuela

G.h.parambae
NW Ecuador

G.h.chaplinae
W Ecuador

Grallaria dignissima (Ochre-striped Antpitta)
E Ecuador, NE Peru

Grallaria eludens (Elusive Antpitta)
E Peru

Grallaria ruficapilla (Chestnut-crowned Antpitta)

G.r.ruficapilla
C Colombia, N Ecuador

G.r.perijana
NW Venezuela

G.r.avilae
N Venezuela

G.r.nigrolineata
W Venezuela

G.r.connectens
SW Ecuador

G.r.albiloris
S Ecuador, NW Peru

G.r.interior
N Peru

Grallaria watkinsi (Watkins' Antpitta)
SW Ecuador, NW Peru

Grallaria bangsi (Santa Marta Antpitta)
NE Colombia

Grallaria andicola (Stripe-headed Antpitta)
C Peru

Grallaria punensis (Puna Antpitta)
SE Peru

Grallaria rufocinerea (Bicoloured Antpitta)
C Colombia

Grallaria nuchalis (Chestnut-naped Antpitta)

G.n.ruficeps
C Colombia

G.n.obsoleta
NW Ecuador

G.n.nuchalis
E Ecuador

Grallaria blakei (Chestnut Antpitta)
Andes of Peru

Grallaria albigula (White-throated Antpitta)
SE Peru, E Bolivia

Grallaria erythroleuca (Chestnut-brown Antpitta)
SE Peru

Grallaria hypoleuca (Bay-backed Antpitta)

G.h.flavotincta
WC Colombia

G.h.castanea
C Colombia, E Ecuador

G.h.hypoleuca
W Colombia

Grallaria griseonucha (Grey-naped Antpitta)

G.g.tachirae
Venezuela

G.g.griseonucha
W Venezuela

Grallaria rufula (Rufous Antpitta)

G.r.spatiator
NE Colombia

G.r.saltuensis
NE Colombia

G.r.rufula
Colombia, W Venezuela, Ecuador

G.r.cajamarcae
N Peru

G.r.obscura
C Peru

G.r.occabambae
SE Peru

G.r.cochabambae
N Bolivia

Grallaria erythrotis (Rufous-faced Antpitta)
Yungas of N Bolivia

Grallaria quitensis (Tawny Antpitta)
  G.q.quitensis
    C Colombia, Ecuador
  G.q.alticola
    E Colombia
  G.q.atuensis
    N Peru
Grallaria milleri (Brown-banded Antpitta)
  C Colombia

HYLOPEZUS
Hylopezus perspicillatus (Streak-chested Antpitta)
  H.p.intermedius
    Caribbean slope from E Costa Rica to W Panama
  H.p.lizanoi
    W Costa Rica, W Panama
  H.p.perspicillatus
    E Panama, NW Colombia
  H.p.periophthalmicus
    Pacific coast of W Colombia, W Ecuador
  H.p.pallidior
    C Colombia
Hylopezus macularius (Spotted Antpitta)
  H.m.diversus
    SE Colombia, NE Peru, S Venezuela
  H.m.macularius
    E Venezuela, the Guianas, N Brazil
  H.m.paraensis
    Brazil
  H.m.auricularis
    Bolivia
Hylopezus fulviventris (Fulvous-bellied Antpitta)
  H.f.dives
    Caribbean slope of Nicaragua, Costa Rica
  H.f.flammulatus
    Caribbean slope of W Panama
  H.f.barbacoae
    E Panama, W Colombia
  H.f.caquetae
    Colombia
  H.f.fulviventris
    E Ecuador
Hylopezus berlepschi (Amazonian Antpitta)
  H.b.yessupi
    Peru
  H.b.berlepschi
    N Bolivia, C & S Brazil
Hylopezus ochroleucus (Speckle-breasted Antpitta)
  H.o.ochroleucus
    E Brazil
  H.o.nattereri
    SE Brazil, E Paraguay, NE Argentina

MYRMOTHERA
Myrmothera campanisona (Thrush-like Antpitta)
  M.c.modesta
    SE Colombia

M.c.dissors
  E Colombia, S Venezuela, NW Brazil
M.c.campanisona
  SE Venezuela, the Guianas, N Brazil
M.c.signata
  E Ecuador, NE Peru
M.c.minor
  E Peru, W Brazil
M.c.subcanescens
  N & NC Brazil
Myrmothera simplex (Brown-breasted Antpitta)
  M.s.guaiquinimae
    SE Venezuela
  M.s.duidae
    S Venezuela
  M.s.simplex
    SE Venezuela

GRALLARICULA
Grallaricula flavirostris (Ochre-breasted Antpitta)
  G.f.costaricensis
    Costa Rica, Panama
  G.f.brevis
    Mt Pirri (E Panama)
  G.f.ochraceiventris
    W Colombia
  G.f.mindoensis
    N Ecuador
  G.f.zarumae
    SW Ecuador
  G.f.flavirostris
    E Colombia, E Ecuador
  G.f.similis
    N Peru
  G.f.boliviana
    N Bolivia
Grallaricula ferrugineipectus (Rusty-breasted Antpitta)
  G.f.rara
    E Colombia, NW Venezuela
  G.f.ferrugineipectus
    NE Colombia, N Venezuela
  G.f.leymebambae
    N Peru
Grallaricula nana (Slate-crowned Antpitta)
  G.n.occidentalis
    W Colombia
  G.n.nana
    E Colombia, W Venezuela
  G.n.olivascens
    N Venezuela
  G.n.cumanensis
    N Venezuela
  G.n.pariae
    NE Venezuela
  G.n.kukenamensis
    SE Venezuela
Grallaricula loricata (Scallop-breasted Antpitta)
  N Venezuela

Grallaricula peruviana (Peruvian Antpitta)
NW Peru

Grallaricula lineifrons (Crescent-faced Antpitta)
N Ecuador

Grallaricula ochraceifrons (Ochre-fronted Antpitta)   103.4
N Peru

Grallaricula cucullata (Hooded Antpitta)
G.c.cucullata
W Colombia
G.c.venezuelana
NW Venezuela

104. CONOPOPHAGIDAE (GNATEATERS)

CONOPOPHAGA

Conopophaga lineata (Rufous Gnateater)
C.l.lineata
E & SE Brazil
C.l.vulgaris
SE Brazil, E Paraguay, NE Argentina
C.l.cearae
E Brazil

Conopophaga aurita (Chestnut-belted Gnateater)
C.a.inexpectata
SE Colombia, NW Brazil
C.a.aurita
the Guianas, N Brazil
C.a.occidentalis
E Ecuador, NE Peru
C.a.australis
E Peru, W Brazil
C.a.snethlageae
C Brazil
C.a.pallida
Brazil

Conopophaga roberti (Hooded Gnateater)
NE Brazil

Conopophaga peruviana (Ash-throated Gnateater)
E Ecuador, E Peru, W Brazil

Conopophaga ardesiaca (Slaty Gnateater)
C.a.saturata
SE Peru
C.a.ardesiaca
Bolivia

Conopophaga castaneiceps (Chestnut-crowned Gnateater)
C.c.chocoensis
W Colombia
C.c.castaneiceps
E Colombia, NE Ecuador
C.c.chapmani
SE Ecuador, NE Peru
C.c.brunneinucha
C Peru

Conopophaga melanops (Black-cheeked Gnateater)
C.m.perspicillata
E Brazil

C.m.melanops
SE Brazil

Conopophaga melanogaster (Black-bellied Gnateater)
N Bolivia, Brazil

105. RHINOCRYPTIDAE (TAPACULOS)

PTEROPTOCHOS

Pteroptochos castaneus (Chestnut-throated Huet-huet)
C Chile

Pteroptochos tarnii (Black-throated Huet-huet)
S Chile, W Argentina

Pteroptochos megapodius (Moustached Turka)
P.m.atacamae
N Chile
P.m.megapodius
C Chile

SCELORCHILUS

Scelorchilus albicollis (White-throated Tapaculo)
S.a.atacamae
N Chile
S.a.albicollis
C Chile

Scelorchilus rubecula (Chucao Tapaculo)
S.r.rubecula
Chile, W Argentina
S.r.mochae
Mocha I

RHINOCRYPTA

Rhinocrypta lanceolata (Crested Gallito)
R.l.saturata
Paraguay
R.l.lanceolata
Argentina

TELEDROMAS

Teledromas fuscus (Sandy Gallito)
Argentina

LIOSCELES

Liosceles thoracicus (Rusty-belted Tapaculo)
L.t.dugandi
SE Colombia, W Brazil
L.t.erithacus
E Ecuador, E Peru
L.t.thoracicus
SE Peru, C Brazil

MELANOPAREIA

Melanopareia torquata (Collared Crescentchest)
M.t.torquata
E Brazil
M.t.rufescens
C Brazil

Melanopareia maximiliani (Olive-crowned Crescentchest)
M.m.maximiliani
W Bolivia
M.m.argentina
W Bolivia, Paraguay, N Argentina
Melanopareia maranonica (Maranon Crescentchest)
N Peru
Melanopareia elegans (Elegant Crescentchest)
M.e.elegans
W Ecuador
M.e.paucalensis
E Peru

PSILORHAMPHUS
Psilorhamphus guttatus (Spotted Bamboowren)
SE Brazil, NE Argentina

MERULAXIS
Merulaxis ater (Slaty Bristlefront)
SE Brazil
Merulaxis stresemanni (Stresemann's Bristlefront)
Brazil

EUGRALLA
Eugralla paradoxa (Ochre-flanked Tapaculo)
Chile, S Argentina

MYORNIS
Myornis senilis (Ash-coloured Tapaculo)
C Colombia, N Ecuador

SCYTALOPUS
Scytalopus unicolor (Unicoloured Tapaculo)
S.u.latrans
Colombia, W Venezuela, E Ecuador, N Peru
S.u.subcinereus
SW Ecuador, W Peru
S.u.intermedius
N Peru
S.u.unicolor
N Peru
S.u.parvirostris
C & S Peru, N Bolivia
Scytalopus speluncae (Mouse-coloured Tapaculo)
SE Brazil, NE Argentina
Scytalopus macropus (Large-footed Tapaculo)
Peru
Scytalopus femoralis (Rufous-vented Tapaculo)
S.f.sanctaemartae
NE Colombia
S.f.atratus
E Colombia
S.f.confusus
C Colombia

S.f.micropterus
E Ecuador, N Peru
S.f.nigricans
NE Colombia
S.f.femoralis
C Peru,
S.f.bolivianus
SE Peru, N Bolivia
Scytalopus argentifrons (Silvery-fronted Tapaculo)
S.a.argentifrons
Costa Rica, W Panama
S.a.chiriquensis
W Panama
Scytalopus panamensis (Pale-throated Tapaculo)
E Panama, W Colombia, W Ecuador
Scytalopus vicinior (Narino Tapaculo)
E Panama, W Colombia
Scytalopus latebricola (Brown-rumped Tapaculo)
S.l.latebricola
NE Colombia
S.l.meridanus
C & E Colombia, W Venezuela
S.l.caracae
N Venezuela
S.l.spillmanni
W Ecuador
Scytalopus novacapitalis (Brasilia Tapaculo)
SE Brazil
Scytalopus indigoticus (White-breasted Tapaculo)
SE Brazil
Scytalopus magellanicus (Andean Tapaculo)
S.m.griseicollis
E Colombia
S.m.fuscicauda
Venezuela
S.m.canus
W Colombia
S.m.opacus
E Ecuador
S.m.altirostris
N Peru
S.m.affinis
W Peru
S.m.acutirostris
C & SE Peru
S.m.urubambae
S Peru
S.m.simonsi
N Bolivia
S.m.zimmeri
C Bolivia
S.m.fuscus
N Chile W Argentina
S.m.magellanicus
S Chile, S Argentina
Scytalopus superciliaris (White-browed Tapaculo)
S.s.superciliaris
NW Argentina

S.s.santabarbarae
Argentina

Acropternis orthonyx (Ocellated Tapaculo)
A.o.orthonyx
E Colombia, W Venezuela
A.o.infuscata
E Ecuador

106. TYRANNIDAE 106.1

PHYLLOMYIAS
Phyllomyias fasciatus (Planalto Tyrannulet)
106.1
P.f.cearae
NE Brazil
P.f.fasciatus
C & E Brazil
P.f.brevirostris
S Brazil, Paraguay, NE Argentina
Phyllomyias burmeisteri (Rough-legged Tyrannulet)
P.b.zeledoni
Costa Rica, W Panama
P.b.viridiceps
N Venezuela
P.b.wetmorei
NW Venezuela
P.b.bunites
SC Venezuela
P.b.leucogonys
Colombia, Ecuador, Peru
P.b.burmeisteri
SE Brazil, E Bolivia, Paraguay, N Argentina
Phyllomyias virescens (Greenish Tyrannulet)
P.v.urichi
NE Venezuela
P.v.virescens
SE Brazil, Paraguay, NE Argentina
P.v.reiseri
E Brazil
Phyllomyias sclateri (Sclater's Tyrannulet)
P.s.subtropicalis
SE Peru
P.s.sclateri
Bolivia, NW Argentina
Phyllomyias griseocapilla (Grey-capped Tyrannulet)
SE Brazil
Phyllomyias griseiceps (Sooty-headed Tyrannulet)
P.g.cristatus
E Panama, N Colombia, N Venezuela
P.g.caucae
WC Colombia
P.g.griseiceps
Ecuador
P.g.pallidiceps
NE Brazil, Peru, SE Venezuela
Phyllomyias plumbeiceps (Plumbeous-crowned Tyrannulet)
Colombia, Ecuador, Peru

Phyllomyias nigrocapillus (Black-capped Tyrannulet)
P.n.flavimentum
N Colombia
P.n.nigrocapillus
W Colombia, Ecuador, N Peru
P.n.aureus
W Venezuela
Phyllomyias cinereiceps (Ashy-headed Tyrannulet)
Colombia, Ecuador, Peru
Phyllomyias uropygialis (Tawny-rumped Tyrannulet)
Venezuela, Colombia to W Bolivia

ZIMMERIUS 106.1
Zimmerius vilissimus (Paltry Tyrannulet)
Z.v.vilissimus
S Mexico to Honduras
Z.v.parvus
Nicaragua to Panama
Z.v.tamae
N Colombia
Z.v.improbus
N Colombia, W Venezuela
Z.v.petersi
N Venezuela
Zimmerius bolivianus (Bolivian Tyrannulet)
Z.b.bolivianus
N Bolivia
Z.b.viridissimus
SE Peru
Zimmerius cinereicapillus (Red-billed Tyrannulet)
C Peru, NE Ecuador
Zimmerius gracilipes (Slender-footed Tyrannulet)
Z.g.acer
S Venezuela, the Guianas, N Brazil
Z.g.gracilipes
S Venezuela, S Guyana, N Brazil, N Bolivia, E Peru
Z.g.gilvus
W Brazil, NW Bolivia
Zimmerius viridiflavus (Golden-faced Tyrannulet)
Z.v.minimus
N Colombia
Z.v.cumanensis
NE Venezuela
Z.v.albigularis
W Ecuador, SW Colombia
Z.v.flavidifrons
SW Ecuador
Z.v.chrysops
N Peru, Ecuador, S Colombia, W Venezuela
Z.v.viridiflavus
C Peru

ORNITHION
Ornithion inerme (White-lored Tyrannulet)
S Venezuela, the Guianas, Ecuador, N Brazil

Ornithion semiflavum (Yellow-bellied Tyrannulet)
S Mexico to Costa Rica

Ornithion brunneicapillum (Brown-capped Tyrannulet)
O.b.brunneicapillum
Costa Rica, Panama, W Colombia, NW Ecuador
O.b.dilutum
NW Venezuela, N Colombia

CAMPTOSTOMA

Camptostoma obsoletum (Southern Beardless Tyrannulet)
C.o.flaviventre
W Costa Rica, Panama
C.o.orphnum
Coiba I, Afuerita I
C.o.major
Pearl Is
C.o.caucae
W Colombia
C.o.bogotensis
C Colombia
C.o.pusillum
N Colombia, NW Venezuela
C.o.napaeum
N Venezuela, the Guianas, N Brazil
C.o.venezuelae
N & C Venezuela, Trinidad
C.o.maranonicum
N Peru
C.o.olivaceum
S Colombia, E Ecuador, NE Peru, W Brazil
C.o.sclateri
W Ecuador, W Peru
C.o.griseum
W Peru
C.o.bolivianum
C Bolivia, NW Argentina
C.o.cinerascens
E Bolivia, E & C Brazil
C.o.obsoletum
SE Brazil, Uruguay, N Argentina

Camptostoma imberbe (Northern Beardless Tyrannulet)
C.i.imberbe
S USA, Mexico to NW Costa Rica
C.i.ridgwayi
SW USA, W Mexico

PHAIOMYIAS

Phaiomyias murina (Mouse-coloured Tyrannulet)
P.m.eremonoma
Panama
P.m.incomta
Trinidad, the Guianas, Colombia, Venezuela, N Brazil
P.m.tumbezana
SW Ecuador, N Peru
P.m.inflava
W Peru

P.m.maranonica
NC Peru
P.m.wagae
E Peru, W Bolivia
P.m.ignobilis
S Bolivia, Paraguay, NW Argentina
P.m.murina
S Brazil

SUBLEGATUS

Sublegatus modestus (Scrub Flycatcher)
S.m.arenarum
SW Costa Rica
S.m.atrirostris
Panama to NE Colombia
S.m.glaber
N Venezuela, Trinidad
S.m.pallens
Aruba, Curacao, Bonaire
S.m.tortugensis
Tortuga
S.m.orinocensis
C Venezuela
S.m.modestus
E Peru to S Brazil
S.m.brevirostris
E Bolivia to W Argentina

Sublegatus obscurior (Dusky Flycatcher)
W Amazonia

SUIRIRI

Suiriri suiriri (Suiriri Flycatcher)
S.s.affinis
C Brazil, Surinam, NW Bolivia
S.s.bahiae
E Brazil
S.s.suiriri
E Bolivia, Brazil, Uruguay, Paraguay, N Argentina

TYRANNULUS

Tyrannulus elatus (Yellow-crowned Tyrannulet)
Panama to N Bolivia & C Brazil

MYIOPAGIS

Myiopagis gaimardii (Forest Elaenia)
M.g.macilvainii
E Panama, N Colombia
M.g.trinitatis
Trinidad
M.g.bogotensis
E Colombia, N Venezuela
M.g.guianensis
the Guianas, N Brazil
M.g.gaimardii
S Venezuela, Brazil, E Peru, N Bolivia
M.g.subcinereus
EC Brazil

Myiopagis caniceps (Grey Elaenia)
M.c.parambae
W Colombia, NW Ecuador

M.c.absita
  Panama
M.c.cinerea
  S Venezuela, E Colombia, E Ecuador, Peru,
  NW Brazil
M.c.caniceps
  E & S Brazil, Paraguay, N Argentina
Myiopagis subplacens (Pacific Elaenia)
  SW Ecuador, NW Peru
Myiopagis flavivertex (Yellow-crowned Elaenia)
  French Guiana, Surinam, S Venezuela, N
  Peru, N Brazil
Myiopagis viridicata (Greenish Elaenia)
M.v.jaliscensis
  SW Mexico
M.v.minima
  Tres Marias Is
M.v.placens
  S Mexico to Honduras
M.v.pacifica
  SE Chiapas, Mexico
M.v.accola
  Nicaragua to Panama, NW Colombia
M.v.pallens
  NW Colombia, N Venezuela
M.v.restricta
  N coast of Venezuela
M.v.zuliae
  N Venezuela
M.v.implacens
  S Colombia, W Ecuador
M.v.viridicata
  SE Peru, E Bolivia, Brazil, Paraguay, N
  Argentina
Myiopagis cotta (Yellow Elaenia)
  Jamaica

PSEUDELAENIA
**Pseudelaenia leucospodia (Grey & White
Tyrannulet)**   106.8
P.l.cinereifrons
  SW Ecuador
P.l.leucospodia
  NW Peru

ELAENIA
Elaenia flavogaster (Yellow-bellied Elaenia)
E.f.subpagana
  Pacific coast of S Mexico to Panama
E.f.flavogaster
  Trinidad, N Sth America
E.f.semipagana
  W Ecuador, NW Peru
E.f.pallididorsalis
  Panama, Pearl Is
Elaenia martinica (Caribbean Elaenia)
E.m.martinica
  Lesser Antilles islands
E.m.chinchorrensis
  Great Key I, E Mexico

E.m.barbadensis
  Barbados
E.m.riisii
  Virgin Is, Antigua I, Curacao I
E.m.caymanensis
  Cayman Is
E.m.cinerescens
  St Andrew I, Old Providence I
E.m.remota
  Yucatan coast islands
Elaenia spectabilis (Large Elaenia)
E.s.spectabilis
  NE Peru, N & C Brazil, W Bolivia, N
  Argentina
E.s.ridleyana
  Fernando de Noronha I
Elaenia albiceps (White-crested Elaenia)
E.a.griseigularis
  Ecuador
E.a.diversa
  NC Peru
E.a.urubambae
  SE Peru
E.a.albiceps
  Bolivia, Brazil
E.a.modesta
  Peru, NW Chile
E.a.chilensis
  S Chile, SW Argentina
Elaenia parvirostris (Small-billed Elaenia)
  C & E Sth America
Elaenia mesoleuca (Olivaceous Elaenia)
  SE Brazil, Paraguay, NE Argentina
Elaenia strepera (Slaty Elaenia)
  Colombia, Venezuela, Peru, Bolivia, NW
  Argentina
Elaenia gigas (Mottle-backed Elaenia)
  C Colombia, E Ecuador, E Peru, NE Bolivia
Elaenia pelzelni (Brownish Elaenia)
  W Brazil, NE Peru, Bolivia
Elaenia cristata (Plain-crested Elaenia)
E.c.alticola
  S Venezuela, N Brazil
E.c.cristata
  French Guiana to E Peru, Venezuela
Elaenia chiriquensis (Lesser Elaenia)
E.c.chiriquensis
  SW Costa Rica, Panama
E.c.brachyptera
  SW Colombia, NW Ecuador
E.c.albivertex
  Trinidad, C & W Nthn S America
Elaenia ruficeps (Rufous-crowned Elaenia)
  the Guianas, N Brazil, S Venezuela
Elaenia frantzii (Mountain Elaenia)
E.f.browni
  N Colombia, N Venezuela
E.f.pudica
  E Colombia, W Venezuela
E.f.ultima
  Guatemala to Nicaragua
E.f.frantzii
  Nicaragua to Panama

Elaenia obscura (Highland Elaenia)
E.o.sordida
  SE Brazil, E Paraguay, NE Argentina
E.o.obscura
  S Brazil, Peru, Bolivia, Paraguay, N
  Argentina
Elaenia dayi (Great Elaenia)
E.d.dayi
  Mt Roraima (S Venezuela)
E.d.auyantepui
  SE Venezuela
E.d.tyleri
  S Venezuela
Elaenia pallatangae (Sierran Elaenia)
E.p.pallatangae
  W Colombia, Ecuador
E.p.davidwillardi
  Venezuela
E.p.olivina
  S Guyana, S Venezuela
E.p.exsul
  Bolivian Andes
E.p.intensa
  Peru
Elaenia fallax (Greater Antillean Elaenia)
E.f.fallax
  Jamaica
E.f.cherriei
  Hispaniola

MECOCERCULUS
Mecocerculus leucophrys (White-throated Tyrannulet)
M.l.montensis
  N Colombia
M.l.chapmani
  S Venezuela
M.l.nigriceps
  N Venezuela
M.l.notatus
  W Colombia
M.l.setophagoides
  E Colombia
M.l.palliditergum
  coast of N Venezuela
M.l.gularis
  E Venezuela
M.l.parui
  S Venezuela
M.l.rufomarginatus
  S Colombia, Ecuador, NW Peru
M.l.roraimae
  C Venezuela
M.l.brunneomarginatus
  C Peru
M.l.pallidior
  W Peru
M.l.leucophrys
  SE Peru, NW Argentina
Mecocerculus poecilocercus (White-tailed Tyrannulet)
  Colombia, Ecuador, Peru

Mecocerculus hellmayri (Buff-banded Tyrannulet)
  SE Peru, Bolivia, Argentina
Mecocerculus calopterus (Rufous-winged Tyrannulet)
  W Ecuador, NW Peru
Mecocerculus minor (Sulphur-bellied Tyrannulet)
  E Colombia, Venezuela, NW Peru
Mecocerculus stictopterus (White-banded Tyrannulet)
M.s.stictopterus
  Colombia, Ecuador, N Peru
M.s.taeniopterus
  SE Peru, W Bolivia
M.s.albocaudatus
  NW Venezuela

SERPOPHAGA
Serpophaga cinerea (Torrent Tyrannulet)
S.c.grisea
  Costa Rica, W Panama
S.c.cinerea
  W Ecuador, W Peru, W Bolivia
Serpophaga hypoleuca (River Tyrannulet)
S.h.venezuelana
  N Venezuela
S.h.hypoleuca
  SE Colombia, E Peru, Venezuela
S.h.pallida
  EC Brazil
Serpophaga nigricans (Sooty Tyrannulet)
  SE Brazil, Uruguay, Paraguay, N Argentina
Serpophaga araguayae (Bananal Tyrannulet)
  Goias (Brazil)
Serpophaga subcristata (White-crested Tyrannulet)
S.s.straminea
  SC Brazil, Uruguay
S.s.subcristata
  Bolivia to E Brazil, Argentina
S.s.munda
  E Bolivia, N Argentina

INEZIA
Inezia tenuirostris (Slender-billed Tyrannulet)
  N Colombia, NW Venezuela
Inezia inornata (Plain Tyrannulet)
  SW Brazil, E Bolivia, W Paraguay, S Peru
Inezia subflava (Pale-tipped Tyrannulet)
I.s.intermedia
  NW Venezuela, N Colombia
I.s.caudata
  the Guianas, Venezuela
I.s.subflava
  EC Brazil
I.s.obscura
  NW Brazil

STIGMATURA
Stigmatura napensis (Lesser Wagtail Tyrant)
S.n.napensis
  SE Colombia, NE Peru, W Brazil

S.n.bahiae
NE Brazil
Stigmatura budytoides (Greater Wagtail Tyrant)
S.b.budytoides
NC Bolivia
S.b.inzonata
SE Bolivia, N Argentina
S.b.flavocinerea
C Argentina
S.b.gracilis
Brazil

ANAIRETES   106.1
Anairetes alpinus (Ash-breasted Tit Tyrant)
A.a.alpinus
N Peru
A.a.bolivianus
C Bolivia
Anairetes agraphia (Unstreaked Tit Tyrant)
A.a.agraphia
SE Peru
A.a.plengei
C Peru
A.a.squamigera
E Peru
Anairetes agilis (Agile Tit Tyrant)
E Colombia, N Ecuador
Anairetes reguloides (Pied-crested Tit Tyrant)
A.r.nigrocristatus
N Peru
A.r.albiventris
W Peru
A.r.reguloides
SW Peru, NW Chile
Anairetes flavirostris (Yellow-billed Tit Tyrant)
A.f.huancabambae
NW Peru
A.f.arequipae
SW Peru, NW Chile
A.f.cuzcoensis
SE Peru
A.f.flavirostris
W Peru, Bolivia, N Chile, W Argentina
Anairetes parulus (Tufted Tit Tyrant)
A.p.aequatorialis
S Colombia, Ecuador, Peru, Bolivia
A.p.patagonicus
C Argentina
A.p.parulus
S & C Chile, W Argentina
Anairetes fernandezianus (Juan Fernandez Tit Tyrant)
Masatierra I

TACHURIS
Tachuris rubrigastra (Many-coloured Rush Tyrant)
T.r.libertatis
W Peru
T.r.alticola
Peru, W Bolivia, NW Argentina

T.r.rubrigastra
SE Brazil, Paraguay, Uruguay, Chile, N Argentina
T.r.loaensis
N Chile

CULICIVORA
Culicivora caudacuta (Sharp-tailed Tyrant)
S Brazil, E Bolivia, Paraguay, NE Argentina

POLYSTICTUS
Polystictus superciliaris (Grey-backed Tachuri)
SE Brazil
Polystictus pectoralis (Bearded Tachuri)
P.p.bogotensis
Colombia
P.p.brevipennis
Guyana, S Venezuela, N Brazil
P.p.pectoralis
E Bolivia, SW Brazil, Uruguay, Paraguay, N Argentina

PSEUDOCOLOPTERYX
Pseudocolopteryx dinellianus (Dinelli's Doradito)
NW Argentina, SE Bolivia, W Paraguay
Pseudocolopteryx sclateri (Crested Doradito)
Trinidad, Guyana, Brazil, Paraguay, E Argentina
Pseudocolopteryx acutipennis (Subtropical Doradito)
Colombia, Ecuador, Peru, Bolivia, W Argentina
Pseudocolopteryx flaviventris (Warbling Doradito)
Uruguay, C Chile, N Argentina, S Brazil

EUSCARTHMUS
Euscarthmus meloryphus (Tawny-crowned Pygmy Tyrant)
E.m.paulus
Venezuela, Colombia
E.m.fulviceps
W Ecuador, W Peru
E.m.meloryphus
Brazil, E Bolivia, Paraguay, N Argentina
Euscarthmus rufomarginatus (Rufous-sided Pygmy Tyrant)
E.r.rufomarginatus
SE Brazil
E.r.savannophilus
Surinam

MIONECTES
Mionectes striaticollis (Streak-necked Flycatcher)
M.s.columbianus
Colombia, E Ecuador
M.s.viridiceps
W Ecuador
M.s.palamblae
N Peru

M.s.poliocephalus
N & C Peru
M.s.striaticollis
Bolivia, SE Peru
M.s.selvae
WC Colombia
Mionectes olivaceus (Olive-striped Flycatcher)
M.o.olivaceus
Costa Rica, W Panama
M.o.hederaceus
E Panama, W Colombia, W Ecuador
M.o.galbinus
N Colombia
M.o.pallidus
E Colombia
M.o.venezuelanus
N Venezuela, Trinidad
M.o.fasciaticollis
E Ecuador, E Peru
M.o.meridae
NW Venezuela, NE Colombia
Mionectes oleagineus (Ochre-bellied Flycatcher)   106.1
M.o.assimilis
S Mexico to Costa Rica
M.o.obscurus
El Salvador
M.o.dyscolus
W Costa Rica, W Panama
M.o.lutescens
W Panama
M.o.parcus
E Panama, N Colombia, NW Venezuela
M.o.chloronotus
W Amazonia
M.o.abdominalis
N Venezuela
M.o.pallidiventris
NE Venezuela, Trinidad
M.o.intensus
E Venezuela, Guyana
M.o.dorsalis
SE Venezuela
M.o.pacificus
W Colombia, W Ecuador
M.o.hauxwelli
E Ecuador, NE Peru
M.o.wallacei
N Brazil, the Guianas
M.o.maynana
E Peru
M.o.oleagineus
SE Brazil
Mionectes macconnelli (MacConnell's Flycatcher)   106.1
M.m.mercedesfosteri
Venezuela
M.m.macconnelli
Guyana, French Guiana, N Brazil, E Venezuela
M.m.roraimae
S Guyana, S Venezuela

M.m.peruanus
C Peru
M.m.amazonus
C Brazil, E Bolivia
Mionectes rufiventris (Grey-hooded Flycatcher)   106.1
SE Brazil, Paraguay, N Argentina

LEPTOPOGON
Leptopogon rufipectus (Rufous-breasted Leptopogon)
L.r.rufipectus
E Ecuador, E Colombia, Peru
L.r.venezuelanus
NE Colombia, NW Venezuela
Leptopogon taczanowskii (Inca Leptopogon)
E Peru
Leptopogon amaurocephalus (Sepia-capped Leptopogon)
L.a.pileatus
S Mexico, Guatemala, Honduras
L.a.faustus
Costa Rica, Panama
L.a.idius
Coiba I
L.a.diversus
N Colombia
L.a.orinocensis
W & SW Venezuela
L.a.obscuritergum
S Venezuela
L.a.peruvianus
N & W Amazonia
L.a.amaurocephalus
SE Brazil, E Bolivia, N Argentina, Paraguay
Leptopogon superciliaris (White-bellied Leptopogon)
L.s.superciliaris
SE Colombia, N Peru
L.s.poliocephalus
N & C Colombia
L.s.hellmayri
Costa Rica to W Panama
L.s.venezuelensis
N Venezuela, N Brazil
L.s.pariae
NE Venezuela, Trinidad
L.s.transandinus
SW Colombia, W Ecuador
L.s.albidiventer
Bolivia, SE Peru

PHYLLOSCARTES
Phylloscartes nigrifrons (Black-fronted Tyrannulet)
S Venezuela
Phylloscartes poecilotis (Variegated Bristle Tyrant)
P.p.poecilotis
N Colombia to Ecuador
P.p.pifanoi
NE Colombia, NW Venezuela

Phylloscartes chapmani (Chapman's Tyrannulet)
P.c.chapmani
S Venezuela
P.c.duidae
SE Venezuela
Phylloscartes ophthalmicus (Marble-faced Bristle Tyrant)
P.o.ophthalmicus
NW Venezuela, Colombia, Ecuador, Peru
P.o.ottonis
SE Peru, W Bolivia
P.o.purus
coast of Venezuela
Phylloscartes eximius (Southern Bristle Tyrant)
SE Brazil, Paraguay, N Argentina
Phylloscartes gualaquizae (Ecuadorean Bristle Tyrant)
E Ecuador, N Peru
Phylloscartes flaviventris (Yellow-bellied Bristle Tyrant)
NW Venezuela
Phylloscartes venezuelanus (Venezuelan Bristle Tyrant)
N Venezuela
Phylloscartes orbitalis (Spectacled Bristle Tyrant)
E Ecuador, Peru, S Colombia
Phylloscartes lanyoni (Antioquia Bristle Tyrant) 106.7
N Colombia
Phylloscartes flaveolus (Yellow Tyrannulet)
P.f.semiflavus
Nicaragua to Panama
P.f.cerulus
Venezuela, Colombia, Ecuador
P.f.amazonus
French Guiana, N Brazil
P.f.leucophrys
Colombia, W Venezuela
P.f.magnirostris
SW Ecuador
P.f.flaveolus
S Colombia, SE Brazil, E Bolivia, Paraguay
Phylloscartes roquettei (Minas Geraes Tyrannulet)
Minas Geraes (Brazil)
Phylloscartes ventralis (Mottle-cheeked Tyrannulet)
P.v.angustirostris
Peru, Bolivia, NW Argentina
P.v.tucumanus
NW Argentina
P.v.ventralis
SE Brazil, Uruguay, Paraguay, NE Argentina
Phylloscartes paulistus (Sao Paulo Tyrannulet)
SE Brazil, Paraguay
Phylloscartes oustaleti (Oustalet's Tyrannulet)
SE Brazil

Phylloscartes difficilis (Ihering's Tyrannulet)
SE Brazil
Phylloscartes ceciliae (Long-tailed Tyrannulet) 106.2
NE Brazil
Phylloscartes flavovirens (Yellow-green Tyrannulet)
Panama
Phylloscartes virescens (Olive-green Tyrannulet)
the Guianas
Phylloscartes superciliaris (Rufous-browed Tyrannulet)
P.s.superciliaris
Costa Rica to W Panama
P.s.griseocapillus
NW Venezuela
P.s.palloris
E Panama
Phylloscartes sylviolus (Bay-ringed Tyrannulet)
SE Brazil, Paraguay

PSEUDOTRICCUS
Pseudotriccus pelzelni (Bronze-olive Pygmy Tyrant)
P.p.berlepschi
E Panama
P.p.annectens
W Colombia, W Ecuador
P.p.pelzelni
E Colombia, E Ecuador
P.p.peruvianus
E Peru
Pseudotriccus simplex (Hazel-fronted Pygmy Tyrant)
Bolivia, SE Peru
Pseudotriccus ruficeps (Rufous-headed Pygmy Tyrant)
Colombia to NW Bolivia

CORYTHOPIS 106.3
Corythopis delalandi (Delalande's Antpipit)
E Bolivia, S Brazil, E Paraguay, NE Argentina
Corythopis torquata (Ringed Antpipit)
C.t.sarayacuensis
SE Colombia, E Ecuador, NE Peru
C.t.torquata
C Peru, W Brazil
C.t.anthoides
S Venezuela, the Guianas, N & E Brazil
C.t.subtorquata
NE Bolivia

MYIORNIS
Myiornis auricularis (Eared Pygmy Tyrant)
SE Brazil, N Argentina
Myiornis ecaudatus (Short-tailed Pygmy Tyrant)
M.e.miserabilis
C Colombia, Trinidad, Guyana, Venezuela, Surinam

M.e.ecaudatus
  E Peru & Bolivia
**Myiornis atricapillus (Black-capped Pygmy
Tyrant)** 106.6
  E Costa Rica to NW Ecuador
**Myiornis albiventris (White-breasted Pygmy
Tyrant)**
  C Peru

LOPHOTRICCUS
**Lophotriccus pileatus (Scale-crested Pygmy
Tyrant)**
  L.p.luteiventris
    Costa Rica, W Panama
  L.p.sanctaeluciae
    NE Colombia, N coast of Venezuela
  L.p.squamaecrista
    S & C Colombia, W Ecuador, W Venezuela
  L.p.pileatus
    Peru, E Ecuador
  L.p.hypochlorus
    SE Peru
**Lophotriccus eulophotes (Long-crested
Pygmy Tyrant)**
  W Brazil, Peru
**Lophotriccus vitiosus (Double-banded
Pygmy Tyrant)**
  L.v.affinis
    SW Colombia to NE Peru, NW Brazil
  L.v.guianensis
    SC Colombia, the Guianas, NE Brazil
  L.v.vitiosus
    E Peru
  L.v.congener
    NW Brazil
**Lophotriccus galeatus (Helmeted Pygmy
Tyrant)**
  the Guianas, N Brazil, Venezuela

ATALOTRICCUS
Atalotriccus pilaris (Pale-eyed Pygmy Tyrant)
  A.p.wilcoxi
    W Panama
  A.p.pilaris
    W Venezuela, N & E Colombia
  A.p.venezuelensis
    N Venezuela
  A.p.griseiceps
    S Venezuela, W Guyana

POECILOTRICCUS
**Poecilotriccus ruficeps (Rufous-crowned
Tody Tyrant)**
  P.r.melanomystax
    C Colombia
  P.r.ruficeps
    E Ecuador, S Colombia, SW Venezuela
  P.r.rufigenis
    W Ecuador, W Colombia
  P.r.peruvianus
    NW Peru

**Poecilotriccus capitale (Black & White Tody
Flycatcher)**
  SE Colombia, E Ecuador, NE Peru,
**Poecilotriccus tricolor (Tricoloured Tody
Flycatcher)**
  SC Brazil, SE Peru
Poecilotriccus andrei (Black-chested Tyrant)
  P.a.andrei
    SE Venezuela, NW Brazil
  P.a.klagesi
    Rio Tapajos (Brazil)

ONCOSTOMA
Oncostoma olivaceum (Southern Bentbill)
  E Panama, N Colombia
Oncostoma cinereigulare (Northern Bentbill)
  S Mexico to W Panama

HEMITRICCUS
Hemitriccus minor (Snethlage's Tody Tyrant)
  H.m.minor
    E Brazil
  H.m.pallens
    WC Brazil
  H.m.minima
    C Brazil
**Hemitriccus josephinae (Boat-billed Tody
Tyrant)**
  Guyana, W Surinam, NE Brazil
**Hemitriccus diops (Drab-breasted Pygmy
Tyrant)**
  SE Brazil, Paraguay
**Hemitriccus obsoletus (Brown-breasted
Pygmy Tyrant)**
  H.o.obsoletus
    SE Brazil
  H.o.zimmeri
    SE Brazil
**Hemitriccus flammulatus (Flammulated
Pygmy Tyrant)**
  H.f.flammulatus
    Peru, Bolivia, W Brazil
  H.f.olivascens
    E Bolivia
**Hemitriccus zosterops (White-eyed Tody
Tyrant)**
  H.z.zosterops
    S Venezuela, S Colombia, the Guianas, NW
    Brazil
  H.z.flaviviridis
    N Peru
  H.z.griseipectus
    SE Peru, N Bolivia, C Brazil
  H.z.naumburgae
    E Brazil
Hemitriccus aenigma (Zimmer's Tody Tyrant)
  NC Brazil
**Hemitriccus orbitatus (Olivaceous Tody
Tyrant)**
  SE Brazil
**Hemitriccus iohannis (Johanne's Pygmy
Tyrant)**
  W Amazonia

Hemitriccus striaticollis (Stripe-necked Tody Tyrant)
H.s.griseiceps
WC Brazil
H.s.striaticollis
C Brazil to N Bolivia, N Peru, E Colombia
Hemitriccus nidipendulus (Hangnest Tody Tyrant)
H.n.nidipendulus
EC Brazil
H.n.paulistus
SE Brazil
Hemitriccus spodiops (Yungas Tody Tyrant)
NE Brazil
Hemitriccus margaritaceiventer (Pearly-vented Tody Tyrant)
H.m.impiger
NE Colombia, N Venezuela
H.m.septentrionalis
S Colombia
H.m.duidae
S Venezuela
H.m.auyantepui
SE Venezuela
H.m.breweri
C Venezuela
H.m.rufipes
C Peru, NW Bolivia
H.m.margaritaceiventer
WC Brazil to N Argentina
H.m.wuchereri
E Brazil
Hemitriccus inornatus (Pelzeln's Tody Tyrant)
NW Brazil
Hemitriccus granadensis (Black-throated Tody Tyrant)
H.g.lehmanni
N Colombia
H.g.intensus
NW Venezuela
H.g.federalis
N Venezuela
H.g.granadensis
Colombia, NE Ecuador
H.g.andinus
E Colombia, W Venezuela
H.g.pyrrhops
SE Ecuador, N Peru
H.g.caesius
SE Peru
Hemitriccus mirandae (Buff-breasted Tody Tyrant)
NE Brazil
Hemitriccus kaempferi (Kaempfer's Tody Tyrant)
S Brazil
Hemitriccus rufigularis (Buff-throated Tody Tyrant)
C & SE Peru, NW Bolivia
Hemitriccus furcatus (Fork-tailed Pygmy Tyrant)
SE Brazil

TODIROSTRUM
Todirostrum senex (Plumbeous-crowned Tody Flycatcher)
C Brazil
Todirostrum russatum (Ruddy Tody Flycatcher)
SE Venezuela, NE Brazil
Todirostrum plumbeiceps (Ochre-faced Tody Flycatcher)
T.p.obscurum
SE Peru, N Bolivia
T.p.viridiceps
Bolivia, NW Argentina
T.p.plumbeiceps
SE Brazil, Paraguay, NE Argentina
T.p.cinereipectum
SC Brazil
Todirostrum fumifrons (Smoky-fronted Tody Flycatcher)
T.f.fumifrons
NE Brazil
T.f.penardi
French Guiana, Surinam
Todirostrum latirostre (Rusty-fronted Tody Flycatcher)
T.l.mituense
Colombia
T.l.caniceps
SE Colombia, E Ecuador, E Peru
T.l.latirostre
C Brazil
T.l.mixtum
SE Peru
T.l.ochropterum
S Brazil, SE Bolivia
T.l.austroriparium
Santarem (Brazil)
T.l.senectum
Lower Amazon, Brazil
Todirostrum sylvia (Slate-headed Tody Flycatcher)
T.s.schistaceiceps
S Mexico to Panama
T.s.superciliare
N Colombia
T.s.griseolum
N Venezuela
T.s.sylvia
Guyana, French Guiana, N Brazil
T.s.schulzi
NE Brazil
Todirostrum maculatum (Spotted Tody Flycatcher)
T.m.amacurense
NE Venezuela, Guyana, Trinidad
T.m.maculatum
E Venezuela, the Guianas, N Brazil
T.m.signatum
E Ecuador, E Peru, W Brazil
T.m.diversum
WC Brazil

T.m.annectens
C Brazil

Todirostrum poliocephalum (Grey-headed Tody Flycatcher)
SE Brazil

Todirostrum cinereum (Common Tody Flycatcher)
T.c.virididorsale
SC Mexico
T.c.finitimum
S Mexico to Panama
T.c.wetmorei
C & E Costa Rica, Panama
T.c.sclateri
SW Colombia, W Ecuador, NW Peru
T.c.cinereum
the Guianas, N Brazil, S Venezuela, S Colombia
T.c.peruanum
E Ecuador, N & E Peru
T.c.coloreum
C Brazil, E Bolivia
T.c.cearae
E Brazil

Todirostrum viridanum (Short-tailed Tody Flycatcher)
NW Venezuela

Todirostrum pictum (Painted Tody Flycatcher)
the Guianas, N Brazil

Todirostrum chrysocrotaphum (Yellow-browed Tody Flycatcher)
T.c.guttatum
SW Venezuela, Colombia, NW Brazil, N Peru
T.c.neglectum
E Peru, N Bolivia, SW Brazil
T.c.similis
NE Brazil
T.c.illigeri
NE Brazil
T.c.chrysocrotaphum
E Peru, N Bolivia, W Brazil

Todirostrum nigriceps (Black-headed Tody Flycatcher)
Costa Rica, Panama, N Colombia, W Ecuador

Todirostrum calopterum (Golden-winged Tody Flycatcher)
T.c.calopterum
S Colombia, E Ecuador
T.c.pulchellum
SE Peru

CNIPODECTES
Cnipodectes subbrunneus (Brownish Flycatcher)
C.s.subbrunneus
W Ecuador, W Colombia
C.s.panamensis
E Panama, N Colombia
C.s.minor
SE Colombia, E Peru, W Brazil

RAMPHOTRIGON
Ramphotrigon megacephala (Large-headed Flatbill)
R.m.pectoralis
S Colombia, S Venezuela
R.m.venezuelensis
W Venezuela
R.m.boliviana
SE Peru, W Brazil, N Bolivia
R.m.megacephala
SE Brazil, Paraguay, N Argentina

Ramphotrigon ruficauda (Rufous-tailed Flatbill)
N Amazonia

Ramphotrigon fuscicauda (Dusky-tailed Flatbill)
E Ecuador, S Colombia, E Peru

RHYNCHOCYCLUS
Rhynchocyclus brevirostris (Eye-ringed Flatbill)
R.b.brevirostris
S Mexico to W Panama
R.b.pallidus
S Mexico
R.b.hellmayri
E Panama, N Colombia
R.b.pacificus
NW Ecuador, W Colombia

Rhynchocyclus olivaceus (Olivaceous Flatbill)
R.o.bardus
E Panama, N Colombia
R.o.mirus
NW Colombia
R.o.tamborensis
Colombia
R.o.flavus
NW Colombia, N Venezuela
R.o.aequinoctialis
S Colombia, E Ecuador, E Peru, NC Bolivia
R.o.guianensis
N & NW Amazonia
R.o.sordidus
C Brazil
R.o.olivaceus
SE Brazil

Rhynchocyclus fulvipectus (Fulvous-breasted Flatbill)
Colombia, E Ecuador, SE Peru

TOLMOMYIAS
Tolmomyias sulphurescens (Yellow-olive Flycatcher)
T.s.cinereiceps
S Mexico to Costa Rica
T.s.flavoolivaceus
Panama
T.s.berlepschi
Trinidad
T.s.exortivus
N Venezuela, N Colombia

T.s.asemus
Colombia
T.s.confusus
C Colombia, SW Venezuela, NE Ecuador
T.s.duidae
SE Venezuela, NW Brazil
T.s.aequatorialis
W Ecuador, NW Peru
T.s.cherriei
E Colombia to the Guianas, N Brazil
T.s.peruvianus
SE Ecuador, N Peru
T.s.insignis
NE Peru, W Brazil
T.s.mixtus
NE Brazil
T.s.inornatus
SE Peru
T.s.pallescens
C Brazil to N Argentina
T.s.grisecens
C Paraguay, N Argentina
T.s.sulphurescens
S Brazil, N Argentina
Tolmomyias assimilis (Yellow-margined Flycatcher)
T.a.flavotectus
Costa Rica, to NW Ecuador
T.a.neglectus
E Colombia, S Venezuela, NE Brazil
T.a.obscuriceps
S Colombia to NE Peru
T.a.examinatus
SE Venezuela, the Guianas, N Brazil
T.a.assimilis
C Brazil
T.a.clarus
C Peru
T.a.paraensis
N Brazil
T.a.calamae
N Bolivia, SW Brazil
Tolmomyias poliocephalus (Grey-crowned Flycatcher)
T.p.poliocephalus
W Amazonia
T.p.klagesi
S Venezuela
T.p.sclateri
the Guianas, N Brazil
Tolmomyias flaviventris (Yellow-breasted Flycatcher)
T.f.aurulentus
N Colombia, N Venezuela
T.f.collingwoodi
C & E Colombia to Guyana, N Brazil, Trinidad
T.f.dissors
NE Brazil, SE Venezuela
T.f.viridiceps
SE Colombia, E Ecuador, E Peru, W Brazil
T.f.zimmeri
NC Peru

T.f.subsimilis
SE Peru, NW Bolivia, SW Brazil
T.f.flaviventris
E Brazil

PLATYRINCHUS
Platyrinchus saturatus (Cinnamon-crested Spadebill)
P.s.saturatus
S Venezuela, the Guianas, N Brazil
P.s.pallidiventris
C Brazil
Platyrinchus cancrominus (Stub-tailed Spadebill)
P.c.cancrominus
S Mexico to E Nicaragua
P.c.timothei
SE Mexico to N Guatemala
P.c.dilutus
El Salvador to NW Costa Rica
Platyrinchus mystaceus (White-throated Spadebill)
P.m.neglectus
E Costa Rica to E Colombia
P.m.perijanus
NW Venezuela
P.m.insularis
Trinidad, N Venezuela
P.m.imatacae
E Venezuela
P.m.ventralis
S Venezuela
P.m.duidae
SE Venezuela, N Brazil
P.m.ptaritepui
SE Venezuela
P.m.albogularis
W Ecuador, W Colombia
P.m.zamorae
E Ecuador, N Peru
P.m.mystaceus
E Brazil, Paraguay, N Argentina
P.m.partridgei
C Bolivia
P.m.bifasciatus
SW Brazil
P.m.cancromus
E Brazil
P.m.niveigularis
NE Brazil
Platyrinchus coronatus (Golden-crowned Spadebill)
P.c.superciliaris
Nicaragua to W Ecuador
P.c.gumia
the Guianas, N Brazil
P.c.coronatus
W Amazonia
Platyrinchus flavigularis (Yellow-throated Spadebill)
P.f.flavigularis
Colombia, Peru

P.f.vividus
  W Venezuela
Platyrinchus platyrhynchos (White-crested Spadebill)
  P.p.platyrhynchos
    S Venezuela, the Guianas, N Brazil
  P.p.senex
    E Ecuador, E Peru, NW Bolivia
  P.p.nattereri
    W Brazil
  P.p.amazonicus
    E Brazil
Platyrinchus leucoryphus (Russet-winged Spadebill)
  SE Brazil, E Paraguay

ONYCHORHYNCHUS
Onychorhynchus coronatus (Royal Flycatcher)
  O.c.mexicanus
    SE Mexico to E Panama
  O.c.fraterculus
    NE Colombia, E Venezuela
  O.c.castelnaui
    Western Amazonia
  O.c.coronatus
    E Venezuela, the Guianas, N Brazil
  O.c.occidentalis
    W Ecuador
  O.c.swainsoni
    SE Brazil

MYIOTRICCUS
Myiotriccus ornatus (Ornate Flycatcher)
  M.o.ornatus
    C Colombia
  M.o.stellatus
    W Colombia, W Ecuador
  M.o.phoenicurus
    SE Colombia, E Ecuador, N Peru
  M.o.aureiventris
    S Peru

TERENOTRICCUS
Terenotriccus erythrurus (Ruddy-tailed Flycatcher)
  T.e.fulvigularis
    Guatemala to Panama, Colombia, Venezuela
  T.e.signatus
    C & S Colombia to NE Peru
  T.e.venezuelensis
    E Colombia, W Venezuela, NW Brazil
  T.e.brunneifrons
    SW Brazil, E Peru, N Bolivia
  T.e.erythrurus
    the Guianas, S Venezuela, N Brazil
  T.e.amazonus
    C Brazil
  T.e.hellmayri
    N Brazil

MYIOBIUS
Myiobius villosus (Tawny-breasted Flycatcher)
  M.v.villosus
    E Panama, Colombia, W Ecuador
  M.v.schaeferi
    NW Venezuela, NE Colombia
  M.v.clarus
    E Ecuador
  M.v.peruvianus
    SE Peru, W Bolivia
Myiobius barbatus (Sulphur-rumped Flycatcher)
  M.b.sulphureipygius
    S Mexico to Honduras
  M.b.aureatus
    S Honduras to W Ecuador
  M.b.semiflavus
    EC Colombia
  M.b.barbatus
    SE Colombia to the Guianas, N Brazil, N Peru
  M.b.amazonicus
    E Peru
  M.b.insignis
    NE Brazil
  M.b.mastacalis
    SE Brazil
Myiobius atricaudus (Black-tailed Flycatcher)
  M.a.atricaudus
    SW Costa Rica, Panama, W Colombia
  M.a.portovelae
    W Ecuador, N Peru
  M.a.modestus
    E Venezuela
  M.a.adjacens
    S Colombia, E Ecuador, E Peru, W Brazil
  M.a.connectens
    NE Brazil
  M.a.snethlagei
    NE Brazil
  M.a.ridgwayi
    SE Brazil

MYIOPHOBUS
Myiophobus flavicans (Flavescent Flycatcher)
  M.f.flavicans
    Colombia, Ecuador, N Peru
  M.f.perijanus
    NW Venezuela
  M.f.venezuelanus
    N Venezuela
  M.f.caripensis
    NE Venezuela
  M.f.superciliosus
    Peru
Myiophobus phoenicomitra (Orange-crested Flycatcher)
  M.p.litae
    NW Ecuador, W Colombia
  M.p.phoenicomitra
    E Ecuador, N Peru

Myiophobus inornatus (Unadorned
Flycatcher)
Peru, Bolivia
Myiophobus roraimae (Roraiman Flycatcher)
M.r.sadiecoatsi
Venezuela
M.r.roraimae
S Venezuela, SE Colombia, W Guyana
M.r.rufipennis
SE Peru
Myiophobus lintoni (Orange-banded
Flycatcher)
SE Ecuador
Myiophobus pulcher (Handsome Flycatcher)
M.p.pulcher
W Colombia, W Ecuador
M.p.bellus
E Colombia, E Ecuador
M.p.oblitus
SE Peru
Myiophobus cryptoxanthus (Olive-crested
Flycatcher)
E Ecuador, E Peru
Myiophobus ochraceiventris (Ochraceous-
breasted Flycatcher)
C Peru, NW Bolivia
Myiophobus fasciatus (Bran-coloured
Flycatcher)
M.f.furfurosus
SW Costa Rica, Panama
M.f.fasciatus
the Guianas, Trinidad, N Venezuela,
Colombia
M.f.crypterythrus
S Colombia, W Ecuador, N Peru
M.f.saturatus
E Peru
M.f.rufescens
W Peru, N Chile
M.f.auriceps
SE Peru, N & E Bolivia, N Argentina W
Paraguay
M.f.flammiceps
Brazil, Bolivia, Uruguay, Paraguay, N
Argentina

LATHROTRICCUS   106.4
Lathrotriccus euleri (Euler's Flycatcher)
L.e.johnstonei
Grenada
L.e.lawrencei
Trinidad, N Venezuela, Surinam
L.e.bolivianus
N & W Amazonia
L.e.argentinus
SW to SE Amazonia >> E Peru, E Brazil
L.e.euleri
SE Brazil, NE Argentina >> Peru, Bolivia

APHANOTRICCUS
Aphanotriccus capitalis (Tawny-chested
Flycatcher)
Nicaragua, Costa Rica

Aphanotriccus audax (Black-billed
Flycatcher)
Panama, NW Colombia

XENOTRICCUS
Xenotriccus callizonus (Belted Flycatcher)
Mexico
Xenotriccus mexicanus (Pileated Flycatcher)
S Mexico

PYRRHOMYIAS
Pyrrhomyias cinnamomea (Cinnamon
Flycatcher)
P.c.assimilis
NW Colombia
P.c.pyrrhoptera
Colombia, Venezuela, Ecuador, N Peru
P.c.vieillotioides
Venezuela
P.c.spadix
NE Venezuela
P.c.pariae
NE Venezuela
P.c.cinnamomea
Bolivia, Peru, NW Argentina

MITREPHANES
Mitrephanes phaeocercus (Tufted Flycatcher)
M.p.tenuirostris
W Mexico
M.p.phaeocercus
S & E Mexico to Honduras
M.p.burleighi
Guerrero, SW Oaxaca
M.p.nicaraguae
S Mexico to Nicaragua
M.p.aurantiiventris
Costa Rica, W Panama
M.p.vividus
C Panama
M.p.eminulus
E Panama, W Colombia
M.p.berlepschi
S Colombia, NE Ecuador
Mitrephanes olivaceus (Olive Flycatcher)
E Peru, Bolivia

CONTOPUS
Contopus borealis (Olive-sided Flycatcher)
W Canada, N & W USA >> C America, N &
W Sth America
Contopus fumigatus (Smoke-coloured
Pewee)
C.f.pallidiventris
S Arizona, N Mexico >> Belize
C.f.pertinax
C & S Mexica, Guatemala
C.f.minor
Honduras, Belize, N Nicaragua
C.f.lugubris
Costa Rica, W Panama
C.f.cineraceus
N Venezuela

C.f.duidae
  SE Venezuela
C.f.ardosiacus
  NE Peru, E Ecuador, Colombia, NW
  Venezuela
C.f.zarumae
  W Ecuador, NE Peru
C.f.fumigatus
  SE Peru, Bolivia
C.f.brachyrhynchus
  NW Argentina
Contopus ochraceus (Ochraceous Pewee)
  Costa Rica, W Panama
Contopus sordidulus (Western Wood Pewee)
C.s.sordidulus
  S Mexico >> Colombia, Ecuador
C.s.siccicola
  NW USA
C.s.veliei
  W Mexico, SW USA >> Panama
C.s.saturatus
  W Nth America >> N Sth America
C.s.peninsulae
  S Baja California, SE Mexico
C.s.griscomi
  Mexico
C.s.amplus
  WC Nth America >> N Sth America
Contopus virens (Eastern Wood Pewee)
  E Canada, E USA >> Central America, N
  Sth America
Contopus cinereus (Tropical Pewee)
C.c.cinereus
  SE Brazil, N Argentina, Paraguay
C.c.pallescens
  S Brazil, N Paraguay
C.c.surinamensis
  the Guianas, S Venezuela, N Brazil
C.c.bogotensis
  Trinidad, N Venezuela, N Colombia
C.c.punensis
  SW Ecuador, Peru
C.c.rhizophorus
  W Costa Rica
C.c.brachytarsus
  SE Mexico to Panama
C.c.aithalodes
  Coiba I (Panama)
Contopus albogularis (White-throated Pewee)
  French Guiana, Surinam
Contopus nigrescens (Blackish Pewee)
C.n.nigrescens
  E Ecuador
C.n.canescens
  NE Peru
**Contopus caribaeus (Greater Antillean
  Pewee)**
C.c.caribaeus
  Cuba
C.c.morenoi
  S Cuba
C.c.nerlyi
  islands off SC Cuba

C.c.tacitus
  Gonave I
C.c.bahamensis
  Bahama Is
C.c.hispaniolensis
  Hispaniola
C.c.pallidus
  Jamaica
Contopus latirostris (Lesser Antillean Pewee)
C.l.latirostris
  St Lucia I
C.l.brunneicapillus
  Domenica, Guadelupe, Martinique Is
C.l.blancoi
  Puerto Rico

EMPIDONAX
**Empidonax flaviventris (Yellow-bellied
  Flycatcher)**
  S Canada, N USA >> S Mexico, C America
Empidonax virescens (Acadian Flycatcher)
  E USA, E Mexico >> C America, Colombia,
  W Ecuador
Empidonax traillii (Traill's Flycatcher)
  Canada, USA >> Mexico, C America, N
  Sth America
Empidonax alnorum (Alder Flycatcher)
  E Canada, E USA >> C America & Sth
  America
**Empidonax griseipectus (Grey-breasted
  Flycatcher)**
  SW Ecuador, NW Peru
**Empidonax albigularis (White-throated
  Flycatcher)**
E.a.timidus
  NW Mexico
E.a.albigularis
  C & S Mexico, Guatemala, Honduras
E.a.australis
  Nicaragua to Panama
Empidonax minimus (Least Flycatcher)
  E Canada, E USA >> S Mexico, C
  America, NW Sth America
**Empidonax hammondii (Hammond's
  Flycatcher)**
  W Canada, W USA >> Mexico, Guatemala
Empidonax oberholseri (Wright's Flycatcher)
  W USA, Mexico
Empidonax wrightii (Grey Flycatcher)
  W USA, California >> Mexico
Empidonax affinis (Pine Flycatcher)
E.a.pulverius
  NW Mexico
E.a.trepidus
  NE Mexico, Guatemala
E.a.affinis
  C Mexico
E.a.bairdi
  C & S Mexico
E.a.vigensis
  E Mexico
Empidonax difficilis (Western Flycatcher)

E.d.difficilis
W Canada, W USA, NW Mexico
E.d.cineritius
Baja California
E.d.insulicola
Channel Is (S California)
E.d.hellmayri
N Mexico
E.d.occidentalis
S & C Mexico
Empidonax flavescens (Yellowish Flycatcher)
E.f.imperturbatus
S Mexico
E.f.salvini
SE Mexico to Nicaragua
E.f.flavescens
Costa Rica, W Panama
Empidonax fulvifrons (Buff-breasted
Flycatcher)
E.f.pygmaeus
SW USA, W Mexico
E.f.fulvifrons
NE Mexico
E.f.rubicundus
C & S Mexico
E.f.fusciceps
SE Mexico, Guatemala
E.f.brodkorbi
S Oaxaca
E.f.inexpectatus
SC Honduras
Empidonax atriceps (Black-capped
Flycatcher)
Costa Rica, W Panama

NESOTRICCUS
Nesotriccus ridgwayi (Cocos I Flycatcher)
Cocos I (Panama)

CNEMOTRICCUS
Cnemotriccus fuscatus (Fuscous Flycatcher)
C.f.cabanisi
Trinidad, Tobago I, Venezuela, Colombia
C.f.fuscatior
SW Venezuela, N Brazil, E Peru
C.f.duidae
SE Venezuela
C.f.fumosus
the Guianas, N Brazil
C.f.bimaculatus
W & C Brazil, E Bolivia, Paraguay, N
Argentina
C.f.beniensis
N Bolivia
C.f.fuscatus
SE Brazil, N Argentina

SAYORNIS
Sayornis phoebe (Eastern Phoebe)
C & SE Canada, E USA, E Mexico
Sayornis nigricans (Black Phoebe)

S.n.nigricans
SW USA, Mexico
S.n.semiatra
W Mexico, W USA
S.n.aquatica
Guatemala, Nicaragua
S.n.amnicola
Costa Rica, W Panama
S.n.angustirostris
C & N Colombia, E Panama, W Venezuela
to Peru
S.n.latirostris
Bolivia, NW Argentina
Sayornis saya (Say's Phoebe)
S.s.pallida
S Mexico
S.s.saya
WC Canada, W USA, N Mexico
S.s.quiescens
NW Baja California

PYROCEPHALUS
Pyrocephalus rubinus (Vermilion Flycatcher)
P.r.mexicanus
SW USA, Mexico
P.r.flammeus
N Mexico
P.r.blatteus
S Mexico to Honduras
P.r.pinicola
Nicaragua
P.r.piurae
W Colombia to NW Peru
P.r.saturatus
N Colomnia to Surinam, N Brazil
P.r.ardens
NC Peru
P.r.cocachacrae
SW Peru, N Chile
P.r.obscurus
W Peru
P.r.rubinus
W & C Amazonia
P.r.major
SE Peru
P.r.nanus
Galapagos Is
P.r.dubius
Chatham I (Galapagos)

OCHTHOECA
Ochthoeca cinnamomeiventris (Slaty-backed
Chat Tyrant)
O.c.nigrita
W Venezuela
O.c.cinnamomeiventris
Colombia, E Ecuador
O.c.angustifasciata
N Peru
O.c.thoracica
Peru, Bolivia

Ochthoeca diadema (Yellow-bellied Chat Tyrant)
O.d.rubellula
NE Colombia, NW Venezuela
O.d.jesupi
N Colombia
O.d.tovarensis
N Venezuela
O.d.diadema
Colombia, W Venezuela
O.d.meridana
NW Venezuela
O.d.gratiosa
NW Peru, W Ecuador, W Colombia
O.d.cajamarcae
Cajamarca (Peru)
Ochthoeca frontalis (Crowned Chat Tyrant)
O.f.albidiadema
E Colombia
O.f.frontalis
W Ecuador, W Colombia
O.f.spodionota
C & SE Peru
Ochthoeca jelskii (Jelski's Chat Tyrant) 106.5
NW Peru, SW Ecuador
Ochthoeca pulchella (Golden-browed Chat Tyrant)
O.p.similis
C Peru
O.p.pulchella
W Bolivia, SE Peru
Ochthoeca rufipectoralis (Rufous-breasted Chat Tyrant)
O.r.poliogastra
N Colombia
O.r.rubicundulus
NE Colombia, NW Venezuela
O.r.obfuscata
SW & W Colombia to Peru
O.r.rufopectus
C Colombia, Ecuador, NW Peru
O.r.centralis
N Peru
O.r.tectricialis
S Peru
O.r.rufipectoralis
Bolivia, SE Peru
Ochthoeca fumicolor (Brown-backed Chat Tyrant)
O.f.superciliosa
W Venezuela
O.f.ferruginea
C Colombia
O.f.fumicolor
E Colombia, W Venezuela
O.f.brunneifrons
Peru, Ecuador, W Colombia
O.f.berlepschi
SE Peru, W Bolivia
Ochthoeca oenanthoides (D'Orbigny's Chat Tyrant)
O.o.polionota
Peru

O.o.oenanthoides
Bolivia, N Chile, NW Argentina
Ochthoeca parvirostris (Patagonian Chat Tyrant)
S Chile, S Argentina
Ochthoeca leucophrys (White-browed Chat Tyrant)
O.l.dissors
N Peru
O.l.interior
C Peru
O.l.urubambae
C & S Peru
O.l.leucometopa
W Peru, NW Chile
O.l.leucophrys
Bolivia
O.l.tucumana
W Argentina
Ochthoeca piurae (Piura Chat Tyrant)
NW Peru
Ochthoeca littoralis (Drab Water Tyrant)
NE Bolivia to S Venezuela, the Guianas

MYIOTHERETES
Myiotheretes striaticollis (Streak-throated Bush Tyrant)
M.s.striaticollis
Venezuela, Colombia, Ecuador, Peru, W Bolivia
M.s.pallidus
NW Argentina, E Peru, Bolivia
Myiotheretes erythropygius (Red-rumped Bush-Tyrant)
M.e.orinomus
N Colombia
M.e.erythropygius
C Colombia to Ecuador, Peru, N Bolivia
Myiotheretes pernix (Santa Marta Bush-Tyrant)
NE Colombia
Myiotheretes fumigatus (Smoky Bush-Tyrant)
M.f olivaceus
N Colombia, W Venezuela
M.f.fumigatus
Colombia, Ecuador, Peru
M.f.lugubris
W Venezuela
M.f.cajamarcae
S Ecuador, N Peru
Myiotheretes fuscorufus (Rufous-bellied Bush-Tyrant)
Bolivia, SE Peru

POLIOXOLMIS 106.10
Polioxolmis rufipennis (Rufous-webbed Bush-Tyrant)
P.r.rufipennis
SE Peru, N Bolivia
P.r.bolivianus
C Bolivia

Xolmis pyrope (Fire-eyed Diucon)
X.p.pyrope
  S Chile, S Argentina
X.p.fortis
  Chiloe I
Xolmis cinerea (Grey Monjita)
X.c.cinerea
  S Brazil, Uruguay, N Argentina
X.c.pepoaza
  E Bolivia, Paraguay, N Argentina
Xolmis coronata (Black-crowned Monjita)
  E Bolivia, Uruguay, Paraguay, N Argentina
Xolmis velata (White-rumped Monjita)
  E Bolivia, S Brazil, Paraguay
Xolmis dominicana (Black & White Monjita)
  S Brazil, Uruguay, Paraguay, E Argentina
Xolmis irupero (White Monjita)
X.i.nivea
  E Brazil
X.i.irupero
  E Bolivia, S Brazil, Uruguay, Paraguay, N Argentina

Neoxolmis rubetra (Rusty-backed Monjita)
  W Argentina
Neoxolmis rufiventris (Chocolate-vented Tyrant)
  Uruguay, Chile, Argentina

Agriornis montana (Black-billed Shrike Tyrant)
A.m.solitaria
  Ecuador, Colombia
A.m.insolens
  Peru
A.m.intermedia
  W Bolivia, N Chile
A.m.montana
  E Bolivia, NW Argentina
A.m.maritima
  NC Chile
A.m.leucura
  C Chile, C Argentina
Agriornis andicola (White-tailed Shrike Tyrant)
A.a.andicola
  Ecuador
A.a.albicauda
  Peru, W Bolivia, N Chile, NW Argentina
Agriornis livida (Great Shrike Tyrant)
A.l.livida
  SC Chile
A.l.fortis
  S Chile, S Argentina
Agriornis microptera (Grey-bellied Shrike Tyrant)
A.m.andecola
  Bolivia, S peru, N Chile, NW Argentina

A.m.microptera
  Uruguay, Argentina
Agriornis murina (Mouse-brown Shrike Tyrant)
  E Bolivia, Paraguay, N Argentina

Muscisaxicola maculirostris (Spot-billed Ground Tyrant)
M.m.niceforoi
  C Colombia
M.m.rufescens
  Ecuador
M.m.maculirostris
  Peru, Bolivia, Chile, W Argentina
Muscisaxicola macloviana (Dark-faced Ground Tyrant)
M.m.mentalis
  Peru, Bolivia, Chile, Argentina
M.m.macloviana
  Falkland I
Muscisaxicola fluviatilis (Little Ground Tyrant)
  E Peru, N Bolivia, W Brazil, N Argentina
Muscisaxicola capistrata (Cinnamon-bellied Ground Tyrant)
  Bolivia, S Peru, Argentina, Chile
Muscisaxicola rufivertex (Rufous-naped Ground Tyrant)
M.r.occipitalis
  Peru, NW Bolivia
M.r.pallidiceps
  SW Bolivia, NW Argentina, N Chile
M.r.rufivertex
  S Chile, SW Argentina
Muscisaxicola albilora (White-browed Ground Tyrant)
  Ecuador, Peru, Bolivia, Chile
Muscisaxicola juninensis (Puna Ground Tyrant)
  S Peru, N Chile, N Argentina
Muscisaxicola alpina (Plain-capped Ground Tyrant)
M.a.columbiana
  WC Colombia
M.a.quesadae
  C Colombia
M.a.alpina
  N Ecuador
M.a.grisea
  Peru, W Bolivia
Muscisaxicola cinerea (Cinereous Ground Tyrant)
M.c.cinerea
  N & C Chile
M.c.argentina
  NW Argentina
Muscisaxicola albifrons (White-fronted Ground Tyrant)
  S Peru, W Bolivia, N Chile
Muscisaxicola flavinucha (Ochre-naped Ground Tyrant)

M.f.flavinucha
Peru, Bolivia, N Argentina, N Chile
M.f.brevirostris
S Argentina, S Chile
Muscisaxicola frontalis (Black-fronted Ground Tyrant)
Bolivia, Peru, Argentina, Chile

LESSONIA
Lessonia rufa (Rufous-backed Negrito)
Chile, Argentina >> Bolivia, S Brazil
Lessonia oreas (Salvin's Negrito)
S Peru, N Chile, W Bolivia, NW Argentina

KNIPOLEGUS
Knipolegus striaticeps (Cinereous Tyrant)
E Bolivia, SC Brazil, N Argentina
Knipolegus hudsoni (Hudson's Black Tyrant)
S Argentina >> Bolivia, S Brazil
Knipolegus poecilocercus (Amazonian Black Tyrant)
S Venezuela to NE Peru, C Brazil
Knipolegus signatus (Jelski's Bush Tyrant)
K.s.signatus
N Peru
K.s.cabanisi
SE Peru, SE Bolivia, N Argentina
Knipolegus cyanirostris (Blue-billed Black Tyrant)
SE Brazil, Uruguay, Paraguay, E Argentina
Knipolegus poecilurus (Rufous-tailed Tyrant)
K.p.poecilurus
Colombia
K.p.peruanus
Peru, Bolivia
K.p.venezuelanus
Trinidad, Venezuela, NW Brazil
K.p.salvini
Guyana
K.p.paraquensis
S Venezuela
Knipolegus orenocensis (Riverside Tyrant)
K.o.orenocensis
Venezuela
K.o.xinguensis
N Brazil
K.o.sclateri
N Peru, C Brazil
Knipolegus aterrimus (White-winged Black Tyrant)
K.a.heterogyna
N Peru
K.a.anthracinus
S Peru, W Bolivia
K.a.aterrimus
E Bolivia, W Argentina
K.a.franciscanus
CE Brazil
Knipolegus lophotes (Crested Black Tyrant)
S Brazil, Uruguay
Knipolegus nigerrimus (Velvety Black Tyrant)
SE Brazil

HYMENOPS
Hymenops perspicillata (Spectacled Tyrant)
H.p.perspicillata
Argentina, Paraguay, Uruguay
H.p.andina
C Chile, NW Argentina, S Bolivia

FLUVICOLA
Fluvicola pica (Pied Water Tyrant)
F.p.pica
Trinidad, Venezuela, the Guianas, Colombia, N Brazil
F.p.albiventer
S Brazil, E Bolivia, Paraguay, N Argentina
Fluvicola nengeta (Masked Water Tyrant)
F.n.atripennis
SW Ecuador, NW Peru
F.n.nengeta
E Brazil
Fluvicola leucocephala (White-headed Marsh Tyrant)
NE Sth America

COLONIA
Colonia colonus (Long-tailed Tyrant)
C.c.leuconotus
S Honduras to Panama, W Colombia, Ecuador
C.c.fuscicapillus
Colombia, E Ecuador, E Peru, W Brazil, Bolivia
C.c.poecilonotus
the Guianas, SE Venezuela
C.c.niveiceps
S Peru, N Bolivia
C.c.colonus
S Brazil, Paraguay, N Argentina

ALECTRURUS
Alectrurus tricolor (Cock-tailed Tyrant)
S Brazil, Paraguay, E Bolivia, N Argentina
Alectrurus risora (Strange-tailed Tyrant)
N Argentina, S Brazil, Paraguay, Uruguay

GUBERNETES
Gubernetes yetapa (Streamer-tailed Tyrant)
S Brazil, Paraguay, E Bolivia, N Argentina

SATRAPA
Satrapa icterophrys (Yellow-browed Tyrant)
Bolivia, Brazil, Paraguay, Uruguay, N Argentina

TUMBEZIA
Tumbezia salvini (Tumbes Tyrant) NW Peru

MUSCIGRALLA
Muscigralla brevicauda (Short-tailed Field Tyrant)
SW Ecuador, W Peru, N Chile

HIRUNDINEA
Hirundinea ferruginea (Cliff Flycatcher)
H.f.ferruginea
Guyana, French Guiana, N Brazil
H.f.sclateri
E Colombia, Peru
H.f.bellicosa
S Brazil, Paraguay, NE Argentina
H.f.pallidior
Bolivia, NW Argentina

MACHETORNIS
Machetornis rixosus (Cattle Tyrant)
M.r.flavigularis
N Colombia, N Venezuela
M.r.obscurodorsalis
SW Venezuela, E Colombia
M.r.rixosus
E Bolivia, S Brazil, Paraguay, Uruguay, N Argentina

MUSCIPIPRA
Muscipipra vetula (Shear-tailed Grey Tyrant)
SE Brazil, Paraguay, Argentina

ATTILA
Attila phoenicurus (Rufous-tailed Attila)
Venezuela to N Argentina
Attila cinnamomeus (Cinnamon Attila)
Sth America
Attila torridus (Ochraceous Attila)
W Ecuador, SW Colombia
Attila citriniventris (Citron-bellied Attila)
Venezuela, Ecuador, N Peru, NW Brazil
Attila bolivianus (Dull-capped Attila)
A.b.nattereri
N Brazil
A.b.bolivianus
SW Brazil, E Bolivia, E Peru
Attila rufus (Grey-hooded Attila)
A.r.hellmayri
E Brazil
A.r.rufus
SE Brazil
Attila spadiceus (Bright-rumped Attila)
A.s.pacificus
NW Mexico
A.s.cozumelae
Cozumel I
A.s.gaumeri
SE Mexico
A.s.flammulatus
SE Mexico to El Salvador
A.s.salvadorensis
El Salvador to NW Nicaragua
A.s.citreopygus
Nicaragua to W Panama
A.s.sclateri
E Panama, NW Colombia
A.s.caniceps
N Colombia
A.s.parvirostris
NE Colombia, NW Venezuela

A.s.parambae
W Ecuador, W Colombia
A.s.spadiceus
the Guianas, Trinidad, N Venezuela, N Brazil NE Peru, N Bolivia
A.s.uropygiatus
SE Brazil

CASIORNIS
Casiornis rufa (Rufous Casiornis)
E Bolivia, C Brazil, Paraguay, N Argentina
Casiornis fusca (Ash-throated Casiornis)
NE Brazil

RHYTIPTERNA
Rhytipterna simplex (Greyish Mourner)
R.s.frederici
N Sth America
R.s.simplex
SE Brazil
Rhytipterna immunda (Pale-bellied Mourner)
Surinam, French Guiana, SE Colombia, N Brazil
Rhytipterna holerythra (Rufous Mourner)
R.h.holerythra
Guatemala to Panama, N Colombia
R.h.rosenbergi
W Colombia, NW Ecuador

LANIOCERA
Laniocera hypopyrra (Cinereous Mourner)
the Guianas, Venezuela, Colombia, Bolivia, N & W Brazil
Laniocera rufescens (Speckled Mourner)
L.r.rufescens
Guatemala to Panama, W Colombia
L.r.tertia
NW Ecuador, SW Colombia
L.r.griseigula
NC Colombia

SIRYSTES
Sirystes sibilator (Sirystes)
S.s.albogriseus
E Panama, NW Colombia
S.s.albocinereus
S Colombia, E Ecuador, E Peru, W Brazil
S.s.subcanescens
NE Brazil, S Surinam
S.s.sibilator
E Brazil, Paraguay, NE Argentina
S.s.atimastus
SW Brazil

MYIARCHUS
Myiarchus semirufus (Rufous Flycatcher)
Peru
Myiarchus yucatanensis (Yucatan Flycatcher)
M.y.yucatanensis
Yucatan peninsula, Guatemala, Belize
M.y.navae
S Mexico

M.y.lanyoni
Cozumel I
Myiarchus barbirostris (Sad Flycatcher)
Jamaica
Myiarchus tuberculifer (Olivaceous Flycatcher)
M.t.olivascens
SW USA, W Mexico
M.t.lawrenceii
E & S Mexico
M.t.querulus
SW Mexico
M.t.platyrhynchus
SE Mexico
M.t.manens
S Yucatan
M.t.connectens
Guatemala to N Nicaragua
M.t.littoralis
Pacific coast from SE Honduras to Costa Rica
M.t.nigricapillus
SE Nicaragua, Costa Rica, W Panama
M.t.brunneiceps
E Panama, W Colombia
M.t.pallidus
N Colombia, W Venezuela
M.t.tuberculifer
Colombia and Surinam to Bolivia and S Brazil
M.t.nigriceps
W Ecuador, W Colombia
M.t.atriceps
C Peru, C Bolivia, NW Argentina
Myiarchus swainsoni (Swainson's Flycatcher)
M.s.swainsoni
E Paraguay, NE Argentina >> E Colombia, N Venezuela
M.s.phaeonotus
S Guyana, S Venezuela
M.s.pelzelni
E Peru, SE Brazil, N Bolivia
M.s.ferocior
SE Bolivia, N Argentina >> SE Colombia
Myiarchus venezuelensis (Venezuelan Flycatcher)
NE Colombia, NE Venezuela, Tobago I
Myiarchus panamensis (Panama Flycatcher)
M.p.actiosus
Costa Rica
M.p.panamensis
Panama, N Colombia, NW Venezuela
Myiarchus ferox (Short-crested Flycatcher)
M.f.brunnescens
Venezuela
M.f.ferox
S Colombia to Peru, N Brazil, the Guianas
M.f.australis
E Bolivia to S Brazil, N Argentina
Myiarchus apicalis (Apical Flycatcher)
SW Colombia
Myiarchus cephalotes (Pale-edged Flycatcher)

M.c.caribbaeus
Venezuela
M.c.cephalotes
Bolivia, Peru, Ecuador, W Colombia
Myiarchus phaeocephalus (Sooty-crowned Flycatcher)
M.p.phaeocephalus
W Ecuador, NW Peru
M.p.interior
NW Peru
Myiarchus cinerascens (Ash-throated Flycatcher)
M.c.cinerascens
W USA, W Mexico >> Guatemala, NW Costa Rica
M.c.pertinax
Baja California
Myiarchus nuttingi (Pale-throated Flycatcher)
M.n.inquietus
W & C Mexico
M.n.nuttingi
mountains of C America
M.n.flavidior
Pacific coast of C America
Myiarchus crinitus (Great Crested Flycatcher)
E Canada, E USA >> Mexico, C America
Myiarchus tyrannulus (Brown Crested Flycatcher)
M.t.magister
SW USA, W Mexico, Tres Marias I
M.t.cooperi
SW USA, W Mexico to Honduras
M.t.cozumelae
Cozumel I
M.t.tyrannulus
Trinidad, Venezuela, the Guianas, N Brazil, N Colombia to N Argentina
M.t.brachyurus
El Salvador to NW Costa Rica
M.t.insularum
Honduras coastal islands
M.t.bahiae
N & E Brazil
Myiarchus magnirostris (Galapagos Flycatcher)
Galapagos Is
Myiarchus nugator (Grenada Flycatcher)
Lesser Antilles Is
Myiarchus validus (Rufous-tailed Flycatcher)
Jamaica
Myiarchus antillarum (Puerto Rican Flycatcher)
Puerto Rico
Myiarchus sagrae (La Sagra's Flycatcher)
M.s.sagrae
Cuba, Grand Cayman I
M.s.lucaysiensis
Bahama Is
Myiarchus stolidus (Stolid Flycatcher)
M.s.dominicensis
Hispaniola
M.s.stolidus
Jamaica

Myiarchus oberi (Wied's Crested Flycatcher)
M.o.oberi
Dominica, Guadeloupe Is
M.o.sanctaeluciae
St Lucia I
M.o.berlepschii
Nevis, St Kitts Is
M.o.sclateri
Martinique I

DELTARHYNCHUS
Deltarhynchus flammulatus (Flammulated Flycatcher)
SW & S Mexico

PITANGUS
Pitangus sulphuratus (Great Kiskadee)
P.s.derbianus
S Texas, C Mexico
P.s.texanus
E Mexico, S Texas
P.s.guatimalensis
S Mexico, Guatemala to Costa Rica
P.s.trinitatis
C Colombia, Trinidad, Venezuela
P.s.caucensis
W Colombia
P.s.rufipennis
N Venezuela, Colombia
P.s.sulphuratus
S Colombia, the Guianas, N Brazil, NE Peru, E Ecuador
P.s.maximiliani
E & C Brazil, E Bolivia
P.s.bolivianus
E Bolivia
P.s.argentinus
Paraguay, Bolivia, N Argentina

PHILOHYDOR   106.9
Philohydor lictor (Lesser Kiskadee)
P.l.panamensis
E Panama, N Colombia
P.l.lictor
Venezuela, the Guianas, N Brazil, E Peru, E Ecuador, E Colombia

MEGARHYNCHUS
Megarhynchus pitangua (Boat-billed Flycatcher)
M.p.mexicanus
SE Mexico to NW Colombia
M.p.tardiusculus
NW Mexico
M.p.caniceps
W Mexico
M.p.deserticola
C Guatemala
M.p.pitangua
Trinidad, N & C Sth America
M.p.chrysogaster
W Ecuador, NW Peru

MYIOZETETES
Myiozetetes cayanensis (Rusty-margined Flycatcher)
M.c.rufipennis
N Venezuela, E Colombia, E Panama
M.c.hellmayri
Colombia, E Ecuador, NW Venezuela
M.c.cayanensis
the Guianas, N Brazil, E Bolivia, S Venezuela
M.c.erythropterus
SE Brazil
Myiozetetes similis (Social Flycatcher)   106.6
M.s.primulus
N Mexico
M.s.hesperis
W Mexico
M.s.texensis
S Mexico to N Costa Rica
M.s.columbianus
SW Costa Rica, Panama, N Colombia, N Venezuela
M.s.connivens
Colombia, SE Peru, Venezuela, N Brazil
M.s.grandis
W Ecuador, NW Peru
M.s.pallidiventris
E Brazil, E Paraguay, NE Argentina
M.s.similis
E Brazil, Paraguay, NE Argentina
Myiozetetes granadensis (Grey-capped Flycatcher)
M.g.granadensis
E Honduras to CE Panama
M.g.occidentalis
E Panama to NW Peru
M.g.obscurior
S Venezuela, S Colombia, E Ecuador, W Brazil SE Peru
Myiozetetes luteiventris (Dusky-chested Flycatcher)
M.l.luteiventris
E Colombia, Ecuador, Peru, Brazil
M.l.septentrionalis
Surinam

PHELPSIA   106.9
Phelpsia inornata (White-bearded Flycatcher)
N Venezuela

CONOPIAS
Conopias parva (White-ringed Flycatcher)
C.p.distincta
E Costa Rica
C.p.albovittata
E Panama, W Colombia, NW Ecuador
C.p.parva
Guyana, French Guiana, N Brazil
Conopias trivirgata (Three-striped Flycatcher)
C.t.berlepschi
N Brazil, SW Venezuela
C.t.trivirgata
SE Brazil, Paraguay, NE Argentina

Conopias cinchoneti (Lemon-browed Flycatcher)
  C.c.icterophrys
    C Colombia, W Venezuela
  C.c.cinchoneti
    E Ecuador, Peru

MYIODYNASTES
Myiodynastes hemichrysus (Golden-bellied Flycatcher) Costa Rica, W Panama
Myiodynastes maculatus (Streaked Flycatcher)
  M.m.insolens
    SE Mexico to Honduras >> N Sth America
  M.m.nobilis
    NE Colombia
  M.m.difficilis
    W Costa Rica to Venezuela
  M.m.chapmani
    SW Colombia to W Peru
  M.m.maculatus
    N Peru, Venezuela, the Guianas, N Brazil
  M.m.tobagensis
    Guyana, Trinidad, Tobago I
  M.m.solitarius
    S Peru to Uruguay, N Argentina
Myiodynastes luteiventris (Sulphur-bellied Flycatcher)
  S USA to Costa Rica >> N Sth America
Myiodynastes bairdi (Baird's Flycatcher)
  SW Ecuador, W Peru
Myiodynastes chrysocephalus (Golden-crowned Flycatcher)
  M.c.cinerascens
    NW Venezueal, N Colombia
  M.c.minor
    Colombia, Ecuador
  M.c.chrysocephalus
    Peru, Bolivia

LEGATUS
Legatus leucophaius (Piratic Flycatcher)
  L.l.variegatus
    SE Mexico, Guatemala, Honduras
  L.l.leucophaius
    Nicaragua to Panama, Trinidad, N & C Sth America

EMPIDONOMUS
Empidonomus varius (Variegated Flycatcher)
  E.v.varius
    E Bolivia, SE Brazil, Paraguay, N Argentina
  E.v.rufinus
    N & E Brazil, E Peru, E Venezuela, Guyana French Guiana, Colombia

GRISEOTYRANNUS
Griseotyrannus aurantioatrocristatus (Crowned Slaty Flycatcher)   106.9
  G.a.aurantioatrocristatus
    E Peru, E Bolivia, S Brazil, Uruguay, Paraguay, N Argentina

  G.a.pallidiventris
    NE Brazil

TYRANNOPSIS
Tyrannopsis sulphurea (Sulphury Flycatcher)
  Trinidad, the Guianas, Venezuela, Brazil, Peru, Ecuador

TYRANNUS
Tyrannus niveigularis (Snowy-throated Kingbird)
  SW Colombia, Ecuador, NW Peru
Tyrannus albogularis (White-throated Kingbird)
  Brazil, Venezuela to E Peru, N Bolivia
Tyrannus melancholicus (Tropical Kingbird)
  T.m.satrapa
    S USA to N Colombia, N Venezuela
  T.m.despotes
    NE Brazil
  T.m.melancholicus
    tropical Sth America
Tyrannus couchii (Couch's Kingbird)
  S USA, E Mexico
Tyrannus vociferans (Cassin's Kingbird)
  T.v.vociferans
    SW USA, W Mexico, Guatemala
  T.v.xenopterus
    SW Mexico
Tyrannus crassirostris (Thick-billed Kingbird)
  T.c.pompalis
    SE Arizona, N Mexico
  T.c.crassirostris
    S Mexico, W Guatemala
Tyrannus verticalis (Western Kingbird)
  SW Canada, W USA, W Mexico, Guatemala
Tyrannus forficata (Scissor-tailed Flycatcher)
  SC USA >> C America
Tyrannus savana (Fork-tailed Flycatcher)
  T.s.monachus
    S Mexico to NC Brazil
  T.s.sanctaemartae
    N Colombia, NW Venezuela
  T.s.circumdatus
    C Brazil
  T.s.savana
    C & S Sth America >> N Sth America, W Indies
Tyrannus tyrannus (Eastern Kingbird)
  S Canada, C & E USA, C America, N Sth America
Tyrannus dominicensis (Grey Kingbird)
  T.d.dominicensis
    SE USA to Colombia >> NC Sth America
  T.d.vorax
    Lesser Antilles Is, Trinidad
Tyrannus caudifasciatus (Loggerhead Kingbird)
  T.c.bahamensis
    Bahama Is
  T.c.caudifasciatus
    Cuba

T.c.flavescens
Isle of Pines, Gtr Antilles Is
T.c.caymanensis
Cayman Is
T.c.jamaicensis
Jamaica
T.c.taylori
Puerto Rico
T.c.gabbii
Haiti
Tyrannus cubensis (Giant Kingbird)
Cuba, Gtr Antilles Is

XENOPSARIS
**Xenopsaris albinucha (White-naped
Xenopsaris)**
X.a.albinucha
Brazil, Paraguay, Argentina
X.a.minor
Venezuela, the Guianas

PACHYRAMPHUS
Pachyramphus viridis (Green-backed Becard)
P.v.griseigularis
Guyana
P.v.xanthogenys
E Ecuador
P.v.peruanus
C Peru
P.v.viridis
E & S Brazil, Bolivia, Paraguay, N Argentina
Pachyramphus versicolor (Barred Becard)
P.v.costaricensis
Costa Rica
P.v.versicolor
W Venezuela, Colombia, Ecuador, Peru
P.v.meridionalis
Peru, N Bolivia
Pachyramphus spodiurus (Slaty Becard)
W Ecuador, NW Peru
Pachyramphus rufus (Cinereous Becard)
P.r.rufus
Panama, N Sth America
P.r.juruanus
E Peru, W Brazil
**Pachyramphus castaneus (Chestnut-crowned
Becard)**
P.c.saturatus
SE Colombia, E Ecuador, N Peru, NW
Brazil
P.c.intermedius
N Venezuela
P.c.parui
S Venezuela
P.c.amazonus
NE Brazil
P.c.castaneus
E Brazil, Paraguay, Argentina
**Pachyramphus cinnamomeus (Cinnamon
Becard)**
P.c.cinnamomeus
E Panama, Colombia, Ecuador

P.c.fulvidior
SE Mexico to W Panama
P.c.magdalenae
N & E Colombia, W Venezuela
P.c.badius
W Venezuela
**Pachyramphus polychopterus (White-winged
Becard)**
P.p.similis
Guatemala to Panama
P.p.cinereiventris
N Colombia
P.p.dorsalis
W Colombia, NW Ecuador
P.p.tenebrosus
S Colombia
P.p.tristis
N & NE Sth America
P.p.nigriventris
W Amazonia
P.p.polychopterus
E Brazil
P.p.spixii
S Brazil, E Bolivia to N Argentina
**Pachyramphus marginatus (Black-capped
Becard)**
P.m.nanus
N Sth America
P.m.marginatus
E Brazil
**Pachyramphus albogriseus (Black & White
Becard)**
P.a.ornatus
W Nicaragua to W Panama
P.a.coronatus
N Colombia, NW Venezuela
P.a.albogriseus
E Colombia, N Venezuela
P.a.guayaquilensis
E Ecuador
P.a.salvini
N Peru, E Ecuador
Pachyramphus major (Grey-collared Becard)
106.6

P.m.uropygialis
W Mexico
P.m.major
E Mexico
P.m.matudae
S Mexico, N Guatemala
P.m.itzensis
SE Mexico, Belize
P.m.australis
Guatemala to E Nicaragua
**Pachyramphus surinamus (Glossy-backed
Becard)**
Surinam, E Brazil, French Guiana
**Pachyramphus aglaiae (Rose-throated
Becard)**
P.a.albiventris
S Arizona, W Mexico
P.a.gravis
NE Mexico

P.a.yucatanensis
E Mexico
P.a.insularis
Tres Marias Is
P.a.aglaiae
NE Mexico
P.a.sumichrasti
SE Mexico to El Salvador
P.a.hypophaeus
Honduras to NE Costa Rica
P.a.latirostris
W Nicaragua, W Costa Rica
Pachyramphus homochrous (One-coloured Becard)
P.h.homochrous
E Panama to N Peru
P.h.quimarinus
NW Colombia
P.h.canescens
N Colombia, NW Venezuela
Pachyramphus minor (Pink-throated Becard)
N Sth America
Pachyramphus validus (Plain Becard)
P.v.audax
S Peru, C Bolivia, NW Argentina
P.v.validus
E Bolivia to C Brazil, N Argentina
Pachyramphus niger (Jamaican Becard)
Jamaica

TITYRA
Tityra cayana (Black-tailed Tityra)
T.c.cayana
N Sth America, Trinidad
T.c.braziliensis
E & S Brazil, Paraguay, NE Argentina
Tityra semifasciata (Masked Tityra)
T.s.hannumi
NW Mexico
T.s.griseiceps
W Mexico
T.s.personata
E Mexico to El Salvador
T.s.costaricensis
S Honduras to W Panama
T.s.columbiana
E Panama, Colombia, W Venezuela
T.s.nigriceps
SW Colombia, W Ecuador
T.s.semifasciata
N & NW Amazonia
T.s.fortis
C & SE Peru, N & E Bolivia, C Brazil
Tityra inquisitor (Black-crowned Tityra)
T.i.fraserii
SE Mexico to W Panama
T.i.albitorques
E Panama to Peru, N Brazil
T.i.buckleyi
SE Colombia, E Ecuador
T.i.erythrogenys
E Colombia, Venezuela, the Guianas

T.i.pelzelni
E Bolivia, W Brazil
T.i.inquisitor
E Brazil, Paraguay, N Argentina

107. PIPRIDAE (MANAKINS)

SCHIFFORNIS 107.2
Schiffornis major (Greater Manakin)
S.m.major
N Brazil, E Peru
S.m.duidae
SE Venezuela
Schiffornis virescens (Greenish Manakin)
SE Brazil to NE Argentina
Schiffornis turdinus (Thrush-like Manakin)
S.t.veraepacis
SE Mexico to W Panama
S.t.dumicola
W & C Panama
S.t.panamensis
E Panama, NW Colombia
S.t.acrolophites
E Panama, NW Colombia
S.t.rosenbergi
W Ecuador, W Colombia
S.t.stenorhynchus
N Venezuela, E Colombia
S.t.aeneus
E Ecuador, N Peru
S.t.amazonus
S Venezuela to E Peru, W Brazil
S.t.olivaceus
E Venezuela, Guyana
S.t.wallacii
N Brazil, French Guiana, Surinam
S.t.steinbachi
N Bolivia, SE Peru
S.t.intermedius
E Brazil
S.t.turdinus
E Brazil

SAPAYOA
Sapayoa aenigma (Broad-billed Manakin)
E Panama to NW Ecuador

PIPRITES
Piprites griseiceps (Grey-hooded Manakin)
Costa Rica, Nicaragua
Piprites chloris (Wing-barred Manakin)
P.c.antioquiae
NC Colombia
P.c.perijanus
N Colombia, W Venezuela
P.c.tschudii
S Colombia to C Peru, NW Brazil
P.c.chlorion
the Guianas, N Brazil
P.c.grisescens
Para (Brazil)
P.c.bolivianus
N Bolivia, C Brazil

P.c.chloris
SE Brazil, Paraguay, NE Argentina
Piprites pileatus (Black-capped Manakin)
SE Brazil

NEOPIPO
Neopipo cinnamomea (Cinnamon Manakin)
N.c.helenae
Guyana, French Guiana, N Brazil
N.c.cinnamomea
E Ecuador, E Peru, W Brazil

CHLOROPIPO
**Chloropipo flavicapilla (Yellow-headed
Manakin)**
Colombia
Chloropipo holochlora (Green Manakin)
C.h.suffusa
E Panama
C.h.litae
E Panama to NW Ecuador
C.h.holochlora
E Colombia, E Ecuador, N Peru
C.h.viridior
SE Peru
Chloropipo uniformis (Olive Manakin)
C.u.duidae
S Venezuela
C.u.uniformis
Guyana, N Brazil
Chloropipo unicolor (Jet Manakin)
Peru

XENOPIPO
Xenopipo atronitens (Black Manakin)
N & NW Amazonia

ANTILOPHIA
Antilophia galeata (Helmeted Manakin)
Brazil, Paraguay

TYRANNEUTES
**Tyranneutes stolzmanni (Dwarf Tyrant
Manakin)**
N & W Amazonia
Tyranneutes virescens (Tiny Tyrant Manakin)
Venezuela to Surinam, N Brazil

NEOPELMA
**Neopelma pallescens (Pale-bellied Tyrant
Manakin)**
WC Brazil
**Neopelma chrysocephalum (Saffron-crested
Tyrant Manakin)**
the Guianas, N Brazil, S Venezuela, E
Colombia
Neopelma aurifrons (Wied's Tyrant Manakin)
N.a.aurifrons
SE Brazil
N.a.chrysolophum
E Brazil

**Neopelma sulphureiventer (Sulphur-bellied
Tyrant Manakin)**
E Peru, N Bolivia, W Brazil

HETEROCERCUS
**Heterocercus aurantiivertex (Orange-crowned
Manakin)**
E Ecuador, NE Peru
**Heterocercus flavivertex (Yellow-crowned
Manakin)**
E Colombia, Venezuela, N Brazil
**Heterocercus linteatus (Flame-crowned
Manakin)**
NE Peru, WC Brazil

MACHAEROPTERUS
Machaeropterus regulus (Striped Manakin)
M.r.antioquiae
W & C Colombia
M.r.striolatus
E Colombia, E Ecuador, NE Peru
M.r.obscurostriatus
W Venezuela
M.r.zulianus
W Venezuela
M.r.aureopectus
SE Venezuela
M.r.regulus
SE Brazil
**Machaeropterus pyrocephalus (Fiery-capped
Manakin)**
M.p.pallidiceps
E Venezuela
M.p.pyrocephalus
N Brazil, E Peru
**Machaeropterus deliciosus (Club-winged
Manakin)**
W Colombia, NW Ecuador

MANACUS
Manacus manacus (White-bearded Manakin)
M.m.candei
SE Mexico to Costa Rica
M.m.vitellinus
E Panama
M.m.milleri
N Colombia
M.m.viridiventris
W Colombia
M.m.cerritus
NW Panama
M.m.aurantiacus
SW Costa Rica, W Panama
M.m.trinitatis
Trinidad, NE Venezuela
M.m.abditivus
N Colombia
M.m.flaveolus
E Colombia
M.m.bangsi
SW Colombia
M.m.interior
S Colombia, E Ecuador, N Peru, NW Brazil

M.m.umbrosus
S Venezuela
M.m.manacus
the Guianas, N Brazil
M.m.leucochlamys
W Ecuador
M.m.maximus
SW Ecuador
M.m.expectatus
W Brazil, NE Peru
M.m.longibarbatus
C Brazil
M.m.purissimus
E Brazil
M.m.gutturosus
SE Brazil, E Paraguay, NE Argentina
M.m.purus
N Brazil
M.m.subpurus
C Brazil

CORAPIPO
Corapipo leucorrhoa (White-ruffed Manakin)
C.l.altera
E Nicaragua, NW Costa Rica, Panama
C.l.heteroleuca
SW Costa Rica, W Panama
C.l.leucorrhoa
Colombia
Corapipo gutturalis (White-throated Manakin)
Guyana, Brazil, Venezuela

ILICURA
Ilicura militaris (Pin-tailed Manakin)
SE Brazil

MASIUS
Masius chrysopterus (Golden-winged Manakin)
M.c.bellus
W Colombia
M.c.pax
SE Colombia
M.c.coronulatus
SW Colombia, W Ecuador
M.c.chrysopterus
E Colombia, NW Venezuela, E Ecuador
M.c.peruvianus
N Peru

CHIROXIPHIA
Chiroxiphia linearis (Long-tailed Manakin)
C.l.linearis
S Mexico, Guatemala
C.l.fastuosa
El Salvador, W Nicaragua, NW Costa Rica
Chiroxiphia lanceolata (Lance-tailed Manakin)
Panama to NW Venezuela
Chiroxiphia pareola (Blue-backed Manakin)

C.p.atlantica
Tobago I
C.p.pareola
the Guianas, N & E Brazil
C.p.regina
N Brazil
C.p.napensis
SE Colombia, E Ecuador, NE Peru
C.p.boliviana
SE Peru, Bolivia
Chiroxiphia caudata (Swallow-tailed Manakin)
Brazil, Paraguay, N Argentina

PIPRA
Pipra pipra (White-crowned Manakin)
P.p.anthracina
Costa Rica, W Panama
P.p.bolivari
NW Colombia
P.p.coracina
E Colombia, E Ecuador, NW Peru
P.p.discolor
NE Peru
P.p.minima
W Colombia
P.p.unica
NC Colombia
P.p.pipra
the Guianas, E Venezuela, N Brazil
P.p.occulta
NC Peru
P.p.pygmaea
NC Peru
P.p.microlopha
E Peru, W Brazil
P.p.comata
Peru
P.p.separabilis
EC Brazil
P.p.cephaleucos
E Brazil
Pipra coronata (Blue-crowned Manakin)
P.c.velutina
SW Costa Rica, W Panama
P.c.minuscula
E Panama to NW Ecuador
P.c.caquetae
C Colombia
P.c.carbonata
SE Colombia, NE Ecuador, NW Brazil
P.c.coronata
NW Brazil
P.c.caelestipileata
W Brazil, SE Peru
P.c.exquisita
C Peru
P.c.regalis
NC Bolivia
Pipra isidorei (Blue-rumped Manakin)
P.i.isidorei
E Colombia, E Ecuador
P.i.leucopygia
N Peru

Pipra caeruleocapilla (Caerulean-capped Manakin)
C & SE Peru

Pipra nattereri (Snow-capped Manakin)
Brazil

Pipra vilasboasi (Golden-crowned Manakin)
Brazil

Pipra iris (Opal-crowned Manakin)
P.i.iris
EC Brazil
P.i.eucephala
C Brazil

Pipra serena (White-fronted Manakin)
P.s.suavissima
Guyana, SE Venezuela
P.s.serena
French Guiana, N Brazil

Pipra aureola (Crimson-hooded Manakin)
P.a.aureola
the Guianas, NE Venezuela, NE Brazil
P.a.aurantiicollis
N Brazil
P.a.flavicollis
N Brazil
P.a.borbae
C Brazil

Pipra fasciicauda (Band-tailed Manakin)
P.f.calamae
C & W Brazil
P.f.saturata
N Peru
P.f.purusiana
W Brazil, E Peru
P.f.fasciicauda
E Bolivia, SE Peru
P.f.scarlatina
Paraguay, S Brazil

Pipra filicauda (Wire-tailed Manakin)
P.f.subpallida
E Colombia, NW Venezuela
P.f.filicauda
NE Peru to SC Venezuela, WC Brazil

Pipra mentalis (Red-capped Manakin)
P.m.mentalis
SE Mexico to E Costa Rica
P.m.ignifera
W Costa Rica, W Panama
P.m.minor
E Panama to W Ecuador

Pipra erythrocephala (Golden-headed Manakin)
P.e.erythrocephala
Panama, N Sth America
P.e.berlepschi
W Amazonia
P.e.flammiceps
E Colombia

Pipra rubrocapilla (Red-headed Manakin)
NE Peru, W Brazil

Pipra chloromeros (Round-tailed Manakin)
Peru, Bolivia

Pipra cornuta (Scarlet-horned Manakin)
Guyana, Venezuela, Brazil

108. COTINGIDAE (COTINGAS)

PHOENICERCUS
Phoenicercus carnifex (Guianian Red Cotinga)
the Guianas, E Venezuela Brazil
Phoenicercus nigricollis (Black-necked Red Cotinga)
NW Sth America

LANIISOMA
Laniisoma elegans (Shrike-like Cotinga)
L.e.venezuelensis
NE Colombia, NW Venezuela
L.e.buckleyi
E Peru, E Ecuador
L.e.elegans
SE Brazil
L.e.cadwaladeri
Bolivia

PHIBALURA
Phibalura flavirostris (Swallow-tailed Cotinga)
P.f.flavirostris
Paraguay, SE Brazil, NE Bolivia, NE Argentina
P.f.boliviana
NE Bolivia

TIJUCA
Tijuca atra (Black & Gold Cotinga)
SE Brazil
Tijuca condita (Grey-winged Cotinga)
SE Brazil

CARPORNIS
Carpornis cucullatus (Hooded Berryeater)
SE Brazil
Carpornis melanocephalus (Black-headed Berryeater)
SE Brazil

AMPELION
Ampelion rubrocristata (Red-crested Cotinga)
Venezuela, Colombia to NW Bolivia
Ampelion rufaxilla (Chestnut-crested Cotinga)
A.r.antioquiae
W Colombia
A.r.rufaxilla
Peru
Ampelion sclateri (Bay-vented Cotinga)
C Peru
Ampelion stresemanni (White-cheeked Cotinga)
W Peru

PIPREOLA
Pipreola riefferii (Green & Black Fruiteater)
P.r.occidentalis
W Colombia, W Ecador
P.r.riefferii
E Colombia, E Ecuador

P.r.melanolaema
NW Venezuela
P.r.confusa
Peru, N Bolivia
P.r.chachapoyas
N Peru
P.r.tallmanorum
C Peru
Pipreola intermedia (Band-tailed Fruiteater)
P.i.intermedia
N & C Peru
P.i.signata
SE Peru, Bolivia
Pipreola arcuata (Barred Fruiteater)
P.a.arcuata
Colombia, W Venezuela
P.a.viridicauda
N Bolivia, C Peru
Pipreola aureopectus (Golden-breasted Fruiteater)
P.a.decora
N Colombia
P.a.festiva
N Venezuela
P.a.aureopectus
E Colombia, W Venezuela
P.a.jucunda
W Colombia, W Ecuador
P.a.lubomirskii
S Colombia, E Ecuador, E Peru
P.a.pulchra
E Peru
Pipreola frontalis (Scarlet-breasted Fruiteater)
P.f.squamipectus
SE Ecuador, N Peru
P.f.frontalis
S Peru, Bolivia
Pipreola chlorolepidota (Fiery-throated Fruiteater)
E Ecuador, E Peru
Pipreola formosa (Handsome Fruiteater)
P.f.formosa
N Venezuela
P.f.rubidior
NE Venezuela
P.f.pariae
NE Venezuela
Pipreola whitelyi (Red-banded Fruiteater)
P.w.kathleenae
SE Venezuela
P.w.whitelyi
Guyana

AMPELIOIDES
Ampelioides tschudii (Scaled Fruiteater)
NW Venezuela to N Peru

IODOPLEURA
Iodopleura pipra (Buff-throated Purpletuft)
I.p.leucopygia
Guyana

I.p.pipra
E Brazil
Iodopleura fusca (Dusky Purpletuft)
the Guianas, Venezuela
Iodopleura isabellae (White-browed Purpletuft)
I.i.isabellae
Colombia, E Ecuador, E Peru, N Brazil
I.i.paraensis
NE Brazil

CALYPTURA
Calyptura cristata (Kinglet Calyptura)
S Brazil

LIPAUGUS
Lipaugus subalaris (Grey-tailed Piha)
E Ecuador, E Peru
Lipaugus cryptolophus (Olivaceus Piha)
L.c.mindoensis
W Ecuador, SW Colombia
L.c.cryptolophus
E Colombia, E Ecuador, E Peru
Lipaugus fuscocinereus (Dusky Piha)
Colombia, E Ecuador
Lipaugus vociferans (Screaming Piha)
the Guianas, N & W Amazonia
Lipaugus unirufus (Rufous Piha)
L.u.unirufus
SE Mexico to N Colombia
L.u.castaneotinctus
SW Colombia to NW Ecuador
Lipaugus lanioides (Cinnamon-vented Piha)
SE Brazil
Lipaugus streptophorus (Rose-coloured Piha)
NW Guyana, Venezuela, N Brazil

CHIROCYLLA
Chirocylla uropygialis (Scimitar-winged Piha)
N Bolivia

PORPHYROLAEMA
Porphyrolaema porphyrolaema (Purple-throated Cotinga)
W Amazonia

COTINGA
Cotinga amabilis (Lovely Cotinga)
SE Mexico to Costa Rica
Cotinga ridgwayi (Ridgway's Cotinga)
SW Costa Rica, Panama
Cotinga nattererii (Blue Cotinga)
NW Venezuela to N & E Peru
Cotinga maynana (Plum-throated Cotinga)
W Amazonia
Cotinga cotinga (Purple-breasted Cotinga)
E Colombia to the Guianas, N Brazil
Cotinga maculata (Banded Cotinga)
SE Brazil
Cotinga cayana (Spangled Cotinga)
N & W Sth America

XIPHOLENA
Xipholena punicea (Pompadour Cotinga)
N & NW Amazonia
Xipholena lamellipennis (White-tailed Cotinga)
E Brazil
Xipholena atropurpurea (White-winged Cotinga)
SE Brazil

CARPODECTES
Carpodectes hopkei (Black-tipped Cotinga)
E Panama to NW Ecuador
Carpodectes nitidus (Snowy Cotinga)
E Honduras to E Panama
Carpodectes antoniae (Yellow-billed Cotinga)
W Costa Rica, W Panama

CONIOPTILON
Conioptilon mcilhennyi (Black-faced Cotinga)
SE Peru

GYMNODERUS
Gymnoderus foetidus (Bare-necked Fruitcrow)
N Sth America

HAEMATODERUS
Haematoderus militaris (Crimson Fruitcrow)
the Guianas, Brazil

QUERULA
Querula purpurata (Purple-throated Fruitcrow)
Costa Rica, Panama, N Sth America

PYRODERUS
Pyroderus scutatus (Red-ruffed Fruitcrow)
P.s.occidentalis
W Colombia
P.s.granadensis
E Colombia, W Venezuela
P.s.orenocensis
N Venezuela, Guyana
P.s.masoni
E Peru
P.s.scutatus
SE Brazil, Paraguay, N Argentina

CEPHALOPTERUS
Cephalopterus glabricollis (Bare-necked Umbrellabird)
Costa Rica, W Panama
Cephalopterus ornatus (Amazonian Umbrellabird)
N & W Sth America
Cephalopterus penduliger (Long-wattled Umbrellabird)
Colombia, Ecuador

PERISSOCEPHALUS
Perissocephalus tricolor (Capuchin Bird)
the Guianas, Venezuela, N Brazil

PROCNIAS
Procnias tricarunculata (Three-wattled Bellbird)
Nigaragua to Panama
Procnias alba (White Bellbird)
Venezuela to Surinam, N Brazil
P.a.wallacei
Belem (Brazil)
Procnias averano (Bearded Bellbird)
P.a.carnobarba
N Venezuela, Trinidad, W Guyana
P.a.averano
NE Brazil
Procnias nudicollis (Bare-throated Bellbird)
S Brazil, Paraguay, NE Argentina

RUPICOLA
Rupicola rupicola (Guianan Cock of the Rock)
E Colombia to the Guianas, N Brazil
Rupicola peruviana (Andean Cock of the Rock)
R.p.sanguinolenta
W Colombia, W Ecuador
R.p.aequatorialis
W Venezuela, Colombia, E Ecuador, N Peru
R.p.peruviana
C Peru
R.p.saturata
SE Peru, N Bolivia

109. OXYRUNCIDAE (SHARPBILL)

OXYRUNCUS
Oxyruncus cristatus (Sharpbill)
O.c.frater
Costa Rica, W Panama
O.c.brooksi
E Panama
O.c.hypoglaucus
Guyana, SE Venezuela
O.c.tocantinsi
C Brazil
O.c.cristatus
SE Brazil, Paraguay

110. PHYTOTOMIDAE (PLANTCUTTERS)

PHYTOTOMA
Phytotoma rara (Chilean Plantcutter)
S Chile, S Argentina, Falkland Is
Phytotoma rutila (Red-breasted Plantcutter)
P.r.angustirostris
NW Argentina, Bolivia
P.r.rutila
N Argentina, Uruguay, Paraguay
Phytotoma raimondii (Peruvian Plantcutter)
NW Peru

111. PITTIDAE (PITTAS)

PITTA
Pitta phayrei (Phayre's Pitta)   111.1
  Burma, Thailand, N Indochina
Pitta nipalensis (Blue-naped Pitta)
  P.n.nipalensis
    Himalayas to Burma, S China
  P.n.hendeei
    N Vietnam
Pitta soror (Blue-backed Pitta)
  P.s.soror
    C Indochina
  P.s.tonkinensis
    C China to N Vietnam
  P.s.douglasi
    Hainan I
  P.s.petersi
    C Vietnam
Pitta oatesi (Fulvous Pitta)
  P.o.oatesi
    Burma, Thailand, NW Laos
  P.o.castaneiceps
    N Laos, N Vietnam
  P.o.bolovenensis
    S Laos
  P.o.deborah
    Malaysia
Pitta schneideri (Schneider's Pitta)
  N Sumatra
Pitta caerulea (Giant Pitta)
  P.c.caerulea
    Malaysia, Sumatra, S Thailand
  P.c.hosei
    Borneo
Pitta kochi (Koch's Pitta)
  N Luzon I
Pitta erythrogaster (Red-breasted Pitta)
  P.e.erythrogaster
    Philippine Is
  P.e.thompsoni
    Culion I
  P.e.propinqua
    Balabac I, Palawan I
  P.e.inspeculata
    Taulaut Is
  P.e.caeruleitorques
    Gr Sanghir I
  P.e.palliceps
    Sanghir Is
  P.e.cyanonota
    Ternate I
  P.e.rubrinucha
    Buru I
  P.e.celebensis
    N Sulawesi
  P.e.dohertyi
    Sula Is
  P.e.piroensis
    Seram
  P.e.rufiventris
    N Moluccas

  P.e.obiensis
    Obi I
  P.e.bernsteini
    Gebe I
  P.e.novaehibernicae
    New Ireland
  P.e.gazellae
    New Britain
  P.e.extima
    New Hanover
  P.e.splendida
    Tabar I
  P.e.macklotii
    W & S New Guinea, N Queensland
  P.e.habenichti
    N New Guinea
  P.e.oblita
    SE New Guinea
  P.e.loriae
    SE New Guinea
  P.e.aruensis
    Aru Is
  P.e.finschii
    D'Entrecasteaux Arch
  P.e.meeki
    Rossel I
  P.e.kuehni
    Kai Is
Pitta arcuata (Blue-banded Pitta)
  Borneo
Pitta granatina (Garnet Pitta)
  P.g.coccinea
    Malaysia, E Sumatra
  P.g.granatina
    S Borneo
Pitta venusta (Black-crowned Garnet Pitta)
  P.v.venusta
    W Sumatra
  P.v.ussheri
    N Borneo
Pitta cyanea (Blue Pitta)
  P.c.cyanea
    Himalayas, Burma, N Thailand
  P.c.aurantiaca
    SE Thailand
  P.c.willoughbyi
    S Vietnam, S Laos
Pitta elliotii (Elliot's Pitta)
  S Indochina
Pitta guajana (Blue-tailed Pitta)
  P.g.irena
    Malaysia, Sumatra
  P.g.ripleyi
    S Thailand
  P.g.bangkae
    Bangka I
  P.g.affinis
    W Java
  P.g.guajana
    E Java, Bali
  P.g.schwaneri
    Borneo

Pitta gurneyi (Gurney's Pitta)
S Burma, S Thailand
Pitta baudi (Blue-headed Pitta)
Borneo
Pitta sordida (Hooded Pitta)
P.s.cucullata
Himalayas to Malaysia, Indochina
P.s.mulleri
Sumatra, Java, Borneo, S Thailand,
Malaysia
P.s.abbotti
Nicobar Is
P.s.bangkana
Bangka, Billiton Is
P.s.novaeguineae
New Guinea
P.s.hebetior
Dampier I
P.s.sanghirana
Sanghir Is
P.s.forsteni
Sulawesi
P.s.goodfellowi
Aru Is
P.s.mefoorana
Numfor I
P.s.rosenbergii
Biak I
P.s.palawanensis
Balabac, Palawan Is
P.s.sordida
Philippine Is
Pitta brachyura (Blue-winged Pitta)
Himalayas, C India >> S India, Sri Lanka
Pitta nympha (Fairy Pitta)
NE Asia, E China >> Indochina, Borneo
Pitta angolensis (African Pitta)
P.a.pulih
Sierra Leone to S Cameroun
P.a.angolensis
Cameroun to N Angola
P.a.longipennis
Uganda, E Zaire to Transvaal
Pitta reichenowi (Green-breasted Pitta)
Cameroun to Uganda
Pitta superba (Superb Pitta)
Admiralty Is
Pitta maxima (Great Pitta)
P.m.maxima
Batjan, Halmahera Is
P.m.morotaiensis
Morotai I
Pitta steerei (Steere's Pitta)
P.s.steerei
Mindanao I
P.s.coelestis
Bohol, Leyte, Samar Is
Pitta moluccensis (Moluccan Pitta)
P.m.moluccensis
S China, Burma >> Borneo, Moluccas
P.m.megarhyncha
coast of Burma, Thailand, Malaysia,
Sumatra

Pitta versicolor (Noisy Pitta)
P.v.intermedia
N Queensland
P.v.simillima
S New Guinea, N Queensland
P.v.versicolor
S Queensland, New South Wales
Pitta elegans (Elegant Pitta)   111.2
P.e.virginalis
Djampea, Kalaotua, Kalao
P.e.vigorsii
Tukangbesi, Tanimbar
P.e.hutzi
S Nasa Penida
P.e.concinna
Lombok, Flores, Sumbawa, Alor Is
P.e.maria
Sumba I
P.e.elegans
Timor, S Moluccas
Pitta iris (Rainbow Pitta)
N Northern Territory
Pitta anerythra (Black-faced Pitta)
P.a.anerythra
Ysabel I
P.a.pallida
Bougaianville I
P.a.nigrifrons
Choiseul I

112. XENICIDAE (NEW ZEALAND WRENS)

ACANTHISITTA
Acanthisitta chloris (Rifleman)
A.c.granti
North Island, New Zealand
A.c.chloris
South Island, New Zealand
A.c.citrina
SW South Island, New Zealand

XENICUS
Xenicus longipes (Bush Wren)
X.l.stokesii
North Island, New Zealand e?
X.l.longipes
South Island, New Zealand
X.l.variabilis
SW of Stewart I, New Zealand
Xenicus gilviventris (Rock Wren)
South Island, New Zealand

113. PHILEPITTIDAE (ASITIES)

PHILEPITTA
Philepitta castanea (Velvet Asity)
E Madagascar
Philepitta schlegeli (Schlegel's Asity)
W Madagascar

NEODREPANIS
Neodrepanis coruscans (Wattled False Sunbird)
E Madagascar
Neodrepanis hypoxantha (Small-billed False Sunbird)
E Madagascar e?

114. MENURIDAE (LYREBIRDS)

MENURA
Menura novaehollandiae (Superb Lyrebird)
M.n.edwardi
E Queensland
M.n.novaehollandiae
New South Wales, Victoria
Menura alberti (Prince Albert's Lyrebird)
E Queensland, New South Wales

115. ATRICHORNITHIDAE (SCRUBBIRDS)

ATRICHORNIS
Atrichornis clamosus (Western Scrubbird)
SW Australia
Atrichornis rufescens (Rufous Scrubbird)
SE Queensland, NE New South Wales

116. ALAUDIDAE (LARKS)

MIRAFRA
Mirafra cantillans (Western Singing Bush Lark) 116.3
M.c.marginata
Somalia, Uganda, Kenya, Tanzania
M.c.chadensis
Senegal to Sudan
M.c.simplex
W Arabia
M.c.cantillans
N India
Mirafra javanica (Eastern Singing Bush Lark)
M.j.williamsoni
C Burma, Thailand, Indochina
M.j.beaulieui
S Vietnam
M.j.philippinensis
Luzon, Mindoro Is
M.j.mindanensis
Mindanao I
M.j.javanica
S Borneo, Java, Bali
M.j.parva
Lombok, Sumbawa, Sumba, Flores Is
M.j.timorensis
Savu Is, Timor I
M.j.sepikiana
N New Guinea
M.j.aliena
NE New Guinea
M.j.woodwardi
WC Australia

M.j.halli
NW Australia
M.j.subrufescens
NW Australia
M.j.soderbergi
W Northern Territory
M.j.melvillensis
Melville I
M.j.rufescens
E Northern Territory, N Queensland
M.j.horsfieldii
C & S Queensland, New South Wales, Victoria
M.j.secunda
South Australia
Mirafra hova (Hova Lark)
Madagascar
Mirafra passerina (Monotonous Lark) 116.2
Namibia, Botswana
Mirafra albicauda (White-tailed Bush Lark)
M.a.albicauda
Chad to Ethiopia and Tanzania
M.a.rukwensis
S Tanzania
Mirafra cheniana (Melodious Lark)
Zimbabwe, Transvaal, Orange Free State, N & E Cape Province
Mirafra cordofanica (Kordofan Lark)
Niger, Chad, W Sudan
Mirafra williamsi (Williams' Lark)
Kenya
Mirafra candida (Nyiro Bush Lark)
Nyiro river, Kenya
Mirafra pulpa (Friedmann's Lark)
Sagon river, Ethiopia
Mirafra africana (Rufous-naped Lark)
M.a.henrici
Guinea, Liberia
M.a.batesi
Niger, N Nigeria
M.a.stresemanni
N Cameroun
M.a.bamendae
Cameroun
M.a.kurrae
E Chad
M.a.tropicalis
E Zaire, Uganda, Kenya, Tanzania
M.a.sharpii
Somalia
M.a.ruwenzoria
W Uganda
M.a.athi
Kenya
M.a.harterti
S Kenya
M.a.malbranti
SC Zaire
M.a.nyikae
E Zambia, N Malawi
M.a.chapini
SE Zaire, NW Zambia

M.a.occidentalis
  Gabon, W & S Angola
M.a.irwini
  Cuando, Angola
M.a.kabalii
  NE Angola
M.a.anchietae
  Huila, Angola
M.a.gomesi
  E Angola
M.a.grisescens
  W Zambia, Zimbabwe
M.a.pallida
  Namibia, Botswana
M.a.ghansiensis
  SW Botswana
M.a.nigrescens
  SW Tanzania
M.a.zuluensis
  SW Tanzania, N Natal, Mozambique
M.a.rostrata
  Natal, E Transkei
M.a.transvaalensis
  Transvaal, Natal, Zimbabwe, Zambia,
  Malawi
M.a.africana
  S Natal, E Cape Province
Mirafra hypermetra (Red-winged Bush Lark)
M.h.kathangorensis
  Sudan
M.h.kidepoensis
  S Sudan, N Uganda
M.h.gallarum
  S Ethiopia
M.h.hypermetra
  S Somalia, Kenya, N Tanzania
Mirafra somalica (Somali Lark)
M.s.somalica
  N Somalia
M.s.rochei
  S Somalia
Mirafra ashi (Ash's Lark)   116.1
  S Somalia
Mirafra angolensis (Angola Lark)
M.a.marungensis
  SE Zaire
M.a.angolensis
  W Angola
M.a.niethammeri
  Cuando District (Angola)
M.a.antonii
  E Angola
M.a.minyanyae
  NW Zambia
Mirafra rufocinnamomea (Flappet Lark)
M.r.buckleyi
  Gambia & Mali to Nigeria & Cameroun
M.r.serlei
  E Nigeria
M.r.tigrina
  NE Cameroun
M.r.furensis

W Sudan
M.r.sobatensis
  E Sudan
M.r.rufocinnamomea
  Ethiopia
M.r.omoensis
  SW Ethiopia
M.r.kawirondensis
  Uganda, Kenya, Tanzania
M.r.fischeri
  SE Zaire & Zambia to Kenya & Mozambique
M.r.torrida
  C Tanzania
M.r.zombae
  S Malawi, SE Tanzania to C Mozambique
M.r.smithersi
  NW Transvaal to SC Zambia
M.r.pintoi
  S Mozambique, NE Transvaal
M.r.mababiensis
  SW Zambia, N Botswana
Mirafra apiata (Clapper Lark)
M.a.reynoldsi
  W Zambia
M.a deserti
  Namibia, W Botswana
M.a.fasciolata
  E Cape Province to W Transvaal
M.a.apiata
  SW Cape Province
M.a.marjoriae
  Cape Town
M.a.damarensis
  N Namibia   116.2
M.a.nata
  E Botswana
Mirafra africanoides (Fawn-coloured Lark)
M.a.intercedens
  Ethiopia, Somalia, Kenya, NE Tanzania
M.a.alopex
  S Somalia, Ethiopia
M.a.macdonaldi
  S Ethiopia
M.a.longonotensis
  E Uganda, W Kenya
M.a.omaruru
  NW Namibia
M.a.trapnelli
  W Zambia
M.a.harei
  C Namibia
M.a.makarikari
  S Zambia, NE Botswana
M.a.sarwensis
  N Namibia, C Botswana, Mozambique
M.a.vincenti
  Zimbabwe, S Mozambique
M.a.africanoides
  S Namibia, N Cape Province
Mirafra collaris (Collared Lark)
  Somalia, N Kenya
Mirafra assamica (Rufous-winged Bush Lark)

M.a.assamica
  N India, Nepal, Assam
M.a.affinis
  S India, Sri Lanka
M.a.microptera
  C Burma
M.a.subsessor
  N Thailand
M.a.marionae
  S Burma, S Thailand, S Indochina
Mirafra rufa (Rusty Bush Lark)
M.r.nigriticola
  Mali
M.r.rufa
  W Sudan
M.r.lynesi
  C Sudan
Mirafra gilletti (Gillett's Lark)
  Ethiopia, Somalia
Mirafra degodiensis (Degodi Lark) 116.3
  Ethiopia
Mirafra poecilosterna (Pink-breasted Lark)
M.p.australoabyssinica
  N Sudan, Ethiopia
M.p.poecilosterna
  N & E Kenya
M.p.massaica
  N Uganda, Kenya, Tanzania
Mirafra sabota (Sabota Lark)
M.s.sabotoides
  W Botswana
M.s.plebeja
  SW Zaire
M.s.ansorgei
  W Angola
M.s.sabota
  S Mozambique, Transvaal, Natal, Cape
  Province
M.s.suffusca
  E Transvaal, C Natal, S Mozambique
M.s.herero
  SW Namibia
M.s.bradfieldi
  E Cape Province
M.s.naevia
  C Namibia 116.2
M.s.waibeli
  N Namibia, NW Botswana
Mirafra erythroptera (Indian Red-winged Lark)
M.e.sindiana
  NW India
M.e.furva
  Kathiawar (NW India)
M.e.erythroptera
  S & C India

HETEROMIRAFRA 116.3
Heteromirafra ruddi (Rudd's Lark)
  SE Transvaal, Orange Free State
Heteromirafra archeri (Archer's Lark) 116.3
  W Somalia
Heteromirafra sidamoensis (Sidamo Lark)
  Ethiopia

CERTHILAUDA
Certhilauda curvirostris (Long-billed Lark)
C.c.damarensis
  WC Namibia
C.c.bradshawi
  S Namibia, N Cape Province
C.c.subcoronata
  E Cape Province
C.c.curvirostris
  S & W Cape Province
C.c.semitorquata
  S Transvaal, Natal, E Cape Province
C.c.falcirostris
  NW Cape Province
C.c.transvaalensis
  N Natal, C Transvaal, N Cape Province
C.c.gilli
  SC Cape Province
C.c.brevirostris
  S Cape Province
C.c.algida
  E Cape Province
Certhilauda chuana (Short-clawed Lark) 116.2
  N Cape Province, SE Botswana, W
  Transvaal
Certhilauda albescens (Karoo Lark)
C.a.codea
  N Cape Province
C.a.albescens
  SW Cape Province
C.a.guttata
  C & W Cape Province
Certhilauda erythrochlamys (Dune Lark)
  116.2
C.e.erythrochlamys
  W Namibia
C.e.barlowi
  S Namibia
C.e.cavei
  S Namibia
Certhilauda burra (Red Lark) 116.3
  S Namibia, W Cape Province

PINAROCORYS
Pinarocorys nigricans (Dusky Lark)
  S Zaire & Angola to Natal & Mozambique
Pinarocorys erythropygia (Rufous-rumped Lark)
  Gambia to Sudan & Uganda

CHERSOMANES 116.3
Chersomanes albofasciata (Spike-heeled Lark)
C.a.beesleyi
  N Tanzania
C.a.obscurata
  C Angola
C.a.longispina
  W Huila, Angola
C.a.erikssoni
  N Namibia
C.a.kalahariae
  S Botswana

C.a.boweni
W Namibia
C.a.arenaria
C Namibia, N Cape Province
C.a.barlowi
S Botswana
C.a.bathoeni
SE Botswana
C.a.subpallida
NE Transvaal
C.a.alticola
N Orange Free State, S Transvaal
C.a.baddeleyi
S Orange Free State, N Cape Province
C.a.albofasciata
Natal, S Orange Free State, C & E Cape Province
C.a.bushmanensis
NW Cape Province
C.a.garrula
W Cape Province
C.a.macdonaldi
S Karoo, Cape Province
C.a.latimerae
Transkei

ALAEMON
Alaemon alaudipes (Hoopoe Lark)
A.a.boavistae
Cape Verde Is
A.a.alaudipes
N Africa, Sahara
A.a.doriae
E Arabia, Iraq to NW India
A.a.desertorum
Red Sea
Alaemon hamertoni (Lesser Hoopoe Lark)
A.h.altera
NE Somalia
A.h.tertia
EC Somalia
A.h.hamertoni
SE Somalia

RAMPHOCORIS
Ramphocoris clotbey (Thick-billed Lark)
N Africa, N Arabia, Syria

MELANOCORYPHA
Melanocorypha calandra (Calandra Lark)
M.c.calandra
S Europe, N Africa, Iran
M.c.psammochroa
Transcaspia, Iran, Afghanistan
M.c.gaza
Jordan
M.c.dathei
Turkey
Melanocorypha bimaculata (Bimaculated Lark)
M.b.bimaculata
SW Asia, Iran, NE Africa

M.b.torquata
E Iran, Afghanistan, NW India
M.b.rufescens
Asia Minor, Jordan, NE Africa
Melanocorypha mongolica (Mongolian Lark)
C Asia, Mongolia, Tsinghai
Melanocorypha maxima (Long-billed Calandra Lark)
M.m.maxima
Sikkim, S Tibet, W China
M.m.holdereri
NE Tibet, N Kashmir
Melanocorypha leucoptera (White-winged Lark)
S Russia, C Asia, Iran
Melanocorypha yeltoniensis (Black Lark)
S Russia, C Asia, Caucasus

AMMOMANES
Ammomanes cincturus (Bar-tailed Lark)
A.c.cincturus
Cape Verde Is
A.c.pallens
Mali to Sudan
A.c.arenicolor
N Africa, Sinai, Arabia
A.c.zarudnyi
E Iran, Afghanistan, NW India
Ammomanes phoenicurus (Rufous-tailed Desert Lark)
A.p.phoenicurus
N India
A.p.testaceus
C India
Ammomanes deserti (Desert Lark)
A.d.payni
Morocco
A.d.algeriensis
Algeria, Tunisia
A.d.whitakeri
NW Libya
A.d.mya
S Algeria, Niger
A.d.janeti
S Algeria
A.d.geyri
S Algeria, Mauretania, N Nigeria
A.d.monodi
C Mauretania
A.d.mirei
Tibesti Mountains
A.d.kollmanspergeri
NE Chad
A.d.deserti
E Libya, Egypt
A.d.erythrochrous
Sudan
A.d.isabellinus
Egypt, Arabia, Israel, Syria
A.d.samharensis
E Sudan, S Arabia
A.d.taimuri
Oman

A.d.assabensis
  Eritrea, N Somalia
A.d.akeleyi
  Somalia
A.d.azizi
  EC Arabia
A.d.saturatus
  S Arabia
A.d.annae
  Transjordan
A.d.insularis
  Bahrain I
A.d.cheesmani
  E Iraq, W Iran
A.d.parvirostris
  Transcaspia
A.d.orientalis
  N Afghanistan
A.d.iranicus
  S Iran, S Afghanistan
A.d.phoenicuroides
  NW India
Ammomanes grayi (Gray's Lark)
A.g.grayi
  W & S Namibia
A.g.hoeschi
  NW Namibia

CALANDRELLA
Calandrella razae (Raza Island Lark)
  Raza I, Cape Verde Is
**Calandrella brachydactyla (Greater Short-
toed Lark)** 116.3
C.b.woltersi
  Turkey
C.b.dukhunensis
  N India, Burma, Mongolia, W China, Tibet
C.b.longipennis
  C Asia, Afghanistan, N India
C.b.orientalis
  C Asia, Mongolia, Manchuria
C.b.artemisiana
  Caucasus, W Siberia, Asia Minor, Iran
C.b.brachydactyla
  S Europe, Russia, N Africa
C.b.rubiginosa
  S Morocca, S Algeria, S Tunisia
C.b.hermonensis
  Lebanon, Israel, Red Sea
C.b.eremica
  SW Arabia
Calandrella cinerea (Red-capped Lark) 116.3
C.c.erlangeri
  Ethiopia
C.c.saturatior
  Uganda to Botswana & Natal
C.c.williamsi
  W Kenya
C.c.alluvia
  S Mozambique
C.c.fulvida
  E Zimbabwe, W Mozambique, S Zambia, S
  Malawi

C.c.ongumaensis
  N Namibia
C.c.witputzi
  S Namibia
C.c.vagilans
  NE Cape Province, e Botswana, Transvaal
C.c.niveni
  SW Transvaal, Orange Free State, Natal
C.c.spleniata
  S Angola, W Namibia
C.c.cinerea
  S Cape Province
Calandrella blanfordi (Blanford's Lark)
  N Ethiopia, Somalia
**Calandrella acutirostris (Hume's Short-toed
Lark)**
C.a.acutirostris
  N Afghanistan, CS Asia, N & C India
C.a.tibetana
  C Asia, S Tibet, N India
Calandrella raytal (Indian Sand Lark)
C.r.raytal
  N India, N Burma
C.r.krishnarkumarsinhji
  Kathiawar (NW India)
C.r.adamsi
  NW India
**Calandrella rufescens (Lesser Short-toed
Lark)**
C.r.rufescens
  Tenerife I
C.r.polatzeki
  Gran Canaria, Lanzarote Is
C.r.apetzii
  S Spain
C.r.minor
  N Sahara, Sinai
C.r.nicolli
  N Egypt
C.r.aharonii
  Asia Minor, Jordan
C.r.pseudobaetica
  W Caspian Sea
C.r.persica
  Iraq to Afghanistan
C.r.heinei
  SE Russia, W Siberia
**Calandrella somalica (Somali Short-toed
Lark)** 116.3
C.s.somalica
  Somalia
C.s.vulpecula
  Somalia
C.s.megaensis
  S Ethiopia
C.s.athensis
  S Kenya, NE Tanzania
**Calandrella cheleensis (Mongolian Short-toed
Lark)**
C.c.cheleensis
  N China, Manchuria
C.c.leucophaea
  Turkestan

C.c.kukunoorensis
Kuku Nor
C.c.seebohmi
C Asia
C.c.biecki
N Kansu
C.c.tangutica
NE Tibet

SPIZOCORYS
Spizocorys conirostris (Pink-billed Lark)
S.c.damarensis
Namibia
S.c.crypta
NE Botswana
S.c.barlowi
Namibia, NW Cape Province
S.c.makawai
W Barotseland
S.c.griseovinacea
W Transvaal
S.c.transiens
C Cape Province to SE Botswana
S.c.conirostris
S Botswana, S Transvaal, N Cape Province
Spizocorys sclateri (Sclater's Lark)
S.s.sclateri
Namibia, Cape Province
S.s.capensis
C & E Cape Province
Spizocorys obbiensis (Obbia Lark)
Somalia
Spizocorys personata (Masked Lark)
S.p.personata
E Ethiopia
S.p.yavelloensis
S Ethiopia
S.p.mcchesneyi
N Kenya
S.p.intensa
C Kenya

BOTHA
Botha fringillaris (Botha's Lark)
S Transvaal, N Orange Free State

EREMALAUDA   116.3
Eremalauda dunni (Dunn's Lark)
E.d.dunni
S Sahara, W Sudan
E.d.eremodites
SW Arabia
Eremalauda starki (Stark's Lark)
E.s.starki
W Angola to Namibia
E.s.gregaria
W Cape Province, C Namibia

CHERSOPHILUS
Chersophilus duponti (DuPont's Lark)
C.d.duponti
S Spain, N Algeria, N Tunisia

C.d.margaritae
S Algeria, S Tunisia, Libya, NW Egypt

PSEUDALAEMON
Pseudalaemon fremantlii (Short-tailed Lark)
P.f.fremantlii
Somalia
P.f.megaensis
S Ethiopia
P.f.delamerei
SE Kenya, NE Tanzania

GALERIDA
Galerida cristata (Crested Lark)
G.c.pallida
Portugal, Spain
G.c.cristata
C Europe, Crimea, N Morocco
G.c.meridionalis
S Italy, SE Europe
G.c.subtaurica
C & S Asia Minor
G.c.caucasica
Caucasus, W Asia Minor, Cyprus, Crete
G.c.riggenbachi
W Morocco
G.c.macrorhyncha
S Algeria
G.c.randoni
C Algeria
G.c.carthaginis
N Algeria, N Tunisia
G.c.arenicola
SE Algeria, S Tunisia
G.c.balsaci
W Mauretania
G.c.festae
Libya
G.c.senegallensis
Senegal & Gambia to Sierra Leone & Mali
G.c.jordonsi
Air Mts (Niger)
G.c.alexanderi
N Cameroun, Mali, S Niger, Chad
G.c.zalingei
W Sudan
G.c.isabellina
N Chad, C Sudan
G.c.somaliensis
Somalia
G.c.altirostris
Egypt, Arabia
G.c.maculata
C Egypt, SW Arabia
G.c.nigricans
N Egypt
G.c.cinnamomina
Lebanon
G.c.zion
Turkey, Syria
G.c.magna
C & S Asia, NW China, N India

G.c.leautungensis
Manchuria, N China
G.c.coreensis
Korea
G.c.lynesi
Kashmir
G.c.chendoola
N & NW India
Galerida malabarica (Thekla Lark)   116.3
G.m.theklae
Portugal, Spain, Balearic Is
G.m.erlangeri
N Morocco
G.m.ruficolor
C Morocco, N Algeria, N Tunisia
G.m.superflua
E Morocco, C Algeria, S Tunisia, Libya
G.m.deichleri
S Algeria, S Tunisia
G.m.harrarensis
Harar (Ethiopia)
G.m.huei
Bale (Ethiopia)
G.m.praetermissa
Ethiopia
G.m.ellioti
Somalia
G.m.mallablensis
S Somalia
G.m.huriensis
Kenya
G.m.malabarica
W India
Galerida deva (Sykes' Crested Lark)
WC & SC India
Galerida modesta (Sun Lark)
G.m.giffardi
Ghana to Sudan
G.m.modesta
S Sudan
G.m.nigrita
Guinea, Sierra Leone
G.m.struempelli
W Cameroun
G.m.bucolica
Central African Republic, N Zaire, SW Sudan

CALENDULA
Calendula magnirostris (Large-billed Lark)
C.m.magnirostris
W & SW Cape Province
C.m.harei
SW Transvaal, Orange Free State, Cape Province
C.m.montivaga
Lesotho

LULLULA
Lullula arborea (Wood Lark)
L.a.arborea
N & NW Europe, N Africa

L.a.pallida
S Europe, N Africa, Caucasus, Iran

ALAUDA
Alauda arvensis (Eurasian Sky Lark)
A.a.arvensis
N & W Europe
A.a.sierrae
N Portugal, Spain
A.a.harterti
W Nth Africa
A.a.cantarella
SE Europe, Iran, E Nth Africa
A.a.dulcivox
SE Russia, C Asia, N India
A.a.kiborti
EC Asia, Manchuria, E China
A.a.intermedia
NE Manchuria, E China
A.a.pekinensis
NC & NE Asia, Japan, N China
A.a.lonnbergi
Sakhalin I, Korea, NE China, Japan
A.a.japonica
Japan, Riukiu Is
Alauda gulgula (Oriental Sky Lark)
A.g.inconspicua
SW Asia, Afghanistan, NW India
A.g.lhamarum
W Himalayas
A.g.inopinata
SE Tibet, W China, N Burma, Nepal
A.g.sala
Hainan I
A.g.herberti
C & S Thailand, S Indochina
A.g.wattersi
Taiwan
A.g.wolfei
Philippine Is
A.g.vernayi
E Bhutan, SE Tibet, N Burma
A.g.weigoldi
C China
A.g.coelivox
SE China, N Vietnam
A.g.gulgula
S India, Sri Lanka, S Burma
A.g.australis
S India, Sri Lanka

EREMOPTERIX
Eremopterix australis (Black-eared Sparrow-Lark)
S Namibia, Botswana, W Transvaal, N Cape Province
Eremopterix leucotis (Chestnut-backed Sparrow-Lark)
E.l.melanocephala
Senegal to Nile Valley
E.l.leucotis
Ethiopia

E.l.madaraszi
  Kenya, Tanzania, Mozambique
E.l.hoeschi
  Namibia, S Angola, SW Zambia
E.l.smithi
  S & SE Africa
Eremopterix signata (Chestnut-headed Sparrow-Lark)
E.s.harrisoni
  SE Sudan, N Kenya
E.s.signata
  Ethiopia, Somalia, Kenya
Eremopterix verticalis (Grey-backed Sparrow-Lark)
E.v.verticalis
  Botswana, W Zimbabwe, Transvaal, W
  Cape Province
E.v.damarensis
  Angola, Namibia, NW Cape Province
E.v.harti
  S Angola, SW Zambia
E.v.khama
  NE Botswana
Eremopterix nigriceps (Black-crowned Sparrow-Lark)
E.n.nigriceps
  Cape Verde Is
E.n.albifrons
  Mauretania to Nile valley
E.n.melanauchen
  Egypt, Ethiopia, Sudan, Socotra I, Arabia,
  Iraq
E.n.affinis
  Pakistan, NW India
Eremopterix grisea (Ashy-crowned Sparrow-Lark)
  Pakistan, India, Sri Lanka
Eremopterix leucopareia (Fischer's Sparrow-Lark)
  Uganda, Kenya, Malawi

EREMOPHILA
Eremophila alpestris (Shore (Horned) Lark)
E.a.flava
  N Europe, N Asia
E.a.balcanica
  SE Europe
E.a.penicillata
  Asia Minor, W Iran
E.a.albigula
  N Iran, Afghanistan, C Asia
E.a.brandti
  EC Asia, N China
E.a.longirostris
  Baluchistan, NW Himalayas
E.a.teleschowi
  C Siking
E.a.przewalskii
  NW Tsinghai
E.a.argalea
  NW India
E.a.elwesi
  N India, Sikkim, Nepal, S Tibet

E.a.nigrifrons
  E Tsinghai, NW China
E.a.khamensis
  C Sikang
E.a.atlas
  C Morocco
E.a.bicornis
  S Asia Minor, Lebanon
E.a.arcticola
  N Alaska, W Canada, NW USA
E.a.alpina
  W Washington State
E.a.hoyti
  N Canada, N USA
E.a.alpestris
  NE Canada, NE USA
E.a.leucolaema
  S Canada, C & S USA, N Mexico
E.a.enthymia
  C Canada, C USA
E.a.praticola
  SE Canada, C & EC USA
E.a.strigata
  NW USA
E.a.merrilli
  W Canada, W USA
E.a.lamprochroma
  W & SW USA
E.a.utahensis
  WC USA
E.a.sierrae
  NE California
E.a.rubea
  C California
E.a.actia
  S California, N Baja California
E.a.insularis
  S California islands
E.a.ammophila
  SW USA, NE Baja California, NW Mexico
E.a.leucansiptila
  SW USA, NE Baja California, NW Mexico
E.a.occidentalis
  S USA, N Mexico
E.a.adusta
  S USA
E.a.giraudi
  S Texas, NC Mexico
E.a.enertera
  WC Baja California
E.a.aphrasta
  NC Mexico
E.a.diaphora
  E Mexico
E.a.lactea
  Coahuila (Mexico)
E.a.chrysolaema
  SC Mexico
E.a.oaxacae
  S Mexico
E.a.peregrina
  Colombia

Eremophila bilopha (Temminck's Horned Lark)
N Africa, N Arabia, Iraq

117. HIRUNDINIDAE (SWALLOWS, MARTINS)

PSEUDOCHELIDONINAE

PSEUDOCHELIDON
Pseudochelidon eurystomina (African River Martin)
Zaire
Pseudochelidon sirintarae (White-eyed River Martin)
C Thailand

HIRUNDININAE

TACHYCINETA   117.3
Tachycineta bicolor (Tree Swallow)
W Canada, W USA, Central America, Cuba
Tachycineta albilinea (Mangrove Swallow)
T.a.rhizophorae
NW Mexico
T.a.albilinea
E & S Mexico, Central America
T.a.stolzmanni
W Peru
Tachycineta albiventer (White-winged Swallow)
NE Sth America, Trinidad
Tachycineta leucorrhoa (White-rumped Swallow)
SC Sth America
Tachycineta leucopyga (Chilean Swallow)
S Sth America
Tachycineta thalassina (Violet-green Swallow)
T.t.lepida
NW Canada, W & SW USA, NW Mexico, C America
T.t.brachyptera
C & S Baja California, NW Mexico
T.t.thalassina
C Mexico
Tachycineta cyaneoviridis (Bahama Swallow)
E Cuba, Bahama Is
Tachycineta euchrysea (Golden Swallow)
T.e.euchrysea
Jamaica
T.e.sclateri
Hispaniola

PHAEOPROGNE
Phaeoprogne tapera (Brown-chested Martin)
P.t.tapera
Colombia, Venezuela, the Guianas, Ecuador, Peru, Brazil

P.t.fusca
C Sth America

PROGNE
Progne subis (Purple Martin)
P.s.subis
S Canada, W USA, E Mexico to N Sth America
P.s.hesperia
Arizona, Baja California, W Mexico to Nicaragua
P.s.arboricola
Utah
Progne cryptoleuca (Cuban Martin)   117.4
Cuba, Isle of Pines
Progne dominicensis (Caribbean Martin)
117.4
Jamaica, Hispaniola, Lesser Antilles
Progne sinaloae (Sinaloa Martin)   117.4
NW Mexico
Progne chalybea (Grey-breasted Martin)
P.c.chalybea
S USA, C America, N Sth America
P.c.warneri
Mexico
P.c.macroramphus
C Sth America
Progne modesta (Southern Martin)
P.m.modesta
C & S Galapagos Is
P.m.elegans
Bolivia, Brazil, Argentina
P.m.murphyi
W Peru, N Chile

NOTIOCHELIDON
Notiochelidon murina (Brown-bellied Swallow)
N.m.murina
Colombia, Ecuador, Peru
N.m.meridensis
W Venezuela
N.m.cyanodorsalis
W Bolivia
Notiochelidon cyanoleuca (Blue & White Swallow)
N.c.cyanoleuca
Costa Rica, Panama, N & C Sth America
N.c.peruviana
W Peru
N.c.patagonica
S Sth America
Notiochelidon flavipes (Pale-footed Swallow)
Peru, Colombia
Notiochelidon pileata (Black-capped Swallow)
S Mexico, Guatemala

ATTICORA
Atticora fasciata (White-banded Swallow)
N & W Amazonia
Atticora melanoleuca (Black-collared Swallow)
N Amazonia

NEOCHELIDON
Neochelidon tibialis (White-thighed Swallow)
  N.t.minimus
    E Panama, Colombia, W Ecuador
  N.t.griseiventris
    S Colombia, SE Venezuela, E Ecuador, E
    Peru W Brazil
  N.t.tibialis
    SE Brazil

ALOPOCHELIDON
Alopochelidon fucata (Tawny-headed Swallow)
    Venezuela to C Argentina

STELGIDOPTERYX
Stelgidopteryx serripennis (Northern Rough-winged Swallow) 117.4
  S.s.serripennis
    S Canada, USA, Mexico, C America
  S.s.psammochroa
    SW USA, Baja California, S Mexico
  S.s.fulvipennis
    C & S Mexico to Costa Rica
  S.s.ridgwayi
    Yucatan
  S.s.stuarti
    SE Mexico, E Guatemala
  S.s.burleighi
    Belize
Stelgidopteryx ruficollis (Southern Rough-winged Swallow) 117.4
  S.r.decolor
    W Costa Rica, W Panama
  S.r.uropygialis
    W Colombia, Ecuador, NW Peru
  S.r.aequalis
    N Colombia, W Venezuela, Trinidad
  S.r.cacabata
    E Venezuela, the Guianas
  S.r.ruficollis
    SE Colombia, E Ecuador, E Peru, Bolivia, N
    Argentina

CHERAMOECA
Cheramoeca leucosterna (White-backed Swallow)
    W, S & C Australia

RIPARIA
Riparia paludicola (Brown Throated Sand Martin)
  R.p.mauretanica
    W Morocco
  R.p.minor
    Niger, Mali, Chad, Sudan, Ethiopia
  R.p.newtoni
    SE Nigeria
  R.p.ducis
    Uganda, E Zaire, Kenya, Tanzania
  R.p.paludicola
    Zambia, Sth Africa

  R.p.cowani
    Madagascar
  R.p.chinensis
    N India, SE Asia
  R.p.tantilla
    Luzon I
Riparia congica (Congo Sand Martin)
    C Zaire
Riparia riparia (Sand Martin) (Bank Swallow)
  R.r.riparia
    Nth America, N Sth America, Europe, Asia,
    N & E Africa
  R.r.ijimae
    NE Asia, Burma, Thailand, Philippine Is
  R.r.tibetana
    C Asia
  R.r.diluta
    NW India, Nepal
  R.r.fokienensis
    C & S China
  R.r.indica
    Afghanistan, NW India
  R.r.shelleyi
    Egypt, Sudan
Riparia cincta (Banded Martin)
  R.c.erlangeri
    Ethiopia
  R.c.suahelica
    Uganda, Kenya, N Tanzania
  R.c.parvula
    S Zaire
  R.c.cincta
    W & S Africa
  R.c.xerica
    W Angola

PHEDINA
Phedina borbonica (Mascarene Martin)
  P.b.borbonica
    Mauritius, Reunion Is
  P.b.madagascariensis
    Malawi, Madagascar
Phedina brazzae (Brazza's Martin) 117.5
    S Zaire, N Angola

PSEUDHIRUNDO 117.5
Pseudhirundo griseopyga (Grey-rumped Swallow)
  P.g.liberiae
    Liberia
  P.g.gertrudis
    NE Nigeria
  P.g.melbina
    Gabon
  P.g.griseopyga
    Ethiopia to Natal

HIRUNDO
Hirundo rupestris (Crag Martin)
  H.r.rupestris
    S Europe, C & SW Asia, NE Africa, India
  H.r.theresae
    S Morocco

Hirundo obsoleta (Pale Crag Martin)
H.o.spatzi
  SC Algeria
H.o.presaharica
  NC Algeria
H.o.buchanani
  N Niger
H.o.obsoleta
  Turkey, Egypt, Sudan to Iran
H.o.arabica
  E Sudan, W Arabia, Somalia, Socotra I
H.o.perpallida
  E Arabia
H.o.pallida
  E Iran, Afghanistan, NW India
Hirundo fuligula (African Rock Martin)
H.f.pusilla
  S Sudan, Ethiopia
H.f.rufigula
  N Nigeria, Chad, S Sudan, Ethiopia, Zaire,
  Uganda, N Tanzania
H.f.birwae
  Sierra Leone, Guinea
H.f.bansoensis
  SE Nigeria
H.f.fusciventris
  S Tanzania, Malawi, N Mozambique
H.f.anderssoni
  S Angola, Namibia, Botswana, W Cape
  Province
H.f.fuligula
  E Cape Province
H.f.pretoriae
  E Transvaal, Natal
Hirundo concolor (Dusky Crag Martin)
H.c.concolor
  India
H.c.sintaungensis
  Burma, N Thailand, N Laos, N
  Vietnam
Hirundo rustica ((Barn) Swallow)
H.r.rustica
  Europe, W Asia, Africa, India
H.r.transitiva
  Asia Minor, Egypt, Kenya, Uganda
H.r.savignii
  Egypt
H.r.gutteralis
  NE Asia, India, S China, SE Asia, New
  Guinea
H.r.tytleri
  Bhutan, Siberia, W Mongolia, W China,
  Burma
H.r.mandschurica
  Manchuria, E China
H.r.saturata
  E Siberia
H.r.erythrogaster
  N & S America, West Indies
Hirundo lucida (Red-chested Swallow)
H.l.lucida
  Senegal to Ghana
H.l.clara

Mali, Upper Volta
H.l.subalaris
  E Zaire
H.l.rothschildi
  Ethiopia
Hirundo angolensis (Angola Swallow)
H.a.arcticincta
  Uganda, W Kenya, NW Tanzania
H.a.angolensis
  SE Zaire, Tanzania, Angola, Malawi,
  Zambia
Hirundo tahitica (Pacific Swallow)
H.t.domicola
  S India, Sri Lanka
H.t.abbotti
  Malaysia, Sumatra, Borneo, Philippine Is
H.t.namiyei
  Riukiu Is
H.t.mallopega
  Andaman Is, E Sumatra, Java
H.t.frontalis
  New Guinea,
H.t.javanica
  Lesser Sunda Is, Sulawesi, Moluccas
H.t.ambiens
  New Britain
H.t.subfusca
  Polynesia, Melanesia, Fiji Is, Tonga I
H.t.tahitica
  Society Is
H.t.carteri
  W Australia
H.t.parsonsi
  NE Queensland
H.t.neoxena
  S Queensland, New South Wales, Victoria
  South Australia
**Hirundo albigularis (White-throated
Swallow)**
H.a.ambigua
  N & E Angola, NW Zambia
H.a.albigularis
  Sthn Africa
Hirundo aethiopica (Ethiopian Swallow)
H.a.fulvipectus
  Nigeria, Cameroun, Sudan
H.a.aethiopica
  Ethiopia, Somalia, E Kenya, Tanzania
Hirundo smithii (Wire-tailed Swallow)
H.s.smithii
  W, C, E & SE Africa
H.s.filifera
  SW Asia, India, Burma, Thailand, Laos, N
  Vietnam
Hirundo atrocaerulea (Blue Swallow)
  Tanzania to Natal
**Hirundo nigrita (White-throated Blue
Swallow)**
  W & C Africa
Hirundo leucosoma (Pied-winged Swallow)
  Senegal to Nigeria
Hirundo megaensis (White-tailed Swallow)
  S Ethiopia

Hirundo nigrorufa (**Black & Rufous Swallow**)
Angola, S Zaire, Zambia
Hirundo dimidiata (**Pearl-breasted Swallow**)
H.d.marwitzi
Angola, Zambia, SW Tanzania, Malawi
H.d.dimidiata
S Africa, Zimbabwe
Hirundo cucullata (**Greater Striped Swallow**)
Sthn Africa
Hirundo abyssinica (**Lesser Striped Swallow**)
H.a.puella
Sierra Leone to Nigeria
H.a.maxima
S Nigeria, Cameroun
H.a.bannermani
S Sudan
H.a.abyssinica
Ethiopia, Uganda, Kenya, Tanzania
H.a.unitatis
W Uganda, E Zaire, SE Africa
H.a.ampliformis
Caprivi Strip to NE Zambia
Hirundo semirufa (**Red-breasted Swallow**)
H.s.gordoni
W & C Africa
H.s.semirufa
SE Africa
Hirundo senegalensis (**Mosque Swallow**)
H.s.senegalensis
W & NC Africa
H.s.saturatior
C & E Africa
H.s.monteiri
SC Africa
Hirundo daurica (**Red-rumped Swallow**)
H.d.daurica
C Asia
H.d.japonica
E Asia, Japan, China, N India
H.d.gephyra
W China
H.d.nipalensis
Himalayas, India, N Burma
H.d.erythropygia
S India, Sri Lanka
H.d.hyperythra
Sri Lanka
H.d.rufula
S Europe, Iran, Afghanistan, NW India
H.d.domicella
NW Africa
H.d.disjuncta
Sierra Leone
H.d.kumboensis
Cameroun
H.d.emini
E Africa
H.d.melanocrissa
N Ethiopia
Hirundo striolata (**Greater Striated Swallow**)
H.s.striolata
Taiwan, Philippine Is, Borneo, Sumatra,
Lesser Sunda Is

H.s.mayri
Assam, N Burma, NW Thailand
H.s.stanfordi
Burma, N Thailand, N Laos
H.s.vernayi
S Thailand
H.s.badia
C Malaysia
Hirundo rufigula (**Red-throated Cliff Swallow**)
Gabon to Zambia & Angola
Hirundo preussi (**Preuss'Cliff Swallow**)
W Africa, N Zaire
Hirundo andecola (**Andean Swallow**)
H.a.oroyae
C Peru
H.a.andecola
S Peru, N Bolivia, N Chile
Hirundo nigricans (**Tree Martin**)
H.n.timoriensis
Timor, Lesser Sunda Is
H.n.neglecta
W & N Australia
H.n.nigricans
E & S Australia, New Guinea, Solomon Is
Hirundo perdita (**Red Sea Cliff Swallow**)   117.1
Red Sea
Hirundo spilodera (**South African Cliff Swallow**)
Southern Africa
Hirundo pyrrhonota (**American Cliff Swallow**)
H.p.pyrrhonota
Nth America, C America, N Sth America
H.p.tachina
SW USA, WC America
H.p.minima
SW USA, N Mexico
H.p.melanogaster
S Mexico, S Brazil, N Argentina
Hirundo fulva (**Cave Swallow**)   117.3
H.f.pelodoma
S USA, Mexico, Guatemala
H.f.citata
SE Mexico
H.f.fulva
Puerto Rico, Cuba, Hispaniola, Jamaica
H.f.chapmani
SW Ecuador
H.f.rufocollaris
W Peru
Hirundo fluvicola (**Indian Cliff Swallow**)
Afghanistan, Himalayas, N India
Hirundo ariel (**Fairy Martin**)
Australia, Tasmania
Hirundo fuliginosa (**Forest Swallow**)
S Cameroun, Gabon

DELICHON
Delichon urbica (**Common House Martin**)
D.u.urbica
Europe, C & W Asia, W & SE Africa
D.u.meridionalis
Mediterranean, N Africa, Iran, N India

D.u.lagopoda
E Asia, S China, Burma, Thailand
Delichon dasypus (Asian House Martin)
D.d.cashmiriensis
Himalayas, India, W China
D.d.nigrimentalis
S China, Taiwan
D.d.dasypus
NE Asia, N China, Malaysia, Borneo,
Philippine Is
Delichon nipalensis (Nepal House Martin)
D.n.nipalensis
Himalayas, Assam
D.n.cuttingi
NE Burma

PSALIDOPROCNE
Psalidoprocne nitens (Square-tailed Saw-wing)
P.n.nitens
W Africa, N Zaire
P.n.centralis
NE Zaire
Psalidoprocne fuliginosa (Mountain Saw-wing)
Cameroun, Fernando Po I
Psalidoprocne albiceps (White-headed Saw-wing)
P.a.albiceps
EC Africa
P.a.suffusa
Angola
Psalidoprocne pristoptera (Black Saw-wing)
P.p.pristoptera
Somalia, N Ethiopia
P.p.blanfordi
S Ethiopia
P.p.mangbettorum
NE Zaire
P.p.ruwenzori
E Zaire, W Uganda
P.p.massaica
Kenya
P.p.holomelaena
Mozambique & Malawi to E Cape Province
Psalidoprocne oleaginea (Kaffa Saw-wing)
SW Ethiopia
Psalidoprocne antinorii (Brown Saw-wing)
S Ethiopia
Psalidoprocne petiti (Petit's Saw-wing)
P.p.petiti
Gabon, Cameroun, Central African
Republic
P.p.chalybea
N Cameroun, NE Zaire
Psalidoprocne orientalis (Eastern Saw-wing)
117.2
P.o.reichenowi
Angola, SW Zaire, Zambia
P.o.orientalis
Tanzania, Malawi, E Zambia,
Mozambique

Psalidoprocne obscura (Fantee Saw-wing)
Portuguese Guinea to Cameroun

118. MOTACILLIDAE (WAGTAILS, PIPITS)

DENDRONANTHUS
Dendronanthus indicus (Forest Wagtail)
NE Asia, China, India, Thailand, Malaysia,
Sumatra, Java, Borneo

MOTACILLA
Motacilla flava (Yellow Wagtail)
M.f.flavissima
NW Europe, Spain, N Africa
M.f.flava
N Europe, S Africa
M.f.iberiae
S France, Spain, NW Africa
M.f.cinereocapilla
Italy, Sardinia, Arabia, NE Africa
M.f.pygmaea
Egypt
M.f.beema
SE Russia, C Asia, India, NE Africa
M.f.leucocephala
Mongolia, C Asia, NW India
M.f.lutea
SE Russia, S Africa, India
M.f.zaissanensis
C Asia, India
M.f.thunbergi
NE Europe, C & S Africa
M.f.plexa
N Siberia, N Asia, India
M.f.angarensis
N & C Asia, E Mongolia, E China, Burma
M.f.macronyx
C Asia, E China, Burma, Malaysia, Sumatra
M.f.simillima
N Siberia, E China, Philippine Is
M.f.tschutschensis
NE Asia, Alaska, E China, Java
M.f.taivana
E Siberia, China, Philippine Is
M.f.feldegg
Balkans, Asia Minor, Iraq, Iran, E Africa
M.f.melanogrisea
SW Asia, NW & C India
Motacilla citreola (Citrine Wagtail)
M.c.citreola
Russia, C Asia, Manchuria, India, SE China
M.c.werae
Siberia, Iran, SW Asia, India
M.c.calcarata
E Iran, Himalayas, Tibet, Burma,
Afghanistan
Motacilla cinerea (Grey Wagtail)
M.c.patriciae
Azores Is
M.c.schmitzi
Madeira I
M.c.canariensis
Canary Is

M.c.cinerea
Europe, N Africa, Iran, India, SE Asia, C & S Africa
M.c.robusta
NE Asia, Japan, E China, Philippine Is
Motacilla alba (Pied (White) Wagtail)
M.a.yarrelli
British Isles, Spain Morocco
M.a.alba
Europe, Russia, N & E Africa, Iran, Arabia
M.a.subpersonata
Morocco
M.a.dukhunensis
S Russia, SW Asia, Afghanistan, India
M.a.personata
Siberia, W Asia, Afghanistan, Iran, N India
M.a.persica
Iran
M.a.baicalensis
C Asia, Iran, India, Thailand, SW China
M.a.ocularis
E Siberia, E India, China, Thailand Philippine Is
M.a.lugens
NE Asia, NE China, Japan, Taiwan
M.a.leucopsis
E Asia, China, Thailand, Himalayas
M.a.alboides
S China, E Himalayas, Burma
Motacilla grandis (Japanese Pied Wagtail)
Japan, Korea, E China, Taiwan
Motacilla maderaspatensis (Large Pied Wagtail)
Pakistan, India
Motacilla aguimp (African Pied Wagtail)
M.a.vidua
W, C & E Africa, S Africa
M.a.aguimp
Orange River, NW Cape Province
Motacilla clara (Mountain Wagtail)
M.c.chapini
Guinea to E Zaire
M.c.clara
Ethiopia
M.c.torrentium
Uganda and E Zaire to W Angola & Natal
Motacilla capensis (Cape Wagtail)
M.c.simplicissima
Angola, S Zaire, Zambia
M.c.bradfieldi
W Cape Province, Namibia
M.c.capensis
Sthn Africa
M.c.wellsi
E Zaire, Uganda, Kenya
Motacilla flaviventris (Madagascar Wagtail)
Madagascar

TMETOTHYLACUS
Tmetothylacus tenellus (Golden Pipit)
Somalia, Kenya, Tanzania

MACRONYX
Macronyx capensis (Cape Longclaw)
M.c.capensis
W Cape Province
M.c.colletti
Botswana, Transvaal, C Cape Province, Natal
M.c.latimerae
N Cape Province, W Transvaal
M.c.stabilior
Zimbabwe
Macronyx croceus (Yellow-throated Longclaw)
M.c.croceus
W & C Africa
M.c.tertius
Kenya to N Natal & N Transvaal
M.c.vulturnus
Transkei, Natal
Macronyx fuelleborni (Fülleborn's Longclaw)
M.f.fuelleborni
C Tanzania
M.f.ascensi
Angola, Zaire, Zambia, SW Tanzania
Macronyx flavicollis (Abyssinian Longclaw)
Ethiopia
Macronyx aurantiigula (Pangani Longclaw)
Somalia, Kenya
Macronyx ameliae (Rosy-breasted Longclaw)
M.a.ameliae
W Kenya to Natal
M.a.altanus
SW Tanzania to E Angola & Zimbabwe
Macronyx grimwoodi (Grimwood's Longclaw)
M.g.grimwoodi
E Angola, NW Zambia
M.g.cuandocubangensis
Cuando, Angola

HEMIMACRONYX 118.1
Hemimacronyx sharpei (Sharpe's Longclaw)
Kenya
Hemimacronyx chloris (Yellow-breasted Pipit)
S & E Sth Africa

ANTHUS
Anthus novaeseelandiae (Richard's Pipit)
A.n.eximius
Yemen
A.n.cameroonensis
Cameroun
A.n.lynesi
E Cameroun, Sudan
A.n.stabilis
Sudan
A.n.cinnamomeus
Ethiopia, E Africa, Zambia
A.n.lacuum
E Africa
A.n.lwenarum
NW Zambia

A.n.rufuloides
Sth Africa
A.n.editus
Lesotho, W Natal
A.n.richardi
C Asia, Pakistan, India, Thailand, Siberia, N
Vietnam
A.n.dauricus
N Mongolia
A.n.centralasiae
C Asia
A.n.sinensis
E Siberia, E China, Malaysia, Sumatra
A.n.rufulus
Nepal, Burma, Thailand, Laos, Vietnam
A.n.waitei
Pakistan, NW India
A.n.malayensis
S India, Sri Lanka, Malaysia, Sumatra,
Java, Borneo
A.n.lugubris
Palawan, Philippine Is
A.n.albidus
Lesser Sunda Is, S Sulawesi
A.n.medius
Savu, Timor Is
A.n.exiguus
C New Guinea
A.n.rogersi
N Australia
A.n.subaustralis
C & W Australia
A.n.bilbali
SW Australia
A.n.australis
SE Australia
A.n.bistriatus
Tasmania
A.n.reischeki
North I, New Zealand
A.n.novaeseelandiae
South I, New Zealand
A.n.chathamensis
Chatham I
A.n.aucklandicus
Auckland Is
A.n.steindachneri
Antipodes Is
Anthus hoeschi (Mountain Pipit)   118.8
Angola, Namibia, Botswana
Anthus godlewskii (Blyth's Pipit)
EC Asia, Tibet, India, Burma, Sri Lanka
Anthus campestris (Tawny Pipit)
A.c.campestris
Europe, N Africa, Iran, Arabia, SW Asia
A.c.griseus
C Asia, Afghanistan, W India
A.c.kastschenkoi
W Siberia, N India
Anthus similis (Long-billed Pipit)
A.s.nicholsoni
S Africa, Zimbabwe

A.s.leucocraspedon
Namibia
A.s.bannermani
Sierra Leone, Guinea
A.s.josensis
C Nigeria
A.s.asbenaicus
S Sahara
A.s.jebelmarrae
W Sudan
A.s.hararensis
Ethiopia, Kenya, N Tanzania
A.s.nivescens
NE Sudan, Red Sea
A.s.sokotrae
Socotra I
A.s.dewittei
E & C Zaire, W Uganda
A.s.chyuluensis
Uganda to N Tanzania   118.2
A.s.moco
C Angola
A.s.frondicolus
N Zimbabwe, S Zambia, N Botswana
A.s.palliditinctus
NW Namibia, SW Angola
A.s.petricola
Lesotho
A.s.captus
Lebanon, Syria, Israel
A.s.arabicus
SW Arabia
A.s.decaptus
Afghanistan, Pakistan, NW India
A.s.jerdoni
E Afghanistan, Himalayas, Burma
A.s.yamethini
C Burma
A.s.similis
C & S India
A.s.travancoriensis
SW India
Anthus nyassae (Woodland Pipit)   118.2
A.n.nyassae
Angola to Malawi
A.n.schoutedeni
S Angola to SW Tanzania
Anthus vaalensis (Buffy Pipit)
A.v.saphiroi
E Ethiopia, Somalia
A.v.goodsoni
Kenya
A.v.neumanni
Angola, S Zaire, Zambia
A.v.chobiensis
W Zambia, Botswana, Zimbabwe
A.v.clanceyi
N Cape Province, SW Transvaal, SE
Botswana
A.v.vaalensis
Namibia, Cape Province, W Zimbabwe

A.v.daviesii
E Cape Province
Anthus leucophrys Plain-backed Pipit)
A.l.ansorgei
Senegal to N Nigeria
A.l.gouldii
Sierra Leone, Ivory Coast
A.l.zenkeri
S Nigeria to N Uganda & Kenya
A.l.omoensis
Ethiopia, Uganda, W Kenya
A.l.bohndorffi
Angola, S Zaire, Zambia, Malawi
A.l.enunciator
Transkei, S Natal, E Transvaal
A.l.leucophrys
S Angola, Botswana, Sth Africa
A.l.tephridorsus
NW Zimbabwe
Anthus pallidiventris (Long-legged Pipit)
A.p.pallidiventris
Gabon NW Angola
A.p.esobe
C Zaire
Anthus pratensis (Meadow Pipit)
A.p.theresae
W Ireland
A.p.pratensis
Greenland, Europe, N Africa, Asia Minor,
Iran
Anthus trivialis (Tree Pipit)
A.t.trivialis
Europe, Asia, India, Africa
A.t.differens
Turkey
A.t.haringtoni
EC Asia, Himalayas, N India
Anthus hodgsoni (Indian Tree Pipit)
A.h.yunnanensis
N & E Asia, India, SE China, Borneo
A.h.hodgsoni
Himalayas, NW China, Japan, India
Anthus roseatus (Hodgson's Pipit)
C & E Asia, N India, Tibet, W China
Anthus cervinus (Red-throated Pipit)
E Europe, E Asia, W & E Africa, India
Anthus gustavi (Petchora Pipit)
A.g.gustavi
NE Asia, China, Philippine Is, Borneo,
Sulawesi
A.g.commandorensis
Commander Is
A.g.menzbieri
S Ussuriland
Anthus rubescens (Buff-bellied Pipit)   118.4
A.r.rubescens
N & NE Asia, Nth America, Mexico
A.r.pacificus
W Canada, W USA, W Mexico
A.r.alticola
SW USA, NW Mexico
A.r.japonicus
E Asia, Japan, N India, China, Burma

A.r.coutellii
C Asia, Tibet, N China, N India, Iran
Anthus spinoletta (Water Pipit)   118.4
A.s.spinoletta
S & E Europe
A.s.kleinschmidti
Faroe Is
Anthus petrosus (Rock Pipit)   118.4
A.p.petrosus
British Isles
A.p.littoralis
NW Europe
Anthus nilghiriensis Nilgiri Pipit)
S India
Anthus sylvanus (Upland Pipit)
Afghanistan, Himalayas, W China
Anthus berthelotii (Canarian Pipit)
A.b.berthelotii
Canary Is
A.b.madeirensis
Madeira I
Anthus lineiventris (Striped Pipit)
A.l.stygium
SE Transvaal & Transkei
A.l.angolensis
S Zaire to Zambia
A.l.lineiventris
Botswana, Transvaal, S Zimbabwe
Anthus brachyurus (Short-tailed Pipit)
A.b.eludens
S Zaire, N Angola, Gabon   118.5
A.b.leggei
Angola, Zaire, Uganda, Tanzania, Zambia
A.b.brachyurus
Natal
Anthus caffer (Bushveld Pipit)
A.c.australoabyssinicus
S Ethiopia
A.c.blayneyi
S Kenya, Tanzania
A.c.mzimbaensis
Malawi
A.c.traylori
E Natal, S Mozambique
A.c.caffer
Botswana, Angola, Zimbabwe,
Mozambique, Transvaal, Natal
Anthus latistriatus (Jackson's Pipit)   118.3
Kavirondo (Kenya)
Anthus sokokensis (Sokoke Pipit)
SE Kenya, NE Tanzania
Anthus melindae (Malindi Pipit)
A.m.melindae
S Somalia, Kenya
A.m.mallablensis
SC Somalia   118.6
Anthus crenatus (African Rock Pipit)
E Transvaal, Cape Province
Anthus gutturalis (New Guinea Pipit)
A.g.gutturalis
SE New Guinea
A.g.rhododendri
EC New Guinea

A.g.wollastoni
  WC New Guinea
Anthus spragueii (Sprague's Pipit)
  NC & S USA, S Mexico
Anthus furcatus (Short-billed Pipit)
A.f.brevirostris
  Peru, Bolivia
A.f.furcatus
  Brazil, Paraguay, Uruguay, Argentina
Anthus lutescens (Yellowish Pipit)
A.l.parvus
  W Panama
A.l.peruvianus
  Peru, N Chile
A.l.lutescens
  Colombia, Venezuela, the Guianas, Brazil,
  Argentina
Anthus chacoensis (Chaco Pipit)
  Paraguay, Argentina
Anthus correndera (Correndera Pipit)
A.c.calcaratus
  Peru
A.c.catamarcae
  Bolivia, N Chile, NW Argentina
A.c.chilensis
  S Chile, S Argentina
A.c.grayi
  Falkland Is
A.c.correndera
  S Brazil, Uruguay, Paraguay, N Argentina
Anthus antarcticus (South Georgia Pipit)
  S Georgia I
Anthus nattereri (Ochre-breasted Pipit)
  SE Brazil, Paraguay
Anthus hellmayri (Hellmayr's Pipit)
A.h.hellmayri
  Peru, Bolivia, NW Argentina
A.h.dabbenei
  Chile, W Argentina
A.h.brasilianus
  SE Brazil, Uruguay, N Argentina
Anthus bogotensis (Paramo Pipit)
A.b.bogotensis
  Colombia, Ecuador
A.b.immaculatus
  Bolivia, Peru
A.b.shiptoni
  Bolivia, NW Argentina
A.b.meridae
  NW Venezuela

119. CAMPEPHAGIDAE (CUCKOO SHRIKES)

PTEROPODOCYS
Pteropodocys maxima (Ground Cuckoo Shrike)
P.m.pallida
  N Australia
P.m.maxima
  S Australia

CORACINA
Coracina novaehollandiae (Black-faced Cuckoo Shrike)
C.n.macei
  C & S India
C.n.nipalensis
  Assam, Himalayas
C.n.rexpineti
  SE China, Taiwan, N Laos
C.n.layardi
  Sri Lanka
C.n.andamana
  Andaman Is
C.n.siamensis
  Burma, Thailand, S Indochina
C.n.larutensis
  N Malaysia
C.n.larvivorus
  Hainan I
C.n.javensis
  Java, Bali
C.n.lettiensis
  Sumba, Leti, Moa Is
C.n.subpallida
  Kai Is, C Western Australia
C.n.melanops
  E Australia, New Guinea, S Moluccas
C.n.novaehollandiae
  Tasmania, Flinders I
Coracina personata (Wallacean Cuckoo Shrike)   119.1.2
C.p.floris
  Sumbawa, Flores
C.p.sumbensis
  Sumba Is
C.p.personata
  Roti to Sermata, Timor, Wetar Is
C.p.alfrediana
  Lomblen, Alor Is
C.p.pollens
  Kai Is
C.p.unimoda
  Tanimbar Is
Coracina fortis (Buru Is Cuckoo Shrike)
  S Moluccas
Coracina atriceps (Moluccan Cuckoo Shrike)
C.a.atriceps
  Seram I
C.a.magnirostris
  Ternate, Halmahera Is
Coracina schistacea (Slaty Cuckoo Shrike)
  Sula, Banggai, Peleng Is
Coracina caledonica (Melanesian Greybird)
C.c.bougainvillei
  Bougainville I
C.c.kulambangrae
  Kulambangara I
C.c.welchmani
  Ysabel I
C.c.amadonis
  Guadalcanal I
C.c.thilenii
  Espiritu Santo, Malekula Is

C.c.seiuncta
  Erromango I
C.c.lifuensis
  Lifu, Loyalty Is
C.c.caledonica
  New Caledonia I
Coracina caeruleogrisea (Stout-billed Greybird)
  C.c.strenua
    W & C New Guinea, Japen I
  C.c.caeruleogrisea
    Aru Is, S New Guinea
  C.c.adamsoni
    SE New Guinea
Coracina temminckii (Caerulean Cuckoo Shrike)
  C.t.temminckii
    N Sulawesi
  C.t.rileyi
    C & SE Sulawesi
  C.t.tonkeana
    E Sulawesi
Coracina larvata (Black-faced Greybird)
  C.l.melanocephala
    Sumatra
  C.l.larvata
    Java
  C.l.normani
    Borneo
Coracina striata (Barred Cuckoo Shrike)
  C.s.dobsoni
    Andaman Is
  C.s.sumatrensis
    Thailand, Malaysia, Sumatra, Borneo
  C.s.bungurensis
    Anamba, Natuna Is
  C.s.simalurensis
    Simalur I (Sumatra)
  C.s.babiensis
    Babi I (Sumatra)
  C.s.kannegieteri
    Nias I
  C.s.enganensis
    Enggano I
  C.s.vordermani
    Kangean I
  C.s.difficilis
    Palawan, Balabac Is
  C.s.striata
    Luzon, Lubang Is
  C.s.mindorensis
    Mindoro I
  C.s.panayensis
    Masbate, Panay, Negros Is
  C.s.boholensis
    Bohol, Leyte, Samar Is
  C.s.kochii
    Mindanao, Basilan Is
  C.s.guillemardi
    Sulu Arch
Coracina bicolor (Pied Cuckoo Shrike)   119.2
  Muna I, Sulawesi
Coracina lineata (Lineated Cuckoo Shrike)

C.l.axillaris
  Waigeu I, C New Guinea
C.l.maforensis
  Numfor I
C.l.sublineata
  New Ireland, New Britain
C.l.nigrifrons
  Bougainville, Ysabel Is
C.l.ombriosa
  Kulambangra, New Georgia, Rendova Is
C.l.pusilla
  Guadalcanal I
C.l.malaitae
  Malaita I
C.l.makirae
  San Cristobal I
C.l.gracilis
  Rennell I
C.l.lineata
  E Queensland, E New South Wales
Coracina boyeri (White-lored Cuckoo Shrike)
  C.b.boyeri
    Japen I, W New Guinea
  C.b.subalaris
    S New Guinea
Coracina leucopygia (White-rumped Cuckoo Shrike)
  Muna I, Sulawesi
Coracina papuensis (Papuan Cuckoo Shrike)
  C.p.papuensis
    Japen I, W New Guinea
  C.p.intermedia
    S New Guinea
  C.p.oriomo
    SE New Guinea, N Queensland
  C.p.angustifrons
    SE New Guinea
  C.p.louisiadensis
    Louisiade Arch
  C.p.ingens
    Admiralty Is
  C.p.sclateri
    Bismarck Arch
  C.p.perpallida
    Bougainville, Choiseul, Ysabel Is
  C.p.elegans
    New Georgia, Rendova, Guadalcanal Is
  C.p.eyerdami
    Malaita I
  C.p.hypoleuca
    Aru, Melville Is, N Australia
  C.p.stalkeri
    N & C Queensland
Coracina robusta (Little Cuckoo Shrike)
  E Australia
Coracina longicauda (Black-hooded Greybird)
  C.l.grisea
    WC New Guinea
  C.l.longicauda
    C & SE New Guinea

Coracina parvula **(Halmahera Cuckoo Shrike)**
119.2
Halmahera
Coracina abbotti **(Pygmy Cuckoo Shrike)**
119.2
C Sulawesi
Coracina analis **(Caledonian Greybird)**
New Caledonia
Coracina caesia **(African Grey Cuckoo Shrike)**
C.c.preussi
E Nigeria, Fernando Po I
C.c.pura
Ethiopia & Sudan to Malawi
C.c.caesia
Zimbabwe, Sth Africa
Coracina pectoralis **(White-breasted Cuckoo Shrike)**
W, C, E & SC Africa
Coracina graueri **(Grauer's Cuckoo Shrike)**
E Zaire
Coracina cinerea **(Madagascar Cuckoo Shrike)**
C.c.cucullata
Great Comoro I
C.c.cinerea
N & E Madagascar
C.c.pallida
C, W & SW Madagascar
Coracina azurea **(African Blue Cuckoo Shrike)**
W & WC Africa
Coracina typica **(Mauritius Greybird)**
Mauritius I
Coracina newtoni **(Reunion Greybird)**
Reunion I
Coracina coerulescens **(Philippine Black Greybird)**
C.c.coerulescens
Luzon I
C.c.deschauenseei
Marinduque I
Coracina dohertyi **(Sumba Cicadabird)** 119.2
Sumba, Flores Is
Coracina dispar **(Kai Cicadabird)** 119.2
Banda, Kai, Tanimbar, Damar Is
Coracina tenuirostris **(Common Cicadabird)**
C.t.timoriensis
Timor, Lomblen Is
C.t.kalaotuae
Kalaotua I
C.t.emancipata
Djampea I
C.t.pererrata
Tukangbesi I
C.t.edithae
S Sulawesi
C.t.amboinensis
Ambon, Ceram Is
C.t.obiensis
Obi, Bisa Is
C.t.pelingi
Peleng I
C.t.tenuirostris
Queensland, New South Wales, Victoria

C.t.melvillensis
N Queensland, Northern Territory, NW Australia
C.t.aruensis
Aru Is, S New Guinea
C.t.muelleri
Kofiau, Misol Is, New Guinea, D'Entrecasteaux Arch
C.t.nehrkorni
Waigeu I
C.t.grayi
N Moluccas
C.t.numforana
Numfor I
C.t.meyeri
Biak I
C.t.tagulana
Tagula I, Louisiade Arch
C.t.rostrata
Rossel I
C.t.admiralitatis
Admiralty Is
C.t.matthiae
Storm, St Matthias Is
C.t.remota
New Ireland, New Hanover
C.t.heinrothi
New Britain
C.t.rooki
Rook I
C.t.monacha
Palau Is
C.t.nesiotis
Yap I
C.t.saturatior
N & C Solomon Is
C.t.nisoria
Russell I
C.t.erythropygia
Guadalcanal, Malaita Is
C.t.salomonis
San Cristobal I
C.t.insperata
Ponapé I
C.t.ultima
Lihir, Tanga Is
Coracina sula **(Sula Cicadabird)** 119.2
Sula Arch
Coracina morio **(Sulawesi Cicadabird)** 119.2
C.m.morio
Muna I, Sulawesi
C.m.talautensis
Talaut Is
C.m.salvadorii
Sangihe Is
Coracina ceramensis **(Pale Cicadabird)** 119.2
C.c.ceramensis
Seram, Buru, Boano Is
C.c.hoogerwerfi
Obi I
Coracina incerta **(Philippine Cicadabird)** 119.2
C.i.incerta
Waigeu, Japen Is, New Guinea

C.i.everetti
Sulu Is
C.i.mindanensis
Mindanao, Basilan Is
C.i.elusa
Mindoro I
C.i.lecroyae
Luzon I
C.i.ripleyi
Bohol, Samar, Leyte Is
Coracina ostenta (White-winged Cuckoo Shrike) 119.3
Panay, Negros Is
Coracina schisticeps (New Guinea Greybird)
C.s.schisticeps
Misol I, NW New Guinea
C.s.reichenowi
N New Guinea
C.s.poliopsa
S New Guinea
C.s.vittata
D'Entrecasteaux Arch
Coracina melaena (Black Greybird)
C.m.waigeuense
Waigeu I
C.m.tommasonis
Japen I
C.m.melaena
W New Guinea
C.m.meeki
E New Guinea
C.m.goodsoni
Aru Is
C.m.batantae
Batanta I
Coracina montana (Black-bellied Greybird)
C.m.montana
New Guinea
C.m.bicinia
Sepik district, New Guinea
Coracina holopolia (Black-bellied Cuckoo Shrike)
C.h.holopolia
Bougainville, Choiesul, Guadalcanal Is
C.h.pygmaea
Kulambangra, Vangunu Is
C.h.tricolor
Malaita I
Coracina mcgregori (Sharp-tailed Greybird)
N Mindanao
Coracina panayensis (Philippines Greybird)
Negros, Panay Is
Coracina polioptera (Indochinese Cuckoo Shrike)
C.p.jabouillei
N Vietnam
C.p.indochinensis
Burma, C Thailand, C Laos, S Vietnam
C.p.polioptera
S Burma, S Thailand, S Laos
Coracina melaschistos (Dark-grey Cuckoo Shrike)

C.m.melaschistos
N India, Himalayas
C.m.avensis
W China, Burma, N Thailand, N Vietnam
C.m.intermedia
C & S China, Burma, S Thailand S Vietnam
C.m.saturata
N Vietnam, Hainan I
Coracina fimbriata (Lesser Cuckoo Shrike)
C.f.neglecta
S Burma, S Thailand
C.f.culminata
S Malaysia
C.f.schierbrandi
Sumatra, Borneo
C.f.compta
W Sumatran islands
C.f.fimbriata
Java, Bali
Coracina melanoptera (Black-headed Cuckoo Shrike)
C.m.melanoptera
N India
C.m.sykesi
S India, Sri Lanka

CAMPOCHAERA
Campochaera sloetii (Orange Cuckoo Shrike)
C.s.sloetii
NW New Guinea
C.s.flaviceps
SE New Guinea

CHLAMYDOCHAERA
Chlamydochaera jefferyi (Black-breasted Triller)
Borneo

LALAGE
Lalage melanoleuca (Black & White Triller)
L.m.melanoleuca
Luzon, Mindoro Is
L.m.minor
Samar, Leyte, Mindanao Is
Lalage nigra (Pied Triller)
L.n.davisoni
Nicobar Is
L.n.nigra
Malaysia, Sumatra, Java
L.n.chilensis
Borneo, Philippine Is
L.n.leucopygialis
Sulawesi, Sula Arch,
Lalage sueurii (White-winged Triller)
L.s.sueurii
E Java, Lesser Sunda Is, S Sulawesi
L.s.tricolor
N & C Australia, SE New Guinea
Lalage aurea (Red-bellied Triller)
N Moluccas

Lalage atrovirens (Black-browed Triller)
  L.a.moesta
    Tenimber Is
  L.a.atrovirens
    Misol, Waigeu Is, N New Guinea
  L.a.leucoptera
    Biak I
Lalage leucomela (White-browed Triller)
  L.l.keyensis
    Kai Is
  L.l.rufiventer
    Melville I, Northern Territory
  L.l.leucomela
    E Queensland, NE New South Wales
  L.l.yorki
    N Queensland
  L.l.polygrammica
    Aru Is, E New Guinea
  L.l.obscurior
    D'Entrecasteaux Arch
  L.l.trobriandi
    Trobriand Is
  L.l.pallescens
    Louisiade Arch
  L.l.falsa
    New Britain, Rook I
  L.l.karu
    New Ireland
  L.l.albidior
    New Hanover
  L.l.ottomeyeri
    Lihir Is
  L.l.tabarensis
    Tabar I
  L.l.conjuncta
    St Matthias Is
  L.l.sumunae
    Dyaul I
Lalage maculosa (Spotted Triller)
  L.m.ultima
    Efate I
  L.m.modesta
    N & C New Hebrides
  L.m.melanopygia
    Santa Cruz I
  L.m.vanikorensis
    Vanikoro I
  L.m.soror
    Kandavu I
  L.m.pumila
    Viti Levu I
  L.m.mixta
    C & NW Fiji Is
  L.m.woodi
    Vanua Levu I
  L.m.rotumae
    Rotuma I
  L.m.nesophila
    Lau Archipelago
  L.m.tabuensis
    Tonga I
  L.m.vauana
    Vavau group, Fiji Is

  L.m.keppeli
    Keppel, Boscawen Is
  L.m.futunae
    Futuna, Horne Is
  L.m.whitmeei
    Niue, Savage Is
  L.m.maculosa
    Upolu, Savaii, Samoa Is
Lalage sharpei (Samoan Triller)
  L.s.sharpei
    Upolu, Samoa Is
  L.s.tenebrosa
    Savaii, Samoa Is
Lalage leucopyga (Long-tailed Triller)
  L.l.affinis
    San Cristobal I
  L.l.deficiens
    Torres, Banks Is
  L.l.albiloris
    C & N New Hebrides
  L.l.simillima
    S New Hebrides, Loyalty Is
  L.l.montrosieri
    New Caledonia I
  L.l.leucopyga
    Norfolk I

CAMPEPHAGA
Campephaga flava (African Black Cuckoo Shrike) 119.4
    Uganda, Somalia to Angola & Sth Africa
Campephaga phoenicea (Red-shouldered Cuckoo Shrike)
    Gambia to Ethiopia, N Zaire, Uganda
Campephaga petiti (Petit's Cuckoo Shrike)
    S Cameroun to Kenya
Campephaga quiscalina (Purple-throated Cuckoo Shrike)
  C.q.quiscalina
    Sierra Leone to Cameroun & N Angola
  C.q.martini
    E Zaire, Uganda, Kenya
  C.q.munzneri
    Tanzania
Campephaga lobata (Western Wattled Cuckoo Shrike)
    Liberia, Ghana
Campephaga oriolina (Eastern Wattled Cuckoo Shrike) 119.4
    S Cameroun, Gabon to E Zaire

PERICROCOTUS
Pericrocotus roseus (Rosy Minivet)
  P.r.càntonensis
    E China, Thailand, Laos
  P.r.stanfordi
    S China, S Thailand, S Laos
  P.r.roseus
    SW China, Burma, Himalayas, N India
Pericrocotus divaricatus (Ashy Minivet)
  P.d.divaricatus
    NE Asia, Japan, E China, SE Asia, Philippine Is

P.d.tegimae
Riukiu Is
Pericrocotus cinnamomeus (Small Minivet)
P.c.malabaricus
W India
P.c.cinnamomeus
S India, Sri Lanka
P.c.pallidus
Pakistan
P.c.peregrinus
N India, Himalayas
P.c.vividus
Andaman Is
P.c.thai
Bhutan, Burma, N Thailand Laos
P.c.sacerdos
Cambodia, S Vietnam
P.c.seperatus
S Burma, S Thailand
P.c.saturatus
Java, Bali
P.c.igneus
Malaysia, Sumatra, Borneo, Palawan I
P.c.trophis
Simalur I
Pericrocotus lansbergei (Flores Minivet)
Sumbawa, Flores Is
Pericrocotus erythropygius (Jerdon's Minivet)
P.e.erythropygius
W Pakistan, C India
P.e.albifrons
C Burma
Pericrocotus solaris (Yellow-throated Minivet)
P.s.solaris
E Himalayas, NW Burma
P.s.rubrolimbatus
S Burma, N Thailand
P.s.montpellieri
SW China
P.s.griseogularis
SE China, Taiwan, N Indochina
P.s.deignani
S Vietnam
P.s.nassovicus
SE Thailand, Cambodia
P.s.montanus
Malaysia, W Sumatra
P.s.cinereigula
N Borneo
Pericrocotus ethologus (Long-tailed Minivet)
P.e.favillaceus
Afghanistan, W Himalayas, W India
P.e.laetus
E Nepal, W Assam
P.e.ethologus
W China, N Thailand, N Indochina
P.e.yvettae
NE Burma
P.e.mariae
SE Assam, W Burma
P.e.ripponi
E Burma, NW Thailand

P.e.annamensis
S Vietnam
Pericrocotus brevirostris (Short-billed Minivet)
P.b.brevirostris
Himalayas, Nepal, W Assam
P.b.affinis
E Assam, NW Burma
P.b.neglectus
N Thailand, N Laos
P.b.anthoides
S China, N Vietnam
Pericrocotus miniatus (Sunda Minivet)
W Sumatra, Java
Pericrocotus flammeus (Scarlet Minivet)
P.f.flammeus
S India, Sri Lanka
P.f.siebersi
Java, Bali
P.f.exul
Lombok I
P.f.andamanensis
Andaman Is
P.f.minythomelas
Simalur I
P.f.modiglianii
Enggano I
P.f.speciosus
Himalayas, N India
P.f.elegans
S Assam, N Burma, N Thailand, N Indochina
P.f.fohkiensis
Fukien
P.f.semiruber
E India, S Burma, Thailand
P.f.flammifer
S Thailand, N & C Malaysia
P.f.xanthogaster
S Malaysia, Sumatra
P.f.insulanus
Borneo
P.f.novus
Luzon, Negros Is
P.f.leytensis
Samar, Leyte Is
P.f.johnstoniae
Mt Apo (Mindanao I)
P.f.marchesae
Jolo I
P.f.gonzalesi
Mt Katanglad (Mindanao I)
P.f.nigroluteus
Mindanao I
P.f.fraterculus
Assam

Hemipus picatus (Bar-winged Flycatcher Shrike)
H.p.capitalis
Himalayas, N Burma, N Thailand, N Indochina

H.p.picatus
India, S Burma, S Thailand, S Indochina
H.p.intermedius
S Thailand, Malaysia, Sumatra, N Borneo
H.p.leggei
Sri Lanka
Hemipus hirundinaceus (Black-winged Flycatcher Shrike)
Malaysia, Sumatra, Java, Bali, Borneo

TEPHRODORNIS
Tephrodornis gularis (Brown-tailed Wood Shrike)
T.g.sylvicola
W India
T.g.pelvica
E Himalayas, N Burma
T.g.jugans
S Burma, N Thailand
T.g.vernayi
SW Thailand
T.g.annectens
S Thailand, N Malaysia
T.g.fretensis
S Malaysia, Sumatra
T.g.gularis
SW Sumatra, Java
T.g.frenata
Borneo
T.g.mekongensis
E Thailand, Cambodia, S Indochina
T.g.hainana
N Indochina, Hainan I
T.g.latouchei
Fukien
Tephrodornis pondicerianus (Common Wood Shrike)
T.p.affinis
Sri Lanka
T.p.pondicerianus
E India, Burma, N Thailand S Laos
T.p.pallidus
Pakistan, NW India
T.p.orientis
Cambodia, S Vietnam

120. PYCNONOTIDAE (BULBULS)

SPIZIXOS
Spizixos canifrons (Crested Finchbill)
S.c.canifrons
S Assam, W Burma
S.c.ingrami
E Burma to S China, N Indochina
Spizixos semitorques (Collared Finchbill)
S.s.semitorques
S China
S.s.cinereicapillus
Taiwan

PYCNONOTUS
Pycnonotus zeylanicus (Straw-crowned Bulbul)
Malaysia to Java, Borneo

Pycnonotus striatus (Striated Green Bulbul)
P.s.striatus
E Himalayas, W Burma
P.s.arctus
NE Assam
P.s.paulus
Burma to N Indochina, S China
Pycnonotus leucogrammicus (Striated Bulbul)
W Sumatra
Pycnonotus tympanistrigus (Olive-crowned Bulbul)
W Sumatra
Pycnonotus melanoleucos (Black & White Bulbul)
Malaysia, Sumatra, Borneo
Pycnonotus priocephalus (Grey-headed Bulbul)
SW India
Pycnonotus atriceps (Black-headed Bulbul)
P.a.fuscoflavescens
Andaman Is
P.a.atriceps
NE India to Bali, Borneo, Palawan I
P.a.hyperemnus
W Sumatra islands
P.a.baweanus
Bawean I
P.a.hodiernus
Maratua I
Pycnonotus melanicterus (Black-crested Bulbul)
P.m.melanicterus
Sri Lanka
P.m.gularis
SW India
P.m.flaviventris
Himalayas, NE India, N Burma
P.m.vantynei
S Burma to N Indochina
P.m.xanthops
SE Burma, N Thailand
P.m.auratus
NE Thailand, W Laos
P.m.johnsoni
SE Thailand, S Indochina
P.m.elbeli
SE Thailand islands
P.m.negatus
SW Thailand
P.m.caecilii
N Malaysia
P.m.dispar
Sumatra, Java
P.m.montis
N Borneo
Pycnonotus squamatus (Scaly-breasted Bulbul)
P.s.weberi
Malaysia, Sumatra
P.s.squamatus
W & C Java

P.s.borneensis
Borneo
Pycnonotus cyaniventris (Grey-bellied Bulbul)
P.c.cyaniventris
Malaysia, Sumatra
P.c.paroticalis
Borneo
Pycnonotus jocosus (Red-whiskered Bulbul)
P.j.fuscicaudatus
W India
P.j.abuensis
N Bombay, SW Rajasthan
P.j.pyrrhotis
Nepal, N India
P.j.emeria
E India, Burma, SW Thailand
P.j.whistleri
Andaman Is
P.j.monticola
E Himalayas to SW China
P.j.pattani
Thailand, N Malaysia, S Indochina
P.j.hainanensis
N Vietnam, SE China
P.j.jocosus
S China
Pycnonotus xanthorrhous (Anderson's Bulbul)
P.x.xanthorrhous
NE Burma to N Vietnam
P.x.andersoni
S China
Pycnonotus sinensis (Chinese Bulbul)
P.s.hoyi
C China
P.s.sinensis
E China
P.s.hainanus
Hainan I, SE China, N Vietnam
P.s.formosae
Taiwan
P.s.orii
S Riukiu Is
Pycnonotus taivanus (Formosan Bulbul)
Taiwan
Pycnonotus leucogenys (White-cheeked Bulbul)
P.l.mesopotamiae
Iraq
P.l.dactylus
E Saudi Arabia
P.l.leucotis
S Iran to NW India
P.l.humii
NW Pakistan
P.l.leucogenys
Himalayas
Pycnonotus cafer (Red-vented Bulbul)
P.c.cafer
Sri Lanka
P.c.pusillus
S India

P.c.humayuni
Pakistan, NW India
P.c.wetmorei
NE India
P.c.intermedius
W Himalayas
P.c.bengalensis
E Himalayas, NE India
P.c.primrosei
S Assam
P.c.stanfordi
N Burma, W Yunnan
P.c.melanchimus
SC Burma
Pycnonotus aurigaster (White-eared Bulbul)
P.a.chrysorrhoides
S China
P.a.resurrectus
SE China, N Vietnam
P.a.dolichurus
C Vietnam
P.a.latouchei
Burma to SW China, N Vietnam
P.a.klossi
SE Burma, N Thailand
P.a.schauenseei
S Burma, SW Thailand
P.a.thais
S Thailand
P.a.germani
SE Thailand, S Indochina
P.a.aurigaster
Sumatra, Java, S Sulawesi
Pycnonotus xanthopygos (White-eyed Bulbul)
Syria to Aden
Pycnonotus nigricans (Red-eyed Bulbul)
P.n.nigricans
Namibia, S Botswana, N Cape Province
P.n.grisescentior
E & N Botswana, S Angola to Zimbabwe
P.n.superior
NE Cape Province, S Transvaal
Pycnonotus capensis (Cape Bulbul)
S & SW Cape Province
Pycnonotus barbatus (Garden Bulbul)
P.b.barbatus
Nth Africa
P.b.inornatus
Senegal to Ghana
P.b.goodi
S Sahara, N Cameroun
P.b.arsinoe
Nile valley, Sudan
P.b.schoanus
E Ethiopia
P.b.somaliensis
SE Ethiopia, Somalia
P.b.nigeriae
C & S Nigeria to Gabon
P.b.gabonensis
W Gabon

P.b.tricolor
N Namibia to S Uganda, S Tanzania
P.b.minor
N Zaire, S Sudan
P.b.spurius
S Ethiopia
P.b.fayi
Kenya, C Tanzania
P.b.naumanni
S Tanzania to N Transvaal & Mozambique
P.b.ngamii
S Zambia, Botswana
P.b.pallidus
SE Tanzania to E Natal
P.b.layardi
SE Africa
P.b.tenebrior
E Cape Province, S Lesotho
P.b.micrus
SE Kenya, E Tanzania
P.b.peasei
S Ethiopia, E Kenya
P.b.dodsoni
N Kenya, S Somalia, E Ethiopia
Pycnonotus eutilotus (Puff-backed Bulbul)
Malaysia, Sumatra, Borneo
Pycnonotus nieuwenhuisii (Blue-wattled Bulbul)
P.n.inexpectatus
Lesten (Sumatra)
P.n.nieuwenhuisii
Kayan river (Borneo)
Pycnonotus urostictus (Yellow-wattled Bulbul)
P.u.ilokensis
N Luzon I
P.u.urostictus
C Luzon I
P.u.atricaudatus
Bohol, Samar, Leyte Is
P.u.philippensis
Mindanao I
P.u.basilanicus
Basilan I
Pycnonotus bimaculatus (Orange-spotted Bulbul)
P.b.snouckaerti
NW Sumatra
P.b.barat
SW Sumatra, W & C Java
P.b.bimaculatus
E Java, Bali
Pycnonotus finlaysoni (Stripe-throated Bulbul)
P.f.davisoni
S Burma
P.f.eous
Thailand, C & S Indochina
P.f.finlaysoni
Malaysia
Pycnonotus xantholaemus (Yellow-throated Bulbul)
S India

Pycnonotus penicillatus (Yellow-tufted Bulbul)
Sri Lanka
Pycnonotus flavescens (Flavescent Bulbul)
P.f.flavescens
S Assam, W Burma
P.f.vividus
Burma, Thailand, N Indochina
P.f.sordidus
S Indochina
P.f.leucops
N Borneo
Pycnonotus goiavier (Yellow-vented Bulbul)
P.g.jambu
SE Thailand, S Indochina
P.g.personatus
Malaysia, Sumatra
P.g.analis
Java, Bali, Lombok I, S Sulawesi
P.g.gourdini
Borneo
P.g.goiavier
N & C Philippine Is
P.g.suluensis
S Philippine Is
Pycnonotus luteolus (White-browed Bulbul)
P.l.luteolus
C & S India
P.l.insulae
Sri Lanka
Pycnonotus plumosus (Olive-brown Bulbul)
P.p.plumosus
Malaysia, E Sumatra, Java
P.p.porphyreus
W Sumatra and islands
P.p.billitonis
Billiton I, W & S Borneo
P.p.hutzi
N & E Borneo
P.p.chiroplethis
Anamba Is
P.p.hachisukae
N Borneo islands
P.p.cinereifrons
Palawan I
P.p.sibergi
Bawean I, Java Sea
Pycnonotus blanfordi (Blanford's Olive Bulbul)
P.b.blanfordi
C & S Burma
P.b.conradi
Thailand, N Malaysia, S Indochina
P.b.robinsoni
C Malaysia
Pycnonotus simplex (White-eyed Brown Bulbul)
P.s.simplex
S Thailand, Malaysia, Sumatra
P.s.prillwitzi
Java
P.s.oblitus
Bangka, Billiton Is, S & W Borneo

P.s.halizonus
Anamba, N Natuna Is
P.s.perplexus
N & E Borneo
Pycnonotus brunneus (Red-eyed Brown Bulbul)
P.b.brunneus
Malaysia, Sumatra, Borneo
P.b.zapolius
Anamba Is
Pycnonotus erythrophthalmus (Lesser Brown Bulbul)
P.e.erythrophthalmus
Malaysia, Billiton I, Sumatra
P.e.salvadorii
Borneo
Pycnonotus masukuensis (Shelley's Greenbul)
P.m.kakamegae
E Zaire, W Kenya, W Tanzania
P.m.roehli
C Tanzania
P.m.masukuensis
SW Tanzania, N Malawi
Pycnonotus montanus (Cameroun Montane Greenbul)
Togo, Ghana, Cameroun
Pycnonotus virens (Little Greenbul)
P.v.erythropterus
Gambia to S Nigeria
P.v.virens
Gabon to Sudan, Uganda, Angola
P.v.holochlorus
W Uganda
P.v.zombensis
E Angola, SE Zaire to E Kenya,
Mozambique
P.v.marwitzi
SE Kenya
P.v.zanzibaricus
Zanzibar I
Pycnonotus hallae (Hall's Greenbul)
E Zaire
Pycnonotus gracilis (Little Grey Greenbul)
P.g.extremus
Sierra Leone to S Nigeria
P.g.gracilis
Cameroun to Uganda & N Angola
P.g.ugandae
E Zaire, Uganda
Pycnonotus ansorgei (Ansorge's Greenbul)
P.a.ansorgei
Sierra Leone to S Nigeria
P.a.muniensis
Cameroun, Gabon, N Zaire
P.a.kavirondensis
Kenya
Pycnonotus curvirostris (Plain Greenbul)
P.c.leoninus
Sierra Leone to Ghana
P.c.curvirostris
S Ghana to W Kenya, N Angola
Pycnonotus importunus (Sombre Greenbul)

P.i.fricki
WC Kenya
P.i.somaliensis
S Somalia
P.i.subalaris
S Kenya
P.i.insularis
E Tanzania, Zanzibar I
P.i.hypoxanthus
S Tanzania, Malawi, Mozambique
P.i.loquax
E Zimbabwe to SE Tanzania
P.i.mentor
S Mozambique, N Natal, NE Transvaal
P.i.errolius
E Natal, E Transvaal
P.i.noomei
E Zimbabwe to Natal, C Cape Province
P.i.importunus
S & E Cape Province
Pycnonotus latirostris (Yellow-whiskered Greenbul)
P.l.congener
Sierra Leone to S Nigeria
P.l.latirostris
SE Nigeria to W Zaire, N Angola
P.l.eugenius
S Sudan, E Zaire to NW Tanzania
P.l.saturatus
E Kenya, N Tanzania
P.l.australis
SW Tanzania
Pycnonotus gracilirostris (Slender-billed Greenbul)
P.g.gracilirostris
Senegal to Nigeria
P.g.congensis
S Cameroun to W Zaire, N Angola
P.g.chagwensis
N Zaire to W Kenya, NW Tanzania
P.g.percivali
C Kenya
Pycnonotus tephrolaemus(Mountain Greenbul)
P.t.tephrolaemus
Cameroun Mt, Fernando Po I
P.t.bamendae
SE Nigeria, Cameroun
P.t.kikuyuensis
E Zaire to Uganda, W Kenya
P.t.kungwensis
W Tanzania
P.t.nigriceps
SW Kenya, N Tanzania
P.t.usambarae
NE Tanzania
P.t.neumanni
E Tanzania
P.t.fusciceps
SW Tanzania, Malawi, NW Mozambique
P.t.chlorigula
C Tanzania

Pycnonotus milanjensis (Stripe-cheeked Greenbul)
P.m.striifacies
  SE Kenya, N Tanzania
P.m.olivaceiceps
  SW Tanzania, N Malawi, N Mozambique
P.m.milanjensis
  Malawi, W Mozambique, Zimbabwe
P.m.disjuncta
  E Zimbabwe

CALYPTOCICHLA
Calyptocichla serina (Golden Greenbul)
  Sierra Leone to N Zaire, Gabon

BAEOPOGON
Baeopogon indicator (Honeyguide Greenbul)
B.i.leucurus
  Sierra Leone, Liberia
B.i.togoensis
  Ghana, Togo
B.i.indicator
  Cameroun to N Angola, W Zaire
B.i.chlorosaturata
  E Zaire to Sudan, Uganda
Baeopogon clamans (Sjöstedt's Honeyguide Greenbul)
  Cameroun to NE Zaire, Gabon

IXONOTUS
Ixonotus guttatus (Spotted Greenbul)
I.g.guttatus
  Ghana, Gabon to C Zaire
I.g.bugoma
  E Zaire, W Uganda

CHLOROCICHLA
Chlorocichla falkensteini (Yellow-necked Greenbul)
C.f.viridescentior
  River Ja, Cameroun
C.f.falkensteini
  SW Zaire, N Angola
Chlorocichla simplex (Simple Greenbul)
  Guinea to Zaire, N Angola
Chlorocichla flavicollis (Yellow-throated Leaf-Love)
C.f.flavicollis
  Senegal to Cameroun
C.f.adamauae
  N Cameroun
C.f.simplicicolor
  E Cameroun
C.f.soror
  Cameroun to Zaire, S Sudan
C.f.flavigula
  Angola to NW Tanzania, Zambia
C.f.pallidigula
  Uganda, W Kenya
Chlorocichla flaviventris (Yellow-bellied Greenbul)
C.f.centralis
  Kenya, Tanzania, N Mozambique

C.f.zambesiae
  N Namibia, S Zambia, Zimbabwe
C.f.occidentalis
  N Namibia, S Angola to Mozambique
C.f.ortiva
  S Mozambique, Malawi, SE Zimbabwe
C.f.flaviventris
  Natal
Chlorocichla laetissima (Joyful Greenbul)
C.l.laetissima
  E Zaire to S Sudan, W Kenya
C.l.schoutedeni
  SE Zaire, SW Tanzania, NE Zambia
Chlorocichla prigoginei (Prigogine's Greenbul)
  Lake Edward, E Zaire

THESCELOCICHLA
Thescelocichla leucopleura (Swamp Palm Bulbul)
  Senegal to Gabon, W Uganda

PHYLLASTREPHUS
Phyllastrephus scandens (Leaf-Love)
P.s.scandens
  Senegal to Cameroun
P.s.acedis
  S Cameroun, Gabon, SW Zaire
P.s.orientalis
  Central African Republic to Sudan & Tanzania
P.s.upembae
  S Zaire, W Tanzania
Phyllastrephus terrestris (Terrestrial Brownbul)
P.t.bensoni
  C Kenya
P.t.katangae
  Katanga, Zaire
P.t.suahelicus
  SE Kenya, E Tanzania, N Mozambique
P.t.rhodesiae
  S Zaire to N Namibia
P.t.intermedius
  S Angola to Mozambique, N Natal
P.t.terrestris
  Transvaal, S Natal, Cape Province
P.t.montanus
  Transkei, Natal, E Transvaal
Phyllastrephus strepitans (Northern Brownbul)
  S Sudan to E Tanzania
Phyllastrephus cerviniventris (Grey Olive Greenbul)
P.c.cerviniventris
  C Kenya to Zambia & Mozambique
P.c.schoutedeni
  Katanga, Zaire
Phyllastrephus fulviventris (Pale Olive Greenbul)
  Central African Republic to Angola

Phyllastrephus poensis (Cameroun Olive Greenbul)
S Nigeria, Cameroun mt
Phyllastrephus hypochloris (Toro Olive Greenbul) 120.4
S Sudan, S Uganda, W Kenya, E Zaire
Phyllastrephus baumanni (Baumann's Greenbul)
Sierra Leone to S Nigeria
Phyllastrephus poliocephalus (Grey-headed Greenbul)
SE Nigeria, Cameroun mt
Phyllastrephus flavostriatus (Yellow-streaked Greenbul)
P.f.graueri
E Zaire
P.f.olivaceogriseus
E Zaire, SW Uganda
P.f.kungwensis
W Tanzania
P.f.uzungwensis
Morogoro (E Tanzania)
P.f.tenuirostris
SE Kenya to NE Mozambique
P.f.alfredi
SW Tanzania, E Zambia, N Malawi
P.f.vincenti
S Malawi, W Mozambique
P.f.dryobates
S Mozambique
P.f.dendrophilus
E Zimbabwe, W Mozambique
P.f.distans
E Transkei
P.f.flavostriatus
E Sth Africa
Phyllastrephus debilis (Tiny Greenbul)
P.d.rabai
E Kenya, E Tanzania
P.d.albigula
N Tanzania
P.d.debilis
S Tanzania, Mozambique
Phyllastrephus lorenzi (Sassi's Olive Greenbul)
E Zaire
Phyllastrephus albigularis (White-throated Greenbul)
P.a.albigularis
Sierra Leone to Sudan, Uganda
P.a.viridiceps
N Angola
Phyllastrephus fischeri (Fischer's Greenbul)
P.f.sucosus
S Sudan to E Zaire, W Tanzania
P.f.nandensis
N Nandi (Kenya)
P.f.ngurumanensis
SW Kenya
P.f.fischeri
E Tanzania, Mozambique
Phyllastrephus cabanisi (Cabanis' Greenbul)
Angola, S Zaire, Zambia, W Tanzania

Phyllastrephus placidus (Olive Mountain Greenbul)
SE Kenya to Malawi, Mozambique
Phyllastrephus icterinus (Icterine Greenbul)
P.i.icterinus
Sierra Leone to Nigeria
P.i.tricolor
Cameroun to W Uganda, C Zaire
Phyllastrephus xavieri (Xavier's Greenbul)
P.x.serlei
Cameroun Mt
P.x.xavieri
Cameroun to S Central African Republic, N Zaire
P.x.sethsmithi
E Zaire, Uganda
Phyllastrephus leucolepis (Liberian Greenbul) 120.2
SE Liberia
Phyllastrephus madagascariensis (Madagascar Tetraka)
P.m.madagascariensis
E Madagascar
P.m.inceleber
N & W Madagascar
Phyllastrephus zosterops (Short-billed Tetraka)
P.z.fulvescens
N Madagascar
P.z.andapae
NE Madagascar
P.z.zosterops
E Madagascar
P.z.ankafanae
SE Madagascar
Phyllastrephus apperti (Appert's Tetraka)
SW Madagascar
Phyllastrephus tenebrosus (Dusky Tetraka)
E Madagascar
Phyllastrephus cinereiceps (Grey-crowned Greenbul)
E Madagascar 120.3

BLEDA
Bleda syndactyla (Common Bristle-Bill)
B.s.syndactyla
Senegal to S Nigeria
B.s.multicolor
S Nigeria to Zambia
B.s.woosnami
Sudan, Uganda, NW Kenya
B.s.nandensis
N Nandi (Kenya)
Bleda eximia (Green-tailed Bristle-Bill)
B.e.eximia
Sierra Leone to Ghana
B.e.notata
S Nigeria to Central African Republic
B.e.ugandae
Sudan, Uganda, N Zaire
Bleda canicapilla (Grey-headed Bristle-Bill)
Gambia to S Nigeria

CRINIGER

Criniger barbatus (Bearded Bulbul)
  C.b.barbatus
    Sierra Leone to Togo
  C.b.ansorgeanus
    S Nigeria
  C.b.chloronotus
    Cameroun to Central African Republic
  C.b.weileri
    E Zaire

Criniger calurus (Red-tailed Bulbul)
  C.c.verreauxi
    Guinea to Nigeria
  C.c.calurus
    S Nigeria to Central African Republic
  C.c.emini
    Zaire, Uganda

Criniger ndussumensis (White-bearded Bulbul)
    Cameroun to N Zaire

Criniger olivaceus (Yellow-bearded Bulbul)
    Senegal to Ghana

Criniger finschii (Finsch's Bearded Bulbul)
    Malaysia, Sumatra, Borneo

Criniger flaveolus (Ashy-fronted Bearded Bulbul)
  C.f.flaveolus
    Himalayas to NE Burma
  C.f.burmanicus
    SE Burma, W Thailand

Criniger pallidus (Olivaceous Bearded Bulbul)
  C.p.griseiceps
    S Burma
  C.p.robinsoni
    Tenasserim, S Burma
  C.p.henrici
    N Thailand, N Indochina
  C.p.pallidus
    Hainan I
  C.p.isani
    NE Thailand
  C.p.annamensis
    C Indochina
  C.p.khmerensis
    S Indochina

Criniger ochraceus (Ochraceous Bearded Bulbul)
  C.o.hallae
    S Vietnam
  C.o.cambodianus
    SE Thailand, SW Cambodia
  C.o.ochraceus
    S Burma, SW Thailand
  C.o.sordidus
    C Malaysia
  C.o.sacculatus
    S Malaysia
  C.o.sumatranus
    W Sumatra
  C.o.fowleri
    N Borneo

  C.o.ruficrissus
    NE Borneo

Criniger bres (Grey-cheeked Bearded Bulbul)
  C.b.tephrogenys
    Malaysia, E Sumatra
  C.b.bres
    W & C Java
  C.b.balicus
    E Java, Bali
  C.b.gutteralis
    Borneo
  C.b.frater
    Palawan I

Criniger phaeocephalus (Grey-headed Bearded Bulbul)
  C.p.phaeocephalus
    Malaysia, Sumatra, Bangka, Billiton Is
  C.p.connectens
    NE Borneo
  C.p.sulphuratus
    C Borneo
  C.p.diardi
    W Borneo

SETORNIS

Setornis criniger (Long-billed Bulbul)
    E Sumatra, Bangka I, Borneo

HYPSIPETES

Hypsipetes viridescens (Blyth's Olive Bulbul)
  H.v.cacharensis
    S Assam
  H.v.myitkyinensis
    NE Burma
  H.v.viridescens
    S Burma, SW Thailand

Hypsipetes propinquus (Grey-eyed Bulbul)
  H.p.aquilonis
    N Vietnam
  H.p.propinquus
    E Burma, N Thailand, N Laos
  H.p.simulator
    SE Thailand, S Indochina
  H.p.innectens
    S Vietnam
  H.p.lekhakuni
    S Burma, SW Thailand
  H.p.cinnamomeoventris
    N Malaysia

Hypsipetes charlottae (Crested Olive Bulbul)
  H.c.cryptus
    Malaysia, Sumatra & islands
  H.c.charlottae
    S & W Borneo
  H.c.perplexus
    N & E Borneo

Hypsipetes palawanensis (Golden-eyed Bulbul)
    Palawan I

Hypsipetes criniger (Hairy-backed Bulbul)
  H.c.criniger
    Malaysia, E Sumatra

H.c.sericeus
W Sumatra
H.c.viridis
Borneo
Hypsipetes philippinus (Rufous-breasted Bulbul)
H.p.philippinus
Luzon, Samar, Leyte, Bohol Is
H.p.guimarasensis
Masbate, Panay, Negros Is
H.p.saturatior
E Mindanao I
H.p.mindorensis
Mindoro I
H.p.rufigularis
W Mindanao, Basilan Is
Hypsipetes siquijorensis (Slaty-crowned Bulbul)
H.s.cinereiceps
Tablas, Romblon Is
H.s.monticola
Cebu I
H.s.siquijorensis
Siquijor I
Hypsipetes everetti (Yellow-washed Bulbul)
H.e.everetti
Dinagat, E & C Mindanao Is
H.e.haynaldi
Sulu Arch
H.e.samarensis
Samar, Leyte Is
H.e.catarmanensis
Camiguin South I
Hypsipetes affinis (Golden Bulbul)
H.a.platenae
Sanghir Is
H.a.aureus
Togian I
H.a.harterti
Peleng, Banggai Is
H.a.longirostris
Sula Is
H.a.chloris
Morotai, Halmahera, Batjan Is
H.a.lucasi
Obi I
H.a.mystacalis
Buru I
H.a.affinis
Ceram I
H.a.flavicaudus
Ambon I
Hypsipetes indicus (Golden-browed Bulbul)
H.i.ictericus
W India
H.i.indicus
SW India, Sri Lanka
H.i.guglielmi
SW Sri Lanka
Hypsipetes mcclellandii (Mountain Streaked Bulbul)
H.m.mcclellandii
E Himalayas, Assam

H.m.ventralis
SW Burma
H.m.tickelli
E Burma, NW Thailand
H.m.similis
NE Burma to N Indochina
H.m.holtii
S China
H.m.loquax
N & E Thailand, S Laos
H.m.griseiventer
S Vietnam
H.m.canescens
SE Thailand
H.m.peracensis
S Thailand, N Malaysia
Hypsipetes malaccensis (Green-backed Bulbul)
S Vietnam, Malaysia, Sumatra, Borneo
Hypsipetes virescens (Sumatran Bulbul)
H.v.sumatranus
W Sumatra
H.v.virescens
Java
Hypsipetes flavalus (Ashy Bulbul)
H.f.flavalus
E Himalayas, NW Burma
H.f.cannipennis
S China, NE Vietnam
H.f.castanotus
Hainan I
H.f.bourdellei
E Thailand, N Laos
H.f.remotus
S Indochina
H.f.hildebrandi
S Burma, NW Thailand
H.f.davisoni
S Burma, SW Thailand
H.f.cinereus
Malaysia, Sumatra
H.f.connectens
N Borneo
Hypsipetes amaurotis (Chestnut-eared Bulbul)
H.a.hensoni
N Japan, S Korea, NE China
H.a.amaurotis
C Japan, Riukiu Is
H.a.matchiae
S Kyushu Is
H.a.squamiceps
Bonin Is
H.a.magnirostris
Volcano Is
H.a.borodinonis
Borodino Is
H.a.ogawae
N Riukiu Is
H.a.pryeri
C Riukiu Is
H.a.insignis
Miyakojima Is (S Riukiu Is)

H.a.stejnegeri
S Riukiu Is
H.a.nagamichii
S Taiwan
H.a.batanensis
Batjan I
H.a.fugensis
Calayan, Fuga Is
H.a.camiguinensis
Camiguin Is
Hypsipetes crassirostris (Thick-billed Bulbul)
Seychelles Is
Hypsipetes borbonicus (Reunion Bulbul)
H.b.borbonicus
Reunion I
H.b.olivaceus
Mauritius I
Hypsipetes madagascariensis (Black Bulbul)
H.m.madagascariensis
Madagascar
H.m.grotei
Glorioso I
H.m.parvirostris
Comoro Is
H.m.rostratus
Aldabra I
H.m.humii
Sri Lanka
H.m.ganeesa
SW India
H.m.psaroides
Himalayas
H.m.nigrescens
E Assam, W Burma
H.m.concolor
E Burma, SW China to S Vietnam
H.m.ambiens
NE Burma
H.m.sinensis
SW China, Thailand, Laos
H.m.stresemanni
SW China, Thailand, Laos
H.m.leucothorax
W China to N Vietnam
H.m.leucocephalus
SE China
H.m.nigerrimus
Taiwan
H.m.perniger
Hainan I
Hypsipetes nicobariensis (Nicobar Bulbul)
Nicobar Is
Hypsipetes thompsoni (Bingham's Bulbul)
S Burma, NW Thailand

MALIA   120.1
Malia grata (Malia)
M.g.recondita
N Sulawesi
M.g.stresemanni
C & SE Sulawesi
M.g.grata
SW Sulawesi

NEOLESTES
Neolestes torquatus (Black-collared Bulbul)
Gabon, Zaire, Angola

TYLAS
Tylas eduardi (Kinkimavo)
T.e.eduardi
E Madagascar
T.e.albigularis
WC Madagascar

121. IRENIDAE (LEAFBIRDS, IORAS)

AEGITHINA
Aegithina tiphia (Common Iora)
A.t.multicolor
S India, Sri Lanka
A.t.deignani
SC India, N & C Burma
A.t.humei
C India
A.t.tiphia
NE India
A.t.septentrionalis
Pakistan, NW India
A.t.philipi
SW China, C Burma, N Thailand, Laos N
Vietnam
A.t.cambodiana
Cambodia, SE Thailand, S Vietnam
A.t.horizoptera
S Burma, S Thailand, Malaysia, Sumatra
A.t.micromelaena
N Malaysia
A.t.singapurensis
S Malaysia
A.t.scapularis
Java, Bali
A.t.viridus
S Borneo
A.t.aequanimis
N Borneo, Palawan I
A.t.trudiae
Brunei Bay
Aegithina nigrolutea (Marshall's Iora)
Pakistan, NC India
Aegithina viridissima (Green Iora)
A.v.viridissima
S Thailand, Malaysia, Sumatra, Borneo
A.v.thapsina
Anamba Is
Aegithina lafresnayei (Great Iora)
A.l.lafresnayei
S Thailand, Malaysia
A.l.innotata
S Burma, N Thailand, N Indochina
A.l.xanthotis
Cambodia, S Indochina

CHLOROPSIS
**Chloropsis flavipennis (Yellow-quilled
Leafbird)**
Cebu, Mindanao Is

Chloropsis palawanensis (Palawan Leafbird)
Palawan I
Chloropsis sonnerati (Greater Green Leafbird)
C.s.sonnerati
Java
C.s.zosterops
S Thailand, Malaysia, Sumatra, Borneo
C.s.parvirostris
Nias I
Chloropsis cyanopogon (Lesser Green Leafbird)
C.c.cyanopogon
S Burma, Thailand, Malaysia, Sumatra, Borneo
C.c.septentrionalis
S Thailand
Chloropsis cochinchinensis (Blue-winged Leafbird)
C.c.cochinchinensis
Burma, SE Thailand, Cambodia, S Indochina
C.c.kinneari
E Thailand, N Indochina
C.c.serithai
S Thailand
C.c.moluccensis
S Thailand, Malaysia
C.c.icterocephala
S Malaysia, Sumatra
C.c.natunensis
Natuna Is
C.c.billitonis
Billiton I
C.c.viridinucha
Borneo
C.c.nigricollis
Java
C.c.jerdoni
S India, Sri Lanka
Chloropsis aurifrons (Golden-fronted Leafbird)
C.a.aurifrons
Himalayas, Burma, NE India
C.a.frontalis
C & S India
C.a.insularis
SW India, Sri Lanka
C.a.pridii
S Burma, N Thailand, N Laos
C.a.inornata
C & S Thailand, Cambodia, S Vietnam
C.a.incompta
SW Thailand, S Indochina
C.a.media
Sumatra
Chloropsis hardwickei (Orange-bellied Leafbird)
C.h.hardwickei
Burma, E Himalayas, N Thailand, N Vietnam
C.h.malayana
Malaysia

C.h.melliana
S China, N Vietnam
C.h.lazulina
Hainan I
Chloropsis venusta (Blue-masked Leafbird)
Sumatra

IRENA
Irena puella (Blue-backed Fairy Bluebird)
I.p.puella
S India
I.p.sikkimensis
Sikkim, Assam, N India, Burma, Thailand
I.p.malayensis
Malaysia
I.p.criniger
Sumatra, Borneo
I.p.turcosa
Java
I.p.tweeddalei
Palawan I
Irena cyanogaster (Black-mantled Fairy Bluebird)
I.c.cyanogaster
Luzon I
I.c.ellae
Samar, Leyte Is
I.c.melanochlamys
Basilan I
I.c.hoogstraali
Mindanao I

122. LANIIDAE (SHRIKES)

PRIONOPINAE

EUROCEPHALUS
Eurocephalus ruppelli (Ruppell's White-crowned Shrike)
S Sudan to Tanzania
Eurocephalus anguitimens (White-crowned Shrike)
E.a.anguitimens
Namibia to E Transvaal
E.a.niveus
E Transvaal

PRIONOPS
Prionops plumata (Long-crested Helmet Shrike)
P.p.plumata
Senegal to Nigeria
P.p.adamauae
N Cameroun
P.p.concinnata
Cameroun to Sudan & Uganda
P.p.cristata
E & S Ethiopia, SE Sudan
P.p.melanoptera
Somalia, E Ethiopia

P.p.vinaceigularis
  E & C Ethiopia, N & E Kenya
P.p.angolica
  Namibia to Zaire, S Kenya
P.p.poliocephala
  E Tanzania to Transvaal, Natal
Prionops poliolopha (Grey-crested Helmet Shrike)
  SW Kenya, W Tanzania
Prionops caniceps (Red-billed Shrike)
P.c.caniceps
  Sierra Leone to Togo
P.c.harterti
  S Nigeria
P.c.rufiventris
  Cameroun, Central African Republic, N Zaire
P.c.mentalis
  E & S Zaire, W Uganda
Prionops alberti (Yellow-crested Helmet Shrike)
  E Zaire
Prionops retzii (Retz's Red-billed Helmet Shrike)
P.r.neumanni
  S Somalia
P.r.graculina
  E Kenya, NE Tanzania
P.r.tricolor
  Tanzania, Zambia to Mozambique
P.r.intermedia
  NW Tanzania
P.r.nigricans
  W Tanzania, S Zaire to Angola
P.r.retzii
  Namibia to NE Transvaal
Prionops gabela (Rand's Red-billed Helmet Shrike)
  Gabela (Angola)
Prionops scopifrons (Chestnut-fronted Helmet Shrike)
P.s.keniensis
  C Kenya
P.s.kirki
  E Kenya, NE Tanzania
P.s.scopifrons
  SE Tanzania, Mozambique

MALACONOTINAE

LANIOTURDUS
Lanioturdus torquatus (Chat-Shrike)
L.t.torquatus
  S Angola, Namibia
L.t.mesicus
  Huila (Angola)

NILAUS
Nilaus afer (Brubru Shrike)
N.a.afer
  Senegal to Ethiopia

N.a.camerunensis
  Cameroun, Central African Republic
N.a.hilgerti
  C Ethiopia
N.a.minor
  E Eritrea to E Kenya
N.a.massaicus
  SW Kenya to Rwanda, E Zaire
N.a.nigritemporalis
  SC Zaire to Tanzania & Natal
N.a.brubru
  S Angola to N Cape Province
N.a.solivagus
  Natal
N.a.affinis
  N Angola
N.a.miombensis
  Sul do Save, Mozambique

DRYOSCOPUS
Dryoscopus pringlii (Pringle's Puffback)
  Somalia to NE Tanzania
Dryoscopus gambensis (Puffback)
D.g.gambensis
  Senegal to Cameroun & Chad
D.g.congicus
  Gabon to SW Zaire
D.g.malzacii
  Central African Republic to Sudan & Kenya
D.g.erythreae
  E Sudan, Ethiopia
D.g.erwini
  E Zaire to N Tanzania
Dryoscopus cubla (Black-backed Puffback)
D.c.affinis
  E Kenya, Zanzibar
D.c.nairobiensis
  S Kenya, N Tanzania
D.c.hamatus
  S Kenya to Mozambique & Angola
D.c.chapini
  S Mozambique to N Transvaal
D.c.okavangensis
  N Botswana to S Angola, Namibia
D.c.cubla
  Natal, Cape Province
Dryoscopus senegalensis (Red-eyed Puffback)
  S Nigeria to Uganda
Dryoscopus angolensis (Pink-footed Puffback)
D.a.boydi
  Cameroun
D.a.angolensis
  SW Zaire, N Angola
D.a.nandensis
  E Zaire to Sudan & W Kenya
D.a.kungwensis
  W Tanzania
Dryoscopus sabini (Sabine's Puffback)
D.s.sabini
  Sierra Leone to S Nigeria

D.s.melanoleucus
Cameroun to C Zaire

Tchagra minuta (Lesser Tchagra)
T.m.minuta
Sierra Leone to Ethiopia & Kenya
T.m.reichenowi
E Kenya, NE Tanzania
T.m.anchietae
Angola to Tanzania, Malawi
T.m.remota
S Zimbabwe
Tchagra senegala (Black-headed Tchagra)
T.s.cucullata
N Africa
T.s.percivali
S Arabia
T.s.remigialis
Sudan
T.s.notha
Mali to Chad
T.s.senegala
Senegal to Sierra Leone
T.s.pallida
Upper Volta, Ivory Coast to N Central
African Republic
T.s.camerunensis
Cameroun to S Sudan, Uganda
T.s.habessinica
E Sudan to Somalia
T.s.armena
S Uganda to SE Zaire, Zambia
T.s.orientalis
S Somalia to E Tanzania
T.s.mozambica
SE Tanzania to E Transvaal
T.s.confusa
E Transvaal, Natal, E Cape Province
T.s.kalahari
Zimbabwe to S Angola, Namibia
T.s.rufofusca
SW Zaire, N Angola
Tchagra tchagra (Levaillant's Tchagra)
T.t.tchagra
S Cape Province
T.t.natalensis
E Cape Province to Natal
T.t.caffrariae
E Cape Province
Tchagra australis (Brown-headed Tchagra)
T.a.ussheri
Sierra Leone to Nigeria
T.a.emini
E Zaire to S Sudan, W Kenya
T.a.frater
Nigeria to Gabon, W Zaire
T.a.minor
Kenya, NW Tanzania
T.a.littoralis
E Kenya to Zimbabwe, Natal
T.a.congener

S Tanzania, Malawi, Zambia
T.a.ansorgei
W Angola
T.a.bocagei
Cuando (Angola)
T.a.souzae
SE Zaire to Angola
T.a.australis
E Botswana, Transvaal, S Zimbabwe
T.a.damarensis
Namibia, W Botswana
Tchagra jamesi (Three-streaked Tchagra)
T.j.jamesi
Uganda to Somalia, N Kenya
T.j.mandana
E Kenya, Manda, Lamu Is
Tchagra cruenta (Rosy-patched Shrike)
T.c.cruenta
Sudan, N Ethiopia
T.c.hilgerti
Somalia, E Ethiopia, N Kenya
T.c.cathemagmena
S Kenya, E Tanzania

**Laniarius ruficeps (Red-crowned Bush
Shrike)**
L.r.ruficeps
Somalia
L.r.rufinuchalis
Ethiopia to SE Kenya
L.r.kismayensis
S Somalia
Laniarius luhderi (Lühder's Bush Shrike)
L.l.luhderi
Cameroun to Sudan, Kenya
L.l.castaneiceps
S Uganda, W Kenya
L.l.brauni
NW Angola
L.l.amboimensis
W Angola
Laniarius turatii (Turati's Boubou)   122.3
Senegal to Sierra Leone
Laniarius aethiopicus (Tropical Boubou)
L.a.major
Sierra Leone to Sudan & Malawi
L.a.aethiopicus
Ethiopia, Somalia, N Kenya
L.a.ambiguus
Kenya, N Tanzania
L.a.erlangeri
S Somalia
L.a.sublacteus
E Kenya, Zanzibar I
L.a.limpopoensis
S Zimbabwe, N Transvaal
L.a.mossambicus
S Zaire, E Zambia, Mozambique
Laniarius bicolor (Gabon Boubou)
L.b.bicolor
Cameroun, Gabon

L.b.guttatus
  W Zaire, Angola
L.b.sticturus
  S Angola, N Botswana, W Zambia
Laniarius ferrugineus (Southern Boubou)
  L.f.transvaalensis
    S & E Transvaal
  L.f.tongensis
    S Mozambique, N Natal
  L.f.natalensis
    W Natal
  L.f.pondoensis
    E Cape Province
  L.f.savensis
    S Mozambique
  L.f.ferrugineus
    W Cape Province
Laniarius barbarus (Common Gonolek)
  L.b.helenae
    Sierra Leone
  L.b.barbarus
    Senegal to N Cameroun
Laniarius erythrogaster (Black-headed Gonolek)
  N Cameroun to Tanzania
Laniarius mufumbiri (Mufumbiri Shrike)
  Uganda, Rwanda
Laniarius atrococcineus (Burchell's Gonolek)
  S Angola to Transvaal
Laniarius atroflavus (Yellow-breasted Boubou)
  L.a.atroflavus
    Cameroun Mt
  L.a.craterum
    highlands of Cameroun
Laniarius fuelleborni (Fülleborn's Black Boubou)
  L.f.camerunensis
    W Cameroun
  L.f.poensis
    E Nigeria, Cameroun
  L.f.holomelas
    E Zaire, Uganda
  L.f.usambaricus
    W Tanzania
  L.f.ulugurensis
    Tanzania
  L.f.fuelleborni
    SW Tanzania, N Malawi, E Zambia
Laniarius funebris (Slate-coloured Boubou)
  L.f.funebris
    S Sudan & Somalia to Tanzania
  L.f.degener
    S Ethiopia, S Somalia, E Kenya
Laniarius leucorhynchus (Sooty Boubou)
  122.3
  Sierra Leone to Sudan, W Kenya

TELOPHORUS
Telophorus bocagei (Grey-green Bush Shrike)
  T.b.bocagei
    C Cameroun to N Angola

T.b.jacksoni
  C Zaire, Uganda, W Kenya
T.b.ansorgei
  SW Zaire, N Angola
Telophorus sulfureopectus (Sulphur-breasted Bush Shrike)
  T.s.sulfureopectus
    Senegal to NE Zaire
  T.s.terminus
    E Cape Province to SE Tanzania
  T.s.similis
    Ethiopia to Angola & Cape Province
Telophorus olivaceus (Olive Bush Shrike)
  T.o.makawa
    C & S Malawi, E Zimbabwe
  T.o.bertrandi
    S Malawi
  T.o.vitorum
    Sol do Save, Mozambique
  T.o.interfluvius
    Mozambique
  T.o.taylori
    E & N Transvaal
  T.o.olivaceus
    S Mozambique, E Transvaal, E Cape Province
Telophorus nigrifrons (Black-fronted Bush Shrike)
  T.n.nigrifrons
    C Kenya, Tanzania, N Malawi
  T.n.manningi
    SE Zaire, N Namibia
  T.n.sandgroundi
    Mozambique, Zimbabwe, NE Transvaal
Telophorus multicolor (Many-coloured Bush Shrike)
  T.m.multicolor
    Sierra Leone to Cameroun Mt
  T.m.batesi
    Cameroun to W Uganda, N Angola
  T.m.graueri
    E Zaire
Telophorus kupeensis (Serle's Bush Shrike)
  Kupe Mt (Cameroun)
Telophorus zeylonus (Bokmakierie Shrike)
  T.z.restrictus
    Zimbabwe
  T.z.phanus
    S Angola, Namibia
  T.z.thermophilus
    Namibia, W Cape Province
  T.z.zeylonus
    Transvaal, Cape Province
Telophorus viridis (Perrin's Bush Shrike)
  S Zaire, N Angola, NW Zambia
Telophorus quadricolor (Four-coloured Bush Shrike)
  T.q.nigricauda
    SE Kenya, E Tanzania
  T.q.quartus
    Mozambique, S Malawi
  T.q.quadricolor
    Zimbabwe to Natal

Telophorus dohertyi (Doherty's Bush Shrike)
W Kenya, W Uganda, E Zaire

Malaconotus cruentus (Fiery-breasted Bush Shrike)
M.c.cruentus
Sierra Leone to Cameroun
M.c.gabonensis
Cameroun, Gabon
M.c.adolfi-friederici
E Zaire
Malaconotus lagdeni (Lagden's Bush Shrike)
M.l.lagdeni
Liberia to Ghana
M.l.centralis
E Zaire
Malaconotus gladiator (Green-breasted Bush Shrike)
Cameroun Mt
Malaconotus blanchoti (Grey-headed Bush Shrike)
M.b.blanchoti
Senegal to N Cameroun
M.b.catharoxanthus
N Zaire to E Ethiopia, W Kenya
M.b.approximans
S Ethiopia to N Tanzania
M.b.hypopyrrhus
Rwanda, Tanzania to Natal
M.b.extremus
E Cape Province
M.b.interpositus
SE Zaire, W Zambia
M.b.citrinipectus
SW Angola
Malaconotus monteiri (Monteiro's Bush Shrike)   122.2
M.m.perspicillatus
Cameroun
M.m.monteiri
S Zaire, Angola
Malaconotus alius (Black-cap Bush Shrike)
Tanzania

Nicator chloris (Western Nicator)
Senegal to Sudan, Zaire, Uganda
Nicator gularis (Eastern Nicator)
Kenya & Zambia to Natal
Nicator vireo (Yellow-throated Nicator)
Cameroun to N Angola, Uganda

LANIINAE

Corvinella corvina (Yellow-billed Shrike)
C.c.corvina
Senegal & Gambia to Mali
C.c.affinis
Guinea to Uganda
C.c.chapini
NE Zaire, W Kenya

C.c.caliginosa
SW Sudan
Corvinella melanoleuca (Magpie Shrike)
C.m.aequatorialis
S Kenya, Tanzania
C.m.angolensis
NE Namibia, S Angola
C.m.expressa
S Malawi, N Mozambique to N Natal, E Transvaal
C.m.melanoleuca
W Sthn Africa

Lanius tigrinus (Tiger Shrike)
E Asia, Sumatra to Philippine Is & Sulawesi
Lanius souzae (Souza's Shrike)
L.s.souzae
Zaire, Angola, Zambia
L.s.tacitus
E Zambia
L.s.burigi
SE Zaire to Tanzania, Mozambique
Lanius bucephalus (Bull-headed Shrike)
L.b.bucephalus
E Asia, Japan, China
L.b.sicarius
NW China
Lanius cristatus (Brown Shrike)
L.c.cristatus
E Asia, India, China
L.c.confusus
NE Asia >> Thailand, Malaysia
L.c.superciliosus
Japan, E China >> Sunda Is
L.c.luconiensis
China, Philippine Is, N Borneo, N Sulawesi N Moluccas
Lanius collurio (Red-backed Shrike)
L.c.juxtus
S Britain >> W Europe
L.c.collurio
Europe, Siberia, W Asia >> S Africa
L.c.pallidifrons
W Siberia, WC Asia
L.c.kobylini
Iran, Arabia, E Africa
L.c.phoenicuroides
S Russia, NW India, NE Africa
L.c.speculigerus
Iran, C Asia
L.c.isabellinus
E Turkestan.NW India >> NE Africa   122.1
L.c.tsaidamensis
W China
Lanius collurioides (Burmese Shrike)
L.c.collurioides
E India to N Vietnam
L.c.nigricapillus
S Vietnam
Lanius gubernator (Emin's Shrike)
Ghana to Sudan
Lanius vittatus (Bay-backed Shrike)

L.v.nargianus
Iran, Afghanistan, Pakistan
L.v.vittatus
N & C India
Lanius schach (Black-headed Shrike)
L.s.stresemanni
E New Guinea
L.s.bentet
Malaysia, Sumatra to Timor I
L.s.suluensis
Sulu Arch
L.s.nasutus
Philippine Is, N Borneo
L.s.schach
E & C China, N Vietnam
L.s.longicaudatus
C Thailand
L.s.tricolor
Himalayas, Burma, N Indochina
L.s.nigriceps
E India
L.s.caniceps
S & W India, Sri Lanka
L.s.erythronotus
C Asia, W India
L.s.lahulensis
NW India
L.s.tephronotus
Himalayas, S & W China >> Gtr Sunda Is
Lanius validirostris (Strong-billed Shrike)
L.v.validirostris
Luzon I
L.v.tertius
Mindoro I
L.v.hachisuka
SE Mindanao I
L.v.quartus
W Mindanao I
Lanius mackinnoni (Mackinnon's Shrike)
Cameroun to Angola & Tanzania
Lanius minor (Lesser Grey Shrike)
L.m.minor
S & E Europe >> E & S Africa
L.m.turanicus
C Asia, Iran >> E Africa
Lanius ludovicianus (Loggerhead Shrike)
L.l.gambeli
W Nth America >> W Mexico
L.l.excubitorides
C Canada, C USA >> E Mexico
L.l.migrans
E Nth America >> NE Mexico
L.l.ludovicianus
SE USA
L.l.miamensis
S Florida
L.l.mexicanus
C Mexico
L.l.sonoriensis
SW USA, NW Mexico
L.l.grinnelli
N Baja California

L.l.nelsoni
S Baja California
L.l.anthonyi
Santa Barbara I
L.l.mearnsi
San Clemente I
Lanius excubitor (Great Grey Shrike)
L.e.borealis
E Canada, E USA
L.e.invictus
W Canada, W USA
L.e.sibiricus
E Siberia, NE Asia, Mongolia
L.e.excubitor
Europe, Asia Minor, W Siberia
L.e.bianchii
Sakhalin I, N Japan
L.e.mollis
C Asia
L.e.funereus
Chinese Turkestan
L.e.leucopterus
W Siberia, Russian Turkestan
L.e.homeyeri
SE Europe, WC Asia
L.e.leucopterus
W Siberia, Russian Turkestan
L.e.meridionalis
S France, Iberia
L.e.koenigi
Canary Is
L.e.algeriensis
NW Africa
L.e.elegans
Sahara, S Egypt
L.e.leucopygos
S Sahara, Chad, W Sudan
L.e.aucheri
NE Africa, Arabia, S Iran
L.e.buryi
SW Arabia
L.e.uncinatus
Socotra I
L.e.lahtora
NW India
L.e.pallidirostris
SC Asia, Iraq, Arabia, NE Africa
Lanius excubitoroides (Grey-backed Fiscal)
L.e.excubitoroides
Sudan to E Zaire, W Kenya
L.e.intercedens
C Ethiopia, W Kenya
L.e.bohmi
Ethiopia to Tanzania
Lanius sphenocercus (Chinese Great Grey Shrike)
L.s.sphenocercus
NE Asia, Korea, N & E China
L.s.giganteus
NW China
Lanius cabanisi (Long-tailed Fiscal)
Somalia, Kenya, Tanzania

Lanius dorsalis (Taita Fiscal)
Ethiopia, E Uganda, NE Tanzania
Lanius somalicus (Somali Fiscal)
Somalia, E & S Ethiopia, N Kenya
Lanius collaris (Fiscal Shrike)
L.c.smithii
Sierra Leone to W Tanzania
L.c.humeralis
Ethiopia to Zambia, N Mozambique
L.c.marwitzi
SW Tanzania
L.c.capelli
W Uganda to Angola
L.c.aridicolus
W Namibia
L.c.predator
Transkei to S Mozambique
L.c.collaris
Zimbabwe to Cape Province, Natal
L.c.subcoronatus
SW Angola, Namibia, W Botswana
Lanius newtoni (Newton's Fiscal)
Sao Thome I
Lanius senator (Woodchat Shrike)
L.s.senator
Europe, N Africa
L.s.badius
W Mediterranean islands, NW Africa
L.s.niloticus
Syria, Iran, NE Africa
Lanius nubicus (Masked Shrike)
SE Europe to Iraq, NE Africa

PITYRIASINAE

PITYRIASIS
Pityriasis gymnocephala (Bornean Bristlehead)
Borneo

123. VANGIDAE (VANGA SHRIKES)

CALICALICUS
Calicalicus madagascariensis (Red-tailed Vanga)
E Madagascar

SCHETBA
Schetba rufa (Rufous Vanga)
S.r.rufa
E Madagascar
S.r.occidentalis
W Madagascar

VANGA
Vanga curvirostris (Hook-billed Vanga)
V.c.curvirostris
E & N Madagascar
V.c.cetera
SW Madagascar

XENOPIROSTRIS
Xenopirostris xenopirostris (Lafresnaye's Vanga)
SW Madagascar
Xenopirostris damii (Van Dam's Vanga)
NW Madagascar
Xenopirostris polleni (Pollen's Vanga)
NW & E Madagascar

FALCULEA
Falculea palliata (Sicklebill)
N & W Madagascar

ARTAMELLA
Artamella viridis (White-headed Vanga)
A.v.viridis
E Madagascar
A.v.annae
W Madagascar

LEPTOPTERUS
Leptopterus chabert (Chabert Vanga)
L.c.chabert
N & E Madagascar
L.c.schistocercus
SW Madagascar
Leptopterus madagascarinus (Blue Vanga)
L.m.madagascarinus
N & C Madagascar
L.m.comorensis
Moheli I
L.m.bensoni
Grand Comoro I

ORIOLIA
Oriolia bernieri (Bernier's Vanga)
E Madagascar

EURYCEROS
Euryceros prevostii (Helmet Bird)
E Madagascar

HYPOSITTA
Hypositta corallirostris (Coral-billed Nuthatch)
Madagascar

124. BOMBYCILLIDAE (WAXWINGS)

BOMBYCILLINAE

BOMBYCILLA
Bombycilla garrulus (Bohemian Waxwing)
B.g.garrulus
N Europe
B.g.centralasiae
N, C & E Asia, China, Japan
B.g.pallidiceps
W Canada, N & W USA
Bombycilla japonica (Japanese Waxwing)
NE Asia, Japan

Bombycilla cedrorum (Cedar Waxwing)
Canada, USA, Mexico, Central America,
Colombia, Venezuela

PTILOGONATINAE

PTILOGONYS
Ptilogonys cinereus (Grey Silky Flycatcher)
P.c.otofuscus
NW Mexico
P.c.cinereus
C & E Mexico
P.c.pallescens
SW Mexico
P.c.molybdophanes
S Mexico, W Guatemala
P.c.schistaceus
S Mexico
Ptilogonys caudatus (Long-tailed Silky Flycatcher)
Costa Rica, W Panama

PHAINOPEPLA
Phainopepla nitens (Phainopepla)
P.n.lepida
SW USA, NW Mexico
P.n.nitens
S Texas, NC Mexico

PHAINOPTILA
Phainoptila melanoxantha (Black & Yellow Silky Flycatcher)
Costa Rica, W Panama

HYPOCOLIINAE

HYPOCOLIUS
Hypocolius ampelinus (Grey Hypocolius)
Iraq, SW Arabia

125. DULIDAE (PALMCHAT)

DULUS
Dulus dominicus (Palm Chat)
Gonave I, Hispaniola

126. CINCLIDAE (DIPPERS)

CINCLUS
Cinclus cinclus (White-throated (Dipper))
C.c.hibernicus
Ireland, W Scotland
C.c.gularis
Scotland, W & C England
C.c.cinclus
N Europe, Turkey, W Spain, Corsica
C.c.aquaticus
W Europe, Balkans
C.c.minor
NW Africa
C.c.caucasicus
Caucasus, Iraq, Iran
C.c.rufiventris
W Syria

C.c.persicus
SW Iran
C.c.leucogaster
C Asia
C.c.cashmeriensis
Himalayas
C.c.przewalskii
S Tibet, W China
Cinclus pallasii (Brown Dipper)
C.p.tenuirostris
C Asia, Himalayas
C.p.dorjei
S Assam, N Burma, N Thailand
C.p.pallasii
NE Asia, Japan, W China, N Thailand, N Vietnam
C.p.marila
Khasia hills, India
Cinclus mexicanus (North American Dipper)
C.m.unicolor
W Canada, W USA
C.m.mexicanus
N & C Mexico
C.m.anthonyi
S Mexico, Guatemala
C.m.dickermani
S Mexico
C.m.ardesiacus
Costa Rica, W Panama
Cinclus leucocephalus (White-capped Dipper)
C.l.rivularis
N Colombia
C.l.leuconotus
W Venezuela, S Colombia, Ecuador
C.l.leucocephalus
Peru, Bolivia
Cinclus schulzi (Rufous-throated Dipper)
NW Argentina

127. TROGLODYTIDAE (WRENS)

DONACOBIUS 127.4
Donacobius atricapillus (Black-capped Mocking Thrush) (Donacobius)
D.a.brachypterus
E Panama, N Colombia
D.a.nigrodorsalis
SE Colombia, E Ecuador, E Peru
D.a.atricapillus
Venezuela, the Guianas to NE Argentina
D.a.albovittatus
E Bolivia

CAMPYLORHYNCHUS
Campylorhynchus jocosus (Boucard's Wren)
SC Mexico
Campylorhynchus gularis (Spotted Wren)
NW & W Mexico
Campylorhynchus yucatanicus (Yucatan Cactus-Wren)
SE Mexico
Campylorhynchus brunneicapillus (Cactus-Wren)

C.b.couesi
  SW USA, N Baja California, NW Mexico
C.b.sandiegense
  California
C.b.bryanti
  W Baja California
C.b.purus
  C Baja California
C.b.seri
  Tiburon I
C.b.affinis
  S Baja California
C.b.brunneicapillus
  NW Mexico
C.b.guttatus
  C Mexico
Campylorhynchus chiapensis (Giant Wren)
  Chiapas
Campylorhynchus griseus (Bicoloured Wren)
C.g.albicilius
  N Colombia, NW Venezuela
C.g.bicolor
  Colombia
C.g.minor
  E Colombia, N Venezuela
C.g.pallidus
  S Venezuela
C.g.griseus
  E Venezuela, Guyana, N Brazil
Campylorhynchus rufinucha (Rufous-naped Wren)
C.r.humilis
  SW Mexico
C.r.rufinucha
  E Mexico
C.r.nigricaudatus
  S Mexico, W Guatemala
C.r.castaneus
  Guatemala to Nicaragua
C.r.capistratus
  El Salvador to NW Costa Rica
C.r.nicoyae
  Costa Rica
Campylorhynchus turdinus (Thrush-like Wren)
C.t.aenigmaticus
  SW Colombia
C.t.hypostictus
  NW & W Amazonia
C.t.turdinus
  EC Brazil
C.t.unicolor
  E Bolivia, W Brazil
Campylorhynchus albobrunneus (White-headed Wren)　　127.4
C.a.albobrunneus
  C Panama
C.a.harterti
  E Panama, W Colombia
Campylorhynchus nuchalis (Stripe-backed Wren)
C.n.pardus
  N Colombia

C.n.brevipennis
  N Venezuela
C.n.nuchalis
  C Venezuela
Campylorhynchus fasciatus (Fasciated Wren)
C.f.pallescens
  SW Ecuador, NW Peru
C.f.fasciatus
  W Peru
Campylorhynchus zonatus (Banded-backed Wren)
C.z.vanblockeri
  Oaxaca (Mexico)
C.z.zonatus
  EC Mexico
C.z.restrictus
  S Mexico, Guatemala
C.z.vulcanius
  S Mexico to Nicaragua
C.z.costaricensis
  E Costa Rica, W Pamama
C.z.curvirostris
  N Colombia
C.z.brevirostris
  N Colombia, NW Ecuador
Campylorhynchus megalopterus (Grey-barred Wren)
C.m.megalopterus
  C Mexico
C.m.nelsoni
  SC Mexico

ODONTORCHILUS
Odontorchilus cinereus (Tooth-billed Wren)
  N Brazil
Odontorchilus branickii (Grey-mantled Wren)
O.b.branickii
  Colombia, Ecuador, Peru
O.b.minor
  N Ecuador

SALPINCTES
Salpinctes obsoletus (Rock Wren)
S.o.obsoletus
  SW Canada, W USA, N & C Mexico
S.o.guadeloupensis
  Guadeloupe I
S.o.tenuirostris
  San Benito I
S.o.exsul
  San Benedicto I
S.o.neglectus
  SE Mexico to Honduras
S.o.guttatus
  El Salvador to Costa Rica

CATHERPES
Catherpes mexicanus (Canyon Wren)
C.m.pallidior
  Wyoming
C.m.conspersus
  W USA, NW Mexico

C.m.albifrons
SW USA, N Mexico
C.m.mexicanus
N, C & S Mexico
C.m.croizati
Mexico
C.m.meliphonus
NW Mexico
C.m.cantator
S Mexico

HYLORCHILUS
Hylorchilus sumichrasti (Slender-billed Wren)
H.s.sumichrasti
Vera Cruz
H.s.navai
Chiapas

CINNYCERTHIA
Cinnycerthia unirufa (Rufous Wren)
C.u.unirufa
NE Colombia
C.u.chakei
NW Venezuela
C.u.unibrunnea
S & C Colombia, Ecuador
Cinnycerthia peruana (Sepia-brown Wren)
C.p.bogotensis
W Colombia
C.p.olivascens
SW Colombia, W Ecuador
C.p.peruana
Peru
C.p.fulva
S Peru, N Bolivia

CISTOTHORUS
Cistothorus platensis (Sedge Wren)
C.p.stellaris
E Canada, E USA, NE Mexico
C.p.tinnulus
W Mexico
C.p.potosinus
San Luis Potosi
C.p.jalapensis
Vera Cruz
C.p.warneri
W Chiapas
C.p.russelli
Belize
C.p.graberi
SE Honduras, NE Nicaragua
C.p.elegans
S Mexico, Guatemala
C.p.lucidus
S Costa Rica, W Panama
C.p.alticola
N Colombia, N Venezuela, N Guyana
C.p.tamae
Colombia, Venezuela

C.p.tolimae
C Colombia
C.p.aequatorialis
S Colombia, Ecuador, Peru
C.p.graminicola
C Peru
C.p.minimus
S Peru
C.p.boliviae
NW Bolivia
C.p.polyglottus
SE Brazil to NE Argentina
C.p.tucumanus
NW Argentina
C.p.platensis
C & E Argentina
C.p.hornensis
S Chile, S Argentina
C.p.falklandicus
Falkland Is
Cistothorus meridae (Paramo Wren)
NW Venezuela
Cistothorus apolinari (Apolinar's Wren)
C Colombia
Cistothorus palustris (Marsh Wren)
C.p.browningi
British Columbia
C.p.palustris
E USA
C.p.waynei
CE USA
C.p.griseus
SE USA
C.p.marianae
SE USA
C.p.thryophilus
S USA
C.p.iliacus
C Canada, NC to SE USA
C.p.laingi
WC Canada, WC USA >> N Mexico
C.p.plesius
SW Canada, W USA >> NW Mexico
C.p.deserticola
California
C.p.paludicola
W USA, N Baja California
C.p.aestuarinus
SW USA, NW Mexico
C.p.tolucensis
C Mexico

THRYOMANES
Thryomanes bewickii (Bewick's Wren)
T.b.bewickii
C & S USA
T.b.altus
EC & SC USA
T.b.cryptus
SC USA, NE Mexico
T.b.eremophilus
WC & SW USA, NW Mexico

T.b.calophonus
SW Canada, NW USA
T.b.drymoecus
W USA
T.b.pulichi
Texas
T.b.marinensis
SW USA
T.b.atrestus
W USA
T.b.spilurus
SW USA
T.b.correctus
SW USA
T.b.nesophilus
Santa Cruz I (S California)
T.b.catalinae
Santa Catalina I (California)
T.b.leucophrys
San Clemente I (California)
T.b.charienturus
NW Baja California
T.b.cerroensis
WC Baja California
T.b.magdalenensis
SW Baja California
T.b.sadai
N Mexico
T.b.murinus
C Mexico
T.b.mexicanus
SC Mexico
Thryomanes sissonii (Revillagigedo Wren)
Socorro I, Revillagigedo Group

FERMINIA
Ferminia cerverai (Zapata Wren)
Cuba

THRYOTHORUS
Thryothorus atrogularis (Black-throated Wren)
T.a.atrogularis
Nicaragua, Costa Rica, W Panama
T.a.xerampelinus
E Panama
Thryothorus spadix (Sooty-headed Wren) 127.2
E Panama, W Colombia
Thryothorus fasciatoventris (Black-bellied Wren)
T.f.melanogaster
SW Costa Rica, W Panama
T.f.albigularis
E Panama, W Colombia
T.f.fasciatoventris
N Colombia
Thryothorus euophrys (Plain-tailed Wren)
T.e.euophrys
Ecuador
T.e.longipes
Ecuador

T.e.atriceps
NW Peru
T.e.schulenbergi
Amazonas (Peru)
Thryothorus eisenmanni (Inca Wren) 127.1
Cuzco (Peru)
Thryothorus genibarbis (Moustached Wren)
T.g.macrurus
Colombia
T.g.amaurogaster
E Colombia
T.g.saltuensis
W Colombia
T.g.yananchae
SW Colombia
T.g.consobrinus
NW Venezuela
T.g.ruficaudatus
N Venezuela
T.g.tachirensis
NW Venezuela
T.g.mystacalis
Ecuador
T.g.genibarbis
NC Brazil
T.g.juruanus
W Brazil, NE Bolivia
T.g.intercedens
C Brazil
T.g.bolivianus
C Bolivia
Thryothorus coraya (Coraya Wren)
T.c.obscurus
E Venezuela
T.c.caurensis
E Colombia, S Venezuela, N Brazil
T.c.ridgwayi
E Venezuela, W Guyana
T.c.coraya
the Guianas, N Brazil
T.c.herberti
N Brazil
T.c.griseipectus
Colombia, Ecuador, W Brazil, N Peru
T.c.amazonicus
E Peru
T.c.albiventris
N Peru
T.c.cantator
Junin (Peru)
Thryothorus felix (Happy Wren)
T.f.sonorae
NW Mexico
T.f.pallidus
W Mexico
T.f.lawrencii
Maria Madre I (Mexico)
T.f.magdalenae
Maria Magdalena I (Mexico)
T.f.felix
C & S Mexico
T.f.grandis
C Mexico

Thryothorus maculipectus (Spot-breasted Wren)
T.m.microstictus
  NE Mexico
T.m.maculipectus
  E Mexico
T.m.umbrinus
  Guatemala, Honduras, El Salvador, E Nicaragua S Mexico
T.m.canobrunneus
  SE Mexico, Guatemala
Thryothorus rutilus (Rufous-breasted Wren)
T.r.hyperythrus
  W Costa Rica, W Panama
T.r.tobagensis
  Tobago I
T.r.rutilus
  Trinidad, N Venezuela
T.r.intensus
  NW Venezuela
T.r.laetus
  NW Venezuela, N Colombia
T.r.interior
  C Colombia
T.r.hypospodius
  NC Colombia
T.r.columbianus
  W Colombia
T.r.paucimaculatus
  W Ecuador, NW Peru
T.r.sclateri
  N Peru
Thryothorus semibadius (Riverside Wren)
  SW Costa Rica, SW Panama   127.4
Thryothorus nigricapillus (Bay Wren)
T.n.costaricensis
  E Nicaragua, Costa Rica, NW Panama
T.n.castaneus
  C Panama
T.n.schottii
  E Panama, NW Colombia
T.n.reditus
  NE Panama
T.n.connectens
  SW Colombia
T.n.nigricapillus
  W Ecuador
Thryothorus thoracicus (Stripe-breasted Wren)
  E Nicaragua, Costa Rica, W Panama
Thryothorus leucopogon (Stripe-throated Wren)
T.l.grisescens
  E Panama
T.l.leucopogon
  SE Panama, W Colombia, NW Ecuador
Thryothorus pleurostictus (Banded Wren)
T.p.nisorius
  W & C Mexico
T.p.oaxacae
  SW Mexico
T.p.acaciarum
  S Mexico

T.p.oblitus
  S Mexico, Guatemala, W El Salvador
T.p.pleurostictus
  Guatemala
T.p.lateralis
  El Salvador, W Honduras
T.p.ravus
  Nicaragua, NW Costa Rica
Thryothorus ludovicianus (Carolina Wren)
T.l.ludovicianus
  C, S, SE USA
T.l.miamensis
  Florida
T.l.nesophilus
  Florida
T.l.burleighi
  Cat I
T.l.lomitensis
  S USA, NE Mexico
T.l.berlandieri
  NC & NE Mexico
T.l.tropicalis
  NC Mexico
T.l.albinucha
  SE Mexico, N Guatemala
T.l.subfulvus
  Guatemala, Nicaragua
Thryothorus rufalbus (Rufous & White Wren)
T.r.transfinis
  S Mexico
T.r.rufalbus
  Guatemala, El Salvador
T.r.sylvus
  Honduras
T.r.castanonotus
  W Honduras to W Panama
T.r.skutchi
  Panama
T.r.cumanensis
  N Colombia, N Venezuela
T.r.minlosi
  EC Colombia, NW Venezuela
Thryothorus nicefori (Niceforo's Wren)
  N Colombia
Thryothorus sinaloa (Bar-vented Wren)
T.s.cinereus
  NW Mexico
T.s.sinaloa
  WC Mexico
T.s.russeus
  SW Mexico
Thryothorus modestus (Plain Wren)
T.m.modestus
  S Mexico, Guatemala, Honduras, El Salvador, Nicaragua
T.m.zeledoni
  E Nicaragua, E Costa Rica, NW Panama
T.m.roberti
  Honduras
T.m.vanrossemi
  El Salvador
T.m.elutus
  W Panama

***Thryothorus leucotis* (Buff-breasted Wren)**
  T.l.galbraithii
    E Panama, NW Colombia
  T.l.conditus
    Islas San Miguel, Panama
  T.l.leucotis
    N Colombia
  T.l.collinus
    NE Colombia
  T.l.venezuelanus
    NE Colombia, NW Venezuela
  T.l.zuliensis
    Colombia, Venezuela
  T.l.hypoleucus
    NC Venezuela
  T.l.bogotensis
    E Colombia, C Venezuela
  T.l.albipectus
    NE Venezuela, the Guianas, NE Brazil
  T.l.peruanus
    E Peru, W Brazil, SE Colombia, E Ecuador
  T.l.rufiventris
    C Brazil
***Thryothorus superciliaris* (Superciliated Wren)**
  T.s.superciliaris
    Ecuador
  T.s.baroni
    S Ecuador, N Peru
***Thryothorus guarayanus* (Fawn-breasted Wren)**
    Bolivia, SW Brazil
***Thryothorus longirostris* (Long-billed Wren)**
  T.l.bahiae
    NE Brazil
  T.l.longirostris
    C Brazil
***Thryothorus griseus* (Grey Wren)**
    W Brazil

TROGLODYTES
***Troglodytes troglodytes* ((Winter) Wren)**
  T.t.hiemalis
    C & S Canada, C & E USA
  T.t.pullus
    SE USA
  T.t.pacificus
    W Canada, W USA
  T.t.muiri
    California
  T.t.obscurior
    California
  T.t.ochroleucus
    Alaska
  T.t.helleri
    Kodiak I
  T.t.semidiensis
    Semidi I
  T.t.kiskensis
    Aleutian Is
  T.t.meligerus
    Attu, Agatta, Aleutian Is

  T.t.alascensis
    Pribilof Is
  T.t.pallescens
    Kamchatka, Commander Is
  T.t.kurilensis
    N Kurile Is
  T.t.fumigatus
    S Kurile Is, Japan, Izu Is
  T.t.mosukei
    Iku, Borodino Is
  T.t.ogawae
    Tanegashima, Yakushima Is
  T.t.taivanus
    Taiwan
  T.t.dauricus
    NE Asia, Korea
  T.t.idius
    N China
  T.t.szetschuanus
    W China
  T.t.talifuensis
    Sikang, NE Burma
  T.t.subpallidus
    Afghanistan
  T.t.nipalensis
    C & E Himalayas
  T.t.neglectus
    W Himalayas
  T.t.magrathi
    NW India
  T.t.zagrossiensis
    Iran
  T.t.tianschanicus
    NE Iran, C Asia
  T.t.hyrcanus
    Caucasus, NW Iran
  T.t.cypriotes
    Crete, Rhodes, Cyprus
  T.t.juniperi
    NW Libya
  T.t.kabylorum
    NW Africa, Balearic Is
  T.t.koenigi
    Corsica, Sardinia
  T.t.troglodytes
    Europe, Asia Minor
  T.t.indigenus
    Ireland, Scotland, England
  T.t.hirtensis
    St Kilda I Scotland
  T.t.hebridensis
    Outer Hebrides, Scotland
  T.t.fridariensis
    Fair Isle, Scotland
  T.t.zetlandicus
    Shetland Is, Scotland
  T.t.borealis
    Faroe Is
  T.t.islandicus
    Iceland
***Troglodytes tanneri* (Clarion Wren)** 127.4
    Isla Clarion
***Troglodytes aedon* (House Wren)**

T.a.aedon
SE Canada, E USA
T.a.parkmanii
SW Canada, C & W USA, N Mexico
T.a.cahooni
SW USA, NW Mexico
T.a.compositus
C & NE Mexico
T.a.brunneicollis
C & S Mexico
T.a.intermedius
S Mexico, Guatemala, Honduras, El
Salvador, Nicaragua, Costa Rica
T.a.beani
Cozumel I, SE Mexico
T.a.inquietus
Panama
T.a.carychrous
Coiba I, Panama
T.a.rufescens
Dominica I
T.a.mesoleucus
St Lucia I
T.a.guadelupensis
Guadeloupe I
T.a.musicus
St Vincent I
T.a.atopus
N Colombia
T.a.striatulus
Colombia, Venezuela
T.a.columbae
E Colombia
T.a.effutitus
NW & W Venezuela
T.a.albicans
Trinidad, Colombia, Venezuela, W Ecuador,
N Peru, the Guianas, Brazil
T.a.tobagensis
Tobago I
T.a.audax
W Peru
T.a.puna
W Bolivia, Peru
T.a.carabayae
Peru
T.a.tecellatus
Peru, N Chile
T.a.rex
Bolivia, Paraguay, Argentina
T.a.atacamensis
N & C Chile
T.a.musculus
C & S Brazil, E Paraguay, N Argentina
T.a.bonariae
S Brazil, Uruguay, NE Argentina
T.a.chilensis
S Chile, S Argentina
T.a.cobbi
Falkland Is
Troglodytes solstitialis (Mountain Wren)
T.s.chiapensis
S Mexico

T.s.rufociliatus
E Guatemala, N El Salvador
T.s.nannoides
W El Salvador
T.s.rehni
Honduras
T.s.ochraceus
Costa Rica
T.s.ligea
W Panama
T.s.festinus
E Panama
T.s.monticola
N Colombia
T.s.solitarius
Colombia, Venezuela
T.s.solstitialis
S Colombia, Ecuador, N Peru
T.s.macrourus
EC Peru
T.s.frater
SE Peru, Bolivia
T.s.auricularis
N Argentina
Troglodytes rufulus (Tepui Wren)
T.r.rufulus
SE Venezuela
T.r.fulvigularis
SE Venezuela
T.r.yavii
S Venezuela
T.r.duidae
S Venezuela
T.r.wetmorei
S Venezuela
T.r.marahuacae
Amazonas (Venezuela)

THRYORCHILUS 127.4
Thryorchilus browni (Timberline Wren)
T.b.ridgwayi
Costa Rica
T.b.basultoi
Costa Rica
T.b.browni
W Panama

UROPSILA
Uropsila leucogastra (White-bellied Wren)
U.l.leucogastra
EC & E Mexico
U.l.centralis
C Mexico
U.l.restricta
Mexico
U.l.pacifica
SW Mexico
U.l.musica
S Mexico
U.l.brachyura
SE Mexico, Guatemala

HENICORHINA
Henicorhina leucosticta (White-breasted Wood Wren)
  H.l.decolorata
    Mexico
  H.l.prostheleuca
    S & E Mexico, Guatemala
  H.l.tropaea
    Guatemala, Honduras, Nicaragua, Costa Rica, NW Panama
  H.l.smithei
    Peten(Guatemala)
  H.l.costaricensis
    Cartago(Costa Rica)
  H.l.pittieri
    SW Costa Rica, W Panama
  H.l.alexandri
    Panama
  H.l.dariensis
    E Panama, NW Colombia
  H.l.albilateralis
    C Colombia
  H.l.leucosticta
    S Venezuela, Guyana, Surinam, N Brazil
  H.l.eucharis
    Colombia
  H.l.inornata
    S Colombia, Ecuador
  H.l.hauxwelli
    S Colombia, E Ecuador, Peru
Henicorhina leucophrys (Grey-breasted Wood Wren)
  H.l.mexicana
    E Mexico
  H.l.festiva
    W Mexico
  H.l.castanea
    Honduras, S Mexico, N Guatemala
  H.l.capitalis
    S Mexico, W Guatemala, El Salvador
  H.l.minuscula
    S Mexico
  H.l.collina
    Costa Rica, W Panama
  H.l.anachoreta
    N Colombia
  H.l.bangsi
    N Colombia
  H.l.manastarae
    NW Venezuela
  H.l.sanluisensis
    NW Venezuela
  H.l.venezuelensis
    N Venezuela
  H.l.meridana
    W Venezuela
  H.l.tamae
    E Colombia, NW Venezuela
  H.l.leucophrys
    W Colombia, Ecuador, Peru
  H.l.brunneiceps
    SW Colombia, N Ecuador

  H.l.hilaris
    SW Ecuador
  H.l.boliviana
    W Bolivia
Henicorhina leucoptera (Bar-winged Wood Wren)   127.2
    NC Peru

MICROCERCULUS
Microcerculus philomela (Nightingale Wren)   127.3
    C & S Costa Rica
Microcerculus marginatus (Scaly-breasted Wren)   127.3
  M.m.taeniatus
    S Colombia, Ecuador
  M.m.corrasus
    N Colombia
  M.m.squamulatus
    NE Colombia, N Venezuela
  M.m.marginatus
    S Mexico to Upper Amazonia
Microcerculus ustulatus (Flutist Wren)
  M.u.duidae
    S Venezuela
  M.u.lunatipectus
    S Venezuela
  M.u.obscurus
    SE Venezuela
  M.u.ustulatus
    SE Venezuela, W Guyana
Microcerculus bambla (Wing-banded Wren)
  M.b.albigularis
    E Ecuador, W Brazil
  M.b.caurensis
    S Venezuela
  M.b.bambla
    SE Venezuela, the Guianas, NE Brazil

CYPHORINUS
Cyphorinus thoracicus (Chestnut-breasted Wren)
  C.t.dichrous
    S Colombia, Ecuador, Peru
  C.t.thoracicus
    E Peru
Cyphorinus aradus (Musician Wren)
  C.a.richardsoni
    Nicaragua, SE Honduras
  C.a.infuscatus
    Costa Rica, NW Panama
  C.a.lawrencii
    Panama, NW Colombia
  C.a.propinquus
    N Colombia
  C.a.chocoanus
    W Colombia
  C.a.phaeocephalus
    SW Colombia, Ecuador
  C.a.urbanoi
    S Venezuela
  C.a.aradus
    S Venezuela, the Guianas, NE Brazil

C.a.faroensis
  N Brazil
C.a.griseolateralis
  N Brazil
C.a.interpositus
  S Brazil
C.a.transfluvialis
  N Brazil, SE Colombia
C.a.salvini
  SE Colombia, E Ecuador, NE Peru
C.a.modulator
  E Peru, W Brazil

128. MIMIDAE (MOCKING BIRDS, THRASHERS)

DUMETELLA
Dumetella carolinensis (Catbird)
  S Canada, C, S & SE USA, C America,
    West Indies

MELANOPTILA
Melanoptila glabrirostris (Black Catbird)
  SE Mexico, N Guatemala, N Honduras

MELANOTIS
Melanotis caerulescens (Blue Mockingbird)
M.c.longirostris
  Tres Marias Is
M.c.caerulescens
  C & S Mexico
Melanotis hypoleucus (Blue & White Mockingbird)
  SE Mexico, Guatemala, Honduras, El
    Salvador

MIMUS
Mimus polyglottos (Northern Mockingbird)
M.p.polyglottos
  C, E & SE USA
M.p.leucopterus
  SW USA, NW Mexico
M.p.orpheus
  Bahama Is, Greater Antilles
Mimus gilvus (Tropical Mockingbird)
M.g.gracilis
  S Mexico, Guatemala, Honduras, El
    Salvador
M.g.leucophaeus
  SE Mexico, Belize
M.g.antillarum
  Martinique, Windward Is
M.g.tobagensis
  Trinidad, Tobago I
M.g.rostratus
  N Venezuela islands
M.g.magnirostris
  St Andrews I
M.g.tolimensis
  W & C Colombia
M.g.melanopterus
  N & E Colombia, Venezuela, Guyana, N
    Brazil

M.g.gilvus
  French Guiana, Surinam
M.g.antelius
  N & E Brazil
Mimus gundlachii (Bahama Mockingbird)
M.g.gundlachii
  N Cuba, Inagua, Caicos I
M.g.hillii
  Jamaica
Mimus thenca (Chilean Mockingbird)
  C Chile
Mimus longicaudatus (Long-tailed Mockingbird)
M.l.platensis
  La Plata I, W Ecuador
M.l.albogriseus
  SW Ecuador, N Peru
M.l.longicaudatus
  W Peru
M.l.maranonicus
  NE Peru
Mimus saturninus (Chalk-browed Mockingbird)
M.s.saturninus
  N Brazil
M.s.arenaceus
  NE Brazil
M.s.frater
  N Bolivia, SW Brazil
M.s.modulator
  SE Bolivia, S Brazil, Uruguay, N Argentina
Mimus patagonicus (Patagonian Mockingbird)
  W & S Argentina, S Chile
Mimus triurus (White-banded Mockingbird)
  E Bolivia, S Brazil, Paraguay, Uruguay,
    Argentina
Mimus dorsalis (Brown-backed Mockingbird)
  Bolivia, NW Argentina

NESOMIMUS
Nesomimus trifasciatus (Galapagos Mockingbird)
N.t.trifasciatus
  Gardner, Champion, Galapagos Is
N.t.macdonaldi
  Hood, Gardner Is
N.t.melanotis
  Chatham I
N.t.parvulus
  Narboro', Albemarle, Daphne, Seymour,
    Indefatigable Is
N.t.barringtoni
  Barrington I
N.t.personatus
  Abingdon, Bindloe, James, Jervis Is
N.t.wenmani
  Wenman I
N.t.hulli
  Culpeper I
N.t.bauri
  Tower I

MIMODES
Mimodes graysoni (Socorro Thrasher)
Socorro, Revillagigedo Is

OREOSCOPTES
Oreoscoptes montanus (Sage Thrasher)
W & SW USA, Baja California

TOXOSTOMA
Toxostoma rufum (Brown Thrasher)
T.r.rufum
SE Canada, NC, E & SE USA
T.r.longicauda
SC Canada, C & S USA
Toxostoma longirostre (Long-billed Thrasher)
T.l.sennetti
S Texas, NE Mexico
T.l.longirostre
E Mexico
Toxostoma guttatum (Cozumel Thrasher)
Cozumel I
Toxostoma cinereum (Grey Thrasher)
T.c.mearnsi
W Baja California
T.c.cinereum
S Baja California
Toxostoma bendirei (Bendire Thrasher)
T.b.bendirei
SW USA, NW Mexico
T.b.candidum
CW Sonora (Mexico)
T.b.rubricatum
SE Sonora (Mexico)
Toxostoma ocellatum (Ocellated Thrasher)
T.o.ocellatum
SC Mexico
T.o.villai
S Mexico
Toxostoma curvirostre (Curve-billed Thrasher)
T.c.palmeri
S Arizona, N Sonora (Mexico)
T.c.insularum
San Esteban, Tiburon Is
T.c.maculatum
NW Mexico
T.c.occidentale
WC Mexico
T.c.celsum
S USA, NC Mexico
T.c.curvirostre
C & SC Mexico
T.c.oberholseri
S Texas, NE Mexico
Toxostoma lecontei (Le Conte Thrasher)
T.l.lecontei
SW USA, NW Mexico, N Baja California
T.l.macmillanorum
S California
T.l.arenicola
W Baja California
Toxostoma redivivum (California Thrasher)

T.r.sonomae
N California
T.r.redivivum
S California, NW Baja California
Toxostoma dorsale (Crissal Thrasher)
T.d.coloradense
SW USA, NW Mexico, Baja California
T.d.dorsale
SW USA, N Mexico
T.d.trinitatis
N Baja California
T.d.dumosum
NC Mexico

CINCLOCERTHIA
Cinclocerthia ruficauda (Brown Trembler)
C.r.pavida
NW Leeward Is
C.r.tremula
Guadeloupe I
C.r.ruficauda
Dominica I
C.r.tenebrosa
St Vincent I
Cinclocerthia gutturalis (Grey Trembler) 128.1
C.g.gutturalis
Martinique I
C.g.macrorhyncha
St Lucia I

RAMPHOCINCLUS
Ramphocinclus brachyurus (White-breasted Trembler)
R.b.brachyurus
Martinique I
R.b.sanctaeluciae
St Lucia I

ALLENIA
Allenia fusca (Scaly-breasted Thrasher)
Lesser Antilles

MARGAROPS
Margarops fuscatus (Pearly-eyed Thrasher)
M.f.fuscatus
S Bahama Is, Hispaniola, Puerto Rico, N Leeward Is
M.f.densirostris
S Leeward Is
M.f.bonairensis
Bonaire, Los Hermanos Is, N Venezuela

129. PRUNELLIDAE (ACCENTORS)

PRUNELLA
Prunella collaris (Alpine Accentor)
P.c.collaris
SW Europe, N Africa, W Mediterranean islands
P.c.subalpina
SE Europe, Crete, W Turkey
P.c.montana
Caucasus, S Iran, N Iraq

P.c.rufilata
Tadzhikistan, W Sinkiang, N Afghanistan
P.c.whymperi
NW India, W Himalayas
P.c.nipalensis
E Sinkiang, E Himalayas, SE Tibet, SW China
P.c.tibetana
E Tibet, NW China
P.c.erythropygia
Altai, N China, Korea, Japan
P.c.fennelli
Taiwan
Prunella himalayana (Himalayan Accentor)
C Asia, Pakistan, N India, Himalayas
Prunella rubeculoides (Robin Accentor)
P.r.rubeculoides
N India, Pakistan, Himalayas, SE Tibet
P.r.fusca
E Tibet, W China
Prunella strophiata (Rufous-breasted Accentor)
P.s.jerdoni
NW Himalayas
P.s.strophiata
E Himalayas, N Burma, W China
Prunella montanella (Mountain Accentor)
Siberia, Mongolia, Korea
Prunella fulvescens (Brown Accentor)
P.f.fulvescens
Tien Shan, Afghanistan, Pakistan
P.f.dahurica
Altai, Mongolia
P.f.dresseri
SW Sinkiang, N Tibet
P.f.nanschanica
Nan Shan Mts
P.f.khamensis
NE Tibet, W China
P.f.sushkini
S & SE Tibet
Prunella ocularis (Radde's Accentor)
P.o.ocularis
Iran, S Russia, NE Turkey
P.o.fagani
Yemen
Prunella atrogularis (Black-throated Accentor)
P.a.atrogularis
Ural Mts, Iran
P.a.huttoni
Altai, Sinkiang, Pakistan, W Himalayas
Prunella koslowi (Koslov's Accentor)
Mongolia, Ningsia
Prunella modularis (Dunnock)(Hedge Accentor)
P.m.hebridium
Ireland, W Scotland
P.m.occidentalis
E Scotland, England, W France
P.m.modularis
Scandinavia, E & C Europe, N Africa, Turkey

P.m.mabbotti
Iberian peninsula, SW France
P.m.obscura
E Caucasus, N Iran, Lebanon
P.m.euxina
N Turkey, W Caucasus
Prunella rubida (Japanese Hedge Sparrow)
P.r.rubida
S Kurile, N Hokkaido Is
P.r.fervida
Honshu, Kyushu Is
Prunella immaculata (Maroon-backed Accentor)
SE Tibet, N Burma, W China

130. TURDIDAE (THRUSHES, CHATS)

BRACHYPTERYX
Brachypteryx stellata (Gould's Shortwing)
B.s.stellata
E Himalayas, SE Tibet, NE Burma
B.s.fusca
N Vietnam
Brachypteryx hyperythra (Rusty-bellied Shortwing)
E Himalayas, Assam
Brachypteryx major (White-bellied Shortwing)
B.m.major
Mysore, W Madras
B.m.albiventris
Kerala, SW Madras
Brachypteryx leucophrys (Lesser Shortwing)
B.l.nipalensis
Himalayas, Burma, W Yunnan
B.l.carolinae
S China, N Thailand, N Indochina
B.l.langbianensis
S Indochina
B.l.wrayi
Malaysia
B.l.leucophrys
Sumatra, Java, Lesser Sunda Is, Timor I
Brachypteryx montana (Blue Shortwing)
B.m.cruralis
E Himalayas, N Burma, W China >> N Indochina
B.m.sinensis
NW Fukien
B.m.goodfellowi
Taiwan
B.m.sillimani
S Palawan I
B.m.poliogyna
Luzon, Mindoro Is
B.m.andersoni
Mt Isarog (S Luzon I)
B.m.brunneiceps
Negros I
B.m.malindangensis
Mt Malindang (Mindanao I)
B.m.mindanensis
Mt Apo (Mindanao I)

B.m.erythrogyna
N Borneo
B.m.saturata
Sumatra
B.m.montana
Java
B.m.floris
Flores I

HEINRICHIA 130.1
Heinrichia calligyna (Sulawesi Shortwing)
H.c.simplex
N Sulawesi
H.c.calligyna
SC Sulawesi
H.c.picta
SE Sulawesi

POGONOCICHLA
Pogonocichla stellata (White Starred Robin)
P.s.ruwenzorii
Kivu area, NE Zaire
P.s.elgonensis
Mt Elgon (Kenya)
P.s.keniensis
N Uganda
P.s.pallidiflava
S Sudan
P.s.friedmanni
Kigezi (Uganda)
P.s.guttifer
S Sudan, Kenya, NE Tanzania
P.s.macarthuri
Chyulu Mts (SE Kenya)
P.s.orientalis
Malawi, Zambia, Tanzania, Mozambique
P.s.hygrica
SC Mozambique
P.s.transvaalensis
NE Transvaal, SW Mozambique
P.s.margaritata
C Natal
P.s.stellata
S Natal, Cape Province

SWYNNERTONIA 130.15
Swynnertonia swynnertoni (Swynnerton's Bush Robin)
S.s.swynnertoni
E Zimbabwe
S.s.rodgersi
Morogoro (E Tanzania)
S.s.umbratica
W Mozambique

STIPHRORNIS 130.15
Stiphrornis erythrothorax (Forest Robin)
S.e.erythrothorax
Sierra Leone to S Nigeria
S.e.gabonensis
Fernando Po I, W Gabon
S.e.xanthogaster
S Cameroun, N Zaire

S.e.mabirae
E Zaire, Uganda

SHEPPARDIA 130.15
Sheppardia polioptera (Grey-winged Akelat)
S.p.polioptera
S Sudan, Uganda, N Angola
S.p.nigriceps
Sierra Leone to W Cameroun
S.p.tessmanni
E Cameroun
S.p.grimwoodi
NW Zambia
Sheppardia bocagei (Bocage's Akelat) 130.15
S.b.insulana
Fernando Po I
S.b.granti
Mt Kupe (Cameroun)
S.b.kaboboensis
Kivu area (E Zaire)
S.b.kungwensis
W Tanzania
S.b.schoutedeni
E Zaire
S.b.chapini
NW Zambia
S.b.bocagei
SW Zaire, W Angola
S.b.hallae
SE Zaire
S.b.ilyae
W Tanzania
Sheppardia cyornithopsis (Lowland Akelat)
S.c.houghtoni
Sierra Leone, Liberia
S.c.cyornithopsis
S Cameroun
S.c.lopezi
NE Zaire, Uganda
S.c.pallidigularis
W Kenya
S.c.acholiensis
S Sudan
Sheppardia aequatorialis (Jackson's Akelat)
E Zaire, S Uganda, W Kenya
Sheppardia sharpei (Sharpe's Akelat)
S.s.usambarae
C Tanzania
S.s.sharpei
SW Tanzania, N Malawi
Sheppardia gunningi (East Coast Akelat)
S.g.sokokensis
E Kenya, E Tanzania
S.g.bensoni
NW Malawi
S.g.gunningi
C Mozambique
Sheppardia gabela (Gabela Akelat)
Angola
Sheppardia montana (Usambara Akelat)
Usambara Mts (NE Tanzania)
Sheppardia lowei (Iringa Akelat)
S Tanzania, N Malawi

ERITHACUS
Erithacus rubecula (European Robin)
  E.r.melophilus
    British Isles
  E.r.rubecula
    W Europe, NW Morocco >> NE Africa
  E.r.superbus
    Teneriffe, Grand Canary Is
  E.r.witherbyi
    E Algeria, Tunisia
  E.r.sardus
    Corsica, Sardinia
  E.r.balcanicus
    Balkans, Turkey
  E.r.hyrcanus
    E Turkey, S Russia >> Iran, Iraq
  E.r.tataricus
    W Siberia >> Iran
Erithacus akahige (Japanese Robin)
  E.a.akahige
    Sakhalin I, N Japan >> S China
  E.a.rishirensis
    Rishiri Is
  E.a.tanensis
    S Japan & islands
Erithacus komadori (Riukiu Robin)
  E.k.komadori
    Tanegashima, N Riukiu Is
  E.k.namiyei
    Okinawa I
  E.k.subrufus
    S Riukiu Is
Erithacus sibilans (Swinhoe's Robin)
    SE Siberia, N China >> S China
Erithacus calliope (Siberian Rubythroat)
    Siberia >> S China, India
Erithacus pectoralis (Himalayan Rubythroat)
  E.p.pectoralis
    S Russia, W Himalayas >> NW India
  E.p.confusus
    E Himalayas >> NE India
  E.p.tschebaiewi
    Tibet, NW China >> Assam, Burma
Erithacus ruficeps (Rufous-headed Robin)
    SW Shensi
Erithacus obscurus (Black-throated Blue Robin)
    SW Shensi, SE Kansu
Erithacus pectardens (David's Rubythroat)
    SE Tibet, SW China
Erithacus brunneus (Indian Bluechat)
  E.b.brunneus
    Pakistan, N India >> S India
  E.b.wickhami
    Chin hills (Burma)
Erithacus cyane (Siberian Blue Robin)
  E.c.cyane
    C Siberia >> SE Asia
  E.c.bochaiensis
    E Siberia, Japan >> Malaysia, Borneo
Erithacus indicus (White-browed Bush Robin)
  E.i.indicus
    E Himalayas, NE India

  E.i.yunnanensis
    W China >> N Burma, N Indochina
  E.i.formosanus
    Taiwan
Erithacus hyperythrus (Rufous-breasted Bush Robin)
    Himalayas, N Burma, SE Tibet
Erithacus johnstoniae (Collared Bush Robin)
    Taiwan

TARSIGER 130.11
Tarsiger cyanurus (Red-flanked Bluetail)
  T.c.cyanurus
    N Russia, N Japan >> S China
  T.c.pallidior
    Pakistan, NW India
  T.c.rufilatus
    Himalayas, W China >> Burma, Indochina
Tarsiger chrysaeus (Golden Bush Robin)
  T.c.whistleri
    Pakistan, NW India
  T.c.chrysaeus
    Nepal, NE India, W China >> N Vietnam

LUSCINIA
Luscinia luscinia (Thrush Nightingale)
    Europe, W Asia >> SC Africa
Luscinia megarhynchos (Nightingale)
  L.m.megarhynchos
    W Europe, N Africa >> W & C Africa
  L.m.africanus
    Syria, SW Iran >> E Africa
  L.m.hafizi
    C Asia >> E Africa
Luscinia svecicus (Bluethroat)
  L.s.svecicus
    N Europe, N Asia >> India, China
  L.s.cyaneculus
    C Europe >> N Africa
  L.s.volgae
    SW Russia
  L.s.luristanicus
    Iran >> Iraq, Sudan
  L.s.pallidogularis
    Turkestan, W Siberia >> India
  L.s.abbotti
    Pakistan, NW India
  L.s.saturatior
    C Asia to E Tibet, Afghanistan

COPSYCHUS
Copsychus saularis (Magpie Robin)
  C.s.saularis
    Pakistan, N & W India
  C.s.ceylonensis
    SE India, Sri Lanka
  C.s.erimelas
    NE India to Thailand & Indochina
  C.s.andamanensis
    Andaman Is

C.s.prosthopellus
S & E China, Hainan I
C.s.musicus
S Thailand, Malaysia, Sumatra Billiton I
C.s.nesiotes
Bangka I, SE Sumatra
C.s.zacnecus
Simalur I
C.s.nesiarchus
Nias I
C.s.masculus
Batu I
C.s.pagiensis
Mentawei, Siberut, Sipora Is
C.s.javensis
W Java
C.s.amoenus
E Java, Bali
C.s.problematicus
S & W Borneo
C.s.adamsi
N Borneo, Banguey I
C.s.pluto
Maratua I, E & SE Borneo
C.s.deuteronymus
Luzon I
C.s.mindanensis
S Philippine Is
Copsychus sechellarum (Seychelles Magpie Robin)
Seychelles Is
Copsychus albospecularis (Madagascar Magpie Robin)
C.a.albospecularis
N Madagascar
C.a.inexpectatus
E Madagascar
C.a.pica
W Madagascar
C.a.winterbottomi
SW Madagascar
Copsychus malabaricus (White-rumped Shama)
C.m.malabaricus
S India
C.m.leggei
Sri Lanka
C.m.indicus
Nepal, Assam, NE India
C.m.albiventris
Andaman Is
C.m.interpositus
Burma, Thailand, Indochina
C.m.minor
Hainan I
C.m.mallopercnus
Malaysia
C.m.tricolor
Sumatra, W Java
C.m.mirabilis
Prinsen I
C.m.melanurus
N West Sumatra islands

C.m.opisthopelus
S West Sumatra islands
C.m.javanus
WC & C Java
C.m.omissus
E Java
C.m.ochroptilus
Anamba Is
C.m.abbotti
Bangka, Billiton Is
C.m.eumesus
Natuna Is
C.m.suavis
Borneo
C.m.nigricauda
Kangean I
Copsychus stricklandii (Strickland's Shama)
C.s.stricklandii
Banguey I, N Borneo
C.s.barbouri
Maratua I, E Borneo
Copsychus luzoniensis (White-browed Shama)
C.l.luzonensis
Catanduanes, Marinduque, Luzon Is
C.l.parvimaculatus
Polillo Is
C.l.superciliaris
Ticao, Masbate, Panay, Negros Is
Copsychus niger (Black Shama)
C.n.niger
Calamianes, Balabac, Palawan Is
C.n.cebuensis
Cebu I
Copsychus pyrrhopygus (Orange-tailed Shama)
Malaysia, Sumatra, Borneo

IRANIA
Irania gutteralis (Irania)
Asia Minor, S Asia >> E Africa

ALETHE
Alethe diademata (Fire-crested Alethe) 130.15
A.d.diademata
Guinea to Togo
A.d.castanea
Nigeria to Zaire
A.d.woosnami
E Zaire, Uganda
Alethe poliophrys (Red-throated Alethe)
A.p.poliophrys
NE Zaire
A.p.kaboboensis
E Zaire
Alethe poliocephala (Brown-chested Alethe)
A.p.castanonota
Sierra Leone to Ghana
A.p.poliocephala
Fernando Po I, S Cameroun to N Angola
A.p.hallae
Gabela, Angola

A.p.giloensis
S Sudan
A.p.carruthersi
S Sudan, NE Zaire, Uganda
A.p.vandewhegei
Ruanda, Burundi
A.p.nandensis
W Kenya
A.p.akeleyae
C Kenya
A.p.kungwensis
W Tanzania
A.p.ufipae
SW Tanzania
Alethe fuelleborni (White-chested Alethe)
A.f.usambarae
E Tanzania
A.f.fuelleborni
SW Tanzania, N Malawi
A.f.xuthera
S Mozambique
Alethe choloensis (Cholo Alethe)
A.c.choloensis
E & S Malawi
A.c.namuli
Namuli Mt (Mozambique)

COSSYPHA
Cossypha isabellae (Mountain Robin Chat)

130.15

C.i.batesi
E Nigeria
C.i.isabellae
Cameroun Mt
Cossypha roberti (White-bellied Robin Chat)
C.r.roberti
S Cameroun, Fernando Po I
C.r.rufescentior
E Zaire
Cossypha archeri (Archer's Robin Chat)
C.a.archeri
NE Zaire, W Uganda
C.a.albimentalis
Kivu area (E Zaire)
C.a.kimbutui
Mt Kabobo (SE Zaire)
Cossypha anomala (Olive-flanked Robin Chat)
C.a.mbuluensis
Mbulu dist (N Tanzania)
C.a.albigularis
S Tanzania
C.a.macclounii
N Malawi, SW Tanzania
C.a.anomala
Milanje (Malawi)
C.a.gurue
N Mozambique
Cossypha caffra (Cape Robin Chat)
C.c.iolaema
S Sudan to Zambia & Mozambique
C.c.kivuensis
Kivu area (E Zaire)

C.c.drakensbergi
E Transvaal
C.c.vespera
E Zimbabwe
C.c.namaquensis
S Namibia, W Transvaal
C.c.caffra
Natal, Swaziland, Cape Province
Cossypha humeralis (White-throated Robin Chat)
C.h.humeralis
Zimbabwe
C.c.crepuscula
E Transvaal, Natal, S Mozambique
Cossypha cyanocampter (Blue-shouldered Robin Chat)
C.c.cyanocampter
Sierra Leone to Cameroun, Gabon
C.c.bartteloti
NE Zaire, Uganda, Kenya
C.c.pallidiventris
N Nandi (Kenya)
Cossypha semirufa (Ruppell's Robin Chat)
C.s.semirufa
S & W Ethiopia, SE Sudan, N Kenya
C.s.donaldsoni
E & SE Ethiopia
C.s.intercedens
C & S Kenya, N Tanzania
Cossypha heuglini (White-browed Robin Chat)
C.h.pallidior
Chad
C.h.heuglini
S Sudan, Ethiopia to Zambia, Malawi
C.h.subrufescens
Gabon, W Zaire
C.h.intermedia
E Somalia, E Kenya, E Tanzania
C.h.orphea
S Angola, SW Zambia, NW Botswana
C.h.euronota
N Natal, E Transvaal, Zimbabwe, Mozambique
Cossypha natalensis (Red-capped Robin Chat)
C.n.intensa
S Sudan & Ethiopia to Zambia & Mozambique
C.n.larischi
N Angola
C.n.garguensis
Mt Gargues (Kenya)
C.n.tennenti
Mt Endau (Kenya)
C.n.natalensis
Natal, Cape Province
C.n.egregior
S Mozambique
C.n.hylophona
Malawi
Cossypha dichroa (Chorister Robin Chat)
E Transvaal, Natal, S Cape Province

Cossypha heinrichii (White-headed Robin Chat)
N Angola, W Zaire
Cossypha niveicapilla (Snowy-crowned Robin Chat)
C.n.niveicapilla
Senegal to Sudan & SW Ethiopia
C.n.melanonota
NW Cameroun, Gabon, Zaire
Cossypha albicapilla (White-crowned Robin Chat)
C.a.albicapilla
Senegal to Guinea
C.a.giffardi
Ghana to N Cameroun
C.a.omoensis
SE Sudan, SW Ethiopia

XENOCOPSYCHUS 130.15
Xenocopsychus ansorgei (Angola Cave Chat)
W Angola

NEOCOSSYPHUS
Neocossyphus rufus (Red-tailed Ant-thrush)
N.r.rufus
Tanzania, N Kenya, Zanzibar
N.r.gabunensis
S Cameroun to Uganda
Neocossyphus poensis (White-tailed Ant-thrush)
N.p.poensis
Sierra Leone to Gabon, Fernando Po I
N.p.praepectoralis
N & E Zaire, W Uganda
N.p.kakamegoes
Kakamega (W Kenya)
N.p.nigridorsalis
N Nandi (Kenya)
N.p.pallidigularis
N Angola
Neocossyphus fraseri (Rufous Flycatcher Thrush) 130.15
N.f.fraseri
Fernando Po I
N.f.rubicunda
Cameroun to W Zaire & Angola
N.f.vulpina
N & E Zaire
Neocossyphus finschii (Finsch's Flycatcher Thrush) 130.15
Sierra Leone to Nigeria

MODULATRIX 130.8
Modulatrix stictigula (Spot Throat)
M.s.stictigula
N Tanzania
M.s.pressa
N Malawi, SW Tanzania

ARCANATOR 130.8.9
Arcanator orostruthus (Dappled Mountain Robin)

A.o.amani
N Tanzania
A.o.orostruthus
N Mozambique
A.o.sanjei
Morogoro (E Tanzania)

PINARORNIS
Pinarornis plumosus (Boulder Chat)
E Zambia, Zimbabwe, Mozambique

CICHLADUSA
Cichladusa arquata (Collared Palm Thrush)
SE Zaire, S Kenya to Zambia & Mozambique
Cichladusa ruficauda (Rufous-tailed Palm Thrush)
Gabon to N Angola
Cichladusa guttata (Spotted Morning Thrush)
C.g.guttata
S Sudan to SE Kenya, N Tanzania
C.g.rufipennis
SE Ethiopia, E Kenya, E Tanzania

CERCOTRICHAS
Cercotrichas leucosticta (Forest Scrub Robin) 130.15
C.l.leucosticta
Sierra Leone to Ghana
C.l.collsi
NE Zaire
C.l.reichenowi
N Angola
Cercotrichas barbata (Miombo Bearded Scrub Robin)
C.b.thamnodytes
NE Angola, S Zaire, N Zambia
C.b.barbata
Angola, NW Zambia
Cercotrichas quadrivirgata (Eastern Bearded Scrub Robin)
C.q.erlangeri
Juba river, Somalia
C.q.quadrivirgata
E Kenya to Transvaal, C Mozambique
C.q.interna
SE Zambia, Zimbabwe
C.q.rovuma
E Transvaal, Malawi, S Tanzania
C.q.wilsoni
SE Transvaal, Natal, S Mozambique
C.q.brunnea
NC Tanzania
C.q.greenwayi
Mafia, Zanzibar Is
Cercotrichas signata (Brown Scrub Robin)
C.s.oatleyi
N Transvaal
C.s.tongensis
N Natal
C.s.reclusa
W Natal

C.s.signata
E Cape Province, S Natal
Certotrichas hartlaubi (Brown-backed Scrub Robin)
C.h.hartlaubi
Cameroun to W Kenya, N Angola
C.h.kenia
C Kenya
Cercotrichas leucophrys (White-browed Scrub Robin) 130.15
C.l.leucoptera
S Sudan to Somalia, N Kenya, N Uganda
C.l.eluta
S Somalia
C.l.brunneiceps
C Kenya, N Tanzania
C.l.vulpina
SC Kenya
C.l.zambesiana
E Kenya to S Mozambique, E Zambia
C.l.munda
N Angola, S Zaire, W Zambia
C.l.ovamboensis
S Angola, SW Zambia, N Namibia
C.l.simulator
S Mozambique
C.l.makalaka
E Botswana
C.l.pectoralis
S Zimbawe, NE Cape Province, Swaziland
C.l.strepitans
Natal, Transvaal
C.l.leucophrys
S & E Cape Province
Certotrichas galactotes (Rufous Scrub Robin)
C.g.galactotes
W Mediterranean, N Africa >> S Sahara
C.g.syriaca
E Mediterranean, Middle East >> E Africa
C.g.familiaris
S Saudi Arabia, Iran to NW India
C.g.minor
Senegal to Sudan & Ethiopia
C.g.hamertoni
N Somalia
Certotrichas paena (Kalahari Scrub Robin)
C.p.benguellensis
S Angola
C.p.paena
Zimbabwe, Botswana, W Transvaal, N Cape Province
C.p.damarensis
Namibia
C.p.oriens
S Transvaal to N Cape Province
Cercotrichas coryphaeus (Karroo Scrub Robin)
C.c.coryphaeus
Namibia, Botswana, W Cape Province
C.c.cinerea
SE Namibia, SW Cape Province
C.c.eurina
Orange Free State

Cercotrichas podobe (Black Scrub Robin)
C.p.podobe
Senegal to N Somalia
C.p.melanoptera
W Saudi Arabia, Yemen, Aden

DRYMODES
Drymodes brunneopygia (Southern Scrub Robin)
D.b.brunneopygia
Interior of New South Wales, Victoria, E South Australia
D.b.pallidus
S & W South Australia
Drymodes superciliaris (Northern Scrub Robin)
D.s.beccarii
NW New Guinea
D.s.nigriceps
WC New Guinea
D.s.brevirostris
Aru Is, S New Guinea
D.s.colcloughi
Northern Territory
D.s.superciliaris
N Queensland

NAMIBORNIS
Namibornis herero (Herero Chat)
Namibia

CHAETOPS
Chaetops frenatus (Cape Rockjumper)
W Cape Province
Chaetops aurantius (Orange-breasted Rockjumper) 130.12
Natal, E Cape Province

PHOENICURUS
Phoenicurus alaschanicus (Przewalski's Redstart)
W China
Phoenicurus erythronotus (Eversmann's Redstart)
C Asia >> Iran, N India
Phoenicurus caeruleocephalus (Blue-headed Redstart)
C Asia, Himalayas
Phoenicurus ochruros (Black Redstart)
P.o.gibraltariensis
Europe, N Africa >> Israel, Egypt
P.o.ochruros
Asia Minor, N Iran >> Israel, Iraq
P.o.semirufus
Syria, Lebanon >> Israel
P.o.phoenicuroides
C Asia >> NE Africa, N India
P.o.rufiventris
Himalayas, Tibet, NW China >> India, N Burma
Phoenicurus phoenicurus (Common Redstart)

P.p.phoenicurus
Europe, N Africa, C Asia >> W & E Africa
P.p.samamisicus
Iran, S Russia >> NE Africa
P.p.algeriensis
SW Iberian peninsula, NW Africa
Phoenicurus hodgsoni (Hodgson's Redstart)
W China >> Burma, NE India
Phoenicurus frontalis (Blue-fronted Redstart)
N India, Tibet, W China >> N Vietnam
Phoenicurus fuliginosus (Plumbeous Water Redstart)
P.f.fuliginosus
Tibet, Himalayas to N Thailand, W China
P.f.affinis
Taiwan
Phoenicurus leucocephalus (White-capped Redstart)
C Asia >> India, E China, Indochina
Phoenicurus schisticeps (White-throated Redstart)
Tibet, W China >> Assam, N Burma
Phoenicurus auroreus (Daurian Redstart)
P.a.leucopterus
W China, Tibet >> NE India, N Indochina
P.a.auroreus
Siberia, N China >> Japan & S China
Phoenicurus moussieri (Moussier's Redstart)
Tunisia, Algeria, Morocco
Phoenicurus erythrogaster (Guldenstadt's Redstart)
P.e.erythrogaster
Caucasus, Iran
P.e.grandis
C Asia, Tibet, Pakistan, N India >> NE China

RHYACORNIS
Rhyacornis bicolor (Philippine Water Redstart)
N Luzon I

HODGSONIUS
Hodgsonius phaenicuroides (White-bellied Redstart)
H.p.phaenicuroides
Himalayas, N India, Tibet, N Burma
H.p.ichangensis
W China >> N Indochina

CINCLIDIUM
Cinclidium leucurum (White-tailed Blue Robin)
C.l.leucurum
India, Indochina, Malaysia, Burma
C.l.cambodianum
Cambodia
Cinclidium diana (Sunda Blue Robin)
C.d.sumatranum
N & WC Sumatra
C.d.diana
Java
Cinclidium frontale (Blue-fronted Callene)

C.f.frontale
Nepal, Sikkim
C.f.orientale
N Indochina

GRANDALA
Grandala coelicolor (Hodgson's Grandala)
N India, Burma, W China

SIALIA
Sialia sialis (Eastern Bluebird)
S.s.sialis
E USA >> N Mexico
S.s.grata
S Florida
S.s.fulva
Arizona, N Mexico >> Guatemala
S.s.guatemalae
SE Mexico, Guatemala
S.s.meridionalis
El Salvador, Nicaragua
S.s.caribaea
Nicaragua
Sialia mexicana (Western Bluebird)
S.m.occidentalis
W Canada, W USA
S.m.bairdi
SW USA, NW Mexico
S.m.anabelae
N Baja California
S.m.amabilis
C Mexico
S.m.mexicana
NE Mexico
S.m.australis
SC Mexico
Sialia currucoides (Mountain Bluebird)
W Canada, W USA >> SW USA, W Mexico

ENICURUS
Enicurus scouleri (Little Forktail)
E.s.scouleri
SE Russia, Himalayas, N India, W China
E.s.fortis
Taiwan
Enicurus velatus (Lesser Forktail)
E.v.sumatranus
Sumatra
E.v.velatus
Java
Enicurus ruficapillus (Chestnut-backed Forktail)
Malaysia, Borneo, Sumatra, S Thailand
Enicurus immaculatus (Black-backed Forktail)
Himalayas, Burma, Thailand
Enicurus schistaceus (Slaty-backed Forktail)
Himalayas, Burma, Thailand, Indochina
Enicurus leschenaulti (White-crowned Forktail)
E.l.indicus
NE India, Burma, Thailand, Indochina

E.l.sinensis
  W & S China, Hainan I
E.l.frontalis
  Malaysia, Sumatra, Nias I, Borneo
E.l.chaseni
  Batu Is, W Sumatra
E.l.leschenaulti
  Java, Bali
E.l.borneensis
  W Borneo
Enicurus maculatus (Spotted Forktail)
E.m.maculatus
  W & C Himalayas
E.m.guttatus
  E Himalayas, SW China
E.m.bacatus
  S China, N Indochina
E.m.robinsoni
  Dalat (Vietnam)

COCHOA
Cochoa purpurea (Purple Cochoa)
  N India to N Vietnam
Cochoa viridis (Green Cochoa)
  N India to S China, Indochina
Cochoa azurea (Malaysian Cochoa)
  W & C Java
Cochoa beccarii (Sumatran Cochoa)
  W Sumatra   130.14

MYADESTES
Myadestes obscurus (Hawaiian Thrush)
                                               130.10
  Hawaii
**Myadestes myadestinus (Large Kauai
Thrush)**
  Kauai I
Myadestes palmeri (Small Kauai Thrush)
  Kauai I
Myadestes lanaiensis (Lanai Thrush)
M.l.lanaiensis
  Lanai I
M.l.rutha
  Molokai I
Myadestes townsendi (Townsend's Solitaire)
M.t.townsendi
  W Canada, W USA, W Mexico
M.t.calophonus
  N Mexico
**Myadestes occidentalis (Brown-backed
Solitaire)**   130.10
M.o.occidentalis
  E Mexico
M.o.cinereus
  C Mexico
M.o.insularis
  Tres Marias Is
M.o.deignani
  Oaxaca, S Chiapas
M.o.oberholseri
  S Mexico, Guatemala, El Salvador
Myadestes elisabeth (Cuban Solitaire)

M.e.elisabeth
  E & W Cuba
M.e.retrusus
  Isle of Pines
**Myadestes genibarbis (Rufous-throated
Solitaire)**
M.g.solitarius
  Jamaica
M.g.montanus
  Hispaniola
M.g.dominicanus
  Dominica I
M.g.genibarbis
  Martinique I
M.g.sanctaeluciae
  St Lucia I
M.g.sibilans
  St Vincent I
Myadestes melanops (Black-faced Solitaire)
                                               130.13
  Costa Rica, W Panama
Myadestes coloratus (Varied Solitaire) 130.13
  E Panama
Myadestes ralloides (Andean Solitaire)
M.r.plumbeiceps
  W Colombia, W Ecuador
M.r.candelae
  NC Colombia
M.r.venezuelensis
  N Venezuela, E Colombia to N Peru
M.r.ralloides
  C & S Peru, W Bolivia
Myadestes unicolor (Slate-coloured Solitaire)
M.u.unicolor
  S Mexico, Guatemala, N Honduras
M.u.pallens
  Nicaragua
**Myadestes leucogenys (Rufous-brown
Solitaire)**
M.l.gularis
  Guyana
M.l.chubbi
  W Ecuador
M.l.peruvianus
  C Peru
M.l.leucogenys
  SE Brazil

ENTOMODESTES
**Entomodestes leucotis (White-eared
Solitaire)**
  Peru, Bolivia
Entomodestes coracinus (Black Solitaire)
  W Colombia, W Ecuador

SAXICOLA
Saxicola rubetra (Whinchat)
  Europe, Asia, N Africa >> W & C Africa
**Saxicola macrorhyncha (Stoliczka's
Bushchat)**
  S Afghanistan, N India
Saxicola insignis (Hodgson's Bushchat)
  Russia, W China, Tibet >> N India

Saxicola dacotiae (Canary Islands Chat)
S.d.dacotiae
Fuerteventura I
S.d.murielae
Allegranza I
Saxicola torquata (Common Stonechat)
S.t.hibernans
British Isles, W France
S.t.rubicola
W Europe, N Africa >> Middle East
S.t.variegata
Asia Minor, NE Africa, Iraq
S.t.armenica
Iran, N Iraq, NE Africa, Saudi Arabia
S.t.maura
E Russia, C Asia >> Iran, Iraq, N India
S.t.indica
Himalayas >> C India
S.t.przewalskii
W China >> Burma, N India
S.t.stejnegeri
E Siberia, Japan >> S China, Burma,
Indochina
S.t.felix
SW Saudi Arabia, Yemen
S.t.albofasciata
Ethiopia
S.t.jebelmarrae
W Sudan
S.t.moptana
S Mali
S.t.nebularum
Sierra Leone, Ivory Coast
S.t.adamauae
N & W Cameroun
S.t.pallidigula
Cameroun Mt, Fernando Po I
S.t.axillaris
E Zaire, Uganda, Kenya, N Tanzania
S.t.promiscua
C Tanzania
S.t.salax
Cameroun to N Angola & Zimbabwe
S.t.altivaga
S Malawi, N Mozambique
S.t.stonei
Angola to Mozambique, N Sth Africa
S.t.clanceyi
W Namibia, NW Cape Province
S.t.torquata
SW Cape Provicne to Natal, Transvaal
S.t.oreobates
S Zimbabwe, W Natal, Orange Free State
S.t.sibilla
Madagascar
S.t.voeltzkowi
Gt Comoro I
S.t.tectes
Reunion I
Saxicola bifasciata (Buff-streaked Chat)130.15
S Transvaal, Natal, Cape Province
Saxicola leucura (White-tailed Stonechat)
Pakistan, N India

Saxicola caprata (Pied Stonechat)
S.c.rossorum
SW Asia >> Iran & Pakistan
S.c.bicolor
Pakistan >> N & C India
S.c.burmanica
C India, Burma, N Thailand, Indochina
S.c.nilgiriensis
S India
S.c.atrata
Sri Lanka
S.c.caprata
Luzon, Mindoro, Cebu Is
S.c.randi
Negros, Bohol, Siquijor Is
S.c.anderseni
C Mindanao I
S.c.fruticola
Java to Flores & Alor Is
S.c.pyrrhonota
Kisser, Wetar, Savu, Timor Is
S.c.francki
Sumba I
S.c.albonotata
Saleyer I, Sulawesi
S.c.cognata
Babar I
S.c.aethiops
New Britain, N New Guinea
S.c.belensis
WC New Guinea
S.c.wahgiensis
EC & E New Guinea
Saxicola jerdoni (Jerdon's Bushchat)
E India, Burma, N Indochina
Saxicola ferrea (Grey Bushchat)
Himalayas to S China >> S Indochina
Saxicola gutturalis (White-bellied Bushchat)
S.g.gutturalis
Timor I
S.g.luctuosa
Semau I

OENANTHE
Oenanthe leucopyga (White-crowned Black Wheatear)
O.l.aegra
Algeria, Tunisia
O.l.ernesti
Egypt, Saudi Arabia, Iraq
O.l.leucopyga
Mali to Ethiopia
Oenanthe monacha (Hooded Wheatear)
Egypt to Pakistan
Oenanthe leucura (Black Wheatear)
O.l.leucura
W Mediterranean
O.l.syenitica
NW Africa
Oenanthe monticola (Mountain Wheatear)
O.m.albipileata
Benguella (Angola)

O.m.nigricauda
Huambo (Angola)
O.m.atmorii
W Namibia
O.m.monticola
E Namibia, Cape Province
O.m.griseiceps
S Botswana, Natal, Transvaal
Oenanthe phillipsi (Somali Wheatear) 130.5
Somalia
Oenanthe oenanthe (Northern Wheatear)
O.o.leucorhoa
Greenland, NE Canada >> W Europe, W
Africa
O.o.oenanthe
N & C Europe, N Asia >> N & C Africa
O.o.nivea
Balearic Is, S Spain
O.o.virago
E Mediterranean islands >> Israel, Egypt
O.o.seebohmi
Morocco, Algeria
O.o.libanotica
S Europe, NW Africa
Oenanthe lugens (Mourning Wheatear)
O.l.halophila
N Africa
O.l.lugens
N Egypt, Israel, Iraq
O.l.persica
S Egypt, S Israel, S Iran, N Sudan
O.l.lugentoides
SW Saudi Arabia, Yemen
O.l.boscaweni
S Saudi Arabia
O.l.vauriei
NE Somalia
O.l.lugubris
N & C Ethiopia
O.l.schalowi
S Kenya, NE Tanzania
Oenanthe finschii (Finsch's Wheatear)
O.f.finschii
Turkey, Saudi Arabia >> Iran, Pakistan
O.f.barnesi
NE Iran, C Asia >> S Iran, Pakistan
Oenanthe picata (Variable Wheatear)
Iran, Pakistan >> N India
Oenanthe moesta (Red-rumped Wheatear)
O.m.moesta
N Africa
O.m.brooksbanki
E Egypt, Jordan, Iraq
Oenanthe pleschanka (Pied Wheatear)
E Europe to W China >> E Africa
Oenanthe cypriaca (Cyprus Wheatear) 130.6
Cyprus >> E & N Africa
Oenanthe alboniger (Hume's Wheatear)
S Iran, Afghanistan, Pakistan
Oenanthe hispanica (Black-eared Wheatear)
O.h.hispanica
S Europe, N Africa >> W Africa

O.h.melanoleuca
E Europe, Asia Minor >> NE & W Africa
Oenanthe xanthoprymna (Red-tailed Wheatear)
O.x.xanthoprymna
SW Iran >> NE Africa
O.x.chrysopygia
E Turkey, N Iran >> Iraq, Saudi Arabia
O.x.kingi
Afghanistan >> Pakistan, NW India
Oenanthe deserti (Desert Wheatear)
O.d.homochroa
N Africa
O.d.deserti
W & C Asia >> Pakistan, N India, NE Africa
O.d.oreophila
Tibet & Sinkiang >> Pakistan, Saudi Arabia
Oenanthe pileata (Capped Wheatear)
O.p.neseri
S Angola, N Namibia
O.p.livingstonii
C Kenya, E Angola to Zimababwe & Malawi
O.p.pileata
S Namibia, Sth Africa
Oenanthe bottae (Red-breasted Wheatear)
O.b.bottae
Yemen
O.b.frenata
Ethiopia
O.b.heuglini
Mali, Central African Republic, Sudan
Oenanthe isabellina (Isabelline Wheatear)
E Europe, W China >> India, C Africa

CERCOMELA
Cercomela sinuata (Sicklewing Chat)
C.s.hypernephela
Lesoto
C.s.ensifera
E Transvaal, Orange Free State, N Cape
Province
C.s.sinuata
S Cape Province
Cercomela schlegelii (Karoo Chat)
C.s.benguellensis
S Angola
C.s.schlegelii
coastal Namibia
C.s.namaquensis
Little Namaqualand
C.s.pollux
Orange Free State, Cape Province
Cercomela fusca (Brown Rockchat)
Pakistan, N & C India
Cercomela tractrac (Tractrac Chat)
C.t.hoeschi
S Angola, N Namibia
C.t.albicans
N coastal Namibia
C.t.barlowi
Gt Namaqualand
C.t.nebulosa
S coastal Namibia

C.t.tractrac
Little Namaqualand

Cercomela familiaris (Familiar Chat)
C.f.falkensteini
Ghana to SW Sudan, N Ethiopia
C.f.omoensis
SE Sudan, SW Ethiopia
C.f.modesta
Uganda to Angola & Mozambique
C.f.angolensis
N Angola, N Namibia
C.f.galtoni
E Namibia, W Botswana, N Cape Province
C.f.hellmayri
E Botswana, Zimbabwe, Transvaal
C.f.actuosa
W Natal
C.f.familiaris
S Mozambique, Natal to S Cape Province

Cercomela scotocerca (Brown-tailed Rockchat)
C.s.furensis
W Sudan
C.s.scotocerca
Sudan coast
C.s.turkana
SW Ethiopia, NW Kenya
C.s.spectatrix
Somalia
C.s.validior
Run (Somalia)

Cercomela dubia (Sombre Rockchat)
Somalia, C Ethiopia

Cercomela melanura (Blackstart)
C.m.melanura
Israel to Saudi Arabia
C.m.neumanni
W Saudi Arabia, Yemen, Aden
C.m.lypura
W Red Sea coast
C.m.aussae
E Ethiopia
C.m.airensis
E Niger, Chad, Sudan
C.m.ultima
E Mali, W Niger

Cercomela sordida (Moorland Chat)
C.s.sordida
Ethiopia
C.s.rudolfi
Mt Elgon (Kenya)
C.s.ernesti
Aberdare Mts, Mt Kenya
C.s.olimotiensis
N Tanzania
C.s.hypospodia
Mt Kilimanjaro (Tanzania)

MYRMECOCICHLA
Myrmecocichla tholloni (Congo Moorchat)
Gabon, N Angola

Myrmecocichla aethiops (Northern Anteater Chat)

M.a.aethiops
Senegal to Chad, N Nigeria, N Cameroun
M.a.sudanensis
W Sudan
M.a.cryptoleuca
C Kenya, N Tanzania

Myrmecocichla formicivora (Southern Anteater Chat)
M.f.formicivora
E Botswana, Natal, C & E Cape Province
M.f.orestes
E Transvaal, W Natal
M.f.minor
W Botswana, Namibia

Myrmecocichla nigra (Sooty Chat)
Nigeria to Sudan, Angola, Tanzania

Myrmecocichla melaena (Ruppell's Black Chat)
Ethiopia

Myrmecocichla albifrons (White-fronted Black Chat)
M.a.frontalis
Senegal to Nigeria & Chad
M.a.limbata
N & E Cameroun, Central African Republic
M.a.albifrons
N Ethiopia
M.a.pachyrhyncha
SW Ethiopia
M.a.clericalis
S Sudan, NE Zaire, N Uganda

Myrmecocichla arnoti (White-headed Black-Chat)
M.a.leucolaema
SE Zaire, W Tanzania
M.a.harterti
SW Zaire, Angola
M.a.arnoti
E Angola & Namibia to Malawi, N Transvaal

Myrmecocichla cinnamomeiventris (Mocking Cliffchat) 130.15
M.c.cavernicola
Fiko (Mali)
M.c.bambarae
Kulikoro (Mali)
M.c.coronata
Togo to N Cameroun, W Sudan
M.c.kordofanensis
C Sudan
M.c.albiscapulata
E & S Ethiopia, E Sudan
M.c.subrufipennis
Sudan & Ethiopia to Zambia & Malawi
M.c.odica
E Botswana, Transvaal, Zimbabwe, Mozambique
M.c.cinnamomeiventris
E Transvaal, Natal, E Cape Province
M.c.autochthones
N Natal, S Mozambique

Myrmecocichla semirufa (White-winged Cliffchat)
Ethiopia

SAXICOLOIDES
Saxicoloides fulicata (Black-backed Robin)
S.f.cambaiensis
Pakistan, N & W India
S.f.erythrura
NE India
S.f.intermedia
C India
S.f.fulicata
S India
S.f.leucoptera
Sri Lanka

PSEUDOCOSSYPHUS
Pseudocossyphus imerinus (Madagascar Robin Chat)
P.i.erythronotus
N Madagascar
P.i.sharpei
C Madagascar
P.i.imerinus
SE Madagascar
P.i.salomonseni
E Madagascar
Pseudocossyphus bensoni (Farkas'Robin Chat)
SW Madagascar

MONTICOLA
Monticola rupestris (Cape Rock Thrush)
South Africa
Monticola explorator (Sentinel Rock Thrush)
M.e.explorator
Transvaal, Natal, Cape Province
M.e.tenebriformis
Lesotho >> N Natal, S Mozambique
Monticola brevipes (Short-toed Rock Thrush)
M.b.niveiceps
Huila (Angola)
M.b.brevipes
S Angola, Namibia, N Cape Province
M.b.pretoriae
N Cape Province, SE Botswana, Transvaal
M.b.leucocapilla
S Botswana, W Transvaal, Orange Free State
Monticola angolensis (Miombo Rock Thrush)
M.a.angolensis
S Zaire, Angola to Tanzania, Mozambique
M.a.niassae
E Zimbabwe, SE Zaire, N Mozambique
M.a.hylophila
NW Zimbabwe
Monticola saxatilis (Mountain Rock Thrush)
M.s.saxatilis
C & S Europe to W China >> N India & E Africa
M.s.coloratus
E Europe
Monticola cinclorhynchus (Blue-capped Rock Thrush)
Himalayas >> W Burma, S India

Monticola gularis (White-throated Rock Thrush)
NE Asia >> Burma, Thailand, Indochina
Monticola rufiventris (Chestnut-bellied Rock Thrush)
Himalayas, W China >> Indochina
Monticola rufocinereus (Little Rock Thrush)
M.r.rufocinereus
Ethiopia, Sudan to NE Tanzania
M.r.sclateri
W Saudi Arabia
Monticola solitarius (Blue Rock Thrush)
M.s.solitarius
S Europe, Middle East >> W & C Africa
M.s.longirostris
N Iraq, Iran, Pakistan >> N India & NE Africa
M.s.pandoo
Himalayas, C Asia >> India, SE Asia, Indonesia
M.s.philippensis
NE Russia, China, Japan >> Philippine Is, Indonesia
M.s.madoci
Malaysia

MYIOPHONEUS
Myiophoneus blighi (Ceylon Whistling Thrush)
Sri Lanka
Myiophoneus melanurus (Shiny Whistling Thrush)
Sumatra
Myiophoneus glaucinus (Sunda Whistling Thrush)
M.g.glaucinus
Java, Bali
M.g.castaneus
Sumatra
M.g.borneensis
Borneo
Myiophoneus robinsoni (Malayan Whistling Thrush)
Malaysia
Myiophoneus horsfieldii (Malabar Whistling Thrush)
S India
Myiophoneus insularis (Formosan Whistling Thrush)
Taiwan
Myiophoneus caeruleus (Himalayan (Blue) Whistling Thrush)
M.c.temminckii
C Asia, Pakistan, N India, Burma
M.c.eugenei
S Burma, Thailand, W China, Indochina
M.c.caeruleus
W China >> C & S China, N Indochina
M.c.crassirostris
SE Thailand, N Malaysia
M.c.dichrorhynchus
C & S Malaysia, W Sumatra

M.c.flavirostris
Java

GEOMALIA 130.1
Geomalia heinrichi (Sulawesi Mountain Thrush)
Sulawesi

ZOOTHERA
Zoothera schistacea (Slaty-backed Ground Thrush)
Tanimbar Is
Zoothera dumasi (Moluccan Ground Thrush)
Z.d.dumasi
Buru I
Z.d.joiceyi
Ceram I
Zoothera interpres (Chestnut-capped Ground Thrush)
Z.i.interpres
S Thailand, Sumatra to Flores I, Borneo, Basilan I
Z.i.leucolaema
Enggano I
Zoothera erythronota (Red-backed Ground Thrush)
Z.e.erythronota
Sulawesi
Z.e.mendeni
Peleng I
Zoothera dohertyi (Chestnut-backed Ground Thrush) 130.1
Lombok to Timor I
Zoothera wardii (Pied Ground Thrush)
N India >> S India, Sri Lanka
Zoothera cinerea (Ashy Ground Thrush)
Mindoro, N Luzon Is
Zoothera peronii (Orange-banded Ground Thrush)
Z.p.peronii
W Timor I
Z.p.audacis
E Timor I, Wetar I to Babar I
Zoothera citrina (Orange-headed Ground Thrush)
Z.c.citrina
Pakistan to N Burma >> S India, Sri Lanka
Z.c.cyanotus
S India
Z.c.innotata
Burma, S China, Indochina >> Malaysia
Z.c.melli
SE China
Z.c.courtoisi
Anhwei
Z.c.aurimacula
S Vietnam, Hainan I
Z.c.andamanensis
Andaman Is
Z.c.albogularis
Nicobar Is

Z.c.gibsonhilli
S Burma, S Thailand
Z.c.aurata
N Borneo
Z.c.rubecula
W Java
Z.c.orientis
E Java, Bali
Zoothera everetti (Everett's Ground Thrush)
N Borneo
Zoothera sibirica (Siberian Ground Thrush)
Z.s.sibirica
NE Asia >> SE Asia, Java
Z.s.davisoni
Japan >> S China, Thailand, Malaysia
Zoothera piaggiae (Abyssinian Ground Thrush)
Z.p.piaggiae
Ethiopia, Sudan, E Zaire, N Kenya
Z.p.hadii
SE Sudan
Z.p.ruwenzorii
Ruwenzori Mts
Z.p.kilimensis
C & S Kenya, N Tanzania
Z.p.rowei
N Tanzania
Zoothera tanganjicae (Kivu Ground Thrush)
SW Uganda
Zoothera crossleyi (Crossley's Ground Thrush)
Z.c.crossleyi
Mt Kupe, Cameroun Mt
Z.c.pilettei
NE Zaire
Zoothera gurneyi (Orange Ground Thrush)
Z.g.chuka
Mt Kenya
Z.g.otomitra
S Kenya to Angola, N Malawi
Z.g.gurneyi
Natal, E Cape Province
Z.g.disruptans
C & S Malawi, E Zimbabwe, N Transvaal, Mozambique
Zoothera oberlaenderi (Oberlaender's Ground Thrush)
NE Zaire, Uganda
Zoothera cameronensis (Black-eared Ground Thrush)
Z.c.cameronensis
Cameroun
Z.c.graueri
NW Zaire
Zoothera kibalensis (Prigogine's Ground Thrush)
W Uganda
Zoothera princei (Grey Ground Thrush)
Z.p.princei
Sierra Leone to Ghana
Z.p.batesi
S Cameroun, N Zaire
Zoothera guttata (Spotted Ground Thrush)

Z.g.lippensi
  E Zaire
Z.g.guttata
  S Malawi, Natal, Cape Province
Z.g.fischeri
  coast of Kenya & Tanzania
Z.g.belcheri
  Thyolo Mt (Malawi)
Z.g.maxis
  S Sudan
Z.g.nateliens
  Cape Province, S Africa
Zoothera spiloptera (Spotted-winged Thrush)
  Sri Lanka
Zoothera andromedae (Sunda Ground Thrush)
  Sumatra to Timor, Mindoro, Mindanao Is
Zoothera mollissima (Plain-backed Mountain Thrush)
Z.m.whiteheadi
  N Pakistan, W Himalayas
Z.m.mollissima
  E Himalayas, SE Tibet >> Burma, Indochina
Z.m.griseiceps
  SW China, N Vietnam
Zoothera dixoni (Long-tailed Mountain Thrush)
  E Himalayas, Tibet >> Burma, Thailand, N Vietnam
Zoothera dauma (White's Thrush)
Z.d.aurea
  N & NE Asia >> S China, Indochina
Z.d.dauma
  Himalayas, Burma, China, Thailand >> S India
Z.d.neilgherriensis
  S India
Z.d.imbricata
  Sri Lanka
Z.d.toratugumi
  Manchuria, Japan >> Taiwan
Z.d.major
  N Riukiu Is
Z.d.hancii
  S Riukiu Is, S Thailand, S Vietnam, Taiwan
Z.d.horsfieldi
  Sumatra, Java, Lombok I
Zoothera machiki (Fawn-breasted Thrush) 130.1
  Tanimbar Is
Zoothera heinei (Heine's Ground Thrush) 130.1
Z.h.papuensis
  SE New Guinea
Z.h.eichhorni
  St Matthias Is
Z.h.choiseuli
  Choiseul I
Z.h.heinei
  S Queensland
Zoothera lunulata (Australian Ground Thrush) 130.2

Z.l.cuneata
  NC Queensland
Z.l.lunulata
  New South Wales, Victoria, South Australia
Z.l.macrorhyncha
  Tasmania
Zoothera talaseae (Melanesian Ground Thrush) 130.3
Z.t.talaseae
  New Britain
Z.t.turipavae
  Guadalcanal I
Z.t.margaretae
  San Cristobal I
Zoothera monticola (Greater Long-billed Thrush)
Z.m.monticola
  Himalayas, Assam, NE Burma
Z.m.atrata
  N Vietnam
Zoothera marginata (Lesser Long-billed Thrush)
  N India to Indochina

IXOREUS 130.4
Ixoreus naevia (Varied Thrush)
I.n.naevia
  SE Alaska, W Canada, NW USA >> SW USA
I.n.meruloides
  N Alaska, NW Canada >> WC USA

RIDGWAYIA 130.4
Ridgwayia pinicola (Aztec Thrush)
  C Mexico

AMALOCICHLA
Amalocichla sclateriana (Greater New Guinea Thrush)
A.s.occidentalis
  W New Guinea
A.s.sclateriana
  SE New Guinea
Amalocichla incerta (Lesser New Guinea Thrush)
A.i.incerta
  Arfak Mts (W New Guinea)
A.i.olivascentior
  WC New Guinea
A.i.brevicauda
  E & SE New Guinea

CATAPONERA
Cataponera turdoides (Cataponera Thrush)
C.t.abditiva
  NC Sulawesi
C.t.tenebrosa
  S Sulawesi
C.t.turdoides
  SW Sulawesi
C.t.heinrichi
  SE Sulawesi

NESOCICHLA
Nesocichla eremita (Tristan Thrush)
  N.e.eremita
    Tristan da Cunha I
  N.e.gordoni
    Inaccessible I
  N.e.procax
    Nightingale I

CICHLHERMINIA
Cichlherminia lherminieri (Forest Thrush)
  C.l.lherminieri
    Guadeloupe I
  C.l.lawrencii
    Montserrat I
  C.l.dominicensis
    Dominica I
  C.l.sanctaeluciae
    St Lucia I

CATHARUS
Catharus gracilirostris (Slender-billed Nightingale Thrush)
  C.g.gracilirostris
    Costa Rica
  C.g.accentor
    W Panama
Catharus aurantiirostris (Orange-billed Nightingale Thrush)
  C.a.aenopennis
    NW Mexico
  C.a.clarus
    NC Mexico
  C.a.melpomene
    C Mexico to Costa Rica
  C.a.russatus
    SW Costa Rica, W Panama
  C.a.griseiceps
    W Panama
  C.a.phaeoplurus
    C Colombia
  C.a.aurantiirostris
    NE Colombia, NW Venezuela
  C.a.birchalli
    NE Venezuela, Trinidad
  C.a.barbaritoi
    C Venezuela
  C.a.sierrae
    Santa Marta Mts (Colombia)
  C.a.inornatus
    EC Colombia
  C.a.insignis
    N Colombia
Catharus fuscater (Slaty-backed Nightingale Thrush)
  C.f.hellmayri
    Costa Rica, W Panama
  C.f.mirabilis
    E Panama
  C.f.sanctaemartae
    N Colombia
  C.f.fuscater
    Ecuador, Colombia, W Venezuela

  C.f.opertaneus
    W Colombia
  C.f.caniceps
    N & C Peru
  C.f.mentalis
    SE Peru, N Bolivia
Catharus occidentalis (Russet Nightingale Thrush)
  C.o.olivascens
    N Mexico
  C.o.durangensis
    Durango
  C.o.lambi
    N Puebla
  C.o.fulvescens
    C Mexico
  C.o.occidentalis
    SE Mexico
Catharus frantzii (Frantzius' Nightingale Thrush)
  C.f.chiapensis
    C Chiapas
  C.f.confusus
    NE Puebla
  C.f.nelsoni
    E Oaxaca
  C.f.waldroni
    N Nicaragua
  C.f.wetmorei
    Chiriqui, Panama
  C.f.juancitonis
    S Mexico to Honduras
  C.f.frantzii
    Costa Rica, W Panama
Catharus mexicanus (Black-headed Nightingale Thrush)
  C.m.mexicanus
    EC Mexico
  C.m.cantator
    S Mexico, E Guatemala, Honduras
  C.m.fumosus
    Nicaragua, Costa Rica, W Panama
Catharus dryas (Spotted Nightingale Thrush)
  C.d.harrisoni
    Oaxaca
  C.d.ovandensis
    Chiapas
  C.d.dryas
    W Guatemala, Honduras, W Ecuador
  C.d.maculatus
    E Colombia, E Ecuador, Peru, Bolivia
  C.d.ecuadoreanus
    W Ecuador
  C.d.blakei
    Jujuy (Argentina)
Catharus fuscescens (Veery)
  C.f.fuscescens
    E Canada, E USA >> E Mexico, Panama
  C.f.fuliginosa
    SE Canada >> E USA
  C.f.salicicola
    W Canada, W USA >> Mexico & N Sth America

C.f.subpallidus
NW USA >> SW USA

Catharus minimus (Grey-cheeked Thrush)
C.m.minimus
E Siberia, Canada >> E USA, N Sth
America
C.m.aliciae
SE Canada >> E USA, West Indies

Catharus ustulatus (Swainson's Thrush)
C.u.almae
S Alaska, W Canada >> S USA
C.u.ustulatus
SE Alaska, W Canada >> W USA, Mexico
C.u.oedicus
W USA >> Mexico
C.u.swainsoni
C & E Canada, E USA >> S Mexico, West
Indies E Sth America

Catharus guttatus (Hermit Thrush)
C.g.guttatus
Alaska, W Canada >> W USA, N & C
Mexico
C.g.nanus
SE Alaska, W Canada >> W USA, Baja
California
C.g.slevini
W USA >> NW Mexico
C.g.sequoiensis
W USA >> N Mexico
C.g.polionotus
W USA >> S Mexico
C.g.auduboni
W & SW USA >> Mexico, Guatemala
C.g.faxoni
Canada, E USA >> SE USA
C.g.crymophilus
Newfoundland >> SE USA

HYLOCICHLA

Hylocichla mustelina (Wood Thrush)
SE Canada, E USA >> E Mexico, Cuba,
Central America

PLATYCICHLA

Platycichla flavipes (Yellow-legged Thrush)
P.f.venezuelensis
Colombia, N & W Venezuela
P.f.melanopleura
NE Venezuela, Trinidad
P.f.xanthoscelus
Tobago I
P.f.polionota
S Venezuela, Guyana
P.f.flavipes
SE Brazil, Argentina, NE Paraguay

Platycichla leucops (Pale-eyed Thrush)
N Sth America

PSOPHOCICHLA

**Psophocichla litsipsirupa (Groundscraper
Thrush)** 130.15
P.l.simensis
Ethiopia

P.l.litsipsirupa
Botswana, Zimbabwe, Zambia, S Africa
P.l.pauciguttatus
NW Botswana, Namibia
P.l.stierlingi
N Angola to W & S Tanzania

TURDUS

Turdus bewsheri (Comoro Thrush)
T.b.comorensis
Great Comoro I
T.b.moheliensis
Moheli I
T.b.bewsheri
Anjouan I

**Turdus olivaceofuscus (Gulf of Guinea
Thrush)**
T.o.olivaceofuscus
Sao Thome I
T.o.xanthorhynchus
Principé I

Turdus olivaceus (Olive Thrush)
T.o.chiguancoides
Senegal to W Ghana
T.o.saturatus
W Ghana to N Zaire
T.o.adamauae
N Cameroun
T.o.nigrilorum
Cameroun Mt
T.o.poensis
Fernando Po I
T.o.bocagei
S Zaire, N Angola, NW Tanzania
T.o.centralis
N Zaire, Uganda, Central African Republic
T.o.pelios
Sudan, Ethiopia
T.o.graueri
S Uganda, NW Tanzania
T.o.stormsi
SE Zaire, NE Angola, NW Zambia
T.o.williami
Zambia
T.o.swynnertoni
E Zimbabwe
T.o.transvaalensis
N Transvaal
T.o.smithi
N Cape Province, Orange Free State,
Transvaal
T.o.olivaceus
SW Cape Province
T.o.pondoensis
Transkei, Natal, Swaziland

**Turdus abyssinicus (Mountain Thrush)
(Northern Olive Thrush)**
T.a.abyssinicus
Ethiopia, W Kenya, NE Tanzania
T.a.baraka
E Zaire, W & S Uganda
T.a.polius
N Kenya

T.a.mwaki
  NW Kenya
T.a.porini
  W Kenya
T.a.elgonensis
  WC Kenya
T.a.deckeni
  NE Tanzania
T.a.oldeani
  NE Tanzania
T.a.bambusicola
  Ruanda, Kivu (E Zaire)
T.a.roehli
  NE Tanzania
T.a.nyikae
  C Tanzania, Malawi, NE Zambia
T.a.milanjensis
  S Malawi, Mozambique
Turdus helleri (Taita Olive Thrush)
  SE Kenya
Turdus libonyanus (Kurrichane Thrush)
T.l.verreauxi
  S Zaire, Angola, N Namibia
T.l.chobiensis
  E Namibia, Zambia, NW Zimbabwe
T.l.libonyanus
  E Botswana, Transvaal, N Natal
T.l.peripheris
  C Natal
T.l.tropicalis
  SE Zaire, Mozambique, Malawi, Tanzania
Turdus tephronotus (African Bare-eyed Thrush)
  Ethiopia, Somalia, Kenya, E Tanzania
Turdus menachensis (Yemen Thrush)
  S Saudi Arabia, Yemen
Turdus ludoviciae (Somali Blackbird)
  N Somalia
Turdus dissimilis (Black-breasted Thrush)
T.d.dissimilis
  NE India to SW China, N Indochina
T.d.hortulorum
  Siberia, Manchuria >> SE China, Vietnam
Turdus unicolor (Tickell's Thrush)
  Pakistan, Nepal, N India
Turdus cardis (Japanese Grey Thrush)
  Japan, China >> Indochina
Turdus albocinctus (White-collared Blackbird)
  Himalayas, SE Tibet, SW Sikang >> N Burma
Turdus torquatus (Ring Ousel)
T.t.torquatus
  N Europe >> S Europe, NW Africa
T.t.alpestris
  S & E Europe >> Asia Minor, N Africa
T.t.amicorum
  Turkey, Caucasus, N Iran >> S Iran
Turdus boulboul (Grey-winged Blackbird)
  Himalayas, S China, N Indochina >> N Burma
Turdus merula (Blackbird)

T.m.merula
  W Europe
T.m.azorensis
  Azores Is
T.m.cabrerae
  Madeira, W Canary Is
T.m.mauritanicus
  Morocco to Tunisia
T.m.aterrimus
  SE Europe, Caucasus >> Mediterranean islands
T.m.insularum
  Crete, Rhodes, Mytilene I
T.m.syriacus
  S Turkey, Middle East, Iran
T.m.intermedius
  C Asia, Afghanistan >> S Iraq
T.m.maximus
  Pakistan, India, SE Tibet
T.m.sowerbyi
  Szechwan
T.m.mandarinus
  Kweichow
T.m.nigropileus
  SC India
T.m.spencei
  E India
T.m.simillimus
  SE India
T.m.bourdilloni
  S India
T.m.kinnisii
  Sri Lanka
Turdus poliocephalus (Island Thrush)
T.p.erythropleurus
  Christmas Is
T.p.loeseri
  N Sumatra
T.p.indrapurae
  C Sumatra
T.p.biesenbachi
  Mt Papandajan (W Java)
T.p.fumidus
  Mt Gedeh (W Java)
T.p.stresemanni
  Mt Lawoe (C Java)
T.p.javanicus
  C Java
T.p.whiteheadi
  E Java
T.p.seebohmi
  N Borneo
T.p.niveiceps
  Taiwan
T.p.thomassoni
  N Luzon I
T.p.mayonensis
  S Luzon I
T.p.mindorensis
  Mindoro I
T.p.nigrorum
  Negros I

T.p.malindangensis
  Mt Malindang (NW Mindanao I)
T.p.katanglad
  Mt Katanglad (C Mindanao I)
T.p.kelleri
  Mt Apo (SE Mindanao I)
T.p.hygroscopus
  S Sulawesi
T.p.celebensis
  SW Sulawesi
T.p.schlegelii
  W Timor I
T.p.sterlingi
  E Timor I
T.p.deningeri
  Seram I
T.p.versteegi
  W New Guinea
T.p.carbonarius
  Bismarck Mts (New Guinea)
T.p.keysseri
  Huon, SE New Guinea
T.p.tolokiwae
  Tolokiwa I (Bismarck Arch)
T.p.papuensis
  SE New Guinea
T.p.canescens
  Goodenough I
T.p.heinrothi
  St Matthias Is
T.p.bougainvillei
  Bougainville I
T.p.kulambangrae
  Kulambangra I
T.p.sladeni
  Guadalcanal I
T.p.rennellianus
  Rennell I
T.p.vanikorensis
  Vanikoro, Santa Cruz, Espiritu Santo Is
T.p.placens
  Ureparapara, Vanue Lava Is
T.p.whitneyi
  Gaua, Banks Is
T.p.malekulae
  Pentecost, Malekula, Ambrim Is
T.p.becki
  Paema, Lopevi, Epi, Mai Is
T.p.efatensis
  Efate, Nguna Is
T.p.albifrons
  Erromanga I
T.p.pritzbueri
  Tana, Lifu Is
T.p.mareensis
  Mare I e?
T.p.xanthopus
  New Caledonia I
T.p.poliocephalus
  Norfolk I
T.p.layardi
  Viti Levu, Ovalau, Yasawa, Koro Is

T.p.ruficeps
  Kandavu I
T.p.vitiensis
  Vanua Levu I
T.p.hades
  Ngau I
T.p.tempesti
  Taveuni I
T.p.samoensis
  Savaii, Upolu Is
Turdus chrysolaus (Red-billed Thrush)
T.c.orii
  N & C Kurile Is >> Japan, Riukiu Is
T.c.chrysolaus
  N Japan >> SE China, N Philippine Is
Turdus celaenops (Seven Islands Thrush)
  Izu, Yakushima Is
Turdus rubrocanus (Grey-headed Thrush)
T.r.rubrocanus
  Pakistan, Himalayas
T.r.gouldi
  SE Tibet, W China
Turdus kessleri (Kessler's Thrush)
  W China >> E Tibet, SW Sikang
Turdus feae (Fea's Thrush)
  N China >> Burma, Assam
Turdus pallidus (Pale Thrush)
  NE Siberia >> China, Japan, Taiwan
Turdus obscurus (Eye-browed Thrush)
  NE Asia >> China, Indonesia
Turdus ruficollis (Black-throated Thrush)
T.r.atrogularis
  W Siberia, C Asia >> N India, China
T.r.ruficollis
  E Siberia >> W China, Burma, NE India
Turdus naumanni (Dusky Thrush)
T.n.eunomus
  N Siberia >> Japan, S China, Burma
T.n.naumanni
  C Siberia, C Asia >> N China
Turdus pilaris (Fieldfare)
  N Europe, N Asia >> S Europe, Caucasus
Turdus iliacus (Redwing)
T.i.coburni
  Iceland, Faroe Is >> NW Europe
T.i.iliacus
  N Europe, C Asia >> N Africa, Caucasus
Turdus philomelos (Song Thrush)
T.p.hebridensis
  Outer Hebrides, Isle of Skye
T.p.clarkei
  British Isles, W Europe
T.p.philomelos
  C Asia, C & E Europe >> N Africa, S Europe, Iran
T.p.nataliae
  C Asia, Iran
Turdus mupinensis (Mongolian Song Thrush)
  W China
Turdus viscivorus (Mistle Thrush)
T.v.viscivorus
  Europe, Asia Minor, S Russia

T.v.bonapartei
Siberia, C Asia, Himalayas >> N India
Turdus aurantius (White-chinned Thrush)
Jamaica
Turdus ravidus (Grand Cayman Thrush)
Grand Cayman e?
Turdus plumbeus (Red-legged Thrush)
T.p.plumbeus
N Bahama Is
T.p.schistaceus
E Cuba
T.p.rubripes
C & W Cuba, Isle of Pines
T.p.coryi
Cayman Brac I
T.p.ardosiaceus
Hispaniola, Puerto Rico, Gonave I
T.p.albiventris
Dominica I
Turdus chiguanco (Chiguanco Thrush)
T.c.chiguanco
coastal Peru, NW Bolivia
T.c.conradi
S Ecuador, C Peru
T.c.anthracinus
S Bolivia, NE Chile, W Argentina
Turdus nigrescens (Sooty Robin)
Costa Rica, W Panama
Turdus fuscater (Great Thrush)
T.f.opertaneus
NW Colombia
T.f.cacozelus
N Colombia
T.f.clarus
E Colombia, W Venezuela
T.f.quindio
S & W Colombia, N Ecuador
T.f.gigas
E Colombia, W Venezuela
T.f.gigantodes
S Ecuador, N Peru
T.f.ockendeni
SE Peru
T.f.fuscater
W Bolivia
Turdus infuscatus (Black Robin) 130.13
SE Mexico to Honduras
Turdus serranus (Glossy-black Thrush)
T.s.cumanensis
NE Venezuela
T.s.atrosericeus
NE Colombia, N Venezuela
T.s.fuscobrunneus
C & S Colombia, Eucador
T.s.serranus
Peru, Bolivia
Turdus nigriceps (Slaty Thrush)
T.n.nigriceps
SE Ecuador, E Peru, E Bolivia, W Argentina
T.n.subalaris
S Brazil, Paraguay, N Argentina
Turdus reevei (Plumbeous-backed Thrush)
W Ecuador, NW Peru

Turdus olivater (Black-hooded Thrush)
T.o.sanctaemartae
N Colombia
T.o.olivater
E Colombia, Venezuela
T.o.paraquensis
S Venezuela
T.o.kemptoni
C Venzuela
T.o.duidae
Mt Duida (S Venezuela)
T.o.roraimae
S Venezuela, S Guyana
T.o.caucae
C Colombia
T.o.ptaritepui
SE Venezuela
Turdus maranonicus (Maranon Thrush)
N Peru
Turdus fulviventris (Chestnut-bellied Thrush)
E Colombia, Venezuela, E Ecuador
Turdus rufiventris (Rufous-bellied Thrush)
T.r.juensis
NE Brazil
T.r.rufiventris
S Brazil, Uruguay, Paraguay, N Argentina
Turdus falcklandii (Austral Thrush)
T.f.falcklandii
Falkland Is
T.f.magellanicus
S Chile, S Argentina
T.f.pembertoni
SC Argentina
Turdus leucomelas (Pale-breasted Thrush)
T.l.leucomelas
S Brazil, E Peru, Paraguay
T.l.albiventer
N Colombia, Venezuela, NE Brazil, the Guianas
T.l.cautor
N Colombia
Turdus amaurochalinus (Creamy-bellied Thrush)
C Sth America
Turdus plebejus (Mountain Robin)
T.p.differens
SE Mexico, Guatemala
T.p.rafaelensis
Nicaragua, El Salvador
T.p.plebejus
Costa Rica, W Panama
Turdus ignobilis (Black-billed Thrush)
T.i.ignobilis
E Colombia
T.i.goodfellowi
W Colombia
T.i.debilis
SE Colombia, Venezuela, W Amazonia
T.i.murinus
SE Venezuela, Guyana
T.i.arthuri
SE Venezuela, Guyana, French Guiana

Turdus lawrencii (Lawrence's Thrush)
Upper Amazonia
Turdus fumigatus (Cocoa Thrush)
T.f.aquilonalis
NE Colombia, N Venezuela, Trinidad
T.f.orinocensis
E Colombia, W Venezuela
T.f.fumigatus
N & E Brazil, the Guianas
Turdus personus (Lesser Antillean Thrush)
T.p.bondi
St Vincent I
T.p.personus
Grenada I
Turdus obsoletus (Pale-vented Thrush)
T.o.obsoletus
Costa Rica, Panama, NW Colombia
T.o.parambanus
W Colombia, W Ecuador
T.o.colombianus
C Colombia
T.o.hauxwelli
Upper Amazonia
Turdus haplochrous (Unicoloured Thrush)
E Bolivia
Turdus grayi (Clay-coloured Thrush)
T.g.tamaulipensis
E Mexico
T.g.microrhynchus
San Luis Potosi
T.g.lanyoni
S Mexico, W Guatemala
T.g.linnaei
S Mexico
T.g.grayi
S Mexico to Guatemala
T.g.megas
W Guatemala to Nicaragua
T.g.casius
Costa Rica to NW Colombia
T.g.incomptus
N Colombia
Turdus nudigenis (Bare-eyed Thrush)
T.n.nudigenis
Lesser Antilles, Trinidad, NE Sth America
T.n.extimus
N Brazil
T.n.maculirostris
W Ecuador, NW Peru
Turdus jamaicensis (White-eyed Thrush)
Jamaica
Turdus albicollis (White-necked Thrush)
T.a.calliphthongus
NW Mexico
T.a.lygrus
C & S Mexico
T.a.assimilis
C Mexico
T.a.renominatus
SC Mexico
T.a.oaxacae
Oaxaca

T.a.leucauchen
S Mexico to Honduras
T.a.rubicundus
W Guatemala, El Salvador
T.a.atrotinctus
E Nicaragua
T.a.oblitus
Costa Rica
T.a.cnephosus
SW Costa Rica, W Panama
T.a.coibensis
Coiba I
T.a.daguae
E Panama to NW Ecuador
T.a.minusculus
NE Colombia
T.a.phaeopygoides
NE Colombia, N Venezuela, Trinidad
T.a.phaeopygus
E Colombia to the Guianas, N Brazil
T.a.berlepschi
C & S Colombia
T.a.spodiolaemus
E Ecuador to N Bolivia, W Brazil
T.a.contemptus
S Bolivia
T.a.crotopezus
E Brazil
T.a.albicollis
SE Brazil
T.a.paraguayensis
SW Brazil, Paraguay, N Argentina
Turdus rufopalliatus (Rufous-backed Robin)
T.r.griseor
NW Mexico
T.r.rufopalliatus
W Mexico
T.r.graysoni
Tres Marias Is
Turdus swalesi (La Selle Thrush)
T.s.dodae
C Dominica
T.s.swalesi
Haiti
Turdus rufitorques (Rufous-collared Robin)
SE Mexico, Guatemala, El Salvador
Turdus migratorius (American Robin)
T.m.migratorius
Canada, C USA >> E USA, E Mexico
T.m.nigrideus
E Canada, EC USA
T.m.achrusterus
S USA >> SE Mexico
T.m.caurinus
SE Alaska, W Canada >> SW USA
T.m.propinquus
W Canada, W USA, SW Mexico >>
Guatemala
T.m.phillipsi
C Mexico
T.m.confinis
S Baja California

T.m.permixtus
SW Mexico

131. ORTHONYCHIDAE (LOGRUNNERS)

ORTHONYX
Orthonyx temminckii (Spine-tailed Logrunner)
O.t.novaeguineae
W New Guinea
O.t.dorsalis
W New Guinea
O.t.victoriana
SE New Guinea
O.t.temminckii
SE Queensland, E New South Wales
Orthonyx spaldingii (Spalding's Logrunner)
N Queensland

ANDROPHOBUS
Androphobus viridis (Green-backed Babbler)
W New Guinea

PSOPHODES
Psophodes olivaceus (Eastern Whipbird)
P.o.lateralis
N Queensland
P.o.magnirostris
C Queensland
P.o.olivaceus
S Queensland, E New South Wales, Victoria
Psophodes nigrogularis (Western Whipbird)
P.n.leucogaster
NW Victoria, SE South Australia
P.n.nigrogularis
SW Western Australia
P.n.pondalowiensis
S South Australia

SPHENOSTOMA
Sphenostoma cristatum (Wedgebill)
C & W Australia

CINCLOSOMA
Cinclosoma punctatum (Spotted Quail Thrush)
C.p.punctatum
Eastern Australia
C.p.dovei
Tasmania
Cinclosoma castanotum (Chestnut Quail Thrush)
C.c.castanotum
SE Australia
C.c.mayri
New South Wales
C.c.morgani
South Australia
C.c.clarum
C Australia
C.c.dundasi
SW Western Australia

Cinclosoma alisteri (Nullarbor Quail Thrush)
Nullarbor Plain
Cinclosoma cinnamomeum (Cinnamon Quail Thrush)
C.c.cinnamomeum
EC Australia
C.c.samueli
South Australia
C.c.marginatum
SW Australia
Cinclosoma castaneothorax (Chestnut-breasted Quail Thrush) 131.1
S Queensland, N New South Wales
Cinclosoma ajax (Ajax Quail Thrush)
C.a.ajax
W New Guinea
C.a.muscale
S New Guinea
C.a.alare
SC New Guinea
C.a.goldiei
SE New Guinea

PTILORRHOA
Ptilorrhoa leucosticta (High Mountain Rail Babbler)
P.l.leucosticta
W New Guinea
P.l.mayri
W New Guinea
P.l.centralis
W New Guinea
P.l.sibilans
N New Guinea
P.l.amabilis
E New Guinea
P.l.loriae
SE New Guinea
P.l.menawa
N coast of New Guinea
Ptilorrhoa caerulescens (Lowland Rail Babbler)
P.c.caerulescens
W New Guinea
P.c.neumanni
N New Guinea
P.c.nigricrissa
S New Guinea
P.c.geislerorum
E New Guinea
Ptilorrhoa castanonota (Mid-mountain Rail Babbler)
P.c.castanonota
W New Guinea
P.c.saturata
W New Guinea
P.c.uropygialis
W New Guinea
P.c.buergersi
C New Guinea
P.c.par
E New Guinea

P.c.pulcher
  SE New Guinea
P.c.gilliardi
  Batanta I

EUPETES
Eupetes macrocerus (Malay Rail Babbler)
  E.m.macrocerus
    Malaysia, Thailand, Sumatra
  E.m.borneensis
    N Borneo

MELAMPITTA
Melampitta lugubris (Lesser Melampitta)
  M.l.lugubris
    W New Guinea
  M.l.rostrata
    W New Guinea
  M.l.longicauda
    NC & E New Guinea
Melampitta gigantea (Greater Melampitta)
  W New Guinea

IFRITA
Ifrita kowaldi (Blue-capped Babbler)
  I.k.kowaldi
    E & C New Guinea
  I.k.brunnea
    WC New Guinea

132. TIMALIIDAE (BABBLERS)

TRICHASTOMA   132.1
Trichastoma rostratum (White-chested Jungle Babbler)
  T.r.rostratum
    Malaysia, Sumatra, Billiton I
  T.r.macropterum
    Banggai I, Borneo
Trichastoma celebense (Sulawesi Jungle Babbler)
  T.c.celebense
    N Sulawesi
  T.c.connectens
    NC Sulawesi
  T.c.rufofuscum
    C Sulawesi
  T.c.finschi
    SW Sulawesi
  T.c.improbatum
    E & S Sulawesi, Pula I
  T.c.togianense
    Togian I
Trichastoma bicolor (Ferruginous Jungle Babbler)
  Malaysia, E Sumatra, Bangka I, Borneo
Trichastoma woodi (Bagobo Babbler)   132.1
  Mindanao

MALACOCINCLA   132.1
Malacocincla abbotti (Abbott's Jungle Babbler)

M.a.abbotti
  Himalayas, Burma, Thailand NW
  Malaysia
M.a.krishnarajui
  E Ghats (India)
M.a.williamsoni
  E Thailand, NW Cambodia
M.a.obscurius
  SE Thailand
M.a.rufescentior
  Thailand
M.a.altera
  C Laos, C Vietnam
M.a.olivacea
  Thailand, Malaysia, E Sumatra
M.a.sirensis
  Pulau Matu Siri, Billiton Is
M.a.baweana
  Bawean I
M.a.finschi
  Borneo
Malacocincla sepiaria (Horsfield's Jungle Babbler)
  M.s.tardinata
    Malaysia
  M.s.liberalis
    NW Sumatra
  M.s.barussana
    SW Sumatra
  M.s.sepiaria
    W & C Java
  M.s.minus
    E Java, Bali
  M.s.rufiventris
    W & S Borneo
  M.s.harterti
    N & E Borneo
Malacocincla perspicillata (Black-browed Jungle Babbler)   132.1
  M.p.perspicillata
    Borneo
  M.p.vanderbilti
    N Sumatra
Malacocincla malaccensis (Short-tailed Jungle Babbler)
  M.m.malaccensis
    N Natuna Is, Malaysia, Sumatra, Anamba Is
  M.m.saturata
    Bangka, Billiton Is, W Borneo
  M.m.poliogenys
    E Borneo
  M.m.feriata
    Sarawak, N Borneo
Malacocincla cinereiceps (Ashy-headed Jungle Babbler)
    Balabac, Palawan Is

PELLORNEUM   132.1
Pellorneum tickelli (Tickell's Jungle Babbler)
  P.t.assamense
    Assam, NW Burma
  P.t.grisescens
    Arakan Yoma (SW Burma)

P.t.fulvum
 SW Yunnan, NE Burma, N Thailand,
 Indochina
P.t.annamense
 S & C Vietnam, S Laos
P.t.tickelli
 N Malaysia, E Burma, W Thailand
P.t.ochracea
 S China
P.t.australis
 N Malaysia
***Pellorneum albiventre* (Plain Brown Babbler)**
P.a.ignotum
 Mishmi hills (NE Assam)
P.a.albiventre
 Bhutan, Assam, W Burma
P.a.nagaense
 Burma
P.a.cinnamomeum
 C Burma, NW Thailand, S Indochina
P.a.pusillum
 NW Vietnam, N Laos
***Pellorneum palustre* (Marsh Spotted Babbler)**
 C & E Assam
***Pellorneum ruficeps* (Spotted Babbler)**
P.r.olivaceum
 SW India
P.r.ruficeps
 W & C India
P.r.punctatum
 W Himalayas
P.r.mandellii
 Sikkim, Bhutan, Nepal, NE India
P.r.chamelum
 S Assam
P.r.pectorale
 Mishmi hills (NE Assam)
P.r.ripleyi
 Lakhimpur (NE Assam)
P.r.vocale
 C Manipur
P.r.stageri
 NE Burma
P.r.shanense
 SW Yunnan, C Burma
P.r.hilarum
 C Burma
P.r.victoriae
 Chin hills (N Burma)
P.r.minus
 S Burma
P.r.subochraceum
 S Burma, SW Thailand
P.r.insularum
 Mergui Arch
P.r.acrum
 C Thailand, N Malaysia
P.r.chthonium
 N Thailand
P.r.indistinctum
 N Thailand
P.r.oreum
 S China, N Indochina

P.r.vividum
 N Vietnam
P.r.elbeli
 E Thailand
P.r.ubonense
 E Thailand, S Laos
P.r.deignani
 S Vietnam
P.r.dilloni
 S Indochina
P.r.euroum
 W Cambodia, C & SE Thailand
P.r.smithi
 coastal islands of SE Thailand & Cambodia
***Pellorneum fuscocapillum* (Brown-capped Jungle Babbler)**
P.f.babaulti
 N & E Sri Lanka
P.f.fuscocapillum
 SW Sri Lanka
P.f.scortillum
 SW Sri Lanka
***Pellorneum pyrrogenys* (Temminck's Jungle Babbler)**
P.p.buettikoferi
 Sumatra
P.p.pyrrogenys
 W Java
P.p.besuki
 E Java
P.p.erythrote
 W Sarawak, N Borneo
P.p.longstaffi
 Sarawak, N Borneo
P.p.canicapillum
 N Borneo
***Pellorneum capistratum* (Black-capped Babbler)**
P.c.nigrocapitatum
 Malaysia, N Natuna, Billiton Is
P.c.nyctilampis
 Sumatra, Bangka I
P.c.capistratoides
 W & S Borneo
P.c.morrelli
 Banggai I, N & E Borneo
P.c.capistratum
 Java

MALACOPTERON
***Malacopteron magnirostre* (Moustached Tree Babbler)**
M.m.magnirostre
 S Burma, Thailand, Malaysia, Sumatra
M.m.cinereocapillum
 Borneo
M.m.flavum
 Anamba Is
***Malacopteron affine* (Sooty-capped Babbler)**
M.a.affine
 S Thailand, Malaysia, Sumatra
M.a.notatum
 Banyak I

M.a.phoeniceum
  Borneo
***Malacopteron cinereum* (Scaly-crowned Babbler)**
M.c.indochinense
  SE Thailand, S Indochina
M.c.rufifrons
  Java
M.c.cinereum
  Malaysia, Sumatra, Bangka I, Borneo
M.c.niasense
  Nias I
M.c.bungurense
  N Natuna Is
***Malacopteron magnum* (Rufous-crowned Tree Babbler)**
M.m.magnum
  S Burma, Malaysia, Sumatra, Borneo, Natuna Is
M.m.saba
  NE Borneo
***Malacopteron palawanense* (Palawan Tree Babbler)**
  Balabac, Palawan Is
***Malacopteron albogulare* (Grey-breasted Babbler)**
M.a.albogulare
  Malaysia, NE Sumatra
M.a.moultoni
  NW Borneo

ILLADOPSIS   132.1
***Illadopsis cleaveri* (Blackcap Thrush Babbler)**
I.c.johnsoni
  Sierra Leone, Liberia
I.c.cleaveri
  Ghana
I.c.marchanti
  S Nigeria
I.c.batesi
  SE Nigeria, Gabon, Cameroun
I.c.poense
  Fernando Po I
***Illadopsis albipectus* (Scaly-breasted Thrush Babbler)**
I.a.barakae
  N Zaire, Uganda, S Sudan, SW Kenya
I.a.albipectus
  N Angola, W Zaire
***Illadopsis rufescens* (Rufous-winged Thrush Babbler)**
  Sierra Leone to Ghana
***Illadopsis puveli* (Puvel's Thrush Babbler)**
I.p.puveli
  Guinea to Sierra Leone
I.p.strenuipes
  S Nigeria to NE Zaire
***Illadopsis rufipennis* (Pale-breasted Thrush Babbler)**
I.r.bocagei
  Fernando Po I
I.r.extremis
  Sierra Leone to Ghana

I.r.rufipennis
  S Nigeria to Uganda, Kenya
I.r.distans
  NE Tanzania, Zanzibar I
***Illadopsis fulvescens* (Brown Thrush Babbler)**
I.f.gularis
  Sierra Leone to Ghana
I.f.moloneyanum
  E Ghana, Togo
I.f.iboensis
  S Nigeria
I.f.fulvescens
  Cameroun, W Zaire
I.f.ugandae
  N Zaire, Uganda
I.f.dilutius
  N Angola
***Illadopsis pyrrhopterus* (Mountain Thrush Babbler)**
I.p.pyrrhopterus
  East Africa, Malawi
I.p.kivuense
  E Zaire, W Uganda, W Tanzania
***Illadopsis abyssinicus* (African Hill Babbler)**
                                                    132.1
I.a.monachus
  Cameroun Mt
I.a.claudi
  Fernando Po I
I.a.poliothorax
  W Kenya
I.a.loima
  NW Kenya
I.a.ansorgei
  C & W Angola, SE Zaire, W Tanzania
I.a.stierlingi
  S & C Tanzania, N Malawi
I.a.abyssinicus
  W Ethiopia, W Kenya, N Tanzania
I.a.atriceps
  Cameroun to NE Zaire, W Uganda
I.a.hildegardae
  SW Tanzania

KAKAMEGA
***Kakamega poliothorax* (Grey-chested Thrush Babbler)**
  S Cameroun, Fernando Po I, E Zaire, SW Kenya

PTYRTICUS
***Ptyrticus turdinus* (African Thrush Babbler)**
P.t.harterti
  C Cameroun
P.t.turdinus
  SW Sudan, NE Zaire
P.t.upembae
  SE Zaire, N Zambia

POMATORHINUS
***Pomatorhinus hypoleucos* (Long-billed Scimitar Babbler)**

P.h.hypoleucos
  Assam, Bangladesh, W Burma
P.h.tickelli
  Thailand, N Indochina
P.h.brevirostris
  S Indochina
P.b.wrayi
  Malaysia
P.b.hainanus
  Hainan I

Pomatorhinus erythrogenys **(Rusty-cheeked Scimitar Babbler)**
P.e.erythrogenys
  W Himalayas
P.e.ferrugilatus
  W & C Nepal
P.e.haringtoni
  W Himalayas
P.e.mcclellandi
  S Assam, W Burma
P.e.imberbis
  Karenni (Burma)
P.e.celatus
  C Burma, NW Thailand
P.e.odicus
  NE Burma, N Laos, SW China
P.e.decarlei
  S Szechwan, N Yunnan
P.e.dedekeni
  E & S Sikang, NW Yunnan
P.e.gravivox
  NW Szechwan, S Kansu
P.e.sowerbyi
  N Shensi
P.e.cowensae
  SW Hupeh, E Szechwan
P.e.swinhoei
  Anhwei, Kiangsi, Kwangsi, Hunan, Fukien
P.e.erythrocnemis
  Taiwan

Pomatorhinus horsfieldii **(Travancore Scimitar Babbler)**
P.h.melanurus
  Sri Lanka
P.h.travancoreensis
  SW India
P.h.horsfieldii
  W India
P.h.obscurus
  NW India
P.h.maderaspatensis
  EC India

Pomatorhinus schisticeps **(Slaty-headed Scimitar Babbler)**
P.s.leucogaster
  NW Himalayas
P.s.schisticeps
  E Himalayas, NE India, NW Burma
P.s.salimalii
  Mishmi hills (NE Assam)
P.s.cryptanthus
  Lakhimpur (NE Assam)

P.s.mearsi
  W Burma
P.s.ripponi
  E Burma, N Thailand, N Laos
P.s.nuchalis
  E Burma
P.s.difficilis
  S Burma, SW Thailand
P.s.olivaceus
  S Burma, SW Thailand
P.s.fastidiosus
  Malaysia
P.s.humilis
  E Thailand, S Laos, C Vietnam
P.s.annamensis
  S Vietnam
P.s.klossi
  SE Thailand, SW Cambodia

Pomatorhinus montanus **(Chestnut-backed Scimitar Babbler)**
P.m.occidentalis
  S Malaysia, Sumatra
P.m.montanus
  W & C Java
P.m.ottolanderi
  E Java, Bali
P.m.bornensis
  Borneo

Pomatorhinus ruficollis **(Streak-breasted Scimitar Babbler)**
P.r.ruficollis
  W & C Nepal
P.r.godwini
  E Himalayas, N Assam
P.r.bakeri
  SE Assam, W Burma
P.r.bhamoensis
  N Burma
P.r.similis
  NE Burma, NW Yunnan
P.r.albipectus
  SW Yunnan, N Laos
P.r.beaulieui
  N Laos
P.r.laurentei
  S Yunnan
P.r.reconditus
  SE Yunnan, N Vietnam
P.r.stridulus
  SE China
P.r.hunonensis
  C China
P.r.eidos
  S Szechwan
P.r.musicus
  Taiwan
P.r.nigrostellatus
  Hainan I

Pomatorhinus ochraceiceps **(Red-billed Scimitar Babbler)**
P.o.stenorhynchus
  NE Assam, N Burma

P.o.austeni
Manipur, E Assam

P.o.ochraceiceps
Burma, N Thailand, N Indochina

P.o.alius
E Thailand, S Indochina

Pomatorhinus ferruginosus (Coral-billed Scimitar Babbler)

P.f.ferruginosus
E Himalayas

P.f.formosus
S Assam

P.f.phayrei
Arakan Yoma (SW Burma)

P.f.stanfordi
NE Burma

P.f.mariae
C Burma

P.f.albogularis
E Burma, NW Thailand

P.f.orientalis
N Indochina

GARRITORNIS

Garritornis isidorei (Isidor's Rufous Babbler)

G.i.isidorei
C & S New Guinea, Misol I

G.i.calidus
N New Guinea

POMATOSTOMUS

Pomatostomus temporalis (Grey-crowned Babbler)

P.t.tregellasi
SE South Australia, Victoria, SE New South Wales

P.t.trivirgatus
E New South Wales, S Queensland

P.t.temporalis
coastal C Queensland

P.t.cornwalli
coastal N Queensland

P.t.strepitans
S New Guinea

P.t.intermedius
C Australia

P.t.mountfordae
N Northern Territory

P.t.browni
NW Northern Territory

P.t.rubeculus
C Northern Territory

P.t.bamba
Melville I

P.t.nigrescens
Western Australia

Pomatostomus superciliosus (White-browed Babbler)

P.s.gilgandra
W New South Wales, Victoria, South Australia

P.s.superciliosus
SE South Australia, Victoria

P.s.ashbyi
SW Western Australia

P.s.gwendolenae
Gascoyne valley, Western Australia

Pomatostomus halli (Hall Babbler)
SW Queensland

Pomatostomus ruficeps (Chestnut-crowned Babbler)
SW Queensland, W New South Wales, NW Victoria, NE South Australia

XIPHIRHYNCHUS

Xiphirhynchus superciliaris (Slender-billed Scimitar Babbler)

X.s.superciliaris
E Himalayas

X.s.intextus
S Assam, W Burma

X.s.forresti
NE Burma, NW Yunnan

X.s.rothschildi
N Vietnam

JABOUILLEIA

Jabouilleia danjoui (Danjou's Babbler)

J.d.danjoui
C Vietnam

J.d.parvirostris
C Vietnam

RIMATOR

Rimator malacoptilus (Long-billed Wren Babbler)

R.m.malacoptilus
E Himalayas, Assam, NE Burma

R.m.pasquieri
N Vietnam

R.m.albostriatus
W Sumatra

PTILOCICHLA

Ptilocichla leucogrammica (Bornean Wren Babbler)
Borneo

Ptilocichla mindanensis (Streaked Ground Babbler)

P.m.minuta
Leyte, Samar Is

P.m.fortichi
Bohol I

P.m.mindanensis
Mindanao I

P.m.basilanica
Basilan I

Ptilocichla falcata (Palawan Wren Babbler)
Balabec, Palawan Is

KENOPIA

Kenopia striata (Striped Wren Babbler)
Malaysia, E Sumatra, Borneo

NAPOTHERA

Napothera rufipectus (Sumatran Wren Babbler)
W Sumatra

Napothera atrigularis (Black-throated Wren Babbler)
Borneo

Napothera macrodactyla (Large Wren Babbler)
N.m.macrodactyla
Malaysia
N.m.beauforti
NE Sumatra
N.m.lepidopleura
Java

Napothera marmorata (Marbled Wren Babbler)
N.m.grandior
C Malaysia
N.m.marmorata
W Sumatra

Napothera crispifrons (Limestone Wren Babbler)
N.c.annamensis
N Indochina
N.c.calcicola
NE Thailand
N.c.crispifrons
N Thailand, S Burma

Napothera brevicaudata (Streaked Wren Babbler)
N.b.striata
S Assam, SW Burma
N.b.venningi
W Yunnan, NE Burma
N.b.brevicaudata
N Thailand
N.b.stevensi
N Indochina
N.b.proxima
C Vietnam, S Laos
N.b.rufiventer
S Vietnam
N.b.griseigularis
SE Thailand, SW Cambodia
N.b.leucosticta
N Malaysia

Napothera crassa (Mountain Wren Babbler)
N Borneo

Napothera rabori (Luzon Wren Babbler)
N.r.rabori
Ilocos Norte (Luzon I)
N.r.mesoluzonica
Laguna (Luzon I)
N.r.sorsogonensis
Sorsogon (Luzon I)

Napothera epilepidota (Lesser Wren Babbler)
N.e.guttaticollis
N Assam
N.e.roberti
S Assam, NW Burma
N.e.bakeri
C Burma

N.e.davisoni
N Thailand
N.e.amyae
N Indochina
N.e.delacouri
Kwangsi
N.e.hainana
Hainan I
N.e.clara
S Vietnam
N.e.granti
N Malaysia
N.e.lucilleae
N Sumatra
N.e.diluta
W Sumatra
N.e.mendeni
SW Sumatra
N.e.epilepidota
W & C Java
N.e.exsul
N Borneo

PNOEPYGA

Pnoepyga albiventer (Scaly-breasted Wren Babbler)
P.a.pallidior
N & C Himalayas
P.a.albiventer
E Himalayas, Assam, N Burma, S China

Pnoepyga pusilla (Pygmy Wren Babbler)
P.p.pusilla
Himalayas, Assam, N Burma, N Thailand, S China
P.p.formosana
Taiwan
P.p.annamensis
S Indochina
P.p.harterti
C Malaysia
P.p.lepida
W Sumatra
P.p.rufa
Java
P.p.everetti
Flores I
P.p.timorensis
Timor I

SPELAEORNIS

Spelaeornis caudatus (Short-tailed Wren Babbler)
Nepal, Sikkim, Bhutan

Spelaeornis badeigularis (Mishmi Wren Babbler)
Mishmi hills (NE Assam)

Spelaeornis troglodytoides (Bar-winged Wren Babbler)
S.t.sherriffi
E Bhutan
S.t.souliei
NE Burma, NW Yunnan

341

S.t.rocki
NW Yunnan
S.t.troglodytoides
Sikang, NW Szechwan
S.t.halsueti
Shensi
Spelaeornis formosus (Spotted Wren Babbler)
E Himalayas, W Burma, S China
Spelaeornis chocolatinus (Godwin-Austin's Wren Babbler)
S.c.chocolatinus
S Assam, Manipur
S.c.oatesi
N Burma
S.c.reptatus
NE Burma, SW Yunnan
S.c.kinneari
N Vietnam
Spelaeornis longicaudatus (Long-tailed Wren Babbler)
S Assam, Manipur

SPHENOCICLA
Sphenocicla humei (Wedge-billed Wren Babbler)
S.h.humei
Sikkim to N Assam
S.h.roberti
S Assam, NE Burma

NEOMIXIS
Neomixis tenella (Northern Jery)
N.t.tenella
N Madagascar
N.t.decaryi
W Madagascar
N.t.orientalis
C & S Madagascar
M.t.debilis
SW Madagascar
Neomixis viridis (Southern Green Jery)
N.v.delacouri
NE Madagascar
N.v.viridis
SE Madagascar
Neomixis striatigula (Stripe-throated Jery)
N.s.sclateri
NE Madagascar
N.s.striatigula
SE Madagascar
N.s.pallidior
SW Madagascar
Neomixis flavoviridis (Wedge-tailed Jery)
SE Madagascar

STACHYRIS
Stachyris rodolphei (Deignan's Babbler)
Thailand
Stachyris rufifrons (Red-fronted Tree Babbler)
S.r.pallescens
Arakan Yoma (SW Burma)

S.r.rufifrons
SE Burma, W Thailand
S.r.obscura
S Thailand
S.r.poliogaster
W Malaysia, Sumatra
S.r.sarawacensis
Borneo
Stachyris ambigua (Buff-chested Babbler)
S.a.ambigua
E Himalayas, Assam
S.a.planicola
NE Burma
S.a.adjuncta
N & E Thailand, N Indochina
S.a.insuspecta
S Laos
Stachyris ruficeps (Red-headed Tree Babbler)
S.r.ruficeps
E Himalayas, N Assam
S.r.rufipectus
NW Burma
S.r.bhamoensis
NE Burma, NW Yunnan
S.r.davidi
C & S China, N Indochina
S.r.praecognita
Taiwan
S.r.goodsoni
Hainan I
S.r.pagana
S Vietnam
Stachyris pyrrhops (Red-billed Tree Babbler)
Pakistan, W Himalayas
Stachyris chrysaea (Golden-headed Tree Babbler)
S.c.chrysaea
E Himalayas, Assam, N Burma
S.c.binghami
SE Assam, SW Burma
S.c.aurata
S Burma, N Indochina
S.c.assimilis
C Burma, NW Thailand
S.c.chrysops
C Malaysia
S.c.frigida
W Sumatra
Stachyris plateni (Pygmy Tree Babbler)
S.p.pygmaea
Samar, Leyte Is
S.p.plateni
Mindanao I
Stachyris capitalis (Rufous-crowned Tree Babbler)
S.c.dennistouni
NE Luzon I
S.c.affinis
S Luzon I
S.c.euroaustralis
Philippine Is
S.c.nigrocapitata
Samar, N Leyte Is

S.c.boholensis
Bohol I
S.c.capitalis
Dinagat, Mindanao Is
S.c.isabelae
Basilan I
Stachyris speciosa (Rough-templed Tree Babbler)
Negros I
Stachyris whiteheadi (Whitehead's Tree Babbler)
N Luzon I
Stachyris striata (Striped Tree Babbler)
N Luzon I
Stachyris nigrorum (Negros Tree Babbler)
Negros I
Stachyris hypogrammica (Palawan Tree Babbler)
Palawan I
Stachyris grammiceps (White-breasted Tree Babbler)
W Java
Stachyris herberti (Sooty Tree Babbler)
Laos
Stachyris nigriceps (Black-throated Tree Babbler)
S.n.nigriceps
E Himalayas
S.n.coei
Mishmi hills (E Assam)
S.n.coltarti
Naga hills, E Assam, N Burma
S.n.spadix
S Assam, S Burma, NW Thailand
S.n.yunnanensis
N Thailand, E Burma, N Indochina, SW Yunnan
S.n.rileyi
S Vietnam
S.n.dipora
N Malaysia
S.n.davisoni
C Malaysia
S.n.larvata
Lingga Arch, Sumatra
S.n.natunensis
N Natuna Is
S.n.tionis
Tioman I
S.n.hartleyi
W Sarawak
S.n.borneensis
Borneo
Stachyris poliocephala (Grey-headed Tree Babbler)
S.p.poliocephala
Malaysia, Sumatra, Borneo
S.p.pulla
NE Sumatra
Stachyris striolata (Spot-necked Tree Babbler)
S.s.swinhoei
Hainan I

S.s.tonkinensis
Kwansi, N Indochina
S.s.helenae
N Thailand, N Laos
S.s.guttata
W Thailand
S.s.nigrescentior
S Thailand
S.s.umbrosa
NE Sumatra
S.s.striolata
W Sumatra
Stachyris oglei (Austen's Spotted Tree Babbler)
E Assam
Stachyris maculata (Chestnut-rumped Tree Babbler)
S.m.pectoralis
C & S Malaysia
S.m.maculata
Sumatra, Borneo
S.m.banjakensis
Banyak I
S.m.hypopyrrha
Batu I
Stachyris leucotis (White-necked Tree Babbler)
S.l.leucotis
S Malaysia
S.l.sumatrensis
Sumatra
S.l.obscurata
Borneo
Stachyris nigricollis (Black-throated Tree Babbler)
S.n.erythronotus
N Malaysia
S.n.nigricollis
S Malaysia, E Sumatra, Borneo
Stachyris thoracica (White-collared Tree Babbler)
S.t.thoracica
S Sumatra, W & C Java
S.t.orientalis
E Java
Stachyris erythroptera (Chestnut-winged Tree Babbler)
S.e.erythroptera
S Malaysia, N Natuna Is
S.e.apega
Bangka, Billiton Is
S.e.pyrrhophaea
Sumatra, Batu I
S.e.fulviventris
Banyak I
S.e.bicolor
Banggai I, N & E Borneo
S.e.rufa
SW Borneo
Stachyris melanothorax (Pearl-cheeked Tree Babbler)
S.m.melanothorax
W Java

S.m.albigula
Java
S.m.mendeni
Java
S.m.intermedia
E Java
S.m.baliensis
Bali
Stachyris zantholeuca (White-bellied Tree Babbler) 132.3
S.z.zantholeuca
E Himalayas to Thailand
S.z.tyrannula
N Thailand, N Indochina, Hainan I
S.z.griseiloris
SE China, Taiwan
S.z.sordida
E Thailand, S Indochina
S.z.canescens
SE Thailand, Cambodia
S.z.interposita
Malaysia
S.z.saani
NW Sumatra
S.z.brunnescens
Borneo

DUMETIA
Dumetia hyperythra (Rufous-bellied Babbler)
D.h.hyperythra
SW Nepal, N & C India
D.h.albogularis
S India
D.h.phillipsi
Sri Lanka
D.h.navarroi
W India

RHOPOCICHLA
Rhopocichla atriceps (Black-headed Babbler)
R.a.atriceps
C India
R.a.bourdilloni
SW India
R.a.siccata
N & E Sri Lanka
R.a.nigrifrons
SW Sri Lanka

MACRONOUS
Macronous gularis (Striped Tit-Babbler)
M.g.rubricapilla
Nepal, NE India, Assam
M.g.ticehursti
W Burma
M.g.sulphureus
E Burma, N Thailand
M.g.lutescens
SE Yunnan, N & E Thailand, Laos, N Vietnam

M.g.kinneari
C Vietnam
M.g.versuricola
E Cambodia, S Vietnam
M.g.saraburiensis
E Thailand, W Cambodia
M.g.connectens
S Thailand
M.g.inveteratus
coastal islands of SE Thailand & Cambodia
M.g.condorensis
Pulau Kondor, S Vietnam
M.g.archipelagicus
Mergui Arch
M.g.chersonesophilus
N Malaysia
M.g.gularis
S Malaysia, Sumatra, Batu I
M.g.zopherus
Anamba Is
M.g.zaperissus
N Natuna Is
M.g.everetti
Bunguran, N Natuna Is
M.g.ruficoma
Bangka, Billiton Is
M.g.javanicus
W & C Java
M.g.flavicollis
E Java
M.g.prillwitzi
Kangean I
M.g.montanus
NE Borneo
M.g.bornensis
Borneo
M.g.cagayenensis
Cagayan Sulu I
M.g.argenteus
N Borneo islands
M.g.woodi
Palawan I
Macronous kelleyi (Grey-faced Tit-Babbler)
S Indochina
Macronous striaticeps (Brown Tit-Babbler)
M.s.mindanensis
Samar, Leyte, Bohol, Mindanao Is
M.s.alcasidi
Dinagat I
M.s.striaticeps
Basilan, Malamaui Is
M.s.kettlewelli
Sulu Arch
Macronous ptilosus (Fluffy-backed Tit-Babbler)
M.p.ptilosus
Malaysia
M.p.trichorrhos
Sumatra, Batu Is
M.p.sordidus
Bangka, Billiton Is
M.p.reclusus
Borneo

MICROMACRONUS
Micromacronus leytensis (Leyte Tit-Babbler)
M.l.leytensis
Leyte I
M.l.sordidus
Mindanao I

TIMALIA
Timalia pileata (Chestnut-capped Babbler)
T.p.bengalensis
E Himalayas, NW Burma
T.p.smithi
N Burma, S China, N Thailand, N Indochina
T.p.intermedia
C & S Burma, SW Thailand
T.p.patriciae
WC Thailand
T.p.dictator
S & E Thailand, S Indochina
T.p.pileata
Java

CHRYSOMMA
Chrysomma sinense (Oriental Yellow-eyed Babbler)
C.s.nasale
Sri Lanka
C.s.hypoleucum
Pakistan, India, Bangladesh, W Burma
C.s.sinense
S China, E Himalayas, Burma, Thailand, Indochina

MOUPINIA
Moupinia altirostris (Jerdon's Babbler)
M.a.scindica
Pakistan
M.a.griseigularis
N India, S Assam, NE Burma
M.a.altirostris
SC Burma
Moupinia poecilotis (Rufous-crowned Babbler)
W China

CHAMAEA
Chamaea fasciata (Wren-Tit)
C.f.phaea
coast of Oregon
C.f.rufula
coast of N California
C.f.intermedia
San Francisco area
C.f.fasciata
coast of C California
C.f.henshawi
SW Oregon, N & C California
C.f.canicauda
NW Baja California

TURDOIDES
Turdoides nipalensis (Spiny Babbler)
W & C Nepal

Turdoides altirostris (Iraq Babbler)
SE Iraq, SW Iran
Turdoides caudatus (Common Babbler)
T.c.salvadorii
SE Iraq, SW Iran
T.c.huttoni
Afghanistan, E Iran, S Pakistan
T.c.eclipes
N Pakistan
T.c.caudatus
SE Pakistan, India
Turdoides earlei (Striated Babbler)
T.e.sonivius
Pakistan, NW India
T.e.earlei
NE India, Assam, Burma
Turdoides gularis (White-throated Babbler)
C & S Burma
Turdoides longirostris (Slender-billed Babbler)
Nepal, Assam
Turdoides malcolmi (Large Grey Babbler)
C India
Turdoides squamiceps (Arabian Babbler)
T.s.squamiceps
coast of W & S Saudi Arabia
T.s.yemensis
S Yemen, Aden
T.s.muscatensis
coast of Oman
Turdoides fulvus (Fulvous Babbler)
T.f.maroccanus
SW Morocco
T.f.fulvus
N Algeria, Tunisia, NW Libya
T.f.buchanani
C Sahara
T.f.acaciae
S Egypt, N Sudan, NE Ethiopia
Turdoides aylmeri (Scaly Chatterer)
T.a.aylmeri
Somalia, SE Ethiopia
T.a.boranensis
SC Ethiopia
T.a.kenianus
C Kenya
T.a.loveridgei
SE Kenya, NE Tanzania
T.a.mentalis
NC Tanzania
Turdoides rubiginosus (Rufous Chatterer)
T.r.bowdleri
SE Ethiopia
T.r.rubiginosus
E Uganda, W Kenya, C & S Ethiopia, S Sudan
T.r.heuglini
East African coast from Somalia to Tanzania
T.r.schnitzeri
NW Tanzania
Turdoides subrufus (Rufous Babbler)

T.s.subrufus
SW India
T.s.hyperythus
SW Madras
Turdoides striatus (White-headed Jungle Babbler)
T.s.malabaricus
SW India
T.s.somervillei
W coast of India
T.s.sindianus
Pakistan, NW India
T.s.striatus
N India, E Assam
T.s.orientalis
C & S India
Turdoides rufescens (Ceylon Jungle Babbler)
Sri Lanka
Turdoides affinis (White-headed Babbler)
T.a.affinis
S India
T.a.taprobanus
Sri Lanka
Turdoides melanops (Black-lored Babbler)
T.m.vepres
S Kenya
T.m.clamosus
C Kenya
T.m.sharpei
W Kenya, S Uganda, NW Tanzania
T.m.melanops
SW Angola, N Namibia, Botswana
T.m.angolensis
Huila (Angola)
T.m.querulus
N Namibia, S Angola
Turdoides tenebrosus (Dusky Babbler)
NE Zaire, S Sudan, SW Ethiopia
Turdoides reinwardtii (Blackcap Babbler)
T.r.reinwardtii
Senegal to Sierra Leone
T.r.stictilaemus
Ghana to N Zaire
T.r.houyi
N Cameroun, Central African Republic
Turdoides plebejus (Brown Babbler)
T.p.platycircus
Senegal to Sierra Leone, Togo, Niger
T.p.uamensis
E & C Cameroun
T.p.plebejus
S Cameroun, Nigeria, Central African
Republic
T.p.leucocephalus
E Sudan, Ethiopia
T.p.cinereus
E Nigeria to Ethiopia, Sudan, W Kenya
T.p.gularis
W Cameroun, N Zaire
Turdoides jardineii (Arrow-marked Babbler)
T.j.hypostictus
S Zaire, N Angola

T.j.tanganjicae
SE Zaire, N Zambia
T.j.emini
Uganda, Tanzania
T.j.kikuyuensis
SW Kenya, NW Tanzania
T.j.kirkii
coastal zone from Kenya to Mozambique,
Malawi
T.j.convergens
S Mozambique, N Angola
T.j.tamalakanei
SW Zambia, N Botswana, S Angola
T.j.jardineii
Zimbabwe, Mozambique, Transvaal, Natal
Turdoides squamulatus (Scaly Babbler)
T.s.jubaensis
S Somalia
T.s.carolinae
River Sheballi (S Somalia)
T.s.squamulatus
coast of Kenya
Turdoides leucopygius (White-rumped Babbler)
T.l.leucopygius
coast of E Ethiopia
T.l.limbatus
NW Ethiopia
T.l.smithii
W Somalia, SE Ethiopia
T.l.lacuum
SW Ethiopia
T.l.omoensis
SW Ethiopia, SE Sudan
T.l.ater
SE Zaire, NE Zambia, SW Tanzania
T.l.hartlaubii
W Zambia, S Angola, N Botswana
T.l.griseosquamatus
N Botswana
Turdoides hindei (Hinde's Pied Babbler)
E Kenya
Turdoides hypoleucus (Northern Pied Babbler)
T.h.hypoleucus
C Kenya
T.h.rufuensis
NE Tanzania
Turdoides bicolor (Pied Babbler)
Namibia, Botswana, W Transvaal
Turdoides gymnogenys (Bare-cheeked Babbler)
T.g.gymnogenys
SW Angola
T.g.kaokensis
N Namibia

BABAX
Babax lanceolatus (Chinese Babax)
B.l.lanceolatus
SW China, NE Burma
B.l.woodi
SE Assam, W Burma

B.l.latouchei
SE China
Babax waddelli (Giant Babax)
B.w.waddelli
SE Tibet
B.w.lumsdeni
NE Tibet
B.w.jomo
SE Tibet
Babax koslowi (Koslow's Babax)
N Sikang

GARRULAX
Garrulax cinereifrons (Ashy-headed Laughing Thrush)
SW Sri Lanka
Garrulax palliatus (Grey & Brown Laughing Thrush)
G.p.palliatus
W Sumatra
G.p.schistochlamys
N Borneo
Garrulax rufifrons (Red-fronted Laughing Thrush)
G.r.rufifrons
W Java
G.r.slamatensis
Java
Garrulax perspicillatus (Spectacled Laughing Thrush)
C & S China, N Vietnam
Garrulax albogularis (White-throated Laughing Thrush)
G.a.whistleri
Pakistan, W Himalayas, NW India
G.a.albogularis
E Himalayas, Bhutan
G.a.eous
SE Sikang, SW China, NW Vietnam
G.a.ruficeps
Taiwan
Garrulax leucolophus (White-crested Laughing Thrush)
G.l.leucolophus
Himalayas, N Assam
G.l.patkaicus
S Assam, W Burma
G.l.belangeri
S Burma, SW Thailand
G.l.diardi
SE Burma, Thailand, Yunnan, Indochina
G.l.bicolor
W Sumatra
Garrulax monileger (Lesser Necklaced Laughing Thrush)
G.m.monileger
E Himalayas, NE Burma
G.m.badius
Mishmi hills (NE Assam)
G.m.stuarti
SE Burma, NW Thailand
G.m.fuscatus
SW Thailand

G.m.mouhoti
SE Thailand, S Indochina
G.m.pasquieri
C Vietnam
G.m.schauenseei
E Burma, NE Thailand, N Laos
G.m.tonkinensis
Kwangsi, N Vietnam
G.m.melli
Kwangtung to Anhwei (SE China)
G.m.schmackeri
Hainan I
Garrulax pectoralis (Greater Necklaced Laughing Thrush)
G.p.pectoralis
Nepal
G.p.melanotis
E Himalayas, Assam, N Burma
G.p.subfusus
SE Burma, W Thailand, NW Laos
G.p.robini
NE Laos, N Vietnam
G.p.picticollis
Kwangtung to Anhwei (SE China)
G.p.semitorquatus
Hainan I
Garrulax lugubris (Black Laughing Thrush)
G.l.lugubris
Malaysia, W Sumatra
G.l.calvus
NE Borneo
Garrulax striatus (Striated Laughing Thrush)
G.s.striatus
NW Himalayas
G.s.vibex
C Himalayas
G.s.sikkimensis
Sikkim, E Himalayas
G.s.cranbrooki
Bhutan, Assam, N & W Burma
Garrulax strepitans (Tickell's Laughing Thrush)
G.s.strepitans
E & S Burma, W Thailand, NW Laos
G.s.ferrarius
SE Thailand
Garrulax milleti (Black-hooded Laughing Thrush)
S Vietnam
Garrulax maesi (Maes' Laughing Thrush)
G.m.grahami
SE Sikang, SW China
G.m.maesi
Kwangsi, N Vietnam
G.m.varennei
NE & C Laos
G.m.castanotis
Hainan I
Garrulax nuchalis (Chestnut-backed Laughing Thrush)
NE Assam, N Burma
Garrulax chinensis (Black-throated Laughing Thrush)

G.c.lochmius
SW Yunnan, SE Burma, N Thailand, N Laos
G.c.propinquus
S Burma, SW Thailand
G.c.germaini
S Vietnam
G.c.chinensis
S China, NE Indochina
G.c.monachus
Hainan I

Garrulax vassali (White-cheeked Laughing Thrush)
S Indochina

Garrulax galbanus (Austen's Laughing Thrush)
G.g.galbanus
SE Assam, W Burma
G.g.courtoisi
NE Kiangsi
G.g.simaoensis
Yunnan

Garrulax delesserti (Rufous-vented Laughing Thrush)
G.d.delesserti
SW India
G.d.gularis
Bhutan, Assam, N Burma, N Laos

Garrulax variegatus (Variegated Laughing Thrush)
G.v.variegatus
W Himalayas
G.v.similis
Pakistan, NW India

Garrulax davidi (David's Laughing Thrush)
G.d.chinganicus
N Manchuria
G.d.davidi
N China, S Mongolia
G.d.experrectus
N Kansu
G.d.concolor
NW Szechwan

Garrulax sukatschewi (Black-fronted Laughing Thrush)
S Kansu

Garrulax cineraceus (Ashy Laughing Thrush)
G.c.cineraceus
S Assam, W Burma
G.c.strenuus
NE Burma, SW China
G.c.cinereiceps
C & SE China

Garrulax rufogularis (Rufous-chinned Laughing Thrush)
G.r.occidentalis
Pakistan, W Himalayas
G.r.grosvenori
W Nepal
G.r.rufogularis
E Himalayas, N Assam
G.r.assamensis
NE Assam

G.r.rufitinctus
S Assam
G.r.rufiberbis
N Burma
G.r.intensior
N Vietnam

Garrulax lunulatus (Bar-backed Laughing Thrush)
S Kansu, S Shensi

Garrulax bieti (Biet's Laughing Thrush)
SE Sikang, W Szechwan

Garrulax maximus (Giant Laughing Thrush)
W China, SE Tibet

Garrulax ocellatus (White-spotted Laughing Thrush)
G.o.griseicauda
W Himalayas
G.o.ocellatus
E Himalayas, S Tibet
G.o.maculipectus
NW Yunnan, NE Burma
G.o.artemisiae
SW Szechwan, E Sikang

Garrulax caerulatus (Grey-sided Laughing Thrush)
G.c.caerulatus
E Himalayas
G.c.subcaerulatus
S Assam
G.c.livingstoni
E Assam, NW Burma
G.c.kaurensis
N Burma
G.c.latifrons
W Yunnan, NE Burma
G.c.ricinus
S Yunnan
G.c.berthemyi
NW Fukien

Garrulax poecilorhynchus (Rufous Laughing Thrush)
Taiwan

Garrulax mitratus (Chestnut-capped Laughing Thrush)
G.m.mitratus
W Sumatra
G.m.major
C Malaysia
G.m.damnatus
E Sarawak
G.m.griswoldi
C Borneo
C.m.treacheri
N Borneo

Garrulax ruficollis (Rufous-necked Laughing Thrush)
E Himalayas, NE Burma

Garrulax merulinus (Spot-breasted Laughing Thrush)
G.m.merulinus
W Yunnan, N Burma, S Assam
G.m.obscurus
SE Yunnan, N Indochina

G.m.annamensis
  S Vietnam
Garrulax canorus (Melodious Laughing Thrush (Hwamei))
  G.c.canorus
    S China, N Indochina
  G.c.owstoni
    Hainan I
  G.c.taewanus
    Taiwan
Garrulax sannio (White-browed Laughing Thrush)
  G.s.albosuperciliaris
    E Assam
  G.s.comis
    Yunnan, SE Sikang, NE Burma, N Indochina
  G.s.sannio
    N Vietnam, S China
  G.s.oblectans
    WC China
Garrulax cachinnans (Nilgiri White-breasted Laughing Thrush)
  Nilgiri Hills, W Madras
Garrulax jerdoni (White-breasted Laughing Thrush)
  G.j.jerdoni
    Coorg, W Mysore
  G.j.fairbanki
    Palni hills (S India)
  G.j.meridionalis
    S Kerala (SW India)
Garrulax lineatus (Himalayan Streaked Laughing Thrush)
  G.l.bilkevitchi
    Tadzhikistan, NW Pakistan
  G.l.gilgit
    NE Pakistan
  G.l.lineatus
    W Himalayas
  G.l.setafer
    Sikkim, Nepal, W Bengal
  G.l.imbricatus
    Bhutan, SE Tibet
Garrulax virgatus (Striped Laughing Thrush)
  S Assam, SW Burma
Garrulax austeni (Brown-capped Laughing Thrush)
  G.a.austeni
    S Assam
  G.a.victoriae
    N Burma
Garrulax squamatus (Blue-winged Laughing Thrush)
  E Himalayas, Burma, Assam, SW China
Garrulax subunicolor (Plain-coloured Laughing Thrush)
  G.s.subunicolor
    E Himalayas, E Assam
  G.s.griseatus
    NE Burma, NW Yunnan
  G.s.fooksi
    NW Vietnam

Garrulax elliotii (Elliot's Laughing Thrush)
  G.e.prjevalskii
    Kansu, E Tsinghai
  G.e.elliotii
    C & SW China
Garrulax henrici (Prince Henry's Laughing Thrush)
  G.h.henrici
    SE Tibet, SW Sikang
  G.h.gucenensis
    W China
Garrulax affinis (Black-faced Laughing Thrush)
  G.a.affinis
    W & C Nepal
  G.a.bethelae
    E Himalayas
  G.a.oustaleti
    NE Assam, SW Sikang, N Burma, NW Yunnan
  G.a.muliensis
    NW Yunnan, SE Sikang
  G.a.blythii
    SW Szechwan, E Sikang
  G.a.saturatus
    N Vietnam
  G.a.morrisonianus
    Taiwan
Garrulax erythrocephalus (Red-headed Laughing Thrush)
  G.e.erythrocephalus
    W Himalayas
  G.e.kali
    W & C Nepal
  G.e.nigrimentum
    Sikkim, Bhutan
  G.e.imprudens
    NE Assam
  G.e.chrysopterus
    S Assam
  G.e.godwini
    SE Assam
  G.e.erythrolaema
    E Manipur, SW Burma
  G.e.woodi
    NE Burma, SW Yunnan
  G.e.connectens
    N Indochina
  G.e.subconnectens
    NW Thailand
  G.e.schistaceus
    E Burma, NW Thailand
  G.e.melanostigma
    SE Burma, NW Thailand
  G.e.ramsayi
    S Burma
  G.e.peninsulae
    S Thailand, N Malaysia
Garrulax yersini (Yersin's Laughing Thrush)
  S Vietnam
Garrulax formosus (Crimson-winged Laughing Thrush)

G.f.formosus
  SW Szechwan, NE-Yunnan
G.f.greenwayi
  NW Vietnam
Garrulax milnei (Red-tailed Laughing Thrush)
G.m.sharpei
  E Burma, Yunnan, NW Thailand, N Indochina
G.m.vitryi
  S Laos
G.m.sinianus
  Kwangsi
G.m.milnei
  NW Fukien

LIOCICHLA
Liocichla phoenicea (Red-faced Liocichla)
L.p.phoenicea
  E Himalayas
L.p.bakeri
  S Assam, NW Burma
L.p.ripponi
  E & S Burma, NW Thailand
L.p.wellsi
  S Yunnan, N Indochina
Liocichla omeiensis (Mount Omei Liocichla)
  Mt Omei (Szechwan)
Liocichla steerii (Steere's Liocichla)
  S Taiwan

LEIOTHRIX
Leiothrix argentauris (Silver-eared Mesia)
L.a.argentauris
  Himalayas, N Assam
L.a.vernayi
  S Assam, Burma, W Yunnan
L.a.galbana
  E Burma, N Thailand
L.a.ricketti
  SE Yunnan, N Indochina
L.a.cunhaci
  S Indochina
L.a.tahanensis
  S Thailand, N Malaysia
L.a.rookmakeri
  NW Sumatra
L.a.laurinae
  W Sumatra
Leiothrix lutea (Pekin Robin)
L.l.kumaiensis
  W Himalayas
L.l.calipyga
  E Himalayas
L.l.luteola
  SW Burma, S Assam
L.l.yunnanensis
  NE Burma, NW Yunnan, SE Sikang
L.l.kwangtungensis
  S China, NE Vietnam
L.l.lutea
  SE & C China

CUTIA
Cutia nipalensis (Nepal Cutia)
C.n.nipalensis
  E Himalayas, Assam, W Burma
C.n.melanchima
  E Burma, NW Thailand, N Indochina
C.n.cervinicrissa
  N Malaysia
C.n.legalleni
  S Vietnam

PTERUTHIUS
Pteruthius rufiventer (Rufous-bellied Shrike Babbler)
P.r.rufiventer
  E Himalayas, Assam, N Burma, Yunnan
P.r.delacouri
  NW Vietnam
Pteruthius flaviscapis (Red-winged Shrike Babbler)
P.f.validirostris
  Himalayas, Assam, NW Burma
P.f.ricketti
  NE Burma, S China, N Indochina
P.f.annamensis
  S Vietnam
P.f.schauenseei
  S Thailand, E Burma
P.f.cameranoi
  Malaysia, W Sumatra
P.f.flaviscapis
  Java
P.f.robinsoni
  N Borneo
Pteruthius xanthochlorus (Green Shrike Babbler)
P.x.occidentalis
  W Himalayas
P.x.xanthochlorus
  E Himalayas
P.x.hybridus
  Naga hills, Assam, W Burma
P.x.pallidus
  NE Burma, SE Sikang.W & S China
Pteruthius melanotis (Black-eared Shrike Babbler)
P.m.melanotis
  E Himalayas, Burma, N Thailand, N Indochina
P.m.tahanensis
  C Malaysia
Pteruthius aenobarbus (Chestnut-fronted Shrike Babbler)
P.a.aenobarbulus
  Garo hills, Assam
P.a.intermedius
  E Burma, NW Thailand, N Indochina
P.a.yaoshanensis
  Kwangsi
P.a.indochinensis
  S Vietnam
P.a.aenobarbus
  W Java

GAMPSORHYNCHUS
Gampsorhynchus rufulus (White-headed Shrike Babbler)
G.r.rufulus
Sikkim, Assam, Burma
G.r.torquatus
SE Burma, Thailand, S Laos, Vietnam
G.r.saturatior
C Malaysia

ACTINODURA
Actinodura egertoni (Rusty-fronted Barwing)
A.e.egertoni
Nepal, Sikkim, Bhutan, N Assam
A.e.lewisi
Mishmi hills, NE Assam
A.e.khasiana
S Assam
A.e.ripponi
E Assam, SW Burma
Actinodura ramsayi (Spectacled Barwing)
A.r.yunnanensis
SE Yunnan, N Vietnam
A.r.radcliffei
E Burma, N Laos
A.r.ramsayi
E Burma, NW Thailand
Actinodura nipalensis (Hoary Barwing)
A.n.nipalensis
W & C Nepal
A.n.vinctura
E Nepal, Sikkim, Bhutan
Actinodura waldeni (Austen's Barwing)
A.w.daflaensis
N Assam
A.w.waldeni
SE Assam, NW Burma
A.w.poliotis
Mt Victoria, N Burma
A.w.saturatior
NE Burma, NW Yunnan
Actinodura souliei (Streaked Barwing)
A.s.souliei
NW Yunnan
A.s.griseinucha
NW Vietnam
Actinodura morrisoniana (Formosan Barwing)
Taiwan

MINLA
Minla cyanouroptera (Blue-winged Minla)
M.c.cyanouroptera
C & E Himalayas, E Assam
M.c.aglae
SE Assam, W Burma
M.c.sordida
E & S Burma, NW Thailand
M.c.wingatei
NE Burma, N Thailand, S China, N Indochina

M.c.croizati
Szechwan
M.c.rufodorsalis
SE Thailand, SW Cambodia
M.c.orientalis
S Vietnam
M.c.sordidior
S Thailand, N Malaysia
Minla strigula (Chestnut-tailed Minla)
M.s.simlaensis
W Himalayas
M.s.strigula
E Himalayas, N Assam
M.s.cinereigenae
Mt Japvo (E Assam)
M.s.yunnanensis
E Assam, W Burma, N Indochina
M.s.castanicauda
S Burma, NW Thailand
M.s.malayana
C Malaysia
Minla ignotincta (Red-tailed Minla)
M.i.ignotincta
E Nepal, Burma, Assam, NW Yunnan
M.i.mariae
SE Yunnan, N Vietnam
M.i.sini
Kwangsi
M.i.jerdoni
SW Szechwan
Minla annectens (Chestnut-backed Minla)
132.3
M.a.annectens
E Himalayas to NW Burma
M.a.mixta
SE Burma to N Indonesia
M.a.saturata
SE Burma, NW Thailand
M.a.eximia
Dalat (S Vietnam)

ALCIPPE
Alcippe chrysotis (Golden-breasted Fulvetta)
A.c.chrysotis
E Himalayas, E Assam
A.c.albilineata
S Assam
A.c.forresti
NE Burma, NW Yunnan
A.c.amoena
NW Vietnam
A.c.swinhoii
SE Sikang, S Szechwan
Alcippe variegaticeps (Variegated Fulvetta)
Kwangsi
Alcippe cinerea (Yellow-throated Fulvetta)
E Himalayas, N Burma, N Laos
Alcippe castaneceps (Chestnut-headed Fulvetta)
A.c.castaneceps
E Himalayas, Assam, Burma, NW Thailand
A.c.exul
N Thailand, Laos, NW Vietnam

A.c.soror
C Malaysia
A.c.klossi
S Vietnam
Alcippe vinipectus (White-browed Fulvetta)
A.v.kangrae
W Himalayas
A.v.vinipectus
W & C Nepal
A.v.chumbiensis
E Nepal, SE Tibet, Sikkim, Bhutan
A.v.austeni
S Assam
A.v.ripponi
Chin hills, W Burma
A.v.perstriata
NE Burma
A.v.valentinae
N Vietnam
A.v.bieti
NW Yunnan, SE Sikang
Alcippe striaticollis (Chinese Mountain Fulvetta)
NW China
Alcippe ruficapilla (Spectacled Fulvetta)
A.r.ruficapilla
S Shensi, Szechwan
A.r.sordidior
NW Yunnan
A.r.danisi
SE Yunnan, N Laos
Alcippe cinereiceps (Streak-throated Fulvetta)
A.c.ludlowi
E Bhutan, SE Tibet
A.c.manipurensis
E Assam, N Burma, NW Yunnan
A.c.tonkinensis
NE Laos, NW Vietnam
A.c.guttaticollis
Fukien, N Kwangtung
A.c.formosana
Taiwan
A.c.fucata
Hupeh, Hunan
A.c.cinereiceps
W Hupeh, Szechwan, SE Sikang
A.c.fessa
SW Kansu
Alcippe rufogularis (Rufous-throated Fulvetta)
A.r.rufogularis
E Himalayas, N Assam
A.r.collaris
E Assam, Bangladesh
A.r.major
E Burma, N & E Thailand, N Laos
A.r.stevensi
N Indochina
A.r.kelleyi
C Vietnam
A.r.khmerensis
SE Thailand, SW Cambodia

Alcippe brunnea (Gould's Fulvetta)
A.b.mandellii
S Assam, W Burma
A.b.intermedia
E Burma
A.b.dubia
S Burma
A.b.genestieri
SW China, N Indochina
A.b.superciliaris
E & SE China
A.b.brunnea
Taiwan
A.b.arguta
Hainan I
A.b.olivacea
W Hupeh, Szechwan
Alcippe brunneicauda (Brown Fulvetta)
A.b.brunneicauda
Malaysia, Sumatra, NW Borneo, N Natuna Is
A.b.eriphaea
Borneo
Alcippe poioicephala (Brown-checked Fulvetta)
A.p.poioicephala
S India
A.p.brucei
C & S India
A.p.fusca
S Assam, NW Burma
A.p.phayrei
SW Burma
A.p.haringtoniae
NE Burma, NW Thailand
A.p.alearis
N & E Thailand, N Indochina
A.p.karenni
SE Burma, SW Thailand
A.p.davisoni
S Thailand, Mergui Arch
Alcippe pyrrhoptera (Javanese Fulvetta)
W & C Java
Alcippe peracensis (Mountain Fulvetta)
A.p.grotei
N & C Vietnam, S Laos
A.p.annamensis
S Indochina
A.p.eremita
SE Thailand
A.p.peracensis
N & C Malaysia
Alcippe morrisonia (Grey-cheeked Fulvetta)
A.m.yunnanensis
SE Sikang, NW Yunnan, NE Burma
A.m.fraterculus
SW Yunnan, SE Burma, N Indochina
A.m.schaefferi
SE Yunnan, NW Vietnam
A.m.rufescentior
Hainan I
A.m.morrisonia
Taiwan

A.m.hueti
Kwangtung to Anhwei (SE China)
A.m.davidi
W Hupeh, Szechwan
Alcippe nipalensis (Nepal Fulvetta)
A.n.nipalensis
E Himalayas
A.n.commoda
Bangladesh, Assam, N Burma
A.n.stanfordi
SW Burma

LIOPTILUS
Lioptilus nigricapillus (Bush Blackcap)
E Cape Province, Natal, N Transvaal

KUPEORNIS   132.2
Kupeornis gilberti (White-throated Mountain Babbler)
Mt Kupe (Cameroun)
Kupeornis rufocinctus (Red-collared Flycatcher Babbler)
E Zaire
Kupeornis chapini (Chapin's Flycatcher Babbler)
K.c.chapini
Ituri river (E Zaire)
K.c.nyombensis
Mt Nyombe (E Zaire)
K.c.kalindei
E Zaire

PAROPHASMA
Parophasma galinieri (Abyssinian Catbird)
C & S Ethiopia

PHYLLANTHUS
Phyllanthus atripennis (Capuchin Babbler)
P.a.atripennis
Senegal to Liberia
P.a.rubiginosus
Ivory Coast to S Nigeria
P.a.bohndorffi
NE Zaire to W Uganda

CROCIAS
Crocias langbianis (Mt Langbian Sibia)
S Vietnam
Crocias albonotatus (Spotted Sibia)
W & C Java

HETEROPHASIA
Heterophasia capistrata (Black-capped Sibia)
H.c.capistrata
W Himalayas
H.c.nigriceps
C Himalayas
H.c.bayleyi
E Himalayas
Heterophasia gracilis (Grey Sibia)
S Assam, N Burma, W Yunnan
Heterophasia melanoleuca (Black-headed Sibia)

H.m.desgodinsi
NE Burma, W China
H.m.castanoptera
SE Burma
H.m.tonkinensis
N Vietnam
H.m.melanoleuca
E Burma, NW Thailand
H.m.engelbachi
S Laos
H.m.robinsoni
S Vietnam
Heterophasia auricularis (White-eared Sibia)
Taiwan
Heterophasia pulchella (Beautiful Sibia)
SE Tibet, Assam, NE Burma
Heterophasia picaoides (Long-tailed Sibia)
H.p.picaoides
E Himalayas, NE Burma
H.p.cana
E & S Burma, N Thailand, N Indochina
H.p.wrayi
C Malaysia
H.p.simillima
W Sumatra

YUHINA
Yuhina castaniceps (Striated Yuhina)
Y.c.rufigenis
W Bengal, Sikkim
Y.c.plumbeiceps
N Assam, N Burma
Y.c.castaniceps
S Assam, SW Burma
Y.c.striata
E Burma, NW Thailand
Y.c.torqueola
N Thailand, S China, N Indochina
Y.c.everetti
N Borneo
Yuhina bakeri (White-naped Yuhina)
E Himalayas, Assam
Yuhina flavicollis (Whiskered Yuhina)
Y.f.albicollis
W Himalayas
Y.f.flavicollis
E Himalayas
Y.f.rouxi
Assam, N Burma, SW China, N Indochina
Y.f.clarki
E Burma
Y.f.humilis
S Burma
Y.f.constantiae
N Laos
Y.f.rogersi
Thailand
Yuhina gularis (Striped-throated Yuhina)
Y.g.vivax
W Himalayas
Y.g.gularis
E Nepal, Burma, Assam, NW Vietnam

Y.g.omeiensis
SW China
Yuhina diademata (White-collared Yuhina)
NE Burma, N Vietnam, SW & S China
Yuhina occipitalis (Rufous-vented Yuhina)
Y.o.occipitalis
E Himalayas, SE Tibet, N Assam
Y.o.obscurior
NE Burma, NW Yunnan
Yuhina brunneiceps (Formosan Yuhina)
Taiwan
Yuhina nigrimenta (Black-chinned Yuhina)
Y.n.nigrimenta
Himalayas, Assam
Y.n.intermedia
NE Burma, SW China, N Indochina
Y.n.pallida
Fukien, SE China

MYZORNIS
Myzornis pyrrhoura (Fire-tailed Myzornis)
E Himalayas, SE Tibet, NE Burma

HORIZORHINUS
Horizorhinus dohrni (Dohrn's Thrush-Babbler)
Principé I

OXYLABES
Oxylabes cinereiceps (Grey-crowned Oxylabes)
E Madagascar
Oxylabes madagascariensis (White-throated Oxylabes)
E Madagascar
Oxylabes xanthophrys (Yellow-browed Oxylabes)
EC Madagascar

MYSTACORNIS
Mystacornis crossleyi (Crossley's Babbler)
Madagascar

133. PANURIDAE (PARROTBILLS)

PANURUS
Panurus biarmicus (Bearded Reedling)
P.b.biarmicus
England, , S & E Europe
P.b.occidentalis
Balkans
P.b.russicus
E Europe, Russia, Iran, Manchuria

CONOSTOMA
Conostoma oemodium (Great Parrotbill)
Himalayas, NE Burma, SE Tibet, SW China

PARADOXORNIS
Paradoxornis paradoxus (Three-toed Parrotbill)
P.p.paradoxus
NW China
P.p.taipaiensis
Shensi
Paradoxornis unicolor (Brown Parrotbill)
E Himalayas, N Burma, SE Tibet, SW China
Paradoxornis flavirostris (Gould's Parrotbill)
E Himalayas, W Burma
Paradoxornis guttaticollis (Spot-breasted Parrotbill)
Assam to N Indochina
Paradoxornis conspicillatus (Spectacled Parrotbill)
P.c.conspicillatus
WC China
P.c.rocki
Hupeh
Paradoxornis ricketti (Yunnan Parrotbill)
NW Yunnan
Paradoxornis webbianus (Vinous-throated Parrotbill)
P.w.mantschuricus
E Manchuria
P.w.suffusus
NW to SE China
P.w.fulvicauda
NE China, S Korea
P.w.webbianus
coast of S Kiangsu, N Chekiang
P.w.bulomachus
Taiwan
P.w.elisabethae
SE Yunnan
P.w.brunneus
NE Burma, NW Yunnan
Paradoxornis alphonsianus (Ashy-throated Parrotbill)
P.a.alphonsianus
WC China
P.a.yunnanensis
SE Yunnan, NW Vietnam
Paradoxornis zappeyi (Zappey's Parrotbill)
W Szechwan
Paradoxornis przewalskii (Grey-crowned Parrotbill)
S Kansu
Paradoxornis fulvifrons (Fulvous-fronted Parrotbill)
P.f.chayulensis
N Assam, SE Tibet
P.f.fulvifrons
Nepal, Sikkim, Bhutan
P.f.albifacies
NW Yunnan, SE Sikang
P.f.cyanophrys
NW Szechwan, SW Shensi
Paradoxornis nipalensis (Blyth's Parrotbill)
P.n.nipalensis
C Nepal

P.n.humii
 E Nepal, Sikkim
P.n.crocotius
 SE Tibet, E Bhutan
P.n.poliotis
 Assam, NE Burma, NW Yunnan
P.n.patriciae
 SE Assam
P.n.ripponi
 Mt Victoria (N Burma)
P.n.feae
 SE Burma, NW Thailand
P.n.verreauxi
 E Sikang, SW Szechwan
P.n.pallidus
 NW Fukien
P.n.morrisonianus
 Taiwan
P.n.beaulieu
 N Laos
P.n.craddocki
 NW Vietnam
Paradoxornis davidianus (David's Parrotbill)
P.d.davidianus
 Fukien
P.d.tonkinensis
 N Vietnam
P.d.thompsoni
 E Burma, NW Laos, E Thailand
Paradoxornis atrosuperciliaris (Lesser Red-headed Parrotbill)
P.a.oatesi
 W Bengal, Sikkim
P.a.atrosuperciliaris
 Assam, N Burma, N Laos, W Yunnan
Paradoxornis ruficeps (Greater Red-headed Parrotbill)
P.r.ruficeps
 E Himalays
P.r.bakeri
 Assam, N & E Burma
P.r.magnirostris
 N Vietnam
Paradoxornis gularis (Grey-headed Parrotbill)
P.g.gularis
 E Himalayas, N Assam
P.g.transfluvialis
 S Assam, N & E Burma, NW Thailand
P.g.laotianus
 E Burma, N Thailand, N Indochina
P.g.fokiensis
 Fukien, Anhwei
P.g.hainanus
 Hainan I
P.g.rasus
 Chin hills, W Burma
P.g.margaritae
 S Vietnam
Paradoxornis heudei (Heude's Parrotbill)
P.h.heudei
 Kiangsu (C China)
P.h.polivanovi
 Lake Khanka (USSR)

134. PICATHARTIDAE (BALD CROWS)

PICATHARTES
Picathartes gymnocephalus (White-necked Bald Crow)(Rockfowl)
 Sierra Leone to Ghana
Picathartes oreas (Grey-necked Bald Crow)(Rockfowl)
 Cameroun

135. POLIOPTILIDAE (GNATCATCHERS)

MICROBATES
Microbates collaris (Collared Gnatwren)
 M.c.paraguensis
  Venezuela
 M.c.collaris
  SE Colombia to French Guiana
 M.c.perlatus
  N Brazil
Microbates cinereiventris (Half-collared Gnatwren)
 M.c.semitorquatus
  S Nicaragua to W Panama
 M.c.magdalenae
  E Panama, N Colombia
 M.c.cinereiventris
  W Colombia, SW Ecuador
 M.c.peruvianus
  Colombia, E Ecuador, Peru

RAMPHOCAENUS
Ramphocaenus melanurus (Long-billed Gnatwren)
 R.m.rufiventris
  SE Mexico, Central America to E Ecuador
 R.m.ardeleo
  Yucatan, N Guatemala
 R.m.sanctaemarthae
  N Colombia, NW Venezuela
 R.m.griseodorsalis
  W Colombia
 R.m.pallidus
  NE Colombia, N Venezuela
 R.m.trinitatis
  E Colombia to Trinidad
 R.m.duidae
  S Venezuela, to NE Ecuador
 R.m.badius
  NE Peru, SE Ecuador
 R.m.obscurus
  Peru
 R.m.amazonum
  E Peru, NW Brazil
 R.m.sticturus
  NW Brazil
 R.m.albiventris
  E Venezuela, the Guianas, NE Brazil
 R.m.austerus
  Brazil

R.m.melanurus
  N & E Brazil

POLIOPTILA
Polioptila caerulea (Blue-grey Gnatcatcher)
  P.c.caerulea
    C & E USA, E Mexico, West Indies
  P.c.amoenissima
    SW USA, NW Mexico, N Baja California
  P.c.obscura
    S Baja California
  P.c.gracilis
    SE Sonora (Mexico)
  P.c.mexicana
    SE Mexico
  P.c.nelsoni
    SW & S Mexico
  P.c.deppei
    Yucatan
  P.c.cozumelae
    Cozumel I
Polioptila melanura (Black-tailed Gnatcatcher)
  P.m.californica
    SW California, NW Baja California
  P.m.lucida
    SW USA, N Mexico, NE Baja California
  P.m.melanura
    S USA
  P.m.pontilis
    C Baja California
  P.m.margaritae
    S Baja California, Santa Margarita I
  P.m.curtata
    Tiburon I
Polioptila lembeyei (Cuban Gnatcatcher)
  E Cuba
Polioptila albiloris (White-lored Gnatcatcher)
  P.a.vanrossemi
    S & W Mexico
  P.a.albiventris
    N Yucatan
  P.a.albiloris
    Guatemala to Costa Rica
Polioptila nigriceps (Black-capped Gnatcatcher)
  P.n.restricta
    Sonora (Mexico)
  P.n.nigriceps
    N Sinaloa (Mexico)
Polioptila plumbea (Tropical Gnatcatcher)
  P.p.brodkorbi
    SE Mexico to Nicaragua
  P.p.superciliaris
    SE Mexico to Panama
  P.p.cinericia
    Coiba I (Panama)
  P.p.bilineata
    Colombia to W Peru
  P.p.plumbiceps
    N Colombia, N Venezuela
  P.p.anteocularis
    Colombia

P.p.daguae
  Colombia
P.p.innotata
  E Colombia, S Venezuela, N Brazil
P.p.plumbea
  the Guianas, NE Brazil
P.p.maior
  N Peru
P.p.parvirostris
  E Peru
P.p.atricapilla
  NE Brazil
Polioptila lactea (Cream-bellied Gnatcatcher)
  SE Brazil to Argentina
Polioptila guianensis (Guianan Gnatcatcher)
  P.g.facilis
    Venezuela, NE Brazil
  P.g.guianensis
    the Guianas
  P.g.paraensis
    Brazil
Polioptila schistaceigula (Slate-throated Gnatcatcher)
  E Panama to W Ecuador
Polioptila dumicola (Masked Gnatcatcher)
  P.d.berlepschi
    Brazil, E Bolivia
  P.d.dumicola
    Bolivia to Uruguay, Argentina
  P.d.saturata
    Bolivia

136. SYLVIIDAE (OLD WORLD WARBLERS)   136.1

OLIGURA   136.1
Oligura castaneocoronata (Chestnut-headed Ground Warbler)
  O.c.castaneocoronata
    Himalayas, Burma, Thailand, S China
  O.c.abadiei
    N Vietnam
  O.c.ripleyi
    SW China

TESIA   136.9
Tesia superciliaris (Java Ground Warbler)
  Java
Tesia olivea (Slaty-bellied Ground Warbler)
  Burma, Thailand, Laos, SW China, N Vietnam
Tesia cyaniventer (Grey-bellied Ground Warbler)
  Himalayas, Burma, Thailand, Laos, S China
Tesia everetti (Russet-capped Stubtail)   136.1.12
  T.s.everetti
    Flores I
  T.s.sumbawana
    Sumbawa I

UROSPHENA   136.2.9
Urosphena subulata (Timor Stubtail)   136.1

U.s.subulata
Timor I
U.s.advena
Babar I
Urosphena whiteheadi (Bornean Stubtail)
NE Borneo
Urosphena squameiceps (Scaly-headed Stubtail)
E Siberia, Japan >> S China, Burma
Indochina

CETTIA   136.1.9
Cettia pallidipes (Pale-footed Stubtail)   136.12
C.p.pallidipes
Himalayas, N Burma
C.p.laurentei
S China >> NW Thailand, N Indochina
C.p.osmastoni
S Andaman
Cettia diphone (Japanese Bush Warbler)
C.d.borealis
Manchuria, Korea, S China
C.d.canturians
N China
C.d.viridis
S Sakhalin >> SE China
C.d.cantans
C & S Japan
C.d.riukiuensis
Riukiu Is
C.d.restricta
Borodino I
C.d.diphone
Bonin Is
C.d.seebohmi
N Luzon
Cettia annae (Palau Bush Warbler)   136.3
Palau Is
Cettia parens (Shade Warbler)   136.3
San Cristobal I
Cettia ruficapilla (Fiji Warbler)   136.3
C.r.ruficapilla
Kandavu I
C.r.badiceps
Viti Levu I
C.r.castaneoptera
Vanua Levu I
C.r.funebris
Taveuni I
Cettia fortipes (Strong-footed Bush Warbler)
C.f.pallida
Pakistan, NW India
C.f.fortipes
Nepal
C.f.davidiana
SW China
Cettia vulcania (Müller's Bush Warbler)   136.4
C.v.sepiaria
N Sumatra
C.v.flaviventris
C & S Sumatra
C.v.vulcania
E Sumatra, Java, Bali, Lombok I

C.v.everetti
Timor I
C.v.oreophila
NE Borneo
C.v.banksi
Borneo
C.v.palawana
Palawan I
Cettia major (Large Bush Warbler)
C.m.major
Nepal, Burma, W China
C.m.vafer
Assam
Cettia carolinae (Yamdena Warbler)   136.5
Tanimber Is
Cettia flavolivacea (Aberrant Bush Warbler)
C.f.flavolivacea
Nepal, S Tibet
C.f.intricata
S China
C.f.weberi
Chin hills (Burma)
C.f.alexanderi
Burma
C.f.oblita
W China, N Vietnam
C.f.stresemanni
Assam
Cettia robustipes (Swinhoe's Bush Warbler)
       136.1
C.r.brunnescens
Nepal, Assam, SE Tibet
C.r.acanthizoides
SE Tibet, S China
C.r.robustipes
S China, Taiwan
Cettia brunnifrons (Rufous-capped Bush Warbler)
C.n.whistleri
NW Himalayas
C.b.brunnifrons
E Himalayas, Nepal
C.b.umbratica
Burma
Cettia cetti (Cetti's Warbler)
C.c.cetti
S Europe, Asia Minor, N Africa
C.c.orientalis
Middle East, N Iran
C.c.albiventris
W Pakistan

BRADYPTERUS
Bradypterus baboecalus (African Sedge Warbler)
B.b.centralis
S Nigeria, Cameroun to E Zaire
B.b.chadensis
L Chad
B.b.sudanensis
S Sudan
B.b.abyssinicus
Ethiopia

B.b.elgonensis
N Kenya
B.b.tongensis
SE Kenya to S Zambia & Transkei
B.b.msiri
Angola, S Zaire, Botswana
B.b.benguellensis
S Angola
B.b.baboecalus
S Cape Province
Bradypterus graueri (Grauer's Warbler)
E Zaire
Bradypterus grandis (Ja River Warbler)
Cameroun, Gabon
Bradypterus carpalis (White-winged Warbler)
NE Zaire, Uganda
Bradypterus alfredi (Bamboo Warbler)
B.a.kungwensis
N Zambia, SW Tanzania
B.a.alfredi
E Zaire, SW Ethiopia, W Uganda
Bradypterus sylvaticus (Knysna Scrub Warbler)
B.s.sylvaticus
Cape Province coast
B.s.pondoensis
Transkei & Natal coast
Bradypterus lopezi (Lopez' Warbler)
B.l.lopezi
Fernando Po I
B.l.camerunensis
Cameroun Mt
B.l.manengubae
Manenguba Mt (Cameroun)
B.l.barakae
E Zaire, W Uganda
Bradypterus mariae (Evergreen Forest Warbler)
B.m.mariae
SE Kenya, NE Tanzania
B.m.usambarae
S Kenya, W Tanzania, E Zambia, N Malawi
B.m.ufipae
S Tanzania, Zambia B.m.granti
W & S Malawi
B.m.priesti
E Zimbabwe, W Mozambique
B.m.boultoni
W Angola
Bradypterus barratti (Scrub Warbler)
B.b.barratti
E Transvaal, Natal
B.b.godfreyi
Lesotho, E Cape Province
B.b.cathkinensis
S Natal
Bradypterus victorini (Victorin's Scrub Warbler)
S Cape Province
Bradypterus cinnamomeus (Cinnamon Bracken Warbler)
B.c.bangwaensis
E Nigeria, Cameroun

B.c.cavei
S Sudan
B.c.cinnamomeus
S Ethiopia, Kenya, NE Zaire
B.c.mildbreadi
E Zaire
B.c.nyassae
E Tanzania, Malawi
Bradypterus brunneus (Brown Feather-tailed Warbler)   136.6
EC Madagascar
Bradypterus thoracicus (Spotted Bush Warbler)
B.t.davidi
NE Asia, S China
B.t.suschkini
C Asia, N Vietnam
B.t.przewalskii
W Himalayas, S & W China
B.t.kashmirensis
NW Himalayas
B.t.thoracicus
Bangladesh, E Himalayas, Nepal
B.t.shanensis
Burma >> Thailand
Bradypterus major (Large-billed Bush Warbler)
B.m.major
W Himalayas, Turkestan, Sinkiang
B.m.innae
Sinkiang
Bradypterus tacsanowskius (Chinese Bush Warbler)
E Siberia, N China >> S Burma, Indochina
Bradypterus luteoventris (Brown Bush Warbler)
B.l.luteoventris
Assam, Nepal, SE Tibet
B.l.ticehursti
Burma
Bradypterus palliseri (Palliser's Warbler)
Sri Lanka
Bradypterus seebohmi (Mountain Scrub Warbler)
B.s.idoneus
Taiwan, N Thailand, S Indochina
B.s.melanorhynchus
NW Fukien
B.s.seebohmi
N Luzon I
B.s.montis
E Java
B.s.timorensis
Timor
Bradypterus caudatus (Long-tailed Ground Warbler)
B.c.caudatus
N Luzon I
B.c.unicolor
Mindanao I
B.c.malindangensis
S Mindanao I

Bradypterus accentor (Kinabalu Scrub Warbler)
N Borneo

Bradypterus castaneus (Chestnut-backed Bush Warbler)
B.c.disturbans
Buru I
B.c.musculus
Seram I
B.c.castaneus
C Moluccas

BATHMOCERCUS
Bathmocercus cerviniventris (Black-capped Rufous Warbler)
Sierra Leone, Ghana

Bathmocercus rufus (Black-faced Rufous Warbler)
B.r.rufus
Cameroun, Gabon
B.r.vulpinus
E Cameroun to Sudan, N Kenya

Bathmocercus winifredae (Mrs Moreau's Warbler)
E Tanzania

AMPHILAIS   136.6
Amphilais seebohmi (Seebohm's Feather-tailed Warbler)
SC Madagascar

NESILLAS
Nesillas typica (Tsikirity Warbler)
N.t.obscura
NW Madagascar
N.t.typica
C & E Madagascar
N.t.lantzii
S & W Madagascar
N.t.longicaudata
Anjouan I
N.t.brevicaudata
Great Comoro I
N.t.moheliensis
Moheli I

Nesillas mariae (Comoro Warbler)
Moheli I

Nesillas aldabranus (Aldabra Warbler)
Aldabra I

THAMNORNIS
Thamnornis chloropetoides (Kiritika Warbler)
S Madagascar

MELOCICHLA
Melocichla mentalis (Moustached Grass Warbler)   136.1
M.m.mentalis
Guinea to Central African Republic & C Zaire
M.m.amauroura
S Sudan to N Zambia

M.m.orientalis
Tanzania to Zimbabwe & N Mozambique
M.m.incanus
Meru (Tanzania)
M.m.luangwae
S Zambia

ACHAETOPS
Achaetops pycnopygius (Damaraland Rock Jumper)   136.1
A.p.pycnopygius
S Angola, N Namibia
A.p.spadix
Huila (Angola)

SPHENOEACUS
Sphenoeacus afer (Cape Grassbird)
S.a.excisus
Zimbabwe
S.a.natalensis
Transvaal, Natal
S.a.intermedius
E Cape Province
S.a.afer
S Cape Province

MEGALURUS
Megalurus pryeri (Japanese Marsh Warbler)
M.p.sinensis
China
M.p.pryeri
Japan

Megalurus timoriensis (Tawny Marshbird)
M.t.timoriensis
Timor I
M.t.stresemanni
NW New Guinea
M.t.mayri
N New Guinea
M.t.wahgiensis
C New Guinea
M.t.montanus
C New Guinea
M.t.macrurus
C & SE New Guinea
M.t.harterti
E New Guinea
M.t.alpinus
SE New Guinea
M.t.muscalis
S New Guinea
M.t.celebensis
Sulawesi
M.t.inquirendus
Sumba I
M.t.amboinensis
Ambon I
M.t.interscapularis
New Britain
M.t.alisteri
N Western Australia, Northern Territory, N Queensland

M.t.oweni
S Queensland, New South Wales
M.t.tweeddalei
Luzon, Panay, Tablas Is
M.t.mindorensis
Mindoro I
M.t.alopex
Bohol, Cebu, Leyte Is
M.t.crex
Mindanao I
Megalurus palustris (Striated Canegrass Warbler)
M.p.toklao
NE India to Indochina, S China
M.p.palustris
Java
M.p.forbesi
Luzon, Mindanao Is
Megalurus albolimbatus (Fly River Grass Warbler)
SE New Guinea
Megalurus gramineus (Little Marshbird)
M.g.papuensis
New Guinea
M.g.gramineus
Tasmania, Sthn Australia
Megalurus punctatus (Fernbird) 136.1
M.p.vealeae
Nth I, New Zealand
M.p.punctatus
Sth I, New Zealand
M.p.stewartianus
Stewart I
M.p.wilsoni
Codfish I
M.p.caudatus
Snares I

CINCLORHAMPHUS
Cinclorhamphus mathewsi (Rufous Songlark)
Australia
Cinclorhamphus cruralis (Brown Songlark)
Australia

EREMIORNIS
Eremiornis carteri (Spinifex Bird)
Central Australia

MEGALURULUS 136.1.7
Megalurulus bivittatus (Buff-banded Bushbird)
Timor I
Megalurulus mariei (New Caledonian Grass Warbler)
New Caledonia

CICHLORNIS
Cichlornis whitneyi (Thicket Warbler)
C.w.whitneyi
Espiritu Santo I
C.w.turipavae
Guadalcanal I

Cichlornis grosvenori (Whiteman Mountains Warbler)
New Britain
Cichlornis llaneae (Bougainville Thicket Warbler)
Bougainville I

ORTYGOCICHLA
Ortygocichla rubiginosa (Rufous-faced Thicket Warbler)
New Britain
Ortygocichla rufa (Long-legged Warbler) 136.7

O.r.rufa
Viti Levu, Fiji Is
O.r.cluniei
Vanua Levu I

CHAETORNIS
Chaetornis striatus (Bristled Grass Warbler)
Pakistan, N India

GRAMINICOLA
Graminicola bengalensis (Large Grass Warbler)
G.b.bengalensis
Himalayas, N India, Bangladesh
G.b.sinica
S China
G.b.striata
N Vietnam

SCHOENICOLA
Schoenicola platyura (Broad-tailed Warbler)
S.p.alexinae
Sierra Leone to Angola, Tanzania
S.p.brevirostris
Sthn Africa
S.p.platyura
SW India, Sri Lanka

LOCUSTELLA
Locustella lanceolata (Lanceolated Warbler)
NC & NE Asia, Sakhalin I >> Philippine Is, SE Asia
Locustella naevia (Grasshopper Warbler)
L.n.naevia
Europe, N Africa, Russia
L.n.obscurior
Caucasus
L.n.straminea
Iran, N India, W Siberia, C Asia
L.n.mongolica
Afghanistan, W Mongolia
Locustella certhiola (Pallas' Grasshopper Warbler)
L.c.certhiola
NC Asia >> NW India
L.c.centralasiae
C Asia, Himalayas >> Burma
L.c.rubescens
E Siberia >> C & S India, Sri Lanka
L.c.minor
E China, Indochina

Locustella ochotensis (Middendorff's Grasshopper Warbler)
 L.o.subcerthiola
  Kamchatka, Kurile Is >> Philippine Is
 L.o.ochotensis
  E Siberia, N Japan >> Philippine Is, Borneo, Sulawesi
Locustella pleskei (Styan's Grasshopper Warbler)
  Japan, Korea >> SE China & Sunda Is
Locustella fluviatilis (River Warbler)
  NE Europe, NW Asia >> SE Africa
Locustella luscinioides (Savi's Warbler)
 L.l.luscinioides
  C & E Europe, Iberia, N Africa
 L.l.sarmatica
  S Russia
 L.l.fusca
  WC Asia
Locustella fasciolata (Gray's Grasshopper Warbler)
  NE Asia, Sakhalin, Japan >> Philippine Is, Sulawesi, New Guinea
Locustella amnicola (Stepanyan's Grasshopper Warbler)
  Sakhalin I, Kurile Is

ACROCEPHALUS
Acrocephalus melanopogon (Moustached Warbler)
 A.m.melanopogon
  Europe, N Africa
 A.m.mimica
  A Asia, S Russia, Iraq, Iran, Afghanistan
 A.m.albiventris
  Krasnodor (USSR)
Acrocephalus paludicola (Aquatic Warbler)
  CS & E Europe, W Russia, Asia Minor, Egypt
Acrocephalus schoenobaenus (Sedge Warbler)
  Europe & W & C Asia >> EC & SE Africa
Acrocephalus sorghophilus (Speckled Reed Warbler)
  N & E China >> Philippine Is
Acrocephalus bistrigiceps (Schrenk's Reed Warbler)
 A.b.bistrigiceps
  E Siberia, N China >> SE China, Thailand
 A.b.tangorum
  N Manchuria >> ?Thailand
Acrocephalus agricola (Paddyfield Warbler)
 A.a.septimus
  SW Russia >> SE Iran, W India
 A.a.agricola
  C Asia >> E Iran, N India
Acrocephalus concinens (Blunt-winged Paddyfield Warbler)
 A.c.haringtoni
  Afghanistan, NW India, Pakistan
 A.c.stevensi
  Bangladesh, Assam, Burma

 A.c.concinens
  N & C China >> S Burma, S Thailand
Acrocephalus scirpaceus (Reed Warbler)
 A.s.scirpaceus
  Europe, W Russia >> WC & E Africa
 A.s.fuscus
  Asia Minor, C Asia >> E Africa
Acrocephalus cinnamomeus (Reichenow's Reed Warbler) 136.1
 A.c.guiersi
  Senegambia
 A.c.cinnamomeus
  NC Africa to SC Africa
 A.c.fraterculus
  NE Zambia to Natal
Acrocephalus baeticatus (African Reed Warbler)
 A.b.avicenniae
  Ethiopia
 A.b.suahelicus
  E Tanzania, Zanzibar
 A.b.baeticatus
  N Botswana, Zimbabwe to Natal & Cape Province
 A.b.hallae
  SW Angola, N Namibia to N & W Cape Province
Acrocephalus palustris (Marsh Warbler)
 A.p.palustris
  Europe, C Russia >> E Africa, Natal
 A.p.laricus
  Iran >> E Africa
Acrocephalus dumetorum (Blyth's Reed Warbler) 136.1
  C Europe, W Siberia >> Iran, India
Acrocephalus arundinaceus (Great Reed Warbler)
 A.a.arundinaceus
  Europe, Asia Minor >> NW, W & SC Africa
 A.a.zarudnyi
  S Russia, C Asia >> C Africa
 A.a.griseldis
  S Iran >> E Africa
Acrocephalus stentoreus (Clamorous Reed Warbler)
 A.s.stentoreus
  Egypt, Sinai
 A.s.brunnescens
  Iran, Afghanistan, NW India
 A.s.amyae
  N India, Thailand, Indochina
 A.s.meridionalis
  Sri Lanka
 A.s.siebersi
  W Java
 A.s.lentecaptus
  Lombok, Sumbawa Is
 A.s.sumbae
  Moluccas, Sumba I
 A.s.celebensis
  Sulawesi
 A.s.harterti
  Philippine Is

A.s.gouldi
Western Australia
A.s.australis
SE Australia, Tasmania
Acrocephalus orinus (Large-billed Reed Warbler)
N India
Acrocephalus orientalis (Oriental Great Reed Warbler)
SE Siberia, N China >> Philippine Is
Acrocephalus luscinia (Nightingale Reed Warbler)
A.l.luscinia
Guam, Saipan, Almagan Is
A.l.syrinx
Caroline Is
A.l.yamashinae
Pagan I
A.l.nijoi
Agiguan I
A.l.astrolabii
Mariana Is
A.l.rehsei
Nauru I
Acrocephalus kingi (Hawaiian Reed Warbler)
Nihoa (Hawaii)
Acrocephalus aequinoctialis (Polynesian Reed Warbler)
A.a.aequinoctialis
Christmas I
A.a.pistor
Fanning I
Acrocephalus caffer (Long-billed Reed Warbler)   136.1
A.c.caffer
Tahiti
A.c.garretti
Huahine I
A.c.longirostris
Moorea I
A.c.percernis
Nukuhiva I
A.c.mendanae
Hivaoa, Tahuata Is
A.c.consobrinus
Motane I
A.c.fatuhivae
Fatuhiva I
A.c.idae
Uahuka I
A.c.dido
Uapu I
A.c.aquilonis
Eiao I
A.c.postremus
Hatutu I
Acrocephalus atyphus (Tuamotu Warbler)
A.a.atyphus
NW Tuamotu Is
A.a.ravus
SE Tuamotu Is
A.a.palmarum
Anaa I

A.a.niauensis
Niau I
A.a.eremus
Makatea I
A.a.flavidus
Napuka I
Acrocephalus vaughani (Pitcairn Warbler)
A.v.vaughanii
Pitcairn I
A.v.rimitarae
Rimitara I
A.v.taiti
Henderson I
Acrocephalus kerearako (Cook Islands Warbler)   136.15
A.k.kerearako
Mangaia I
A.k.kaoko
Mitiaro I
Acrocephalus rufescens (Rufous Swamp Warbler)
A.r.rufescens
W & WC Africa
A.r.chadensis
Chad
A.r.ansorgei
S Sudan to N Angola, Botswana
Acrocephalus brevipennis (Cape Verde Swamp Warbler)
Cape Verde Is
Acrocephalus gracilirostris (Swamp Warbler)
A.g.neglectus
L Chad
A.g.tsanae
N Ethiopia
A.g.jacksoni
NE Zaire, Uganda, S Sudan
A.g.parvus
S Kenya
A.g.leptorhynchus
Tanzania, S Zaire, Angola to Mozambique
A.g.winterbottomi
E Angola, W Zambia
A.g.cunenensis
S Angola
A.g.gracilirostris
Namibia, Botswana, Sth Africa
A.g.zuluensis
SE Zimbabwe to E Natal
Acrocephalus newtoni (Madagascar Swamp Warbler)
Madagascar
Acrocephalus aedon (Thick-billed Warbler)
A.a.aedon
NC Asia >> Burma, Thailand, Indonesia
A.a.stegmanni
E Siberia, Mongolia >> SE China, Thailand
Indochina   136.8

BEBRORNIS
Bebrornis rodericanus (Rodriguez Brush Warbler)
Rodriguez I

***Bebrornis sechellensis* (Seychelles Brush Warbler)**
Cousin I

HIPPOLAIS
***Hippolais caligata* (Booted Warbler)**
H.c.caligata
C & E Russia >> S & W India
H.c.annectens
Altai, NW Mongolia >> N India
H.c.rama
E Iran to Pakistan
***Hippolais pallida* (Olivaceous Warbler)**
H.p.opaca
S Spain, NW Africa
H.p.elaeica
SE Europe, Iran, SE Asia >> Ethiopia
H.p.reiseri
S Algeria, S Tunisia
H.p.pallida
Egypt, Sudan
H.p.laeneni
L Chad
***Hippolais languida* (Upcher's Warbler)**
H.l.languida
SE Turkey, Syria >> E Africa
H.l.magnirostris
E Iran, S Russia, Afghanistan >> E Africa
***Hippolais olivetorum* (Olive-tree Warbler)**
SE Europe, Asia Minor, N Africa
***Hippolais polyglotta* (Melodious Warbler)**
SW Europe >> NW & W Africa
***Hippolais icterina* (Icterine Warbler)**
H.i.icterina
C & E Europe, W Siberia >> Sth Africa
H.i.alaria
Asia Minor, Iran

CHLOROPETA
***Chloropeta natalensis* (Yellow Warbler)**
C.n.batesi
E Nigeria, Cameroun, Gabon, E Zaire
C.n.massaica
Zaire to Ethiopia & Tanzania
C.n.major
Angola, S Zaire, Zambia
C.n.natalensis
S Tanzania, Malawi, Zimbabwe, N Sth Africa
***Chloropeta similis* (Mountain Yellow Warbler)**
Zaire to Sudan, Malawi, Tanzania
***Chloropeta gracilirostris* (Yellow Swamp Warbler)**
C.g.gracilirostris
E Zaire, W Uganda
C.g.bensoni
L Mweru (NE Zambia)

CISTICOLA
***Cisticola erythrops* (Red-faced Cisticola)**
C.e.erythrops
Gambia to Zaire

C.e.pyrrhomitra
Ethiopia, S Sudan
C.e.nilotica
SW Sudan
C.e.sylvia
E Zaire, Sudan to Tanzania
C.e.arcana
Zambia
C.e.elusa
Zimbabwe
C.e.nyasa
SE Africa
***Cisticola lepe* (Angolan Cisticola)**   136.1
C Angola to SE Zaire, Botswana
***Cisticola cantans* (Singing Cisticola)**
C.c.swanzii
Gambia to S Nigeria
C.c.concolor
Niger to Sudan
C.c.cantans
N Ethiopia
C.c.belli
E Zaire & S Sudan to Tanzania
C.c.pictipennis
Kenya, N Tanzania
C.c.munzneri
Zambia to Mozambique
***Cisticola lateralis* (Whistling Cisticola)**
C.l.lateralis
Gambia to Cameroun
C.l.antinorii
NE Zaire, Sudan, Uganda
C.l.vincenti
Angola, S Zaire
***Cisticola woosnami* (Trilling Cisticola)**
C.w.woosnami
E Zaire, Uganda to N Malawi
C.w.lufira
SE Zaire, Zambia, W Tanzania
***Cisticola anonyma* (Chattering Cisticola)**
Ghana to Zaire, Angola
***Cisticola bulliens* (Bubbling Cisticola)**
N & C Angola
***Cisticola chubbi* (Chubb's Cisticola)**
C.c.discolor
Cameroun Mt
C.c.adametzi
highlands of Cameroun
C.c.chubbi
NE Zaire, Kenya, Tanzania
C.c.marungensis
Marungu highland (E Zaire)
***Cisticola hunteri* (Hunter's Cisticola)**
W Kenya, N Tanzania
***Cisticola nigriloris* (Black-browed Cisticola)**
136.1
S Tanzania, N Malawi, E Zambia
***Cisticola aberrans* (Lazy Cisticola)**
C.a.petrophila
N Nigeria to Sudan & NE Zaire
C.a.admiralis
Mali & Sierra Leone to Ghana

C.a.emini
E Zaire, NW Tanzania
C.a.teitensis
SE Kenya, NE Tanzania
C.a.bailunduensis
W Angola
C.a.lurio
N Mozambique, E Malawi
C.a.nyika
NE Zambia, Botswana to Mozambique
C.a.aberrans
Transvaal, Natal
C.a.minor
E Cape Province, S Natal
Cisticola bodessa (Boran Cisticola)
C.b.bodessa
E Sudan, S Ethiopia, N Kenya
C.b.kaffensis
Kaffa (S Ethiopia)
Cisticola chiniana (Rattling Cisticola)
C.c.simplex
S Sudan, N Kenya
C.c.fricki
Ethiopia, N Kenya
C.c.humilis
C Kenya
C.c.fischeri
E Zaire, Tanzania
C.c.ukamba
SW Kenya, Tanzania
C.c.heterophrys
E Kenya, E Tanzania
C.c.keithi
Tanzania
C.c.mbeya
Tanzania
C.c.fortis
S Zaire, Angola, N Zambia
C.c.procera
S Tanzania, Malawi, E Zambia, N
Mozambique
C.c.huilensis
C Angola, N Namibia
C.c.frater
N Namibia, SW Angola
C.c.bensoni
Zambia
C.c.smithersi
NE Botswana, NE Angola
C.c.chiniana
S Zambia, S Zimbabwe to Cape Province
C.c.campestris
Natal, Mozambique, E Transvaal
Cisticola cinereola (Ashy Cisticola)
E Ethiopia, N Somalia to NE Tanzania
Cisticola ruficeps (Red-pate Cisticola)
C.r.guinea
Gambia to N Nigeria
C.r.ruficeps
Central African Republic to Sudan
C.r.scotoptera
E Sudan, N Ethiopia

C.r.mongalla
S Sudan, N Uganda
Cisticola rufilata (Grey Cisticola)
C.r.ansorgei
Angola to S Zaire
C.r.venustula
SE Angola to N Mozambique
C.r.rufilata
N Namibia to W Transvaal
C.r.vicinior
C Zimbabwe, N Transvaal, SE Botswana
Cisticola subruficapilla (Red-headed Cisticola)
C.s.newtoni
Mossamedes (Angola)
C.s.windhoekensis
Namibia
C.s.karasensis
Namibia
C.s.namaqua
NW Cape Province
C.s.subruficapilla
SW Cape Province
C.s.jamesi
E Cape Province
C.s.euroa
W Orange Free State, NE Cape Province
Cisticola lais (Wailing Cisticola)
C.l.distincta
E Uganda, N Kenya
C.l.mashona
E Zimbabwe, N Transvaal
C.l.namba
C Angola
C.l.semifasciata
S Tanzania, Malawi, N Mozambique
C.l.lais
SE Sthn Africa
C.l.monticola
S Transvaal
C.l.maculata
SW & S Cape Province
C.l.oreobates
Mozambique
Cisticola restricta (Tana River Cisticola)
Tana river, Kenya
Cisticola njombe (Churring Cisticola)
Zambia, Malawi, Tanzania
Cisticola galactotes (Winding Cisticola)
C.g.amphilecta
Senegal to Cameroun, N Zaire
C.g.zalingei
N Cameroun to W Sudan
C.g.grisea
Gabon, NW Angola
C.g.lugubris
Ethiopia
C.g.marginata
S Sudan, N Uganda
C.g.haematocephala
S Somalia, E Kenya, E Tanzania
C.g.nyansae
S Uganda, Kenya, NW Tanzania

C.g.suahelica
E Zaire, Zambia
C.g.schoutedeni
NW Zambia
C.g.luapula
Angola, Namibia to Mozambique
C.g.galactotes
Malawi, Natal
C.g.stagnans
N Botswana, SW Zambia
C.g.isodactyla
S Malawi, SE Zimbabwe, W Mozambique
Cisticola carruthersi (Carruthers' Cisticola)
NE Zaire, Kenya
Cisticola pipiens (Chirping Cisticola)
C.p.congo
E Angola, S Zaire, Zambia, N Botswana
C.p.pipiens
S Angola
C.p.arundinicola
Botswana, SE Angola
Cisticola tinniens (Levaillant's Cisticola)
C.t.dyleffi
E Zaire
C.t.oreophila
C Kenya
C.t.shiwae
NE Zambia, SE Zaire
C.t.perpulla
Angola
C.t.tinniens
SE Africa
Cisticola robusta (Stout Cisticola)
C.r.robusta
Ethiopia
C.r.omo
SW Ethiopia
C.r.santae
Cameroun
C.r.nuchalis
NE Zaire, S Sudan, N Kenya, NW Tanzania
C.r.ambigua
C & S Kenya, NE Tanzania
C.r.angolensis
N & C Angola, NW Zambia
C.r.awemba
SW Tanzania, NE Zambia
Cisticola aberdare (Aberdare Mountain Cisticola)
C Kenya
Cisticola natalensis (Croaking Cisticola)
C.n.strangei
Senegal to W & C Zaire
C.n.inexpectata
Ethiopia
C.n.argentea
S Ethiopia, N Kenya
C.n.tonga
Sudan
C.n.kapitensis
C Kenya
C.n.littoralis
E Kenya, E Tanzania

C.n.katanga
NE Angola, S Zaire, Zambia, SW Tanzania
C.n.huambo
N & C Angola
C.n.natalensis
SE Africa
C.n.holubii
NE Botswana, SW Zimbabwe
C.n.matengorum
S Tanzania, E Zambia, Mozambique
Cisticola fulvicapilla (Piping Cisticola)
C.f.dispar
Angola
C.f.muelleri
Zambia to Mozambique
C.f.hallae
S Angola, NE Namibia, NW Botswana
C.f.dextra
E Botswana, S Zimbabwe to S Mozambique
C.f.ruficapilla
Transvaal, N Cape Province
C.f.lebombo
N Natal, S Mozambique
C.f.fulvicapilla
S Natal, Cape Province
C.f.silberbaueri
SW Cape Province
C.f.dumicola
Botswana, S Zambia
Cisticola angusticauda (Tabora Cisticola)
SW Kenya to Zambia
Cisticola melanura (Slender-tailed Cisticola)
136.1
NE Angola, S Zaire
Cisticola brachyptera (Siffling Cisticola)
C.b.brachyptera
Senegal to Sudan, Uganda
C.b.zedlitzi
Ethiopia
C.b.reichenowi
N Kenya
C.b.ankole
SW Uganda, NW Tanzania, E Zaire
C.b.kericho
C & S Kenya
C.b.katonae
S Kenya, N Tanzania
C.b.loanda
C Angola, S Zaire, Zambia
C.b.isabellina
S Tanzania, Malawi, Mozambique, Zimbabwe
Cisticola rufa (Rufous Cisticola)
Gambia to N Cameroun
Cisticola troglodytes (Foxy Cisticola)
C.t.troglodytes
Niger to Sudan, Kenya
C.t.ferruginea
E Sudan, W Ethiopia
Cisticola nana (Tiny Cisticola)
S Ethiopia, E Kenya, N Tanzania
Cisticola incana (Incana) 136.1
Socotra

***Cisticola juncidis* (Zitting Cisticola) (Fan-tailed Warbler)**
  C.j.juncidis
    S Europe, Asia Minor, Egypt
  C.j.cisticola
    Iberian peninsula, NW Africa
  C.j.neurotica
    Syria, Iraq
  C.j.uropygialis
    W & NC Africa
  C.j.perennia
    E Africa
  C.j.terrestris
    S Africa
  C.j.cursitans
    Pakistan, India, Sri Lanka
  C.j.salimalii
    SW India
  C.j.omalura
    Sri Lanka
  C.j.malaya
    Malaysia
  C.j.brunniceps
    Japan
  C.j.tinnabulans
    E China, Philippine Is
  C.j.nigrostriata
    Palawan I
  C.j.fuscicapilla
    E Java, Lesser Sunda Is, Moluccas
  C.j.constans
    Buton, Peleng Is, Sulawesi
  C.j.leanyeri
    Northern Territory (Australia)
  C.j.normani
    W Queensland
  C.j.laveryi
    E Queensland coast
***Cisticola haesitata* (Socotra Cisticola)**
  Socotra I
***Cisticola cherina* (Madagascar Cisticola)**
  Madagascar
***Cisticola aridula* (Desert Cisticola)**
  C.a.aridula
    Niger, Chad, Sudan
  C.a.lavendulae
    Ethiopia, Somalia
  C.a.tanganyika
    Kenya, Tanzania
  C.a.lobito
    S Angola
  C.a.perplexa
    E Angola, S Zambia
  C.a.kalahari
    Namibia
  C.a.traylori
    NE Angola, W Zambia
  C.a.caligena
    E Zimababwe to S Mozambique & Natal
  C.a.eremica
    NW Namibia
***Cisticola textrix* (Tink-tink Cisticola)**

  C.t.bulubulu
    W & C Angola, W Zambia
  C.t.anselli
    E Angola, NW Zambia
  C.t.major
    Transvaal, E Cape Province
  C.t.marleyi
    N Natal
  C.t.textrix
    SW Cape Province
***Cisticola eximia* (Black-necked Cisticola)**
  C.e.occidens
    Guinea to Nigeria
  C.e.winneba
    S Ghana
  C.e.eximia
    NE Zaire to Sudan, Ethiopia
***Cisticola dambo* (Cloud-scraping Cisticola)**
  C.d.kasai
    C Zaire
  C.d.dambo
    Angola, S Zaire, Zambia
***Cisticola brunnescens* (Pectoral-patch Cisticola)**
  C.b.mbangensis
    Cameroun
  C.b.lynesi
    Cameroun
  C.b.wambera
    W Ethiopia
  C.b.brunnescens
    E Ethiopia, Somalia
  C.b.midcongo
    W Zaire
  C.b.cinnamomea
    E Zaire, Tanzania, Zambia, Zimbabwe
  C.b.hindii
    W Kenya, N Tanzania
  C.b.nakuruensis
    Kenya, Tanzania
  C.b.egregia
    S Zimbabwe, E Transvaal, Natal
***Cisticola ayresii* (Wing-snapping Cisticola)**
  C.a.gabun
    Gabon
  C.a.imatong
    S Sudan
  C.a.entebbe
    E Zaire, Uganda, NW Tanzania
  C.a.itombwensis
    E Zaire
  C.a.mauensis
    Kenya
  C.a.ayresii
    C, E & SE Africa
***Cisticola exilis* (Gold-capped Cisticola)**
  C.e.tytleri
    Bangladesh, Assam, Nepal, N India, China
  C.e.erythrocephala
    S India
  C.e.equicaudata
    Burma, Thailand, S Vietnam

C.e.lineocapilla
Java, Bali, Northern Territory (Australia)
C.e.rustica
Sulawesi, Moluccas, Philippine Is
C.e.exilis
E Australia
C.e.diminuta
Solomon Is, N Queensland, New Guinea
C.e.alexandrae
NW Australia, W Queensland
C.e.mixta
S Queensland
C.e.volitans
Taiwan, SW China
C.e.courtoisi
Kiangsi, Fukien
C.e.semirufa
Philippine Is, Sulu Arch
C.e.polionota
Bismarck Arch

SCOTOCERCA
Scotocerca inquieta (Streaked Scrub Warbler)
S.i.theresae
S Morocco
S.i.saharae
SE Morocco, Algeria
S.i.harterti
Cyrenaica
S.i.grisea
W Arabia
S.i.buryi
S Arabia
S.i.inquieta
E Egypt, N Arabia
S.i.platyura
Transcaspia, Afghanistan
S.i.striata
Iran, Baluchistan, Pakistan, NW India

RHOPOPHILUS
Rhopophilus pekinensis (White-browed Chinese Warbler)
R.p.albosuperciliaris
Tarim Basin to Lop Nor
R.p.leptorhynchus
Kansu, NE Tsinghai
R.p.pekinensis
S Manchuria, Korea, Shansi, Ningsia

PRINIA   136.1
Prinia burnesii (Long-tailed Prinia)
P.b.burnesii
Pakistan, NW India
P.b.cinerascens
N Bangladesh, Assam
Prinia criniger (Hill Prinia)
P.c.striatula
W Pakistan
P.c.criniger
Himalayas, Assam

P.c.catharia
Assam to NW China
P.c.parvirostris
SE Yunnan
P.c.parumstriata
C & SE China
P.c.striata
Taiwan
Prinia polychroa (Brown Hill Prinia)
P.p.cooki
S Burma to S Laos, Cambodia
P.p.bangsi
Taiwan, SE Yunnan
P.p.rocki
S Vietnam
P.p.polychroa
Java
Prinia atrogularis (White-browed Prinia)
P.a.atrogularis
Nepal, N India, SE Tibet
P.a.khasiana
Assam, Burma
P.a.erythropleura
Burma, N Thailand
P.a.superciliaris
SE China, Burma
P.a.waterstradti
Malaysia
P.a.dysancrita
W Sumatra
P.a.klossi
N Indochina
Prinia cinereocapilla (Hodgson's Long-tailed Warbler)
N India
Prinia buchanani (Rufous-fronted Prinia)
Pakistan, India
Prinia rufescens (Lesser Brown Prinia)
P.r.rufescens
Bhutan, Nepal to Burma
P.r.beavani
SE Burma to N Indochina
P.r.extrema
Malaysia
P.r.peninsularis
S Thailand
P.r.objurgans
S China, N Vietnam
P.r.dalatensis
S Vietnam
Prinia hodgsonii (Franklin's Prinia)
P.h.hodgsonii
India, W Burma
P.h.albogularis
SW India
P.h.leggei
Sri Lanka
P.h.rufula
Kashmir to NW Yunnan, N Burma
P.h.confusa
S China, N Indochina
P.h.erro
E & SE Burma to S Indochina

Prinia gracilis (Graceful Warbler)
P.g.akyildizi
S Turkey coast
P.g.palaestinae
Syria, Israel, Sinai
P.g.deltae
Nile delta, Israel coast
P.g.natronensis
Natrun (Egypt)
P.g.gracilis
S Egypt to S Sudan
P.g.carlo
Red Sea to S Somalia, E Ethiopia
P.g.yemenensis
S Arabia
P.g.hufufae
E Saudi Arabia, Bahrein
P.g.carpenteri
Oman coast
P.g.lepida
S Iran to N India
P.g.stevensi
S Nepal to NE India

Prinia sylvatica (Jungle Prinia)
P.s.insignis
NW India
P.s.gangetica
Nepal, N India, Bangladesh
P.s.mahendrae
Orissa
P.s.sylvatica
S India
P.s.valida
Sri Lanka

Prinia familiaris (Bar-winged Prinia)
P.f.prinia
S Sumatra, W Java
P.f.familiaris
E Java, Bali

Prinia flaviventris (Yellow-bellied Prinia)
P.f.sindiana
NW India, Pakistan
P.f.flaviventris
Bhutan, Himalayas, Bangladesh
P.f.delacouri
Thailand
P.f.sonitans
SE China, Taiwan, Hainan I
P.f.rafflesi
Malaysia, Sumatra, Java
P.f.halistona
Nias I
P.f.latrunculus
Borneo

Prinia socialis (Ashy Prinia)
P.s.stewarti
Pakistan, India
P.s.inglisi
NE India, Nepal, Bhutan, Assam,
Bangladesh
P.s.socialis
S India

P.s.brevicauda
Sri Lanka

Prinia subflava (Tawny-flanked Prinia)
P.s.terricolor
Pakistan, N India
P.s.inornata
SE & C India
P.s.franklinii
SW India
P.s.insularis
Sri Lanka
P.s.fusca
Nepal, Assam, Bangladesh
P.s.blanfordi
Burma
P.s.herberti
S Thailand, C Indochina
P.s.extensicauda
W & S China, N Indochina
P.s.flavirostris
Taiwan
P.s.subflava
Senegal to Ethiopia
P.s.melanorhyncha
Sierra Leone to Gabon
P.s.pallescens
Mali, Niger, Sudan
P.s.tenella
C Uganda, Kenya, Tanzania
P.s.affinis
SE Zaire, Zambia to Mozambique
P.s.kasokae
N Angola
P.s.graueri
C Angola, S Zaire, W Zambia
P.s.bechuanae
N Botswana, S Angola, N Namibia
P.s.pondoensis
E Cape Province, Natal

Prinia somalica (Pale Prinia)
P.s.somalica
N Somalia
P.s.erlangeri
S Ethiopia, S Somalia, SE Sudan, N Kenya

Prinia fluviatilis (Lake Chad Prinia)　　136.1
Mali, L Chad area

Prinia maculosa (Karroo Prinia)
P.m.psammophila
Namibia coast
P.m.maculosa
Namibia, Botswana, W Cape Province
P.m.hypoxantha
E Cape Province, Natal, E Transvaal

Prinia flavicans (Black-chested Prinia)
P.f.bihe
E Angola, W Zambia
P.f.ansorgei
S Angola
P.f.flavicans
Namibia, Botswana, W Cape Province
P.f.nubilosa
E Botswana, W Zimbabwe, N Transvaal

P.f.ortleppi
NE Cape Province, SW Transvaal, W Orange Free State

Prinia substriata (White-breasted Prinia)
W & C Cape Province

Prinia molleri (Sao Thome Prinia)
Sao Thome I

Prinia robertsi (Roberts' Prinia)
E Zimbabwe, W Mozambique

Prinia leucopogon (White-chinned Prinia)
P.l.leucopogon
SE Nigeria to Zambia
P.l.reichenowi
NE Zaire, S Sudan, Kenya, Tanzania

Prinia leontica (Sierra Leone Prinia)
Sierra Leone to Ghana

Prinia bairdii (Banded Prinia)
P.b.bairdii
Cameroun, Gabon
P.b.melanops
E Zaire, W Kenya
P.b.obscura
Kivu, Ruwenzori
P.b.heinrichi
N Angola

Prinia erythroptera (Red-winged Warbler)
136.1
P.e.erythroptera
Guinea to Nigeria
P.e.jodoptera
Cameroun to Sudan
P.e.major
Ethiopia
P.e.rhodoptera
W Kenya, Zambia to Mozambique

Prinia pectoralis (Rufous-eared Prinia)
P.p.etoshae
N Namibia
P.p.ocularia
Namibia, Botswana, N Cape Province
P.p.pectoralis
S Namibia, Cape Province, W Transvaal

DRYMOCICHLA
Drymocichla incana (Red-winged Grey Warbler)
Cameroun to Uganda

UROLAIS
Urolais epichlora (Green Longtail)
U.e.epichlora
N Cameroun, E Nigeria
U.e.mariae
Fernando Po I

SPILOPTILA
Spiloptila clamans (Cricket Warbler)
Mali to N Ethiopia

APALIS
Apalis thoracica (Bar-throated Apalis)
A.t.fuscigularis
SE Kenya

A.t.griseiceps
SE Kenya, NE Tanzania
A.t.uluguru
E Tanzania
A.t.lynesi
N Mozambique
A.t.quarta
Mozambique
A.t.pareensis
NE Tanzania
A.t.iringae
S Tanzania
A.t.youngi
SW Tanzania, N Malawi
A.t.murina
NE Zambia, N Malawi
A.t.whitei
Zambia, Tanzania, Malawi
A.t.flavigularis
Malawi
A.t.rhodesiae
W Zimbabwe
A.t.arnoldi
E Zimbabwe
A.t.drakensbergensis
E Transvaal, N Natal
A.t.flaviventris
W & C Transvaal
A.t.spelonkensis
E Transvaal
A.t.lebomboensis
NE Natal
A.t.darglensis
C Natal
A.t.capensis
W Cape Province
A.t.venusta
E Cape Province, Nata,
A.t.thoracica
S Cape Province
A.t.claudei
S Cape Province
A.t.griseopyga
coast of SW Cape Province

Apalis pulchra (Black-collared Apalis)
A.p.pulchra
Cameroun to S Sudan, Kenya
A.p.murpheyi
SE Zaire

Apalis ruwenzorii (Collared Apalis)
A.r.catiodes
SW Lake Kivu (E Zaire)
A.r.ruwenzorii
NW Lake Kivu

Apalis nigriceps (Black-capped Apalis)
A.n.nigriceps
Sierra Leone to Cameroun
A.n.collaris
NE Zaire, W Uganda

Apalis jacksoni (Black-throated Apalis)
A.j.bambuluensis
N Cameroun

A.j.minor
S Cameroun, Central African Republic
A.j.jacksoni
SE Sudan, Uganda, Kenya, Angola
Apalis chariessa (White-winged Apalis)
A.c.chariessa
Tana R (Kenya)
A.c.macphersoni
N Tanzania, N Mozambique
Apalis binotata (Masked Apalis)
Cameroun to W Uganda
Apalis personata (Mountain Masked Apalis)
        136.14
A.p.personata
SW Uganda, E Zaire
A.p.marungensis
SE Zaire
Apalis flavida (Yellow-chested Apalis)
A.f.caniceps
Sierra Leone to Sudan, NW Kenya
A.f.viridiceps
NE Ethiopia, N Somalia
A.f.abyssinica
Illubabor (Ethiopia)
A.f.flavocincta
SW Uganda, Kenya, N Tanzania
A.f.pugnax
Mt Kenya
A.f.tenerrima
E Kenya, Tanzania
A.f.golzi
SW Kenya, N Tanzania, Rwanda
A.f.niassae
E Angola to SW Tanzania
A.f.neglecta
Angola to Mozambique & Natal
A.f.flavida
S Angola, Namibia, Botswana
A.f.florisuga
E Cape Province, Natal
A.f.renata
Sol do Save (Mozambique), N Natal
Apalis ruddi (Rudd's Apalis)
A.r.ruddi
S Mozambique
A.r.caniviridis
S Malawi
A.r.fumosa
S Mozambique, N Natal
Apalis rufogularis (Buff-throated Apalis)
A.r.rufogularis
Fernando Po I, S Nigeria to Gabon
A.r.sanderi
SW Nigeria
A.r.nigrescens
S Sudan to Zambia
A.r.angolensis
N Angola
A.r.brauni
C & S Angola
A.r.kigezi
W Uganda

A.r.eidos
Lake Kivu (Zaire)
A.r.argentea
W Tanzania
Apalis sharpii (Sharpe's Apalis)
Sierra Leone to Ghana
Apalis goslingi (Gosling's Apalis)   136.1
S Cameroun to Angola
Apalis bamendae (Bamenda Apalis)   136.1
Bamenda Mts (Cameroun)
Apalis porphyrolaema (Chestnut-throated Apalis)
A.p.porphyrolaema
Uganda, Kenya, NE Tanzania
A.p.kaboboensis
E Zaire
A.p.chapini
C Tanzania
A.p.strausae
SW Tanzania, N Malawi, NE Zambia
Apalis melanocephala (Black-headed Apalis)
A.m.nigrodorsalis
C Kenya
A.m.moschi
SE Kenya, NE Tanzania
A.m.muhuluensis
S Tanzania
A.m.melanocephala
E Tanzania
A.m.lightoni
Malawi, Mozambique
A.m.fuliginosa
S Malawi
A.m.tenebricosa
Mozambique
A.m.adjacens
N Mozambique
A.m.addenda
Sol do Save, Mozambique
Apalis chirindensis (Chirinda Apalis)
A.c.vumbae
E Zimbabwe, W Mozambique
A.c.chirindensis
SE Zimbabwe, Mt Gorongoza
(Mozambique)
Apalis cinerea (Grey Apalis)
A.c.cinerea
Cameroun to N Tanzania
A.c.sclateri
Fernando Po I
A.c.grandis
Angola
Apalis alticola (Brown-headed Apalis)
A.a.dowsetti
E Zaire
A.a.alticola
Tanzania & SE Zaire to NE Angola & N Malawi
Apalis karamojae (Karamoja Apalis)
A.k.karamojae
NE Uganda, N Tanzania
A.k.stronachi
NC Tanzania

Apalis rufifrons (Red-faced Warbler) 136.1
A.r.rufifrons
Sudan, Ethiopia, Somalia
A.r.smithi
SE Ethiopia to N Tanzania
A.r.rufidorsalis
SE Kenya

STENOSTIRA
Stenostira scita (Fairy Flycatcher) 136.1
S.s.scita
S Namibia, NW Cape Province
>>Zimbabwe
S.s.saturatior
C, E & S Cape Province >> N Cape
Province
S.s.rudebecki
W Orange Free State, Lesotho >> Natal, S
Mozambique

PHYLLOLAIS
Phyllolais pulchella (Buff-bellied Warbler)
Chad, Ethiopia to Tanzania

ORTHOTOMUS
**Orthotomus metopias (Red-capped Forest
Warbler)**
O.m.metopias
NE Tanzania
O.m.altus
N Tanzania
O.m.pallidus
Itanga (Tanzania)
**Orthotomus moreaui (Long-billed Forest
Warbler)**
O.m.moreaui
NE Tanzania
O.m.sousae
N Mozambique
Orthotomus cucullatus (Mountain Tailor Bird)
O.c.coronatus
E Himalayas, Assam
O.c.thais
S Thailand
O.c.cucullatus
Sumatra, Java, Bali
O.c.cinereicollis
NE Borneo
O.c.viridicollis
Palawan I
O.c.heterolaemus
Mindanao I
O.c.philippinus
N Luzon I
O.c.everetti
Flores I
O.c.riedeli
N Sulawesi
O.c.meisei
SC Sulawesi
O.c.stentor
SE Sulawesi

O.c.hedymeles
S Sulawesi
O.c.dumasi
Buru, Seram Is
O.c.batjanensis
Batjan I
Orthotomus sutorius (Long-tailed Tailor Bird)
O.s.guzuratus
NW Himalayas, N India
O.s.patia
Nepal, NE India, Burma
O.s.sutorius
Sri Lanka
O.s.fernandonis
Sri Lanka
O.s.luteus
NE Assam, N Burma
O.s.inexpectatus
W China
O.s.longicauda
S China
O.s.maculicollis
Malaysia
O.s.edela
Java
**Orthotomus atrogularis (Black-necked Tailor
Bird)**
O.a.nitidus
Assam, Burma
O.a.atrogularis
Borneo, Malaysia, Sumatra
O.a.humphreysi
N Borneo
O.a.anambensis
Tioman, Anamba, Natuna Is
O.a.chloronotos
N Luzon I
O.a.castaneiceps
Masbate, Panay, Ticao Is
O.a.rabori
Negros I
O.a.frontalis
Samar, Leyte, Bohol Mindanao Is
O.a.mearnsi
Basilan I
Orthotomus derbianus (Luzon Tailor Bird)
O.d.derbianus
S Catanduanes , C Luzon Is
O.d.nilesi
C Philippine Is
**Orthotomus sericeus (Red-headed Tailor
Bird)**
O.s.hesperius
Burma, Malaysia, Sumatra
O.s.sericeus
Borneo, Palawan I
O.s.rubicundulus
Sirhassen, Natuna Is
O.s.nuntius
Balabac, Cagayan Sulu Is
Orthotomus ruficeps (Grey Tailor Bird)
O.r.cineraceus
S Burma, Malaysia, Sumatra

O.r.baeus
Nias, N & S Pagai Is
O.r.concinnus
Siberut, Sipura Is
O.r.ruficeps
coast of Java
O.r.palliolatus
Kangean, Karimon Is
O.r.baweanus
Bawean I
O.r.borneoensis
Borneo
O.r.cagayanensis
Cagayan Sulu I
Orthotomus sepium (Ashy Tailor Bird)
O.s.sundaicus
Panaitan I (Java)
O.s.sepium
Java, Bali, Lombok I
Orthotomus cinereiceps (White-eared Tailor Bird)
O.c.obscurior
Mindanao I
O.c.cinereiceps
Basilan I
Orthotomus nigriceps (Black-headed Tailor Bird)
Mindanao I
Orthotomus samarensis (Samar Tailor Bird)
Samar, Bohol, Leyte Is

CAMAROPTERA
Camaroptera brachyura (Green-backed Camaroptera)
C.b.bororensis
SE Zaire to Mozambique
C.b.pileata
E Kenya, E Tanzania, Zanzibar
C.b.fugglescouchmani
S Tanzania, N Malawi
C.b.constans
S Mozambique, E Transvaal, N Natal
C.b.brachyura
SE Transvaal, S Cape Province, Natal
Camaroptera brevicaudata (Grey-backed Camaroptera)
C.b.brevicaudata
Senegal to Ethiopia
C.b.tincta
Sierra Leone to Tanzania, S Zaire
C.b.abessinica
NE Sudan to Kenya
C.b.insulata
Ethiopia
C.b.erlangeri
E Somalia, E Kenya, E Tanzania
C.b.griseigula
S Kenya, N Tanzania
C.b.aschani
E Zaire
C.b.intercalata
S Zaire, Angola, S Zambia

C.b.beirensis
S Mozambique
C.b.transitiva
S Zimbabwe to N Transvaal
C.b.sharpei
Sthn Africa
Camaroptera harterti (Hartert's Cameroptera)
NW Angola
Camaroptera superciliaris (Yellow-browed Camaroptera)
NW Angola
Camaroptera chloronota (Olive-green Camaroptera)
C.c.kelsalli
Senegal to Ghana
C.c.chloronota
Togo to Cameroun, W Zaire
C.c.granti
Fernando Po I
C.c.kamitugaensis
E Zaire
C.c.toroensis
S Uganda, E Zaire

CALAMONASTES   136.1.10
Calamonastes simplex (Grey Wren Warbler)
C.s.simplex
Ethiopia, Somalia, N Kenya
C.s.undosus
S Kenya, Tanzania, Zambia
C.s.cinerea
S Zaire, N Angola
C.s.katangae
SE Zaire, W Zambia
C.s.huilae
S Angola
Calamonastes stierlingi (Stierling's Wren Warbler)
C.s.buttoni
N Malawi, NW Zambia
C.s.stierlingi
S Malawi, Tanzania, Mozambique
C.s.irwini
W Malawi, Zambia, S Zimbabwe
C.s.pintoi
S Mozambique, E Transvaal
C.s.olivascens
Mozambique
Calamonastes fasciolata (Barred Camaroptera)
C.f.pallidior
S Angola
C.f.fasciolata
SW Africa
C.f.europhila
W Transvaal

EURYPTILA
Euryptila subcinnamomea (Kopje Warbler)
W Namibia, NW Cape Province

POLIOLAIS
Poliolais lopezi (White-tailed Warbler)
P.l.lopezi
Fernando Po I
P.l.alexanderi
Cameroun Mt
P.l.manengubae
Manenguba, Kupe Mt (Cameroun)

GRAUERIA
Graueria vittata (Grauer's Warbler)
E Zaire

EREMOMELA
**Eremomela icteropygialis (Yellow-bellied
Eremomela)**
E.i.alexanderi
Niger, Chad, Sudan
E.i.griseoflava
Somalia, Ethiopia, N Kenya
E.i.karamojensis
Somalia to NE Uganda
E.i.crawfurdi
SW Kenya, NW Tanzania, Rwanda
E.i.abdominalis
S Kenya, N Tanzania
E.i.polioxantha
S Zaire, S Tanzania to N Transvaal, N Natal
E.i.helenorae
SW Zambia to W Mozambique
E.i.salvadorii
S Zaire to C Angola, W Zambia   136.1
E.i.puellula
W Angola
E.i.sharpei
Namibia
E.i.icteropygialis
S Namibia, W Cape Province
E.i.saturatior
E Cape Province, S Transvaal
**Eremomela flavicrissalis (Yellow-vented
Eremomela)**
Somalia, NE Kenya
Eremomela scotops (Green-cap Eremomela)
E.s.congensis
W & C Zaire
E.s.kikuyuensis
C Kenya
E.s.citriniceps
E Zaire, Uganda, Tanzania
E.s.occipitalis
E Kenya, E Tanzania
E.s.angolensis
N Angola
E.s.pulchra
E Angola to Malawi
E.s.scotops
Botswana to Tanzania, Mozambique
E.s.chlorochlamys
S Mozambique to Transvaal & Natal
**Eremomela pusilla (Smaller Green-backed
Eremomela)**
Senegal to N Cameroun

**Eremomela canescens (Green-backed
Eremomela)**
E.c.elegans
Chad, W Sudan
E.c.abyssinica
Ethiopia, E Sudan
E.c.canescens
Cameroun to S Sudan & N Zaire
E.c.elgonensis
W Kenya, Uganda
**Eremomela gregalis (Yellow-rumped
Eremomela)**
E.g.gregalis
Namibia, W Cape Province
E.g.albigularis
SW Cape Province
**Eremomela badiceps (Brown-crowned
Eremomela)**
E.b.fantiensis
Sierra Leone to Ghana
E.b.badiceps
Fernando Po I, Cameroun to N Angola & S
Sudan
Eremomela turneri (Turner's Eremomela)
E.t.kalindei
E & C Zaire
E.t.turneri
Uganda, W Kenya
**Eremomela atricollis (Black-necked
Eremomela)**
E.a.venustula
NW Zambia, SW Zaire
E.a.atricollis
W Angola, SE Zaire, Zambia
**Eremomela usticollis (Burnt-neck
Eremomela)**
E.u.usticollis
C Sthn Africa
E.u.rensi
Mozambique, Malawi
E.u.baumgarti
S Angola & Namibia to W Transvaal

RANDIA
**Randia pseudozosterops (Marvantsetra
Warbler)**
Madagascar

NEWTONIA
Newtonia amphichroa (Tulear Newtonia)
NE Madagascar
Newtonia brunneicauda (Common Newtonia)
N.b.brunneicauda
Madagascar
N.b.monticola
Mt Ankarata
Newtonia archboldi (Tabity Newtonia)
SW Madagascar
Newtonia fanovanae (Fanovana Newtonia)
E Madagascar

SYLVIETTA
Sylvietta virens (Green Crombec)
  S.v.flaviventris
    Sierra Leone to W Nigeria
  S.v.virens
    SE Nigeria to W Zaire
  S.v.baraka
    S Sudan, Uganda
  S.v.tando
    W Angola
  S.v.meridionalis
    C Angola
Sylvietta denti (Lemon-bellied Crombec)
  S.d.hardyi
    Sierra Leone to Ghana
  S.d.denti
    S Cameroun
Sylvietta leucophrys (White-browed Crombec)
  S.l.arileuca
    Uganda
  S.l.leucophrys
    Uganda, N Kenya
  S.l.chloronata
    E Zaire, SW Uganda, W Tanzania
  S.l.chapini
    E Zaire
Sylvietta brachyura (Northern Crombec)
  S.b.brachyura
    Senegal to Ethiopia
  S.b.carnapi
    N Cameroun to W Kenya
  S.b.leucopsis
    Ethiopia, Sudan to N Tanzania
Sylvietta philippae (Somali Short-billed Crombec)
    Somalia
Sylvietta whytii (Red-faced Crombec)
  S.w.loringi
    S Ethiopia to N Tanzania
  S.w.jacksoni
    S & E Uganda, Kenya, Tanzania
  S.w.minima
    E Kenya, E Tanzania
  S.w.whytii
    S Malawi, Mozambique, E Zimbabwe
  S.w.nemorivaga
    W Zimbabwe
Sylvietta ruficapilla (Red-capped Crombec)
  S.r.schoutedeni
    W Tanzania
  S.r.rufigenis
    W & S Zaire
  S.r.chubbi
    E Zaire to Mozambique
  S.r.makayii
    N Angola
  S.r.ruficapilla
    S Angola, W Zambia
  S.r.gephyra
    SE Zaire, SW Zambia
Sylvietta rufescens (Long-billed Crombec)

  S.r.flecki
    N Angola
  S.r.adelphe
    E & SE Zaire
  S.r.ansorgei
    S Angola
  S.r.pallida
    Zambia, Botswana to Mozambique
  S.r.ochrocara
    Namibia
  S.r.resurga
    Natal
  S.r.rufescens
    Namibia, S Botswana, W Cape Province
  S.r.diverga
    E Cape Province
Sylvietta isabellina (Somali Long-billed Crombec)
    Ethiopia, Somalia, N Kenya

HEMITESIA
Hemitesia neumanni (Neumann's Short-tailed Warbler)
    E Zaire, W Uganda

MACROSPHENUS
Macrosphenus kempi (Kemp's Longbill)
  M.k.kempi
    Sierra Leone
  M.k.flammeus
    S Nigeria
Macrosphenus flavicans (Yellow Longbill)
  M.f.flavicans
    SW Nigeria to NW Angola
  M.f.hypochondriacus
    Zaire, Uganda to SW Sudan
Macrosphenus concolor (Grey Longbill)
  M.c.concolor
    Sierra Leone to Uganda
  M.c.griscens
    Kamituga (Zaire)
Macrosphenus pulitzeri (Pulitzer's Longbill)
    Angola
Macrosphenus kretschmeri (Kretschmer's Longbill)
  M.k.kretschmeri
    SE Kenya, NE Tanzania
  M.k.griseiceps
    SE Tanzania, NE Mozambique

AMAUROCICHLA
Amaurocichla bocagei (Bocage's Longbill)
    Sao Thome I

HYPERGERUS
Hypergerus atriceps (Oriole Warbler)
    Senegal to Central African Republic
Hypergerus lepida (Grey-capped Warbler)
    Uganda, Kenya, Tanzania

HYLIOTA
Hyliota flavigaster (Yellow-bellied Flycatcher)
  H.f.flavigaster
    Senegal to Ethiopia, Kenya
  H.f.barbozae
    Angola to Tanzania, Mozambique
  H.f.marginalis
    S Tanzania, Mozambique
Hyliota australis (Southern Yellow-bellied Flycatcher)
  H.a.slatini
    Zaire, W Kenya
  H.a.usambara
    NE Tanzania
  H.a.pallidipectus
    SE Zaire, Angola, Zambia
  H.a.inornata
    Malawi, Mozambique, S Zimbabwe
  H.a.australis
    C Zimababwe
Hyliota violacea (Violet-backed Flycatcher)
  H.v.nehrkorni
    Liberia to Ghana
  H.v.violacea
    Togo to S Cameroun, W Zaire

HYLIA
Hylia prasina (Green Hylia)
  H.p.poensis
    Fernando Po I
  H.p.prasina
    Guinea to N Angola & Tanzania

PHYLLOSCOPUS
Phylloscopus ruficapilla (Yellow-throated Woodland Warbler)
  P.r.ochrogularis
    W Tanzania
  P.r.minullus
    SE Kenya, E Tanzania
  P.r.johnstoni
    SW Tanzania, Malawi
  P.r.quelimanensis
    N Mozambique
  P.r.alacris
    Manica e Sofala (Mozambique)
  P.r.ruficapilla
    Natal, S Cape Province
  P.r.ochraceiceps
    Transvaal
  P.r.voelckeri
    coastal Cape Province
Phylloscopus laurae (Mrs Boulton's Woodland Warbler)
  P.l.eustacei
    SE Zaire, NE Zambia
  P.l.laurae
    Angola, W Zambia
Phylloscopus laetus (Red-faced Woodland Warbler)
  P.l.schoutedeni
    E Zaire

  P.l.laetus
    Rwanda, W Uganda
Phylloscopus budongoensis (Uganda Woodland Warbler)
    E Zaire, Uganda, W Kenya
Phylloscopus herberti (Black-capped Woodland Warbler)
  P.h.herberti
    Fernando Po I
  P.h.camerunensus
    S Nigeria, Cameroun Mt
Phylloscopus umbrovirens (Brown Woodland Warbler)
  P.u.yemenensis
    SW Arabia
  P.u.williamsi
    N Somalia
  P.u.umbrovirens
    N & C Ethiopia, Somalia
  P.u.mackenzianus
    S Sudan, Uganda, Kenya
  P.u.wilhelmi
    E Zaire
  P.u.alpinus
    W Uganda
  P.u.dorcadichrous
    NE Tanzania
  P.u.fugglescouchmani
    E Tanzania
Phylloscopus trochilus (Willow Warbler)
  P.t.trochilus
    Britain, W Europe >> W & WC Africa
  P.t.acredula
    Scandinavia, C & E Europe, W Siberia >> C Africa
  P.t.yakutensis
    N & NE Asia >> E Africa
Phylloscopus collybitus (Chiff-chaff)
  P.c.abietinus
    N & E Europe >> Asia Minor, Arabia, Somalia
  P.c.collybitus
    W & S Europe >> N Africa
  P.c.canariensis
    Canary Is
  P.c.exsul
    Lanzarote I
  P.c.brehmii
    W Pyrenees to N Africa
  P.c.brevirostris
    N & W Turkey
  P.c.tristis
    C Asia, N India, Bangladesh
Phylloscopus sindianus (Eastern Chiff-chaff)
136.11
  P.s.lorenzii
    Caucasus to Armenia
  P.s.sindianus
    C Asia, Himalayas >> N India, Pakistan
Phylloscopus neglectus (Plain Willow Warbler)
    NE Iran, N Afghanistan >> NW India
Phylloscopus bonelli (Bonelli's Warbler)

P.b.bonelli
S Europe, N Africa >> SW Sahara
P.b.orientalis
Balkans, Asia Minor >> NE Africa
Phylloscopus sibilatrix (Wood Warbler)
WC & NE Europe >> C Africa
Phylloscopus fuscatus (Dusky Warbler)
P.f.fuscatus
C & NE Asia >> India, China, Burma,
Thailand
P.f.weigoldi
S Tibet, E Himalayas, W China >> NE
India
P.f.robustus
N China >> S China, Indochina
Phylloscopus fuligiventer (Smoky Warbler)
P.f.fuligiventer
Bhutan, Sikkim, Himalayas >> NE India
P.f.tibetanus
SE Tibet, SW Sikiang >> N Assam
**Phylloscopus affinis (Tickell's Willow
Warbler)**
P.a.affinis
Himalayas to W China >> India, Burma
P.a.subaffinis
W & S China >> SE Asia
**Phylloscopus griseolus (Sulphur-bellied
Willow Warbler)**
C Asia, W Himalayas >> N India
**Phylloscopus armandii (Milne-Edwards'
Willow Warbler)**
P.a.armandii
Mongolia, E China >> Burma, N Laos
P.a.perplexus
Sikiang, W China >> S Burma, Thailand
**Phylloscopus schwarzi (Radde's Bush
Warbler)**
NC & NE Asia >> Burma, Thailand,
Indochina
**Phylloscopus pulcher (Orange-barred Willow
Warbler)**
P.p.kangrae
NW Himalayas
P.p.pulcher
N Burma, E Nepal, S Tibet >> N Thailand
**Phylloscopus maculipennis (Grey-faced
Willow Warbler)**
P.m.virens
? >> NW Himalayas
P.m.maculipennis
E Himalayas, SW China, N Burma, N
Indochina
**Phylloscopus proregulus (Pallas' Leaf
Warbler)**
P.p.proregulus
NE & C Asia, Sakhalin I, N China >> S
China, Indochina
P.p.simlaensis
Afghanistan, NW India, W Himalayas
P.p.chloronotus
E Himalayas, W China >> Thailand,
Malaysia

**Phylloscopus subviridis (Brooks's Willow
Warbler)**
Afghanistan, Pakistan >> N India
**Phylloscopus inornatus (Yellow-browed
Warbler)**
P.i.inornatus
N India, Assam, Bangladesh >> SE Asia
P.i.humei
C Asia, Himalayas, Altai >> N India
P.i.mandellii
Burma, Sikkim, S Tibet, W China >> N
Thailand
Phylloscopus borealis (Arctic Warbler)
P.b.talovka
N Russia, N Siberia, NW Mongolia >> SE
Asia, Philippine Is
P.b.transbaicalicus
E Siberia, N Mongolia >> SE Asia
P.b.borealis
NE Europe, N & NE Asia >> SE China,
Philippine Is
P.b.hylebata
S Asia, Mongolia, Sakhalin I >> Indochina
P.b.xanthodryas
Kamchatka, N Kurile Is >> Philippine Is,
Indochina
P.b.kennikotti
W Alaska >> Philippine Is
Phylloscopus trochiloides (Greenish Warbler)
P.t.viridanus
NE Europe, W & C Asia >> Pakistan, W
India
P.t.trochiloides
Himalayas, Tibet, W China >> N India to
Indochina
P.t.ludlowi
Ladakh, NW Himalayas >> S India
P.t.obscuratus
NW China >> Thailand, Indochina
Phylloscopus nitidus (Green Willow Warbler)
Caucasus, Iran, W India
**Phylloscopus plumbeitarsus (Two-barred
Greenish Warbler)** 136.1
E Asia >> Thailand, Indochina
**Phylloscopus tenellipes (Pale-legged Willow
Warbler)**
NE Asia, Japan >> SE Asia
**Phylloscopus borealoides (Japanese Pale-
legged Willow Warbler)**
Sakhalin, S Kurile Is, N Japan 136.13
**Phylloscopus magnirostris (Large-billed
Willow Warber)**
Himalayas, India, W China, E Tibet
Phylloscopus tytleri (Tytler's Willow Warbler)
Pakistan, India
**Phylloscopus occipitalis (Large Crowned
Willow Warbler)**
C Asia, Afghanistan >> W India
**Phylloscopus coronatus (Temminck's
Crowned Willow Warbler)**
NE Asia, Korea, Japan >> Indochina,
Malaysia

Phylloscopus ijimae (Ijima's Willow Warbler)
Izu Is >> N Philippine Is
Phylloscopus reguloides (Blyth's Crowned
Willow Warbler)
P.r.kashmiriensis
NW India >> N India
P.r.reguloides
NE India, Nepal, Bangladesh >> Burma
P.r.assamensis
Assam, N Burma, SE Tibet >> S Burma
P.r.claudiae
W China >> S China, SE Asia
P.r.fokiensis
S China
P.r.ticehursti
S Vietnam
Phylloscopus davisoni (White-tailed Willow
Warbler)
P.d.davisoni
E & S Sikang >> Burma, N Indochina
P.d.disturbans
SW China
P.d.ogilviegranti
SE China, N Vietnam
P.d.intensior
SE Thailand, N Cambodia
P.d.klossi
S Indochina
Phylloscopus cantator (Yellow-faced Leaf
Warbler)
P.c.cantator
N India, Burma >> NW Thailand
P.c.pernotus
N Laos
Phylloscopus ricketti (Black-browed Leaf
Warbler)
P.r.ricketti
S China, N Vietnam >> Laos, S Vietnam
P.r.goodsoni
S Hainan I
Phylloscopus olivaceus (Philippine Leaf
Warbler)
Philippine Is
Phylloscopus cebuensis (Dubois' Leaf
Warbler)
P.c.cebuensis
Cebu, Negros Is
P.c.luzonensis
N Luzon I
P.c.sorsogonensis
S Luzon I
Phylloscopus trivirgatus (Mountain Leaf
Warbler)
P.t.parvirostris
Malaysia
P.t.trivirgatus
Sumatra, Java, Bali, NW Borneo
P.t.kinabaluensis
NE Borneo
P.t.sarawacensis
W Borneo
P.t.nigrorum

Luzon, Negros, Mindoro Is
P.t.diuatae
NE Mindanao I
P.t.mindanensis
Mindanao I
P.t.malindangensis
Mt Malindang (Mindanao I)
P.t.flavostriatus
Mindanao I
P.t.peterseni
Palawan I
Phylloscopus sarasinorum (Sulawesi Leaf
Warbler)   136.9
P.s.sarasinorum
S Sulawesi
P.s.nesophilus
N & C Sulawesi
Phylloscopus presbytes (Timor Leaf Warbler)
P.p.presbytes
Timor I
P.p.floris
Flores I
Phylloscopus poliocephalus (Island Leaf
Warbler)   136.9
P.p.henrietta
N Halmahera I
P.p.waterstradti
Moluccas
P.p.everetti
Buru I
P.p.matthiae
St Matthias Is
P.p.avicola
Great Kai Is
P.p.ceramensis
Seram I
P.p.maforensis
Numfor I
P.p.misoriensis
Biak I
P.p.poliocephala
NW New Guinea
P.p.albigularis
WC New Guinea
P.p.paniaiae
W New Guinea
P.p.cyclopum
N New Guinea
P.p.giulianettii
C & SE New Guinea
P.p.hamlini
Goodenough I
P.p.becki
Guadalcanal, Malaita Is
P.p.bougainvillei
Bougainville I
P.p.pallescens
Kulambangra I
P.p.moorhousei
New Britain
P.p.leletensis
New Ireland

Phylloscopus makirensis **(San Cristobal Leaf Warbler)** 136.1
   San Cristobal I
Phylloscopus amoenus **(Kulambangra Warbler)**
   Kulambagra I

SEICERCUS
Seicercus burkii **(Yellow-eyed Flycatcher Warbler)**
  S.b.whistleri
   W Himalayas >> S India
  S.b.burkii
   Assam, SE Tibet, Nepal >> C India
  S.b.tephrocephalus
   N Thailand, Burma >> Indochina
  S.b.distinctus
   NW Thailand, W China >> S Vietnam
  S.b.valentini
   E Tibet, W & C China >> SW China
Seicercus xanthoschistos **(Grey-headed Flycatcher Warbler)**
  S.x.xanthoschistos
   W Himalayas
  S.x.jerdoni
   E Himalayas
  S.x.tephrodiras
   E Assam, Burma
  S.x.flavogularis
   Assam, Burma
Seicercus affinis **(Allied Flycatcher Warbler)**
  S.a.affinis
   E Himalayas to S Vietnam
  S.a.intermedius
   NW Fukien >> S China, Indochina
Seicercus poliogenys **(Grey-cheeked Flycatcher Warbler)**
   N India, N Burma >> N Vietnam
Seicercus castaniceps **(Chestnut-headed Flycatcher Warbler)**
  S.c.castaniceps
   Assam, Nepal
  S.c.collinsi
   N Thailand
  S.c.laurentei
   SE Yunnan
  S.c.stresemanni
   S Laos
  S.c.sinensis
   N Indochina, S & W China
  S.c.youngi
   S Thailand
  S.c.annamensis
   S Vietnam
  S.c.butleri
   Malaysia
  S.c.muelleri
   Sumatra
Seicercus montis **(Yellow-breasted Flycatcher Warbler)**
  S.m.davisoni
   Malaysia

  S.m.inornatus
   Sumatra
  S.m.montis
   Borneo
  S.m.xanthopygius
   Palawan I
  S.m.floris
   S Flores I
  S.m.paulinae
   Timor I
Seicercus grammiceps **(Sunda Flycatcher Warbler)**
  S.g.sumatrensis
   Sumatra
  S.g.grammiceps
   Java, Bali

TICKELLIA 136.1
Tickellia hodgsoni **(Broad-billed Flycatcher Warbler)**
  T.h.hodgsoni
   Darjeeling to W Burma
  T.h.tonkinensis
   SE Yunnan, N Laos, N Vietnam

ABROSCOPUS
Abroscopus albogularis **(White-throated Flycatcher Warbler)**
  A.a.albogularis
   Assam, Nepal, W Burma
  A.a.hugonis
   N Thailand
  A.a.fulvifacies
   S China, N Indochina
Abroscopus schisticeps **(Black-faced Flycatcher Warbler)**
  A.s.schisticeps
   Nepal to Sikkim
  A.s.flavimentalis
   SE Tibet, Burma
  A.s.ripponi
   W China, N Vietnam
Abroscopus superciliaris **(Yellow-bellied Flycatcher Warbler)**
  A.s.flaviventris
   Nepal to Assam
  A.s.superciliaris
   N India, Burma, W Thailand
  A.s.drasticus
   N India, SW Thailand
  A.s.smythiesi
   Burma
  A.s.euthymus
   N & C Vietnam
  A.s.bambusarum
   S Thailand
  A.s.sakaiorum
   Malaysia
  A.s.papilio
   Sumatra
  A.s.schwaneri
   Borneo

A.s.vordermani
Java

PARISOMA
Parisoma buryi (Yemen Tit Warbler)
S Arabia
Parisoma lugens (Brown Tit Warbler)
P.l.lugens
N Ethiopia
P.l.griseiventris
C Ethiopia
P.l.jacksoni
E Zaire, Tanzania, Zambia, Malawi
P.l.prigoginei
SE Zaire
Parisoma boehmi (Banded Tit Warbler)
P.b.somalicum
Ethiopia, N Somalia
P.b.marsabit
N Kenya
P.b.boehmi
E & S Kenya, N Tanzania
Parisoma layardi (Layard's Tit Warbler)
P.l.aridicola
Namibia, S Botswana, N & C Cape Province
P.l.barnesi
Lesotho
P.l.subsolana
interior Cape Province
P.l.layardi
SW Cape Province
Parisoma subcaeruleum (Southern Tit Warbler)
P.s.ansorgei
S Angola coast
P.s.cinerascens
Namibia to W Zimbabwe
P.s.orpheanum
C Natal, Zimbabwe, Orange Free State
P.s.subcaeruleum
Transvaal, S Natal, E Cape Province

SYLVIA
Sylvia atricapilla (Blackcap)
S.a.atlantis
Azores
S.a.gularis
Cape Verde Is
S.a.heineken
Madeira
S.a.atricapilla
W Europe, W Russia >> N & E Africa
S.a.dammholzi
Caucasus, Asia Minor, W Iran >> E Africa
S.a.koenigi
Balearic Is
S.a.pauluccii
Sardinia
Sylvia borin (Garden Warbler)
S.b.borin
Europe, W Russia >> WC Africa
S.b.woodwardi
W Siberia >> E & C Africa

Sylvia communis (Whitethroat)
S.c.communis
Europe, N Africa, Russia >> C & S Africa
S.c.icterops
Iran, Asia Minor >> E Africa
S.c.rubicola
C Siberia, Mongolia >> E Africa
Sylvia curruca (Lesser Whitethroat)
S.c.curruca
Europe, W Russia >> N & C Africa
S.c.telengitica
SW Altai, C Asia >> N India
S.c.halimodendri
Kazakhstan, SC Asia >> N India
S.c.minula
W Mongolia, Tibet, Pakistan, NW India
S.c.caucasica
Caucasus, Iran
S.c.margelanica
NE Tsinghai, NW China
S.c.jaxartica
C Asian deserts
S.c.chuancheica
W China
S.c.althaea
Transcaspia, Iran
S.c.monticola
S Russia, Turkestan
Sylvia nana (Desert Whitethroat)
S.n.nana
Iran, Afghanistan, Tibet, N India >> NE Africa
S.n.deserti
N & NE Africa
Sylvia nisoria (Barred Warbler)
S.n.nisoria
C & E Europe, W Russia >> E Africa
S.n.merzbacheri
N Iran, Afghanistan, C Asia >> E Africa
Sylvia hortensis (Orphean Warbler)
S.h.hortensis
S Europe, N & W Africa >> S Sahara
S.h.crassirostris
Asia Minor, Iran, Afghanistan, Arabia >> E Africa
S.h.jerdoni
SE Iran, Pakistan >> N India
S.h.balchanica
Iraq, W Iran >> ? NW India
Sylvia leucomelaena (Red Sea Warbler)
S.l.leucomelaena
Saudi Arabia coast
S.l.negevensis
S Israel
S.l.blanfordi
Eritrea, S Israel coast
S.l.somaliensis
NE Somalia
Sylvia rueppelli (Ruppell's Warbler)
SE Europe, NE Africa >> Sudan
Sylvia melanocephala (Sardinian Warbler)
S.m.melanocephala
S Europe, N Africa

S.m.leucogastra
  Canary Is
S.m.pasiphae
  Rhodes, Crete, Aegean Is
S.m.norrisae
  Egypt
S.m.momus
  Syria, Israel, Sinai
Sylvia melanothorax (Cyprus Warbler)
  Cyprus
Sylvia mystacea (Ménétries' Warbler)
  S Turkey, Iran, Afghanistan >> NE Africa
Sylvia cantillans (Subalpine Warbler)
S.c.cantillans
  SW Europe, NW Africa >> W Africa
S.c.albistriata
  SE Europe, Asia Minor, NE Africa >>
  Sahara oases
S.c.inornata
  N Africa
Sylvia conspicillata (Spectacled Warbler)
S.c.conspicillata
  S Europe, N Africa >> W Africa, Egypt
S.c.orbitalis
  Canary, Cape Verde Is
S.c.bella
  Madeira
Sylvia deserticola (Tristram's Warbler)
S.d.deserticola
  Algeria, Tunisia >> Sahara oases
S.d.maroccana
  W Morocco
S.d.ticehursti
  E Morocco, W Algeria
Sylvia undata (Dartford Warbler)
S.u.dartfordiensis
  S England, NW France
S.u.undata
  SW Europe
S.u.toni
  Portugal, S Spain, N Africa
Sylvia sarda (Marmora's Warbler)
S.s.balearica
  Balearic Is
S.s.sarda
  Corsica, Sardinia, S France, N Africa

REGULUS
Regulus ignicapillus (Firecrest)
R.i.ignicapillus
  C & S Europe, Asia Minor
R.i.madeirensis
  Madeira
R.i.balearicus
  Balearic Is, N Africa
R.i.teneriffae
  Canary Is
Regulus goodfellowi (Taiwan Firecrest)
  Taiwan
Regulus regulus (Goldcrest)
R.r.anglorum
  British Isles

R.r.regulus
  Europe, Asia Minor, Russia, WC Asia >>
  Mediterranean islands
R.r.azoricus
  San Miguel (Azores)
R.r.sanctaemariae
  Santa Maria (Azores)
R.r.inermis
  Azores Is
R.r.interni
  Corsica, Sardinia
R.r.buturlini
  Crimea, Caucasus >> N Iran
R.r.hyrcanus
  N Iran, E Turkey
R.r.coatsi
  W Siberia, Altai
R.r.tristis
  C Asia >> W Iran
R.r.himalayensis
  W Himalayas
R.r.sikkimensis
  Nepal, E Himalayas, NW China
R.r.japonensis
  NE Asia, , Japan, N China, Manchuria >>
  E China
R.r.yunnanensis
  SE Tibet, SW China
Regulus satrapa (Golden-crowned Kinglet)
R.s.amoenus
  W Canada, W USA >> SW USA
R.s.satrapa
  SE Canada, E USA >> NE Mexico, SE
  USA
R.s.olivaceus
  SE Alaska to Oregon >> S California
R.s.apache
  E & S Arizona >> Texas, New Mexico
R.s.aztecus
  NW Mexico
R.s.clarus
  S Mexico, Central America
Regulus calendula (Ruby-crowned Kinglet)
R.c.calendula
  N & E Canada, E USA >> Central America,
  S USA
R.c.grinnelli
  coast of Alaska, British Columbia >> W
  USA
R.c.obscurus
  Guadelupe I (Mexico)

LEPTOPOECILE
**Leptopoecile sophiae (Severtzov's Tit
Warbler)**
L.s.sophiae
  Pakistan, NW India, C Asia
L.s.stoliczkae
  SC Asia, W Gobi Desert
L.s.major
  E Tien Shan
L.s.obscura
  NW China, Tibet, E Himalayas

Leptopoecile elegans (Crested Tit Warbler)
  L.e.meissneri
    SE Tibet
  L.e.elegans
    S China

137. MUSCICAPIDAE (OLD WORLD FLYCATCHERS)

MELAENORNIS   137.1
Melaenornis semipartitus (Silverbird)   137.1
  M.s.semipartitus
    N Ethiopia, Sudan, NW Uganda
  M.s.kavirondensis
    NE Uganda, W Kenya, W Tanzania
Melaenornis pallidus (Pale Flycatcher) 137.1
  M.p.pallidus
    Senegal to Ethiopia, N Zaire
  M.p.parvus
    SW Ethiopia, NW Uganda
  M.p.bowdleri
    Eritrea, C Ethiopia
  M.p.bafirawari
    NE Kenya
  M.p.duyerali
    C Somalia
  M.p.subalaris
    E Kenya, NE Tanzania
  M.p.erlangeri
    S Somalia
  M.p.modestus
    Guinea to S Zaire
  M.p.murinus
    Congo, Angola to W Kenya & NW
    Zimbabwe
  M.p.aquaemontis
    Namibia
  M.p.griseus
    C Tanzania to E Zambia & N Malawi
  M.p.divisus
    SE Zambia to Mozambique & N Transvaal
  M.p.sibilans
    S Mozambique, N Natal
Melaenornis infuscatus (African Brown
  Flycatcher)   137.1
  M.i.benguellensis
    S Angola, NW Namibia
  M.i.namaquensis
    Namibia, W Botswana, W Cape Province
  M.i.placidus
    Botswana, W Transvaal, N Cape Province
  M.i.seimundi
    N Cape Province, W Orange Free State
  M.i.infuscatus
    W Cape Province
Melaenornis mariquensis (Mariqua
  Flycatcher)   137.1
  M.m.acaciae
    SW Angola, Namibia, NW Cape Province
  M.m.mariquensis
    SW Zambia, S Botswana, NE Cape
    Province

  M.m.territinctus
    NE Namibia, NW Botswana
Melaenornis microrhynchus (Grey
  Flycatcher)   137.1
  M.m.microrhynchus
    SW Kenya, W Tanzania, NE Zambia
  M.m.pumilus
    C Ethiopia, N Somalia
  M.m.neumanni
    SE Sudan & C Somalia to NE Uganda & N
    Kenya
  M.m.burae
    SE Somalia, N Kenya
  M.m.taruensis
    SE Kenya
Melaenornis chocolatinus (Abyssinian Slaty
  Flycatcher)
  M.c.chocolatinus
    N Ethiopia, S Eritrea
  M.c.reichenowi
    W Ethiopia
Melaenornis fischeri (White-eyed Slaty
  Flycatcher)   137.1
  M.f.toruensis
    E Zaire, W Uganda
  M.f.semicinctus
    E Zaire
  M.f.nyikensis
    N Tanzania, N Malawi, E Zambia
  M.f.fischeri
    SE Sudan to Tanzania
  M.f.ufipae
    SW Tanzania, E Zaire
Melaenornis brunneus (Angolan Flycatcher)
                                                   137.1
  M.b.brunneus
    W Angola lowlands
  M.b.bailundensis
    W Angola highlands
Melaenornis edolioides (Black Flycatcher)
  M.e.edolioides
    Senegal to Cameroun
  M.e.lugubris
    Ethiopia, Sudan to N Tanzania
  M.e.schistacea
    E Ethiopia, N Kenya
Melaenornis pammelaina (South African
  Black Flycatcher)
  M.p.pammelaina
    Kenya, S Zaire to Natal & E Cape Province
  M.p.diabolicus
    S Zaire to Namibia & W Transvaal
Melaenornis ardesiaca (Berlioz' Black
  Flycatcher)
    E Zaire
Melaenornis annamarulae (Liberian Black
  Flycatcher)
    Liberia
Melaenornis ocreata (Forest Flycatcher) 137.1
  M.o.kelsalli
    Sierra Leone
  M.o.prosphora
    Liberia to Ghana

M.o.ocreata
Cameroun to Zaire, Uganda
Melaenornis cinerascens (White-browed Forest Flycatcher)   137.1
Senegal to W Zaire
Melaenornis silens (Fiscal Flycatcher)
M.s.silens
S Cape Province to S Mozambique
M.s.lawsoni
N Cape Province, W Transvaal, S Botswana

Rhinomyias addita (Buru Jungle Flycatcher)
Buru I
Rhinomyias oscillans (Flores Jungle Flycatcher)
R.o.oscillans
Flores
R.o.stresemanni
Sumba
Rhinomyias brunneata (White-gorgetted Jungle Flycatcher)
R.b.brunneata
SE China >> Thailand, Malaysia
R.b.nicobarica
?S China >> Nicobar Is
Rhinomyias olivacea (Olive-backed Jungle Flycatcher)
R.o.olivacea
Malaysia, Burma, Sumatra, Java, N Borneo
R.o.perolivacea
Banguey I
Rhinomyias umbratilis (White-throated Jungle Flycatcher)
Malaysia
Rhinomyias ruficauda (Rufous-tailed Jungle Flycatcher)
R.r.isola
Sarawak
R.r.ruficrissa
Kinabalu
R.r.ocularis
Sulu, Pangamican, Tawitawi Is
R.r.ruficauda
Basilan I
R.r.zamboanga
W Mindanao I
R.r.boholensis
Bohol I
R.r.samarensis
Leyte, Samar, E Mindanao Is
Rhinomyias colonus (Sula Jungle Flycatcher)
R.c.colonus
Sula Arch
R.c.pelingensis
Peleng I
R.c.subsolanus
E Sulawesi
Rhinomyias gularis (White-browed Jungle Flycatcher)
R.g.gularis
Borneo

R.g.albigularis
Negros I
Rhinomyias insignis (Lepanto Jungle Flycatcher)   137.1
N Luzon
Rhinomyias goodfellowi (Mindanao Jungle Flycatcher)   137.1
Mindanao

Muscicapa striata (Spotted Flycatcher)
M.s.striata
Europe, Asia Minor >> S Africa
M.s.neumanni
W & S Asia >> E Africa
M.s.tyrrhenica
Corsica, Sardinia
M.s.balearica
Balearic Is >> W & S Africa
M.s.sarudnyi
N Iran, Caucasus, Afghanistan >> ?E Africa
M.s.inexpectata
Crimea
M.s.mongola
N Mongolia, SE Transbaikalia
Muscicapa gambagae (Gambaga Spotted Flycatcher)
Ghana to Somalia, Arabia
Muscicapa griseisticta (Grey-streaked Flycatcher)
NE Asia >> New Guinea, Philippine Is
Muscicapa sibirica (Siberian Flycatcher)
M.s.sibirica
E Asia, Japan >> Indochina, Gtr Sunda Is
M.s.gulmergi
Pakistan, W Himalayas
M.s.cacabata
E Himalayas, Tibet >> S Burma, S Thailand
M.s.rothschildi
SW China >> Indochina, Malaysia
Muscicapa dauurica (Brown Flycatcher) 137.2
M.d.dauurica
NE Asia, Japan >> India, S China, Gtr Sunda Is
M.d.williamsoni
? >> SE Asia
M.d.siamensis
N Thailand
M.d.randi
Philippine Is
M.d.umbrosa
NE Borneo
M.d.segregata
Sumba I
Muscicapa muttui (Brown-breasted Flycatcher)
NE India >> SW India
Muscicapa ruficauda (Rufous-tailed Flycatcher)
W Himalayas >> W India

Muscicapa ferruginea (Ferruginous
Flycatcher)
    Himalayas, SE Asia >> Sumatra,
      Philippine Is
Muscicapa sordida (Sri Lankan Dusky Blue
Flycatcher)   137.1
    Sri Lanka
Muscicapa thalassina (Indian Verditer
Flycatcher)   137.1
    M.t.thalassina
      Himalayas to Indochina >> SW India
    M.t.thalassoides
      S Thailand, Malaysia
Muscicapa panayensis (Philippine Verditer
Flycatcher)   137.1
    M.p.septentrionalis
      N Sulawesi
    M.p.meridionalis
      S Sulawesi
    M.p.obiensis
      Obi I, Moluccas
    M.p.harterti
      Seram I
    M.p.panayensis
      Negros, Panay Is
    M.p.nigrimentalis
      Luzon, Mindoro Is
    M.p.nigriloris
      Mindanao I
Muscicapa albicaudata (Nilgiri Verditer
Flycatcher)   137.1
    SW India
Muscicapa indigo (Indigo Flycatcher)   137.1
    M.i.ruficrissa
      Sumatra
    M.i.indigo
      Java
    M.i.cerviniventris
      Borneo
Muscicapa infuscata (African Sooty
Flycatcher)
    S Nigeria to NW Zambia & SW Sudan
Muscicapa ussheri (Ussher's Dusky
Flycatcher)
    Sierra Leone to Nigeria
Muscicapa boehmi (Böhm's Flycatcher) 137.1
    Angola to W Tanzania, N Mozambique
Muscicapa aquatica (Swamp Flycatcher)
    M.a.aquatica
      Senegal to Sudan
    M.a.infulatus
      S Sudan, NE Zaire, W Kenya
    M.a.lualabae
      SE Zaire
    M.a.grimwoodi
      C Zambia
Muscicapa olivascens (Olivaceous
Flycatcher)
    Liberia to E Zaire
Muscicapa lendu (Chapin's Flycatcher)
    M.l.lendu
      E Zaire to E Uganda

    M.l.itombwensis
      Itombwe Mts (Zaire)
Muscicapa adusta (Dusky Flycatcher)
    M.a.albiventris
      Cameroun
    M.a.poensis
      Fernando Po I
    M.a.kumboensis
      W Cameroun
    M.a.okuensis
      Oku (Cameroun)
    M.a.grotei
      Central African Republic
    M.a.pumilis
      NE Zaire, S Sudan, Uganda
    M.a.minima
      NE Ethiopia
    M.a.subadusta
      S Zaire to Zimbabwe, Mozambique
    M.a.angolensis
      Angola, SW Zaire
    M.a.marsabit
      N Kenya
    M.a.murina
      SE Kenya
    M.a.fuelleborni
      Tanzania
    M.a.roehli
      E Tanzania
    M.a.mesica
      C & E Zimbabwe, SE Zambia, W
      Mozambique
    M.a.fuscula
      coast of Natal, E Cape Province
    M.a.adusta
      Cape Province to Transvaal
Muscicapa epulata (Little Grey Flycatcher)
    Liberia to Gabon, E Zaire
Muscicapa sethsmithi (Yellow-footed
Flycatcher)
    Cameroun, Gabon to Uganda
Muscicapa comitata (Dusky Blue Flycatcher)
    M.c.aximensis
      Sierra Leone to S Nigeria
    M.c.camerunensis
      Cameroun
    M.c.comitata
      Gabon to Uganda, Angola
Muscicapa tessmanni (Tessman's Flycatcher)
    Ivory Coast to N Zaire
Muscicapa cassini (Cassin's Grey Flycatcher)
    Sierra Leone to Uganda, Zambia
Muscicapa coerulescens (Ashy Flycatcher)
    M.c.nigrorum
      Guinea to Togo
    M.c.brevicauda
      S Nigeria to Gabon, Sudan, N Angola
    M.c.cinereolus
      S Angola to Kenya, Mozambique
    M.c.impavida
      N Namibia to N Mozambique, W Transvaal
    M.c.vulturna
      Mozambique to E Transvaal, Swaziland

M.c.coerulescens
  E Transvaal, Natal, E Cape Province
Muscicapa griseigularis (Grey-throated Flycatcher)   137.1
M.g.parelii
  Ivory Coast
M.g.griseigularis
  SE Nigeria, E Zaire, Uganda, NW Angola

MYIOPARUS
Myioparus plumbeus (Grey Tit Flycatcher)
M.p.plumbeus
  Senegal to Ethiopia, Uganda, Tanzania
M.p.orientalis
  Kenya to Angola, Natal
M.p.catoleucum
  C Tanzania to Natal & Angola

HUMBLOTIA
Humblotia flavirostris (Humblot's Flycatcher)
  Comoro Is

FICEDULA
Ficedula hypoleuca (Pied Flycatcher)
F.h.hypoleuca
  N Europe, W Siberia >> NC Africa
F.h.tomensis
  SW Asia >> C Africa
F.h.speculigera
  N Africa
Ficedula albicollis (Collared Flycatcher)   137.1
F.a.albicollis
  C & E Europe >> W, C & E Africa
F.a.semitorquata
  SE Europe to NE Iran >> E Africa
Ficedula zanthopygia (Yellow-rumped Flycatcher)
  NE Asia >> Malaysia, Sumatra, Java
Ficedula narcissina (Narcissus Flycatcher)
F.n.narcissina
  Sakhalin, Japan >> Philippine Is, N Borneo
F.n.elisae
  C China >> Thailand, Malaysia
F.n.owstoni
  S Riukiu Is
Ficedula mugimaki (Mugimaki Flycatcher)
  C & E Asia >> Philippine Is
Ficedula hodgsonii (Rusty-breasted Blue Flycatcher)
  Nepal to W China
Ficedula dumetoria (Orange-breasted Flycatcher)
F.d.muelleri
  Malaysia, Sumatra, Borneo
F.d.dumetoria
  Java, Lombok, Sumbawa, Flores Is
F.d.riedeli
  Tanimber Is
Ficedula strophiata (Orange-gorgetted Flycatcher)
F.s.strophiata
  Himalayas to N Thailand, S China >> N Indochina

F.s.fuscogularis
  S Vietnam
Ficedula parva (Red-breasted Flycatcher)
F.p.parva
  C & E Europe, W Himalayas >> N India
F.p.albicilla
  C & NE Asia >> Burma, S China
Ficedula subrubra (Kashmir Flycatcher)   137.1
  Kashmir >> Sri Lanka
Ficedula monileger (White-gorgetted Flycatcher)
F.m.monileger
  Himalayas
F.m.leucops
  NE India to N Vietnam
F.m.gularis
  Arakan (Burma)
Ficedula solitaris (Rufous-browed Flycatcher)
F.s.submoniliger
  SE Burma to S Vietnam   137.1
F.s.malayana
  S Thailand, Malaysia
F.s.solitaris
  Sumatra
Ficedula hyperythra (Thicket Flycatcher)
F.h.hyperythra
  Himalayas, Burma, Thailand
F.h.annamensis
  C Vietnam
F.h.sumatrana
  Malaysia, Sumatra, Borneo
F.h.mjobergi
  W Borneo
F.h.vulcani
  Java, Bali
F.h.mindorensis
  Mindoro I
F.h.calayensis
  Calayan I
F.h.nigrorum
  Negros I
F.h.montigena
  Mindanao I
F.h.daggayana
  N Mindanao I
F.h.malindangensis
  SE Mindanao I
F.h.jugosae
  C & S Sulawesi
F.h.annalisa
  N Sulawesi
F.h.clarae
  Timor I
F.h.audacis
  Babar I
F.h.negroides
  Seram I
F.h.pallidipectus
  Batjan, Buru Is
F.h.alifurus
  Buru I
F.h.luzoniensis
  Luzon I

F.h.rara
Palawan I
Ficedula rufigula (White-vented Flycatcher)
Sulawesi
Ficedula basilanica (Little Slaty Flycatcher)
F.b.basilanica
Basilan, Mindanao Is
F.b.samarensis
Leyte, Samar Is
Ficedula buruensis (Cinnamon-chested Flycatcher)
F.b.buruensis
Buru I
F.b.ceramensis
Seram I
F.b.siebersi
Kai Is
Ficedula henrici (Damar Flycatcher)
Damar I
Ficedula harterti (Hartert's Flycatcher)
Sumba I
Ficedula platenae (Palawan Flycatcher)
Palawan I
Ficedula crypta (Vaurie's Flycatcher)
F.c.crypta
Mindanao I
F.c.disposita
Luzon I
Ficedula bonthaina (Lompobattang Flycatcher)
Sulawesi
Ficedula westermanni (Little Pied Flycatcher)
Himalayas to W China, SE Asia, Moluccas
Ficedula superciliaris (White-browed Blue Flycatcher)
F.s.superciliaris
W Himalayas, C India
F.s.aestigma
E Himalayas, Tibet, W China
Ficedula tricolor (Slaty Blue Flycatcher)
F.t.tricolor
W Himalayas
F.t.minuta
E Himalayas, SE Tibet, S China
F.t.cerviniventris
Assam, Burma
F.t.diversa
W & C China >> N Indochina
Ficedula sapphira (Sapphire-headed Flycatcher)
F.s.sapphira
E Himalayas, N Laos, S China
F.s.laotiana
S Laos
F.s.tienchuanensis
NW China
Ficedula nigrorufa (Black & Orange Flycatcher)
SW India
Ficedula timorensis (Black-banded Flycatcher)
Timor I

CYANOPTILA
Cyanoptila cyanomelaena (Blue & White Flycatcher)
C.c.cumatilis
NE Asia >> S Burma, S Thailand, Gtr Sunda Is
C.c.cyanomelaena
Japan >> Philippine Is, Borneo, Indochina

NILTAVA
Niltava grandis (Large Niltava)
N.g.grandis
E Himalayas to N Vietnam
N.g.decipiens
Sumatra, S Thailand, Malaysia
N.g.griseiventris
SE Yunnan
N.g.decorata
SE Vietnam
Niltava macgregoriae (Small Niltava)
N.m.macgregoriae
W & C Himalayas
N.m.signata
E Himalayas, Assam
Niltava davidi (Fukien Niltava)
S China >> Indochina, SE Thailand
Niltava sundara (Rufous-bellied Niltava)
N.s.whistleri
W Himalayas
N.s.sundara
E Himalayas, SW China, N Laos >> Bangladesh, N Thailand
N.s.denotata
S China >> N Thailand, N Laos
Niltava sumatrana (Sumatran Niltava)
Malaysia, Sumatra
Niltava vivida (Vivid Niltava)
N.v.oatesi
NE India to N Vietnam
N.v.vivida
Taiwan
Niltava hyacinthina (Blue-backed Niltava)
N.h.hyacinthina
Timor, Semau Is
N.h.kuhni
Wetar I
Niltava hoevelli (Blue-fronted Niltava)
C Sulawesi
Niltava sanfordi (Matinan Niltava)
N Sulawesi
Niltava concreta (White-tailed Niltava)
N.c.cyanea
NE India, Burma
N.c.concreta
Malaysia, Sumatra
N.c.everetti
Borneo
Niltava ruecki (Rueck's Niltava)
Malaysia
Niltava herioti (Blue-breasted Niltava)
N.h.herioti
N & C Luzon I

N.h.camarinensis
  S Luzon I
Niltava hainana (Grant's Niltava)
  S China, Thailand, Indochina
Niltava pallipes (White-bellied Niltava)
  SW India
Niltava poliogenys (Brooks' Niltava)
N.p.poliogenys
  C Himalayas, NE India
N.p.vernayi
  EC India
N.p.cachariensis
  E Himalayas, NW Burma
N.p.laurentei
  SE Yunnan
Niltava unicolor (Pale Niltava)
N.u.unicolor
  Himalayas, Burma, N Laos
N.u.diaoluoensis
  Hainan I
N.u.harterti
  Borneo, Malaysia, Sumatra, Java
Niltava rubeculoides (Blue-throated Niltava)
N.r.rubeculoides
  Himalayas, Burma, India >> S India
N.r.dialilaema
  Burma, N Thailand
N.r.rogersi
  Burma
N.r.glaucicomans
  S Thailand, S & C China >> Malaysia
N.r.klossi
  S Vietnam, S Laos, E Thailand
Niltava banyumas (Hill Blue Niltava)
N.b.magnirostris
  E Himalayas, Assam, Burma >> S
  Thailand
N.b.whitei
  Burma to N Vietnam
N.b.lekhakuni
  E Thailand
N.b.deignani
  SE Thailand
N.b.coerulifrons
  Malaysia
N.b.liga
  W Java
N.b.banyumas
  C & E Java
N.b.coeruleata
  Borneo
N.b.lemprieri
  Palawan, Balabac Is
Niltava superba (Bornean Niltava)
  Borneo
Niltava caerulata (Large-billed Niltava)
N.c.albiventer
  Sumatra
N.c.rufifrons
  W Borneo
N.c.caerulata
  N & E Borneo
Niltava turcosa (Malaysian Niltava)

N.t.rupatensis
  Malaysia, Sumatra, W Borneo
N.t.turcosa
  C & E Borneo
Niltava tickelliae (Tickell's Niltava)
N.t.tickelliae
  NE, C & S India, S Yunnan
N.t.jerdoni
  Sri Lanka
N.t.sumatrensis
  Malaysia, Sumatra
N.t.indochina
  Indochina, S Burma, Thailand
N.t.lampra
  Anambas Is
Niltava rufigastra (Mangrove Niltava)
N.r.rufigastra
  Malaysia, Sumatra, Borneo
N.r.lepidula
  Karimon, Java
N.r.rhizophorae
  W Java
N.r.karimatensis
  Karimata Is
N.r.blythi
  Luzon, Polillo Is
N.r.marinduquensis
  Marinduque I
N.r.philippinensis
  C & S Philippine Is, N Sulu Arch
N.r.mindorensis
  Mindoro I
N.r.litoralis
  Palawan I, S Sulu Arch
N.r.omissa
  Sulawesi
N.r.peromissa
  Salayer I
N.r.djampeana
  Djampea I
N.r.kalaoensis
  Kalai I
Niltava hodgsoni (Pygmy Blue Flycatcher)
                                                    137.1

N.h.hodgsoni
  C Himalayas to Thailand, SW China
N.h.sondaica
  Malaysia, Sumatra, Borneo

CULICICAPA
**Culicicapa ceylonensis (Grey-headed Canary
  Flycatcher)**
C.c.calochrysea
  N India, Burma to Indochina, Malaysia
C.c.ceylonensis
  S India, Sri Lanka to Sumatra, Java, Borneo
C.c.connectens
  Sumba Is
C.c.sejuncta
  Flores, Lombok Is
**Culicicapa helianthea (Citrine Canary
  Flycatcher)**

C.h.helianthea
  Banggai I, Sulawesi, Salayer I
C.h.panayensis
  Leyte, Mindanao, Palawan Is
C.h.mayri
  Bongao, Tawitawi Is
C.h.septentrionalis
  NW Luzon I
C.h.zimmeri
  C & S Luzon I

138. PLATYSTEIRIDAE (WATTLE-EYES, PUFF-BACK FLYCATCHERS)

BIAS
Bias flammulatus (African Shrike Flycatcher)
                                                    138.1
  B.f.flammulatus
    Sierra Leone to Gabon, Fernando Po I
  B.f.carolathi
    N Angola
  B.f.aequatorialis
    NE Angola to Sudan, Uganda
Bias musicus (Black & White Flycatcher)
  B.m.musicus
    Sierra Leone to C Zaire, N Angola
  B.m.changamwensis
    E Kenya to Mozambique
  B.m.clarens
    Mozambique, S Malawi, E Zimbabwe

PSEUDOBIAS
Pseudobias wardi (Ward's Flycatcher)
  E Madagascar

BATIS
Batis diops (Ruwenzori Puff-back Flycatcher)
  E Zaire, S Uganda
Batis margaritae (Boulton's Puff-back Flycatcher)
  B.m.margaritae
    S Angola
  B.m.kathleenae
    NW Zambia
Batis mixta (Short-tailed Puff-back Flycatcher)
  B.m.ultima
    coast of Kenya, NE Tanzania
  B.m.mixta
    N Tanzania to N Malawi
  B.m.reichenowi
    SE Tanzania
Batis dimorpha (Malawi Puff-back Flycatcher)
                                                    138.1
  B.d.sola
    N Malawi, E Zambia
  B.d.dimorpha
    C & S Malawi
Batis capensis (Cape Puff-back Flycatcher)
  B.c.kennedyi
    SW Zimbabwe

B.c.erythrophthalma
  E Zimbabwe, W Mozambique
B.c.hollidayi
  S Mozambique, N Natal, N & E Transvaal
B.c.capensis
  SW Cape Province to S Natal
Batis fratrum (Zululand Puff-back Flycatcher)
  B.f.sheppardi
    S Malawi, N Mozambique
  B.f.fratrum
    S Mozambique, N Natal
Batis molitor (Chin Spot Puff-back Flycatcher)
  B.m.pintoi
    Angola, SW Zaire, NW Zambia
  B.m.puella
    E Zaire, Uganda, W Kenya, W Tanzania
  B.m.palliditergum
    S Zaire to Namibia, N Cape Province
  B.m.molitor
    E Cape Province to S Mozambique
Batis soror (Paler Chin Spot Puff-back Flycatcher)
  E Kenya to Mozambique
Batis pirit (Pirit Puff-back Flycatcher)
  B.p.affinis
    W Angola, N Namibia, W Botswana
  B.p.pirit
    C Botswana, SW Transvaal, W Cape Province
Batis senegalensis (Senegal Puff-back Flycatcher)
  Senegal to Nigeria
Batis orientalis (Grey-headed Puff-back Flycatcher)
  B.o.chadensis
    Niger to W Sudan
  B.o.lynesi
    N Sudan
  B.o.orientalis
    Ethiopia, E Sudan, N Kenya
Batis minor (Black-headed Puff-back Flycatcher)
  B.m.erlangeri
    Cameroun & Angola to Somalia
  B.m.minor
    Somalia
Batis perkeo (Pygmy Puff-back Flycatcher)
  S Ethiopia, Kenya, Somalia
Batis minulla (Angola Puff-back Flycatcher)
  Cameroun to Angola
Batis minima (Verreaux's Puff-back Flycatcher)
  Gabon
Batis ituriensis (Chapin's Puff-back Flycatcher)   138.3
  E Zaire
Batis poensis (Fernando Po Puff-back Flycatcher)
  Fernando Po I
Batis occultus (Lawson's Puff-back Flycatcher)   138.2
  Ivory Coast to Gabon

PLATYSTEIRA

Platysteira cyanea (Brown-throated Wattle-eye)
P.c.cyanea
Senegal to Gabon, Central African Republic
P.c.nyanzae
N Zaire, Uganda, S Sudan, Kenya
P.c.aethiopica
E Ethiopia
Platysteira albifrons (White-fronted Wattle-eye)
N Angola, SW Zaire
Platysteira peltata (Black-throated Wattle-eye)
P.p.cryptoleuca
Somalia to N Mozambique
P.p.mentalis
Angola to Zambia & Uganda
P.p.peltata
SE Africa
Platysteira laticincta (Bamenda Wattle-eye)
138.1
Bamenda Mts
Platysteira tonsa (White-spotted Wattle-eye)
S Nigeria to N Zaire
Platysteira castanea (Chestnut Wattle-eye)
P.c.hormophora
Sierra Leone to Togo
P.c.castanea
S Nigeria to SE Sudan, N Angola
Platysteira blissetti (Red-cheeked Wattle-eye)
Guinea to W Cameroun
Platysteira chalybea (Reichenow's Wattle-eye)
138.1
Cameroun to W Angola
Platysteira jamesoni (Jameson's Wattle-eye)
138.1
E Zaire to Uganda, W Kenya, NW Tanzania
Platysteira concreta (Yellow-bellied Wattle-eye)
P.c.concreta
Sierra Leone to Ghana
P.c.harterti
Cameroun, Gabon
P.c.kumbaensis
SW Cameroun
P.c.ansorgei
Angola
P.c.graueri
E Zaire, Uganda
P.c.silvae
W Kenya
P.c.kungwensis
E Zaire

139. MALURIDAE (AUSTRALASIAN WRENS)

CLYTOMIAS
Clytomias insignis (Rufous Wren Warbler)
C.i.insignis
NW New Guinea
C.i.oorti
C & SE New Guinea

MALURUS
Malurus wallacii (Wallace's Wren Warbler)
139.1
M.w.wallacii
Misol, Japen Is, N New Guinea
M.w.capillatus
Aru Is, S New Guinea
Malurus grayi (Broad-billed Wren Warbler)
139.1
M.g.grayi
N New Guinea
M.g.campbelli
C New Guinea
139.3
Malurus alboscapulatus (Black & White Wren)
M.a.alboscapulatus
NW New Guinea
M.a.aida
NW New Guinea
M.a.tappenbecki
NE New Guinea
M.a.randi
W New Guinea
M.a.balim
WC New Guinea
M.a.lorentzi
SW New Guinea
M.a.dogwa
S New Guinea
M.a.moretoni
SE New Guinea
M.a.naimii
SE New Guinea
M.a.mafulu
SE New Guinea
M.a.kutubu
S New Guinea
Malurus melanocephalus (Red-backed Wren)
M.m.cruentatus
N Australia
M.m.melanocephalus
S Queensland, N New South Wales
Malurus leucopterus (White-winged Wren)
M.l.leucopterus
Dirk Hartog I
M.l.edouardi
Barrow I
M.l.leuconotus
SW, S & EC Australia
Malurus cyaneus (Blue Wren)
M.c.cyanochlamys
S Queensland to Victoria, E South Australia
M.c.cyaneus
Tasmania
Malurus splendens (Banded Wren)
M.s.splendens
S Western Australia
M.s.aridus
EC Western Australia
M.s.callainus
S Northern Territory, N South Australia
M.s.whitei
S Queensland

M.s.melanotus
SE South Australia, NW Victoria
Malurus lamberti (Variegated Wren)
M.l.dulcis
NW Western Australia, Northern Territory
M.l.rogersi
W Western Australia
M.l.assimilis
all interior of Australia
M.l.lamberti
E Queensland, E New South Wales
Malurus pulcherrimus (Blue-breasted Wren)
Western & South Australia
Malurus elegans (Red-winged Wren)
SW Western Australia
Malurus coronatus (Purple-crowned Wren)
M.c.coronatus
N Northern Territory, N Western Australia
M.c.macgillivrayi
NE Northern Territory, NW Queensland
Malurus cyanocephalus (Blue Wren Warbler)
139.1
M.c.cyanocephalus
W New Guinea, Salawati, Japen Is
M.c.mysorensis
Biak I
M.c.bonapartii
Aru Is, S New Guinea

STIPITURUS
Stipiturus malachurus (Southern Emu Wren)
S.m.hartogi
Dirk Hartog I
S.m.westernensis
S Western Australia
S.m.malachurus
SE South Australia, E New South Wales,
Victoria
S.m.littleri
Tasmania
S.m.intermedius
S Mt Lofty Mts
S.m.halmaturinus
Kangaroo I
S.m.parimeda
S Eyre peninsula
Stipiturus mallee (Mallee Emu Wren) 139.1
NW Victoria, E South Australia
Stipiturus ruficeps (Rufous-crowned Emu Wren)
W & C Australia

AMYTORNIS
Amytornis textilis (Thick-billed Grass Wren)
A.t.textilis
Western Australia, W South Australia
A.t.myall
SC South Australia
A.t.modestus
Central Australia
Amytornis purnelli (Dusky Grass Wren)
A.p.purnelli
Northern Territory

A.p.ballarae
NW Queensland
Amytornis housei (Black Grass Wren)
N Western Australia
Amytornis woodwardi (White-throated Grass Wren)
Northern Territory
Amytornis dorotheae (Red-winged Grass Wren)
Northern Territory
Amytornis striatus (Striped Grass Wren)
A.s.whitei
NW Western Australia, S Northern Territory
A.s.merrotsyi
N South Australia
A.s.striatus
NW New South Wales, C Victoria
A.s.owensi
NW Victoria
Amytornis barbatus (Grey Grass Wren)
A.b.barbatus
New South Wales
A.b.diamantina
Sth Australia
Amytornis goyderi (Eyrean Grass Wren)
South Australia

140. ACANTHIZIDAE (THORNBILLS, FLYEATERS) 140.1

DASYORNIS
Dasyornis brachypterus (Common Bristlebird)
D.d.brachypterus
SE Queensland to E Victoria
D.b.longirostris
SW Australia
Dasyornis broadbenti (Rufous Bristlebird)
D.b.litoralis
SW Western Australia
D.b.broadbenti
South Australia, Victoria
D.b.whitei
South Australia

PYCNOPTILUS
Pycnoptilus floccosus (Pilot Bird)
SE Australia

ORIGMA
Origma solitaria (Rock Warbler)
E New South Wales

CRATEROSCELIS
Crateroscelis gutturalis (Fern Wren)
NE Queensland
Crateroscelis murina (Lowland Mouse Warbler)
C.m.murina
Japen I, New Guinea
C.m.monacha
Aru Is
C.m.pallida
SE New Guinea

C.m.capitalis
  Waigeu I
C.m.fumosa
  Misol I
Crateroscelis nigrorufa (Mid-mountain Mouse Warbler)
C.n.blissi
  N New Guinea
C.n.nigrorufa
  SE New Guinea
Crateroscelis robusta (Mountain Mouse Warbler)
C.r.peninsularis
  NW New Guinea
C.r.ripleyi
  NW New Guinea
C.r.bastille
  coast of N New Guinea
C.r.deficiens
  N New Guinea
C.r.sanfordi
  C New Guinea
C.r.robusta
  SE New Guinea

SERICORNIS
Sericornis citreogularis (Yellow-throated Sericornis)
S.c.cairnsi
  N Queensland
S.c.citreogularis
  NE Australia
Sericornis maculatus (Spotted Scrub Wren)
140.1
S.m.balstoni
  C coast of W Australia
S.m.maculatus
  SW Australia
S.m.mondraini
  Recherche Arch
S.m.osculans
  coast of Western & South Australia
Sericornis humilis (Tasmanian Sericornis)
140.1
S.h.humilis
  Tasmania
S.h.tregellasi
  King Is
Sericornis frontalis (White-browed Sericornis) 140.1
S.f.longirostris
  coast from SE South Australia to Melbourne
S.f.gularis
  Kent , Flinders Is
S.f.insularis
  Forsyth I
S.f.frontalis
  SE Victoria to C New South Wales
S.f.laevigaster
  E New South Wales to SE Queensland
S.f.herbertoni
  NE Queensland
Sericornis beccarii (Little Sericornis)

S.b.wondiwoi
  NW New Guinea
S.b.cyclopum
  coast of N New Guinea
S.b.weylandi
  Weyland Mts
S.b.imitator
  Arfak Mts
S.b.idenburgi
  NC New Guinea
S.b.jobiensis
  Japen I
S.b.boreonesioticus
  N New Guinea
S.b.virgatus
  NC New Guinea   140.1
S.b.pontifex
  NC New Guinea
S.b.randi
  S New Guinea
S.b.beccarii
  Aru Is
S.b.minimus
  Cape York
S.b.dubius
  N Queensland
Sericornis nouhuysi (Large Mountain Sericornis)
S.n.cantans
  NW New Guinea
S.n.adelberti
  Adelbert Mts
S.n.nouhuysi
  Weyland & Snowy Mts
S.n.stresemanni
  Schrader Mts
S.n.oorti
  Huon Peninsula
S.n.monticola
  SE New Guinea
Sericornis magnirostris (Large-billed Sericornis)
S.m.viridior
  Queensland
S.m.magnirostris
  New South Wales, Victoria
Sericornis keri (Atherton Sericornis)
  NE Queensland
Sericornis spilodera (Pale-billed Sericornis)
S.s.spilodera
  Japen I, NW New Guinea
S.s.granti
  WC New Guinea
S.s.wuroi
  S New Guinea
S.s.guttatus
  SE New Guinea
S.s.ferrugineus
  Waigeu I
S.s.aruensis
  Aru Is
S.s.batantae
  Batanta I

Sericornis perspicillatus (Buff-faced Sericornis)
    C & SE New Guinea
Sericornis rufescens (Arfak Buff-faced Sericornis)
    NW New Guinea
Sericornis papuensis (Papuan Sericornis)
    S.p.meeki
        W New Guinea
    S.p.burgersi
        C New Guinea
    S.p.papuensis
        SE New Guinea
Sericornis arfakianus (Grey-green Sericornis)
    Mts of New Guinea
Sericornis magnus (Scrub Tit)
    Tasmania, King I

PYRRHOLAEMUS   140.2
Pyrrholaemus brunneus (Redthroat)
    W, C & S Australia

CHTHONICOLA   140.2
Chthonicola sagittata (Little Field Wren)
    C Queensland to W Victoria

CALAMANTHUS   140.2
Calamanthus fuliginosus (Striated Field Wren)
    SE South Australia, Victoria, Tasmania
Calamanthus campestris (Rufous Field Wren)
    C.c.rubiginosus
        coast of NW Western Australia
    C.c.isabellinus
        interior W & S Australia
    C.c.campestris
        WC Australia
    C.c.winiam
        NW Victoria
    C.c.ethelae
        York & Eyre peninsulae
    C.c.montanellus
        SW Australia
    C.c.dorrie
        Shark Bay islands

HYLACOLA   140.2
Hylacola pyrrhopygius (Chestnut-rumped Heath Wren)
    W Queensland to SE South Australia
Hylacola cautus (Shy Heath Wren)
    S Western Australia to NW Victoria

ACANTHIZA
Acanthiza murina (De Vis' Tree Warbler)
    C & SE New Guinea
Acanthiza inornata (Western Thornbill)
    A.i.mastersi
        SW Western Australia
    A.i.inornata
        SC Western Australia
Acanthiza reguloides (Buff-tailed Thornbill)
    A.r.squamata
        NE Queensland

A.r.reguloides
    New South Wales, Victoria, SE South Australia
Acanthiza iredalei (Samphire Thornbill)
    A.i.iredalei
        Western Australia
    A.i.hedleyi
        South Australia
    A.i.rosinae
        S South Australia
Acanthiza katharina (Mountain Thornbill)
    Queensland
Acanthiza pusilla (Brown Thornbill)
    A.p.bunya
        S Queensland
    A.p.pusilla
        S Queensland, E New South Wales
    A.p.mcgilli
        E Queensland
    A.p.archibaldi
        King I
    A.p.zietzi
        Kangaroo I
    A.p.diemenensis
        Tasmania
Acanthiza apicalis (Broad-tailed Thornbill)
    A.a.whitlocki
        S Northern Territory, N Western Australia
    A.a.apicalis
        S Western Australia
    A.a.leeuwinensis
        S Western Australia
    A.a.tanami
        C Australia
    A.a.albiventris
        S Queensland, New South Wales, SE South Australia
Acanthiza ewingii (Tasmanian Thornbill)
    Tasmania, King, Flinders Is
Acanthiza chrysorrhoa (Yellow-tailed Thornbill)
    A.c.multi
        S Western Australia
    A.c.pallida
        WC Australia
    A.c.ferdinandi
        C Australia
    A.c.addenda
        EC Australia
    A.c.normantoni
        N Queensland
    A.c.chrysorrhoa
        S Queensland, New South Wales
    A.c.sandlandi
        South Australia, Victoria, Tasmania
Acanthiza uropygialis (Chestnut-tailed Thornbill)
    A.u.augusta
        WC Australia, W New South Wales
    A.u.uropygialis
        C & S Australia
Acanthiza robustirostris (Robust Thornbill)
    W & C Australia

Acanthiza nana (Little Thornbill)
A.n.flava
N Queensland
A.n.modesta
C Queensland to South Australia
A.n.nana
S Queensland, New South Wales
Acanthiza lineata (Striated Thornbill)
A.l.alberti
S Queensland
A.l.lineata
New South Wales
A.l.clelandi
South Australia
A.l.chandleri
Victoria

SMICRORNIS
Smicrornis brevirostris (Weebill)
S.b.cairns
NE Queensland
S.b.flavescens
W Northern Territory
S.b.pallescens
C Queensland
S.b.stirlingi
S Western Australia
S.b.brevirostris
New South Wales
S.b.mallee
Victoria

GERYGONE
Gerygone cinerea (Grey Flyeater)
New Guinea
Gerygone chloronota (Green-backed Flyeater)
G.c.cinereiceps
New Guinea
G.c.aruensis
Waigeu, Aru Is
G.c.chloronota
N Western Australia, N Northern Territory
Gerygone palpebrosa (Black-headed Flyeater)
G.p.palpebrosa
Aru Is, NW New Guinea
G.p.wahnesi
Japen I, NW New Guinea
G.p.inconspicua
SE New Guinea
G.p.tarara
S New Guinea
G.p.personata
N Queensland
G.p.flavida
NE Queensland
Gerygone olivacea (White-throated Flyeater)
G.o.cinerascens
SE New Guinea
G.o.rogersi
NW Australia

G.o.olivacea
S Queensland, New South Wales, Victoria
Gerygone dorsalis (Rufous-sided Flyeater)
140.1.6
G.d.senex
Kalaotua, Madu Is
G.d.kuehni
Damar I
G.d.fulvescens
Roma, Moa, Kisar, Babar Is
G.d.keyensis
Little Kai I
G.d.dorsalis
Tamimbar Is
Gerygone chrysogaster (Yellow-bellied Flyeater)
G.c.leucothorax
NW New Guinea
G.c.notata
NW New Guinea, Misol, Batanta Is
G.c.neglecta
Waigeu I
G.c.dohertyi
SW New Guinea
G.c.chrysogaster
S & E New Guinea, Japen, Aru Is
Gerygone ruficauda (Rufous-tailed Flyeater)
140.4.5
E Queensland
Gerygone magnirostris (Large-billed Flyeater)
G.m.mimikae
S & SE New Guinea
G.m.conspicillata
NW New Guinea
G.m.affinis
Japen, Dampier Is, N New Guinea
G.m.brunneipectus
Aru Is
G.m.cobana
Waigeu, Batanta Is
G.m.hypoxantha
Biak I
G.m.occasa
Kofiau I
G.m.proxima
D'Entrecasteaux Arch
G.m.onerosa
Misima I
G.m.tagulana
Tagula I
G.m.rosseliana
Sudest I
G.m.magnirostris
W Queensland to NE Western Australia
G.m.cairnsensis
N Queensland
Gerygone sulphurea (Golden-bellied Flyeater)
140.1
G.s.flaveola
Sulawesi, Peleng I
G.s.sulphurea
Malaysia, Philippine Is, Gtr Sunda Is to
Alor I

G.s.rhizophorae
Mindanao I, Sulu Arch
Gerygone inornata (Plain Flyeater)
Timor, Wetar, Roti Is
Gerygone ruficollis (Treefern Flyeater)
G.r.ruficollis
NW New Guinea
G.r.insperata
C & SE New Guinea
Gerygone fusca (White-tailed Flyeater)
G.f.fusca
W, C & EC Australia
G.f.mungi
N, NC Australia, Queensland
Gerygone tenebrosa (Dusky Flyeater)
G.t.tenebrosa
NW Western Australia
G.t.whitlocki
coast of W Australia
G.t.christophori
Shark Bay
Gerygone laevigaster (Mangrove Flyeater)
G.l.laevigaster
N Australian coast
G.l.cantator
NE Australian coast
G.l.pallida
S New Guinea
Gerygone flavolateralis (Fan-tailed Flyeater)
G.f.flavolateralis
New Caledonia
G.f.lifuensis
Loyalty Is
G.f.correiae
Banks Is
G.f.citrina
Rennell I
G.f.rouxi
Uvea I
Gerygone mouki (Brown Flyeater)
G.m.mouki
N Queensland
G.m.amalia
S Queensland
G.m.richmondi
N New South Wales
Gerygone modesta (Norfolk Island Flyeater)
140.1
Norfolk I
Gerygone igata (New Zealand Grey Flyeater)
New Zealand
Gerygone albofrontata (Chatham Is Flyeater)
Chatham I

APHELOCEPHALA
Aphelocephala leucopsis (Southern Whiteface)
A.l.castaneiventris
Western Australia
A.l.whitei
C Australia
A.l.leucopsis
South Australia, New South Wales, Victoria

Aphelocephala pectoralis (Chestnut-breasted Whiteface)
C South Australia
Aphelocephala nigricincta (Banded Whiteface)
C Australia

MOHOUA   140.1
Mohoua ochrocephala (Yellowhead)
M.o.albicilla
North I, New Zealand
M.o.ochrocephala
South I, New Zealand

FINSCHIA
Finschia novaeseelandiae (New Zealand Creeper)
South I, New Zealand

EPHTHIANURA
Ephthianura albifrons (White-faced Chat)
E.a.albifrons
S Western Australia to Victoria, New South Wales
E.a.tasmanica
Tasmania
Ephthianura tricolor (Crimson Chat)
C & E Australia
Ephthianura aurifrons (Orange Chat)
all Australia except SW
Ephthianura crocea (Yellow Chat)
E.c.boweri
Fitzroy river (NW Australia)
E.c.tunneyi
Alligator river (Northern Territory)
E.c.crocea
Norman river (W Queensland)
E.c.macgregori
Fitzroy river (Queensland)

ASHBYIA
Ashbyia lovensis (Desert Chat)
N South Australia

141. MONARCHIDAE (MONARCHS, FANTAILS)

ERYTHROCERCUS
Erythrocercus holochlorus (Little Yellow Flycatcher)
Kenya, Tanzania
Erythrocercus mccallii (Chestnut-capped Flycatcher)
E.m.nigeriae
Guinea to Nigeria
E.m.mccallii
Cameroun, Gabon

E.m.congicus
  E Zaire, Uganda
Erythrocercus livingstonei (Livingstone's Flycatcher)
  E.l.thomsoni
    S Tanzania, N Mozambique
  E.l.livingstonei
    S Zambia, S Malawi, C Mozambique
  E.l.francisi
    N Malawi

ELMINIA
Elminia longicauda (Blue Flycatcher)   141.1
  E.l.longicauda
    Sierra Leone to Central African Republic
  E.l.teresita
    Cameroun to S Sudan & Kenya, Angola
Elminia albicauda (White-tailed Blue Flycatcher)   141.1
    Angola to Tanzania
Elminia nigromitrata (Dusky Crested Flycatcher)   141.4
    Liberia to Uganda
Elminia albiventris (White-bellied Crested Flycatcher)   141.4
  E.a.albiventris
    Cameroun, Fernando Po I
  E.a.toroensis
    NE Zaire, Uganda
Elminia albonotata (White-tailed Crested Flycatcher)   141.4
  E.a.albonotata
    Uganda to Tanzania & Malawi
  E.a.swynnertoni
    E Zimbabwe
  E.a.subcaerulea
    N Tanzania, NE Zambia, N Mozambique

PHILENTOMA
Philentoma pyrhopterum (Chestnut-winged Monarch)
  P.p.pyrhopterum
    S Vietnam, Malaysia, Sumatra, Borneo
  P.p.dubium
    Natuna Is
Philentoma velatum (Maroon-breasted Monarch)
  P.v.caesium
    Malaysia, Borneo, Sumatra
  P.v.velatum
    Java

HYPOTHYMIS
Hypothymis azurea (Black-naped Blue Monarch)
  H.a.styani
    India to S China & Vietnam
  H.a.forrestia
    Burma
  H.a.oberholseri
    Taiwan
  H.a.montana
    N & C Thailand

H.a.galerita
  S & SE Thailand
H.a.ceylonensis
  Sri Lanka
H.a.prophata
  Malaysia, Sumatra, Borneo
H.a.tytleri
  Andaman, Cocos Is
H.a.idiochroa
  Car Nicobar I
H.a.nicobarica
  S Nicobar Is
H.a.javana
  Java
H.a.penidae
  Penida I
H.a.karimatensis
  Karimata I
H.a.opisthocyanea
  Anamba Is
H.a.gigantoptera
  Bunguran, Natuna Is
H.a.aeria
  Maratua I
H.a.consobrina
  Simalur I
H.a.leucophila
  Siberut I
H.a.richmondi
  Enggano I
H.a.abbotti
  Babi, Masia Is
H.a.symmixta
  Lesser Sunda Is
H.a.azurea
  Philippine Is
H.a.catarmanensis
  Camiguin I (South)
H.a.puella
  Sulawesi   141.2
H.a.blasii
  Sula Is
Hypothymis helenae (Short-crested Blue Monarch)
  H.h.helenae
    Luzon, Samar Is
  H.h.agusanae
    NE Mindanao
  H.h.personata
    Camiguin I (North)
Hypothymis coelestis (Celestial Blue Monarch)
    Philippine Is

EUTRICHOMYIAS
Eutrichomyias rowleyi (Rowley's Flycatcher)
    Gt Sangai, Peleng, Banggai Is

TERPSIPHONE   141.1
Terpsiphone cyanomelas (Cape Crested Flycatcher)   141.4
  T.c.vivax
    Uganda to S Zaire, Zambia

T.c.bivittatus
  Somalia to E Tanzania
T.c.megalolophus
  Malawi, E Zimbabwe, N Mozambique, N
  Natal
T.c.segregus
  E Transvaal
T.c.cyanomelas
  E Sth Africa
**Terpsiphone nitens (Blue-headed Crested
Flycatcher)**   141.4
T.n.reichenowi
  Sierra Leone to Togo
T.n.nitens
  Nigeria to E Zaire & Angola
**Terpsiphone rufiventer (Red-bellied Paradise
Flycatcher)**
T.r.rufiventer
  Senegal to Guinea
T.r.nigriceps
  Guinea to Togo
T.r.fagani
  S Nigeria
T.r.tricolor
  Fernando Po I
T.r.smithii
  Annobon I
T.r.neumanni
  Cameroun, Gabon
T.r.schubotzi
  Central African Republic
T.r.mayombe
  W Zaire
T.r.somereni
  W Uganda
T.r.emini
  SW Uganda, NW Tanzania
T.r.ignea
  S Zaire, N Angola, Zambia
**Terpsiphone bedfordi (Bedford's Paradise
Flycatcher)**   141.1
  NW Zaire
**Terpsiphone rufocinerea (Rufous-vented
Paradise Flycatcher)**
T.r.batesi
  Cameroun to C Zaire
T.r.rufocinerea
  Gabon to N Angola
T.r.bannermani
  N Angola
**Terpsiphone viridis (African Paradise
Flycatcher)**
T.v.viridis
  Senegal to N Nigeria & Central African
  Republic
T.v.speciosa
  S Ivory Coast to Gabon & W Sudan
T.v.ferretti
  Ethiopia to Tanzania
T.v.restricta
  N Kenya
T.v.kivuensis
  Rwanda, Kivu, NW Tanzania

T.v.suahelica
  E Tanzania
T.v.ungujaensis
  E Tanzania, Zambia
T.v.plumbeiceps
  Angola, W Zaire, Namibia
T.v.granti
  E Sth Africa
T.v.harterti
  Arabia
**Terpsiphone paradisi (Asiatic Paradise
Flycatcher)**
T.p.leucogaster
  Afghanistan, Pakistan, W India
T.p.paradisi
  C & S India, Sri Lanka
T.p.ceylonensis
  Sri Lanka
T.p.incei
  China, Manchuria, Korea
T.p.saturatior
  E Himalayas, Assam, Bangladesh
T.p.burmae
  Burma
T.p.affinis
  S Burma, Thailand, Laos, Malaysia
T.p.nicobarica
  Nicobar, Andaman Is
T.p.indochinensis
  S Thailand, Cambodia
T.p.borneensis
  Borneo
T.p.madzoedi
  N Sumatra
T.p.australis
  Java, S Sumatra
T.p.procera
  Simalur I
T.p.insularis
  Nias I
T.p.sumbaensis
  Sumba I
T.p.floris
  Sumbawa, Alor, Flores Is
**Terpsiphone atrocaudata (Black Paradise
Flycatcher)**
T.a.atrocaudata
  Japan, SE Asia
T.a.illex
  S Japan, Riukiu Is
T.a.periophthalmica
  Mindoro, Botel Tobago Is
**Terpsiphone cyanescens (Blue Paradise
Flycatcher)**
  Palawan I
**Terpsiphone cinnamomea (Rufous Paradise
Flycatcher)**
T.c.cinnamomea
  Mindanao, Basilan Is, Sulu Arch
T.c.talautensis
  Talaut I
T.c.unirufa
  N Philippine Is

Terpsiphone atrochalybea (Sao Thome Paradise Flycatcher)
Sao Thome I

Terpsiphone mutata (Madagascar Paradise Flycatcher)
T.m.mutata
E Madagascar
T.m.singetra
W Madagascar
T.m.pretiosa
Mayotte I
T.m.vulpina
Anjouan I
T.m.voeltzkowiana
Moheli I
T.m.comoroensis
Grand Comoro I

Terpsiphone corvina (Seychelles Paradise Flycatcher)
Seychelles Is

Terpsiphone bourbonnensis (Mascarene Paradise Flycatcher)
T.m.bourbonnensis
Reunion
T.m.desolata
Mauritius

Chasiempis sandwichensis (Elepaio)
C.s.sandwichensis
Hawaii (drier areas)
C.s.bryani
Mauna Kea I
C.s.sclateri
Kauai I
C.s.gayi
Oahu I
C.s.ridgwayi
Hawaii (wetter areas)

Pomarea dimidiata (Rarotonga Flycatcher)
Rarotonga I

Pomarea nigra (Society Is Flycatcher)
P.n.nigra
Tahiti
P.n.pomarea
Maupiti I

Pomarea mendozae (Marquesas Flycatcher)
P.m.mendozae
Tahuata, Hivaoa Is
P.m.motanensis
Motane I
P.m.mira
Huapu I
P.m.nukuhivae
Nukuhiva I

Pomarea iphis (Allied Flycatcher)
P.i.iphis
Huahuna I
P.i.fluxa
Eioa I

Pomarea whitneyi (Large Flycatcher)
Fatuhiva I

Mayrornis schistaceus (Small Slaty Flycatcher)
Vanikoro I

Mayrornis versicolor (Ogea Flycatcher) 141.3
Ogea Levu I

Mayrornis lessoni (Slaty Flycatcher)
M.l.lessoni
NW Fiji Is
M.l.orientalis
S Fiji Is

Neolalage banksiana (Buff-bellied Flycatcher)
Banks Is, New Hebrides

Clytorhynchus pachycephaloides (Southern Shrikebill)
C.p.pachycephaloides
New Caledonia
C.p.grisescens
Banks Is, New Hebrides

Clytorhynchus vitiensis (Fiji Shrikebill)
C.v.powelli
Manua I
C.v.compressirostris
Kandavu I
C.v.vitiensis
Viti Levu, Ovalau Is
C.v.buensis
Vanua Levu I
C.v.layardi
Taviuni I
C.v.pontifex
Ngambia, Rambi Is
C.v.vatuana
N Lau Arch
C.v.heinei
Tonga I
C.v.wiglesworthi
Rotuma I
C.v.nesiotes
S Lau Arch
C.v.fortunae
Fotuna, Alofa Is
C.v.keppeli
Keppel, Boscawen Is

Clytorhynchus nigrogularis (Black-throated Shrikebill)
C.n.nigrogularis
Fiji Is
C.n.sanctaecrucis
Santa Cruz I

Clytorhynchus hamlini (Rennell Shrikebill)
Rennell I

Metabolus rugiensis (Truk Monarch)
Truk I

MONARCHA
Monarcha axillaris (Black Monarch)
  M.a.axillaris
    NW New Guinea
  M.a.fallax
    SE New Guinea
Monarcha rubiensis (Rufous Monarch)
    N & SW New Guinea
Monarcha cinerascens (Island Grey-headed Monarch)
  M.c.cinerascens
    Timor I, Sulawesi, Moluccas
  M.c.perpallidus
    St Matthias, Emirau Is, New Hanover
  M.c.inornatus
    NW New Guinea, Aru Is
  M.c.steini
    Numfor I
  M.c.geelvinkianus
    Japen, Biak Is
  M.c.fuscescens
    N New Guinea islands
  M.c.nigrirostris
    NE New Guinea
  M.c.rosselianus
    Rossel I, D'Entrecasteaux & Louisiade Arch
  M.c.impediens
    Solomon Is
  M.c.fulviventris
    Ninigo, Admiralty Is
Monarcha melanopsis (Pearly-winged Monarch)
    N & E Australia >> SE New Guinea
Monarcha frater (Black-winged Monarch)
  M.f.frater
    NW New Guinea
  M.f.kunupi
    C New Guinea
  M.f.periophthalmicus
    SE New Guinea
  M.f.canescens
    N Queensland
Monarcha erythrosticta (Bougainville Monarch)
    Bougainville I
Monarcha castaneiventris (Chestnut-bellied Monarch)
  M.c.castaneiventris
    Guadalcanal, Malaita, Choiseul Is
  M.c.obscurior
    Rossel I
  M.c.megarhyncha
    San Cristobal I
  M.c.ugiensis
    Ugi Is
Monarcha richardsii (Richard's Monarch)
    C Solomon Is
Monarcha leucotis (White-eared Monarch)
  M.l.castus
    Tanimbar Is
  M.l.buruensis
    Buru I

M.l.pileatus
    Halmahera I
M.l.leucotis
    E Queensland coast
Monarcha guttula (Spot-winged Monarch)
    E New Guinea, Aru Is
Monarcha mundus (Tanimbar Monarch)
    Tanimbar, Babar, Wetar Is
Monarcha sacerdotum (Mees' Monarch)
    Flores I
Monarcha boanensis (Black-chinned Monarch)   141.2
    Boano I
Monarcha trivirgatus (Spectacled Monarch)
  M.t.trivirgatus
    Timor, Lesser Sunda Is
  M.t.bernsteini
    NW New Guinea, Sulawesi
  M.t.melanopterus
    Louisiade Arch
  M.t.diadematus
    Obi I
  M.t.bimaculatus
    Halmahara, Batjan, Morotai Is
  M.t.nigrimentum
    Seram, Ambon Is
  M.t.albiventris
    N Queensland, S New Guinea
  M.t.gouldi
    S Queensland, New South Wales
Monarcha leucurus (Kai Monarch)
    Kai Is
Monarcha everetti (White-tipped Monarch)   141.2
    Djampea I
Monarcha loricatus (Black-tipped Monarch)   141.2
    Buru I
Monarcha julianae (Kofiau Monarch)
    Kofiau I
Monarcha brehmii (Biak Monarch)
    Biak, Misol Is
Monarcha manadensis (Black & White Monarch)
    New Guinea
Monarcha infelix (Unhappy Monarch)
  M.i.infelix
    Manus I
  M.i.coultasi
    Rambutyo I
Monarcha menckei (St Matthias Monarch)
    St Matthias Is
Monarcha verticalis (Bismarck Monarch)   141.1
  M.v.ateralbus
    Dyaul I
  M.v.verticalis
    Bismarck Arch
Monarcha barbatus (Pied Monarch)
  M.b.barbatus
    Bougainville, Guadalcanal, Choiseul, Ysabel Is

M.b.malaitae
  Malaita I
Monarcha browni (Kulambangra Monarch)
M.b.browni
  Kulambangra I
M.b.ganongae
  Ganonga I
M.b.nigrotectus
  Vellalavella I
M.b.meeki
  Rendova I
Monarcha viduus (San Cristobal Monarch)
M.v.viduus
  San Cristobal I
M.v.squamulatus
  Ugi I
Monarcha godeffroyi (Yap Monarch)
  Yap I
Monarcha takatsukasae (Tinian Monarch)
  Tinian I
Monarcha chrysomela (Black & Yellow Monarch)
M.c.chrysomela
  New Hanover, New Ireland
M.c.kordensis
  Biak, Misol Is
M.c.melanonotus
  New Guinea
M.c.aurantiacus
  NE New Guinea
M.c.nitida
  E & S New Guinea, Fergusson I
M.c.aruensis
  Aru Is, S New Guinea
M.c.pulcherrima
  Dyaul I
M.c.whitneyorum
  Lihir Is
M.c.tabarensis
  Tabar I

ARSES
Arses kaupi (Pied Flycatcher)
  N Queensland
Arses telescophthalmus (Frilled Flycatcher)
A.t.telescophthalmus
  Misol I, NW New Guinea
A.t.insularis
  Japen I, N New Guinea
A.t. batantae
  Batanta, Waigeu Is
A.t.aruensis
  Aru Is
A.t.lauterbachi
  NE New Guinea
A.t.harterti
  SW & S New Guinea
A.t.henkei
  SE New Guinea
A.t.lorealis
  N Queensland

MYIAGRA   141.1
Myiagra oceanica (Micronesian Myiagra Flycatcher)   141.1
M.o.erythrops
  Palau I
M.o.freycineti
  Guam I
M.o.oceanica
  Truk I
M.o.pluto
  Ponapé I
Myiagra galeata (Helmet Flycatcher)
M.g.galeata
  Batjan I
M.g.goramensis
  Goram, Little Kai, Ambon, Seram Is
M.g.buruensis
  Buru I
Myiagra atra (Black Myiagra Flycatcher) 141.1
  Numfor, Biak Is
Myiagra rubecula (Leaden Flycatcher)
M.r.rubecula
  S Queensland to Victoria >> S New Guinea, N Australia
M.r.concinna
  N Western Australia, Northern Territory, N Queensland
M.r.yorki
  NE Queensland
M.r.papuana
  New Guinea
M.r.sciurorum
  Louisiade Arch
Myiagra ferrocyanea (Steel-blue Flycatcher)
M.f.ferrocyanea
  Ysabel, Choiseul, Guadalcanal Is
M.f.feminina
  Kulambangra I
M.f.cinerea
  Bougainville I
M.f.malaitae
  Malaita I
Myiagra cervinicauda (San Cristobal Myiagra Flycatcher)   141.1
  San Cristobal I
Myiagra caledonica (New Caledonian Myiagra Flycatcher)
M.c.caledonica
  New Caledonia
M.c.melanura
  Tanna, Erromanga, Mare Is
M.c.marinae
  N & C New Hebrides
M.c.viridinitens
  Lifu, Urea Is
M.c.occidentalis
  Rennell I
Myiagra vanikorensis (Red-bellied Flycatcher)
M.v.vanikorensis
  Santa Cruz, Vanikoro Is
M.v.rufiventris
  Navigator I, NW Fiji Is

M.v.townsendi
S Lau Is
M.v.kandavensis
Kandavu I
M.v.dorsalis
SC Fiji Is, N Lau Is
Myiagra albiventris (White-vented Flycatcher)
Samoa Is
Myiagra azureocapilla (Blue-headed Flycatcher)
M.a.azureocapilla
Taviuni I
M.a.castaneigularis
Vanua Levu, Kambara Is
M.a.whitneyi
Viti Levu I
Myiagra ruficollis (Broad-billed Flycatcher)
M.r.ruficollis
Alor, Timor, Kalao, Djampea Is
M.r.mimikae
New Guinea, N Queensland
M.r.fulviventris
Tanimbar Is
Myiagra cyanoleuca (Satin Flycatcher)
Tasmania, E Australia >> S New Guinea & islands
Myiagra alecto (Shining Flycatcher)   141.2
M.a.alecto
Ternate I, Moluccas
M.a.chlaybeocephalus
New Ireland
M.a.lucidus
Woodlark I
M.a.manumudari
Vulcan I
M.a.rufolateralis
Aru Is
M.a.longirostris
Tanimbar Is
M.a.tormenti
N Western Australia
M.a.wardelli
N Queensland
Myiagra hebetior (Dull Flycatcher)   141.2
M.h.hebetior
St Matthais Is
M.h.eichhorni
New Hanover
M.h.cervinicolor
Dyaul I
Myiagra inquieta (Restless Flycatcher)   141.2
M.i.nana
N Western Australia, Northern Territory
M.i.westralensis
S Western Australia
M.i.inquieta
E Australia

LAMPROLIA
Lamprolia victoriae (Silktail)
L.v.victoriae
Taviuni I

L.v.kleinschmidti
Vanua Levu I

MACHAERIRHYNCHUS
Machaerirhynchus flaviventer (Yellow-breasted Flatbill Flycatcher)
M.f.albifrons
Waigeu, Misol Is, N New Guinea
M.f.albigula
NW New Guinea
M.f.novus
NE New Guinea
M.f.xanthogenys
S New Guinea, Aru Is
M.f.secundus
N Queensland
M.f.flaviventer
Queensland
Machaerirhynchus nigripectus (Black-breasted Flatbill Flycatcher)
M.n.nigripectus
NW New Guinea
M.n.saturatus
C New Guinea
M.n.harterti
SE New Guinea

PELTOPS
Peltops blainvillii (Lowland Peltops Flycatcher)
New Guinea
Peltops montanus (Mountain Peltops Flycatcher)
New Guinea

RHIPIDURA
Rhipidura hypoxantha (Yellow-bellied Fantail)
Himalayas, S China
Rhipidura superciliaris (Blue Fantail)
R.s.superciliaris
Basilan, N Mindanao Is
R.s.apo
SE Mindanao
R.s.samarensis
Bohol, Samar, Leyte Is
Rhipidura cyaniceps (Blue-headed Fantail)
R.c.cyaniceps
Luzon I
R.c.pinicola
NW Luzon I
R.c.albiventris
Masbate, Negros, Panay Is
R.c.sauli
Tablas I
Rhipidura phoenicura (Red-tailed Fantail)
Java
Rhipidura nigrocinnamomea (Black & Cinnamon Fantail)
R.n.hutchinsoni
N Mindanao I
R.n.nigrocinnamomea
SE Mindanao I
Rhipidura albicollis (White-throated Fantail)

R.a.canescens
  W Himalayas
R.a.albicollis
  N India, Nepal
R.a.orissae
  NE India
R.a.stanleyi
  E Himalayas, Assam, Burma
R.a.albogularis
  SW India
R.a.vernayi
  SE India
R.a.celsa
  SE Tibet, S China, Thailand, N Indochina
R.a.atrata
  Sumatra
R.a.sarawacensis
  N Borneo
R.a.kinabalu
  Borneo
R.a.cinerascens
  S Indochina
Rhipidura euryura (White-bellied Fantail)
  Java
Rhipidura aureola (White-browed Fantail)
R.a.aureola
  N India
R.a.compressirostris
  S India, Sri Lanka
R.a.burmanica
  Assam, Burma
Rhipidura javanica (Pied Fantail)
R.j.longicauda
  Sumatra
R.j.javanica
  Java
R.j.nigritorquis
  Philippine Is
Rhipidura perlata (Pearlated Fantail)
  Malaysia
Rhipidura leucophrys (Black & White Fantail)
  (Willy Wagtail)
  R.l.picata
    N Western Australia, Northern Territory
  R.l.leucophrys
    S Australia
  R.l.melaleuca
    Bismarck Arch, Moluccas, New Guinea,
    Solomon Is
Rhipidura rufiventris (Northern Fantail)
R.r.rufiventris
  Timor I
R.r.perneglecta
  S Moluccas
R.r.finitima
  Watubela I
R.r.assimilis
  Kai Is
R.r.bouruensis
  Buru I
R.r.obiensis
  Obi I

R.r.pallidiceps
  Wetar I
R.r.gigantea
  Lihir, Tabar Is
R.r.tangensis
  Boang, Tanga Is
R.r.cinerea
  Seram , Ambon Is
R.r.buttikoferi
  Damar, Moa, Leti Is
R.r.tenkatei
  Roti I
R.r.niveiventris
  Admiralty Is
R.r.mussai
  St Matthias Is
R.r.setosa
  New Ireland, New Hanover, Dyaul I
R.r.finschii
  New Britain
R.r.vidua
  Kofiau I
R.r.gularis
  Waigeu, Japen Is, New Guinea
R.r.nigromentalis
  Sudest, Tagula, Misima Is
R.r.kordensis
  Biak I
R.r.superciliosa
  N Australia
R.r.isura
  coast of N Western Australia
Rhipidura cockerelli (Cockerell's Fantail)
R.c.cockerelli
  Guadalcanal I
R.c.coultasi
  N Solomon Is
R.c.septentrionalis
  Bougainville I
R.c.interposita
  Ysabel I
R.c.floridana
  Florida, Tulagi I
R.c.lavellae
  Vellalavella I
R.c.albina
  Kulambangra, Rendova Is
Rhipidura albolimbata (Friendly Fantail)
R.a.albolimbata
  NW New Guinea
R.a.lorentzi
  SW New Guinea
Rhipidura threnothorax (Sooty Thicket Fantail)
R.t.threnothorax
  New Guinea, Aru, Waigeu Is
R.t.fumosa
  Japen I
Rhipidura hyperythra (Chestnut-bellied Fantail)
R.h.hyperythra
  Aru Is

R.h.mulleri
SW New Guinea
R.h.castaneothorax
SE New Guinea
Rhipidura maculipectus (Black Thicket Fantail)
W & S New Guinea, Aru Is
Rhipidura leucothorax (White-breasted Fantail)
R.l.leucothorax
NW New Guinea
R.l.episcopalis
SE New Guinea
R.l.clamosa
EC New Guinea
Rhipidura atra (Black Fantail)
R.a.atra
NW New Guinea
R.a.vulpes
N New Guinea
Rhipidura fuliginosa (Collared Grey Fantail)
R.f.bulgeri
Lifu I, New Caledonia
R.f.phasiana
New Guinea
R.f.brenchleyi
New Hebrides, Banks Is
R.f.preissi
S Western Australia
R.f.albicauda
C Australia
R.f.alisteri
S Queensland
R.f.albiscapa
King I, Tasmania
R.f.placabilis
North I (New Zealand)
R.f.fuliginosa
South I (New Zealand)
R.f.penitus
Chatham I
R.f.pelzelni
Norfolk I
Rhipidura drownei (Mountain Fantail)
R.d.drownei
Bougainville I
R.d.ocularis
Guadalcanal I
Rhipidura tenebrosa (Dusky Fantail)
San Cristobal I
Rhipidura rennelliana (Rennell Fantail)
Rennell I
Rhipidura spilodera (Spotted Fantail)
R.s.spilodera
N & C New Hebrides, Banks Is
R.s.layardi
Ovalau, Viti Levu Is
R.s.erythronota
Yanganga, Vanua Levu Is
R.s.rufilateralis
Taviuni I
R.s.verreauxi
Lifu, Mare Is, New Caledonia

Rhipidura nebulosa (Samoan Fantail)
R.n.nebulosa
Upolu I
R.n.altera
Savaii I
Rhipidura brachyrhyncha (Dimorphic Rufous Fantail)
R.b.brachyrhyncha
NW New Guinea
R.b.devisi
SE New Guinea
Rhipidura personata (Kandavu Fantail)
Kandavu I (Fiji Is)
Rhipidura dedemi (Seram Rufous Fantail)
Seram I
Rhipidura superflua (Moluccan Fantail)
Buru I
Rhipidura teysmanni (Sulawesi Rufous Fantail) 141.2
R.t.teysmanni
SW Sulawesi
R.t.toradja
N, C & SE Sulawesi
R.t.sulaensis
Sula Is
Rhipidura fuscorufa (Cinnamon-tailed Fantail) 141.2
Tanimbar, Babar Is
Rhipidura diluta (Brown-capped Fantail) 141.2
R.d.diluta
Flores, Lomblen Is
R.d.sumbawensis
Sumbawa I
Rhipidura opistherythra (Long-tailed Fantail)
Tanimbar Is
Rhipidura lepida (Palau Fantail)
Palau Is
Rhipidura rufidorsa (Grey-breasted Rufous Fantail)
R.r.rufidorsa
Misol, Japen Is, NW New Guinea
R.r.kumusi
SE New Guinea
R.r.kubuna
S New Guinea
Rhipidura dahli (Island Rufous Fantail)
R.d.dahli
New Britain
R.d.antonii
New Ireland
Rhipidura matthiae (St Matthias Rufous Fantail)
St Matthias Is
Rhipidura malaitae (Malaita Rufous Fantail)
Malaita I
Rhipidura rufifrons (Rufous Fantail)
R.r.uraniae
Mariana, Guam Is
R.r.saipanensis
Saipan, Tinian Is
R.r.mariae
Rota I

R.r.kubaryi
  Ponapé I
R.r.versicolor
  Yap I
R.r.torrida
  Halmahera, Ternate Is
R.r.semicollaris
  Timor I
R.r.sumbensis
  Sumba I
R.r.agilis
  Santa Cruz I
R.r.melaenolaema
  Vanikoro I
R.r.utupuae
  Utupua I
R.r.commoda
  Bougainville I
R.r.rufofronta
  Guadalcanal I
R.r.granti
  Rendova I
R.r.russata
  San Cristobal I
R.r.semirubra
  Admiralty Is
R.r.hamadryas
  Tanimbar Is
R.r.brunnea
  Malaita I
R.r.kuperi
  Santa Anna I
R.r.ugiensis
  Ugi I
R.r.streptophora
  SW New Guinea
R.r.mimosae
  Kalao I
R.r.celebensis
  Djampea I
R.r.elegantula
  Roma, Leti, Moa, Damar Is
R.r.reichenowi
  Babar I
R.r.squamata
  Banda I
R.r.louisiadensis
  Louisiade Arch
R.r.dryas
  N Western Australia, Northern Territory
R.r.rufifrons
  Queensland to Victoria

142. EOPSALTRIDAE (AUSTRALASIAN ROBINS)

MONACHELLA
Monachella muelleriana (River Flycatcher)
  M.m.muelleriana
    New Guinea
  M.m.coultasi
    New Britain

MICROECA
Microeca leucophaea (Australian Brown Flycatcher)
  M.l.zimmeri
    New Guinea
  M.l.pallida
    Northern Territory, N Queensland
  M.l.leucophaea
    S Queensland, New South Wales
  M.l.barcoo
    C Australia
  M.l.assimilis
    S Western Australia
Microeca flavigaster (Lemon-breasted Flycatcher)
  M.f.tarara
    New Guinea
  M.f.laeta
    New Guinea
  M.f.terraereginae
    N Queensland
  M.f.flavigaster
    Northern Territory
  M.f.tormenti
    N Western Australia
Microeca hemixantha (Tanimbar Microeca Flycatcher)
    Tanimbar Is
Microeca griseoceps (Yellow-footed Flycatcher)
  M.g.occidentalis
    NW New Guinea
  M.g.griseoceps
    SE New Guinea, S Queensland
Microeca flavovirescens (Olive Microeca Flycatcher)
  M.f.flavovirescens
    Aru Is, New Guinea
  M.f.cuicui
    New Guinea
Microeca papuana (Papuan Microeca Flycatcher)
New Guinea & islands

EUGERYGONE
Eugerygone rubra (Red-backed Warbler)
  E.r.rubra
    NW New Guinea
  E.r.saturatior
    C New Guinea

PETROICA
Petroica bivittata (Forest Robin)
  P.b.caudata
    C New Guinea
  P.b.bivittata
    SE New Guinea
Petroica archboldi (Rock Robin)
    C New Guinea
Petroica multicolor (Scarlet Robin)
  P.m.kleinschmidti
    Fiji Is

P.m.pusilla
Samoa Is
P.m.feminina
Efate, Mai I
P.m.similis
Tanna I
P.m.soror
Vanua Leva
P.m.taveunensis
Taveuni
P.m.ambrynensis
New Hebrides, Banks Is
P.m.cognata
Erromanga I
P.m.becki
Kandavu I
P.m.polymorpha
San Cristobal I
P.m.septentrionalis
Bougainville I
P.m.kulambangrae
Kulambangra I
P.m.campbelli
S Western Australia
P.m.boodang
E Australia, Tasmania
P.m.multicolor
Norfolk I
P.m.dennisi
Guadalcanal I
Petroica goodenovii (Red-capped Robin)
W, C & E Australia
Petroica phoenicea (Flame Robin)
New South Wales, SE Australia, Tasmania
Petroica rodinogaster (Pink Robin)
SE Australia, Tasmania
Petroica rosea (Rose Robin)
coastal S Queensland to Victoria
Petroica cucullata (Hooded Robin)
P.c.picata
Western Australia
P.c.cucullata
S Queensland, New South Wales, Victoria
Petroica macrocephala (New Zealand Tit)
P.m.toitoi
North I (New Zealand)
P.m.macrocephala
South I (New Zealand)
P.m.marrineri
Auckland Is
P.m.chathamensis
Chatham I
P.m.dannefaerdi
Snares I
Petroica vittata (Dusky Robin)
Tasmania
Petroica australis (New Zealand Robin)
P.a.longipes
North I (New Zealand)
P.a.australis
South I (New Zealand)
P.a.rakiura
Stewart I

Petroica traversi (Chatham I Robin)
Chatham I

TREGELLASIA
Tregellasia leucops (White-faced Robin)
T.l.leucops
NW New Guinea
T.l.mayri
WC New Guinea
T.l.nigroorbitalis
NC New Guinea
T.l.heurni
C New Guinea
T.l.nigriceps
EC New Guinea
T.l.melanogenys
N New Guinea
T.l.wahgiensis
C & E New Guinea
T.l.albifacies
SE New Guinea
T.l.auricularis
S New Guinea
T.l.albigularis
N Queensland
Tregellasia capito (Pale Yellow Robin)
T.c.nana
N Queensland
T.c.capito
S Queensland, New South Wales, South
Australia

EOPSALTRIA
Eopsaltria australis (Yellow Robin)
E.a.magnirostris
N Queensland
E.a.chrysorrhoa
C Queensland
E.a.coomooboolaroo
EC Queensland
E.a.australis
S Queensland, New South Wales
E.a.austina
C & N New South Wales
E.a.viridior
Victoria
E.a.griseogularis
S Western Australia
E.a.rosinae
South Australia
Eopsaltria georgiana (White-breasted Robin)
coastal S Western Australia
Eopsaltria flaviventris (Yellow-bellied Robin)
New Caledonia

PENEOENANTHE
Peneoenanthe pulverulenta (Mangrove Robin)
P.p.pulverulenta
S New Guinea
P.p.leucura
Aru Is, N Queensland

P.p.cinereiceps
  N Western Australia
P.p.alligator
  NC Australia

POECILODRYAS
Poecilodryas brachyura (White-breasted Robin)
  P.b.brachyura
    NW New Guinea
  P.b.albotaeniata
    Japen I, New Guinea
  P.b.dumasi
    N New Guinea
Poecilodryas hypoleuca (Black & White Robin)
  P.h.hypoleuca
    NW New Guinea
  P.h.steini
    Misol I, W New Guinea
  P.h.hermani
    N New Guinea
Poecilodryas placens (Olive-yellow Robin)
    SE New Guinea
Poecilodryas albonotata (Black-throated Robin)
  P.a.albonotata
    NW New Guinea
  P.a.griseiventris
    SW New Guinea
  P.a.correcta
    SE New Guinea
Poecilodryas superciliosa (White-browed Robin)
  P.s.superciliosa
    N Queensland
  P.s.cerviniventris
    N Western Australia, Northern Territory

PENEOTHELLO
Peneothello sigillatus (White-winged Thicket Flycather)
  P.s.saruwagedi
    NE New Guinea
  P.s.quadrimaculatus
    W New Guinea
  P.s.hagenensis
    EC New Guinea
  P.s.sigillatus
    SE New Guinea
Peneothello cryptoleucus (Grey Thicket Flycatcher)
  P.c.cryptoleucus
    NW New Guinea
  P.c.albidior
    WC New Guinea
  P.c.maximus
    Kumawa Mts
Peneothello cyanus (Slaty Thicket Flycatcher)
  P.c.cyanus
    NW New Guinea
  P.c.atricapillus
    NE New Guinea

P.c.subcyanus
  SE & C New Guinea
Peneothello bimaculatus (White-rumped Thicket Flycatcher)
  P.b.bimaculatus
    NW New Guinea
  P.b.vicarius
    SE New Guinea

HETEROMYIAS
Heteromyias albispecularis (Ground Thicket Robin)
  H.a.albispecularis
    NW New Guinea
  H.a.atricapillus
    NW New Guinea
  H.a.rothschildi
    WC New Guinea
  H.a.centralis
    C New Guinea
  H.a.armiti
    SE New Guinea
Heteromyias cinereifrons (Grey-headed Thicket Robin)
    N Queensland

PACHYCEPHALOPSIS
Pachycephalopsis hattamensis (Green Thicket Flycatcher)
  P.h.hattamensis
    NW New Guinea
  P.h.ernesti
    NW New Guinea
  P.h.axillaris
    WC New Guinea
  P.h.insulanus
    W New Guinea
  P.h.lecroyae
    EC New Guinea
Pachycephalopsis poliosoma (White-throated Thicket Flycatcher)
  P.p.idenburgi
    N New Guinea
  P.p.hypopolia
    NE New Guinea
  P.p.albigularis
    WC New Guinea
  P.p.balim
    WC New Guinea
  P.p.approximans
    C New Guinea
  P.p.hunsteini
    EC New Guinea
  P.p.poliosoma
    SE New Guinea

143. PACHYCEPHALIDAE (WHISTLERS)

EULACESTOMA
Eulacestoma nigropectus (Wattled Shrike Tit)
  E.n.clara
    C New Guinea

E.n.nigropectus
SE New Guinea

Falcunculus frontatus (Crested Shrike Tit)
F.f.leucogaster
S Western Australia
F.f.whitei
Northern Territory, NW Western Australia
F.f.frontatus
E Australia

Oreoica gutturalis (Crested Bellbird)
O.g.pallescens
N Western Australia
O.g.gutturalis
WC & E Australia

Pachycare flavogrisea (Golden-faced Pachycare)
P.f.flavogrisea
W New Guinea
P.f.subaurantia
C New Guinea
P.f.randi
N New Guinea
P.f.subpallida
SE New Guinea

Rhagologus leucostigma (Mottled Whistler)
R.l.leucostigma
NW New Guinea
R.l.novus
N New Guinea
R.l.obscurus
C & SE New Guinea

Hylocitrea bonensis (Buff-throated Thickhead)
H.b.bonensis
NC & SE Sulawesi
H.b.bonthaina
S Sulawesi

143.1
Coracornis raveni (Maroon-backed Whistler)
C & SE Sulawesi

143.1
Aleadryas rufinucha (Rufous-naped Whistler)
A.r.rufinucha
NW New Guinea
A.r.niveifrons
C New Guinea
A.r.lochmia
E New Guinea
A.r.gamblei
SE New Guinea

A.r.prasinonota
SE New Guinea

Pachycephala tenebrosa (Sooty Whistler)
P.t.atra
N New Guinea
P.t.tenebrosa
S New Guinea
Pachycephala olivacea (Olive Whistler)
P.o.macphersoniana
New South Wales, S Queensland
P.o.olivacea
S Victoria, SE South Australia, Tasmania
Pachycephala rufogularis (Red-lored Whistler)
NW Victoria, E South Australia
Pachycephala inornata (Gilbert Whistler)
P.i.gilbertii
S Western Australia
P.i.inornata
SE Australia
Pachycephala grisola (Mangrove Whistler)
143.1
Bangladesh to Indochina, Gtr Sunda , Philippine Is
Pachycephala plateni (Palawan Whistler)
143.1
Palawan
Pachycephala phaionota (Island Whistler)
143.1
Aru Is, N Moluccas, NW New Guinea
Pachycephala hyperythra (Rufous-breasted Whistler)
P.h.hyperythra
W New Guinea
P.h.sepikiana
C New Guinea
P.h.reichenowi
NE New Guinea
P.h.salvadorii
SE New Guinea
Pachycephala modesta (Brown-backed Whistler)
P.m.modesta
SE New Guinea
P.m.hypoleuca
NE New Guinea
P.m.telefolminensis
C New Guinea
Pachycephala philippinensis (Yellow-bellied Whistler)
P.p.fallax
Calayan I
P.p.illex
Camiguin I (North)
P.p.philippinensis
Luzon I
P.p.siquijorensis
Siquijor I
P.p.apoensis
Samar, Leyte, Mindanao Is

P.p.basilanica
Basilan I
P.p.boholensis
Bohol I
Pachycephala sulfuriventer (Sulphur-vented Whistler)
Sulawesi
Pachycephala hypoxantha (Bornean Mountain Whistler)
P.h.hypoxantha
N Borneo
P.h.sarawacensis
Poi Mts, NW Borneo
Pachycephala meyeri (Vogelkop Whistler)
NW New Guinea
Pachycephala soror (Sclater's Whistler)
P.s.soror
W New Guinea
P.s.klossi
C & E New Guinea
P.s.bartoni
SE New Guinea & Goodenough I
P.s.octogenarii
Kumawa Mts
Pachycephala simplex (Grey Whistler)
P.s.simplex
Northern Territory, Melville I
P.s.rufipennis
Kai Is
P.s.gagiensis
Gagi I
P.s.waigeuensis
Waigeu I
P.s.griseiceps
Aru Is, NE New Guinea
P.s.miosnomensis
Meos Num I
P.s.jobiensis
Japen I, N New Guinea
P.s.perneglecta
S New Guinea
P.s.peninsulae
N Queensland
P.s.dubia
SE New Guinea, D'Entrecasteaux Arch
P.s.sudestensis
Tagula I, Louisiade Arch
Pachycephala orpheus (Fawn-breasted Whistler)
Timor, Wetar Is
Pachycephala pectoralis (Golden Whistler)
P.p.teysmanni
Saleyer I
P.p.everetti
Djampea, Kalao Tua, Madu Is
P.p.javana
E Java, Bali
P.p.fulvotincta
Sumbawa I to Alor I
P.p.fulviventris
Sumba I
P.p.calliope
Timor, Semau, Wetar Is

P.p.sharpei
Babar i
P.p.dammeriana
Damar I
P.p.par
Roma I
P.p.compar
Leti, Moa Is
P.p.fuscoflava
Larat I (Tanimbar Is)
P.p.macrorhynchus
Ambon, Seram Is
P.p.buruensis
Buru I
P.p.clio
Sula Is
P.p.pelengensis
Banggai, Peleng Is
P.p.obiensis
Obi I
P.p.tidorensis
Tidore, Ternate Is
P.p.mentalis
Batjan, Halmahera, Morotai Is
P.p.occidentalis
S Western Australia
P.p.fuliginosa
South Australia, W Victoria
P.p.glaucura
Tasmania, King I
P.p.youngi
E Victoria
P.p.pectoralis
New South Wales
P.p.ashbyi
N New South Wales, S Queensland
P.p.queenslandica
N Queensland
P.p.contempta
Lord Howe I
P.p.xanthoprocta
Norfolk I
P.p.collaris
Louisiade Arch
P.p.rosseliana
Rossel I
P.p.fergussonis
Fergusson I
P.p.misimae
Misima I
P.p.citreogaster
New Hanover, New Britain, New Ireland
P.p.sexuvaria
St Matthias Is
P.p.goodsoni
Admiralty Is
P.p.tabarensis
Tabar I
P.p.ottomeyeri
Lihir I
P.p.whitneyi
Whitney I

P.p.bougainvillei
  Buka, Bougainville Is
P.p.orioloides
  Choiseul, Ysabel, Florida Is
P.p.cinnamomea
  Beagle, Guadalcanal Is
P.p.sanfordi
  Malaita I
P.p.pavuvu
  Pavuvu Is
P.p.centralis
  E New Georgia Is
P.p.feminina
  Rennell I
P.p.melanoptera
  S New Georgia Is
P.p.melanonota
  Ganonga, Vellalavella Is
P.p.christophori
  Santa Ana, San Cristobal Is
P.p.littayei
  Loyalty Is
P.p.cucullata
  Aneiteum I
P.p.chlorura
  Erromango I
P.p.intacta
  Banks Is, N & C New Hebrides
P.p.vanikorensis
  Vanikoro, Santa Cruz Is
P.p.utupuae
  Utupua I
P.p.ornata
  N Santa Cruz I
P.p.kandavensis
  Kandavu Is
P.p.lauana
  S Lau Arch
P.p.vitiensis
  Ngau I
P.p.bella
  Vatu Vara I
P.p.koroana
  Karo I
P.p.torquata
  Taviuni I
P.p.ambigua
  Rambi, Kio Is
P.p.optata
  Ovalau, SE Viti Levu Is
P.p.graeffii
  Waia, Viti Levu Is
P.p.aurantiiventris
  Yanganga, Vanua Levu Is
Pachycephala melanops (Tonga Whistler)
  Vavau, Tonga Is
Pachycephala melanura (Mangrove Golden Whistler)
P.m.balim
  N New Guinea
P.m.dahli
  islands off S New Guinea

P.m.bynoei
  Western Australia
P.m.hilli
  N Western Australia
P.m.melanura
  N Western Australia
P.m.violatae
  Northern Territory, Melville I
P.m.spinicauda
  N Queensland, Torres Straits islands
Pachycephala flavifrons (Yellow-fronted Whistler)
  Samoa Is
Pachycephala caledonica (New Caledonian Whistler)
  New Caledonia
Pachycephala implicata (Mountain Whistler)
P.i.implicata
  Guadalcanal I
P.i.richardsi
  Bougainville I
Pachycephala nudigula (Bare-throated Whistler)
P.n.nudigula
  Flores I
P.n.ilsa
  Sumbawa I
Pachycephala lorentzi (Lorentz's Whistler)
  C & E New Guinea
Pachycephala schlegelii (Schlegel's Whistler)
P.s.schlegelii
  W New Guinea
P.s.obscurior
  C & E New Guinea
P.s.cyclopum
  WC New Guinea
Pachycephala aurea (Yellow-backed Whistler)
  SE New Guinea
Pachycephala rufiventris (Rufous Whistler)
  143.1
P.r.falcata
  Melville I, Northern Territory
P.r.colletti
  N Western Australia
P.r.pallida
  NW Queensland
P.r.dulcior
  N Queensland
P.r.rufiventris
  South Australia
P.r.maudeae
  C Australia
P.r.xanthetraea
  New Caledonia
Pachycephala monacha (Aru Whistler) 143.1
  Aru Is
Pachycephala leucogastra (White-bellied Whistler) 143.1
P.l.dorsalis
  C & E New Guinea
P.l.leucogastra
  SE New Guinea

P.l.meeki
Rossel I
**Pachycephala arctitorquis (Wallacean
Whistler)** 143.1
P.a.kebirensis
Moa, Roma, Damar, Wetar Is
P.a.arctitorquis
Tanimbar Is
P.a.tianduana
Tiandou I (W Kai Is)
Pachycephala griseonota (Drab Whistler)
143.1
P.g.cinerascens
N Moluccas, Ternate I
P.g.johni
Obi Major I
P.g.lineolata
Sula Is
P.g.examinata
Buru I
P.g.griseonota
Seram I
P.g.kuehni
Kai Is
**Pachycephala lanioides (White-breasted
Whistler)**
P.l.carnaroni
Shark Bay (Western Australia)
P.l.bulleri
coast of Western Australia
P.l.lanioides
N Western Australia
P.l.fretorum
Northern Territory, NW Queensland

COLLURICINCLA
**Colluricincla megarhyncha (Rufous Shrike
Thrush)**
C.m.sanghirensis
Sanghir Is
C.m.affinis
Waigeu I
C.m.batantae
Batanta I
C.m.misoliensis
Misol I
C.m.megarhyncha
W New Guinea
C.m.ferruginea
NW New Guinea
C.m.aruensis
Aru Is
C.m.goodsoni
S New Guinea
C.m.wuroi
S New Guinea
C.m.palmeri
S New Guinea
C.m.despecta
SE New Guinea
C.m.superflua
SE New Guinea

C.m.nea
E New Guinea
C.m.madaraszi
E New Guinea
C.m.tappenbecki
NE New Guinea
C.m.maeandrina
NE New Guinea
C.m.idenburgi
N New Guinea
C.m.hybrida
N New Guinea
C.m.obscura
Japen I
C.m.melanorhyncha
Biak I
C.m.fortis
D'Entrecasteaux Arch
C.m.trobriandi
Trobriand Is
C.m.discolor
Tagula I
C.m.parvula
Northern Territory, Melville I
C.m.conigravi
N Western Australia
C.m.griseata
Cape York islands
C.m.normani
N Queensland
C.m.parvissima
N Queensland
C.m.gouldii
C & S Queensland
C.m.rufogaster
N New South Wales
**Colluricincla boweri (Stripe-breasted Shrike-
Thrush)**
N Queensland
Colluricincla harmonica (Grey Shrike-Thrush)
C.h.roebucki
Roebuck Bay, N Western Australia
C.h.parryi
Kimberley, W Australia
C.h.julietae
N Western Australia
C.h.kolichisi
W Australia
C.h.brunnea
Northern Territory, Melville I
C.h.superciliosa
N Queensland
C.h.tachycrypta
SE New Guinea
C.h.pallescens
NC Queensland
C.h.harmonica
S Queensland to E Victoria
C.h.strigata
Tasmania, Bass Strait islands
C.h.halmaturina
SW New South Wales, NW Victoria, SE
South Australia

C.h.anda
  NE South Australia
C.h.whitei
  C South Australia
C.h.rufiventris
  W & C Australia
Colluricincla woodwardi (Sandstone Shrike-Thrush)
  C.w.woodwardi
    Northern Territory
  C.w.assimilis
    N Western Australia

PITOHUI
Pitohui kirhocephalus (Variable Pitohui)
  P.k.kirhocephalus
    NW New Guinea
  P.k.salvadorii
    NW New Guinea
  P.k.dohertyi
    NW New Guinea
  P.k.rubiensis
    NW New Guinea
  P.k.stramineipectus
    SW New Guinea
  P.k.decipiens
    SW New Guinea
  P.k.adiensis
    Adi I
  P.k.carolinae
    SW New Guinea
  P.k.brunneivertex
    W New Guinea
  P.k.jobiensis
    Kurudu, Japen Is
  P.k.meyeri
    N New Guinea
  P.k.senex
    N New Guinea
  P.k.brunneicaudus
    N New Guinea
  P.k.meridionalis
    SE New Guinea
  P.k.brunneiceps
    S New Guinea
  P.k.nigripectus
    S New Guinea
  P.k.aruensis
    Aru I
  P.k.uropygialis
    Salawati, Misol Is
  P.k.tibialis
    NW New Guinea
  P.k.pallidus
    Sagewin, Batanta Is
  P.k.cerviniventris
    Waigeu I
Pitohui dichrous (Black-headed Pitohui)
  P.d.dichrous
    N New Guinea
  P.d.monticola
    C New Guinea

Pitohui incertus (Mottle-breasted Pitohui)
  S New Guinea
Pitohui ferrugineus (Rusty Pitohui)
  P.f.leucorhynchus
    Waigeu I
  P.f.fuscus
    Batanta I
  P.f.brevipennis
    Aru I
  P.f.ferrugineus
    Misol I, NW New Guinea
  P.f.holerythrus
    Japen I, N New Guinea
  P.f.clarus
    SE New Guinea
Pitohui cristatus (Crested Pitohui)
  P.c.cristatus
    W New Guinea
  P.c.arthuri
    N & S New Guinea
  P.c.kodonophonos
    SE New Guinea
Pitohui nigrescens (Black Pitohui)
  P.n.nigrescens
    NW New Guinea
  P.n.wandamensis
    N New Guinea
  P.n.meeki
    C New Guinea
  P.n.burgersi
    N & C New Guinea
  P.n.schistaceus
    SE New Guinea
  P.n.harterti
    E New Guinea
Pitohui tenebrosus (Morning Bird)
  Palau Is

TURNAGRA
Turnagra capensis (New Zealand Thrush)
  T.c.turnagra
    North I, (New Zealand)
  T.c.capensis
    South I, (New Zealand)

144. AEGITHALIDAE (LONG-TAILED TITS)

AEGITHALOS
Aegithalos caudatus (Long-tailed Tit)
  A.c.caudatus
    N Europe, N Asia, N Korea
  A.c.rosaceus
    British Isles
  A.c.europaeus
    C Europe
  A.c.aremoricus
    NW & C France
  A.c.taiti
    N Iberia, S France

A.c.irbii
S Iberia, Corsica
A.c.italiae
Italy
A.c.siculus
Sicily
A.c.macedonicus
Albania, Greece
A.c.tauricus
S Russia
A.c.major
Caucasus
A.c.tephronotus
Asia Minor
A.c.alpinus
N Iran
A.c.passekii
SW Iran
A.c.vinaceus
N & W China
A.c.glaucogularis
C China
A.c.trivirgatus
Honshu I
A.c.kiusiuensis
S Japan
A.c.magnus
S Korea
Aegithalos leucogenys (White-cheeked Tit)
Afghanistan to NW India
Aegithalos concinnus (Red-headed Tit)
A.c.iredalei
Pakistan, W Himalayas
A.c.rubricapillus
E Himalayas, Assam
A.c.manipurensis
S Assam, NE India, W Burma
A.c.talifuensis
NE Burma, W China, N Vietnam
A.c.pulchellus
E Burma
A.c.concinnus
C & E China, Taiwan
A.c.annamensis
S Indochina
Aegithalos iouschistos (Blyth's Long-tailed Tit)
A.i.niveogularis
Pakistan, W Himalayas
A.i, iouschistos
E Himalayas, SE Tibet
A.i.bonvaloti
NE Burma, SW China
A.i.obscuratus
W Szechwan
A.i.sharpei
E Burma
Aegithalos fuliginosus (Sooty Long-tailed Tit)
W & C China

PSALTRIA
Psaltria exilis (Pygmy Tit)
W & C Java

PSALTRIPARUS
Psaltriparus minimus (Common Bushtit)
P.m.minimus
W USA
P.m.californicus
S Oregon, California
P.m.sociabilis
S California
P.m.melanurus
NW Baja California
P.m.grindae
S Baja California
P.m.plumbeus
WC & S USA, N Mexico
P.m.providentialis
SE California, S Nevada
P.m.cecaumenorum
NW Mexico
Psaltriparus melanotis (Black-eared Bushtit)
P.m.lloydi
S USA, N Mexico
P.m.dimorphicus
N Mexico
P.m.iulus
W & C Mexico
P.m.melanotis
S Mexico, Guatemala

145. REMIZIDAE (PENDULINE TITS)

REMIZ
Remiz pendulinus (Penduline Tit)
R.p.pendulinus
S & E Europe, Asia Minor, W Siberia
R.p.caspius
N & W Caspian Sea
R.p.coronatus
C Asia, NW India
R.p.macronyx
WC Asia, N Iran
R.p.nigricans
E Iran
R.p.stoliczkae
N Mongolia
R.p.consobrinus
Manchuria, N China, Korea

ANTHOSCOPUS
Anthoscopus punctifrons (Sennar Kapok Tit)
S Sahara
Anthoscopus parvulus (Yellow Penduline Tit)
Senegal to Sudan
Anthoscopus musculus (Mouse-coloured Tit)
NE & E Africa
Anthoscopus flavifrons (Yellow-fronted Tit)
A.f.waldroni
Ghana
A.f.flavifrons
Gabon, Cameroun, N Zaire
A.f.ruthae
E Zaire

Anthoscopus caroli (African Penduline Tit)
  A.c.roccatii
    S Uganda
  A.c.taruensis
    S Kenya, N Tanzania
  A.c.pallescens
    Tanzania
  A.c.ansorgei
    Angola, Zaire
  A.c.rhodesiae
    SE Zaire, NE Zambia, S Tanzania
  A.c.robertsi
    NE Zambia to Mozambique
  A.c.caroli
    Namibia to S Mozambique
  A.c.winterbottomi
    NW Zambia
  A.c.rankinei
    Zambia
  A.c.hellmayri
    Natal to E Zimbabwe
Anthoscopus sylviella (Rungwe Penduline Tit)
    S Kenya, Tanzania
Anthoscopus minutus (Southern Kapok Tit)
  A.m.damarensis
    Zimbabwe, Namibia, W Transvaal
  A.m.gigi
    S Karroo
  A.m.minutus
    W Cape Province

AURIPARUS
Auriparus flaviceps (Verdin)
  A.f.flaviceps
    Baja California, NW Mexico
  A.f.acaciarum
    SW USA, NW Mexico
  A.f.hidalgensis
    N Mexico
  A.f.sinaloae
    N Mexico
  A.f.ornatus
    SW USA, N Mexico
  A.f.fraterculus
    N Mexico

CEPHALOPYRUS
Cephalopyrus flammiceps (Fire-capped Tit Warbler)
  C.f.flammiceps
    Pakistan, W Himalayas, N India
  C.f.olivaceus
    E Himalayas, W China

146 PARIDAE (TITS, CHICKADEES)

PARUS
Parus palustris (Marsh Tit)
  P.p.palustris
    Italy, NW Europe
  P.p.brandtii
    N Caucasus

P.p.brevirostris
    C Asia, Manchuria, N China
P.p.hensoni
    S Kurile Is, N Japan
P.p.hellmayri
    S Korea, China
P.p.hypermelaena
    W China, E Burma
Parus lugubris (Sombre Tit)
  P.l.lugubris
    Hungary, N Greece
  P.l.lugens
    C & S Greece
  P.l.anatoliae
    Asia Minor
Parus hyrcanus (Iranian Sombre Tit)
  P.h.hyrcanus
    N Iran
  P.h.dubius
    W Iran
  P.h.kirmanensis
    SE Iran
  P.h.talischensis
    Azerbaidjan (USSR)
Parus montanus (Willow Tit)
  P.m.loennbergi
    Lapland, N Russia
  P.m.borealis
    NE & EC Europe, Siberia
  P.m.montanus
    SE Europe
  P.m.salicarius
    NW Europe
  P.m.kamtschatkensis
    N Kurile Is, Kamchatka
  P.m.sachalinensis
    S Kurile, Sakhalin Is
  P.m.restrictus
    N Japan
  P.m.songarus
    C Asia
  P.m.affinis
    NW China
  P.m.stoetzneri
    N China, SW Manchuria
  P.m.weigoldicus
    W China
Parus atricapillus (Black-capped Chickadee)
  P.a.turneri
    NW Canada
  P.a.occidentalis
    W Canada, W USA
  P.a.septentrionalis
    WC Canada, C USA
  P.a.nevadensis
    WC USA
  P.a.atricapillus
    E Canada, NE USA
  P.a.bartletti
    Newfoundland
  P.a.practicus
    NE USA
Parus carolinensis (Carolina Chickadee)

P.c.atricapilloides
  SC USA
P.c.agilis
  S USA
P.c.carolinensis
  SE USA
P.c.extimus
  E USA
P.c.impiger
  Florida
Parus sclateri (Mexican Chickadee)
P.s.eidos
  S USA, N Mexico
P.s.garzai
  Mexico
P.s.sclateri
  SC Mexico
P.s.rayi
  S Mexico
Parus gambeli (Mountain Chickadee)
P.g.abbreviatus
  W Canada, NW USA
P.g.inyoensis
  W USA
P.g.gambeli
  SW USA
P.g.baileyae
  SW California
P.g.atratus
  N Baja California
Parus superciliosus (White-browed Tit)
  W China
Parus davidi (Père David's Tit)
  W China
Parus cinctus (Siberian Tit)
P.c.lapponicus
  Lapland
P.c.cinctus
  Siberia
P.c.sayanus
  C Asia
P.c.lathami
  NW Alaska
Parus hudsonicus (Boreal Chickadee)
P.h.columbianus
  W Canada
P.h.cascadensis
  NW USA
P.h.hudsonicus
  Canada
P.h.littoralis
  SE Canada, NE USA
**Parus rufescens (Chestnut-backed
Chickadee)**
P.r.rufescens
  W Canada, W USA
P.r.neglectus
  California
P.r.barlowi
  S California
Parus wollweberi (Bridled Titmouse)
P.w.vandevenderi
  Arizona

P.w.phillipsi
  SW USA, NW Mexico
P.w.wollweberi
  C & S Mexico
P.w.caliginosus
  SW Mexico
Parus rubidiventris (Black-crested Tit)
P.r.rubidiventris
  C Himalayas
P.r.beavani
  E Himalayas, W China, NE Burma
P.r.saramatii
  NW Burma
Parus rufonuchalis (Rufous-naped Tit)
  W & C Asia, N India
Parus melanolophus (Vigors' Crested Tit)
  Pakistan, W Himalayas
Parus ater (Coal Tit)
P.a.ater
  Europe, Siberia
P.a.britannicus
  Great Britain
P.a.hibernicus
  Ireland
P.a.vieirae
  Spain, Portugal
P.a.sardus
  Corsica, Sardinia
P.a.atlas
  N Morocco
P.a.ledouci
  N Africa
P.a.cypriotes
  Cyprus
P.a.moltchanovi
  S Russia
P.a.michalowskii
  Caucasus
P.a.derjugini
  N Armenia
P.a.gaddi
  Iran
P.a.chorassanicus
  NE Iran
P.a.phaeonotus
  SW Iran
P.a.rufipectus
  C Asia
P.a.aemodius
  E Himalayas, W China, N Burma
P.a.pekinensis
  N China
P.a.insularis
  Japan
P.a.kuatunensis
  SE China
P.a.ptilosus
  Taiwan
Parus venustulus (Yellow-bellied Tit)
  S & W China
Parus elegans (Elegant Tit)
P.e.edithae
  Babuyan I

P.e.montigenus
 N Luzon I
P.e.gilliardi
 Batjan I (Luzon I)
P.e.elegans
 S Luzon, Panay, Mindoro Is
P.e.visayanus
 Cebu I
P.e.albescens
 Guimares, Masbate, Negros Is
P.e.mindanensis
 Mindanao I
P.e.suluensis
 Tawitawi, Sulu Is
P.e.bongaoensis
 Bongao I
Parus amabilis (Palawan Tit)
 Balabac, Palawan Is
Parus cristatus (Crested Tit)
P.c.cristatus
 N & E Europe, Alps
P.c.scoticus
 NC Scotland
P.c.abadiei
 NW France
P.c.weigoldi
 S & W Iberia
P.c.mitratus
 C & W Europe
P.c.baschkirikus
 C Russia
Parus dichrous (Brown Crested Tit)
P.d.kangrae
 Pakistan, NW Himalayas
P.d.dichrous
 C & E Himalayas
P.d.dichroides
 NW China
P.d.wellsi
 SW China, NE Burma
Parus afer (Southern Grey Tit)   146.1
P.a.thruppi
 Ethiopia, Somalia
P.a.barakae
 Uganda, Kenya, Tanzania
P.a.benguelae
 SW Angola
P.a.arens
 Lesotho, E Cape Province
P.a.afer
 W Cape Province
Parus cinerascens (Acacia Grey Tit)   146.1
P.c.cinerascens
 Angola to E Cape Province
P.c.orphnus
 Leostho to E Transvaal
Parus griseiventris (Miombo Grey Tit)
 Zambia, Tanzania, Zimbabwe
Parus niger (Southern Black Tit)
P.n.ravidus
 S Zimbabwe, Transvaal, E Zambia
P.n.niger
 Southern Africa

P.n.xanthostomus
 S Zambia
Parus leucomelas (White-winged Black Tit)
P.l.guineensis
 W & WC Africa
P.l.leucomelas
 Ethiopia
P.l.insignis
 C & SC Africa
Parus carpi (Carp's Tit)   146.1
 SW Angola, N Namibia
Parus albiventris (White-breasted Tit)
 Nigeria to Sudan, Tanzania
Parus leuconotus (White-backed Black Tit)
 Ethiopia
Parus funereus (Dusky Tit)
P.f.funereus
 Cameroun to Kenya
P.f.gabela
 Angola
Parus fasciiventer (Stripe-breasted Tit)
P.f.fasciiventer
 Rwanda (E Zaire)
P.f.tanganjicae
 S Kivu (E Zaire)
P.f.kaboboensis
 Mt Kabobo (SE Zaire)
Parus fringillinus (Red-throated Tit)
 S Kenya, N & C Tanzania
Parus rufiventris (Rufous-bellied Tit)
P.r.rufiventris
 WC Africa
P.r.masukuensis
 Zambia, Malawi
P.r.pallidiventris
 Tanzania, Malawi
P.r.diligens
 NE Namibia, S Angola
P.r.stenotopicus
 E Zimbabwe, W Mozambique
Parus major (Great Tit)
P.m.newtoni
 British Isles
P.m.major
 Europe, Asia Minor, C Asia
P.m.excelsus
 NW Africa
P.m.corsus
 Corsica, Sardinia
P.m.aphrodite
 S Greece, Mediterranean islands
P.m.terraesanctae
 Lebanon, Israel, Jordan, Syria
P.m.blanfordi
 Iran
P.m.karelini
 NW Iran
P.m.intermedius
 W Iran
P.m.kapustini
 C Asia
P.m.caschmirensis
 NW India, Pakistan

P.m.decolorans
  E Afghanistan
P.m.ziaratensis
  N Baluchistan, S Afghanistan
P.m.mahrattarum
  S India, Sri Lanka
P.m.stupae
  C & W India
P.m.nipalensis
  Nepal, N India, W Burma
P.m.vauriei
  E Assam
P.m.templorum
  W Thailand, S Indochina
P.m.cinereus
  Java, Lesser Sunda Is
P.m.ambiguus
  SE Burma, Malaysia, Sumatra
P.m.sarawacensis
  W Sarawak
P.m.hainanus
  Hainan I
P.m.nigriloris
  S Riukiu Is
P.m.commixtus
  S China, N Vietnam
P.m.okinawae
  C Riukiu Is
P.m.amamiensis
  N Riukiu Is
P.m.kagoshimae
  S Kyushu, Goto Is
P.m.dageletensis
  Dagelet I (Japan)
P.m.minor
  Japan, E Asia, N China, E Tibet
P.m.tibetanus
  Tibet, SW China, N Burma
P.m.nubicolus
  E Burma, N Thailand, W Indochina
Parus bokharensis (Turkestan Tit)
P.b.bokharensis
  Russia, C Asia
P.b.turkestanicus
  W Mongolia
Parus monticolus (Green-backed Tit)
P.m.monticolus
  W Himalayas
P.m.yunnanensis
  E Himalayas, Burma, W China
P.m.legendrei
  S Vietnam
P.m.insperatus
  Taiwan
Parus nuchalis (White-naped Tit)
  NW India
Parus xanthogenys (Black-spotted Yellow Tit)
P.x.xanthogenys
  W Himalayas
P.x.aplonotus
  C India

P.x.travancoreensis
  S India
Parus spilonotus (Chinese Yellow Tit)
P.s.spilonotus
  E Himalayas
P.s.subviridis
  Burma, Thailand, Assam
P.s.rex
  S China, N Vietnam
P.s.basileus
  S Indochina
Parus holsti (Formosan Yellow Tit)
  Taiwan
Parus caeruleus (Blue Tit)
P.c.obscurus
  British Isles
P.c.caeruleus
  N & E Europe
P.c.ogliastrae
  S Iberia, Corsica, Sardinia
P.c.balearicus
  Majorca I
P.c.orientalis
  E & C Russia
P.c.satunini
  Caucasus, NW Iran
P.c.raddei
  N Iran
P.c.persicus
  SW Iran
P.c.ultramarinus
  NW Africa
P.c.cyrenaicae
  Libya
P.c.ombriosus
  Hierro I (Canary Is)
P.c.palmensis
  Palma I
P.c.teneriffae
  Grand Canary, Tenerife Is
P.c.degener
  Fuerteventura, Lanzarote Is
Parus cyanus (Azure Tit)
P.c.cyanus
  W Russia, WC Asia
P.c.yenisseensis
  C Asia
P.c.tianschanicus
  C & E Asia, Manchuria
P.c.kotkalensis
  S Russia
P.c.flavipectus
  WC Asia
P.c.carruthersi
  N Iran
P.c.berezowskii
  NW China
Parus varius (Varied Tit)
P.v.varius
  Japan, Korea
P.v.sunsunpi
  S Japanese islands

P.v.amamii
Amami I
P.v.orii
C Riukiu Is
P.v.olivaceus
S Riukiu Is
P.v.castaneoventris
Taiwan
P.v.namiyei
N Izu Is
P.v.owstoni
S Izu Is
Parus semilarvatus (White-fronted Tit)
P.s.snowi
N Luzon I
P.s.semilarvatus
C & S Luzon, Negros Is
P.s.nehrkorni
Mindanao I
Parus inornatus (Plain Titmouse)
P.i.sequestratus
SW Oregon, NW California
P.i.zaleptus
SE Oregon, E California, W Nevada
P.i.inornatus
WC California
P.i.kernensis
SC California
P.i.mohavensis
SE California
P.i.transpositus
SW California
P.i.affabilis
N Baja California
P.i.cineraceus
S Baja California
P.i.ridgwayi
WC USA
P.i.plumbescens
SW New Mexico, SW Arizona
Parus bicolor (Tufted Titmouse)
P.b.bicolor
E, C & SE USA
P.b.sennetti
C & S Texas
P.b.paloduro
N Texas
P.b.dysleptus
W Texas, N Mexico
P.b.atricristatus
S Texas, NE Mexico

MELANOCHLORA
Melanochlora sultanea (Sultan Tit)
M.s.sultanea
E Himalayas, Assam, Burma, N Thailand
M.s.flavocristata
S Burma, Malaysia, Sumatra
M.s.seorsa
S China, Hainan I, N Indochina
M.s.gayeti
C Vietnam

SYLVIPARUS
Sylviparus modestus (Yellow-browed Tit)
S.m.simlaensis
NW Himalayas
S.m.modestus
C & E Himalayas, N Burma, SW China, N Laos
S.m.klossi
S Vietnam

147. SITTIDAE (NUTHATCHES)

SITTINAE

SITTA
Sitta europaea (Eurasian Nuthatch)
S.e.europaea
NW Europe
S.e.asiatica
Russia, N Asia, N Japan
S.e.seorsa
W Sinkiang
S.e.amurensis
Manchuria, Korea, C Japan
S.e.arctica
N Siberia
S.e.albifrons
Kamchatka
S.e.roseilia
S Japan
S.e.bedfordi
Quelpart I
S.e.caesia
WC Europe, N Mediterranean
S.e.hispaniensis
Spain, NW Africa
S.e.levantina
Israel, Lebanon, Turkey
S.e.persica
W Iran
S.e.caucasica
Caucasus
S.e.rubiginosa
N Iran, SE Russia
S.e.sinensis
W & S China, Taiwan
S.e.montium
SE Tibet
S.e.nagaensis
N India, Assam, N Burma, N Thailand, SW China
S.e.griseiventris
S Burma, S Vietnam
S.e.nebulosa
C China
S.e.whistleri
S Thailand
Sitta castanea (Chestnut-bellied Nuthatch)
S.c.cashmirensis
Pakistan, NW India
S.c.almorae
W Himalayas

S.c.cinnamoventris
  E Himalayas, Bangladesh
S.c.koelzi
  E Assam, N Burma
S.c.neglecta
  Burma, S Thailand, S Laos, S Vietnam
S.c.castanea
  C India
S.c.prateri
  EC India
S.c.tonkinensis
  N Thailand, N Laos, N Vietnam
Sitta himalayensis (White-tailed Nuthatch)
S.h.himalayensis
  Himalayas, Assam, Nepal
S.h.australis
  S Assam, Burma, N Vietnam
Sitta victoriae (White-browed Nuthatch)
  W Burma
Sitta pygmaea (Pygmy Nuthatch)
S.p.pygmaea
  W California
S.p.melanotis
  SW Canada, W USA, NW Mexico
S.p.canescens
  Nevada
S.p.leuconucha
  S California, Baja California
S.p.chihuahuae
  NW Mexico
S.p.elii
  N Mexico
S.p.brunnescens
  W Mexico
S.p.flavincuha
  E Mexico
Sitta pusilla (Brown-headed Nuthatch)
S.p.pusilla
  S USA
S.p.caniceps
  Florida
S.p.insularis
  Gd Bahama I
Sitta whiteheadi (Corsican Nuthatch)
  Corsica
Sitta yunnanensis (Yunnan Nuthatch)
  W China
Sitta canadensis (Red-breasted Nuthatch)
  Canada, USA
Sitta villosa (Chinese Nuthatch)
S.v.bangsi
  W China
S.v.villosa
  NE China
Sitta leucopsis (White-cheeked Nuthatch)
S.l.leucopsis
  Pakistan, W Himalayas
S.l.przewalskii
  SE Tibet, NW China
Sitta carolinensis (White-breasted Nuthatch)
S.c.aculeata
  W USA

S.c.tenuissima
  SW Canada, NW & W USA
S.c.atkinsi
  Florida
S.c.lagunae
  S Baja California
S.c.nelsoni
  C & S USA, N Mexico
S.c.alexandrae
  N Mexico
S.c.umbrosa
  N Mexico
S.c.mexicana
  C Mexico
S.c.oberholseri
  C Mexico
S.c.kinneari
  SW Mexico
S.c.carolinensis
  E Canada, E USA
Sitta krueperi (Krüper's Nuthatch)
  Turkey, Caucasus
Sitta ledanti (Kabylie Nuthatch)
  N Algeria
Sitta neumayer (Rock Nuthatch)
S.n.neumayer
  SE Europe
S.n.syriaca
  Turkey, N Israel
S.n.rupicola
  Caucasus, Iran
S.n.tschitscherini
  Iraq, Iran
S.n.plumbea
  SE Iran
Sitta tephronota (Eastern Rock Nuthatch)
S.t.tephronota
  C Asia, N Iran, Afghanistan, Pakistan
S.t.obscura
  N & E Iran
S.t.dresseri
  SW Asia
Sitta frontalis (Velvet-fronted Nuthatch)
S.f.frontalis
  India, Burma, N Thailand, N Vietnam, S
  Sumatra, Java
S.f.saturatior
  Malaysia, N Sumatra
S.f.corallipes
  Borneo
S.f.palawana
  Palawan I
S.f.isarog
  NE, E & S Luzon I
S.f.mesoleuca
  N Luzon I
S.f.oenochlamys
  Cebu, Panay, Negros Is
S.f.lilacea
  Samar, Leyte Is
S.f.apo
  SE Mindanao I

S.f.zamboanga
 Mindanao, Basilan Is
Sitta solangiae (Lilac Nuthatch)
 S.s.solangiae
 N Vietnam
 S.s.fortior
 C & S Vietnam
Sitta azurea (Azure Nuthatch)
 S.a.expectata
 Malaysia, Sumatra
 S.a.nigriventer
 W Java
 S.a.azurea
 E Java
Sitta magna (Giant Nuthatch)
 S.m.ligea
 SW China
 S.m.magna
 C Burma, N Thailand
Sitta formosa (Beautiful Nuthatch)
 E Himalayas to N Laos

TICHODROMADINAE

TICHODROMA
Tichodroma muraria (Wallcreeper)
 T.m.muraria
 S & E Europe, Turkey, NW Iran
 T.m.nepalensis
 C Asia, Pakistan, Himalayas, China

DAPHOENOSITTINAE

NEOSITTA
Neositta chrysoptera (Varied Sitella)
 N.c.pileata
 C Australia
 N.c.lathami
 E Victoria
 N.c.chrysoptera
 E New South Wales
 N.c.leucocephala
 C & SE Queensland
 N.c.lumholzi
 E Queensland
 N.c.albata
 Bowen, E Queensland
 N.c.magnirostris
 NE Queensland
 N.c.rothschildi
 N Queensland
 N.c.striata
 N & NW Queensland
 N.c.leucoptera
 Northern Territory, NC Australia
Neositta papuensis (Papuan Sitella)
 N.p.toxopeusi
 NW New Guinea
 N.p.intermedia
 NW New Guinea
 N.p.wahgiensis
 W New Guinea
 N.p.papuensis
 W New Guinea

N.p.alba
 C New Guinea
N.p.albifrons
 SE New Guinea

DAPHOENOSITTA
Daphoenositta miranda (Pink-faced Nuthatch)
 D.m.miranda
 SE New Guinea
 D.m.kuboriensis
 NE New Guinea
 D.m.frontalis
 NW New Guinea

148. CERTHIIDAE (TREECREEPERS)

CERTHIINAE

CERTHIA
Certhia familiaris (Common Treecreeper)
 C.f.brittannica
 Britain, Ireland
 C.f.macrodactyla
 C & S Europe
 C.f.pyrenaica
 Pyrenees
 C.f.familiaris
 N & E Europe, W Siberia
 C.f.corsa
 Corsica
 C.f.persica
 Caucasus, N Iran
 C.f.tianschanica
 Russian & Chinese Turkestan
 C.f.hodgsoni
 Pakistan, W Himalayas
 C.f.mandellii
 E Himalayas
 C.f.bianchii
 W China
 C.f.khamensis
 SE Tibet, SW China, N Burma
 C.f.daurica
 E Siberia, N Mongolia, N Korea, N Japan
Certhia americana (American Treecreeper)
 C.a.montana
 W Canada, W USA
 C.a.occidentalis
 NW Canada, W USA
 C.a.alascensis
 Alaska
 C.a.stewarti
 British Columbia
 C.a.zelotes
 S California
 C.a.leucosticta
 S Nevada, Utah
 C.a.idahoensis
 Idaho
 C.a.albescens
 SW USA, NW Mexico

C.a.molinensis
  C Mexico
C.a.jaliscensis
  SW Mexico
C.a.guerrerensis
  SW Mexico
C.a.alticola
  SE Mexico
C.a.pernigra
  S Mexico, Guatemala
C.a.extima
  Nicaragua
C.a.americana
  C & E Canada, CE & SE USA
C.a.nigrescens
  EC USA
Certhia brachydactyla (Short-toed Treecreeper)
  C & E Europe, Caucasus
Certhia himalayana (Himalayan Treecreeper)
C.h.taeniura
  SW Asia, Afghanistan
C.h.himalayana
  N Pakistan, W Himalayas
C.h.limes
  Pakistan, NW India
C.h.intima
  W Nepal
C.h.yunnanensis
  W China
C.h.ripponi
  N Burma
Certhia nipalensis (Stoliczka's Treecreeper)
  SE Tibet, C Nepal, NE Burma
Certhia discolor (Brown-throated Treecreeper)
C.d.discolor
  E Himalayas, Nepal, Assam
C.d.manipurensis
  E Assam, W Burma
C.d.shanensis
  N Burma, N Thailand
C.d.laotiana
  Laos
C.d.meridionalis
  S Vietnam

SALPORNITHINAE

SALPORNIS
Salpornis spilonotus (Spotted Grey Creeper)
S.s.emini
  Portuguese Guinea to Sudan, Uganda
S.s.erlangeri
  SW Ethiopia
S.s.salvadori
  Angola to Tanzania & Mozambique
S.s.rajputanae
  NW India
S.s.spilonotus
  N & C India
S.s.xylodromus
  E Zimbabwe, W Mozambique

149. RHABDORNITHIDAE (PHILIPPINE CREEPERS)

RHABDORNIS
Rhabdornis mysticalis (Stripe-headed Creeper)
R.m.mysticalis
  Luzon, Masbate, Negros, Panay Is
R.m.minor
  Samar, Leyte, Mindanao Is
Rhabdornis inornatus (Plain-headed Creeper)
R.i.grandis
  N Luzon I
R.i.inornatus
  Samar I
R.i.rabori
  Negros I
R.i.alaris
  Mindanao I
R.i.zamboanga
  Mt Malindang (Mindanao I)
R.i.leytensis
  Leyte I

150. CLIMACTERIDAE (AUSTRALIAN CREEPERS)

CLIMACTERIS
Climacteris erythrops (Red-browed Treecreeper)
C.e.erythrops
  E New South Wales, E & S Victoria
C.e.olinda
  S Victoria
Climacteris affinis (White-browed Treecreeper)
C.a.superciliosus
  WC Australia
C.a.affinis
  C Australia
Climacteris picumnus (Brown Treecreeper)
C.p.melanota
  N Queensland
C.p.picumnus
  S & E Australia
Climacteris rufa (Rufous Treecreeper)
  SW Western Australia
Climacteris melanura (Black-tailed Treecreeper)
C.m.melanura
  N Western Australia, Northern Territory, NW Queensland
C.m.wellsi
  NW Western Australia
Climacteris leucophaea (White-throated Treecreeper)
C.l.minor
  N Queensland
C.l.intermedia
  E Queensland
C.l.leucophaea
  E Australia

C.l.grisescens
S South Australia
Climacteris placens **(Papuan Treecreeper)**
C.p.placens
NW New Guinea
C.p.steini
W New Guinea
C.p.inexpectata
N New Guinea
C.p.meridionalis
SE New Guinea

151. DICAEIDAE (FLOWERPECKERS)

MELANOCHARIS
Melanocharis arfakiana **(Obscure Berrypecker)**
New Guinea
Melanocharis nigra **(Black Berrypecker)**
M.n.pallida
Waigeu I
M.n.nigra
Misol I, W New Guinea
M.n.unicolor
Japen I, N & E New Guinea
M.n.chloroptera
Aru Is, S New Guinea
Melanocharis longicauda **(Mid-mountain Berrypecker)**
M.l.longicauda
NW New Guinea
M.l.chloris
NW New Guinea
M.l.umbrosa
NW New Guinea
M.l.captata
C & E New Guinea
M.l.orientalis
SE New Guinea
Melanocharis versteri **(Fan-tailed Berrypecker)**
M.v.versteri
NW New Guinea
M.v.meeki
NW New Guinea
M.v.virago
N & NE New Guinea
M.v.maculiceps
SE New Guinea
Melanocharis striativentris **(Streaked Berrypecker)**
M.s.axillaris
NW New Guinea
M.s.striativentris
C & SE New Guinea
M.s.prasina
SE New Guinea
M.s.chrysocome
E New Guinea

RHAMPHOCHARIS
Rhamphocharis crassirostris **(Spotted Berrypecker)**

R.c.crassirostris
NW & C New Guinea
R.c.piperata
SE New Guinea
R.c.viridescens
SE New Guinea

PRIONOCHILUS
Prionochilus olivaceus **(Olive-backed Flowerpecker)**
P.o.parsoni
NE Luzon I
P.o.olivaceus
Basilan, Mindanao, Bohol Is
P.o.samarensis
Samar, Leyte Is
Prionochilus maculatus **(Yellow-throated Flowerpecker)**
P.m.septentrionalis
S Burma, S Thailand
P.m.oblitus
Malaysia
P.m.maculatus
Sumatra, Billiton I, Nias I, Borneo
P.m.natunensis
Great Natuna I
Prionochilus percussus **(Crimson-breasted Flowerpecker)**
P.p.ignicapilla
S Burma to Sumatra & Borneo
P.p.regulus
Batu I
P.p.percussus
Java
Prionochilus plateni **(Palawan Yellow-rumped Flowerpecker)**
Palawan, Culion Is
Prionochilus xanthopygius **(Borneo Yellow-rumped Flowerpecker)**
Borneo
Prionochilus thoracicus **(Scarlet-breasted Flowerpecker)**
Malaysia, Billiton I, Borneo

DICAEUM
Dicaeum annae **(Sunda Flowerpecker)** 151.1
Flores, Sumbawa
Dicaeum agile **(Thick-billed Flowerpecker)**
D.a.agile
N India
D.a.zeylonicum
Sri Lanka
D.a.modestum
Bangladesh, Burma, Thailand, N Vietnam
D.a.remotum
S Burma, S Thailand, Malaysia
D.a.atjehense
N Sumatra
D.a.finschi
W Java
D.a.tinctum
Sumba, Flores, Alor Is

D.a.obsoletum
Timor I
D.a.striatissimum
Sibuyan, Luzon Is
D.a.aeruginosum
Cebu, Negros, Mindoro, Mindanao Is
D.a.affine
Palawan I
**Dicaeum everetti (Brown-backed
Flowerpecker)**   151.2
D.e.everetti
Labuan I, Borneo, Malaysia
D.e.bungurense
Great Natuna I
**Dicaeum proprium (Grey-breasted
Flowerpecker)**
Mt Mayo (Mindanao I)
**Dicaeum chrysorrheum (Yellow-vented
Flowerpecker)**
D.c.chrysoclore
E Himalayas to SW China, Indochina
D.c.chrysorrheum
S Thailand to Sumatra, Borneo, Java
**Dicaeum melanoxanthum (Yellow-bellied
Flowerpecker)**
E Himalayas to SW China
Dicaeum vincens (Legge's Flowerpecker)
Sri Lanka
**Dicaeum aureolimbatum (Yellow-sided
Flowerpecker)**
D.a.aureolimbatum
Muna, Buton Is, Sulawesi
D.a.laterale
Great Sanghir I
**Dicaeum nigrilore (Olive-capped
Flowerpecker)**
Mindanao I
**Dicaeum anthonyi (Yellow-crowned
Flowerpecker)**
D.a.anthonyi
Cagayan, Luzon Is
D.a.masawan
NW Mindanao I
D.a.kampalili
SE Mindanao I
Dicaeum bicolor (Bicoloured Flowerpecker)
D.b.inexpectatum
Luzon, Mindanao, Leyte, Samar Is
D.b.bicolor
Mindanao I
D.b.viridissimum
Negros I
Dicaeum australe (Philippine Flowerpecker)
D.a.australe
Philippine Is
D.a.haematostictum
Panay, Negros Is
**Dicaeum retrocinctum (Mindoro
Flowerpecker)**
Mindoro I
**Dicaeum trigonostigma (Orange-bellied
Flowerpecker)**

D.t.rubropygium
Assam, S Burma, S Thailand
D.t.trigonostigma
S Thailand
D.t.melanostigma
Bangka, Billiton Is, Malaysia, Sumatra
D.t.antioproctum
Simalur I
D.t.megastoma
Great Natuna I
D.t.flaviclunis
Java, Bali
D.t.dayakanum
Borneo, N Bornean islands
D.t.sibutuense
Sibutu I
D.t.assimile
Tawitawi, Jolo, Siasi Is
D.t.cinereigulare
Mindanao, Samar, Leyte, Bohol Is
D.t.besti
Siquijor I
D.t.dorsale
Masbate, Panay, Negros Is
D.t.intermedium
Romblon, Tablas Is
D.t.sibuyanicum
Sibuyan I
D.t.isidroi
Camiguin I (South)
D.t.xanthopygium
Marinduque, Mindoro, Luzon Is
**Dicaeum hypoleucum (White-bellied
Flowerpecker)**
D.h.lagunae
N & C Luzon I
D.h.pontifex
Bohol, Samar, Leyte, Mindanao Is
D.h.hypoleucum
W Mindanao, Basilan, Sulu Is
D.h.cagayanensis
NE Luzon I
**Dicaeum erythrorhynchos (Tickell's
Flowerpecker)**
D.e.erythrorhynchos
W Burma, Bangladesh, India
D.e.ceylonense
Sri Lanka
Dicaeum concolor (Plain Flowerpecker)
D.c.olivaceum
Himalayas to S China, N Indochina
D.c.concolor
SW India
D.c.virescens
S Andaman Is
D.c.minullum
Hainan I
D.c.uchidai
Taiwan
D.c.borneanum
Malaysia, Sumatra, Borneo
D.c.sollicitans
Java, Bali

Dicaeum pygmaeum (Palawan Flowerpecker)
D.p.salomonseni
N Luzon I
D.p.fugaensis
Fuga I (N Luzon I)
D.p.pygmaeum
S Luzon, Mindoro, Negros, Leyte, Cebu Is
D.p.davao
Mindanao I
D.p.palawanorum
Balabac, Palawan Is
Dicaeum nehrkorni (Red-headed
Flowerpecker)
Sulawesi
Dicaeum vulneratum (Ashy-fronted
Flowerpecker)
S Moluccas
Dicaeum erythrothorax (White-throated
Flowerpecker)
D.e.schistaceiceps
Halmahera, Morotai, Obi, Batjan Is
D.e.erythrothorax
Buru I
Dicaeum pectorale (Olive-crowned
Flowerpecker)
D.p.ignotum
Gebe I
D.p.pectorale
Misol, Waigeu Is, NW New Guinea
Dicaeum geelvinkianum (Red-capped
Flowerpecker)
D.g.maforense
Numfor I
D.g.misoriense
Biak I
D.g.geelvinkianum
Japen I
D.g.obscurifrons
W New Guinea
D.g.setekwa
SW New Guinea
D.g.diversum
N New Guinea
D.g.centrale
C New Guinea
D.g.albopunctatum
SC New Guinea
D.g.rubrigulare
S New Guinea
D.g.rubrocoronatum
SE New Guinea
D.g.violaceum
D'Entrecasteaux Arch
Dicaeum nitidum (Louisiade Flowerpecker)
D.n.nitidum
Tagula, Misima Is
D.n.rosseli
Rossel I
Dicaeum eximium (New Ireland Flowerpecker)
D.e.layardorum
New Britain
D.e.eximium
New Ireland, New Hanover

D.e.phaeopygium
Dyaul I
Dicaeum aeneum (Solomon Is Flowerpecker)
D.a.aeneum
N Solomon Is
D.a.becki
Guadalcanal I
D.a.malaitae
Malaita I
Dicaeum tristrami (San Cristobal
Flowerpecker)
San Cristobal I
Dicaeum igniferum (Black-banded
Flowerpecker) 151.1
Sumbawa, Flores I to Alor Is
Dicaeum maugei (Blue-cheeked
Flowerpecker)
D.m.maugei
Samau, Timor, Sawu, Roma, Damar Is
D.m.salvadorii
Babar, Moa Is
D.m.splendidum
Saleyer, Djampea Is
D.m.neglectum
Lombok I
Dicaeum hirundinaceum (Mistletoe
Flowerpecker)
D.h.hirundinaceum
Australia
D.h.ignicolle
Aru Is
D.h.keiense
Kai Is
D.h.fulgidum
Tanimbar Is
Dicaeum celebicum (Black-sided
Flowerpecker)
D.c.kuehni
Tukangbesi I
D.c.sulaense
Sula, Banguey Is
D.c.celebicum
Muna, Buton Is, Sulawesi
D.c.sanghirense
Sanghir Is
D.c.talautense
Talaut I
Dicaeum monticolum (Bornean Fire-breasted
Flowerpecker)
Borneo
Dicaeum ignipectus (Green-backed
Flowerpecker)
D.i.ignipectus
Himalayas to S China, Indochina
D.i.dolichorhynchum
S Thailand, Malaysia
D.i.cambodianum
Cambodia, SE Thailand
D.i.formosum
Taiwan
D.i.luzoniense
N Luzon I

D.i.apo
Negros, Mindanao Is
D.i.bonga
Samar I
D.i.beccarii
N Sumatra
Dicaeum sanguinolentum (Blood-breasted
Flowerpecker) 151.1
D.s.sanguinolentum
Java, Bali
D.s.rhodopygiale
Flores I
D.s.wilhelminae
Sumba I
D.s.hanieli
Timor I
Dicaeum cruentatum (Scarlet-backed
Flowerpecker)
D.c.cruentatum
NE India to S China, Indochina
D.c.siamense
E Thailand
D.c.ignitum
Malaysia
D.c.sumatranum
Sumatra
D.c.batuense
Mentawai Is
D.c.simalurense
Simalur I
D.c.nigrimentum
Borneo
D.c.niasense
Nias I
Dicaeum trochileum (Scarlet-headed
Flowerpecker)
D.t.trochileum
Java, Bali, SE Borneo, Kangean Is
D.t.stresemanni
Lombok I

OREOCHARIS
Oreocharis arfaki (Tit Berrypecker)
New Guinea

PARAMYTHIA
Paramythia montium (Crested Berrypecker)
P.m.olivaceum
C New Guinea
P.m.montium
C & SE New Guinea
P.m.brevicauda
SE New Guinea

PARDALOTUS
Pardalotus quadragintus (Forty-spotted
Pardalote)
Tasmania
Pardalotus punctatus (Spotted Pardalote)
W Western Australia, E Australia, Tasmania

Pardalotus xanthopygus (Yellow-tailed
Pardalote)
Western Australia to NW Victoria
Pardalotus rubricatus (Red-browed
Pardalote)
P.r.parryi
N Australia
P.r.rubricatus
C Australia
P.r.carpenteriae
NW Queensland
P.r.yorki
NE Queensland
Pardalotus striatus (Yellow-tipped Pardalote)
E Australia, Tasmania
Pardalotus ornatus (Red-tipped Pardalote)
S Queensland to S Victoria
Pardalotus substriatus (Striated Pardalote)
Southern Australia
Pardalotus melanocephalus (Black-headed
Pardalote)
P.m.uropygialis
N Western Australia to NW Queensland
P.m.melvillensis
Melville I
P.m.restrictus
N Queensland
P.m.barroni
NC Queensland
P.m.bowensis
E Queensland
P.m.melanocephalus
SE Queensland, NE New South Wales

152. NECTARINIIDAE (SUNBIRDS)

ANTHREPTES
Anthreptes gabonicus (Brown Sunbird)
Gambia to Gabon
Anthreptes fraseri (Scarlet-tufted Sunbird)
A.f.cameroonensis
S Nigeria, Cameroun, Central African
Republic, N Angola
A.f.idius
Sierra Leone to Ghana
A.f.fraseri
Fernando Po I
A.f.axillaris
NE Zaire, Uganda
Anthreptes reichenowi (Plain-backed
Sunbird)
A.r.yokanae
S Kenya, NE Tanzania
A.r.reichenowi
SE Zimbabwe, Mozambique
Anthreptes anchietae (Anchieta's Sunbird)
Angola, N Zambia, SW Tanzania, Malawi,
W Mozambique
Anthreptes simplex (Plain-coloured Sunbird)
S Burma, S Thailand, Malaysia, Sumatra,
Borneo
Anthreptes malacensis (Plain-throated
Sunbird)

A.m.malacensis
S Burma to Indochina, Sumatra, S Borneo
A.m.mjobergi
Maratua Is
A.m.borneensis
N Borneo
A.m.birgitae
Luzon I
A.m.chlorigaster
WC Philippine Is, SW Mindanao I
A.m.griseigularis
Samar, Leyte, NE Mindanao Is
A.m.heliolusius
W Mindanao, Basilan Is
A.m.cagayanensis
Cagayan I
A.m.paraguae
Palawan I
A.m.wiglesworthi
Sulu Arch (except Sibutu)
A.m.iris
Sibutu I
A.m.heliocalus
Sangi Is
A.m.celebensis
S & C Sulawesi
A.m.extremus
Sula Is
A.m.convergens
Lesser Sunda Is
A.m.rubrigena
Sumba I
Anthreptes rhodolaema (Shelley's Sunbird)
S Burma, S Thailand, Malaysia, Sumatra,
Borneo
**Anthreptes singalensis (Ruby-cheeked
Sunbird)**
A.s.assamensis
E Nepal, Bangladesh, N Burma, N Thailand
A.s.internotus
S Burma, S Thailand
A.s.koratensis
E Thailand, Laos, Vietnam
A.s.interpositus
S Thailand
A.s.singalensis
Malaysia
A.s.panopsius
W Sumatra islands, Nias I
A.s.sumatranus
Sumatra, Billiton I
A.s.pallidus
N Natuna Is
A.s.borneanus
Banguey I, Borneo
A.s.phoenicotis
E & C Java
A.s.bantenensis
W Java
**Anthreptes longuemarei (Violet-backed
Sunbird)**
A.l.longuemarei
Senegal to Guinea

A.l.haussarum
Liberia to Cameroun, N Zaire, Sudan,
Uganda
A.l.angolensis
S Zaire, Angola, Zambia, Malawi, W
Tanzania
A.l.nyassae
SE Tanzania, N Mozambique, E Zimbabwe
**Anthreptes orientalis (Kenya Violet-backed
Sunbird)**
A.o.orientalis
S Sudan, Ethiopia, N Uganda, Kenya, E
Tanzania
A.o.neumanni
NE Kenya, Somalia
**Anthreptes neglectus (Uluguru Violet-backed
Sunbird)**
SE Kenya, NE Tanzania, N Mozambique
Anthreptes aurantium (Violet-tailed Sunbird)
S Nigeria, Gabon, Central African Republic,
NE Angola
Anthreptes pallidigaster (Amani Sunbird)
E Kenya, NE Tanzania
Anthreptes rectirostris (Green Sunbird)
A.r.rectirostris
Sierra Leone to Ghana
A.r.tephrolaema
Fernando Po I, S Nigeria to Angola &
Uganda
**Anthreptes rubritorques (Banded Green
Sunbird)**
NE Tanzania
Anthreptes collaris (Collared Sunbird)
A.c.subcollaris
Guinea to Nigeria
A.c.hypodilus
Fernando Po I
A.c.somereni
SE Nigeria, N & W Zaire, N Angola, SW
Sudan
A.c.jubaensis
S Ethiopia, Somalia, N Kenya
A.c.djamdjamensis
SW Ethiopia
A.c.garguensis
C & E Zaire, Uganda
A.c.elachior
E Kenya, NE Tanzania, Zanzibar I
A.c.philipsi
E Angola, SE Zaire, N Zambia
A.c.zambesianus
S Tanzania, SE Zambia, Botswana
A.c.chobiensis
N Namibia to S Mozambique
A.c.patersonae
Zimbabwe, W Mozambique
A.c.collaris
E Cape Province, S Natal, Swaziland
Anthreptes platurus (Pygmy Sunbird)
A.p.platurus
Senegal to NW Kenya
A.p.metallicus
NE Africa, SW Arabia

HYPOGRAMMA
Hypogramma hypogrammicum (Blue-naped Sunbird)
H.h.lisettae
N Burma, N Thailand, N & C Indochina
H.h.mariae
Cambodia, S Indochina
H.h.nuchale
S Burma, S Thailand, Malaysia
H.h.hypogrammicum
Sumatra, Borneo
H.h.natunense
N Natuna Is

NECTARINIA
Nectarinia seimundi (Little Green Sunbird)
N.s.kruensis
Sierra Leone to Ghana
N.s.seimundi
Fernando Po I
N.s.traylori
Nigeria to Zaire, Uganda, N Angola
Nectarinia batesi (Bates's Olive Sunbird)
Fernando Po I, S Nigeria to Zaire, Zambia
Nectarinia olivacea (Olive Sunbird)
N.o.guineensis
Guinea to W Ghana
N.o.cephaelis
E Ghana to Zaire, N Angola
N.o.obscura
Principé, Fernando Po Is
N.o.vincenti
S Sudan, NW Kenya, Uganda
N.o.ragazzii
Sudan, Ethiopia to N Zambia, N Malawi
N.o.neglecta
C Kenya, N Tanzania
N.o.changamwensis
E Kenya, E Tanzania
N.o.granti
Pemba, Zanzibar Is
N.o.lowei
W Tanzania, N Zambia
N.o.alfredi
S Tanzania, Malawi, Zambia
N.o.sclateri
E Zimbabwe
N.o.olivacina
E Mozambique, N Natal
N.o.olivacea
C Natal
Nectarinia ursulae (Fernando Po Sunbird)
Fernando Po I, Cameroun Mt
Nectarinia veroxii (Mouse-coloured Sunbird)
N.v.fischeri
Somalia, E Kenya, E Tanzania, Mozambique, E Natal
N.v.zanzibarica
Zanzibar I
N.v.veroxii
E Natal, E Cape Province
Nectarinia balfouri (Socotra Sunbird)
Socotra I

Nectarinia reichenbachii (Reichenbach's Sunbird)
Ghana to N Zaire
Nectarinia hartlaubii (Principé Sunbird)
Principé I
Nectarinia newtonii (Newton's Yellow-breasted Sunbird)
Sao Thome I
Nectarinia thomensis (Sao Thome Giant Sunbird)
Sao Thome I
Nectarinia oritis (Cameroun Blue-headed Sunbird)
N.o.poensis
mountains of Fernando Po I
N.o.oritis
Cameroun Mt
N.o.bansoensis
W Cameroun
Nectarinia alinae (Blue-headed Sunbird)
N.a.alinae
E Zaire, SW Uganda
N.a.tanganjicae
SE Zaire
Nectarinia bannermani (Bannerman's Sunbird)
Angola, S Zaire, NW Zambia
Nectarinia verticalis (Green-headed Sunbird)
N.v.verticalis
Senegal to Nigeria
N.v.bohndorffi
Cameroun to Zaire, Angola
N.v.cyanocephala
W Gabon
N.v.viridisplendens
S Sudan, E Zaire, W Kenya to NE Zambia
Nectarinia cyanolaema (Blue-throated Sunbird)
N.c.magnirostrata
Sierra Leone to Ghana
N.c.cyanolaema
Fernando Po I
N.c.octaviae
Ghana to Uganda & N Angola
Nectarinia fuliginosa (Carmelite Sunbird)
N.f.aurea
Liberia to Gabon
N.f.fuliginosa
Zaire, Angola
Nectarinia rubescens (Green-throated Sunbird)
N.r.stangerii
Fernando Po I
N.r.crossensis
Cameroun
N.r.rubescens
Cameroun to Sudan, Angola, Zambia, Kenya
Nectarinia amethystina (Amethyst Sunbird)
N.a.kalckreuthi
Somalia, E Kenya, NE Tanzania
N.a.doggetti
W Kenya, Uganda, NW Tanzania

N.a.kirkii
SW Tanzania, SE Zaire, Zimbabwe, E Zambia
N.a.deminuta
S Zaire, W Zambia, Angola, W Botswana
N.a.adjuncta
E Transvaal, N Natal, S Mozambique
N.a.amethystina
S Natal, S Transvaal, Cape Province
Nectarinia senegalensis (Scarlet-chested Sunbird)
N.s.senegalensis
Senegal to N Nigeria
N.s.adamauae
NE Cameroun
N.s.acik
Cameroun to S Sudan, Uganda
N.s.cruentata
SE Sudan, Ethiopia
N.s.lamperti
E Zaire, Kenya, Tanzania
N.s.inaestimata
S Somalia to S Zimbabwe
N.s.saturatior
Angola, W Zambia, Namibia
N.s.gutteralis
SE Africa
Nectarinia hunteri (Hunter's Sunbird)
Somalia, Kenya, Tanzania
Nectarinia adelberti (Buff-throated Sunbird)
N.a.adelberti
Sierra Leone to Ghana
N.a.eboensis
Togo to SE Nigeria
Nectarinia zeylonica (Purple-rumped Sunbird)
N.z.flaviventris
Bangladesh, India
N.z.sola
S India
N.z.zeylonica
Sri Lanka
Nectarinia minima (Small Sunbird)
W & S India
Nectarinia sperata (Van Hasselt's Sunbird)
N.s.phayrei
Burma
N.s.brasiliana
Assam, Bangladesh, Thailand, Malaysia, Borneo, Java, Sumatra
N.s.emmae
Cambodia, S Laos, S Vietnam
N.s.mecynorhyncha
Simalur I
N.s.eumecis
Anamba Is
N.s.axantha
Natuna Is
N.s.henkei
N Luzon I
N.s.theresae
C Luzon I
N.s.davaoensis
SE Mindanao I

N.s.juliae
W & S Mindanao, Basilan Is
N.s.marinduquensis
Marinduque I
N.s.sperata
Maratua, Palawan, C Philippine Is
Nectarinia sericea (Black Sunbird)
N.s.talautensis
Talaut I
N.s.sangirensis
Sangihe Is
N.s.grayi
N Sulawesi
N.s.porphyrolaema
C & S Sulawesi
N.s.auriceps
Peleng I, N Moluccas
N.s.auricapilla
Kajoa I (W Moluccas)
N.s.proserpina
Buru I
N.s.aspasioides
S Moluccas
N.s.chlorolaema
Kai Is
N.s.sericea
New Guinea, except SE
N.s.vicina
SE New Guinea
N.s.mariae
Kofiau I
N.s.cochrani
Misol, Waigeu Is
N.s.maforensis
Numfor I
N.s.salvadorii
W Japen I
N.s.chlorocephala
Aru Is
N.s.nigriscapularis
Meos Num, Rani Is
N.s.mysorensis
Biak I
N.s.veronica
Liki I
N.s.cornelia
Tarawai I
N.s.christianae
D'Entrecasteaux & Louisiade Archipelagos
N.s.caeruleogula
New Britain, Rook I
N.s.corinna
Bismarck Arch
N.s.eichhorni
Feni I (Bismarck Arch)
Nectarinia calcostetha (Macklot's Sunbird)
Burma to Malaysia, Indochina, Sumatra, Borneo, Java, Philippine Is
Nectarinia dussumieri (Seychelles Sunbird)
Seychelles Is
Nectarinia lotenia (Loten's Sunbird)
N.l.hindustanica
S India

N.l.lotenia
Sri Lanka
Nectarinia jugularis (Olive-backed Sunbird)
N.j.andamanica
Andaman Is
N.j.klossi
N Nicobar Is
N.j.proselia
Car Nicobar Is
N.j.flammaxillaris
Burma, Thailand, Cambodia, N Malaysia
N.j.pectoralis
C Malaysia
N.j.microleuca
S Malaysia, Singapore
N.j.rhizophorae
N Vietnam, Hainan I
N.j.ornata
Sumatra, Java, Borneo, Lesser Sunda Is
N.j.polyclysta
Enggano
N.j.obscurior
N Luzon I
N.j.jugularis
S Luzon I & C & S Philippine Is
N.j.aurora
Palawan I
N.j.woodi
Sulu Arch
N.j.plateni
Sulawesi, Salayer I
N.j.infrenata
Tukangbesi I
N.j.robustirostris
Sula Is
N.j.teijsmanni
Djampea, Kalao Is
N.j.buruensis
Buru I
N.j.clementiae
S Moluccas
N.j.keiensis
Kai Is
N.j.idenburgi
N New Guinea
N.j.frenata
N Moluccas, Aru Is, New Guinea, N Queensland
N.j.flavigaster
Solomon Is, Bismarck Arch
Nectarinia buettikoferi (Sumba I Sunbird)
Sumba I
Nectarinia solaris (Timor Sunbird)
Timor, Lesser Sunda Is   152.1
Nectarinia asiatica (Purple Sunbird)
N.a.brevirostris
SE Arabia, SE Iran, Afghanistan, Pakistan, N India
N.a.asiatica
S India, Sri Lanka
N.a.intermedia
Bangladesh, Assam, Burma, Thailand, N Vietnam

Nectarinia souimanga (Souimanga Sunbird)
N.s.souimanga
Glorioso I, Madagascar
N.s.apolis
SW Madagascar
N.s.aldabrensis
Aldabra I
N.s.abbotti
Assumption I (Aldabra)
N.s.buchenorum
Cosmoledo I (Aldabra)
Nectarinia humbloti (Humblot's Sunbird)
N.h.humbloti
Great Comoro I
N.h.mohelica
Moheli I (Comoro Is)
Nectarinia comorensis (Anjouan Sunbird)
Anjouan I (Comoro Is)
Nectarinia coquerellii (Mayotte Sunbird)
Mayotte I (Comoro Is)
Nectarinia venusta (Variable Sunbird)
N.v.venusta
Senegal to Cameroun
N.v.falkensteini
Gabon, Angola, Zaire, Zambia, Zimbabwe Tanzania
N.v.igneiventris
Uganda, E Zaire
N.v.fazoglensis
Sudan, Ethiopia
N.v.albiventris
Somalia, E Ethiopia, N Kenya
N.v.blicki
S Ethiopia, S Sudan, NW Kenya
N.v.niassae
SE Tanzania, Malawi to E Zimbabwe
Nectarinia talatala (Southern White-bellied Sunbird)
N.t.anderssoni
SE Tanzania & N Namibia
N.t.talatala
N Cape Province, SW Transvaal, E Botswana
N.t.arresta
Natal
Nectarinia oustaleti (Oustalet's White-bellied Sunbird)
N.o.oustaleti
C Angola
N.o.rhodesiae
N Zambia
Nectarinia fusca (Dusky Sunbird)
N.f.fusca
S Angola to W Cape Province
N.f.indusa
Mossamedes, Angola
Nectarinia chalybea (Lesser Double-collared Sunbird)
N.c.pintoi
Angola, S Zaire, W Zambia
N.c.gertrudis
Tanzania, Malawi

N.c.capricornensis
   W Swaziland, E & N Transvaal
N.c.subalaris
   Transvaal, Natal, E Cape Province
N.c.chalybea
   S Cape Province
N.c.albilateralis
   W Cape Province
Nectarinia manoensis (Miombo Double-collared Sunbird)   152.5
   N.m.manoensis
      S Tanzania, SE Zambia, Zimbabwe,
      Mozambique
   N.m.amicorum
      S Mozambique
Nectarinia afra (Greater Double-collared Sunbird)
   N.a.stuhlmanni
      W Uganda
   N.a.graueri
      Rwanda, SW Uganda
   N.a.chapini
      E Zaire, S Burundi
   N.a.prigoginei
      SE Zaire
   N.a.whytei
      Zambia, Malawi
   N.a.saliens
      Transkei, Transvaal, W Natal
   N.a.afra
      South Africa
   N.a.amicorum
      S Mozambique
Nectarinia prigoginei (Double-breasted Sunbird)   152.2
      SE Zaire
Nectarinia preussi (Northern Double-collared Sunbird)
   N.p.preussi
      Fernando Po I, Cameroun Mt
   N.p.eriksoni
      S Sudan, Uganda, W Kenya, NE Zaire
   N.p.ludovicensis
      Angola
Nectarinia mediocris (Eastern Double-collared Sunbird)
   N.m.mediocris
      Kenya, Zambia
   N.m.usambarica
      SE Kenya, NE Tanzania
   N.m.fuelleborni
      Tanzania, N Malawi, NE Zambia
   N.m.bensoni
      Malawi, Zambia, Mozambique
Nectarinia neergaardi (Neergaard's Sunbird)
      S Mozambique, N Natal
Nectarinia chloropygia (Olive-bellied Sunbird)
   N.c.kempi
      Sierra Leone to Ivory Coast
   N.c.chloropygia
      Ghana to Nigeria
   N.c.insularis
      Fernando Po I

N.c.luhderi
   Cameroun, Zaire, Angola
N.c.bineschensis
   SW Ethiopia
N.c.orphogaster
   NE Angola, E Zaire, S Sudan, Uganda W
   Tanzania
Nectarinia minulla (Tiny Sunbird)
   N.m.amadoni
      Fernando Po I
   N.m.minulla
      Ghana to W Uganda
Nectarinia regia (Regal Sunbird)
   N.r.regia
      Uganda
   N.r.kivuensis
      E Zaire, SW Uganda
   N.r.anderseni
      W Tanzania
Nectarinia loveridgei (Loveridge's Sunbird)
      E Tanzania
Nectarinia moreaui (Moreau's Sunbird)
      NE Tanzania
Nectarinia rockefelleri (Rockefeller's Sunbird)
      E Zaire
Nectarinia violacea (Orange-breasted Sunbird)
      Cape Province
Nectarinia habessinica (Shining Sunbird)
   N.h.kinneari
      W Saudi Arabia
   N.h.hellmayri
      S Arabia
   N.h.habessinica
      NE Sudan, W Ethiopia
   N.h.alter
      E Ethiopia, N Somalia
   N.h.turkanae
      S Ethiopia, S Sudan, S Somalia, N Kenya,
      Uganda
Nectarinia bouvieri (Southern Orange-tufted Sunbird)
      Cameroun to W Kenya & N Angola
Nectarinia osea (Northern Orange-tufted Sunbird)
   N.o.osea
      Syria, Israel, Arabia
   N.o.decorsei
      Mali to S Sudan
Nectarinia cuprea (Coppery Sunbird)
   N.c.cuprea
      Senegal to Zaire, Uganda, Tanzania
   N.c.chalcea
      Malawi, Zimbabwe, Angola W Zambia
Nectarinia tacazze (Tacazze Sunbird)
   N.t.tacazze
      Ethiopia
   N.t.jacksoni
      S Sudan, Uganda, W Kenya, N Tanzania
Nectarinia bocagii (Bocage's Sunbird)
      Angola, Zaire

Nectarinia purpureiventris **(Purple-breasted Sunbird)**
Uganda, E Zaire
Nectarinia shelleyi **(Shelley's Sunbird)**
N.s.hofmanni
E Tanzania
N.s.shelleyi
SE Zaire, E Zambia to N Mozambique
Nectarinia mariquensis **(Mariqua Sunbird)**
N.m.osiris
Ethiopia, S Sudan, N Kenya, N Uganda
N.m.suahelica
S Uganda, E Zaire to NE Zambia
N.m.mariquensis
S Angola to Zimbabwe
N.m.lucens
E Zimbabwe, S Mozambique, Natal
Nectarinia pembae **(Violet-breasted Sunbird)**
N.p.chalcomelas
Somalia, E Kenya
N.p.pembae
Pemba I
Nectarinia bifasciata **(Purple-banded Sunbird)**
N.b.bifasciata
Gabon to C Angola
N.b.microrhyncha
Uganda to Angola, N Malawi, Mozambique
N.b.tsavoensis
E Kenya, NE Tanzania
N.b.strophium
SE Zambia, S Mozambique, N Natal
Nectarinia coccinigastra **(Splendid Sunbird)**
Senegal to NE Zaire
Nectarinia erythrocerca **(Red-chested Sunbird)**
Sudan, Uganda, NW Tanzania
Nectarinia congensis **(Congo Black-bellied Sunbird)**
Zaire
Nectarinia pulchella **(Beautiful Sunbird)**
N.p.pulchella
Senegal, Mali, Niger, W Sudan
N.p.aegra
S Niger, Air Massif
N.p.lucidipectus
S Sudan, Ethiopia, NE Zaire, Uganda, NW Kenya
N.p.melanogastra
S Kenya, Tanzania
Nectarinia nectarinioides **(Smaller Black-bellied Sunbird)**
N.n.erlangeri
S Somalia
N.n.nectarinioides
E Kenya, NE Tanzania
Nectarinia famosa **(Yellow-tufted Malachite Sunbird)**
N.f.cupreonitens
Ethiopia, SE Sudan
N.f.aeneigularis
Kenya, Uganda, E Zaire, N Malawi
N.f.major
Lesotho, Natal

N.f.famosa
Zimbabwe, South Africa
Nectarinia johnstoni **(Red-tufted Malachite Sunbird)**
N.j.johnstoni
W Kenya, N Tanzania
N.j.dartmouthi
E Zaire, W Uganda
N.j.nyikensis
S Tanzania, Zambia, Malawi
Nectarinia notata **(Noted Sunbird)**
N.n.notata
Madagascar
N.n.moebii
Gt Comoro I
N.n.voeltzkowi
Moheli I (Comoro)
Nectarinia rufipennis **(Rufous-winged Sunbird)** 152.3
Morogoro (Tanzania)
Nectarinia johannae **(Madame Verreaux's Sunbird)**
N.j.fasciata
Sierra Leone to Benin
N.j.johannae
Cameroun, Zaire
Nectarinia superba **(Superb Sunbird)**
N.s.ashantiensis
Sierra Leone to Ghana
N.s.nigeriae
S Nigeria
N.s.superba
S Cameroun, W Zaire, Angola
N.s.buvuma
E Zaire, Uganda
Nectarinia kilimensis **(Bronze Sunbird)**
N.k.kilimensis
E Zaire, Uganda, W Kenya, Tanzania
N.k.arturi
S Tanzania, Malawi, NE Zambia, E Zimbabwe
N.k.gadowi
C Angola
Nectarinia reichenowi **(Golden-winged Sunbird)**
N.r.shellyae
E Zaire
N.r.lathburyi
N Kenya
N.r.reichenowi
C & S Kenya, NE Tanzania, Mozambique

AETHOPYGA
Aethopyga primigenius **(Hachisuka's Sunbird)**
A.p.diuatae
NE Mindanao
A.p.primigenius
C & E Mindanao
Aethopyga boltoni **(Apo Sunbird)**
A.b.malindangensis
C & W Mindanao

A.b.boltoni
  E Mindanao
Aethopyga flagrans (Flaming Sunbird)
A.f.decolor
  NE Luzon I
A.f.flagrans
  W & S Luzon I
A.f.guimarasensis
  Panay I, Guimaras I
A.f.daphoenonota
  Negros I
Aethopyga pulcherrima (Mountain Sunbird)
A.p.jeffreyi
  Luzon I
A.p.pulcherrima
  Basilan, Samar, Leyte, Mindanao Is
A.p.decorosa
  Bohol I
Aethopyga duyvenbodei (Sanghir Yellow-backed Sunbird)
  Sangihe Is
Aethopyga shelleyi (Palawan Sunbird)
A.s.flavipectus
  Luzon, Mindoro Is
A.s.rubrinota
  Lubang Is
A.s.bella
  Samar, Leyte, Mindanao Is
A.s.bonita
  Ticao, Masbate, Panay, Negros, Cebu Is
A.s.arolasi
  Sulu Arch
A.s.shelleyi
  Balabac, Palawan Is
Aethopyga gouldiae (Mrs Gould's Sunbird)
A.g.gouldiae
  N Assam, Himalayas, SE Tibet
A.g.isolata
  S Assam, Bangladesh, Burma
A.g.dabryii
  E Sikang, SW China, N Vietnam
A.g.annamensis
  S Laos, S Vietnam
Aethopyga nipalensis (Green-tailed Sunbird)
A.n.horsfieldii
  W Himalayas
A.n.nipalensis
  C Nepal, Sikkim
A.n.koelzi
  E Himalayas, NE Burma, S Assam, N Vietnam
A.n.victoriae
  W Burma
A.n.karenensis
  SE Burma
A.n.angkanensis
  N Thailand
A.n.australis
  S Thailand
A.n.blanci
  Laos
A.n.ezrai
  S Vietnam

Aethopyga eximia (Kühl's Sunbird)
  Java
Aethopyga christinae (Fork-tailed Sunbird)
  152.4
A.c.latouchii
  SE China, N Vietnam
A.c.christinae
  Hainan I
A.c.sokolovi
  S Vietnam
Aethopyga saturata (Black-throated Sunbird)
A.s.saturata
  W Himalayas
A.s.assamensis
  Bangladesh, Assam, N Burma, W China
A.s.galenae
  NW Thailand
A.s.petersi
  E Burma, Laos, N Vietnam, SE Yunnan
A.s.sanguinipectus
  SE Burma
A.s.anomala
  S Thailand
A.s.wrayi
  Malaysia
A.s.ochra
  S Laos, C Vietnam
A.s.cambodiana
  SW Cambodia
A.s.johnsi
  S Vietnam
Aethopyga siparaja (Yellow-backed Sunbird)
A.s.vigorsii
  N India
A.s.seheriae
  Nepal, Assam, Bangladesh, NE India, Burma, N Thailand
A.s.labecula
  NE India, S Bangladesh
A.s.owstoni
  Nauchow I (Hainan)
A.s.tonkinensis
  NE Vietnam
A.s.mangini
  SE Thailand, C & S Indochina
A.s.insularis
  Phuquoc I (Cambodia)
A.s.cara
  S Burma, Thailand
A.s.trangensis
  S Thailand
A.s.siparaja
  Malaysia, Sumatra, Borneo
A.s.nicobarica
  Nicobar Is
A.s.heliogona
  Java
A.s.natunae
  N Natuna Is
A.s.magnifica
  WC Philippine Is
A.s.flavostriata
  N Sulawesi

A.s.beccarii
  S Sulawesi
Aethopyga mystacalis (Scarlet Sunbird)
A.m.temminckii
  Malaysia, Sumatra, Borneo
A.m.mystacalis
  Java
Aethopyga ignicauda (Fire-tailed Sunbird)
A.i.ignicauda
  Himalayas, Sikang, N Burma, Yunnan
A.i.flavescens
  NW Burma

ARACHNOTHERA
**Arachnothera longirostra (Little
Spiderhunter)**
A.l.longirostra
  SW India, Nepal, Assam, Burma, W
  Thailand
A.l.sordida
  S Yunnan, NE Thailand, N Indochina
A.l.pallida
  SE Thailand, C Indochina
A.l.cinireicollis
  S Thailand, Malaysia, Sumatra
A.l.niasensis
  Nias I
A.l.prillwitzi
  Java
A.l.buettikoferi
  Borneo
A.l.atita
  S Natuna Is
A.l.rothschildi
  N Natuna Is
A.l.dilutior
  Palawan I
A.l.flammifera
  Samar, Leyte, Bohol, Mindanao Is
A.l.randi
  Basilan I
**Arachnothera crassirostris (Thick-billed
Spiderhunter)**
  S Thailand, Malaysia, Sumatra, Borneo
**Arachnothera robusta (Long-billed
Spiderhunter)**
A.r.robusta
  Malaysia, Sumatra, Borneo
A.r.armata
  Java
**Arachnothera flavigaster (Greater Yellow-
eared Spiderhunter)**
  S Thailand, Malaysia, Sumatra, Borneo
**Arachnothera chrysogenys (Lesser Yellow-
eared Spiderhunter)**
A.c.chrysogenys
  S Burma, S Thailand, Malaysia, Sumatra,
  Java, W Borneo
A.c.harrissoni
  E Borneo
**Arachnothera clarae (Naked-faced
Spiderhunter)**

A.c.philippinensis
  Samar, Leyte Is
A.c.clarae
  E Mindanao I
A.c.malindangensis
  C & W Mindanao I
A.c.luzonensis
  C Luzon, Laguna Is
**Arachnothera affinis (Grey-breasted
Spiderhunter)**
A.a.caena
  S Burma, Thailand
A.a.modesta
  S Thailand, Malaysia, W Borneo
A.a.pars
  E Borneo
A.a.affinis
  Java, Bali
A.a.concolor
  Sumatra
Arachnothera magna (Streaked Spiderhunter)
A.m.magna
  Himalayas, N Burma, Yunnan
A.m.aurata
  EC Burma
A.m.musarum
  SE Burma, N Thailand, N Laos
A.m.pagodarum
  S Burma, SW Thailand
A.m.remota
  S Vietnam
**Arachnothera everetti (Everett's
Spiderhunter)**
  N & C Borneo
**Arachnothera juliae (Whitehead's
Spiderhunter)**
  N Borneo

153. ZOSTEROPIDAE (WHITE EYES)

ZOSTEROPS
**Zosterops erythropleura (Chestnut-flanked
White-eye)**
  Manchuria, Amur, N Korea, China
Zosterops japonica (Japanese White-eye)
Z.j.yesoensis
  Hokkaido I
Z.j.japonica
  Honshu I, S Japan
Z.j.stejnegeri
  Bonin, Izu Is
Z.j.alani
  Iwo Jima I
Z.j.insularis
  Tanegashima, Yakushima Is
Z.j.loochooensis
  Riukiu Is
Z.j.daitoensis
  Borodino Is
Z.j.simplex
  Sikang, China, Burma, N Vietnam, Taiwan
Z.j.hainana
  Hainan I

Zosterops meyeni (Philippine White-eye)
 Z.m.batanis
  Botel Tobago, Kashoto, Batan Is
 Z.m.meyeni
  Luzon, Calayan, Lubang Is
Zosterops palpebrosa (Oriental White-eye)
 Z.p.occidentis
  NW India
 Z.p.palpebrosa
  C India, Sri Lanka, Bangladesh, W Assam,
  Nepal, Bhutan
 Z.p.nilgiriensis
  SW India
 Z.p.salimalii
  SE India
 Z.p.siamensis
  SE Tibet, Burma, N Thailand, SW China,
  Indochina
 Z.p.nicobarica
  Andaman, Nicobar Is
 Z.p.williamsoni
  S Thailand, Malaysia
 Z.p.joannae
  W China
 Z.p.auriventer
  S Burma, Malaysia, Bangka I, W Borneo
 Z.p.sumatrana
  W Sumatra
 Z.p.buxtoni
  E Sumatra, W Java
 Z.p.melanura
  E & C Java, Bali
 Z.p.unica
  Sumbawa, Flores Is
Zosterops ceylonensis (Large Sri Lanka
 White-eye)
  Sri Lanka
Zosterops conspicillata (Bridled White-eye)
 Z.c.saypani
  Tinian, Saipan Is
 Z.c.conspicillata
  Guam I
 Z.c.rotensis
  Rota I
 Z.c.semperi
  Palau Is
 Z.c.owstoni
  Truk I
 Z.c.takatsukasai
  Ponapé I
 Z.c.hypolais
  Yap I
Zosterops salvadorii (Enggano White-eye)
  Enggano I
Zosterops atricapilla (Black-capped White-
 eye)
 Z.a.viridicata
  N Sumatra
 Z.a.atricapilla
  C & S Sumatra, N Borneo
Zosterops everetti (Everett's White-eye)
 Z.e.everetti
  Cebu I

Z.e.basilanica
  Samar, Leyte, Mindanao, Basilan Is
Z.e.boholensis
  Bohol I
Z.e.siquijorensis
  Siquijor I
Z.e.mandibularis
  Sulu Arch
Z.e.babelo
  Talaut I
Z.e.tahanensis
  N Borneo, Malaysia, S Thailand
Z.e.wetmorei
  S Thailand
Zosterops nigrorum (Philippine Yellow White-
 eye)
 Z.n.meyleri
  Camiguin Is
 Z.n.aureiloris
  N Luzon I
 Z.n.sierramadrensis
  Cagayan Province, Luzon I
 Z.n.luzonica
  SE Luzon I
 Z.n.nigrorum
  Masbate, Negros, Panay Is
 Z.n.richmondi
  Cagayancillo I (Sulu Sea)
 Z.n.mindorensis
  Mindoro I
 Z.n.catarmanensis
  Camiguin I (South)
Zosterops montanus (Mountain White-eye)
  153.1
 Z.m.obstinatus
  Batjan, Ternate, Seram Is
 Z.m.whiteheadi
  N Luzon I
 Z.m.halconensis
  Mindoro I
 Z.m.gilli
  Marinduque I
 Z.m.parkesi
  Palawan I
 Z.m.diuatae
  N Mindanao
 Z.m.vulcani
  Mt Apo & Mt Katanglad (Mindanao)
 Z.m.pectoralis
  N Negros I
 Z.m.montanus
  Gtr & Lesser Sunda Is, Timor I, Sulawesi
 Z.m.difficilis
  S Sumatra
Zosterops wallacei (Yellow-spectacled White-
 eye)
  Sumbawa, Sumba, Flores Is
Zosterops flava (Javan White-eye)
  NW Java, S Borneo
Zosterops chloris (Lemon-bellied White-eye)
  153.1
 Z.c.maxi
  Lombok I

Z.c.intermedius
Sumbawa , Flores Is, SW Sulawesi
Z.c.mentoris
NC Sulawesi
Z.c.flavissimus
Tukangbesi I
Z.c.solombensis
Solombo Besar I
Z.c.zachlorus
Kalambau I
Z.c.chloris
Aru, Kai, Seram, Halmahera, Misol Is
Zosterops citrinella (Ashy-bellied White-eye)
153.1
Z.c.albiventris
S Moluccan islands, Tanimbar Is, Torres Straits islands
Z.c.citrinella
Timor, Sumba Is
Z.c.harterti
Alor I
Zosterops consobrinorum (Pale-bellied White-eye) 153.1
SE Sulawesi
Zosterops grayi (Pearl-bellied White-eye)
153.1
Gt Kai I
Zosterops uropygialis (Golden-bellied White-eye) 153.1
Little Kai I
Zosterops anomala (Lemon-throated White-eye) 153.1
S Sulawesi
Zosterops atriceps (Creamy-throated White-eye) 153.1
Z.a.dehaani
Morotai I
Z.a.fuscifrons
Halmahera I
Z.a.atriceps
Batjan I
Zosterops atrifrons (Moluccan Black-fronted White-eye)
Z.a.nehrkorni
Sangihe Is
Z.a.atrifrons
Banggai I, N & C Sulawesi
Z.a.sulaensis
Sula Is
Z.a.stalkeri
Seram I
Zosterops minor (New Guinea Black-fronted White-eye)
Z.m.minor
Japen I, New Guinea
Z.m.chrysolaema
NW New Guinea
Z.m.rothschildi
C New Guinea
Z.m.gregaria
E New Guinea
Z.m.tenuifrons
SE New Guinea

Z.m.delicatula
SE New Guinea
Z.m.pallidogularis
Fergusson, Goodenough Is
Zosterops meeki (White-throated White-eye)
Z.m.meeki
Tagula I, Louisiade Arch
Z.m.hypoxantha
New Britain
Z.m.ultima
New Hanover, New Ireland
Z.m.admiralitatis
Manus I (Admiralty Is)
Zosterops mysorensis (Biak White-eye)
Biak I, New Guinea
Zosterops fuscicapilla (Yellow-bellied Mountain White-eye)
Z.f.fuscicapilla
C & W New Guinea
Z.f.crookshanki
Goodenough I
Zosterops buruensis (Buru I White-eye)
Buru I
Zosterops kuehni (Ambon White-eye)
Ambon I
Zosterops novaeguineae (New Guinea Mountain White-eye)
Z.n.novaeguineae
NW New Guinea
Z.n.aruensis
Aru Is
Z.n.wuroi
S New Guinea
Z.n.wahgiensis
C New Guinea
Z.n.crissalis
SE New Guinea
Z.n.oreophila
E New Guinea
Z.n.magnirostris
NW New Guinea
Zosterops metcalfii (Yellow-throated White-eye)
Z.m.exigua
Buka, Bougainville, Choiseul Is
Z.m.metcalfii
Ysabel, St George Is
Z.m.floridana
Florida I
Zosterops natalis (Christmas Island White-eye)
Christmas I
Zosterops lutea (Yellow Silver-eye)
Z.l.balstoni
NW Western Australia
Z.l.lutea
Northern Territory, N Queensland
Zosterops griseotincta (Louisiades White-eye)
Z.g.pallidipes
Rossel I
Z.g.aignani
Louisiade Arch

Z.g.griseotincta
Louisiade Arch
Z.g.longirostris
Bonvouloir I
Z.g.eichhorni
Nauna, Nissan, Long Is (New Britain)
Zosterops rennelliana (Rennell I White-eye)
Rennell I
Zosterops rendovae (Solomon Is White-eye)
Z.r.vallalavella
Bagga, Vellalavalla Is
Z.r.luteirostris
Gizo I
Z.r.splendida
Ganonga I
Z.r.kulambangrae
Kulambangra, Vangunu, New Georgia Is
Z.r.rendovae
Rendova I
Z.r.tetiparia
Tetipari I
Zosterops murphyi (Kulambangra Mountain White-eye)
Kulambangra I
Zosterops ugiensis (Grey-throated White-eye)
Z.u.ugiensis
San Cristobal I
Z.u.oblita
Guadalcanal I
Z.u.hamlini
Bougainville I
Zosterops stresemanni (Malaita White-eye)
Malaita I
Zosterops sanctaecrucis (Santa Cruz White-eye)
Santa Cruz I
Zosterops samoensis (Savaii White-eye)
Savaii (Samoa Is)
Zosterops explorator (Layard's White-eye)
Fiji Is
Zosterops flavifrons (Yellow-fronted White-eye)
Z.f.gauensis
Gaua I (Banks Is)
Z.f.perplexa
N New Hebrides, Vanua Levu I
Z.f.brevicauda
Malo, Espiritu Santo Is
Z.f.macgillivrayi
Malekula I
Z.f.efatensis
Nguna, Efate, Erromanga Is
Z.f.flavifrons
Tanna I
Z.r.majuscula
Aneitum I
Zosterops minuta (Small Lifu White-eye)
Lifu, Loyalty Is
Zosterops xanthochroa (New Caledonia White-eye)
New Caledonia
Zosterops gouldi (Western Silver-eye)
S Western Australia

Zosterops lateralis (Grey-backed White-eye)
Z.l.halmaturina
Tasmania, W Victoria, SE South Australia
Z.l.lateralis
Victoria, E New South Wales, SE Queensland, New Zealand
Z.l.familiaris
E New South Wales
Z.l.ramsayi
E Queensland
Z.l.tephropleura
Lord Howe I e?
Z.l.chlorocephala
Capricorn I
Z.l.griseonota
New Caledonia
Z.l.nigrescens
Mare, Uvea Is (Loyalty Is)
Z.l.melanops
Lifu I (Loyalty Is)
Z.l.macmillani
Tanna, Aniwa Is (New Hebrides)
Z.l.tropica
Espiritu Santo I
Z.l.vatensis
N New Hebrides, Banks Is, Torres Is
Z.l.valuensis
Valua I (Banks Is)
Z.l.flaviceps
Fiji Archipelago
Zosterops tenuirostris (Slender-billed White-eye)
Norfolk I
Zosterops albogularis (White-chested White-eye)
Norfolk I
Zosterops inornata (Large Lifu White-eye)
Lifu I
Zosterops cinerea (Grey-brown White-eye)
Z.c.finschii
Palau Is
Z.c.ponapensis
Ponapé I
Z.c.cinerea
Kusaie I
Zosterops abyssinica (White-breasted White-eye)
Z.a.abyssinica
E Ethiopia, SE Sudan
Z.a.socotrana
Socotra I, N Somalia
Z.a.arabs
Yemen, Aden
Z.a.omoensis
SW Ethiopia
Z.a.jubaensis
SE Ethiopia, Somalia, N Kenya
Z.a.flavilateralis
E Kenya, E Tanzania
Zosterops pallida (Cape White-eye)   153.2
Z.p.pallida
Namibia, SW Transvaal, NW Cape Province

Z.p.sundevalli
 N Cape Province
Z.p.caniviridis
 W Transvaal, E Botswana
Z.p.capensis
 W Cape Province
Z.p.virens
 S Mozambique to C & E Cape Province
Zosterops senegalensis (African Yellow White-eye)
Z.s.senegalensis
 Senegal to Ethiopia, Uganda
Z.s.demeryi
 Sierra Leone, Liberia, Ivory Coast
Z.s.stenocricota
 Fernando Po I, SE Nigeria to Gabon
Z.s.stuhlmanni
 E Zaire, Uganda
Z.s.reichenowi
 E Zaire
Z.s.toroensis
 NE Zaire
Z.s.jacksoni
 W Kenya, N Tanzania
Z.s.kasaica
 SW Zaire, NE Angola
Z.s.heinrichi
 N Angola
Z.s.quanzae
 C Angola
Z.s.anderssoni
 S Angola to Mozambique, Natal
Z.s.tongensis
 S Zimbabwe, N Natal, S Mozambique
Z.s.stierlingi
 S Tanzania, Zambia, Malawi
Z.s.kirki
 Gt Comoro I
Z.s.poliogastra
 C Ethiopia
Z.s.kaffensis
 W Ethiopia
Z.s.kulalensis
 N Kenya
Z.s.kikuyuensis
 W Kenya
Z.s.silvana
 SE Kenya
Z.s.eurycricota
 N Tanzania
Z.s.mbuluensis
 N Tanzania
Z.s.winifredae
 NE Tanzania
Zosterops borbonica (Bourbon White-eye)
Z.b.mauritiana
 Mauritius I
Z.b.borbonica
 Reunion I
Z.b.alopekion
 Cilaos, Reunion Is
Z.b.xerophila
 Etang les Bains, Reunion I

Zosterops ficedulina (Principé White-eye)
Z.f.ficedulina
 Principé I
Z.f.feae
 Sao Thome I
Zosterops griseovirescens (Annobon White-eye)
 Annobon I
Zosterops hovarum (Hova Grey-backed White-eye)
 Madagascar
Zosterops maderaspatana (Madagascar White-eye)
Z.m.aldabrensis
 Aldabra I
Z.m.maderaspatana
 Madagascar, Glorioso I
Z.m.anjouanensis
 Anjouan I
Z.m.comorensis
 Moheli I
Z.m.voeltzkowi
 Europa I
Z.m.menaiensis
 Cosmoledo Atoll
Zosterops mayottensis (Chestnut-sided White-eye)
 Mayotte I
Zosterops modesta (Seychelles Brown White-eye)
 Mahe I
Zosterops mouroniensis (Grand Comoro White-eye)
 Gt Comoro I
Zosterops olivacea (Olive White-eye)
 Reunion I
Zosterops chloronothos (Mauritius Olive White-eye)
 Mauritius I
Zosterops vaughani (Pemba White-eye)
 Pemba I

WOODFORDIA
Woodfordia superciliosa (Woodford's White-eye)
 Rennell I
Woodfordia lacertosa (Sanford's White-eye)
 Santa Cruz I

RUKIA
Rukia palauensis (Palau White-eye)
 Palau Is
Rukia oleaginea (Yap White-eye)
 Yap I
Rukia ruki (Truk White-eye)
 Truk I
Rukia longirostra (Ponapé White-eye)
 Ponapé I

TEPHROZOSTEROPS
Tephrozosterops stalkeri (Bicoloured White-eye)
 Seram I

Madanga ruficollis (Rufous-throated White-eye)
    NW Buru I

LOPHOZOSTEROPS
Lophozosterops pinaiae (Grey-hooded White-eye)
    C Seram I
Lophozosterops goodfellowi (Goodfellow's White-eye)
    L.g.goodfellowi
        Mt Apo (Mindanao I)
    L.g.malindangensis
        Mt Malindang (NW Mindanao I)
    L.g.gracilis
        NE Mindanao I
Lophozosterops squamiceps (Streaky-headed White-eye)   153.1
    L.s.stresemanni
        N Sulawesi
    L.s.heinrichi
        N Sulawesi
    L.s.striaticeps
        NC Sulawesi
    L.s.stachyrina
        SC Sulawesi
    L.s.squamiceps
        S Sulawesi
    L.s.analoga
        SE Sulawesi
Lophozosterops javanicus (Javan Grey-throated White-eye)
    L.j.frontalis
        W Java
    L.j.javanicus
        C & E Java
    L.j.elongatus
        E Java, Bali
Lophozosterops superciliaris (White-browed White-eye)
    L.s.hartertianus
        W Sumbawa I
    L.s.superciliaris
        Flores I
Lophozosterops dohertyi (Crested White-eye)
    L.d.dohertyi
        Sumbawa I
    L.d.subcristatus
        Flores I

OCULOCINCTA
Oculocincta squamifrons (Pygmy White-eye)
    N & W Borneo

HELEIA
Heleia muelleri (Timor White-eye)
    W Timor I
Heleia crassirostris (Stripe-headed White-eye)   153.1
    Flores, Sumbawa Is

CHLOROCHARIS
Chlorocharis emiliae (Olive Black-eye)
    C.e.emiliae
        Mt Kinabalu (N Borneo)
    C.e.trinitae
        Mt Trus Madi (N Borneo)
    C.e.fusciceps
        NE Sarawak
    C.e.moultoni
        Sarawak

HYPOCRYPTADIUS
Hypocryptadius cinnamomeus (Cinnamon White-eye)
    Mindanao I

SPEIROPS
Speirops brunnea (Fernando Po Speirops)
    Fernando Po I
Speirops leucophaea (Prince's I Speirops)
    Principé I
Speirops lugubris (Black-capped Speirops)
    S.l.melanocephala
        Mt Cameroun
    S.l.lugubris
        Sao Thome I

154. MELIPHAGIDAE (HONEYEATERS)

TIMELIOPSIS
Timeliopsis fulvigula (Mountain Straight-billed Honeyeater)
    T.f.fulvigula
        NW New Guinea
    T.f.meyeri
        C & SE New Guinea
    T.f.fuscicapilla
        E New Guinea
Timeliopsis griseigula (Lowland Straight-billed Honeyeater)
    T.g.griseigula
        W New Guinea
    T.g.fulviventris
        SE New Guinea

MELILESTES
Melilestes megarhynchus (Long-billed Honeyeater)
    M.m.vagans
        Batanta, Waigeu Is
    M.m.brunneus
        NW New Guinea
    M.m.megarhynchus
        Aru Is, S & SE New Guinea
    M.m.stresemanni
        N New Guinea, Japen I
Melilestes bougainvillei (Bougainville Honeyeater)
    Bougainville I

TOXORHAMPHUS
Toxorhamphus novaeguineae (Yellow-bellied Longbill)

T.n.novaeguineae
W New Guinea & islands
T.n.flaviventris
Aru Is, S New Guinea
Toxorhamphus poliopterus (Slaty-chinned Longbill)
T.p.maximus
NC New Guinea
T.p.poliopterus
C & SE New Guinea

OEDISTOMA
Oedistoma iliolophum (Grey-bellied Honeyeater)
O.i.cinerascens
Waigeu I
O.i.affine
NW New Guinea
O.i.iliolophum
Japen I, N New Guinea
O.i.flavum
S & SE New Guinea
O.i.fergussonis
D'Entrecasteaux Arch
Oedistoma pygmaeum (Pygmy Honeyeater)
O.p.waigeuense
Waigeu I
O.p.pygmaeum
Misol I, W New Guinea
O.p.flavipectus
S New Guinea
O.p.olivascens
SE New Guinea
O.p.meeki
D'Entrecasteaux Arch

GLYCICHAERA
Glycichaera fallax (White-eyed Honeyeater)
G.f.pallida
Batanta, Waigeu Is
G.f.poliocephala
Misol, Aru Is, NW New Guinea
G.f.fallax
Japen I, E & S New Guinea
G.f.sylvia
N New Guinea
G.f.claudi
N Queensland

LICHMERA
Lichmera lombokia (Lombok Honeyeater)
154.1
Lombok, Flores, Sumbawa Is
Lichmera argentauris (Plain Olive Honeyeater)
154.1
W New Guinea islands, Halmahera, Seram Is
Lichmera indistincta (Brown Honeyeater)
L.i.limbata
Bali, Lombok, Timor, Lesser Sunda Is
L.i.indistincta
Western Australia, Northern Territory

L.i.ocularis
NE Australia, S New Guinea
L.i.melvillensis
Melville I
L.i.nupta
Aru Is
Lichmera incana (Silver-eared Honeyeater)
L.i.incana
New Caledonia
L.i.poliotis
Loyalty Is
L.i.mareensis
Mare I (Loyalty Is)
L.i.griseoviridis
C New Hebrides
L.i.flavotincta
Erromanga I (New Hebrides)
Lichmera alboauricularis (White-eared Honeyeater)
L.a.alboauricularis
SE New Guinea
L.a.olivacea
N New Guinea
Lichmera squamata (White-tufted Honeyeater)
154.1
Kai, Tanimbar Is, S Banda Sea islands
Lichmera deningeri (Buru Honeyeater)
Buru I
Lichmera monticola (Spectacled Honeyeater)
Seram I
Lichmera flavicans (Timor Honeyeater)
Timor I
Lichmera notabilis (Black-chested Honeyeater)
Wetar I
Lichmera cockerelli (White-streaked Honeyeater)
N Queensland

MYZOMELA
Myzomela blasii (Ambon Honeyeater)
Seram, Ambon Is
Myzomela albigula (White-chinned Honeyeater)
M.a.albigula
Rossel I
M.a.pallidior
W Louisiade Arch
Myzomela cineracea (Bismarck Honeyeater)
M.c.cineracea
New Britain
M.c.rooki
Umboi I (Bismarck Arch)
Myzomela eques (Red-spot Honeyeater)
M.e.eques
NW New Guinea & islands
M.e.primitiva
N New Guinea
M.e.nymani
S & E New Guinea
M.e.karimuiensis
E New Guinea
Myzomela obscura (Dusky Honeyeater)

M.o.harterti
E Queensland
M.o.munna
N Queensland, Torres Strait
M.o.obscura
Northern Territory, Melville I
M.o.fumata
S New Guinea
M.o.aruensis
Aru Is
M.o.simplex
Damar, Ternate, Batjan, Halmahera Is
M.o.rubrotincta
Obi Is
M.o.mortyana
Morotai I
M.o.rubrobrunnea
Biak I
Myzomela cruentata (Red Honeyeater)
M.c.cruentata
Japen I, New Guinea
M.c.coccinea
New Britain, Duke of York Is
M.c.erythrina
New Ireland
M.c.lavongai
New Hanover
M.c.cantans
Tabar I (Bismarck Arch)
M.c.vinacea
Dyaul I (Bismarck Arch)
Myzomela nigrita (Black Honeyeater)
M.n.steini
Waigeu I
M.n.nigrita
Aru Is, S New Guinea
M.n.meyeri
Japen I, New Guinea
M.n.pluto
Meos Num I
M.n.forbesi
D'Entrecasteaux Arch
M.n.louisiadensis
Louisiade Arch
M.n.hades
St Matthias Is (Bismarck Arch)
M.n.ramsayi
Tingwon I (Bismarck Arch)
M.n.pammelaena
Admiralty Is
M.n.ernstmayri
Manus, Admiralty Is
M.n.nigerrima
Long I (NE New Guinea)
Myzomela pulchella (New Ireland Honeyeater)
New Ireland
Myzomela kuehni (Crimson-hooded Honeyeater)
Wetar I
Myzomela erythrocephala (Mangrove Red-headed Honeyeater)
M.e.erythrocephala
coast of N Western Australia

M.e.infuscata
NE Australia, Aru Is, S New Guinea
M.e.dammermanni
Sumba I
Myzomela adolphinae (Mountain Red-headed Honeyeater)
New Guinea
Myzomela sanguinolenta (Scarlet Honeyeater)   154.1
M.s.chloroptera
N, C & SE Sulawesi
M.s.juga
SW Sulawesi
M.s.eva
Djampea, Saleyer Is
M.s.batjanensis
Batjan I
M.s.elisabethae
Seram I
M.s.wakoloensis
Buru I
M.s.annabellae
Babar, Tanimbar Is
M.s.boiei
Banda I
M.s.caledonica
New Caledonia
M.s.sanguinolenta
coast of E Queensland, New South Wales
Myzomela cardinalis (Cardinal Honeyeater)
M.c.lifuensis
Loyalty Is
M.c.cardinalis
S New Hebrides
M.c.tenuis
N New Hebrides
M.c.tucopiae
Tikopia I
M.c.nigriventris
Samoa Is
M.c.sanctaecrucis
Torres, Santa Cruz Is
M.c.sanfordi
Rennell I
M.c.pulcherrima
San Cristobal, Ugi Is
M.c.kobayashii
Palau Is
M.c.kurodai
Yap I
M.c.saffordi
S Marianas Is
M.c.asuncionis
N Marianas Is
M.c.major
Truk I
M.c.dichromata
Ponapé I
M.c.rubratra
Kusaie I (Caroline Is)
M.c.chermesina
Rotuma Is

Myzomela sclateri (Scarlet-throated
Honeyeater)
  N New Guinea islands & New Britain
Myzomela lafargei (Small Bougainville
Honeyeater)
  Bougainville Group, Solomon Is
Myzomela melanocephala (Black-headed
Honeyeater)
  Guadalcanal Group, Solomon Is
Myzomela eichhorni (Yellow-vented
Honeyeater)
  M.e.eichhorni
  Solomon Is
  M.e.ganongae
  Ganonga I
  M.e.atrata
  Vellalavella, Baga Is
Myzomela malaitae (Malaita Honeyeater)
  Malaita I
Myzomela tristrami (Tristram's Honeyeater)
  San Cristobal, Santa Ana Is
Myzomela jugularis (Orange-breasted
Honeyeater)
  Fiji Is
Myzomela erythromelas (Black-bellied
Honeyeater)
  New Britain
Myzomela vulnerata (Sunda Honeyeater)
  Timor I
Myzomela rosenbergii (Black & Red
Honeyeater)
  M.r.rosenbergii
  NW New Guinea
  M.r.longirostris
  Goodenough I (D'Entrecasteaux Arch)
  M.r.wahgiensis
  W & C New Guinea

CERTHIONYX
Certhionyx niger (Black Honeyeater)
  C Australia
Certhionyx variegatus (Pied Honeyeater)
  C Australia

MELIPHAGA
Meliphaga mimikae (Large Spot-breasted
Honeyeater)
  M.m.rara
  N New Guinea
  M.m.mimikae
  C New Guinea
  M.m.bastille
  E New Guinea
  M.m.granti
  SE New Guinea
Meliphaga montana (White-eared Mountain
Honeyeater)
  M.m.montana
  NW New Guinea
  M.m.margaretae
  Batanta I

M.m.sepik
  C New Guinea
M.m.steini
  Japen I
M.m.germanorum
  N New Guinea
M.m.huonensis
  NE New Guinea
M.m.aicora
  SE New Guinea
Meliphaga orientalis (Small Spot-breasted
Honeyeater)
  M.o.facialis
  N & E New Guinea
  M.o.becki
  NE New Guinea
  M.o.orientalis
  SE New Guinea
  M.o.citreola
  N New Guinea
Meliphaga albonotata (White-marked
Honeyeater)
  M.a.setekwa
  SC New Guinea
  M.a.albonotata
  S New Guinea
Meliphaga aruensis (Puff-backed Honeyeater)
  M.a.sharpei
  W, N & E New Guinea, Waigeu I
  D'Entrecasteaux Arch
  M.a.aruensis
  S New Guinea, Aru Is
Meliphaga analoga (Mimic Meliphaga)
  M.a.papuae
  S New Guinea
  M.a.analoga
  S New Guinea, W Papuan islands
  M.a.longirostris
  Aru Is
  M.a.flavida
  N New Guinea, Japen I
  M.a.connectens
  N New Guinea
Meliphaga vicina (Louisiades Honeyeater)
  Tagula I
Meliphaga gracilis (Graceful Honeyeater)
  M.g.stevensi
  SE New Guinea
  M.g.cinereifrons
  SE New Guinea
  M.g.gracilis
  S New Guinea, Aru Is, N Queensland
  M.g.imitatrix
  NE Queensland
Meliphaga notata (Lesser Lewin Honeyeater)
  M.n.notata
  Torres Strait (N Queensland)
  M.n.mixta
  NE Queensland
Meliphaga flavirictus (Yellow-gaped
Honeyeater)
  M.f.flavirictus
  SE New Guinea

M.f.crockettorum
 N & W New Guinea
***Meliphaga lewinii* (Lewin Honeyeater)**
M.l.lewinii
 E Queensland, E New South Wales
M.l.nea
 E Victoria
***Meliphaga flava* (Yellow Honeyeater)**
 E Queensland
***Meliphaga albilineata* (White-striped Honeyeater)**
 Northern Territory
***Meliphaga virescens* (Singing Honeyeater)**
M.v.virescens
 SC, S & W Australia and islands
M.v.insularis
 Rottnest I (Western Australia)
M.v.westwoodia
 S Queensland
M.v.forresti
 NW & C Australia
M.v.cooperi
 N Northern Territory, Melville I
***Meliphaga versicolor* (Varied Honeyeater)**
M.v.sonoroides
 W Papuan islands, NW New Guinea
M.v.vulgaris
 Japen I, New Guinea, Fergusson I
M.v.intermedia
 Samarai I (E New Guinea)
M.v.versicolor
 S New Guinea, Torres Strait, NE Queensland
***Meliphaga fasciogularis* (Mangrove Honeyeater)**
 E Queensland, N New South Wales
***Meliphaga inexpectata* (Guadalcanal Honeyeater)**
 Guadalcanal I
***Meliphaga fusca* (Fuscous Honeyeater)**
M.f.fusca
 SE South Australia, Victoria, E New South Wales
M.f.dawsoni
 SE Queensland
M.f.subgermana
 E Queensland
M.f.zanda
 NW Queensland, E Northern Territory
M.f.flavescens
 N Western Australia, N Northern Territory
M.f.deserticola
 N Western Australia
M.f.melvillensis
 Melville I
M.f.germana
 SE New Guinea
***Meliphaga plumula* (Yellow-fronted Honeyeater)**
M.p.planasi
 N Western Australia
M.p.plumula
 C Australia

M.p.ethelae
 E South Australia, NW Victoria, W New South Wales
***Meliphaga chrysops* (Yellow-faced Honeyeater)**
M.c.samueli
 SE South Australia
M.c.chrysops
 E Victoria, E New South Wales, E Queensland
***Meliphaga cratitia* (Purple-gaped Honeyeater)**
M.c.cratitia
 W Victoria, SE South Australia, SW Western Australia
M.c.halmaturina
 Kangaroo I
***Meliphaga keartlandi* (Grey-headed Honeyeater)**
 C Australia
***Meliphaga penicillata* (White-plumed Honeyeater)**
M.p.carteri
 NW Western Australia
M.p.geraldtonensis
 W Western Australia
M.p.ladasi
 C Western Australia
M.p.centralia
 C Australia
M.p.leilavalensis
 W Queensland, NE South Australia
M.p.interioris
 NW New South Wales, SC Queensland
M.p.penicillata
 E & N Victoria, E South Australia W New South Wales, SE Queensland
M.p.mellori
 SW Victoria, SE South Australia
***Meliphaga hindwoodi* (Eungella Honeyeater)**
154.2
 EC Queensland
***Meliphaga ornata* (Mallee Honeyeater)**
 S & W Australia
***Meliphaga reticulata* (Reticulated Honeyeater)**
 Timor I
***Meliphaga leucotis* (White-eared Honeyeater)**
M.l.novaenorciae
 S Western Australia
M.l.leucotis
 SE South Australia, Victoria, New South Wales, Kangaroo I
***Meliphaga flavicollis* (Yellow-throated Honeyeater)**
 Tasmania, King I
***Meliphaga melanops* (Yellow-tufted Honeyeater)**
M.m.melanops
 E New South Wales, E & C Victoria, SE Queensland
M.m.cassidix
 S Victoria
***Meliphaga unicolor* (White-gaped Honeyeater)**
 Northern Australia

Meliphaga flaviventer (Tawny-breasted Honeyeater)
M.f.fusciventris
Waigeu, Batanta Is
M.f.flaviventer
W Papuan islands, NW New Guinea
M.f.rubiensis
WC New Guinea
M.f.saturatior
Aru Is, S New Guinea
M.f.tararae
S New Guinea
M.f.giulianettii
SE New Guinea
M.f.visi
SE New Guinea
M.f.kumusii
SE New Guinea
M.f.madaraszi
NE New Guinea
M.f.philemon
N New Guinea
M.f.meyeri
Japen I
M.f.spilogaster
Trobriand Is (D'Entrecasteaux Arch)
M.f.filigera
N Queensland
Meliphaga polygramma (Spotted Honeyeater)
M.p.polygramma
Waigeu I
M.p.keuhni
Misol I
M.p.poikilosternos
Salawati I, NW & C New Guinea
M.p.septentrionalis
N New Guinea
M.p.lophotis
SE New Guinea
M.p.candidior
S New Guinea
Meliphaga macleayana (Yellow-streaked Honeyeater)
NE Queensland
Meliphaga frenata (Bridled Honeyeater)
NE Queensland
Meliphaga subfrenata (Black-throated Honeyeater)
M.s.subfrenata
NW New Guinea
M.s.melanolaema
C New Guinea
M.s.utakwensis
SC New Guinea
M.s.salvadorii
SE New Guinea
Meliphaga obscura (Obscure Honeyeater)
M.o.viridifrons
NW New Guinea
M.o.obscura
C & SE New Guinea

OREORNIS
Oreornis chrysogenys (Orange-cheeked Honeyeater)
C New Guinea

FOULEHAIO
Foulehaio carunculata (Carunculated Honeyeater)
F.c.carunculata
Samoan Is, Tonga, E Fiji Is
F.c.taviunensis
Taveuni, Vanua Levu Is
F.c.procerior
W Fiji Is
Foulehaio provocator (Yellow-faced Honeyeater)
Kandavu I

CLEPTORNIS
Cleptornis marchei (Golden Honeyeater)
Saipan I (Mariana Is)

APALOPTERON
Apalopteron familiare (Bonin Island Honeyeater)
A.f.familiare
N Bonin Is
A.f.hahasima
S Bonin Is

MELITHREPTUS
Melithreptus brevirostris (Brown-headed Honeyeater)
M.b.augustus
S Western Australia, S South Australia, NW Victoria
M.b.brevirostris
SE Queensland, E New South Wales, Victoria
M.b.magnirostris
Kangaroo I
Melithreptus lunatus (White-naped Honeyeater)
M.l.lunatus
E Queensland, SE New South Wales, SE South Australia
M.l.chloropsis
SW Western Australia
Melithreptus albogularis (White-throated Honeyeater)
M.a.subalbogularis
N Western Australia
M.a.albogularis
Northern Territory, N Queensland, NE New South Wales, S New Guinea
Melithreptus affinis (Black-headed Honeyeater)
M.a.alisteri
King I, Furneaux Group
M.a.affinis
Tasmania

Melithreptus gularis (Black-chinned Honeyeater)
SE Queensland, New South Wales, Victoria, SE South Australia
Melithreptus laetior (Golden-backed Honeyeater)
M.l.normantoniensis
N Queensland
M.l.carpentarianus
C Queensland
M.l.laetior
N Western Australia, S Northern Territory, NW South Australia
M.l.parus
WC Western Australia
Melithreptus validirostris (Strong-billed Honeyeater)
M.v.kingi
King I, Furneaux Group
M.v.validirostris
Tasmania

ENTOMYZON
Entomyzon cyanotis (Blue-faced Honeyeater)
E.c.albipennis
N Western Australia, Northern Territory
E.c.apsleyi
Melville I
E.c.cyanotis
E South Australia, Victoria, New South Wales, E & C Queensland
E.c.harterti
S New Guinea, N Queensland

NOTIOMYSTIS
Notiomystis cincta (Stitch-bird)
N.c.hautura
Little Barrier I (New Zealand)
(N.c.cinta - extinct)

PYCNOPYGIUS
Pycnopygius ixoides (New Guinea Brown Honeyeater)
P.i.simplex
N New Guinea
P.i.proximus
N New Guinea
P.i.unicus
NE New Guinea
P.i.ixoides
NW New Guinea
P.i.cinereifrons
S New Guinea
P.i.finschi
SE New Guinea
Pycnopygius cinereus (Grey-fronted Honeyeater)
P.c.cinereus
NW New Guinea
P.c.dorsalis
WC New Guinea

P.c.marmoratus
SE New Guinea
Pycnopygius stictocephalus (Streak-capped Honeyeater)
Aru Is, New Guinea

PHILEMON
Philemon meyeri (Meyer's Friarbird)
E New Guinea
Philemon brassi (Brass's Friarbird)
NW New Guinea
Philemon citreogularis (Little Friarbird)
P.c.papuanus
S New Guinea
P.c.kisserensis
S Banda Sea islands
P.c.occidentalis
N Western Australia
P.c.breda
Melville I
P.c.sordidus
N Northern Territory
P.c.carpentariae
NW Queensland
P.c.johnstoni
NE Queensland
P.c.citreogularis
SE South Australia, Victoria, New South Wales, E & C Queensland
Philemon inornatus (Plain Friarbird)
Timor I
Philemon gilolensis (Striated Friarbird)
N Moluccas
Philemon fuscicapillus (Dusky Friarbird)
N Moluccas
Philemon subcorniculatus (Grey-necked Friarbird)
Seram I
Philemon moluccensis (Moluccas Friarbird)
P.m.moluccensis
Buru I
P.m.plumigenis
Kai Is, Tanimbar Is
Philemon buceroides (Timor Helmeted Friarbird)
P.b.neglectus
Lombok, Sumbawa, Flores Is to Alor I
P.b.buceroides
N Western Australia, Timor, Savu, Wetar Is
Philemon gordoni (Melville I Friarbird)
Melville I, N Northern Territory
Philemon novaeguineae (New Guinea Friarbird)
P.n.novaeguineae
NW & S New Guinea, West Papuan islands
P.n.aruensis
Aru Is
P.n.jobiensis
Japen I, N New Guinea
P.n.brevipennis
S New Guinea

P.n.trivialis
SE New Guinea
P.n.subtuberosus
Trobriand Is (D'Entrecasteaux Arch)
P.n.tagulanus
Tagula I
P.n.yorki
Torres Strait, N Queensland
P.n.confusus
NE Queensland
Philemon cockerelli (New Britain Friarbird)
P.c.umboi
Rook I
P.c.cockerelli
New Britain
Philemon eichhorni (New Ireland Friarbird)
New Ireland
Philemon albitorques (White-naped Friarbird)
Manus I
Philemon argenticeps (Silver-crowned Friarbird)
P.a.argenticeps
N Western Australia
P.a.melvillensis
Melville I
P.a.alexis
N Northern Territory
P.a.kempi
N Queensland
Philemon corniculatus (Noisy Friarbird)
P.c.ellioti
S New Guinea, NE Queensland
P.c.clamans
SE Queensland
P.c.corniculatus
NE Victoria, E New South Wales
Philemon diemenensis (New Caledonian Friarbird)
Loyalty Is, New Caledonia

PTILOPRORA
Ptiloprora plumbea (Leaden Honeyeater)
P.p.granti
C New Guinea
P.p.plumbea
SE New Guinea
Ptiloprora meekiana (Meek's Streaked Honeyeater)
P.m.occidentalis
C New Guinea
P.m.meekiana
SE New Guinea
Ptiloprora erythropleura (Red-sided streaked Honeyeater)
P.e.erythropleura
NW New Guinea
P.e.dammermani
C New Guinea
Ptiloprora guisei (Red-backed Honeyeater)
P.g.acrophila
N coast of New Guinea
P.g.umbrosa
N New Guinea

P.g.guisei
SE New Guinea
Ptiloprora mayri (Mayr's Streaked Honeyeater)
N New Guinea
Ptiloprora perstriata (Black-backed Streaked Honeyeater)
P.p.praedicta
NW New Guinea
P.p.incerta
WC New Guinea
P.p.perstriata
C & E New Guinea

MELIDECTES
Melidectes fuscus (Sooty Honeyeater)
M.f.occidentalis
C New Guinea
M.f.gilliardi
E New Guinea
M.f.fuscus
E & SE New Guinea
Melidectes whitemanensis (Gilliard's Honeyeater)
New Britain
Melidectes princeps (Long-bearded Honeyeater)
EC New Guinea
Melidectes nouhuysi (Short-bearded Honeyeater)
WC New Guinea
Melidectes ochromelas (Mid-mountain Honeyeater)
M.o.ochromelas
W New Guinea
M.o.batesi
C & SE New Guinea
M.o.lucifer
NE New Guinea
Melidectes leucostephes (White-fronted Melidectes)
NW New Guinea
Melidectes belfordi (Belford's Melidectes)
M.b.brassi
NW New Guinea
M.b.joiceyi
W New Guinea
M.b.kinneari
S New guinea
M.b.belfordi
SE New Guinea
M.b.schraderensis
Schrader mountains, New Guinea
Melidectes rufocrissalis (Reichenow's Melidectes)
M.r.rufocrissalis
C New Guinea
M.r.thomasi
E New Guinea
M.r.gilliardi
EC New Guinea
Melidectes foersteri (Foerster's Melidectes)
NE New Guinea

Melidectes torquatus (Cinnamon-breasted Wattle Bird)
M.t.torquatus
   NW New Guinea
M.t.nuchalis
   C New Guinea
M.t.mixtus
   C New Guinea
M.t.cahni
   NE New Guinea
M.t.polyphonus
   NE New Guinea
M.t.emilii
   SE New Guinea

MELIPOTES
Melipotes gymnops (Arfak Melipotes)
   NW New Guinea
Melipotes ater (Huon Melipotes)
   NE New Guinea
Melipotes fumigatus (Common Melipotes)
M.f.goliathi
   C New Guinea
M.f.fumigatus
   SE New Guinea
M.f.kumawa
   Kumawa Mts

MYZA
Myza celebensis (Brown Honeysucker)
M.c.celebensis
   N, C & SE Sulawesi
M.c.meridionalis
   S Sulawesi
Myza sarasinorum (Spot-headed Honeysucker)
M.s.sarasinorum
   N Sulawesi
M.s.chionogenys
   SC Sulawesi
M.s.pholidota
   SE Sulawesi

MELIARCHUS
Meliarchus sclateri (San Cristobal Honeyeater)
   San Cristobal I

GYMNOMYZA
Gymnomyza viridis (Green Honeyeater)
G.v.viridis
   Taveuni, Vanua Levu Is
G.v.brunneirostris
   Viti Levu I
Gymnomyza samoensis (Black-breasted Honeyeater)
   Samoa Is
Gymnomyza aubryana (Red-faced Honeyeater)
   New Caledonia

MOHO
Moho braccatus (Kauai O-o)
   Kauai I (Hawaii Is)

Moho bishopi (Bishop's O-o)
   Maui I (Hawaii Is)   154.4

GLYCIPHILUS   154.3
Glyciphilus melanops (Tawny-crowned Honeyeater)
G.m.melanops
   New South Wales, Victoria, South Australia,
   SW Western Australia
G.m.braba
   Kangaroo I
G.m.crassirostris
   King I (Tasmania)

PHYLIDONYRIS
Phylidonyris pyrrhoptera (Crescent Honeyeater)
P.p.pyrrhoptera
   E New South Wales, S Victoria
P.p.indistincta
   SE South Australia
P.p.halmaturina
   Kangaroo I
P.p.rex
   King I, Furneaux Group
P.p.inornata
   Tasmania
Phylidonyris novaehollandiae (Yellow-winged Honeyeater)
P.n.longirostris
   SW Western Australia
P.n.novaehollandiae
   SE Australian coast, S Queensland to South
   Australia
P.n.campbelli
   Kangaroo I
P.n.caudata
   King I, Furneaux Group
P.n.canescens
   Tasmania
Phylidonyris nigra (White-cheeked Honeyeater)
P.n.nigra
   E Queensland, E New South Wales
P.n.gouldii
   SW Western Australia
Phylidonyris albifrons (White-fronted Honeyeater)
   C Australia
Phylidonyris undulata (Barred Honeyeater)
   New Caledonia
Phylidonyris notabilis (White-bellied Honeyeater)
P.n.notabilis
   Banks Is, NW New Hebrides
P.n.superciliaris
   N New Hebrides

RAMSAYORNIS
Ramsayornis fasciatus (Bar-breasted Honeyeater)
R.f.fasciatus
   N Northern Territory, N Queensland

R.f.apsleyi
Melville I
R.f.broomei
N Western Australia
Ramsayornis modestus (Brown-backed Honeyeater)
New Guinea, N Queensland

PLECTORHYNCHA
Plectorhyncha lanceolata (Striped Honeyeater)
E Queensland, New South Wales, Victoria, SE South Australia

CONOPOPHILA
Conopophila whitei (Grey Honeyeater)
C Western Australia, S Northern Territory
Conopophila albogularis (Rufous-banded Honeyeater)
C.a.mimikae
NW & S New Guinea, Aru Is
C.a.albogularis
N Queensland, N Northern Territory
Conopophila rufogularis (Red-throated Honeyeater)
C.r.rufogularis
N Western Australia, N Northern Territory
C.r.queenslandica
N Queensland
Conopophila picta (Painted Honeyeater)
E Australia

XANTHOMYZA
Xanthomyza phrygia (Regent Honeyeater)
S Queensland, New South Wales, Victoria, SE South Australia

CISSOMELA
Cissomela pectoralis (Banded Honeyeater)
N Australia

ACANTHORHYNCHUS
Acanthorhynchus tenuirostris (Eastern Spinebill)
A.t.cairnsensis
E Queensland
A.t.trochiloides
SE Queensland
A.t.tenuirostris
E New South Wales, E & S Victoria SE South Australia
A.t.loftyi
S South Australia
A.t.halmaturinus
Kangaroo I
A.t.regius
King I, Furneaux Group
A.t.dubius
Tasmania

Acanthorhynchus superciliosus (Western Spinebill)
SW Western Australia

MANORINA
Manorina melanophrys (Bell Miner)
SE Australia
Manorina melanocephala (Noisy Miner)
M.m.melanocephala
New South Wales, Victoria, Tasmania
M.m.crassirostris
E Queensland
Manorina obscura (Dusky Miner)
M.o.obscura
S Western Australia
M.o.clelandi
SW Western Australia
Manorina flavigula (Yellow-throated Miner)
M.f.casuarina
N Western Australia
M.f.lutea
C Western Australia
M.f.alligator
Northern Territory
M.f.melvillensis
Melville I
M.f.pallida
C Australia
M.f.flavigula
W Queensland, New South Wales, Victoria, E & S South Australia, SE Western Australia
Manorina melanotis (Black-eared Miner)
NW Victoria, E South Australia

ANTHORNIS
Anthornis melanura (New Zealand Bellbird)
A.m.obscura
Three Kings I
A.m.dumerilii
North I (New Zealand)
A.m.oneho
C New Zealand
A.m.melanura
South I (New Zealand), Stewart I
A.m.incoronata
Auckland Is

ACANTHAGENYS
Acanthagenys rufogularis (Spiny-cheeked Honeyeater)
A.r.rufogularis
C Australia
A.r.parker
Friday I, Torres Strait

ANTHOCHAERA
Anthochaera chrysoptera (Little Wattle Bird)
A.c.chrysoptera
S Queensland, New South Wales, Victoria SE South Australia
A.c.halmaturina
Kangaroo I

A.c.tasmanica
  Tasmania
A.c.lunulata
  SW Western Australia
Anthochaera carunculata (Red Wattle Bird)
A.c.carunculata
  S & SE coast of Australia
A.c.woodwardi
  SW Western Australia
Anthochaera paradoxa (Yellow Wattle Bird)
  Tasmnania, King I

PROSTHEMADURA
Prosthemadura novaeseelandiae (Parson Bird)(Tui)
P.n.novaeseelandiae
  New Zealand, Auckland, Stewart Is
P.n.kermadecensis
  Kermadec Is
P.n.chathamensis
  Chatham I

PROMEROPS
Promerops cafer (Cape Sugarbird)
  Cape Province, South Africa
Promerops gurneyi (Gurney's Sugarbird)
P.g.gurneyi
  Cape Province, Natal, E Transvaal
P.g.ardens
  E Zimbabwe

155. EMBERIZIDAE

A. EMBERIZINAE (BUNTINGS)

MELOPHUS
Melophus lathami (Crested Bunting)
  Pakistan to S China, Indochina

LATOUCHEORNIS
Latoucheornis siemsseni (Fokien Blue Bunting)
  C China

EMBERIZA
Emberiza calandra (Corn Bunting)
  Europe to Sinkiang >> S Iran
Emberiza citrinella (Yellowhammer)(Yellow Bunting)
E.c.caliginosa
  N & W British Isles
E.c.citrinella
  NW Europe, C Russia >> N Africa
E.c.erythrogenys
  E Europe to Siberia >> Mongolia & Iraq
Emberiza leucocephala (Pine Bunting)
E.l.leucocephala
  Tibet, Siberia >> Iraq, India, China

E.l.fronto
  NE China
Emberiza cia (Rock Bunting)
E.c.cia
  S Europe, Asia Minor
E.c.africana
  N Africa
E.c.prageri
  Caucasus, NW Iran
E.c.par
  C Asia, Pakistan, N India
E.c.stracheyi
  W Himalayas
E.c.decolorata
  Sinkiang
E.c.godlewskii
  Mongolia, NW China
E.c.khamensis
  NE Tibet, W China
E.c.yunnanensis
  SE Tibet, SW China
E.c.omissa
  NE China
E.c.flemingorum
  Nepal
Emberiza cioides (Siberian Meadow Bunting)
E.c.tarbagataica
  C Asia >> N Mongolia
E.c.cioides
  NC Asia
E.c.weigoldi
  NE Asia >> Shensi, C Korea
E.c.castaneiceps
  S Korea, E China
E.c.ciopsis
  N Japan >> S Japan
Emberiza jankowskii (Jankowski's Bunting)
  NE Manchuria
Emberiza buchanani (Grey-necked Bunting)
E.b.cerrutii
  E Turkey, SW Russia, Iran
E.b.buchanani
  Afghanistan, W Pakistan >> SE India
E.b.neobscura
  C Asia, W Mongolia
Emberiza stewarti (White-capped Bunting)
  S Russia, Afgahanistan >> Pakistan, NW India
Emberiza cineracea (Cinereous Bunting)
E.c.cineracea
  SW Turkey >> Eritrea
E.c.semenowi
  Yemen, SW Iran >> NE Africa
Emberiza hortulana (Ortolan Bunting)
  Europe, N Africa, C Asia >> Senegal, Sudan, Iran
Emberiza caesia (Cretzschmar's Bunting)
  E Europe, NE Africa >> Iran, Sudan
Emberiza cirlus (Cirl Bunting)
E.c.cirlus
  S British Isles, S Europe, N Africa
E.c.nigrostriata
  Corsica, Sardinia

Emberiza striolata (House Bunting)
E.s.sahari
  NW Africa
E.s.sanghae
  S Mali
E.s.saturatior
  W Sudan, Ethiopia, NW Kenya
E.s.jebelmarrae
  Darfur, Kordofan
E.s.striolata
  NE Africa, Iran, Pakistan, N & C India
Emberiza impetuani (Larklike Bunting)
E.i.impetuani
  Botswana, W Cape Province
E.i.eremica
  N Namibia, S Angola, NW Cape Province
E.i.sloggetti
  C Cape Province
Emberiza tahapisi (Cinnamon-breasted Rock Bunting)
E.t.arabica
  S Arabia
E.t.insularis
  Socotra I
E.t.septemstriata
  E Sudan, NW Ethiopia,
E.t.tahapisi
  Gabon, Zaire, E & S Africa
E.t.goslingi
  Sierra Leone to Sudan, N Zaire
E.t.nivenorum
  N Namibia
Emberiza socotrana (Socotra Mountain Bunting)
  Socotra I
Emberiza capensis (Cape Bunting)
E.c.vincenti
  C Malawi, E Zambia
E.c.smithersii
  E Zimbabwe, Mozambique
E.c.plowesi
  Zimbabwe, NE Botswana
E.c.reidi
  S Transvaal, Natal, Orange Free State, N Lesotho
E.c.limpopoensis
  C & SW Transvaal
E.c.cinnamomea
  W Cape Province, W Orange Free State
E.c.basutoensis
  Lesotho, W Natal
E.c.vinacea
  N Cape Province
E.c.media
  S Transvaal, C Cape Province
E.c.capensis
  S Namibia, W Cape Province
E.c.nebularum
  SW Angola
E.c.bradfieldi
  N Namibia
Emberiza yessoensis (Japanese Reed Bunting)

E.y.yessoensis
  N Japan >> S Japan
E.y.continentalis
  E Manchuria >> S Korea, E China
Emberiza tristrami (Tristram's Bunting)
  Ussuri region >> SW China, Burma
Emberiza fucata (Grey-hooded Bunting)
E.f.arcuata
  Pakistan, W Himalayas >> N Burma, S China
E.f.fucata
  E Asia, N Japan, SE China >> Indochina
E.f.kuatunensis
  S China
Emberiza pusilla (Little Bunting)
  N Asia >> N India, Burma, S China
Emberiza chrysophrys (Yellow-browed Bunting)
  Siberia, NE Asia >> SE China
Emberiza rustica (Rustic Bunting)
E.r.rustica
  N Europe, N Asia >> E China, Japan
E.r.latifascia
  NE Siberia >> E China & Japan
Emberiza elegans (Yellow-headed Bunting)
E.e.elegans
  Manchuria >> S Japan, E China
E.e.ticehursti
  E Amur >> S Manchuria
E.e.elegantula
  SW China
Emberiza aureola (Yellow-breasted Bunting)
E.a.aureola
  N Asia >> India, Malaysia, Indochina
E.a.ornata
  NE Asia
Emberiza poliopleura (Somali Golden-breasted Bunting)
  EC Africa
Emberiza flaviventris (Golden-breasted Bunting)
E.f.flavigaster
  Mali to Ethiopia
E.f.flaviventris
  C, E & Sthn Africa
E.f.kalaharica
  Mozambique, Zimbabwe to N Cape Province
E.f.carychroa
  Nairobi, Kenya
E.f.princeps
  S Angola, N Namibia
Emberiza affinis (Brown-rumped Bunting)
E.a.affinis
  S Sudan, SW Ethiopia, N Uganda, NE Zaire
E.a.vulpecula
  Cameroun, Central African Republic
E.a.nigeriae
  Gambia to W Cameroun
Emberiza cabanisi (Cabanis's Yellow Bunting)
E.c.cabanisi
  W & NC Africa

E.c.cognominata
SW Zaire, N Angola
E.c.orientalis
Zambia, Tanzania, Zimbabwe, Mozambique
Emberiza rutila (Chestnut Bunting)
NE Asia >> SE China, Burma, Indochina
Emberiza koslowi (Koslow's Bunting)
Tibet, Tsinghai
Emberiza melanocephala (Black-headed Bunting)
SE Europe, Iran, Caucasus >> N & C India
Emberiza bruniceps (Red-headed Bunting)
C Asia, Altai >> S India
Emberiza sulphurata (Japanese Yellow Bunting)
C Japan >> S Japan, SE China, N Philippine Is
Emberiza spodocephala (Black-faced Bunting)
E.s.spodocephala
C & E Asia >> E China, Taiwan
E.s.personata
Sakhalin I, N Japan >> S Japan
E.s.sordida
W China >> E India, N Burma
Emberiza variabilis (Japanese Grey Bunting)
Sakhalin I, N Japan
Emberiza pallasi (Pallas' Reed Bunting)
E.p.pallasi
C & E Asia >> Sinkiang & Mongolia
E.p.polaris
NE Asia >> Manchuria & E China
E.p.lydiae
C Mongolia
Emberiza schoeniclus (Reed Bunting)
E.s.schoeniclus
NW Europe, C Russia >> Turkey, N Africa
E.s.passerina
NW Siberia >> Mongolia, N Iran
E.s.parvirostris
C Siberia >> Mongolia
E.s.pyrrhulina
NE Asia >> Japan
E.s.minor
SE Siberia, Manchuria
E.s.pallidior
SW Siberia >> NW India, Mongolia
E.s.ukrainae
S Russia >> N Caucasus
E.s.incognita
C Russia >> Sinkiang
E.s.pyrrhuloides
W & C Asia, W Mongolia, Sinkiang
E.s.zaidamensis
N Tsinghai
E.s.witherbyi
W Spain, Sardinia, Balearic Is
E.s.canetti
SE Europe, N Turkey
E.s.reiseri
S Yugoslavia, N Greece
E.s.caspia
E Caucasus, W & S Iran

E.s.korejewi
E Iran

CALCARIUS
Calcarius mccownii (McCown's Longspur)
SC Canada, WC USA >> S USA, N Mexico
Calcarius lapponicus (Lapland Bunting)
C.l.lapponicus
N Canada, S Greenland >> N Europe, E USA, N Asia
C.l.coloratus
E Siberia >> N China
C.l.alascensis
Alaska, W Canada >> W USA
Calcarius pictus (Smith's Longspur)
N Canada >> SC USA
Calcarius ornatus (Chestnut-coloured Longspur)
S Canada >> S USA, N Mexico

PLECTROPHENAX
Plectrophenax nivalis (Snow Bunting)
P.n.nivalis
N North America, N Europe >> S USA, S Europe
P.n.insulae
Iceland >> N Scotland
P.n.vlasowae
NE Asia >> C Asia, Manchuria
P.n.townsendi
W Aleutian, Commander Is
Plectrophenax hyperboreus (McKay's Bunting)
Bering Sea islands >> W Alaska

CALAMOSPIZA
Calamospiza melanocorys (Lark Bunting)
S Canada >> C & S USA, N Mexico

PASSERELLA
Passerella iliaca (Fox Sparrow)
P.i.chilcatensis
Alaska
P.i.iliaca
E Canada >> E USA
P.i.zaboria
W Canada >> C & S USA
P.i.altivagans
SW Canada >> California
P.i.unalaschensis
Aleutian Is >> S California
P.i.ridgwayi
Alaska >> S California
P.i.sinuosa
Alaska >> S California
P.i.annectens
Alaska >> S California
P.i.townsendi
SE Alaska >> C California
P.i.fuliginosa
W Canada >> S California

P.i.olivacea
 SW Canada >> NW Mexico
P.i.schistacea
 NW USA >> SW USA
P.i.swarthi
 NW Utah, SE Idaho
P.i.fulva
 C Oregon, California >> NW Mexico
P.i.megarhyncha
 SW Oregon >> NW Mexico
P.i.brevicauda
 N California >> S California
P.i.monoensis
 C California >> S California
P.i.canescens
 E California, Nevada >> S California
P.i.stephensi
 S California

MELOSPIZA
Melospiza melodia (Song Sparrow)
M.m.melodia
 S Canada >> SE USA
M.m.atlantica
 NE USA >> E USA
M.m.euphonia
 NC USA >> S USA
M.m.juddi
 WC Canada >> WC USA
M.m.montana
 WC USA >> SW USA.N Mexico
M.m.fallax
 SE Nevada, SW Utah, Arizona, New Mexico
M.m.saltonis
 S Nevada, SE California
M.m.inexpectata
 SE Alaska >> S Oregon
M.m.rufina
 SE Alaska >> W Washington
M.m.merrilli
 SW Canada >> S California
M.m.morphna
 SW Canada >> N California
M.m.fisherella
 Oregon >> S California, W Nevada
M.m.maxima
 Aleutian Is
M.m.sanaka
 Seguam, Sanak, Unimak Is
M.m.amaka
 Amak I
M.m.insignis
 Kodiak I
M.m.kenaiensis
 S Alaska
M.m.caurina
 SE Alaska >> N California
M.m.cleonensis
 NW California
M.m.gouldii
 WC California
M.m.mailliardi
 C California

M.m.samuelis
 NW California
M.m.maxillaris
 NW California
M.m.pusillula
 NW California
M.m.heermani
 C California
M.m.cooperi
 SW California, N Baja California
M.m.micronyx
 San Miguel I
M.m.clementae
 Santa Roza, Santa Cruz Is
M.m.coronatorum
 Coronados I
M.m.rivularis
 SC Baja California
M.m.goldmani
 WC Mexico
M.m.niceae
 EC Mexico
M.m.mexicana
 SC Mexico
M.m.azteca
 SC Mexico
M.m.villai
 W Mexico
M.m.yuriria
 C Mexico
M.m.adusta
 SW Mexico
M.m.zacapu
 W Mexico
Melospiza lincolnii (Lincoln Sparrow)
M.l.lincolnii
 Canada >> SW USA, S Mexico,
 Guatemala
M.l.gracilis
 S Alaska, W Canada >> California
M.l.alticola
 NW USA >> Mexico, Guatemala
Melospiza georgiana (Swamp Sparrow)
M.g.ericrypta
 W & C Canada >> SW USA, NW Mexico
M.g.georgiana
 NE USA >>SE USA

ZONOTRICHIA
**Zonotrichia capensis (Rufous-collared
Sparrow)**
Z.c.septentrionalis
 S Mexico, Guatemala, El Salvador,
 Honduras
Z.c.antillarum
 Dominica I
Z.c.costaricensis
 Costa Rica, Panama, Venezuela, Colombia
Z.c.orestera
 W Panama
Z.c.insularis
 Curacao, Aruba Is

Z.c.venezuelae
N Venezuela
Z.c.inaccessibilis
C Venezuela
Z.c.perezchinchillae
Amazonas (Venezuela)
Z.c.roraimae
S Colombia, E Venezuela, Guyana, N Brazil
Z.c.macconelli
Venezuela
Z.c.capensis
French Guiana
Z.c.tocantinsi
E Brazil
Z.c.novaesi
Para (Brazil)
Z.c.matutina
NE Brazil, E Bolivia
Z.c.subtorquata
E & C Brazil, Paraguay, Uruguay
Z.c.mellea
C Paraguay, N Argentina
Z.c.arenalensis
N Argentina
Z.c.hypoleuca
E & S Bolivia, NE Argentina
Z.c.choraules
W Argentina
Z.c.australis
S Chile, S Argentina >> N Bolivia
Z.c.chilensis
Chile, S Argentina
Z.c.sanborni
Chile, W Argentina
Z.c.antofagastae
Chile
Z.c.pulacayensis
Peru, W Bolivia, N Argentina
Z.c.peruviensis
W Peru
Z.c.carabayae
Peru, Bolivia
Z.c.huancabambae
N Peru
Z.c.illescasensis
N Peru
Zonotrichia querula (Harris' Sparrow)
W Canada >> W USA
Zonotrichia leucophrys (White-crowned Sparrow)
Z.l.leucophrys
C & E Canada >> SE USA, Cuba
Z.l.gambelii
NW & W Canada >> W USA, N Mexico
Z.l.oriantha
SW Canada >> WC USA, N Mexico
Z.l.pugetensis
SW Canada >> SW California
Z.l.nuttalli
WC California
Zonotrichia albicollis (White-throated Sparrow)
Canada, N & E USA >> S USA, E Mexico

Zonotrichia atricapilla (Golden-crowned Sparrow)
Alaska, W Canada >> W USA, NW Mexico

JUNCO
Junco vulcani (Volcano Junco)
Costa Rica, W Panama
Junco hyemalis (Dark-eyed Junco)
J.h.hyemalis
N Canada, NC USA >> S USA, N Mexico
J.h.carolinensis
EC USA
J.h.aikeni
WC USA >> S USA
J.h.oreganus
S Alaska, W Canada >> C California
J.h.cismontanus
W Canada >> W USA
J.h.shufeldti
NW USA >> S California
J.h.montanus
W Canada, W USA >> NW Mexico
J.h.mearnsi
SW Canada >> WC USA, NW Mexico
J.h.thurberi
S Oregon >> California, N Baja California
J.h.pinosus
C & S California
J.h.pontilus
N Baja California
J.h.townsendi
N Baja California
J.h.insularis
Guadeloupe I
Junco caniceps (Grey-headed Junco)
J.c.caniceps
C & SC USA >> S USA & N Mexico
J.c.dorsalis
New Mexico, N Arizona
J.c.mutabilis
S Nevada, SE California
Junco phaeonotus (Mexican Junco)
J.p.palliatus
SW USA, N Mexico
J.p.phaeonotus
C & S Mexico
J.p.bairdi
S Baja California
J.p.fulvescens
S Mexico
J.p.alticola
S Mexico, W Guatemala

PASSERCULUS
Passerculus sandwichensis (Savannah Sparrow)
P.s.labradorius
E Canada >> SE USA
P.s.savanna
SE Canada >> SE USA, SE Mexico
P.s.princeps
Sable I >> SE USA

P.s.mediogriseus
SE Canada, NE USA >> SE USA
P.s.oblitus
C Canada, C USA, NE Mexico
P.s.nevadensis
SW Canada, WC USA >> SC USA, N
Mexico
P.s.brooksi
SW Canada >> California, Baja California
P.s.athinus
Alaska, W Canada, W USA >> SW USA,
W Mexico
P.s.sandwichensis
Alaska >> W USA
P.s.crassus
Aleutian Is, W Alaska >> C California
P.s.alaudinus
N & C California
P.s.beldingi
S California, N Baja California
P.s.anulus
WC Baja California
P.s.sanctorum
San Benito I, Baja California
P.s.guttatus
W & S Baja California
P.s.magdalenae
S Baja California
P.s.rostratus
S California, Baja California, W Mexico
P.s.rufofuscus
Arizona, New Mexico, N Mexico
P.s.atratus
NW Mexico
P.s.brunnescens
NC Mexico
P.s.wetmorei
SW Guatemala

AMMODRAMUS
Ammodramus maritimus (Seaside Sparrow)
A.m.maritimus
NE USA >> SE USA
A.m.macgillivraii
SE USA
A.m.pelonota
SE USA
A.m.mirabilis
SE USA
A.m.peninsulae
SE USA
A.m.junicola
SE USA
A.m.nigrescens
E Florida
A.m.fisheri
SW & S USA
A.m.sennetti
S USA
**Ammodramus caudacutus (Sharp-tailed
Sparrow)**
A.c.nelsoni
WC Canada >> SE USA

A.c.alterus
E Canada >> SE USA
A.c.subvirgatus
E Canada >> SE USA
A.c.caudacutus
NE USA >> SE USA
A.c.diversus
NE USA >> SE USA
Ammodramus leconteii (Le Conte's Sparrow)
WC Canada >> C & SE USA
Ammodramus bairdii (Baird's Sparrow)
WC Canada, W USA >> N Mexico
Ammodramus henslowii (Henslow's Sparrow)
A.h.susurrans
NE USA >> SE USA
A.h.henslowii
C USA >> SE USA
**Ammodramus savannarum (Grasshopper
Sparrow)**
A.s.pratensis
S Canada, E USA >> SE Mexico, West
Indies
A.s.floridanus
Florida
A.s.perpallidus
SW Canada, C & SW USA >> Mexico,
Guatemala
A.s.ammolegus
S Arizona, NW Mexico >> Guatemala
A.s.bimaculatus
S Mexico, Honduras, Nicaragua, NW Costa
Rica W Panama
A.s.cracens
Guatemala, E Honduras, NE Nicaragua
A.s.caucae
Colombia
A.s.savannarum
Jamaica
A.s.intricatus
Hispaniola
A.s.borinquensis
Puerto Rico
A.s.caribaeus
Bonaire, Curacao Is

XENOSPIZA
Xenospiza baileyi (Sierra Madre Sparrow)
N & C Mexico

MYOSPIZA
Myospiza humeralis (Grassland Sparrow)
M.h.humeralis
Colombia, Venezuela, Guyana, French
Guiana, Brazil
M.h.pallidulus
Colombia, Venezuela
M.h.xanthornus
Bolivia, Brazil, Paraguay, Uruguay,
Argentina
M.h.tarijensis
E Bolivia
Myospiza aurifrons (Yellow-browed Sparrow)

M.a.apurensis
NE Colombia, W Venezuela
M.a.cherriei
E Colombia
M.a.tenebrosus
Venezuela, Colombia
M.a.aurifrons
SE Colombia, E Ecuador, Peru, Bolivia, Brazil

SPIZELLA
Spizella arborea (American Tree Sparrow)
S.a.arborea
N Canada, N USA >> C USA
S.a.ochracae
NW & W Canada >> W USA
Spizella passerina (Chipping Sparrow)
S.p.passerina
SE Canada, C USA >> S USA
S.p.arizonae
W Canada >> SW USA, W Mexico
S.p.atremaeus
NC Mexico
S.p.mexicana
C & S Mexico, Guatemala
S.p.repetens
Guerrero, Oaxaca
S.p.comparanda
Nayarit to Vera Cruz
S.p.pinetorum
S Guatemala to NE Nicaragua
Spizella pusilla (Field Sparrow)
S.p.pusilla
SE Canada >> C & SE USA
S.p.arenacea
C USA >> SE USA, NE Mexico
Spizella wortheni (Worthen's Sparrow) 155.A.2
NE & E Mexico
Spizella atrogularis (Black-chinned Sparrow)
S.a.evura
SW USA >> NW Mexico
S.a.caurina
C California
S.a.cana
SW California >> S Baja California
S.a.atrogularis
NC Mexico
Spizella pallida (Clay-coloured Sparrow)
SC Canada, C & SC USA >> W Mexico
Spizella breweri (Brewer's Sparrow)
S.b.taverneri
SW Canada >> SW USA
S.b.breweri
SW Canada, W & SW USA >> NW Mexico

POOECETES
Pooecetes gramineus (Vesper Sparrow)
P.g.gramineus
SE Canada, E USA >> S Mexico
P.g.confinis
SW Canada, WC USA >> SW USA, W Mexico

P.g.affinis
W USA >> NW Baja California

CHONDESTES
Chondestes grammacus (Lark Sparrow)
C.g.grammacus
N & C USA >> SE USA, E & S Mexico
C.g.strigatus
SW Canada, W USA >> Mexico, Guatemala

AMPHISPIZA
Amphispiza bilineata (Black-throated Sparrow)
A.b.bilineata
NC Texas, NE Mexico
A.b.opuntia
SC USA, N Mexico
A.b.deserticola
WC USA >> NW Mexico, Baja California
A.b.bangsi
S Baja California
A.b.tortugae
Tortuga I
A.b.belvederei
Cerralvo I
A.b.pacifica
NW Mexico
A.b.cana
San Esteban I
A.b.grisea
C Mexico
Amphispiza belli (Sage Sparrow)
A.b.nevadensis
W USA >> SW USA, N Baja California, NW Mexico
A.b.canescens
SW California, W Nevada, NE Baja California
A.b.belli
S California, NW Baja California
A.b.clementeae
San Clemente I
A.b.cinerea
C Baja California

AIMOPHILA
Aimophila mystacalis (Bridled Sparrow)
SC Mexico
Aimophila humeralis (Black-chested Sparrow)
WC Mexico
Aimophila ruficauda (Stripe-headed Sparrow)
A.r.acuminata
WC Mexico
A.r.lawrencii
S Mexico
A.r.connectens
E Guatemala
A.r.ibarrorum
Guatemala

A.r.ruficauda
  SE Guatemala, El Salvador, Honduras,
  Nicaragua
**Aimophila sumichrasti (Cinnamon-tailed
Sparrow)**
  S Mexico
Aimophila strigiceps (Stripe-capped Sparrow)
  A.s.strigiceps
    E Argentina
  A.s.dabbenei
    NW Argentina
Aimophila aestivalis (Bachman's Sparrow)
  A.a.bachmani
    SC USA >> SE USA
  A.a.illinoensis
    NC USA >> S USA
  A.a.aestivalis
    S California, Georgia, Florida
Aimophila botterii (Botteri's Sparrow)
  A.b.arizonae
    SE Arizona, NW Mexico
  A.b.texana
    S Texas, NE Mexico
  A.b.mexicana
    NC Mexico
  A.b.goldmani
    W Mexico
  A.b.botterii
    SC & S Mexico
  A.b.petenica
    SE Mexico, Guatemala, Honduras
  A.b.tabascensis
    E coast of Mexico
  A.b.spadiconigrescens
    N Honduras, NE Nicaragua
  A.b.vantynei
    C Guatemala
  A.b.vulcanica
    Nicaragua, N Costa Rica
Aimophila cassinii (Cassin's Sparrow)
  SC USA >> C Mexico
**Aimophila quinquestriata (Five-striped
Sparrow)**
  A.q.septentrionalis
    NW Mexico
  A.q.quinquestriata
    W Mexico
Aimophila carpalis (Rufous-winged Sparrow)
  A.c.carpalis
    SC Arizona, NW Mexico
  A.c.distinguenda
    NW Mexico
  A.c.cohaerens
    NW Mexico
**Aimophila ruficeps (Rufous-crowned
Sparrow)**
  A.r.eremoeca
    SC USA >> E Mexico
  A.r.scottii
    Arizona, New Mexico
  A.r.ruficeps
    C California

A.r.canescens
  SW California, NE Baja California
A.r.pallidissima
  S Nuevo Leon
A.r.extima
  S Oaxaca
A.r.obscura
  Santa Catalina I
A.r.sanctorum
  Todos Santos I
A.r.sororia
  S Baja California
A.r.rupicola
  SW Arizona
A.r.simulans
  NW Mexico
A.r.fusca
  W Mexico
A.r.boucardi
  E Mexico
A.r.australis
  S Mexico
Aimophila notosticta (Oaxaca Sparrow)
  S Mexico
Aimophila rufescens (Rusty Sparrow)
  A.r.antonensis
    NW Mexico
  A.r.mcleodii
    NW Mexico
  A.r.disjuncta
    Guerrero
  A.r.rufescens
    W & SW Mexico
  A.r.pyrgitoides
    Guatemala, Honduras, El Salvador, E & SE
    Mexico
  A.r.discolor
    N Honduras, NE Nicaragua
  A.r.pectoralis
    SE Mexico, Guatemala, El Salvador
  A.r.hypaethrus
    NW Costa Rica
  A.r.brodkorbi
    SW Chiapas
  A.r.newmani
    NE Puebla

RHYNCHOSPIZA
Rhynchospiza stolzmanni (Tumbes Sparrow)
  SW Ecuador, N Peru

TORREORNIS
Torreornis inexpectata (Zapata Sparrow)
  T.i.inexpectata
    SW Cuba
  T.i.sigmani
    S Cuba

ORITURUS
Oriturus superciliosus (Striped Sparrow)
  O.s.palliatus
    NW & W Mexico

O.s.superciliosus
  C & SW Mexico

Phrygilus atriceps (Black-hooded Sierra Finch)
P.a.chloronotus
  Peru
P.a.punensis
  Peru, Bolivia
P.a.atriceps
  Peru, Bolivia, Chile, Argentina
Phrygilus gayi (Grey-hooded Sierra Finch)
P.g.gayi
  N Chile
P.g.minor
  C Chile
P.g.caniceps
  S Chile, Argentina
Phrygilus patagonicus (Patagonian Sierra Finch)
  Chile, Argentina
Phrygilus fruticeti (Mourning Sierra Finch)
P.f.peruvianus
  Peru, Bolivia
P.f.fruticeti
  SW Bolivia, Chile, Argentina
Phrygilus unicolor (Plumbeous Sierra Finch)
P.u.nivarius
  N Colombia, NW Venezuela
P.u.geospizopsis
  S Colombia, Ecuador
P.u.inca
  Peru, Bolivia
P.u.unicolor
  SW Peru, Chile, W Argentina
P.u.tucumanus
  Bolivia, NW Argentina
P.u.ultimus
  Argentina
Phrygilus dorsalis (Red-backed Sierra Finch)
  Bolivia, Chile, Argentina
Phrygilus erythronotus (White-throated Sierra Finch)
  Peru, Bolivia, Chile
Phrygilus plebejus (Ash-breasted Sierra Finch)
P.p.ocularis
  Ecuador, N Peru
P.p.plebejus
  Peru, Chile, Bolivia, Argentina
Phrygilus carbonarius (Carbonated Sierra Finch)
  C Argentina
Phrygilus alaudinus (Band-tailed Sierra Finch)
P.a.bipartitus
  W Ecuador, Peru
P.a.humboldti
  S Ecuador, N Peru
P.a.excelsus
  S Peru, Bolivia

P.a.alaudinus
  C Chile
P.a.venturii
  Argentina

Melanodera melanodera (Black-throated Finch)
M.m.princetoniana
  Chile, Argentina
M.m.melanodera
  Falkland Is
Melanodera xanthogramma (Yellow-bridled Finch)
M.x.barrosi
  Chile, W Argentina
M.x.xanthogramma
  S Argentina

Haplospiza rustica (Slaty Finch)
H.r.uniformis
  S Mexico
H.r.barrilesensis
  Honduras, Costa Rica, W Panama
H.r.arcana
  Venezuela
H.r.rustica
  N Venezuela, Colombia, Ecuador, Peru, Bolivia
Haplospiza unicolor (Uniform Finch)
  SE Brazil, E Paraguay, NE Argentina

Acanthidops bairdii (Peg-billed Sparrow)
  Costa Rica

Lophospingus pusillus (Black-crested Finch)
  S Bolivia, Paraguay, Argentina
Lophospingus griseocristatus (Grey-crested Finch)
  Bolivia, N Argentina

Donacospiza albifrons (Long-tailed Reed Finch)
  Brazil, Paraguay, Uruguay, Argentina

Rowettia goughensis (Gough I Finch)
  Gough I

Nesospiza acunhae (Nightingale Finch)
N.a.acunhae
  Inaccessible I
N.a.questi
  Nightingale I, Tristan da Cunha I
Nesospiza wilkinsi (Wilkins's Finch)
N.w.wilkinsi
  Nightingale I, Tristan da Cunha I

N.w.dunnei
Inaccessible I

DIUCA
Diuca speculifera (White-winged Diuca Finch)
D.s.magnirostris
Peru
D.s.speculifera
SE Peru, N Chile, N Bolivia
Diuca diuca (Common Diuca Finch)
D.d.crassirostris
N Chile, Argentina
D.d.diuca
NC Chile, Argentina
D.d.chiloensis
SC Chile
D.d.minor
Argentina

IDIOPSAR
Idiopsar brachyurus (Short-tailed Finch)
Peru, Bolivia, Argentina

PIEZORHINA
Piezorhina cinerea (Cinereous Finch)
NW Peru

XENOSPINGUS
Xenospingus concolor (Slender-billed Finch)
S Peru, N Chile

INCASPIZA
Incaspiza pulchra (Great Inca Finch)
Peru
Incaspiza personata (Rufous-backed Inca Finch)
Peru
Incaspiza ortizi (Grey-winged Inca Finch)
Peru
Incaspiza laeta (Buff-bridled Inca Finch)
Peru
Incaspiza watkinsi (Little Inca Finch)
Peru

POOSPIZA
Poospiza thoracica (Bay-chested Warbling Finch)
SE Brazil
Poospiza boliviana (Bolivian Warbling Finch)
C Bolivia
Poospiza alticola (Plain-tailed Warbling Finch)
N Peru
Poospiza hypochondria (Rufous-sided Warbling Finch)
P.h.hypochondria
Bolivia
P.h.affinis
Argentina
Poospiza erythrophrys (Rusty-browed Warbling Finch)
P.e.cochabambae
Bolivia

P.e.erythrophrys
Bolivia, NW Argentina
Poospiza ornata (Cinnamon Warbling Finch)
NW Argentina
Poospiza nigrorufa (Black & Rufous Warbling Finch)
P.n.nigrorufa
S Brazil, Uruguay, Paraguay
P.n.whitii
Bolivia, NW Argentina
P.n.wagneri
Bolivia
Poospiza lateralis (Red-rumped Warbling Finch)
P.l.lateralis
SE Brazil
P.l.cabanisi
SE Brazil, Uruguay, Paraguay, NE Argentina
Poospiza rubecula (Rufous-breasted Warbling Finch)
N Peru
Poospiza caesar (Chestnut-breasted Mountain Finch)
SE Peru
Poospiza hispaniolensis (Collared Warbling Finch)
SW Ecuador, Peru
Poospiza torquata (Ringed Warbling Finch)
P.t.torquata
Bolivia
P.t.pectoralis
SE Bolivia, W Paraguay, N & C Argentina
Poospiza cinerea (Grey & White Warbling Finch)
P.c.cinerea
C Brazil
P.c.melanoleuca
Bolivia, Paraguay, Uruguay, SW Brazil, N Argentina

SICALIS
Sicalis citrina (Stripe-tailed Yellow Finch)
S.c.browni
Colombia, Venezuela, Guyana, NE Brazil
S.c.citrina
E Brazil
S.c.occidentalis
Peru
Sicalis lutea (Puna Yellow Finch)
Peru, Bolivia, Argentina
Sicalis uropygialis (Bright-rumped Yellow Finch)
S.u.sharpei
N Peru
S.u.connectens
Peru
S.u.uropygialis
S Peru, Bolivia, N Chile, NW Argentina
Sicalis luteocephala (Citron-headed Yellow Finch)
C Bolivia

Sicalis auriventris (Greater Yellow Finch)
Chile, Argentina
Sicalis olivascens (Greenish Yellow Finch)
S.o.salvini
N Peru
S.o.chloris
C Peru, N Chile
S.o.olivascens
SE Peru, W Bolivia, NW Argentina
S.o.mendozae
W Argentina
Sicalis lebruni (Patagonian Yellow Finch)
S Argentina, S Chile
Sicalis colombiana (Orange-fronted Yellow Finch)
S.c.colombiana
Venezuela, E Colombia
S.c.leopoldinae
E Brazil
S.c.goeldii
E Peru, E Brazil
Sicalis flaveola (Saffron Finch)
S.f.flaveola
Colombia, Venezuela, the Guianas, Trinidad
S.f.valida
Ecuador, NW Peru
S.f.brasiliensis
NE Brazil
S.f.pelzelni
SE Brazil, E Bolivia, Paraguay, Uruguay, N Argentina
Sicalis luteola (Grassland Yellow Finch)
S.l.chrysops
S Mexico, Guatemala, E Honduras, Nicaragua
S.l.mexicana
S Mexico
S.l.eisenmanni
Panama
S.l.bogotensis
Colombia, Ecuador, Peru, Venezuela
S.l.luteola
Colombia, Venezuela, Guyana, Brazil
S.l.flavissima
N Brazilian islands
S.l.chapmani
NE Brazil
S.l.luteiventris
C & SC South America
Sicalis raimondii (Raimondi's Yellow Finch)
Peru
Sicalis taczanowskii (Sulphur-breasted Finch)
SW Ecuador, N Peru

COMPOSPIZA
Compospiza garleppi (Cochabamba Mountain Finch)
Bolivia
Compospiza baeri (Tucuman Mountain Finch)
NW Argentina

EMBERIZOIDES
Emberizoides herbicola (Wedge-tailed Grass Finch)
E.h.lucaris
SW Costa Rica
E.h.hypochondriacus
W & C Panama
E.h.floresae
Panama
E.h.apurensis
E Colombia, W Venezuela
E.h.sphenurus
N Colombia to the Guianas, N Brazil
E.h.herbicola
E & S Brazil, E Bolivia, NE Argentina
Emberizoides ypiranganus (Lesser Grass Finch)
S Venezuela
Emberizoides duidae (Mt Duida Grass Finch)
SE Venezuela

EMBERNAGRA
Embernagra platensis (Great Pampa Finch)
E.p.platensis
SE Brazil, Paraguay, Uruguay, E Argentina
E.p.catamarcanus
NW Argentina
E.p.olivascens
SE Bolivia, W Paraguay, NW Argentina
Embernagra longicauda (Buff-throated Pampa Finch)
Brazil

VOLATINIA
Volatinia jacarina (Blue-black Grassquit)
V.j.splendens
Central America, N South America, Trinidad
V.j.jacarina
E & C Brazil, SE Peru, E Bolivia, N Argentina
V.j.peruviensis
Ecuador, Peru, N Chile

SPOROPHILA
Sporophila frontalis (Buffy-throated Seedeater)
SE Brazil, Paraguay, N Argentina
Sporophila falcirostris (Temminck's Seedeater)
SE Brazil
Sporophila schistacea (Slate-coloured Seedeater)
S.s.subconcolor
S Mexico
S.s.schistacea
Costa Rica, Panama, N Colombia
S.s.incerta
W Colombia, Ecuador
S.s.longipennis
Venezuela, E Colombia, N Brazil
Sporophila intermedia (Grey Seedeater)
S.i.intermedia
Colombia, N Venezuela, Guyana, Trinidad

S.i.bogotensis
W Colombia
S.i.agustini
N Colombia
S.i.anchicayae
Colombia
Sporophila plumbea (Plumbeous Seedeater)
S.p.colombiana
N Colombia
S.p.whiteleyana
E Colombia, S Venezuela, the Guianas, N
Brazil
S.p.plumbea
C & S Brazil, Paraguay, NW Bolivia, N
Argentina
Sporophila aurita (Variable Seedeater)
S.a.corvina
Central America from E Mexico to Panama
S.a.aurita
Costa Rica, Panama
S.a.chocoana
Panama, W Colombia
**Sporophila americana (Wing-barred
Seedeater)**
S.a.ophthalmica
SW Colombia, Ecuador, Peru
S.a.murallae
SE Colombia
S.a.americana
NE Venezuela, the Guianas, Brazil Tobago I
S.a.dispar
Brazil
**Sporophila torqueola (White-collared
Seedeater)**
S.t.sharpei
S Texas, NE Mexico
S.t.torqueola
C & SW Mexico
S.t.morelleti
Atlantic slopes of Central America, S
Mexico to Panama
S.t.mutanda
Pacific slopes from SW Mexico to El
Salvador
**Sporophila collaris (Rusty-collared
Seedeater)**
S.c.ochrascens
N Bolivia, W Brazil
S.c.collaris
E Brazil
S.c.melanocephala
Brazil, Paraguay, N Argentina
Sporophila lineola (Lined Seedeater)
S.l.bouvronides
Trinidad, Tobago I
S.l.restricta
Colombia
S.l.lineola
NW & C South America to Argentina
**Sporophila luctuosa (Black & White
Seedeater)**
NW South America

**Sporophila nigricollis (Yellow-bellied
Seedeater)**
S.n.nigricollis
Costa Rica, Panama & N South America to
Bolivia
S.n.vivida
SW Colombia, W Ecuador
S.n.inconspicua
W Peru
Sporophila ardesiaca (Dubois' Seedeater)
Brazil
Sporophila melanops (Hooded Seedeater)
Brazil
**Sporophila obscura (Dull-coloured
Seedeater)**
S.o.haplochroma
N Colombia, NW Venezuela
S.o.pauper
S Colombia, Ecuador, NW Peru
S.o.obscura
C Peru, Bolivia, Argentina
S.o.pacifica
W Peru
**Sporophila caerulescens (Double-collared
Seedeater)**
S.c.caerulescens
Brazil, Bolivia, Paraguay, Uruguay,
Argentina
S.c.hellmayri
Brazil
S.c.yungae
N Bolivia
**Sporophila albogularis (White-throated
Seedeater)**
NE Brazil
**Sporophila leucoptera (White-bellied
Seedeater)**
S.l.mexianae
Mexiana I, Brazil
S.l.cinereola
E Brazil
S.l.leucoptera
C & SW Brazil, Paraguay, N Argentina
S.l, bicolor
E Bolivia
**Sporophila peruviana (Parrot-billed
Seedeater)**
S.p.devronis
C Ecuador, N Peru
S.p.peruviana
Peru
Sporophila simplex (Drab Seedeater)
Peru
**Sporophila nigrorufa (Black & Tawny
Seedeater)**
Brazil, E Bolivia
Sporophila bouvreuil (Capped Seedeater)
S.b.bouvreuil
E Brazil
S.b.crypta
Rio de Janeiro, Brazil
S.b.pileata
S Brazil, Paraguay, N Argentina

S.b.saturata
Brazil
***Sporophila insulata* (Tumaco Seedeater)**
SW Colombia
***Sporophila minuta* (Ruddy-breasted Seedeater)**
S.m.parva
Central America from SW Mexico to Nicaragua
S.m.centralis
SW Costa Rica, W Panama
S.m.minuta
Trinidad and N South America to S Brazil
***Sporophila hypoxantha* (Tawny-bellied Seedeater)**
Paraguay, E & C Brazil, Uruguay
***Sporophila hypochroma* (Rufous-naped Seedeater)**
E Bolivia
***Sporophila ruficollis* (Dark-throated Seedeater)**
S Brazil, Bolivia, Uruguay, Argentina
***Sporophila palustris* (Marsh Seedeater)**
SE Brazil, Paraguay, Uruguay, N Argentina
***Sporophila castaneiventris* (Chestnut-bellied Seedeater)**
N South America
***Sporophila cinnamomea* (Chestnut Seedeater)**
Brazil, E Paraguay
***Sporophila melanogaster* (Black-bellied Seedeater)**
SE Brazil
***Sporophila telasco* (Chestnut-throated Seedeater)**
W Ecuador, Peru, N Chile

ORYZOBORUS
***Oryzoborus nuttingi* (Nicaraguan Seed Finch)**
155.A.1
Nicaragua to Panama
***Oryzoborus crassirostris* (Large-billed Seed Finch)** 155.A.2
O.c.crassirostris
Colombia, Venezuela, the Guianas, N Brazil
O.c.magnirostris
Trinidad, E Venezuela
***Oryzoborus maximiliani* (Great-billed Seed Finch)** 155.A.2
O.m.maximiliani
Brazil
O.m.occidentalis
Colombia, NW Ecuador
***Oryzoborus atrirostris* (Black-billed Seed Finch)** 155.A.2
O.a.atrirostris
N Peru
O.a.gigantirostris
N Bolivia
***Oryzoborus angolensis* (Lesser Seed Finch)**
O.a.funereus
S Mexico, Central America, Colombia, Ecuador

O.a.torridus
Trinidad, Peru, Ecuador, S Colombia, Venezuela, the Guianas, Brazil
O.a.angolensis
S Brazil, Bolivia, Paraguay, N Argentina

AMAUROSPIZA
***Amaurospiza concolor* (Blue Seedeater)**
A.c.relicta
S Mexico
A.c.concolor
Honduras, Nicaragua, Costa Rica, Panama
A.c.aequatorialis
SW Colombia, Ecuador
***Amaurospiza moesta* (Blackish-blue Seedeater)**
E Brazil, N Argentina

MELOPYRRHA
***Melopyrrha nigra* (Cuban Bullfinch)**
M.n.nigra
Cuba, Isle of Pines
M.n.taylori
Grand Cayman I

DOLOSPINGUS
***Dolospingus fringilloides* (White-naped Seedeater)**
Venezuela, Brazil

CATAMENIA
***Catamenia analis* (Band-tailed Seedeater)**
C.a.alpica
N Colombia
C.a.schistaceifrons
C Colombia
C.a.soederstromi
N Ecuador
C.a.insignis
Peru
C.a.analoides
W Peru
C.a.griseiventris
SE Peru
C.a.analis
N Chile, Bolivia, NW Argentina
***Catamenia inornata* (Plain-coloured Seedeater)**
C.i.mucuchiesi
Venezuela
C.i.minor
W Venezuela, Colombia, Ecuador, Peru
C.i.inornata
SE Peru, Bolivia, NW Argentina
***Catamenia homochroa* (Paramo Seedeater)**
C.h.homochroa
W Venezuela, Colombia, Ecudor, Peru, Bolivia
C.h.duncani
Venezuela, NE Brazil
***Catamenia oreophila* (Colombian Seedeater)**
N Colombia

TIARIS
Tiaris canora (Cuban Grassquit)
  Cuba
Tiaris olivacea (Yellow-faced Grassquit)
  T.o.pusilla
    E Mexico, Central America, Colombia,
    Venezuela
  T.o.intermedia
    Cozumel, Holbox Is, E Mexico
  T.o.ravida
    Panama
  T.o.olivacea
    Cuba, Jamaica, Cayman Is
  T.o.bryanti
    Puerto Rico
Tiaris bicolor (Black-faced Grassquit)
  T.b.bicolor
    Bahama Is, Cuba
  T.b.marchii
    Jamaica, Hispaniola
  T.b.omissa
    Puerto Rico, Tobago I, Colombia,
    Venezuela
  T.b.huilae
    Colombia
  T.b.grandior
    Old Providence, St Andrew Is
  T.b.johnstonei
    La Blanquilla I
  T.b.sharpei
    Aruba, Curacao I
  T.b.tortugensis
    La Tortuga I
Tiaris fuliginosa (Sooty Grassquit)
  T.f.fumosa
    Trinidad, Venezuela
  T.f.zuliae
    Venezuela
  T.f.fuliginosa
    NE & C Brazil

LOXIPASSER
Loxipasser anoxanthus (Yellow-shouldered Grassquit)
  Jamaica

LOXIGILLA
Loxigilla portoricensis (Puerto Rican Bullfinch)
  Puerto Rico
Loxigilla violacea (Greater Antillean Bullfinch)
  L.v.violacea
    Bahama Is
  L.v.maurella
    Tortue, Gonave, Saona Is
  L.v.affinis
    Hispaniola
  L.v.parishi
    Ile-a-vache, Beata I
  L.v.ruficollis
    Jamaica
Loxigilla noctis (Lesser Antillean Bullfinch)

  L.n.coryi
    St Kitts, Monserrat Is
  L.n.ridgwayi
    Anguilla, Antigua, Barbuda Is
  L.n.desiradensis
    Desirade I
  L.n.dominicana
    Guadeloupe, Dominica Is
  L.n.noctis
    Martinique I
  L.n.sclateri
    St Lucia I
  L.n.crissalis
    St Vincent I
  L.n.grenadensis
    Grenada I
  L.n.barbadensis
    Barbados I

MELANOSPIZA
Melanospiza richardsoni (St Lucia Black Finch)
  St Lucia I

GEOSPIZA
Geospiza magnirostris (Large Ground Finch)
  Galapagos Is
Geospiza fortis (Medium Ground Finch)
  Galapagos Is
Geospiza fuliginosa (Small Ground Finch)
  Galapagos Is
Geospiza difficilis (Sharp-beaked Ground Finch)
  G.d.difficilis
    Tower, Abingdon Is
  G.d.debilirostris
    James, Albemarle, Narborough Is
  G.d.septentrionalis
    Culpepper, Wenman Is
Geospiza scandens (Cactus Ground Finch)
  G.s.scandens
    James, Jervis Is
  G.s.intermedia
    Barrington, Charles, Duncan, Indefatigable,
    Albemarle Is
  G.s.abingdoni
    Abingdon I
  G.s.rothschildi
    Bindloe I
Geospiza conirostris (Large Cactus Ground Finch)
  G.c.conirostris
    Hood I
  G.c.propinqua
    Tower I
  G.c.darwini
    Culpepper I

CAMARHYNCHUS
Camarhynchus crassirostris (Vegetarian Tree Finch)
  Galapagos Is

Camarhynchus psittacula (Large Insectivorous Tree Finch)
  C.p.habeli
    Abingdon, Bindloe Is
  C.p.affinis
    Albemarle, Narborough Is
  C.p.psittacula
    Seymour, Barrington, Indefatigable,
    Charles, Duncan, Jervis, James Is
Camarhynchus pauper (Charles Insectivorous Tree Finch)
    Charles I
Camarhynchus parvulus (Small Insectivorous Tree Finch)
  C.p.parvulus
    James, Jervis, Indefatigable, Seymour,
    Barrington, Albemarle, Duncan, Charles,
    Narborough Is
  C.p.salvini
    Chatham I
Camarhynchus pallidus (Woodpecker Finch)
  C.p.pallidus
    James, Jervis, Seymour, Duncan,
    Indefatigable, Charles Is
  C.p.productus
    Albemarle, Narborough Is
  C.p.striatipectus
    Chatham I
Camarhynchus heliobates (Mangrove Finch)
    Albemarle, Narborough Is

CERTHIDEA
Certhidea olivacea (Warbler Finch)
  C.o.becki
    Culpepper, Wenman Is
  C.o.mentalis
    Tower I
  C.o.fusca
    Abingdon, Bindloe Is
  C.o.olivacea
    James, Jervis, Seymour, Duncan,
    Albemarle Narborough Is
  C.o.bifasciata
    Barrington I
  C.o.luteola
    Chatham I
  C.o.cinerascens
    Hood I
  C.o.ridgwayi
    Charles I

PINAROLOXIAS
Pinaroloxias inornata (Cocos Finch)
    Cocos Is

PIPILO
Pipilo ocai (Collared Towhee)
  P.o.alticola
    W Mexico
  P.o.nigrescens
    W Mexico
  P.o.guerrerensis
    SW Mexico

P.o.brunnescens
  S Mexico
P.o.ocai
  EC Mexico
Pipilo erythrophthalmus (Rufous-sided Towhee)
  P.e.erythrophthalmus
    S Canada, E USA >> S USA
  P.e.rileyi
    SE USA
  P.e.alleni
    Florida
  P.e.canaster
    SC USA >> SE USA
  P.e.arcticus
    S Canada, NC, C & SC USA >> N Mexico
  P.e.montanus
    SW USA >> N Mexico
  P.e.gaigei
    Texas
  P.e.curtatus
    SW Canada, W USA >> SE California
  P.e.oregonus
    W USA >> S California
  P.e.falcinellus
    W USA
  P.e.falcifer
    SW USA
  P.e.megalonyx
    SW USA, NW Baja California
  P.e.clementae
    San Clemente I
  P.e.umbraticola
    NW Baja California
  P.e.magnirostris
    S Baja California
  P.e.griseipygius
    W Mexico
  P.e.orientalis
    EC Mexico
  P.e.maculatus
    E Mexico
  P.e.macronyx
    SC Mexico
  P.e.vulcanorum
    SC Mexico
  P.e.oaxacae
    S Mexico
  P.e.repetens
    S Mexico, W Guatemala
  P.e.chiapensis
    S Mexico
  P.e.socorroensis
    Socorro I, Revillagigedo Group
  P.e.sympatricus
    Vera Cruz
Pipilo fuscus (Brown Towhee)
  P.f.bullatus
    SW Oregon, NC California
  P.f.carolae
    SC California
  P.f.petulans
    NC California

P.f.crissalis
 WC California
P.f.eremophilus
 EC California
P.f.senicula
 S California, NW Baja California
P.f.aripolius
 C Baja California
P.f.albigula
 S Baja California
P.f.mesoleucus
 SW USA, N Mexico
P.f.intermedius
 N Mexico
P.f.jamesi
 Tiburon I
P.f.mesatus
 SW USA
P.f.texanus
 W & C Texas, N Mexico
P.f.perpallidus
 N & C Mexico
P.f.fuscus
 E & C Mexico
P.f.potosinus
 N & C Mexico
P.f.campoi
 EC Mexico
P.f.toroi
 SC Mexico
Pipilo aberti (Abert's Towhee)
P.a.aberti
 SW USA
P.a.vorhiesi
 Arizona
P.a.dumeticolus
 NE Baja California, NW Mexico
Pipilo albicollis (White-throated Towhee)
P.a.albicollis
 SC Mexico
P.a.marshalli
 Puebla, Mexico

CHLORURUS
Chlorurus chlorurus (Green-tailed Towhee)
 W USA >> C Mexico

MELOZONE
Melozone kieneri (Rusty-crowned Ground Sparrow)
M.k.grisior
 NW Mexico
M.k.kieneri
 W Mexico
M.k.rubricatum
 C & SW Mexico
M.k.obscurior
 SW Oaxaca
Melozone biarcuatum (Prevost's Ground Sparrow)

M.b.biarcuatum
 S Mexico to W Honduras
M.b.cabanisi
 C Costa Rica
Melozone leucotis (White-eared Ground Sparrow)
M.l.occipitalis
 S Mexico, Guatemala, El Salvador
M.l.nigrior
 Nicaragua
M.l.leucotis
 Costa Rica

ARREMON
Arremon taciturnus (Pectoral Sparrow)
A.t.axillaris
 Colombia, W Venezuela
A.t.taciturnus
 SE Venezuela, the Guianas, Brazil, Bolivia
A.t.semitorquatus
 EC Brazil
A.t.nigrirostris
 SE Peru, N Bolivia
Arremon flavirostris (Saffron-billed Sparrow)
A.f.flavirostris
 EC Brazil
A.f.dorbignii
 E Bolivia, NW Argentina
A.f.devillii
 Brazil, E Bolivia
A.f.polionotus
 Brazil, Paraguay, Argentina
Arremon aurantiirostris (Orange-billed Sparrow)
A.a.saturatus
 SE Mexico, Guatemala, Belize
A.a.rufidorsalis
 Honduras, Nicaragua, Costa Rica
A.a.aurantiirostris
 W Costa Rica, Panama
A.a.strictocollaris
 E Panama, NW Colombia
A.a.occidentalis
 W Colombia, NW Ecuador
A.a.erythrorhynchus
 N Colombia
A.a.spectabilis
 SE Colombia, E Ecuador, Peru
A.a.santarosae
 SW Ecuador
Arremon schlegeli (Golden-winged Sparrow)
A.s.fratruelis
 N Colombia
A.s.canidorsum
 Colombia
A.s.schegeli
 E Colombia, Venezuela
Arremon abeillei (Black-capped Sparrow)
A.a.abeillei
 NW Peru, SW Ecuador
A.a.nigriceps
 NW Peru

Arremonops rufivirgatus (Olive Sparrow)
  A.r.rufivirgatus
    S Texas, NE Mexico
  A.r.ridgwayi
    E Mexico
  A.r.crassirostris
    SE Mexico
  A.r.verticalis
    SE Mexico, Guatemala, Belize
  A.r.sinaloae
    W Mexico
  A.r.sumichrasti
    SW Mexico
  A.r.rhyptothorax
    Yucatan, Mexico
  A.r.chiapensis
    S Mexico
  A.r.superciliosus
    W Costa Rica
Arremonops tocuyensis (Tocuyo Sparrow)
  NE Colombia, NW Venezuela
Arremonops chloronotus (Green-backed Sparrow)
  A.c.chloronotus
    SE Mexico, Guatemala, NW Honduras
  A.c.twomeyi
    NC Honduras
Arremonops conirostris (Black-striped Sparrow)
  A.c.richmondi
    E Honduras, Nicaragua, Costa Rica, Panama
  A.c.striaticeps
    C & E Panama, Colombia, W Ecuador
  A.c.inexpectatus
    Colombia
  A.c.conirostris
    E Colombia, N Venezuela, N Brazil
  A.c.umbrinus
    E Colombia, W Venezuela

Atlapetes albinucha (White-naped Brush Finch)   155.A.2
  E Mexico
Atlapetes gutturalis (Yellow-throated Brush Finch)   155.A.2
  A.g.griseipectus
    S Mexico, W Guatemala, El Salvador
  A.g.fuscipygius
    Honduras, El Salvador, NW Nicaragua
  A.g.parvirostris
    Costa Rica
  A.g.brunnescens
    Panama
  A.g.coloratus
    Panama
  A.g.azuerensis
    S Panama
  A.g.gutturalis
    N Colombia

Atlapetes pallidinucha (Pale-naped Brush Finch)
  A.p.pallidinucha
    E Colombia, SW Venezuela
  A.p.papallactae
    Colombia, Ecuador
Atlapetes rufinucha (Rufous-naped Brush Finch)
  A.r.phelpsi
    Venezuela, Colombia
  A.r.elaeoprorus
    Colombia
  A.r.simplex
    Colombia
  A.r.caucae
    Colombia
  A.r.spodionotus
    S Colombia, N Ecuador
  A.r.comptus
    SW Ecuador, Peru
  A.r.latinuchus
    SE Ecuador, NE Peru
  A.r.chugurensis
    NW Peru
  A.r.baroni
    N Peru
  A.r.melanolaemus
    SE Peru
  A.r.rufinucha
    Bolivia
  A.r.carrikeri
    E Bolivia
Atlapetes leucopis (White-rimmed Brush Finch)
  S Colombia, N Ecuador
Atlapetes melanocephalus (Santa Marta Brush Finch)
  N Colombia
Atlapetes pileatus (Rufous-capped Brush Finch)
  A.p.dilutus
    N & C Mexico
  A.p.pileatus
    S Mexico
Atlapetes flaviceps (Olive-headed Brush Finch)
  Colombia
Atlapetes fuscoolivaceus (Dusky-headed Brush Finch)
  Colombia
Atlapetes tricolor (Tricoloured Brush Finch)
  A.t.crassus
    Colombia, Ecuador
  A.t.tricolor
    C Peru
Atlapetes albofrenatus (Moustached Brush Finch)
  A.a.meridae
    Venezuela
  A.a.albofrenatus
    Colombia
Atlapetes schistaceus (Slaty Brush Finch)

A.s.castaneifrons
Venezuela
A.s.tamae
Venezuela, Colombia
A.s.fumidus
Venezuela, Colombia
A.s.schistaceus
Colombia, Ecuador
A.s.taczanowskii
C Peru
A.s.canigenis
EC Peru
Atlapetes nationi (Rusty-bellied Brush Finch)
A.n.celicae
S Ecuador
A.n.nationi
W Peru
A.n.brunneiceps
SW Peru
A.n.simonsi
S Ecuador
A.n.seebohmi
NW Peru
Atlapetes leucopterus (White-winged Brush Finch)
A.l.leucopterus
Ecuador
A.l.dresseri
SW Ecuador, NW Peru
Atlapetes albiceps (White-headed Brush Finch)
SE Ecuador, NW Peru
Atlapetes pallidiceps (Pale-headed Brush Finch)
S Ecuador
Atlapetes rufigenis (Rufous-eared Brush Finch)
A.r.rufigenis
NW Peru
A.r.forbesi
SC Peru
Atlapetes semirufus (Ochre-breasted Brush Finch)
A.s.denisei
Venezuela
A.s.benedettii
Venezuela
A.s.albigula
Venezuela
A.s.zimmeri
Venezuela, NE Colombia
A.s.majusculus
Colombia
A.s.semirufus
Colombia
Atlapetes personatus (Tepui Brush Finch)
A.p.personatus
Venezuela
A.p.collaris
Venezuela
A.p.duidae
Venezuela

A.p.parui
Venezuela
A.p.paraquensis
Venezuela
A.p.jugularis
Venezuela, N Brazil
Atlapetes fulviceps (Fulvous-headed Brush Finch)
Bolivia, NW Argentina
Atlapetes citrinellus (Yellow-striped Brush Finch)
Argentina
Atlapetes apertus (Plain-breasted Brush Finch)
E Mexico
Atlapetes brunneinucha (Chestnut-capped Brush Finch)
A.b.brunneinucha
E Mexico
A.b.suttoni
S Mexico
A.b.nigrilatera
Oaxaca (Mexico)
A.b.parkesi
S Vera Cruz
A.b.macrourus
S Mexico, SW Guatemala
A.b.alleni
El Salvador, Honduras, W Nicaragua
A.b.elsae
Costa Rica, W & C Panama
A.b.frontalis
E Panama, Colombia, Venezuela, Ecuador, Peru
A.b.allinornatus
NW Venezuela
A.b.inornatus
WC Ecuador
Atlapetes virenticeps (Green-striped Brush Finch)   155.A.2
W & C Mexico
Atlapetes torquatus (Stripe-headed Brush Finch)
A.t.basilicus
N Colombia
A.t.perijanus
E Colombia, W Venezuela
A.t.larensis
Venezuela
A.t.phaeopleurus
N Venezuela
A.t.phygas
NE Venezuela
A.t.assimilis
Venezuela, Ecuador, Colombia, Peru
A.t.nigrifrons
SW Ecuador, NW Peru
A.t.poliophrys
C & SE Peru
A.t.torquatus
NW Bolivia
A.t.fimbriatus
Bolivia

A.t.borelli
Bolivia, Argentina
Atlapetes atricapillus (Black-headed Brush Finch)
A.a.atricapillus
N Colombia
A.a.costaricensis
SW Costa Rica, W Panama
A.a.tacarcunae
E Panama

PEZOPETES
Pezopetes capitalis (Big-footed Sparrow)
Costa Rica, W Panama

OREOTHRAUPIS
Oreothraupis arremonops (Tanager Finch)
SW Colombia, NW Ecuador

PSELLIOPHORUS
Pselliophorus tibialis (Yellow-thighed Sparrow)
Costa Rica, W Panama
Pselliophorus luteoviridis (Yellow-green Sparrow)
E Panama

LYSURUS
Lysurus castaneiceps (Olive Finch)
Colombia, Ecuador, SE Peru
Lysurus crassirostris (Sooty-faced Finch)
155.A.2
Costa Rica, Panama

UROTHRAUPIS
Urothraupis stolzmanni (Black-backed Bush Tanager)
C Colombia to C Ecuador

CHARITOSPIZA
Charitospiza eucosma (Coal-crested Finch)
C & E Brazil, NE Argentina

SALTATRICULA
Saltatricula multicolor (Many-coloured Chaco Finch)
Bolivia, Paraguay, Uruguay, N Argentina

CORYPHASPIZA
Coryphaspiza melanotis (Black-masked Finch)
C.m.marajoara
Marajo I (Brazil)
C.m.melanotis
Brazil, Bolivia, Paraguay, NE Argentina

CORYPHOSPINGUS
Coryphospingus pileatus (Pileated Finch)
C.p.rostratus
Colombia
C.p.brevicaudus
N Colombia, N Venezuela

C.p.pileatus
EC Brazil
Coryphospingus cucullatus (Red-crested Finch)
C.c.cucullatus
the Guianas, Brazil
C.c.rubescens
S Brazil, E Paraguay, Uruguay, Argentina
C.c.fargoi
Peru, Bolivia, N Argentina, W Paraguay

RHODOSPINGUS
Rhodospingus cruentus (Crimson Finch)
Ecuador, Peru

155. EMBERIZIDAE

B. CATAMBLYRHYNCHINAE (PLUSH-CAPPED FINCH)

CATAMBLYRHYNCHUS
Catamblyrhynchus diadema (Plush-capped Finch)
C.d.federalis
N Venezuela
C.d.diadema
NW Venezuela, Colombia, Ecuador
C.d.citrinifrons
Peru, Bolivia, NW Argentina

EMBERIZIDAE

C. CARDINALINAE (CARDINALS, GROSBEAKS)

GUBERNATRIX
Gubernatrix cristata (Yellow Cardinal)
Uruguay, N & E Argentina

PAROARIA
Paroaria coronata (Red-crested Cardinal)
Bolivia, Paraguay, Uruguay, Argentina
Paroaria dominicana (Red-cowled Cardinal)
NE Brazil
Paroaria gularis (Red-capped Cardinal)
P.g.nigrogenis
Trinidad, E Colombia, Venezuela
P.g.gularis
Colombia, Venezuela, the Guianas, Ecuador, Peru, W Brazil
P.g.cervicalis
E Bolivia, Brazil
Paroaria baeri (Crimson-fronted Cardinal)
P.b.baeri
Brazil
P.b.xinguensis
Brazil
Paroaria capitata (Yellow-billed Cardinal)
P.c.capitata
Brazil, Paraguay, N Argentina
P.c.fuscipes
SE Bolivia

SPIZA
Spiza americana (Dickcissel)
  E North America >> Central America,
  Trinidad, Colombia, Venezuela

PHEUCTICUS
Pheucticus chrysopeplus (Yellow Grosbeak)
  P.c.dilutus
    NW Mexico
  P.c.chrysopeplus
    W Mexico
  P.c.rarissimus
    SC Mexico
  P.c.aurantiacus
    S Mexico, Guatemala
  P.c.laubmanni
    N Colombia, N Venezuela
Pheucticus tibialis (Black-thighed Grosbeak)
  C Costa Rica, W Panama
**Pheucticus chrysogaster (Yellow-bellied
Grosbeak)**
  SW Colombia, Ecuador, Peru
**Pheucticus aureoventris (Black-backed
Grosbeak)**
  P.a.meridensis
    Venezuela
  P.a.uropygialis
    Colombia
  P.a.crissalis
    SW Colombia, Ecuador
  P.a.terminalis
    Peru
  P.a.aureoventris
    S Peru, Bolivia, Brazil, Paraguay, N
    Argentina
**Pheucticus ludovicianus (Rose-breasted
Grosbeak)**
  S Canada, C & SE USA >> Mexico,
  Central America, N Sth America
**Pheucticus melanocephalus (Black-headed
Grosbeak)**
  P.m.melanocephalus
    S Canada, WC USA, Mexico
  P.m.maculatus
    SW Canada, W USA, Mexico, Baja
    California

CARDINALIS
Cardinalis cardinalis (Common Cardinal)
  C.c.cardinalis
    E USA
  C.c.floridanus
    SE Georgia, Florida
  C.c.magnirostris
    SE Texas, Louisiana
  C.c.canicaudus
    SC USA, C & E Mexico
  C.c.coccineus
    E Mexico
  C.c.littoralis
    E Mexico
  C.c.yucatanicus
    SE Mexico

  C.c.flammigerus
    SE Mexico, Guatemala, Belize
  C.c.sinaloensis
    W Mexico
  C.c.saturatus
    Cozumel I
  C.c.superbus
    SW USA, NW Mexico
  C.c.townsendi
    Tiburon I, NW Mexico
  C.c.affinis
    WC Mexico
  C.c.mariae
    Tres Marias Is
  C.c.carneus
    W & S Mexico
  C.c.seftoni
    C Baja California
  C.c.igneus
    S Baja California
  C.c.clintoni
    Cerralvo I

PYRRHULOXIA
Pyrrhuloxia phoeniceus (Vermilion Cardinal)
  Colombia, Venezuela
Pyrrhuloxia sinuatus (Pyrrhuloxia)
  P.s.sinuatus
    S USA, N & C Mexico
  P.s.fulvescens
    S Arizona, NW Mexico
  P.s.peninsulae
    Baja California

CARYOTHRAUSTES
**Caryothraustes canadensis (Yellow-green
Grosbeak)**
  C.c.canadensis
    Colombia, Venezuela, the Guianas, Brazil
  C.c.frontalis
    NE Brazil
  C.c.brasiliensis
    EC Brazil
**Caryothraustes poliogaster (Black-faced
Grosbeak)**
  C.p.poliogaster
    SE Mexico, Guatemala, Honduras
  C.p.scapularis
    Nicaragua, Costa Rica, W Panama
  C.p.simulans
    E Panama
**Caryothraustes humeralis (Yellow-
shouldered Grosbeak)**
  Colombia, Ecuador, Brazil

RHODOTHRAUPIS
**Rhodothraupis celaeno (Crimson-collared
Grosbeak)**
  NE Mexico

PERIPORPHYRUS
**Periporphyrus erythromelas (Red & Black
Grosbeak)**
  Venezuela, Guyana, French Guiana, Brazil

PITYLUS
Pitylus grossus (Slate-coloured Grosbeak)
P.g.saturatus
Nicaragua to Ecuador
P.g.grossus
Venezuela, Guyana, Brazil, W Colombia, W
Ecuador Peru, Bolivia
Pitylus fuliginosus (Black-throated Grosbeak)
Brazil, Paraguay, N Argentina

SALTATOR
Saltator atriceps (Black-headed Saltator)
S.a.atriceps
E Mexico, Guatemala, Honduras, Costa
Rica
S.a.suffuscus
SE Vera Cruz
S.a.flavicrissus
Guerrero
S.a.peeti
S Mexico
S.a.raptor
SE Mexico
S.a.lacertosus
W Costa Rica, Panama
Saltator maximus (Buff-throated Saltator)
S.m.gigantodes
E & S Mexico
S.m.magnoides
S Mexico to Panama
S.m.intermedius
SW Costa Rica, NW Panama
S.m.iungens
E Panama, NW Colombia
S.m.maximus
Colombia, Venezuela, the Guianas,
Ecuador, Peru Bolivia, Paraguay
Saltator atripennis (Black-winged Saltator)
S.a.atripennis
Colombia, NW Ecuador
S.a.caniceps
Colombia, W Ecuador
Saltator similis (Green-winged Saltator)
S.s.similis
Brazil, Bolivia, Paraguay, Uruguay,
Argentina
S.s.ochraceiventris
SE Brazil
Saltator coerulescens (Greyish Saltator)
S.c.vigorsii
NW Mexico
S.c.richardsoni
WC Mexico
S.c.grandis
E Mexico, Guatemala, Honduras,
Nicaragua, Costa Rica
S.c.yucatanensis
SE Mexico
S.c.hesperis
Guatemala, El Salvador, Honduras,
Nicaragua
S.c.brevicaudus
W Costa Rica

S.c.plumbeus
N Colombia
S.c.brewsteri
NE Colombia, Venezuela, Trinidad
S.c.olivascens
Venezuela, the Guianas, N Brazil
S.c.azarae
E Colombia, Ecuador, E Peru, Bolivia, Brazil
S.c.mutus
N Brazil
S.c.superciliaris
NE Brazil
S.c.coerulescens
E Bolivia, SW Brazil, Paraguay, N Argentina
Saltator orenocensis (Orinocan Saltator)
S.o.rufescens
NE Colombia, NW Venezuela
S.o.orenocensis
Venezuela
Saltator maxillosus (Thick-billed Saltator)
SE Brazil, Paraguay, N Argentina
**Saltator aurantiirostris (Golden-billed
Saltator)**
S.a.nigriceps
SW Ecuador, NW Peru
S.a.iteratus
N Peru
S.a.albociliaris
Peru, N Chile
S.a.hellmayri
Bolivia
S.a.aurantiirostris
Bolivia, N Argentina, Paraguay, Uruguay, S
Brazil
S.a.parkesi
S Brazil, Uruguay, NE Argentina
S.a.nasica
W Argentina
Saltator cinctus (Masked Saltator)
E Ecuador
Saltator atricollis (Black-throated Saltator)
E Bolivia, Paraguay, S Brazil
Saltator rufiventris (Rufous-bellied Saltator)
Bolivia
Saltator albicollis (Streaked Saltator)
S.a.albicollis
Martinique, St Lucia Is
S.a.guadelupensis
Guadeloupe, Dominica Is
S.a.furax
SW Costa Rica, Panama
S.a.isthmicus
W Panama
S.a.scotinus
Coiba I
S.a.melicus
Taboga I
S.a.speratus
Pearl Is
S.a.striatipectus
E Panama, W Colombia
S.a.perstriatus
NE Colombia, Venezuela, Trinidad

S.a.flavidicollis
SW Colombia, Ecuador, NW Peru
S.a.immaculatus
W Peru
S.a.peruvianus
N Peru

CYANOLOXIA
Cyanoloxia glaucocaerulea (Indigo Grosbeak)
S Brazil, Uruguay, N Argentina

CYANOCOMPSA
Cyanocompsa cyanoides (Blue-black Grosbeak)
C.c.concreta
SE Mexico, Guatemala, Honduras
C.c.toddi
Nicaragua, Costa Rica, W Panama
C.c.cyanoides
E Panama, Colombia, W Venezuela, Ecuador
C.c.rothschildii
Upper Amazonia, W Brazil
Cyanocompsa brissonii (Ultramarine Grosbeak)
C.b.caucae
W Colombia
C.b.minor
N Venezuela
C.b.brissonii
NE Brazil
C.b.sterea
E & S Brazil, NE Argentina, W Paraguay
C.b.argentina
W Brazil, E Bolivia, Paraguay, N Argentina
Cyanocompsa parellina (Blue Bunting)
C.p.beneplacita
NE Mexico
C.p.indigotica
W & SW Mexico
C.p.lucida
NE Mexico
C.p.parellina
E & S Mexico to Nicaragua

GUIRACA
Guiraca caerulea (Blue Grosbeak)
G.c.caerulea
SE USA >> E Mexico & Central America
G.c.interfusa
SW USA >> W Mexico, Guatemala, Honduras
G.c.salicaria
SW USA >> W Mexico, Baja California
G.c.eurhyncha
C & S Mexico
G.c.chiapensis
S Mexico
G.c.deltarhyncha
W coast of Mexico
G.c.lazula
Honduras, Nicaragua, Costa Rica

PASSERINA
Passerina cyanea (Indigo Bunting)
S Canada, E USA >> Central America, Cuba, Jamaica, Colombia, Venezuela
Passerina amoena (Lazuli Bunting)
W USA >> W Mexico, Baja California
Passerina versicolor (Varied Bunting)
P.v.versicolor
S USA, C & S Mexico
P.v.dickeyae
S Arizona, W Mexico
P.v.pulchra
S Baja California, NW Mexico
P.v.purpurascens
S Mexico, Guatemala
Passerina ciris (Painted Bunting)
P.c.ciris
SE USA >> SE Mexico, Bahama Is
P.c.pallidior
S USA >> Mexico, Central America
Passerina rositae (Rose-bellied Bunting)
S Mexico
Passerina leclancherii (Orange-breasted Bunting)
P.l.grandior
Oaxaca
P.l.leclancherii
SW Mexico

PORPHYROSPIZA
Porphyrospiza caerulescens (Blue Finch)
Brazil, SE Bolivia

155. EMBERIZIDAE

D. THRAUPINAE (TANAGERS)

ORCHESTICUS
Orchesticus abeillei (Brown Tanager)
SE Brazil

SCHISTOCLAMYS
Schistoclamys ruficapillus (Cinnamon Tanager)
S.r.capistrata
NE Brazil
S.r.sicki
E Mato Grosso
S.r.ruficapillus
SE Brazil
Schistoclamys melanopis (Black-faced Tanager)
S.m.aterrima
NE Colombia, Venezuela, W Guyana
S.m.melanopis
the Guianas, NE Brazil
S.m.grisea
EC Peru
S.m.olivina
E Bolivia, SC Brazil

S.m.amazonica
SE Brazil

Neothraupis fasciata (White-banded Tanager)
E & S Brazil, E Bolivia, NE Paraguay

Cypsnagra hirundinacea (White-rumped Tanager)
C.h.pallidigula
C Brazil, NE Bolivia
C.h.hirundinacea
S Brazil, E Bolivia, NE Paraguay

Conothraupis speculigera (Black & White Tanager)
S Ecuador, N & E Peru
Conothraupis mesoleuca (Cone-billed Tanager)
Mato Grosso

Lamprospiza melanoleuca (Red-billed Pied Tanager)
the Guianas, N Brazil, SE Peru, N Bolivia

Cissopis leveriana (Magpie Tanager)
C.l.leveriana
Upper Amazonia
C.l.major
Paraguay, SE Brazil, N Argentina

Chlorornis riefferii (Grass-green Tanager)
C.r.riefferii
Colombia, Ecuador
C.r.diluta
N Peru
C.r.elegans
C Peru
C.r.celata
SE Peru
C.r.boliviana
W Bolivia

Compsothraupis loricata (Scarlet-throated Tanager)
E Brazil

Sericossypha albocristata (White-capped Tanager)
SW Venezuela, Colombia, Ecuador, E Peru

Nesospingus speculiferus (Puerto Rican Tanager)
Puerto Rica

Chlorospingus ophthalmicus (Common Bush Tanager)
C.o.albifrons
SW Mexico
C.o.wetmorei
E Mexico
C.o.persimilis
S Oaxaca
C.o.ophthalmicus
SE Mexico
C.o.dwighti
S Mexico, E Guatemala
C.o.postocularis
S Mexico, W Guatemala
C.o.honduratius
El Salvador, Honduras
C.o.regionalis
Nicaragua, E Costa Rica
C.o.novicius
SW Costa Rica, W Panama
C.o.punctulatus
W Panama   155.D.1
C.o.jaqueti
NE Colombia, N Venezuela
C.o.falconensis
NW Venezuela
C.o.venezuelanus
SW Venezuela
C.o.ponsi
W Venezuela
C.o.eminens
NE Colombia
C.o.trudis
Colombia
C.o.exitelus
Colombia
C.o.flavopectus
C Colombia
C.o.macarenae
E Colombia
C.o.nigriceps
C Colombia
C.o.phaeocephalus
Ecuador
C.o.cinereocephalus
C Peru
C.o.hiaticolus
C Peru
C.o.peruvianus
S Peru
C.o.bolivianus
WC Bolivia
C.o.fulvigularis
C Bolivia
C.o.argentinus
C Bolivia, N Argentina

Chlorospingus tacarcunae (Tacarcuna Bush
Tanager)
  E Panama
Chlorospingus inornatus (Pirre Bush
Tanager)
  E Panama
Chlorospingus semifuscus (Dusky-bellied
Bush Tanager)
  C.s.livingstoni
    W Colombia
  C.s.semifuscus
    SW Colombia, W Ecuador
Chlorospingus pileatus (Pileated Bush
Tanager)   155.D.2
  C.p.pileatus
    Costa Rica, W Panama
  C.p.diversus
    W Panama
Chlorospingus parvirostris (Short-billed Bush
Tanager)
  C.p.huallagae
    S Colombia, Peru
  C.p.medianus
    EC Peru
  C.p.parvirostris
    SE Peru, W Bolvia
Chlorospingus flavigularis (Yellow-throated
Bush Tanager)
  C.f.hypophaeus
    W Panama
  C.f.marginatus
    SW Colombia, W Ecuador
  C.f.flavigularis
    S Colombia, E Ecuador, E Peru
Chlorospingus flavovirens (Yellow-green
Bush Tanager)
  W Ecuador
Chlorospingus canigularis (Ash-throated
Bush Tanager)
  C.c.olivaceiceps
    W Costa Rica
  C.c.canigularis
    C Colombia, SW Venezuela
  C.c.conspicillatus
    W Colombia
  C.c.paulus
    SW Ecuador
  C.c.signatus
    E Ecuador, NW Peru

CNEMOSCOPUS
Cnemoscopus rubrirostris (Grey-hooded
Bush Tanager)
  C.r.rubrirostris
    SW Venezuela, Colombia, E Ecuador
  C.r.chrysogaster
    N & C Peru

HEMISPINGUS
Hemispingus atropileus (Black-capped
Hemispingus)
  H.a.atropileus
    SW Venezuela, Colombia, Ecuador

H.a.auricularis
  E Peru
H.a.calophrys
  W Bolivia   155.D.3
Hemispingus parodii (Parodi's Tanager)
  Cuzco
Hemispingus superciliaris (Superciliaried
Hemispingus)
  H.s.chrysophrys
    SW Venezuela
  H.s.superciliaris
    C Colombia
  H.s.nigrifrons
    Colombia, Ecuador
  H.s.maculifrons
    SW Ecuador, NW Peru
  H.s.insignis
    N Peru
  H.s.leucogaster
    C Peru
  H.s.urubambae
    S Peru, W Bolivia
Hemispingus reyi (Grey-capped
Hemispingus)
  SW Venezuela
Hemispingus frontalis (Oleaginous
Hemispingus)
  H.f.frontalis
    Colombia, E Ecuador, E Peru
  H.f.ignobilis
    W Venezuela
  H.f.flavidorsalis
    W Venezuela
  H.f.hanieli
    N Venezuela
  H.f.iteratus
    NE Venezuela
Hemispingus melanotis (Black-eared
Hemispingus)
  H.m.melanotis
    SW Venezuela, C & E Colombia, E Ecuador
  H.m.ochraceus
    SW Colombia, W Ecuador
  H.m.piurae
    NW Peru
  H.m.macrophrys
    W Peru
  H.m.berlepschi
    C Peru
  H.m.castaneicollis
    SE Peru, W Bolivia
Hemispingus goeringi (Slaty-backed
Hemispingus)
  SW Venezuela
Hemispingus rufosuperciliaris (Rufous-
browed Hemispingus)
  C Peru
Hemispingus verticalis (Black-headed
Hemispingus)
  S & C Colombia, E Ecuador
Hemispingus xanthophthalmus (Drab
Hemispingus)
  C Peru

***Hemispingus trifasciatus* (Three-striped Hemispingus)**
SE Peru, W Bolivia

PYRRHOCOMA
***Pyrrhocoma ruficeps* (Chestnut-headed Tanager)**
SE Brazil, E Paraguay, N Argentina

THLYPOPSIS
***Thlypopsis fulviceps* (Fulvous-headed Tanager)**
T.f.fulviceps
NE Colombia, NW Venezuela
T.f.obscuriceps
W Venezuela
T.f.meridensis
W Venezuela
T.f.intensa
NE Colombia
***Thlypopsis ornata* (Rufous-chested Tanager)**
T.o.ornata
SW Colombia, W Ecuador
T.o.media
S Ecuador, N & C Peru
T.o.macropteryx
C & S Peru
***Thlypopsis pectoralis* (Brown-flanked Tanager)**
C Peru
***Thlypopsis sordida* (Orange-headed Tanager)**
T.s.orinocensis
EC Venezuela
T.s.chrysopis
E Ecuador, E Peru, W Brazil
T.s.sordida
E & S Brazil, E Bolivia, Paraguay, N Argentina
***Thlypopsis inornata* (Buff-bellied Tanager)**
N Peru
***Thlypopsis ruficeps* (Rust & Yellow Tanager)**
SE Peru, NW Argentina

HEMITHRAUPIS
***Hemithraupis guira* (Guira Tanager)**
H.g.nigrigula
NC Colombia to NE Brazil
H.g.roraimae
SE Venezuela, Guyana
H.g.guirina
W Colombia to NW Peru
H.g.huambina
SE Colombia to NE Peru, W Brazil
H.g.boliviana
NE Bolivia, NW Argentina
H.g.amazonica
C Brazil
H.g.guira
E Brazil
H.g.fosteri
SE Brazil, Paraguay, NE Argentina

***Hemithraupis ruficapilla* (Rufous-headed Tanager)**
H.r.ruficapilla
SE Brazil
H.r.bahiae
E Brazil
***Hemithraupis flavicollis* (Yellow-backed Tanager)**
H.f.ornata
E Panama, NW Colombia
H.f.albigularis
Colombia
H.f.peruana
SC Colombia to NE Peru
H.f.sororia
N Peru
H.f.centralis
SE Peru, N Bolivia, C Brazil
H.f.aurigularis
SE Colombia, S Venezuela, N Brazil
H.f.hellmayri
SE Venezuela, W Guyana
H.f.flavicollis
Surinam, French Guiana, NE Brazil
H.f.obidensis
N Brazil
H.f.melanoxantha
E Brazil
H.f.insignis
SE Brazil

CHRYSOTHLYPIS
***Chrysothlypis chrysomelas* (Black & Yellow Tanager)**
C.c.titanota
E Costa Rica, W Panama
C.c.chrysomelas
E Costa Rica, W Panama
C.c.ocularis
E Panama
***Chrysothlypis salmoni* (Scarlet & White Tanager)**
W Colombia, NW Ecuador

NEMOSIA
***Nemosia pileata* (Hooded Tanager)**
N.p.hypoleuca
N Colombia, N Venezuela
N.p.surinamensis
Guyana, Surinam
N.p.pileata
French Guiana, Brazil, N Bolivia
N.p.interna
N Brazil
N.p.nana
NE Peru, W Brazil
N.p.caerulea
S & E Brazil, E Bolivia, Paraguay, N Argentina
***Nemosia rourei* (Cherry-throated Tanager)**
SE Brazil

PHAENICOPHILUS
Phaenicophilus palmarum (Black-crowned Palm Tanager)
  Hispaniola, Soana I
Phaenicophilus poliocephalus (Grey-crowned Palm Tanager)
  P.p.poliocephalus
    S Haiti
  P.p.coryi
    Gonave I

CALYPTOPHILUS
Calyptophilus frugivorus (Chat-Tanager)
  C.f.frugivorus
    Dominica I
  C.f.neibei
    C Dominica
  C.f.tertius
    S Haiti
  C.f.abbotti
    Gonave I

RHODINOCICHLA
Rhodinocichla rosea (Rose-breasted Thrush Tanager)
  R.r.schistacea
    W Mexico
  R.r.eximia
    SW Costa Rica, W Panama
  R.r.harterti
    C Colombia
  R.r.beebei
    NE Colombia, NW Venezuela
  R.r.rosea
    NW Venezuela

MITROSPINGUS
Mitrospingus cassinii (Dusky-faced Tanager)
  M.c.costaricensis
    E Costa Rica, W Panama
  M.c.cassinii
    E Panama, W Colombia, W Ecuador
Mitrospingus oleagineus (Olive-backed Tanager)
  M.o.obscuripectus
    SE Venezuela, N Brazil
  M.o.oleagineus
    SE Venezuela, Guyana

CHLOROTHRAUPIS
Chlorothraupis carmioli (Carmiol's Tanager)
  C.c.carmioli
    Nicaragua to NW Panama
  C.c.magnirostris
    W Panama
  C.c.lutescens
    E Panama, NW Colombia
  C.c.frenata
    S Colombia, SE Peru
Chlorothraupis olivacea (Lemon-browed Tanager)
  E Panama to NW Ecuador

Chlorothraupis stolzmanni (Ochre-breasted Tanager)
  C.s.dugandi
    SW Colombia
  C.s.stolzmanni
    W Ecuador

ORTHOGONYS
Orthogonys chloricterus (Olive-green Tanager)
  SE Brazil

EUCOMETIS
Eucometis penicillata (Grey-headed Tanager)
  E.p.pallida
    SE Mexico to E Guatemala
  E.p.spodocephala
    Nicaragua, W Costa Rica
  E.p.stictothorax
    SW Costa Rica, W Panama
  E.p.cristata
    E Panama to W Venezuela
  E.p.affinis
    N Venezuela
  E.p.penicillata
    SE Colombia, E Ecuador, E Peru, the Guianas, N Brazil
  E.p.albicollis
    E Bolivia, SC Brazil, N Paraguay

LANIO
Lanio fulvus (Fulvous Shrike-Tanager)
  L.f.peruvianus
    S Colombia to NE Peru
  L.f.fulvus
    S Venezuela, the Guianas, N Brazil
Lanio versicolor (White-winged Shrike-Tanager)
  L.v.versicolor
    E Peru, N Bolivia, W Brazil
  L.v.parvus
    S Brazil
Lanio aurantius (Black-throated Shrike-Tanager)
  SE Mexico to Honduras
Lanio leucothorax (White-throated Shrike-Tanager)
  L.l.leucothorax
    E Honduras to E Costa Rica
  L.l.reversus
    NW Costa Rica
  L.l.melanopygius
    SW Costa Rica, W Panama
  L.l.ictus
    NW Panama

CREURGOPS
Creurgops verticalis (Rufous-crested Tanager)
  SW Venezuela to Peru

Creurgops dentata (Slaty Tanager)
SE Peru, N Bolivia

HETEROSPINGUS
Heterospingus xanthopygius (Scarlet-browed Tanager)
H.x.rubrifrons
E Costa Rica, Panama
H.x.xanthopygius
E Panama, N Colombia
H.x.berliozi
W Colombia, NW Ecuador

TACHYPHONUS
Tachyphonus cristatus (Flame-crested Tanager)
T.c.cristatus
French Guiana, NE Brazil
T.c.intercedens
E Venezuela, Guyana, Surinam
T.c.orinocensis
E Colombia, S Venezuela
T.c.cristatellus
S Venezuela to N Peru
T.c.fallax
S Colombia to NE Peru
T.c.huarandosae
N Peru
T.c.madeirae
C Brazil
T.c.pallidigula
NE Brazil
T.c.brunneus
E Brazil
Tachyphonus nattereri (Natterer's Tanager)
SW Brazil
Tachyphonus rufiventer (Yellow-crested Tanager)
E Peru, N Bolivia, W Brazil
Tachyphonus surinamus (Fulvous-crested Tanager)
T.s.surinamus
E & S Venezuela, the Guianas, N Brazil
T.s.brevipes
S Venezuela to NE Peru
T.s.napensis
E Peru, NW Brazil
T.s.insignis
N Brazil
Tachyphonus luctuosus (White-shouldered Tanager)
T.l.axillaris
E Honduras to W Panama
T.l.nitidissimus
SW Costa Rica, W Panama
T.l.panamensis
E Panama to W Ecuador, W Venezuela
T.l.luctuosus
tropical South America
T.l.flaviventris
NE Venezuela, Trinidad

Tachyphonus delatrii (Tawny-crested Tanager)
Nicaragua to W Ecuador
Tachyphonus coronatus (Ruby-crowned Tanager)
SE Brazil to NW Argentina
Tachyphonus rufus (White-lined Tanager)
Costa Rica, Panama, N Sth America
Tachyphonus phoenicius (Red-shouldered Tanager)
N Sth America

TRICHOTHRAUPIS
Trichothraupis melanops (Black-goggled Tanager)
SW Amazonia

HABIA
Habia rubica (Red-crowned Ant-Tanager)
H.r.holobrunnea
E Mexico
H.r.rosea
SW Mexico
H.r.affinis
S Mexico
H.r.nelsoni
SE Mexico
H.r.rubicoides
S Mexico to El Salvador
H.r.vinacea
W Costa Rica, W Panama
H.r.alfaroana
NW Costa Rica
H.r.rubra
Trinidad
H.r.crissalis
NE Venezuela
H.r.mesopotamia
E Bolivar (Venezuela)
H.r.perijana
NE Colombia, NW Venezuela
H.r.coccinea
NC Colombia, W Venezuela
H.r.rhodinolaema
SE Colombia to NE Peru, NW Brazil
H.r.peruviana
E Peru, NC Bolivia
H.r.hesterna
C Brazil
H.r.bahiae
E Brazil
H.r.rubica
SE Brazil, Paraguay, N Argentina
Habia fuscicauda (Red-throated Ant-Tanager)
H.f.salvini
SE Mexico to El Salvador
H.f.insularis
SE Mexico, N Guatemala
H.f.discolor
Nicaragua
H.f.fuscicauda
S Nicaragua to W Panama

H.f.willisi
C Panama
H.f.erythrolaema
N Colombia
Habia atrimaxillaris (Black-cheeked Ant-Tanager)
SW Costa Rica
Habia gutturalis (Sooty Ant-Tanager)
NW Colombia
Habia cristata (Crested Ant-Tanager)
W Colombia

PIRANGA
Piranga bidentata (Flame-coloured Tanager)
P.b.bidentata
W Mexico
P.b.flammea
Tres Marias Is
P.b.sanguinolenta
E Mexico to El Salvador
P.b.citrea
Costa Rica, W Panama
Piranga flava (Hepatic Tanager)
P.f.hepatica
SW USA, W Mexico
P.f.dextra
SW USA, E Mexico >> W Guatemala
P.f.figlina
E Guatemala, Belize
P.f.savannarum
Honduras, NE Nicaragua
P.f.albifacies
W Guatemala to N Nicaragua
P.f.testacea
Costa Rica, Panama
P.f.desidiosa
SW Colombia
P.f.lutea
W Ecuador to NW Bolivia
P.f.haemalea
S Venezuela, W Guyana, N Brazil
P.f.faceta
N Colombia, N Venezuela, Trinidad
P.f.toddi
Magdelena (Colombia)
P.f.macconnelli
S Guyana, N Brazil
P.f.saira
E Brazil
P.f.rosacea
E Bolivia
P.f.flava
S Bolivia to Uruguay, N Argentina
Piranga rubra (Summer Tanager)
P.r.cooperi
SW USA >> C Mexico
P.r.rubra
SE USA >> Central & Sth America
P.r.ochracea
Arizona & W Mexico
Piranga roseogularis (Rose-throated Tanager)

P.r.roseogularis
SE Mexico
P.r.tincta
SE Mexico, N Guatemala
P.r.cozumelae
Cozumel I
Piranga olivacea (Scarlet Tanager)
SE Canada, NE USA >> NW Sth America
Piranga ludoviciana (Western Tanager)
W North America >> W Mexico, W Central America
Piranga leucoptera (White-winged Tanager)
P.l.leucoptera
E Mexico to Nicaragua
P.l.latifasciata
Costa Rica, W Panama
P.l.venezuelae
Colombia, Venezuela, N Brazil
P.l.ardens
SW Colombia to Bolivia
Piranga erythrocephala (Red-headed Tanager)
P.e.candida
NW Mexico
P.e.erythrocephala
SC & S Mexico
Piranga rubriceps (Red-hooded Tanager)
W Colombia to N Peru

CALOCHAETES
Calochaetes coccineus (Vermilion Tanager)
S Colombia to E Peru

RAMPHOCELUS
Ramphocelus sanguinolentus (Crimson-collared Tanager)
R.s.sanguinolentus
SE Mexico to Honduras
R.s.apricus
E Honduras to NW Panama
Ramphocelus nigrogularis (Masked Crimson Tanager)
SE Colombia to E Peru, N Brazil
Ramphocelus dimidiatus (Crimson-backed Tanager)
R.d.isthmicus
W & C Panama
R.d.arestus
Coiba I
R.d.limatus
Pearl Arch
R.d.dimidiatus
E Panama, N Colombia, W Venezuela
R.d.molochinus
N Colombia
Ramphocelus melanogaster (Black-bellied Tanager)
R.m.melanogaster
N Peru
R.m.transitus
EC Peru
Ramphocelus carbo (Silver-beaked Tanager)

R.c.unicolor
E Colombia
R.c.capitalis
NE Venezuela
R.c.magnirostris
Trinidad
R.c.carbo
E Peru to Surinam
R.c.venezuelensis
E Colombia, W Venezuela
R.c.connectens
SE Peru, NW Bolivia
R.c.atrosericeus
N & E Bolivia
R.c.centralis
EC Brazil, N Paraguay
Ramphocelus bresilius (Brazilian Tanager)
R.b.bresilius
NE Brazil
R.b.dorsalis
SE Brazil
Ramphocelus passerinii (Scarlet-rumped Tanager)
R.p.passerinii
SE Mexico to W Panama
R.p.costaricensis
W Costa Rica
Ramphocelus flammigerus (Flame-rumped Tanager)
W Colombia
Ramphocelus icteronotus (Yellow-rumped Tanager)
Panama to W Ecuador

SPINDALIS
Spindalis zena (Stripe-headed Tanager)
S.z.townsendi
N Bahama Is
S.z.zena
C Bahama Is
S.z.pretrei
Cuba, Isle of Pines
S.z.salvini
Grand Cayman I
S.z.benedicti
Cozumel I
S.z.dominicensis
Hispaniola
S.z.portoricensis
Puerto Rico
S.z.nigricephala
Jamaica

THRAUPIS
Thraupis episcopus (Blue-grey Tanager)
T.e.cana
SE Mexico to N Venezuela
T.e.caesitia
W Panama
T.e.cumatilis
Coiba I
T.e.nesophilus
E Colombia to Trinidad

T.e.berlepschi
Tobago I
T.e.mediana
SE Colombia, NW Brazil, N Bolivia
T.e.episcopus
the Guianas, N Brazil
T.e.leucoptera
C Colombia
T.e.quaesita
SW Colombia, W Ecuador, NW Peru
T.e.caerulea
SE Ecuador, N Peru
T.e.major
C Peru
T.e.urubambae
SE Peru
T.e.coelestis
SE Colombia to C Peru, W Brazil
Thraupis sayaca (Sayaca Tanager)
T.s.boliviana
NW Bolivia
T.s.obscura
C & S Bolivia, W Argentina
T.s.sayaca
E & S Brazil, Paraguay to Uruguay
Thraupis glaucocolpa (Glaucous Tanager)
155.D.3
N Colombia, NW Venezuela
Thraupis cyanoptera (Azure-shouldered Tanager)
E Paraguay, SE Brazil
Thraupis ornata (Golden-chevroned Tanager)
SE Brazil
Thraupis abbas (Yellow-winged Tanager)
E Mexico to Nicaragua
Thraupis palmarum (Palm Tanager)
T.p.atripennis
E Nicaragua to NW Venezuela
T.p.violilavata
SW Colombia, W Ecuador
T.p.melanoptera
Amazonia, Trinidad
T.p.palmarum
E Bolivia, Paraguay, E & S Brazil
Thraupis cyanocephala (Blue-capped Tanager)
T.c.cyanocephala
W Ecuador to N Bolivia
T.c.annectens
C Colombia
T.c.auricrissa
NC Colombia, W Venezuela
T.c.margaritae
N Colombia
T.c.hypophaea
NW Venezuela
T.c.olivicyanea
N Venezuela
T.c.subcinerea
NE Venezuela
T.c.buesingi
NE Venezuela, Trinidad
Thraupis bonariensis (Blue & Yellow Tanager)

T.b.darwinii
Ecuador to N Chile
T.b.composita
E & C Bolivia
T.b.schulzei
Paraguay, NW Argentina
T.b.bonariensis
S Brazil to EC Argentina

CYANICTERUS
Cyanicterus cyanicterus (Blue-backed Tanager)
E Venezuela, the Guianas

BUTHRAUPIS
Buthraupis arcaei (Blue & Gold Tanager)
B.a.caeruleigularis
E Costa Rica
B.a.arcaei
W Panama
Buthraupis melanochlamys (Black & Gold Tanager)
W Colombia
Buthraupis rothschildi (Golden-chested Tanager)
SW Colombia, NW Ecuador
Buthraupis edwardsi (Moss-backed Tanager)
SW Colombia, NW Ecuador
Buthraupis aureocincta (Gold-ringed Tanager)
W Colombia
Buthraupis montana (Hooded Mountain Tanager)
B.m.gigas
NC Colombia, Venezuela
B.m.cucullata
W Colombia, Ecuador
B.m.cyanonota
N & C Peru
B.m.saturata
SE Peru
B.m.montana
N Bolivia
Buthraupis eximia (Black-chested Mountain Tanager)
B.e.eximia
NC Colombia, SW Venezuela
B.e.zimmeri
WC Colombia
B.e.chloronota
SE Colombia, NW Ecuador
B.e.cyanocalyptra
SC Ecuador
Buthraupis aureodorsalis (Golden-backed Mountain Tanager)
C Peru
Buthraupis wetmorei (Masked Mountain Tanager)
SW Colombia, SC Ecuador

WETMORETHRAUPIS
Wetmorethraupis sterrhopteron (Orange-throated Tanager)
N Peru

ANISOGNATUS
Anisognathus melanogenys (Black-cheeked Mountain Tanager)
N Colombia
Anisognathus lacrymosus (Lacrimose Mountain Tanager)
A.l.pallididorsalis
E Colombia, Venezuela
A.l.melanops
W Venezuela
A.l.tamae
NC Colombia, SW Venezuela
A.l.intensus
SW Colombia
A.l.olivaceiceps
W Colombia
A.l.palpebrosus
SW Colombia, E Ecuador
A.l.caerulescens
S Ecuador, N Peru
A.l.lacrymosus
C Peru
Anisognathus igniventris (Scarlet-bellied Mountain Tanager)
A.i.lunulatus
NC Colombia, W Venezuela
A.i.erythrotus
S Colombia, Ecuador
A.i.ignicrissus
NC Peru
A.i.igniventris
SE Peru, Bolivia
Anisognathus flavinuchus (Blue-winged Mountain Tanager)
A.f.venezuelanus
N Venezuela
A.f.virididorsalis
Venezuela
A.f.antioquiae
Colombia
A.f.victorini
C Colombia, SW Venezuela
A.f.cyanopterus
SW Colombia, W Ecuador
A.f.baezae
S Colombia, E Ecuador
A.f.alamoris
SW Ecuador
A.f.somptuosus
SE Ecuador, E Peru
A.f.flavinuchus
SE Peru, Bolivia
Anisognathus notabilis (Black-chinned Mountain Tanager)
SW Colombia, NW Ecuador

STEPHANOPHORUS
Stephanophorus diadematus (Diademed Tanager)
SE Brazil, N Argentina

IRIDOSORNIS
Iridosornis porphyrocephala (Purplish-mantled Tanager)
W Colombia, W Ecuador
Iridosornis analis (Yellow-throated Tanager)
E Ecuador, E Peru
Iridosornis jelskii (Golden-collared Tanager)
I.j.jelskii
Peru
I.j.bolivianus
SE Peru, W Bolivia
Iridosornis rufivertex (Golden-crowned Tanager)
I.r.rufivertex
W Venezuela to E Ecuador
I.r.caeruleoventris
NW Colombia
I.r.ignicapillus
SW Colombia
I.r.subsimilis
W Ecuador
Iridosornis reinwardti (Yellow-scarfed Tanager)
E Peru

DUBUSIA
Dubusia taeniata (Buff-breasted Mountain Tanager)
D.t.carrikeri
N Colombia
D.t.taeniata
W Venezuela to Ecuador
D.t.stictocephala
SE Peru

DELOTHRAUPIS
Delothraupis castaneoventris (Chestnut-bellied Mountain Tanager)
D.c.peruviana
E Peru
D.c.castaneoventris
W Bolivia

PIPRAEIDEA
Pipraeidea melanonota (Fawn-breasted Tanager)
P.m.venezuelensis
Venezuela to W Bolivia, N Argentina
P.m.melanonota
Paraguay, SE Brazil to NE Argentina

EUPHONIA
Euphonia jamaica (Jamaican Euphonia)
Jamaica
Euphonia plumbea (Plumbeous Euphonia)
S Venezuela to Surinam, N Brazil

Euphonia affinis (Scrub Euphonia)
E.a.godmani
W Mexico
E.a.olmecorum
S Mexico
E.a.affinis
E Mexico to Costa Rica
Euphonia luteicapilla (Yellow-crowned Euphonia)
E Nicaragua to Panama
Euphonia chlorotica (Purple-throated Euphonia)
E.c.cynophora
E Colombia, S Venezuela, N Brazil
E.c.chlorotica
the Guianas, N & NE Brazil
E.c.serrirostris
SE Bolivia to Uruguay, S Brazil
E.c.taczanowskii
E Peru, N Bolivia
E.c.amazonica
C Brazil
Euphonia trinitatis (Trinidad Euphonia)
N Colombia to Trinidad
Euphonia concinna (Velvet-fronted Euphonia)
C Colombia
Euphonia saturata (Orange-crowned Euphonia)
W Colombia to NW Peru
Euphonia finschi (Finsch's Euphonia)
E Venezuela, the Guianas
Euphonia violacea (Violaceous Euphonia)
E.v.rodwayi
E Venezuela, Trinidad
E.v.violacea
the Guianas, N Brazil
E.v.aurantiicollis
SE Brazil, Paraguay
Euphonia laniirostris (Thick-billed Euphonia)
E.l.crassirostris
Costa Rica to N Venezuela
E.l.melanura
Colombia to N Peru, W Brazil
E.l.hypoxantha
E Ecuador, NW Peru
E.l.zopholega
EC Peru
E.l.laniirostris
E Bolivia, SW Brazil
Euphonia hirundinacea (Yellow-throated Euphonia)
E.h.suttoni
E Mexico
E.h.russelli
SE Mexico
E.h.caribbaea
SE Mexico
E.h.hirundinacea
E Mexico to E Nicaragua
E.h.gnatho
NW Nicaragua to W Panama

Euphonia chalybea (Green-throated Euphonia)
SE Brazil, Paraguay
Euphonia elegantissima (Blue-hooded Euphonia) 155.D.3
E.e.rileyi
NW Mexico
E.e.elegantissima
C & S Mexico to Honduras
Euphonia aureata (Golden-rumped Euphonia) 155.D.3
E.a.pelzelni
S Colombia, W Ecuador
E.a.insignis
S Ecuador
E.a.aureata
N Sth America
Euphonia musica (Antillean Euphonia) 155.D.3
E.m.musica
Hispaniola
E.m.sclateri
Puerto Rico
E.m.flavifrons
Lesser Antilles
Euphonia fulvicrissa (Fulvous-vented Euphonia)
E.f.fulvicrissa
Panama, NW Colombia
E.f.omissa
C Colombia
E.f.purpurascens
SW Colombia, NW Ecuador
Euphonia imitans (Spot-crowned Euphonia)
W Costa Rica, W Panama
Euphonia gouldi (Olive-backed Euphonia)
E.g.loetscheri
E Mexico
E.g.gouldi
SE Mexico to Honduras
E.g.praetermissa
E Honduras to Panama
Euphonia chrysopasta (Golden-bellied Euphonia)
E.c.chrysopasta
Western Amazonia
E.c.nitida
E Colombia to French Guiana, N Brazil
Euphonia mesochrysa (Bronze-green Euphonia)
E.m.mesochrysa
C Colombia, E Ecuador
E.m.media
N Peru
E.m.tavarae
SE Peru, C Bolivia
Euphonia minuta (White-vented Euphonia)
E.m.humilis
S Mexico to W Ecuador
E.m.minuta
the Guianas to C Bolivia, W Brazil
Euphonia anneae (Tawny-capped Euphonia)
E.a.anneae
W Costa Rica, W Panama

E.a.rufivertex
W Panama, NW Colombia
Euphonia xanthogaster (Orange-bellied Euphonia)
E.x.oressinoma
W Panama to NW Colombia
E.x.chocoensis
E Panama to NW Ecuador
E.x.badissima
N Venezuela
E.x.quitensis
W Ecuador
E.x.dilutior
S Colombia, NE Peru
E.x.cyanonota
W Brazil
E.x.brunneifrons
SE Peru
E.x.ruficeps
W Bolivia
E.x.brevirostris
N & W Amazonia
E.x.exsul
NE Colombia, N Venezuela
E.x.xanthogaster
S & E Brazil
Euphonia rufiventris (Rufous-bellied Euphonia)
E.r.rufiventris
Western Amazonia
E.r.carnegiei
Venezuela
Euphonia pectoralis (Chestnut-bellied Euphonia)
SE Brazil, Paraguay
Euphonia cayennensis (Golden-sided Euphonia)
SE Venezuela, the Guianas, N Brazil

CHLOROPHONIA
Chlorophonia flavirostris (Yellow-collared Chlorophonia) 155.D.3
SW Colombia, Ecuador
Chlorophonia cyanea (Blue-naped Chlorophonia)
C.c.psittacina
N Colombia
C.c.frontalis
N Venezuela
C.c.minuscula
NE Venezuela
C.c.roraimae
S Venezuela, Guyana
C.c.intensa
W Colombia
C.c.longipennis
W Venezuela to W Bolivia
C.c.cyanea
SE Brazil, Paraguay, NE Argentina
Chlorophonia pyrrhophrys (Chestnut-breasted Chlorophonia)
W Venezuela to E Ecuador

Chlorophonia occipitalis (Blue-crowned Chlorophonia)
SE Mexico to Nicaragua
Chlorophonia callophrys (Golden-browed Chlorophonia)   155.D.3
Costa Rica, W Panama

CHLOROCHRYSA
Chlorochrysa phoenicotis (Glistening-green Tanager)
W Colombia, W Ecuador
Chlorochrysa calliparaea (Orange-eared Tanager)
C.c.bourcieri
Colombia to NE Peru
C.c.calliparaea
EC Peru
C.c.fulgentissima
SE Peru, N Bolivia
Chlorochrysa nitidissima (Multicoloured Tanager)
W Colombia

TANGARA
Tangara phillipsi (Sira Tanager)   155.D.6
E Peru
Tangara inornata (Plain-coloured Tanager)
T.i.rava
Costa Rica, W Panama
T.i.languens
Panama, NW Colombia
T.i.inornata
N Colombia
Tangara cabanisi (Azure-rumped Tanager)
S Mexico, SW Guatemala
Tangara palmeri (Grey & Gold Tanager)
E Panama to W Ecuador
Tangara mexicana (Turquoise Tanager)
T.m.vieilloti
Trinidad
T.m.media
S & E Venezuela, NW Brazil
T.m.mexicana
the Guianas
T.m.boliviana
Western Amazonia
T.m.brasiliensis
SE Brazil
Tangara chilensis (Paradise Tanager)
T.c.paradisea
the Guianas, N Brazil
T.c.coelicolor
E Colombia, S Venezuela, NW Brazil
T.c.chlorocorys
NC Peru
T.c.chilensis
Western Amazonia
Tangara fastuosa (Seven-coloured Tanager)
E Brazil
Tangara seledon (Green-headed Tanager)
SE Brazil, Paraguay, N Argentina
Tangara cyanocephala (Red-necked Tanager)

T.c.cearensis
NE Brazil
T.c.corallina
E Brazil
T.c.cyanocephala
SE Brazil, E Paraguay, N Argentina
Tangara desmaresti (Brassy-breasted Tanager)
SE Brazil
Tangara cyanoventris (Gilt-edged Tanager)
SE Brazil
Tangara johannae (Blue-whiskered Tanager)
W Colombia, NW Ecuador
Tangara schrankii (Green & Gold Tanager)
T.s.venezuelana
S Venezuela
T.s.schrankii
Upper Amazonia
Tangara florida (Emerald Tanager)
T.f.florida
Costa Rica, W Panama
T.f.auriceps
W Colombia, E Panama
Tangara arthus (Golden Tanager)
T.a.arthus
N & E Venezuela
T.a.palmitae
Magdalena (E Colombia)
T.a.sclateri
E Colombia
T.a.aurulenta
C Colombia, NW Venezuela
T.a.occidentalis
W Colombia
T.a.goodsoni
W Ecuador
T.a.aequatorialis
E Ecuador, N Peru
T.a.pulchra
C Peru
T.a.sophiae
SE Peru, W Bolivia
Tangara icterocephala (Silver-throated Tanager)
T.i.frantzii
Costa Rica, W Panama
T.i.oresbia
WC Panama
T.i.icterocephala
E Panama, W Colombia, W Ecuador
Tangara xanthocephala (Saffron-crowned Tanager)
T.x.venusta
W Venezuela to N & C Peru
T.x.xanthocephala
C Peru
T.x.lamprotis
SE Peru
Tangara chrysotis (Golden-eared Tanager)
S Colombia to N Bolivia
Tangara parzudakii (Flame-faced Tanager)
T.p.parzudakii
SW Venezuela to Peru

T.p.urubambae
S Peru
T.p.lunigera
W Colombia, W Ecuador
Tangara xanthogastra (Yellow-bellied Tanager)
T.x.xanthogastra
S Venezuela to N Bolivia
T.x.phelpsi
S Venezuela, N Brazil
Tangara punctata (Spotted Tanager)
T.p.punctata
S Venezuela, the Guianas, N Brazil
T.p.zamorae
E Ecuador, N Peru
T.p.perenensis
E Peru
T.p.annectens
SE Peru
T.p.punctulata
N Bolivia
Tangara guttata (Speckled Tanager)
T.g.eusticta
Costa Rica, W Panama
T.g.tolimae
Tolima (Colombia)
T.g.bogotensis
E Colombia, W Venezuela
T.g.chrysophrys
Venezuela, NW Brazil
T.g.guttata
SE Venezuela, N Brazil
T.g.trinitatis
N Trinidad
Tangara varia (Dotted Tanager)
S Venezuela, the Guianas, N Brazil
Tangara rufigula (Rufous-throated Tanager)
W Colombia, NW Ecuador
Tangara gyrola (Bay-headed Tanager)
T.g.bangsi
Costa Rica, W Panama
T.g.deleticia
E Panama, W Colombia
T.g.nupera
SW Colombia, W Ecuador
T.g.toddi
N Colombia, NW Venezuela
T.g.viridissima
Trinidad, NE Venezuela
T.g.catharinae
E Colombia to C Bolivia
T.g.parva
S Venezuela to NE Peru, NW Brazil
T.g.gyrola
S Venezuela, the Guianas, N Brazil
T.g.albertinae
C Brazil
Tangara lavinia (Rufous-winged Tanager)
T.l.cara
E Guatemala to Costa Rica
T.l.dalmasi
W Panama

T.l.lavinia
E Panama to NW Ecuador
Tangara cayana (Burnished-buff Tanager)
T.c.fulvescens
C Colombia
T.c.cayana
the Guiana to E Peru, N Brazil
T.c.huberi
NE Brazil
T.c.flava
NE Brazil
T.c.sincipitalis
C Brazil
T.c.chloroptera
SE Brazil, Paraguay
T.c.margaritae
C Brazil
Tangara cucullata (Lesser Antillean Tanager)
T.c.versicolor
St Vincent I
T.c.cucullata
Grenada I
Tangara peruviana (Black-backed Tanager)
SE Brazil
Tangara preciosa (Chestnut-backed Tanager)
Paraguay to Uruguay, SE Brazil
Tangara vitriolina (Scrub Tanager)
W Colombia, NW Ecuador
Tangara meyerdeschauenseei (Green-capped Tanager) 155.D.3.4
S Peru
Tangara rufigenis (Rufous-cheeked Tanager)
N Venezuela
Tangara ruficervix (Golden-naped Tanager)
T.r.ruficervix
Colombia
T.r.leucotis
W Ecuador
T.r.taylori
SE Colombia, E Ecuador
T.r.amabilis
N Peru
T.r.inca
S Peru
T.r.fulvicervix
N Bolivia
Tangara labradorides (Metallic-green Tanager)
T.l.labradorides
W Colombia, W Ecuador
T.l.chaupensis
NW Peru
Tangara cyanotis (Blue-browed Tanager)
T.c.lutleyi
S Colombia, Ecuador, E Peru
T.c.cyanotis
NW Bolivia
Tangara cyanicollis (Blue-necked Tanager)
T.c.granadensis
W Colombia
T.c.caeruleocephala
C Colombia to N Peru

T.c.cyanicollis
  E Peru, E Bolivia
T.c.cyanopygia
  W Ecuador
T.c.hannahiae
  E Colombia, W Venezuela
T.c.melanogaster
  C Brazil
T.c.albotibialis
  Goias (Brazil)
Tangara larvata (Golden-masked Tanager)
T.l.larvata
  S Mexico to N Costa Rica
T.l.centralis
  E Costa Rica, W Panama
T.l.franciscae
  W Costa Rica, W Panama
T.l.fanny
  E Panama to NW Ecuador
Tangara nigrocincta (Masked Tanager)
  N & W Amazonia
Tangara dowii (Spangle-cheeked Tanager)
                                   155.D.3
  Costa Rica, W Panama
Tangara fucosa (Green-naped Tanager)
  E Panama   155.D.3
**Tangara nigroviridis (Beryl-spangled
Tanager)**
T.n.cyanescens
  NW Venezuela to W Ecuador
T.n.consobrina
  C Colombia
T.n.nigroviridis
  E Colombia, E Ecuador
T.n.berlepschi
  E Peru, Bolivia
Tangara vassorii (Blue & Black Tanager)
T.v.vassorii
  NW Venezuela to NW Peru
T.v.branickii
  N Peru
T.v.atrocoerulea
  S Peru, Bolivia
Tangara heinei (Black-capped Tanager)
  NW Venezuela to E Ecuador
Tangara viridicollis (Silvery Tanager)
T.v.fulvigula
  S Ecuador, N Peru
T.v.viridicollis
  C & S Peru
**Tangara argyrofenges (Green-throated
Tanager)**
T.a.caeruleigularis
  N Peru
T.a.argyrofenges
  WC Bolivia
Tangara cyanoptera (Black-headed Tanager)
T.c.whitelyi
  S Venezuela, Guyana
T.c.cyanoptera
  N Colombia, N & W Venezuela
**Tangara pulcherrima (Yellow-collared
Tanager)**

T.p.pulcherrima
  Colombia to E Peru
T.p.aureinucha
  W Ecuador
Tangara velia (Opal-rumped Tanager)
T.v.velia
  the Guianas, N Brazil
T.v.iridina
  NW Amazonia
T.v.signata
  NE Brazil
T.v.cyanomelaena
  SE Brazil
Tangara callophrys (Opal-crowned Tanager)
  SE Colombia to E Peru, W Brazil

DACNIS
Dacnis albiventris (White-bellied Dacnis)
  S Venezuela to NE Peru
Dacnis lineata (Black-faced Dacnis)
D.l.egregia
  C Colombia
D.l.aequatorialis
  W Ecuador
D.l.lineata
  N & W Amazonia
Dacnis flaviventer (Yellow-bellied Dacnis)
  N & W Amazonia
Dacnis hartlaubi (Turquoise Dacnis)
  W Colombia
Dacnis nigripes (Black-legged Dacnis)
  SE Brazil
Dacnis venusta (Scarlet-thighed Dacnis)
D.v.vensuta
  Costa Rica, W Panama
D.v.fuliginata
  E Panama to NW Ecuador
Dacnis cayana (Blue Dacnis)
D.c.callaina
  W Costa Rica, W Panama
D.c.ultramarina
  E Nicaragua to NW Colombia
D.c.napaea
  N Colombia
D.c.baudoana
  SW Colombia, W Ecuador
D.c.coerebicolor
  C Colombia
D.c.cayana
  E Colombia to Franch Guiana, N & C Brazil
D.c.glaucogularis
  S Colombia to N & E Bolivia
D.c.paraguayensis
  S & E Brazil, Paraguay, NE Argentina
Dacnis viguieri (Viridian Dacnis)
  E Panama, NE Colombia
Dacnis berlepschi (Scarlet-breasted Dacnis)
  SW Colombia, NW Ecuador

CHLOROPHANES
Chlorophanes spiza (Green Honeycreeper)
C.s.guatemalensis
  S Mexico to Honduras

C.s.arguta
E Honduras to NW Colombia
C.s.exsul
SW Colombia, W Ecuador
C.s.subtropicalis
Colombia, W Venezuela
C.s.spiza
Venezuela, Trinidad, the Guianas, N Brazil
C.s.caerulescens
SE Colombia to Bolivia
C.s.axillaris
E Brazil

CYANERPES
**Cyanerpes nitidus (Short-billed
Honeycreeper)** 155.D.3
NW Amazonia
Cyanerpes lucidus (Shining Honeycreeper)
C.l.lucidus
S Mexico to N Nicaragua
C.l.isthmicus
Costa Rica to NW Colombia
Cyanerpes caeruleus (Purple Honeycreeper)
C.c.chocoanus
W Colombia, W Ecuador
C.c.caerι.aus
Colombia to the Guianas, NE Brazil
C.c.hellmayri
Guyana
C.c.longirostris
Trinidad
C.c.microrhynchus
W & C Amazonia
**Cyanerpes cyaneus (Red-legged
Honeycreeper)**
C.c.carneipes
E & S Mexico
C.c.gemmeus
N Colombia
C.c.eximius
N Colombia, N Venezuela
C.c.tobagensis
Tobago I
C.c.cyaneus
SE Venezuela, Trinidad, the Guianas, N
Brazil
C.c.brevipes
C Brazil
C.c.dispar
S Venezuela to NE Peru, W Brazil
C.c.holti
E Brazil
C.c.violaceus
C Bolivia, W Brazil
C.c.pacificus
W Colombia, W Ecuador
C.c.gigas
Gorgona I (W Colombia)

XENODACNIS
Xenodacnis parina (Tit-like Dacnis)
X.p.bella
N Peru

X.p.petersi
WC Peru
X.p.parina
SC Peru

OREOMANES
Oreomanes fraseri (Giant Conebill) 155.D.3
SW Colombia, Peru, W Bolivia

DIGLOSSA
Diglossa baritula (Slaty Flowerpiercer)
D.b.baritula
C Mexico
D.b.montana
S Mexico to El Salvador
D.b.parva
E Guatemala, Honduras
D.b.plumbea
Costa Rica, W Panama
D.b.veraguensis
W Panama
D.b.hyperythra
NE Colombia, N Venezuela
D.b.mandeli
NE Venezuela
D.b.coelestis
W Venezuela
D.b.dorbignyi
E Colombia, W Venezuela
D.b.decorata
Ecuador, Peru
D.b.sittoides
Bolivia, NW Argentina
Diglossa lafresnayii (Glossy Flowerpiercer)
D.l.gloriosissima
W Colombia
D.l.lafresnayii
W Venezuela to Ecuador, N Peru
D.l.unicincta
N Peru
D.l.pectoralis
C Peru
D.l.albilinea
SE Peru
D.l.mystacalis
W Bolivia
**Diglossa carbonaria (Coal-black
Flowerpiercer)**
D.c.gloriosa
W Venezuela
D.c.nocticolor
N Colombia, W Venezuela
D.c.humeralis
C Colombia, SW Venezuela
D.c.aterrima
W Colombia, Ecuador, NW Peru
D.c.vuilleumieri
NW Colombia, Ecuador
D.c.brunneiventris
NW Colombia to N Chile
D.c.carbonaria
Bolivia

Diglossa venezuelensis **(Venezuelan Flowerpiercer)**
NE Venezuela
Diglossa albilatera **(White-sided Flowerpiercer)**
D.a.federalis
N Venezuela
D.a.albilatera
W Venezuela to Ecaudor
D.a.schistacea
SW Ecuador to NW Peru
D.a.affinis
NC Peru
Diglossa duidae **(Scaled Flowerpiercer)**
D.d.hitchcocki
S Venezuela
D.d.duidae
S Venezuela, N Brazil
D.d.georgebarrowcloughi
Cerre Jime (S Venezuela)
Diglossa major **(Greater Flowerpiercer)**
D.m.gilliardi
SE Venezuela
D.m.disjuncta
SE Venezuela
D.m.chimantae
SE Venezuela
D.m.major
SE Venezuela, N Brazil
Diglossa indigotica **(Indigo Flowerpiercer)**
SW Colombia, W Ecuador

DIGLOSSOPIS   155.D.5
Diglossopis glauca **(Deep-blue Flowerpiercer)**
D.g.tyrianthina
S Colombia, E Ecuador
D.g.glauca
SE Peru, NW Bolivia
Diglossopis caerulescens **(Bluish Flowerpiercer)**
D.c.caerulescens
N Venezuela
D.c.ginesi
NW Venezuela
D.c.saturata
SW Venezuela, Colombia
D.c.media
S Ecuador, NW Peru
D.c.pallida
C Peru
D.c.mentalis
SE Peru, NW Bolivia
Diglossopis cyanea **(Masked Flowerpiercer)**
D.c.tovarensis
N Venezuela
D.c.obscura
NW Venezuela
D.c.cyanea
W Venezuela to Ecuador
D.c.dispar
SW Ecuador, NW Peru
D.c.melanopis
Peru, NW Bolivia

EUNEORNIS
Euneornis campestris **(Orangequit)**
Jamaica

155. EMBERIZIDAE

E. TERSININAE (SWALLOW TANAGER)

TERSINA
Tersina viridis **(Swallow Tanager)**
T.v.grisescens
N Colombia
T.v.occidentalis
E Panama, Colombia, Venezuela, the Guianas, Ecuador, NE Peru, N Bolivia, N Brazil
T.v.viridis
E & S Brazil, E Bolivia, Paraguay, NE Argentina

156. COEREBIDAE (BANANAQUIT)   156.1

COEREBA
Coereba flaveola **(Bananaquit)**
C.f.mexicana
SE Mexico, Central America
C.f.cerinoclunis
Pearl Is
C.f.columbiana
E Panama, C Colombia, SC Venezuela
C.f.gorgonae
Gorgona I, W Colombia
C.f.caucae
W Colombia
C.f.intermedia
SW Venezuela to Ecuador, W Brazil
C.f.magnirostris
N Peru
C.f.pacifica
NW Peru
C.f.dispar
SE Peru, NW Bolivia
C.f.caboti
Cozumel, Holbox Is
C.f.tricolor
Old Providence I
C.f.oblita
St Andrew I
C.f.sharpei
Cayman Is
C.f.bahamensis
Bahama Is
C.f.flaveola
Jamaica
C.f.bananivora
Hispaniola
C.f.nectarea
Tortue I, Haiti
C.f.portoricensis
Puerto Rico
C.f.sanctithomae
Virgin Is

C.f.newtoni
St Croix I
C.f.bartholemica
N Lesser Antilles
C.f.martinicana
Martinique, St Lucia Is
C.f.barbadensis
Barbados I
C.f.atrata
St Vincent I
C.f.aterrima
Grenada I
C.f.uropygialis
Aruba, Curacao Is
C.f.bonairensis
Bonaire I
C.f.melanornis
Cayo Sal I
C.f.lowii
Los Roques I
C.f.ferryi
La Tortuga I
C.f.frailensis
Los Frailes, Los Hermanos Is
C.f.laurae
Los Testigos I
C.f.luteola
N Colombia to Trinidad & Tobago I
C.f.obscura
NE Colombia, W Venezuela
C.f.montana
W Venezuela
C.f.bolivari
E Venezuela
C.f.guianensis
E Venezuela, Guyana
C.f.roraimae
SE Venezuela, NW Brazil, SW Guyana
C.f.minima
N Brazil, French Guiana, Surinam
C.f.chloropyga
S Peru, Bolivia to NE Argentina
C.f.alleni
C Brazil, E Bolivia

157. PARULIDAE (NEW WORLD WARBLERS)

S = 115

MNIOTILTA
Mniotilta varia (Black & White Warbler)
NW, C & SE Canada, C & E USA >>
Central America, West Indies, Venezuela,
Colombia

VERMIVORA
Vermivora bachmanii (Bachman's Warbler)
C & SE USA >> Cuba
Vermivora chrysoptera (Golden-winged Warbler)
E USA >> Central America, Colombia,
Venezuela
Vermivora pinus (Blue-winged Warbler)
E USA >> E Mexico, Central America

Vermivora peregrina (Tennessee Warbler)
NW, C & SE Canada, E USA >> S Mexico,
Colombia, Venezuela
Vermivora celata (Orange-crowned Warbler)
V.c.celata
N & NW Canada, S USA >> Mexico,
Guatemala
V.c.lutescens
W Canada, W USA >> W Mexico
V.c.orestera
WC Canada, WC USA >> C Mexico
V.c.sordida
S California, N Baja California and islands
Vermivora ruficapilla (Nashville Warbler)
V.r.ridgwayi
W USA >> W Mexico, Guatemala
V.r.ruficapilla
S Canada, C & E USA >> Mexico,
Guatemala
Vermivora virginiae (Virginia's Warbler)
SW USA >> W Mexico
Vermivora crissalis (Colima Warbler)
S USA >> EC Mexico
Vermivora luciae (Lucy's Warbler)
SW USA >> W Mexico
Vermivora gutteralis (Flame-throated Warbler)
Costa Rica, W Panama
Vermivora superciliosa (Crescent-chested Warbler)
V.s.sodalis
NC Mexico
V.s.mexicana
E Mexico
V.s.palliata
SW Mexico
V.s.superciliosa
S Mexico, Guatemala, W Honduras
V.s.parva
E Honduras, Nicaragua

PARULA
Parula americana (Parula Warbler)
SE Canada, E USA, E Mexico >> Central
America, West Indies
Parula pitiayumi (Olive-backed Warbler)(Tropical Parula)
P.p.graysoni
Socorro I, Revillagigedo Is
P.p.insularis
Tres Marias Is
P.p.pulchra
NW Mexico
P.p.nigrilora
S Texas, NE Mexico
P.p.inornata
S Mexico, E Guatemala, N Honduras
P.p.speciosa
S Honduras, Nicaragua, Costa Rica, W
Panama
P.p.cirrha
Coiba I (Panama)

P.p.nana
  E Panama, NW Colombia
P.p.elegans
  Colombia, N Venezuela, N Brazil, Trinidad
P.p.roraimae
  S Venezuela, N Brazil
P.p.alarum
  E Ecuador, N Peru
P.p.pacifica
  SW Colombia, W Ecuador, NW Peru
P.p.melanogenys
  S Peru, W Bolivia
P.p.pitiayumi
  E Bolivia, C & S Brazil, Uruguay, Paraguay,
  N Argentina

DENDROICA
Dendroica petechia (Yellow Warbler)
D.p.amnicola
  Canada >> Mexico, Central America, N
  South America
D.p.rubiginosa
  W Canada >> Mexico, Central America
D.p.aestiva
  S Canada, C USA >> Cental America, N
  Sth America
D.p.morcomi
  W USA >> Central America, N Sth
  America
D.p.sonorana
  SW USA >> Central America, Colombia,
  Ecuador
D.p.brewsteri
  Baja California
D.p.hueyi
  C Baja California
D.p.inedita
  NE Mexico
D.p.dugesi
  C Mexico
D.p.rufivertex
  Cozumel I
D.p.flavida
  St Andrew I
D.p.armouri
  Old Providence I
D.p.eoa
  Jamaica, Cayman Is
D.p.gundlachi
  Cuba, Bahama Is
D.p.albicollis
  Hispaniola
D.p.cruciana
  Puerto Rico, Virgin Is
D.p.bartholemica
  N Lesser Antilles
D.p.melanoptera
  C Lesser Antilles
D.p.ruficapilla
  Martinique I
D.p.babad
  St Lucia I

D.p.petechia
  Barbados I
D.p.alsiosa
  Grenadine Is
D.p.rufopileata
  Curacao, Bonaire Is
D.p.obscura
  Los Roques I
D.p.chrysendeta
  NE Colombia, NW Venezuela
D.p.paraguanae
  NW Venezuela
D.p.cienagae
  NC Venezuela
D.p.aurifrons
  NC Venezuela and islands
D.p.castaneiceps
  S Baja California
D.p.rhizophorae
  NW Mexico
D.p.oraria
  E Mexico
D.p.bryanti
  Caribbean coast, SE Mexico to Costa Rica
D.p.xanthotera
  Pacific coast, W Guatemala to Costa Rica
D.p.aureola
  Cocos, Galapagos Is
D.p.aequatorialis
  Pearl Is
D.p.erithachorides
  E Panama, N Colombia
D.p.peruviana
  SW Colombia, W Ecuador, N Peru
Dendroica pensylvanica (Chestnut-sided Warbler)
  S Canada, E USA >> Central America
Dendroica cerulea (Cerulean Warbler)
  Venezuela, Ecuador, Peru, Bolivia, E USA
  >> Colombia
Dendroica caerulescens (Black-throated Blue Warbler)
D.c.caerulescens
  SE Canada, NE USA >> Bahama Is, Gtr
  Antilles
D.c.cairnsi
  EC USA >> Gtr Antilles
Dendroica plumbea (Plumbeous Warbler)
  Dominica, Guadeloupe Is
Dendroica pharetra (Arrow-headed Warbler)
  Jamaica
Dendroica angelae (Elfin Woods Warbler)
  Puerto Rico
Dendroica pinus (Pine Warbler)
D.p.pinus
  SE Canada >> SE USA
D.p.florida
  S Florida
D.p.achrustera
  Bahama Is
D.p.chrysoleuca
  Hispaniola

Dendroica graciae (Grace's Warbler)
D.g.graciae
SW USA >> W Mexico
D.g.yaegeri
W Mexico
D.g.remota
S Mexico, Guatemala, El Salvador, W
Honduras
D.g.decora
Belize, E Honduras, Nicaragua
Dendroica adelaidae (Adelaide's Warbler)
D.a.adelaidae
Puerto Rico
D.a.subita
Barbuda I
D.a.delicata
St Lucia I
Dendroica pityophila (Olive-capped Warbler)
Cuba, Bahama Is
**Dendroica dominica (Yellow-throated
Warbler)**
D.d.albilora
EC & SE USA >> E Mexico, Central
America, Cuba, Jamaica
D.d.dominica
E & SE USA >> Gtr Antilles
D.d.stoddardi
SE USA
D.d.flavescens
Bahama Is
**Dendroica nigrescens (Black-throated Grey
Warbler)**
D.n.nigrescens
SW Canada, W USA >> N Mexico,
Guatemala
D.n.halseii
SW USA, NW Mexico, N Baja California
Dendroica townsendi (Townsend's Warbler)
W Canada, W USA >> Mexico,
Guatemala, Honduras, Nicaragua
Dendroica occidentalis (Hermit Warbler)
. SW USA >> W Mexico, Guatemala,
Honduras, Nicaragua
**Dendroica chrysopareia (Golden-cheeked
Warbler)**
S USA >> Mexico, Guatemala, Honduras,
Nicaragua
**Dendroica virens (Black-throated Green
Warbler)**
C & SE Canada, E USA >> Mexico,
Central America, West Indies
Dendroica discolor (Prairie Warbler)
D.d.discolor
E USA, West Indies
D.d.paludicola
SE USA >> Gtr Antilles
Dendroica vitellina (Vitelline Warbler)
D.v.crawfordi
Little Cayman I
D.v.vitellina
Grand Cayman I
D.v.nelsoni
Swan I

Dendroica tigrina (Cape May Warbler)
C & SE Canada, NC & E USA >> West
Indies, E Central America
Dendroica fusca (Blackburnian Warbler)
SE Canada, E USA >> Central America,
Venezuela, Colombia, Ecuador, Peru
Dendroica magnolia (Magnolia Warbler)
S Canada, E USA >> Mexico, Central
America, Gtr Antilles
Dendroica coronata (Yellow-rumped Warbler)
D.c.coronata
Canada, C & E USA >> Central America,
West Indies
D.c.auduboni
SW Canada, W USA >> Mexico,
Guatemala, W Honduras
D.c.nigrifrons
NC Mexico
D.c.goldmani
W Guatemala
D.c.hooveri
SW USA, NW Mexico
Dendroica palmarum (Palm Warbler)
D.p.palmarum
C & E Canada, E USA >> Gtr Antilles,
Central America
D.p.hypochrysea
SE Canada, NE USA >> SE USA
Dendroica kirtlandii (Kirtland's Warbler)
C Michigan >> Bahama Is
Dendroica striata (Blackpoll Warbler)
Canada, C & E USA >> West Indies, N & C
Sth America
Dendroica castanea (Bay-breasted Warbler)
C & SE Canada, E USA >> Central
America, Colombia, Venezuela

CATHAROPEZA
Catharopeza bishopi (Whistling Warbler)
St Vincent

ISETOPHAGA
Setophaga ruticilla (American Redstart)
S Canada, C & E USA >> Mexico, Central
America, West Indies, N Sth America

SEIURUS
Seiurus aurocapillus (Ovenbird)
S.a.aurocapillus
C & SE Canada, E USA >> West Indies,
Mexico to Colombia and Venezuela
S.a.cinereus
WC USA >> S Mexico, El Salvador,
Honduras, Costa Rica
S.a.furvior
Newfoundland >> Bahamas, Cuba, E
Central America
**Seiurus noveboracensis (Northern Water-
thrush)**
S.n.noveboracensis
Canada, E USA >> West Indies, Central
America, N Sth America

S.n.limnaeus
  W USA, NW Mexico
S.n.notabilis
  SW USA, W Mexico
Seiurus motacilla (Louisiana Water-thrush)
  E USA >> Mexico, West Indies, Central
  America, Colombia, Venezuela

LIMNOTHLYPIS
**Limnothlypis swainsonii (Swainson's
Warbler)**
  SE USA >> E Mexico, West Indies

HELMITHEROS
**Helmitheros vermivorus (Worm-eating
Warbler)**
  E USA >> E Central America, West Indies

PROTONOTARIA
Protonotaria citrea (Prothonotary Warbler)
  E USA >> Central America, West Indies, N
  Sth America

GEOTHLYPIS
Geothlypis trichas (Common Yellowthroat)
  G.t.trichas
    SE Canada, EC USA >> Mexico, West
    Indies, N Sth America
  G.t.typhicola
    SC USA, NE Mexico
  G.t.ignota
    SE USA
  G.t.insperata
    S Texas
  G.t.campicola
    W Canada, NW USA, SW USA >> N
    Mexico
  G.t.arizela
    W Canada, W USA & NW Mexico
  G.t.occidentalis
    WC USA >> Mexico to Honduras
  G.t.sinuosa
    N California
  G.t.scirpicola
    S California, N Baja California
  G.t.chryseola
    W Texas, NW Mexico
  G.t.modesta
    W Sonora, Mexico
  G.t.melanops
    C Mexico
  G.t.chapalensis
    Jalisco
  G.t.riparia
    S Sonora
  G.t.brachydactyla
    E USA, E & S Mexico
Geothlypis beldingi (Peninsular Yellowthroat)
  G.b.goldmani
    C Baja California

G.b.beldingi
  S Baja California
**Geothlypis flavovelata (Yellow-crowned
Yellowthroat)**
  E Mexico
Geothlypis rostrata (Bahama Yellowthroat)
  G.r.tanneri
    N Bahama Is
  G.r.rostrata
    W Bahama Is
  G.r.coryi
    Eleuthera, Cat Is
**Geothlypis semiflava (Olive-crowned
Yellowthroat)**
  G.s.bairdi
    S Honduras, Nicaragua, Costa Rica, NW
    Panama
  G.s.semiflava
    W Colombia, W Ecuador
**Geothlypis speciosa (Black-polled
Yellowthroat)**
  G.s.speciosa
    C Mexico
  G.s.limnatis
    Guanajuata (Mexico)
Geothlypis nelsoni (Hooded Yellowthroat)
  G.n.nelsoni
    E Mexico
  G.n.karlenae
    SW Mexico
**Geothlypis chiriquensis (Chiriqui
Yellowthroat)**
  W Panama
**Geothlypis aequinoctialis (Masked
Yellowthroat)**
  G.a.aequinoctialis
    NE Colombia, Venezuela, the Guianas,
    Surinam, N Brazil
  G.a.auricularis
    W Ecuador, W Peru
  G.a.peruviana
    N Peru
  G.a.velata
    S Peru, Bolivia, Brazil, Paraguay, Uruguay,
    N Argentina
**Geothlypis poliocephala (Grey-crowned
Yellowthroat)**
  G.p.poliocephala
    N & W Mexico
  G.p.ralphi
    NE Mexico
  G.p.palpebralis
    E & S Mexico, Guatemala, Honduras,
    Nicaragua, Costa Rica
  G.p.caninucha
    SW Mexico, W Guatemala, S Honduras, El
    Salvador
  G.p.icterotis
    W Nicaragua, W Costa Rica
  G.p.pontilis
    W Mexico
  G.p.ridgwayi
    SW Costa Rica, W Panama

Geothlypis formosa (Kentucky Warbler)
SE USA >> E Mexico, Central America,
Colombia, Venezuela
Geothlypis agilis (Connecticut Warbler)
EC Canada, NC USA >> Venezuela, NE
Brazil, Colombia
Geothlypis philadelphia (Mourning Warbler)
C & E Canada, NE USA >> Nicaragua,
Costa Rica, Colombia, Venezuela
Geothlypis tolmei (MacGillivray's Warbler)
SW Canada, W USA >> Central America

MICROLIGEA
**Microligea palustris (Green-tailed Ground
Warbler)**
M.p.palustris
Hispaniola
M.p.vasta
SW Dominica I

XENOLIGEA
**Xenoligea montana (White-winged Ground
Warbler)**
Hispaniola

TERETISTRIS
**Teretistris fernandinae (Yellow-headed
Warbler)**
W Cuba
Teretistris fornsi (Oriente Warbler)
E Cuba

LEUCOPEZA
Leucopeza semperi (Semper's Warbler)
St Lucia I

WILSONIA
Wilsonia citrina (Hooded Warbler)
E USA >> E Mexico, Central America
Wilsonia pusilla (Wilson's Warbler)
W.p.pileolata
W Canada, WC USA >> C Mexico, Central
America
W.p.chryseola
SW USA >> W & S Mexico, Guatemala
W.p.pusilla
S & E Canada, NE USA >> E Mexico,
Central America
Wilsonia canadensis (Canada Warbler)
SE Canada, NE USA >> Central America,
N Sth America

CARDELLINA
Cardellina rubifrons (Red-faced Warbler)
SW USA >> W & S Mexico, Guatemala, W
Honduras

ERGATICUS
Ergaticus ruber (Red Warbler)
E.r.melanauris
NW Mexico
E.r.ruber
W & SW Mexico

E.r.rowleyi
Oaxaca
Ergaticus versicolor (Pink-headed Warbler)
S Mexico, W Guatemala

MYIOBORUS
Myioborus pictus (Painted Redstart)
M.p.pictus
SW USA, N Mexico
M.p.guatemalae
S Mexico to N Nicaragua
Myioborus miniatus (Slate-throated Redstart)
M.m.miniatus
W & SW Mexico
M.m.molochinus
E Mexico
M.m.intermedius
S Mexico, E Guatemala
M.m.hellmayri
W Guatemala, El Salvador
M.m.connectens
El Salvador, Honduras
M.m.comptus
W Costa Rica
M.m.aurantiacus
E Costa Rica, W Panama
M.m.ballux
E Panama, Colombia, W Venezuela, NW
Ecuador
M.m.sanctaemartae
N Colombia
M.m.pallidiventris
N Venezuela
M.m.subsimilis
SW Ecuador, NW Peru
M.m.verticalis
SE Ecuador, Peru, Bolivia, SE Venezuela,
Guayana NW Brazil
**Myioborus brunniceps (Brown-capped
Redstart)**
M.b.castaneocapillus
SE Venezuela, W Guyana, N Brazil
M.b.duidae
SE Venezuela
M.b.maguirei
SE Venezuela
M.b.brunniceps
Bolivia, N Argentina
Myioborus pariae (Yellow-faced Redstart)
NE Venezuela
**Myioborus cardonai (Saffron-breasted
Redstart)**
SE Venezuela
Myioborus torquatus (Collared Redstart)
Costa Rica, W Panama
Myioborus ornatus (Golden-fronted Redstart)
M.o.ornatus
E Colombia, SW Venezuela
M.o.chrysops
W Colombia
**Myioborus melanocephalus (Spectacled
Redstart)**

M.m.ruficoronatus
SW Colombia, S Ecuador
M.m.griseonuchus
NW Peru
M.m.malaris
N Peru
M.m.melanocephalus
E Peru
M.m.bolivianus
S Peru, W Bolivia
Myioborus albifrons (White-fronted Redstart)
W Venezuela
Myioborus flavivertex (Yellow-crowned Redstart)
N Colombia
Myioborus albifacies (White-faced Redstart)
S Venezuela

EUTHLYPIS
Euthlypis lachrymosa (Neotropic Fan-tailed Warbler)
E.l.tephra
W Mexico
E.l.schistacea
W Chiapas
E.l.lachrymosa
S Mexico to N Nicaragua

BASILEUTERUS
Basileuterus fraseri (Grey & Gold Warbler)
B.f.ochraceicrista
W Ecuador
B.f.fraseri
C Ecuador, NW Peru
Basileuterus bivittatus (Two-banded Warbler)
B.b.roraimae
Guyana, SE Venezuela, N Brazil
B.b.bivittatus
SE Peru, W Bolivia
B.b.argentinae
SE Bolivia, NW Argentina
Basileuterus chrysogaster (Golden-bellied Warbler)
B.c.chlorophrys
SW Colombia, NW Ecuador
B.c.chrysogaster
E Peru
Basileuterus flaveolus (Flavescent Warbler)
B.f.pallidirostris
NE Colombia, N Venezuela
B.f.flaveolus
Colombia, Venezuela to Bolivia
Basileuterus luteoviridis (Citrine Warbler)
B.l.luteoviridis
SW Venezuela, E Colombia, E Ecuador
B.l.quindianus
C Colombia
B.l.richardsoni
W Colombia
B.l.striaticeps
N Peru
B.l.euophrys
SW Peru, W Bolivia

Basileuterus signatus (Pale-legged Warbler)
B.s.signatus
C Peru
B.s.flavovirens
SE Peru, W Bolivia, NW Argentina
Basileuterus nigrocristatus (Black-crested Warbler)
Venezuela, Colombia, Ecuador
Basileuterus griseiceps (Grey-headed Warbler)
NE Venezuela
Basileuterus basilicus (Santa Marta Warbler)
NE Colombia
Basileuterus cinereicollis (Grey-throated Warbler)
B.c.pallidulus
W Venezuela, NE Colombia
B.c.cinereicollis
E Colombia
Basileuterus coronatus (Russet-crowned Warbler)
B.c.conspicillatus
N Colombia
B.c.regulus
Venezuela, Colombia
B.c.elatus
SW Colombia, W Ecuador
B.c.orientalis
E Ecuador
B.c.castaneiceps
SW Ecuador, NW Peru
B.c.chapmani
NW Peru
B.c.inaequalis
N Peru
B.c.coronatus
SE Peru, W Bolivia
B.c.notius
C Bolivia
Basileuterus culicivorus (Golden-crowned Warbler)
B.c.flavescens
W Mexico
B.c.brasherii
E Mexico
B.c.culicivorus
S Mexico to Costa Rica
B.c.godmani
S Costa Rica, W Panama
B.c.occultus
W Colombia
B.c.austerus
C Colombia
B.c.indignus
N Colombia
B.c.cabanisi
NW Venezuela, NE Colombia
B.c.olivascens
Venezuela, Colombia, Trinidad
B.c.segrex
SE Venezuela, W Guyana, N Brazil
B.c.auricapillus
C Brazil

B.c.azarae
S Brazil, Paraguay, Uruguay, NE Argentina
B.c.viridescens
E Bolivia
Basileuterus rufifrons (Rufous-capped Warbler)
B.r.caudatus
NW Mexico
B.r.dugesi
W & C Mexico
B.r.jouyi
E Mexico
B.r.rufifrons
S Mexico, N Guatemala
B.r.salvini
SW Mexico, N Guatemala
B.r.delattrii
W Guatemala to N Costa Rica
B.r.mesochrysus
S Costa Rica, Panama, N Colombia, W Venezuela
B.r.actuosus
Coiba I (Panama)
Basileuterus belli (Golden-browed Warbler)
B.b.bateli
W Mexico
B.b.belli
C & E Mexico
B.b.clarus
SW Mexico
B.b.scitulus
SE Mexico, Guatemala, W Honduras
B.b.subobscurus
C Honduras
Basileuterus melanogenys (Black-cheeked Warbler)
B.m.melanogenys
Costa Rica
B.m.eximus
Panama
B.m.bensoni
Panama
Basileuterus ignotus (Pirre Warbler)   157.2
Panama
Basileuterus tristriatus (Three-striped Warbler)
B.t.chitrensis
W Panama
B.t.tacarcunae
E Panama, NW Colombia
B.t.daedalus
W Colombia, W Ecuador
B.t.auricularis
E Colombia, SW Venezuela
B.t.meridanus
W Venezuela
B.t.bessereri
N Venezuela
B.t.pariae
NE Venezuela
B.t.baezae
E Ecuador

B.t.tristriatus
SE Ecuador, C Peru
B.t.inconspicuus
SE Peru, NW Bolivia
B.t.punctipectus
C Bolivia
B.t.canens
E Bolivia
Basileuterus trifasciatus (Three-banded Warbler)
B.t.nitidior
SW Ecuador, NW Peru
B.t.trifasciatus
NW Peru
Basileuterus hypoleucus (White-bellied Warbler)
C Brazil, E Paraguay
Basileuterus leucoblepharus (White-browed Warbler)
B.l.leucoblepharus
S Brazil to NE Argentina
B.l.lemurum
Uruguay
Basileuterus leucophrys (White-striped Warbler)
SC Brazil
Basileuterus rivularis (River Warbler)
B.r.leucopygia
Honduras to W Panama
B.r.veraguensis
SW Costa Rica, C Panama
B.r.semicervina
E Panama to NW Peru
B.r.motacilla
N Colombia
B.r.fulvicauda
E Colombia, E Ecuador, NE Peru, W Brazil
B.r.significans
SE Peru
B.r.mesoleuca
E Venezuela, the Guianas, N Brazil
B.r.rivularis
SE Brazil, E Paraguay, NE Argentina
B.r.boliviana
E Bolivia

NEPHELORNIS
Nephelornis oneilli (Pardusco)   157.1
C Peru

ZELEDONIA
Zeledonia coronata (Wren Thrush)
Costa Rica, W Panama

PEUCEDRAMUS
Peucedramus taeniatus (Olive Warbler)
P.t.arizonae
SW USA, N Mexico
P.t.jaliscensis
NW Mexico
P.t.giraudi
C Mexico

P.t.aurantiacus
Chiapas (S Mexico)
P.t.taeniatus
S Mexico, W Guatemala
P.t.micrus
El Salvador, Honduras, N Nicaragua

Granatellus venustus (Red-breasted Chat)
G.v.francescae
Tres Marias Is
G.v.venustus
W & SW Mexico
G.v.melanotis
W coast of Mexico
Granatellus sallaei (Grey-throated Chat)
G.s.sallaei
E Mexico
G.s.boucardi
SE Mexico, E Guatemala, Belize
Granatellus pelzelni (Rose-breasted Chat)
G.p.pelzelni
SE Venezuela, Guyana, Surinam, NW Brazil
G.p.paraensis
N Brazil

ICTERIA
Icteria virens (Yellow-breasted Chat)
I.v.auricollis
SW Canada, W USA >> W Mexico, Guatemala
I.v.virens
E USA >> E Mexico, Central America
I.v.tropicalis
S Sonora (Mexico)

CONIROSTRUM
Conirostrum speciosum (Chestnut-vented Conebill)
C.s.guaricola
C Venezuela
C.s.amazonum
the Guianas to Ecuador >> N Peru
C.s.speciosum
SE Peru & Bolivia to N Argentina
Conirostrum leucogenys (White-eared Conebill)
C.l.panamense
E Panama, NW Colombia
C.l.leucogenys
N Colombia, NE Venezuela
C.l.cyanochrous
W Venezuela
Conirostrum bicolor (Bicoloured Conebill)
C.b.bicolor
N Colombia to the Guianas, N Brazil
C.b.minor
W Brazil, E Ecuador, E Peru
Conirostrum margaritae (Pearly-breasted Conebill)
N Brazil, NE Peru
Conirostrum cinereum (Cinereous Conebill)

C.c.fraseri
SW Colombia, E Ecuador
C.c.littorale
W Peru, N Chile
C.c.cinereum
SE Peru, W Bolivia
Conirostrum tamarugensis (Tamarugo Conebill) 157.3
SC Peru, N Bolivia
Conirostrum ferrugineiventre (White-browed Conebill)
S Peru, W Bolivia
Conirostrum rufum (Rufous-browed Conebill)
N Colombia
Conirostrum sitticolor (Blue-backed Conebill)
C.s.intermedium
W Venezuela
C.s.sitticolor
S Colombia, Ecuador, NW Peru
C.s.cyaneum
Peru, W Bolivia
Conirostrum albifrons (Capped Conebill)
C.a.cyanonotum
N Venezuela
C.a.albifrons
W Venezuela, E Colombia
C.a.centralandium
C Colombia
C.a.atrocyaneum
SW Colombia, Ecuador, N Peru
C.a.sordidum
S Peru, W Bolivia
C.a.lugens
E Bolivia

158. DREPANIDIDAE (HAWAIIAN HONEYCREEPERS) 158.1

TELESPYZA 158.1
Telespyza cantans (Laysan Finch)
Laysan
Telespyza ultima (Nihoa Finch)
Nihoa

PSITTIROSTRA
Psittirostra psittacea (Ou)
Maui, Hawaii

LOXIOIDES
Loxioides bailleui (Palila)
Hawaii

PSEUDONESTOR
Pseudonestor xanthophrys (Maui Parrotbill)
Maui

HEMIGNATHUS
Hemignathus virens (Amakihi)
H.v.stejnegeri
Kauai
H.v.chloris
Oahu

H.v.wilsoni
Maui, Molokai
H.v.virens
Hawaii
Hemignathus parvus (Lesser Amakihi)
Kauai
Hemignathus procerus (Kauai Akialoa)
Kauai
Hemignathus lucidus (Nukupuu)
H.l.affinis
Maui
H.l.hanapepe
Kauai
Hemignathus munroi (Akiapolaau)
Hawaii

OREOMYSTIS
Oreomystis bairdi (Kauai Creeper)
Kauai
Oreomystis mana (Hawaii Creeper)
Hawaii

PAROREOMYZA
Paroreomyza montana (Maui Creeper)
Maui
Paroreomyza flammea (Molokai Creeper)
Molokai
Paroreomyza maculata (Oahu Creeper)
P.m.maculata
Oahu
P.m.newtoni
Maui

LOXOPS
Loxops coccineus (Akepa)
L.c.coccineus
Hawaii
L.c.caerulirostris
Kauai
L.c.ochracea
Maui

VESTIARIA
Vestiaria coccinea (Iiwi)
Kauai, Oahu, Molokai, Maui, Lanai, Hawaii

PALMERIA
Palmeria dolei (Crested Honeycreeper)
Maui

HIMATIONE
Himatione sanguinea (Apapane)
Kauai, Oahu, Molokai, Maui, Hawaii

MELAMPROSOPS   158.2
Melamprosops phaeosoma (Po'ouli)
Maui

159. VIREONIDAE (VIREOS)

CYCLARHINAE

CYCLARHIS
Cyclarhis gujanensis (Rufous-browed Pepper Shrike)
C.g.flaviventris
C Mexico, E Guatemala, N Honduras
C.g.yucatanensis
SE Mexico
C.g.insularis
Cozumel I
C.g.nicaraguae
S Mexico, Guatemala, El Salvador, Honduras, Nicaragua
C.g.subflavescens
Costa Rica, W Panama
C.g.perrygoi
WC Panama
C.g.flavens
E Panama
C.g.coibae
Coiba I (Panama)
C.g.canticus
N & E Colombia
C.g.flavipectus
NE Venezuela, Trinidad
C.g.parvus
E Colombia, N Venezuela
C.g.gujanensis
E Colombia, S Venezuela, the Guianas, Brazil, E Peru, NE Bolivia
C.g.cearensis
E Brazil
C.g.ochrocephala
SE Brazil, Paraguay, Uruguay, NE Argentina
C.g.viridis
Paraguay, N Argentina
C.g.virenticeps
Ecuador, NW Peru
C.g.contrerasi
N Peru
C.g.saturatus
C Peru
C.g.pax
EC Bolivia
C.g.dorsalis
C Bolivia
C.g.tarijae
SE Bolivia, NW Argentina
Cyclarhis nigrirostris (Black-billed Pepper Shrike)
C.n.nigrirostris
C Colombia, E Ecuador
C.n.atrirostris
SW Colombia, W Ecuador

VIREOLANIINAE

VIREOLANIUS
Vireolanius melitophrys (Chestnut-sided Shrike Vireo)

V.m.goldmani
 SC Mexico
V.m.melitophrys
 S Mexico, W Guatemala
Vireolanius pulchellus (Green Shrike Vireo)
V.p.pulchellus
 SE Mexico to Honduras
V.p.verticalis
 Nicaragua, Costa Rica
V.p.viridiceps
 W Costa Rica, W Panama
Vireolanius eximius (Yellow-browed Shrike Vireo) 159.3
V.e.mutabilis
 E Panama, NW Colombia
V.e.eximius
 N Colombia, NW Venezuela
Vireolanius leucotis (Slaty-capped Shrike Vireo)
V.l.mikettae
 W Colombia, NW Ecuador
V.l.leucotis
 N & W Amazonia
V.l.simplex
 N Brazil, S Peru
V.l.bolivianus
 SE Peru, N Bolivia

VIREONINAE 4 |

VIREO
Vireo brevipennis (Slaty Vireo)
V.b.browni
 Guerrero
V.b.brevipennis
 S Mexico
Vireo huttoni (Hutton's Vireo)
V.h.insularis
 Vancouver I
V.h.huttoni
 SW Canada, W USA, N Baja California
V.h.cognatus
 S Baja California
V.h.stephensi
 SW USA, NW Mexico
V.h.carolinae
 S USA, NE Mexico
V.h.pacificus
 W & SW Mexico
V.h.mexicanus
 C &S Mexico
V.h.vulcani
 S Mexico, W Guatemala
Vireo atricapillus (Black-capped Vireo)
 C & S USA >> N & W Mexico
Vireo griseus (White-eyed Vireo)
V.g.noveboracensis
 C & E USA >> E Mexico, Guatemala, Cuba
V.g.griseus
 SE USA >> E Mexico, N Honduras, W Cuba

V.g.maynardi
 S Florida
V.g.bermudianus
 Bermuda I
V.g.micrus
 S Texas, E Mexico
V.g.perquisitor
 EC Mexico
Vireo pallens (Mangrove Vireo)
V.p.paluster
 NW Mexico
V.p.ochraceus
 W Guatemala, W El Salvador
V.p.pallens
 W Honduras, W Nicaragua, W Costa Rica
V.p.semiflavus
 E Mexico, E Guatemala, E Honduras, Nicaragua
Vireo caribaeus (St Andrew Vireo)
 St Andrew I
Vireo bairdi (Cozumel Vireo)
 Cozumel I
Vireo gundlachii (Cuban Vireo)
V.g.magnus
 W Cuba
V.g.sanfelipensis
 W Cuba
V.g.gundlachii
 C & E Cuba
Vireo crassirostris (Thick-billed Vireo)
V.c.crassirostris
 Bahama Is
V.c.stalagmium
 Caicos Is
V.c.tortugae
 Tortue I, Haiti
V.c.approximans
 Old Providence, St Catalina Is
Vireo vicinior (Grey Vireo)
 SW USA >> NW Mexico
Vireo bellii (Bell's Vireo)
V.b.pusillus
 S California >> S Baja California
V.b.arizonae
 SW USA >> NW Mexico
V.b.medius
 S USA >> NC Mexico
V.b.bellii
 C & S USA >> Mexico, Guatemala, El Salvador, Honduras, N Nicaragua
Vireo nelsoni (Dwarf Vireo)
 S Mexico
Vireo hypochryseus (Golden Vireo)
V.h.nitidus
 S Sonora, Mexico
V.h.hypochryseus
 W & SW Mexico
V.h.sordidus
 Tres Marias Is
Vireo modestus (Jamaican White-eyed Vireo)
 Jamaica
Vireo nanus (Flat-billed Vireo)
 Hispaniola

Vireo latimeri (Puerto Rican Vireo)
W Puerto Rico

Vireo osburni (Blue Mountain Vireo)
Jamaica

Vireo carmioli (Carmiol's Vireo)
Costa Rica, W Panama

Vireo solitarius (Solitary Vireo)
V.s.solitarius
Canada, NC & E USA >> E Mexico,
Central America, W Cuba
V.s.alticola
EC USA >> SE USA
V.s.plumbeus
WC USA >> NW Mexico
V.s.cassinii
W USA >> Mexico, Guatemala
V.s.lucasanus
S Baja California
V.s.pinicolus
N Mexico
V.s.repetens
C Mexico
V.s.notius
Belize
V.s.montanus
S Mexico, Guatemala, Honduras, El
Salvador

Vireo flavifrons (Yellow-throated Vireo)
S Canada, E & C USA >> Colombia,
Venezuela, Central America

Vireo philadelphicus (Philadelphia Vireo)
W Canada, N USA >> Mexico, Central
America, Colombia

Vireo olivaceus (Red-eyed Vireo)
V.o.olivaceus
Canada, WC & E USA >> Cuba, C Sth
America
V.o.forreri
Tres Marias Is, N Mexico >> Upper
Amazonia
V.o.hypoleucus
NW Mexico
V.o.insulanus
Pearl Is (Panama)
V.o.caucae
W Colombia
V.o.griseobarbatus
W Ecuador, NW Peru
V.o.pectoralis
N Peru
V.o.solimoensis
E Ecuador, NE Peru
V.o.vividior
Colombia, Venezuela, the Guianas, N
Brazil, Trinidad
V.o.tobagensis
Tobago I
V.o.agilis
NE Brazil
V.o.gracilirostris
Fernando de Noronha I
V.o.diversus
SE Brazil, E Paraguay

V.o.chivi
W & SW Amazonia

Vireo flavoviridis (Yellow-green Vireo)   159.2
S Mexico to Upper Amazonia

Vireo magister (Yucatan Vireo)
V.m.magister
SE Mexico, Belize
V.m.caymanensis
Grand Cayman I

Vireo altiloquus (Black-whiskered Vireo)
V.a.barbatulus
S Florida, Cuba, Haiti >> Colombia,
Venezuela, Brazil, Peru
V.a.altiloquus
Gtr Antilles >> N Sth America
V.a.barbadensis
St Croix, Barbados Is
V.a.bonairensis
Aruba, Curacao, Bonaire Is
V.a.grandior
Old Providence, St Catalina Is
V.a.canescens
St Andrew I

Vireo gilvus (Warbling Vireo)
V.g.swainsonii
W Canada, W USA >> W Mexico,
Guatemala, Honduras, Nicaragua
V.g.victoriae
S Baja California
V.g.leucopolius
WC USA >> N Mexico
V.g.gilvus
SW Canada, C, S & NE USA >> S Mexico,
El Salvador
V.g.brewsteri
NW Mexico
V.g.eleanorae
NE Mexico
V.g.bulli
Oaxaca
V.g.amauronotus
EC Mexico
V.g.connectens
SC Mexico
V.g.strenuus
S Mexico, Guatemala, Honduras
V.g.chiriquensis
Costa Rica, W Panama
V.g.disjunctus
NC Colombia
V.g.mirandae
N Colombia, NW Venezuela
V.g.dissors
W Colombia
V.g.josephae
SW Colombia, W Ecuador
V.g.maranonicus
N Peru
V.g.laetissimus
SE Peru, N Bolivia

Vireo leucophrys (Brown-capped Vireo)   159.3
C Colombia, Ecuador, N Peru

HYLOPHILUS
Hylophilus poicilotis (Rufous-crowned Greenlet)
H.p.amaurocephalus
E Brazil
H.p.poicilotis
SE Brazil, Paraguay, NE Argentina
Hylophilus thoracicus (Lemon-chested Greenlet)
H.t.aemulus
Colombia, Ecuador, Peru, N Bolivia
H.t.griseiventris
E Venezuela, the Guianas, N Brazil
H.t.throacicus
SE Brazil
Hylophilus semicinereus (Grey-chested Greenlet)
H.s.viridiceps
S Venezuela, the Guianas, N Brazil
H.s.semicinereus
N Brazil
H.s.juruanus
NW Brazil
Hylophilus pectoralis (Ashy-headed Greenlet)
the Guianas, N Brazil
Hylophilus sclateri (Tepui Greenlet)
S Venezuela, Guyana, NC Brazil
Hylophilus muscicapinus (Buff-chested Greenlet)
H.m.muscicapinus
S Venezuela, the Guianas, N Brazil
H.m.griseifrons
N Brazil
Hylophilus brunneiceps (Brown-headed Greenlet)
H.b.brunneiceps
E Colombia, S Venezuela, NW Brazil
H.b.inornatus
N Brazil
Hylophilus semibrunneus (Rufous-naped Greenlet)
N Colombia, NW Venezuela, E Ecuador
Hylophilus aurantiifrons (Golden-fronted Greenlet)
H.a.aurantiifrons
E Panama, N Colombia
H.a.helvinus
NW Venezuela
H.a.saturatus
E Colombia, N Venezuela, Trinidad
Hylophilus hypoxanthus (Dusky-capped Greenlet)
H.h.hypoxanthus
SE Colombia
H.h.fuscicapillus
E Ecuador, N Peru
H.h.flaviventris
C Peru
H.h.ictericus
W Brazil, NE Peru, N Bolivia
H.h.albigula
N Brazil
Hylophilus flavipes (Scrub Greenlet)

H.f.viridiflavus
SW Costa Rica, W Panama
H.f.xuthus
Coiba I (Panama)
H.f.flavipes
C & N Colombia
H.f.melleus
N Colombia
H.f.galbanus
NE Colombia, NW Venezuela
H.f.acuticauda
N Venezuela
H.f.insularis
Tobago I
Hylophilus olivaceus (Olivaceous Greenlet)
159.1
E Ecuador, N Peru
Hylophilus ochraceiceps (Tawny-crowned Greenlet)
H.o.ochraceiceps
S Mexico, Guatemala
H.o.pallidipectus
Honduras, El Salvador, Nicaragua, Costa Rica
H.o.nelsoni
E Panama
H.o.bulunensis
E Panama, W Colombia, W Ecuador
H.o.ferrugineifrons
SE Colombia, S Venezuela, Guyana, Ecuador, Peru, NW Brazil
H.o.viridior
S Peru, N Bolivia
H.o.luteifrons
E Venezuela, the Guianas, N Brazil
H.o.lutescens
N Brazil
H.o.rubrifrons
NE Brazil
Hylophilus decurtatus (Grey-headed Greenlet)
H.d.decurtatus
E Mexico, Central America
H.d.darienensis
E Panama, N Colombia
H.d.minor
SW Colombia, W Ecuador

160. ICTERIDAE (NEW WORLD BLACKBIRDS)

ICTERINAE

PSAROCOLIUS
Psarocolius oseryi (Casqued Oropendola)
E Ecuador, E Peru
Psarocolius latirostris (Band-tailed Oropendola)
E Ecuador, N Peru, W Brazil
Psarocolius decumanus (Crested Oropendola)

P.d.melanterus
Panama, N Colombia
P.d.insularis
Trinidad, Tobago I
P.d.decumanus
N Sth America
P.d.maculosus
E Peru to Paraguay, N Argentina
Psarocolius viridis (Green Oropendola)
N Amazonia
Psarocolius atrovirens (Dusky-green Oropendola)
SE Peru, E Bolivia
Psarocolius angustifrons (Russet-backed Oropendola)
P.a.salmoni
C Colombia
P.a.atrocastaneus
W Ecuador
P.a.sincipitalis
NC Colombia
P.a.neglectus
E Colombia, NW Venezuela
P.a.oleagineus
NC Venezuela
P.a.angustifrons
W Amazonia
P.a.alfredi
SE Ecuador, E Peru, E Bolivia
Psarocolius wagleri (Chestnut-headed Oropendola)
P.w.wagleri
SE Mexico to NE Nicaragua
P.w.ridgwayi
S Nicaragua to Panama, W Ecuador
Psarocolius montezuma (Montezuma Oropendola)
S Mexico to Panama
Psarocolius cassini (Chestnut-mantled Oropendola)
NW Colombia
Psarocolius bifasciatus (Para Oropendola)
N Brazil
Psarocolius guatimozinus (Black Oropendola)
E Panama, NW Colombia
Psarocolius yuracares (Olive Oropendola)
P.y.yuracares
W Amazonia
P.y.neivae
N Brazil

CACICUS
Cacicus cela (Yellow-rumped Cacique)
C.c.vitellinus
Panama, N Colombia
C.c.flavicrissus
W Ecuador, NW Peru
C.c.cela
N Sth America, Trinidad
Cacicus haemorrhous (Red-rumped Cacique)
C.h.haemorrhous
SE Colombia, E Ecuador, N Brazil

C.h.affinis
E & S Brazil, Paraguay, NE Argentina
Cacicus uropygialis (Scarlet-rumped Cacique)
C.u.microrhynchus
S Honduras to Panama
C.u.pacificus
E Panama, Colombia, E Ecuador
C.u.uropygialis
S Venezuela to N Peru
Cacicus chrysopterus (Golden-winged Cacique)
E Bolivia to Uruguay
Cacicus koepckeae (Selva Cacique)
Peru
Cacicus leucoramphus (Mountain Cacique)
C.l.leucoramphus
NW Venezuela to E Ecuador
C.l.peruvianus
N Peru
C.l.chrysonotus
S Peru, Bolivia
Cacicus sclateri (Ecuadorian Black Cacique)
E Ecuador, N Peru
Cacicus solitarius (Solitary Cacique)
Central & Northern Sth America
Cacicus melanicterus (Yellow-winged Cacique)
W & SW Mexico
Cacicus holosericeus (Yellow-billed Cacique)
C.h.holosericeus
SE Mexico to Colombia
C.h.flavirostris
W Colombia to NW Peru
C.h.australis
Western Amazonia

ICTERUS
Icterus chrysocephalus (Moriche Oriole)
160.1
N Peru, S Colombia, Venezuela, the Guianas
Icterus cayanensis (Epaulet Oriole)
I.c.cayanensis
Surinam, French Guiana
I.c.tibialis
E Brazil
I.c.valenciobuenoi
SE Brazil
I.c.periporphyrus
NE Bolivia, W Brazil
I.c.pyrrhopterus
SE Bolivia to Uruguay
Icterus chrysater (Yellow-backed Oriole)
I.c.chrysater
S Mexico to Nicaragua
I.c.mayensis
SE Mexico
I.c.hondae
Panama, N Colombia
I.c.giraudii
C Colombia, N Venezuela

Icterus nigrogularis (Yellow Oriole)
  I.n.nigrogularis
    NE Sth America
  I.n.curasoensis
    Aruba, Curacao, Bonaire Is
  I.n.helioeides
    Margarita I
  I.n.trinitatis
    NE Venezuela, Trinidad
Icterus leucopteryx (Jamaican Oriole)
  I.l.bairdi
    Grand Cayman I e?
  I.l.leucopteryx
    Jamaica
  I.l.lawrencii
    St Andrew I
Icterus auratus (Orange Oriole)
  SE Mexico
Icterus mesomelas (Yellow-tailed Oriole)
  I.m.mesomelas
    SE Mexico to Honduras
  I.m.salvinii
    Nicaragua to Panama
  I.m.carrikeri
    N & W Colombia, NW Venezuela
  I.m.taczanowskii
    W Ecuador, NW Peru
Icterus auricapillus (Orange-crowned Oriole)
  E Panama to N Venezuela
Icterus graceannae (White-edged Oriole)
  SW Ecuador, NW Peru
Icterus xantholaemus (Yellow-throated Oriole)
  Ecuador
Icterus pectoralis (Spotted-breasted Oriole)
  I.p.pectoralis
    S Mexico to N Nicaragua
  I.p.espinachi
    S Nicaragua to NW Costa Rica
Icterus gularis (Lichtenstein's Oriole)
  I.g.tamaulipensis
    S Texas, E Mexico
  I.g.yucatanensis
    SE Mexico
  I.g.flavescens
    SW Mexico
  I.g.gularis
    S Mexico
  I.g.troglodytes
    S Mexico, W Guatemala
  I.g.gigas
    S Guatemala, Honduras
Icterus pustulatus (Streak-backed Oriole)
  I.p.microstictus
    W Mexico
  I.p.graysonii
    Tres Marias Is
  I.p.pustulatus
    SW & C Mexico
  I.p.formosus
    S Mexico, NW Guatemala
  I.p.alticola
    Guatemala, E Honduras

  I.p.sclateri
    El Salvador to NW Costa Rica
Icterus cucullatus (Hooded Oriole)
  I.c.nelsoni
    SW USA, NW Mexico
  I.c.sennetti
    S Texas, E Mexico
  I.c.cucullatus
    SW Texas, NC & C Mexico
  I.c.californicus
    N Baja California
  I.c.trochiloides
    S Baja California
  I.c.restrictus
    S Sonora
  I.c.igneus
    SE Mexico, Belize
  I.c.cozumelae
    Cozumel I
  I.c.duplexus
    Mujeres, Holbox Is
  I.c.masoni
    SE Quintana Roo
Icterus icterus (Troupial)
  I.i.ridgwayi
    N Colombia, NW Venezuela, Aruba,
    Curacao Is
  I.i.icterus
    E Colombia, NW Venezuela
  I.i.metae
    SW Venezuela
  I.i.croconotus
    SW Guyana, N Brazil, E Ecuador, E Peru
  I.i.jamaicaii
    E Brazil
  I.i.strictifrons
    N & E Bolivia, SW Brazil
Icterus galbula (Northern Oriole)
  I.g.galbula
    Canada, E USA >> Colombia
  I.g.bullockii
    SW Canada, W USA >> W Mexico to
    Nicaragua
  I.g.parvus
    SW USA >> NW Mexico
  I.g.abeillei
    SC Mexico
Icterus spurius (Orchard Oriole)
  I.s.spurius
    C Canada, E USA >> Colombia, Cuba
  I.s.phillipsi
    C Mexico
  I.s.fuertesi
    E Mexico
Icterus dominicensis (Black-cowled Oriole)
  I.d.prosthemelas
    SE Mexico to Nicaragua
  I.d.praecox
    E Costa Rica, W Panama
  I.d.northropi
    Andros I (Bahamas)
  I.d.melanopsis
    Cuba

I.d.dominicensis
Hispaniola
I.d.portoricensis
Puerto Rico
Icterus wagleri (Black-vented Oriole)
I.w.castaneopectus
NW Mexico
I.w.wagleri
W & S Mexico to Nicaragua
Icterus laudabilis (St Lucia Oriole)
St Lucia I
Icterus bonana (Martinique Oriole)
Martinique I
Icterus oberi (Montserrat Oriole)
Monserrat I
Icterus graduacauda (Audubon's Oriole)
I.g.audubonii
N Mexico
I.g.nayaritensis
WC Mexico
I.g.richardsoni
Oaxaca
I.g.dickeyae
Guerrero
I.g.graduacauda
C & S Mexico
Icterus maculialatus (Bar-winged Oriole)
S Mexico to El Salvador
Icterus parisorum (Scott's Oriole)
SC USA, C & W Mexico

NESOPSAR
Nesopsar nigerrimus (Jamaican Blackbird)
Jamaica

XANTHOPSAR
Xanthopsar flavus (Saffron-cowled Blackbird)
Paraguay, NE Argentina

GYMNOMYSTAX
Gymnomystax mexicanus (Oriole Blackbird)
N Sth America

XANTHOCEPHALUS
Xanthocephalus xanthocephalus (Yellow-headed Blackbird)
SW Canada, W USA, W Mexico

AGELAIUS
Agelaius xanthophthalmus (Yellow-eyed Blackbird)
Peru
Agelaius thilius (Yellow-winged Blackbird)
A.t.alticola
SE Peru, NW Bolivia
A.t.thilius
S Chile, SW Argentina
A.t.petersii
SE Brazil, Uruguay, N Argentina

Agelaius phoeniceus (Red-winged Blackbird)
A.p.arctolegus
Canada, E USA >> SC USA
A.p.fortis
WC USA
A.p.nevadensis
SW Canada, SW USA
A.p.caurinus
W USA
A.p.mailliardorum
WC California
A.p.californicus
C California
A.p.aciculatus
SC California
A.p.neutralis
S California, NW Baja California
A.p.sonoriensis
SW USA, NW Mexico
A.p.nyaritensis
SW Mexico
A.p.gubernator
NC Mexico
A.p.pallidulus
N Yucatan
A.p.nelsoni
SC Mexico
A.p.arthuralleni
N Guatemala
A.p.grinnelli
W Guatemala to NW Costa Rica
A.p.phoeniceus
SE Canada, E & S USA
A.p.littoralis
SE USA
A.p.mearnsi
SE USA
A.p.floridanus
S Florida
A.p.megapotamus
S Texas, NE Mexico
A.p.richmondi
S & SE Mexico, N Guatemala
A.p.matudae
SE Mexico
A.p.brevirostris
E Honduras, SE Nicaragua
A.p.bryanti
NW Bahama Is
A.p.assimilis
W Cuba
A.p.subniger
Isle of Pines
Agelaius tricolor (Tricoloured Blackbird)
W USA
Agelaius icterocephalus (Yellow-hooded Blackbird)
A.i.bogotensis
E Colombia
A.i.icterocephalus
Surinam to NE Peru
Agelaius humeralis (Tawny-shouldered Blackbird)

A.h.humaralis
  Hispaniola
A.h.scopulus
  Cuba
Agelaius xanthomus (Yellow-shouldered Blackbird)
A.x.xanthomus
  Puerto Rico
A.x.monensis
  Mona I (Puerto Rico)
Agelaius cyanopus (Unicoloured Blackbird)
A.c.xenicus
  NE Brazil
A.c.atroolivaceus
  E Brazil
A.c.beniensis
  N Bolivia
A.c.cyanopus
  E Bolivia, Paraguay, N Argentina
Agelaius ruficapillus (Chestnut-capped Blackbird)
A.r.frontalis
  French Guiana, E Brazil
A.r.ruficapillus
  SE Bolivia to Uruguay, N Argentina

STURNELLA
Sturnella superciliaris (Bonaparte's Blackbird)
  S Peru to Uruguay
Sturnella militaris (Red-breasted Blackbird)
  Sth America
Sturnella bellicosa (Peruvian Red-breasted Meadowlark)
S.b.bellicosa
  Ecuador, N Peru
S.b.albipes
  SW Peru, N Chile
S.b.catamarcanus
  NW Argentina
Sturnella defilippi (Lesser Red-breasted Meadowlark)
  SE Brazil, Uruguay, NE Argentina
Sturnella loyca (Long-tailed Meadowlark)
S.l.loyca
  S Chile, S Argentina
S.l.falklandicus
  Falkland Is
Sturnella magna (Eastern Meadowlark)
S.m.magna
  SE Canada, C & E USA
S.m.argutula
  SC & SE USA
S.m.hippocrepis
  Cuba
S.m.hoopesi
  S Texas, NE Mexico
S.m.lilianae
  SW USA, NW Mexico
S.m.auropectoralis
  C & SW Mexico
S.m.saundersi
  Oaxaca, Mexico

S.m.alticola
  S Mexico to Nicaragua
S.m.mexicana
  SE Mexico
S.m.griscomi
  SE Mexico
S.m.inexpectata
  E Guatemala, Honduras
S.m.subulata
  W Panama
S.m.meridionalis
  N Colombia, NW Venezuela
S.m.paralios
  N Colombia, W Venezuela
S.m.monticola
  Mt Roraima
S.m.praticola
  S Venezuela, N Guyana
S.m.quinta
  NE Brazil, Surinam
Sturnella neglecta (Western Meadowlark)
  SW Canada, W USA >> NW Mexico

PSEUDOLEISTES
Pseudoleistes guirahuro (Yellow-rumped Marshbird)
  SE Brazil, N Argentina, Paraguay, Uruguay
Pseudoleistes virescens (Brown-yellow Marshbird)
  SE Brazil, Uruguay, NE Argentina

AMBLYRAMPHUS
Amblyramphus holosericeus (Scarlet-headed Blackbird)
  Bolivia, Brazil, Paraguay, Uruguay, Argentina

HYPOPYRRHUS
Hypopyrrhus pyrohypogaster (Red-bellied Grackle)
  Colombia

CURAEUS
Curaeus curaeus (Austral Blackbird)
C.c.curaeus
  S Argentina, Chile
C.c.recurvirostris
  Megellanes, Chile
C.c.reynoldsi
  Tierra del Fuego
Curaeus forbesi (Forbes's Blackbird)
  E Brazil

GNORIMOPSAR
Gnorimopsar chopi (Chopi Blackbird)
G.c.sulcirostris
  E Bolivia, NW Argentina, E Brazil
G.c.chopi
  SE Bolivia to Uruguay, N Argentina

OREOPSAR
Oreopsar bolivianus (Bolivian Blackbird)
  Bolivia

LAMPROPSAR
Lampropsar tanagrinus (Velvet-fronted Grackle)
 L.t.guianensis
  NE Venezuela, NW Guyana
 L.t.tanagrinus
  Ecuador, N Peru, W Brazil
 L.t.macropterus
  W Brazil
 L.t.boliviensis
  N Bolivia
 L.t.violaceus
  W Brazil

MACROAGELAIUS
Macroagelaius subalaris (Mountain Grackle)
  C Colombia
Macroagelaius imthurni (Golden-tufted Grackle)   160.1
  S Venezuela, N Brazil, W Guyana

DIVES
Dives atroviolacea (Cuban Blackbird)
  Cuba
Dives dives (Melodious Blackbird)
 D.d.dives
  E Mexico to Nicaragua
 D.d.kalinowskii
  W Peru
Dives warszewiczi (Scrub Blackbird)   160.1
  SW Ecuador, NW Peru

QUISCALUS
Quiscalus mexicanus (Great-tailed Grackle)
 Q.m.nelsoni
  SW USA, NW Mexico
 Q.m.graysoni
  NW Mexico
 Q.m.obscurus
  W Mexico
 Q.m.monsoni
  S USA, C Mexico
 Q.m.prosopidicola
  S USA, NE Mexico
 Q.m.mexicanus
  C &S Mexico to N Nicaragua
 Q.m.loweryi
  Belize
 Q.m.peruvianus
  Costa Rica to Peru, Venezuela
Quiscalus major (Boat-tailed Grackle)
 Q.m.torreyi
  E USA
 Q.m.major
  SE USA
Quiscalus nicaraguensis (Nicaraguan Grackle)
  Nicaragua
Quiscalus quiscula (Common Grackle)
 Q.q.versicolor
  C & SE Canada, NE & C USA >> S USA

 Q.q.stonei
  NC USA >> SE USA
 Q.q.quiscula
  SE USA
Quiscalus niger (Antillean Grackle)
 Q.n.caribaeus
  W Cuba
 Q.n.gundlachii
  C & E Cuba
 Q.n.caymanensis
  Grand Cayman I
 Q.n.bangsi
  Little Cayman I
 Q.n.crassirostris
  Jamaica
 Q.n.niger
  Hispaniola
 Q.n.brachypterus
  Puerto Rico
Qusicalus lugubris (Carib Grackle)
 Q.l.guadeloupensis
  Monserrat, Guadeloupe, Martinique Is
 Q.l.inflexirostris
  St Lucia I
 Q.l.contrusus
  St Vincent I
 Q.l.luminosus
  Grenada I
 Q.l.fortirostris
  Barbados, Antigua Is
 Q.l.orquillensis
  Los Hermanos I (Venezuela)
 Q.l.insularis
  Margarita I
 Q.l.lugubris
  Trinidad, N Venezuela, the Guianas, NE Brazil

EUPHAGUS
Euphagus carolinus (Rusty Blackbird)
 E.c.carolinus
  Canada, NE & C USA >> SE USA
 E.c.nigrans
  Newfoundland >> SE USA
Euphagus cyanocephalus (Brewer's Blackbird)
  SW Canada, W USA >> Mexico

MOLOTHRUS
Molothrus badius (Bay-winged Cowbird)
 M.b.fringillarius
  NE Brazil
 M.b.badius
  Bolivia to Uruguay, N Argentina
 M.b.bolivianus
  S Bolivia
Molothrus rufoaxillaris (Screaming Cowbird)
  S Bolivia to Uruguay
Molothrus bonariensis (Shiny Cowbird)
 M.b.cabanisii
  E Panama, Colombia

M.b.aequatorialis
SW Colombia, W Ecuador
M.b.occidentalis
SW Ecuador, W Peru
M.b.venezuelensis
E Colombia, N Venezuela
M.b.minimus
S Lesser Antilles, the Guianas, N Brazil
M.b.riparius
E Peru
M.b.bonariensis
Central Sth America
Molothrus aeneus (Bronzed Cowbird)
M.a.loyei
SW USA, NW Mexico
M.a.assimilis
S & SW Mexico
M.a.aeneus
S Texas, E Mexico to Panama
M.a.armenti
N Colombia
Molothrus ater (Brown-headed Cowbird)
M.a.artemisiae
W Canada, W USA >> C Mexico
M.a.obscurus
SW USA >> S Mexico
M.a.ater
C & S USA >> SE USA
M.a.californicus
SW USA, Los Coronados Is

SCAPHIDURA
Scaphidura oryzivora (Giant Cowbird)
S.o.impacifa
S Mexico to W Panama
S.o.oryzivora
E Panama, Trinidad, N Sth America

DOLICHONYCHINAE

DOLICHONYX
Dolichonyx oryzivorus (Bobolink)
S Canada >> N Sth America, West Indies

161. FRINGILLIDAE (FINCHES)

FRINGILLINAE

FRINGILLA
Fringilla coelebs (Chaffinch)
F.c.moreletti
Azores Is
F.c.maderensis
Madeira I
F.c.canariensis
Gran Canaria, Tenerife Is
F.c.ombriosa
Hierro I
F.c.palmae
Las Palmas I
F.c.africana
NW Africa
F.c.spodiogenys
Tunisia

F.c.coelebs
continent of Europe, Siberia, C Asia, N
Africa
F.c.gengleri
British Isles
F.c.sarda
Sardinia
F.c.schiebeli
Crete
F.c.solomkoi
Crimea
F.c.alexandrovi
N Iran
F.c.transcaspica
S Transcaspia
Fringilla teydea (Blue Chaffinch)
F.t.teydea
Tenerife I
F.t.polatzeki
Gran Canaria I
Fringilla montifringilla (Brambling)
Europe to Japan >> N Africa, N India,
China

CARDUELINAE

SERINUS
Serinus pusillus (Red-fronted Serin)
Asia Minor to Tibet >> Israel
Serinus serinus (European Serin)
W & C Europe, Asia Minor, N Africa
Serinus syriacus (Syrian Serin)
Lebanon, Syria >> Iraq, Egypt
Serinus canaria (Island Canary)
Canary, Azores, Madeira Is
Serinus citrinella (Citril Finch)
S.c.citrinella
S Europe
S.c.corsicana
Corsica, Sardinia
Serinus thibetanus (Tibetan Siskin)
Nepal, SE Tibet >> NE Burma & W China
Serinus canicollis (Yellow-crowned Canary)
S.c.flavivertex
Ethiopia to N Tanzania
S.c.sassii
S Zaire to N Malawi
S.c.huillensis
C Angola
S.c.griseitergum
E Zimbabwe
S.c.thompsonae
Transvaal, N Cape Province
S.c.canicollis
Cape Province
Serinus nigriceps (Black-headed Siskin)
N Ethiopia
Serinus citrinelloides (African Citril Finch)
S.c.citrinelloides
Ethiopia, SE Sudan
S.c.kikuyensis
W Kenya
S.c.brittoni
Kapenguria (Kenya)

S.c.frontalis
W Uganda, E Zaire, NW Tanzania
S.c.hypostictus
S Kenya, E Zambia to Mozambique
S.c.martinsi
Angola
Serinus capistratus (Black-faced Canary)
S.c.capistratus
Gabon to N Angola, Zambia
S.c.hildegardae
S Angola
Serinus koliensis (Papyrus Canary)
Uganda, W Kenya, Rwanda
Serinus scotops (Forest Canary)
S.s.transvaalensis
N & E Transvaal
S.s.umbrosus
SE Transvaal, Natal, S Cape Province
S.s.scotops
S Natal, E Cape Province
Serinus leucopygius (White-rumped Seedeater)
S.l.riggenbachi
Senegal to Chad, Central African Republic
S.l.pallens
Air to N Nigeria
S.l.leucopygius
E Sudan, N Ethiopia
Serinus atrogularis (Yellow-rumped Seedeater)
S.a.rothschildi
E Arabia
S.a.xanthopygius
N Ethiopia
S.a.reichenowi
S Sudan to NE Tanzania
S.a.somereni
E Zaire, W Uganda, W Kenya
S.a.lwenarum
S Zaire, Angola, Zambia
S.a.atrogularis
Zimbabwe, W Transvaal
S.a.impiger
SE Transvaal, W Natal, N Cape Province
S.a.semideserti
S Angola, N Namibia, S Zambia
S.a.deserti
SW Angola, NW Namibia
Serinus citrinipectus (Lemon-breasted Seedeater)
S Malawi, SE Zimbabwe, S Mozambique
Serinus mozambicus (Yellow-fronted Canary)
S.m.caniceps
Senegal to N Cameroun
S.m.punctigula
Cameroun
S.m.barbatus
N Zaire, Sudan to Kenya
S.m.santhome
Sao Thome I
S.m.tando
SW Zaire, N Angola
S.m.samaliyae

SE Zaire, Zambia
S.m.vansoni
SE Angola, Namibia, SW Zambia
S.m.mozambicus
Kenya to Zambia, Mozambique
S.m.granti
S Mozambique, Sth Africa
S.m.grotei
E Sudan, W Ethiopia
S.m.gommaensis
W Ethiopia
Serinus donaldsoni (Grosbeak Canary)
S.d.donaldsoni
Ethiopia, N Kenya
S.d.buchanani
S Kenya, N Tanzania
Serinus flaviventris (Yellow Canary)
S.f.damarensis
Namibia, Botswana
S.f.flaviventris
W Cape Province
S.f.hesperus
W Cape Province, SW Namibia
S.f.aurescens
N Cape Province
S.f.quintoni
S & C Cape Province
S.f.marshalli
NW Cape Province, Transvaal
S.f.guillarmodi
Lesotho
Serinus dorsostriatus (White-bellied Canary)
161.2
S.d.maculicollis
S Ethiopia, Kenya, Somalia
S.d.dorsostriatus
N Tanzania
Serinus sulphuratus (Brimstone Canary)
S.s.loveridgei
SE Tanzania, N Mozambique
S.s.sharpii
Angola to Kenya, Mozambique
S.s.languens
E Natal, Zimbabwe, S Mozambique
S.s.wilsoni
Sth Africa
S.s.sulphuratus
S Cape Province
Serinus albogularis (White-throated Seedeater)
S.a.crocopygius
SW Angola, N Namibia
S.a.sordahlae
S Namibia, NW Cape Province
S.a.albogularis
W Cape Province
S.a.hewitti
C Cape Province
S.a.orangensis
Orange Free State
Serinus reichardi (Stripe-breasted Seedeater)
161.2
S.r.striatipectus

S Sudan, S Ethiopia, N Kenya
S.r.reichardi
  S Zaire, Zambia, Tanzania
Serinus gularis (Streaky-headed Seedeater)
S.g.canicapilla
  Senegal to N Cameroun
S.g.montanorum
  Cameroun
S.g.uamensis
  W Central African Republic
S.g.elgonensis
  N Zaire, S Sudan, W Kenya
S.g.benguellensis
  C Angola, W Zambia
S.g.mendosus
  NE Botswana, NW Transvaal
S.g.gularis
  Zimbabwe to N Cape Province
S.g.endemion
  S Mozambique, E Sth Africa
S.g.humilis
  SW Cape Province
Serinus mennelli (Black-eared Seedeater)
  E Angola to Mozambique
Serinus tristriatus (Brown-rumped Seedeater)
  E Ethiopia
Serinus ankoberensis (Ankober Serin) 161.4
  C Ethiopia
Serinus menachensis (Yemen Serin)
  Saudi Arabia
Serinus striolatus (Streaky Seedeater)
S.s.striolatus
  Ethiopia, N Kenya
S.s.affinis
  Kenya, N Tanzania
S.s.graueri
  Uganda, W Kenya, W Tanzania
S.s.whytii
  S Tanzania, N Malawi
Serinus burtoni (Thick-billed Seedeater)
S.b.burtoni
  Cameroun
S.b.tanganjicae
  E Zaire, W Uganda
S.b.kilimensis
  N Kenya, N Tanzania
S.b.albifrons
  E Kenya
S.b.melanochrous
  S Tanzania
Serinus rufobrunneus (Principé Seedeater)
S.r.rufobrunneus
  Principé I
S.r.thomensis
  Sao Thome I
Serinus leucopterus (White-winged Seedeater)
  SW Cape Province
Serinus tottus (Cape Siskin)
  Cape Province
Serinus symonsi (Drakensburg Siskin)
  Lesotho, W Natal 161.3
Serinus alario (Black-headed Canary)

S.a.leucolaema
  Namibia, Botswana, W Cape Province
S.a.alario
  N & C Cape Province
Serinus estherae (Malay Goldfinch)
S.e.vanderbilti
  N Sumatra
S.e.estherae
  W Java
S.e.orientalis
  E Java
S.e.renatae
  Sulawesi
S.e.mindanensis
  Mindanao I

NEOSPIZA
Neospiza concolor (Sao Thome Grosbeak-Weaver)
  Sao Thome I e?

LINURGUS
Linurgus olivaceus (Oriole-Finch)
L.o.olivaceus
  SE Nigeria, Cameroun, Fernando Po I
L.o.prigoginei
  E Zaire
L.o.elgonensis
  SE Sudan, N Kenya
L.o.kilimensis
  Tanzania, N Malawi

RHYNCHOSTRUTHUS
Rhynchostruthus socotranus (Golden-winged Grosbeak)
R.s.louisae
  N Somalia
R.s.percivali
  SW Arabia
R.s.socotranus
  Socotra I

CARDUELIS
Carduelis chloris (Western Greenfinch)
C.c.chloris
  N Europe >> S Europe
C.c.aurantiiventris
  S Europe, N Africa
C.c.chlorotica
  Syria, Lebanon >> Egypt
C.c.turkestanica
  Caucasus >> Iran, Afghanistan & Iraq
Carduelis sinica (Oriental Greenfinch)
C.s.sinica
  E & C China
C.s.chabarovi
  Manchuria, Mongolia
C.s.ussuriensis
  E Manchuria
C.s.kawarahiba
  Sakhalin I, NE Asia >> Japan
C.s.minor
  S Japan

C.s.kittlitzi
Bonin Is
Carduelis spinoides (Black-headed Greenfinch)
C.s.spinoides
Pakistan, N India, E Himalayas
C.s.heinrichi
S Assam, W Burma
C.s.monguilloti
S Vietnam
Carduelis ambigua (Yunnan Greenfinch)
C.a.taylori
SE Tibet, Sikang
C.a.ambigua
SW China, N Burma
Carduelis spinus (Spruce Siskin)
N Asia, N Europe >> Japan, N Africa & China
Carduelis pinus (Pine Siskin)
C.p.pinus
Canada, USA >> C Mexico
C.p.macroptera
N Baja California, NW & C Mexico
C.p.perplexa
S Mexico, W Guatemala
Carduelis atriceps (Black-capped Siskin)
S Mexico, W Guatemala
Carduelis spinescens (Andean Siskin)
C.s.spinescens
Colombia, W Venezuela
C.s.capitanea
N Colombia
C.s.nigricauda
N Colombia
Carduelis yarrellii (Yellow-faced Siskin)
N Venezuela, N Brazil
Carduelis cucullata (Red Siskin)
NE Colombia, N Venezuela
Carduelis crassirostris (Thick-billed Siskin)
C.c.amadoni
SE Peru
C.c.crassirostris
S Bolivia, C Chile, W Argentina
Carduelis magellanica (Hooded Siskin)
C.m.capitalis
S Colombia, Ecuador, NW Peru
C.m.paula
S Ecuador, W Peru
C.m.peruana
C Peru
C.m.urubambensis
S Peru, N Chile
C.m.boliviana
S Bolivia
C.m.tucumana
NW Argentina
C.m.santaecrucis
EC Bolivia
C.m.alleni
SE Bolivia, Paraguay, NE Argentina
C.m.icterica
SE Brazil, E & S Paraguay

C.m.magellanica
Uruguay, E Argentina
C.m.longirostris
SE Venezuela, Guyana, N Brazil
Carduelis dominicensis (Antillean Siskin)
Hispaniola
Carduelis siemiradzkii (Saffron Siskin)
SW Ecuador
Carduelis olivacea (Olivaceous Siskin)
SE Ecuador, Peru, Bolivia
Carduelis notata (Black-headed Siskin)
C.n.notata
E & C Mexico, N Guatemala
C.n.forreri
W Mexico
C.n.oleacea
Belize to N Nicaragua
Carduelis xanthogastra (Yellow-bellied Siskin)
C.x.xanthogastra
Costa Rica to Colombia, Venezuela
C.x.stejnegeri
C Bolivia
Carduelis atrata (Black Siskin)
S Peru to N Chile, W Argentina
Carduelis uropygialis (Yellow-rumped Siskin)
S Peru to Chile, W Argentina
Carduelis barbata (Black-chinned Siskin)
S Chile, W Argentina
Carduelis tristis (American Goldfinch)
C.t.tristis
C USA >> SE USA, E Mexico
C.t.pallida
W Canada, WC USA >> N Mexico
C.t.jewetti
SW Canada, NW USA
C.t.salicamans
SW USA, N Baja California
Carduelis psaltria (Dark-backed Goldfinch)
C.p.hesperophila
W USA, NW Mexico
C.p.witti
Tres Marias Is
C.p.psaltria
SC USA, N Mexico
C.p.jouyi
SE Mexico
C.p.columbiana
S Mexico to Peru, Venezuela
Carduelis lawrencei (Lawrence's Goldfinch)
SW USA >> NW Mexico
Carduelis carduelis (Eurasian Goldfinch)
C.c.carduelis
W & C Europe
C.c.britannica
British Isles, Netherlands
C.c.parva
W Mediterranean, Azores, Canary Is
C.c.tschusii
Corsica, Sardinia, Sicily
C.c.balcanica
E Mediterranean

C.c.niediecki
Cyprus, Asia Minor, Iraq, Iran, Egypt
C.c.major
SW Siberia
C.c.brevirostris
Caucasus
C.c.loudoni
N Iran
C.c.paropanisi
Central Asia >> S Iran
C.c.subulata
NC Asia >> Turkestan
C.c.caniceps
Pakistan, W Himalayas, Nepal

ACANTHIS
Acanthis flammea (Redpoll)
A.f.flammea
N Europe, Asia, Nth America >> S Europe
& N China
A.f.rostrata
NE Canada >> NE USA, NW Europe
A.f.islandica
Iceland
A.f.cabaret
British Isles, Switzerland
A.f.exilipes
N Eurasia, N Nth America >> C Europe
A.f.hornemanni
Greenland, C & E Canada >> British Isles,
S Canada   161.1
Acanthis flavirostris (Twite)
A.f.flavirostris
NE & C Europe >> S Europe
A.f.pipilans
N British Isles, Ireland
A.f.brevirostris
Caucasus, NW Iran
A.f.korejevi
C Asia
A.f.altaica
EC Asia
A.f.montanella
N Pakistan, Altai, W Sinkiang
A.f.miniakensis
E Sinkiang, NW China
A.f.rufostrigata
Pakistan, Tibet, Himalayas, N India
Acanthis cannabina (Linnet)
A.c.cannabina
Europe, NW Asia >> N Africa
A.c.autochthona
Scotland
A.c.nana
Madeira I
A.c.meadewaldoi
W Canary Is
A.c.harterti
E Canary Is
A.c.bella
Asia Minor, SW Asia >> Egypt, N India
Acanthis yemensis (Yemeni Linnet)
SW Arabia

Acanthis johannis (Warsangli Linnet)
NE Somalia

LEUCOSTICTE
Leucosticte nemoricola (Hodgson's Rosy Finch)
L.n.altaica
W Pakistan, W Sinkiang, Altai
L.n.nemoricola
Himalayas, W China >> N Burma
Leucosticte brandti (Brandt's Rosy Finch)
L.b.margaritacea
W Mongolia
L.b.brandti
W Tien Shan, W Sinkiang
L.b.pamirensis
W Tien Shan, NE Afghanistan
L.b.haematopygia
N Pakistan, Himalayas, Tibet
L.b.pallidior
SW Sinkiang, NE Tsinghai
Leucosticte arctoa (Rosy Finch)
L.a.arctoa
Altai
L.a.cognata
Tannu Tuva
L.a.sushkini
N Mongolia
L.a.gigliolii
Transbaicalia
L.a.brunneonucha
E Siberia, Kurile Is
L.a.griseonucha
Aleutian, Kodiak Is, Alaska
L.a.umbrina
St Matthew, Pribilof Is
L.a.irvingi
N Alaska
L.a.littoralis
E Alaska, W Canada >> SW USA
L.a.tephrocotis
WC Canada >> WC USA
L.a.dawsoni
E California
L.a.wallowa
NE Oregon >> W Nevada
L.a.atrata
WC USA >> SC USA
L.a.australis
SW USA

CALLACANTHIS
Callacanthis burtoni (Red-browed Rosefinch)
W Pakistan, Himalayas

RHODOPECHYS
Rhodopechys sanguinea (Crimson-winged Finch)
R.s.aliena
Morocco

R.s.sanguinea
Caucasus, Iran, SC Asia
Rhodopechys githaginea (Trumpeter Finch)
R.g.amantum
Canary Is
R.g.zedlitzi
N Africa
R.g.githaginea
S Egypt, N Sudan
R.g.crassirostris
Arabia, Iran >> NW India
Rhodopechys mongolica (Mongolian Trumpeter Finch)
E Asia, India >> E China
Rhodopechys obsoleta (Lichtenstein's Desert Finch)
Asia Minor >> N Pakistan

URAGUS
Uragus sibiricus (Long-tailed Rosefinch)
U.s.sibiricus
S Siberia, N Manchuria >> Turkestan
U.s.ussuriensis
C Manchuria, Korea >> NE China
U.s.sanguinolentus
Sakhalin, S Kurile Is >> S Japan
U.s.lepidus
NW China
U.s.henrici
Sikang

UROCYNCHRAMUS
Urocynchramus pylzowi (Przewalski's Rosefinch)
W China

CARPODACUS
Carpodacus rubescens (Blanford's Rosefinch)
Himalayas, W China
Carpodacus nipalensis (Dark Rosefinch)
C.n.kangrae
W Himalayas
C.n.nipalensis
C Himalayas, N Assam
C.n.intensicolor
Sikang, W China >> N Burma
Carpodacus erythrinus (Common Rosefinch)
C.e.erythrinus
E Europe, W Asia >> India, Indochina
C.e.grebnitskii
E Siberia, Manchuria >> SE China
C.e.kubanensis
Caucasus, Iran, W India
C.e.ferghanensis
C Asia >> NW India
C.e.roseatus
Himalayas, Tibet, China >> S India, Indochina
Carpodacus purpureus (Purple Finch)

C.p.purpureus
Canada, NE USA >> SE USA
C.p.californicus
SW Canada >> SW USA, Baja California
Carpodacus cassinii (Cassin's Finch)
SW Canada, W USA >> N Mexico
Carpodacus mexicanus (House Finch)
C.m.frontalis
SW Canada, W USA, NW Mexico
C.m.clementis
San Clemente, Los Coronados Is
C.m.mcgregori
San Benito I
C.m.amplus
Guadeloupe I
C.m.ruberrimus
S Baja California, NW Mexico
C.m.rhodopnus
C Sinaloa
C.m.coccineus
SW Mexico
C.m.potosinus
NC Mexico
C.m.centralis
C Mexico
C.m.mexicanus
SC Mexico
C.m.griscomi
Guerrero
Carpodacus pulcherrimus (Beautiful Rosefinch)
C.p.pulcherrimus
Himalayas
C.p.waltoni
SE Tibet, SW Sikang
C.p.argyrophrys
E Tsinghai, W China
C.p.davidianus
C Mongolia
Carpodacus eos (Stresemann's Rosefinch)
W China, E Sikang
Carpodacus rhodochrous (Pink-browed Rosefinch)
Himalayas
Carpodacus vinaceus (Vinaceous Rosefinch)
C.v.vinaceus
W China, E Sikang
C.v.formosanus
Taiwan
Carpodacus edwardsii (Large Rosefinch)
C.e.edwardsii
W China, E Sikang
C.e.rubicunda
Himalayas, SE Tibet >> N Burma
Carpodacus synoicus (Sinai Rosefinch)
C.s.synoicus
Sinai
C.s.salimalii
NE Afghanistan
C.s.stoliczkae
SW Sinkiang
C.s.beicki
NE Tsinghai, NW China

***Carpodacus roseus* (Pallas's Rosefinch)**
C.r.roseus
Altai, E Asia >> N China, C Japan
C.r.portenko
Sakhalin I >> S Korea, N Japan
***Carpodacus trifasciatus* (Three-banded Rosefinch)**
W China >> SE Tibet
***Carpodacus rhodopeplus* (Spot-winged Rosefinch)**
C.r.rhodopeplus
Himalayas
C.r.verreauxii
W China >> N Burma
***Carpodacus thura* (White-browed Rosefinch)**
C.t.blythi
NE Afghanistan, Pakistan, W Himalayas
C.t.thura
C Himalayas
C.t.femininus
SE Tibet, W China
C.t.dubius
SE Tsinghai, NW China
C.t.deserticolor
NE Tsinghai
***Carpodacus rhodochlamys* (Red-mantled Rosefinch)**
C.r.rhodochlamys
C Asia
C.r.kotschubeii
SC Asia
C.r.grandis
Pakistan, W Himalayas
***Carpodacus rubicilloides* (Eastern Great Rosefinch)**
C.r.lucifer
Ladakh, Himalayas
C.r.rubicilloides
E Sikang, E Tsinghai >> SW China
***Carpodacus rubicilla* (Caucasian Great Rosefinch)**
C.r.rubicilla
Caucasus
C.r.diabolica
NE Afghanistan
C.r.kobdensis
Altai, W Mongolia
C.r.severtzovi
Pakistan to W China
***Carpodacus puniceus* (Rose-breasted Rosefinch)**
C.p.kilianensis
SC Asia
C.p.humii
Pakistan, N India, W Himalayas
C.p.puniceus
C Himalayas, SW Sikang, SE Tibet
C.p.sikangensis
Sikang
C.p.longirostris
E Tsinghai, W China
***Carpodacus roborowskii* (Tibet Rosefinch)**
Tsinghai

PINICOLA
***Pinicola enucleator* (Pine Grosbeak)**
P.e.enucleator
Scandinavia, Russia
P.e.pacatus
Siberia, Altai, Manchuria
P.e.kamschatkensis
NE Asia, Kamchatka
P.e.sakhalinensis
Sakhalin, Kurile Is
P.e.alascensis
Alaska, W Canada, NW USA
P.e.flammulus
W Canada >> NW USA
P.e.carlottae
Queen Charlotte, Vancouver Is
P.e.montanus
SW Canada, WC USA
P.e.californicus
E California
P.e.leucurus
C & E Canada >> NE USA
P.e.eschatosus
SE Canada >> NE USA
***Pinicola subhimachalus* (Red-headed Finch)**
Himalayas, S Sikang

HAEMATOSPIZA
***Haematospiza sipahi* (Scarlet Finch)**
Himalayas to N Vietnam

LOXIA
***Loxia pytyopsittacus* (Parrot Crossbill)**
NE Europe, W Siberia
***Loxia scotica* (Scottish Crossbill)** 161.5
Scotland
***Loxia curvirostra* (Red Crossbill)**
L.c.curvirostra
N Europe, N & NE Asia
L.c.corsicana
Corsica
L.c.balearica
Balearic Is
L.c.poliogyna
Algeria, Tunisia
L.c.guillemardi
Cyprus
L.c.mariae
SW Crimea
L.c.altaiensis
Altai
L.c.tianschanica
Sinkiang
L.c.himalayensis
Himalayas, W China >> N Burma
L.c.meridionalis
S Vietnam
L.c.japonica
NE Asia >> EC China, S Japan
L.c.luzoniensis
N Luzon I
L.c.pusilla
Newfoundland >> NE USA

L.c.minor
S Canada, N USA >> E USA
L.c.bendirei
SW Canada, W USA >> S USA
L.c.grinnelli
SW USA
L.c.stricklandi
S USA, Mexico
L.c.mesamericana
Guatemala to N Nicaragua
Loxia leucoptera (White-winged Crossbill)
L.l.bifasciata
E Europe, N Asia, Japan
L.l.leucoptera
Canada, N USA
L.l.megaplaga
Hispaniola

PYRRHULA
Pyrrhula nipalensis (Brown Bullfinch)
P.n.nipalensis
Pakistan, N India
P.n.ricketti
Tibet to N Vietnam
P.n.victoriae
Burma
P.n.waterstradti
Malaysia
P.n.uchidai
Taiwan
Pyrrhula leucogenys (Philippine Bullfinch)
P.l.leucogenys
N Luzon I
P.l.steerei
W Mindanao I
P.l.coriaria
C Mindanao I
P.l.apo
SE Mindanao I
Pyrrhula aurantiaca (Orange Bullfinch)
Pakistan, NW Himalayas
Pyrrhula erythrocephala (Red-headed Bullfinch)
Himalayas, SE Tibet
Pyrrhula erythaca (Beavan's Bullfinch)
P.e.erythaca
Himalayas to W China
P.e.wilderi
NE China
P.e.owstoni
Taiwan
Pyrrhula pyrrhula (Northern Bullfinch)
P.p.pyrrhula
N Europe to W Mongolia >> S Europe, Iran
P.p.pileata
British Isles
P.p.europoea
NW Europe
P.p.iberiae
Azores Is, N Iberia
P.p.murina
San Miguel (Azores)

P.p.rossikowi
Caucasus, W Turkey
P.p.caspica
NE & N Iran
P.p.cineracea
N Altai >> Amur, Manchuria
P.p.cassinii
Kamchatka >> Japan, N China
P.p.griseiventris
Ussuri, Sakhalin I to Korea, S Japan

COCCOTHRAUSTES
Coccothraustes coccothraustes (Hawfinch)
C.c.coccothraustes
N Europe, W Asia >> N Africa
C.c.burryi
NW Africa
C.c.nigricans
Ukraine, N Iran >> S Iran
C.c.humii
C Russia >> NW India
C.c.japonicus
Sakhalin I, Japan >> E China & Bonin I
Coccothraustes migratorius (Black-tailed Hawfinch)
C.m.migratorius
S Ussuri >> N Korea & E China
C.m.sowerbyi
E China
Coccothraustes personatus (Masked Hawfinch)
C.p.personatus
N Japan >> S Japan, E China
C.p.magnirostris
NE Asia
Coccothraustes icterioides (Black & Yellow Grosbeak)
Afghanistan to N India
Coccothraustes affinis (Allied Grosbeak)
Pakistan to W China, N Burma
Coccothraustes melanozanthos (Spotted-wing Grosbeak)
Pakistan to W China, Thailand
Coccothraustes carnipes (White-winged Grosbeak)
C.c.speculigerus
NE Iran to Pakistan
C.c.carnipes
Pakistan to W China
Coccothraustes vespertinus (Evening Grosbeak)
C.v.vespertinus
C & E Canada >> NE USA
C.v.brooksi
W Canada >> SW USA
C.v.montanus
W & SW Mexico
Coccothraustes abeillei (Hooded Grosbeak)
C.a.pallidus
NW Mexico
C.a.saturatus
W Mexico

C.a.abeillei
C & S Mexico
C.a.cobanensis
S Mexico, Guatemala

Pyrrhoplectes epauletta (Gold-headed Finch)
Himalayas, SE Tibet >> N Burma

162. ESTRILDIDAE (WAXBILLS)

PARMOPTILA
Parmoptila woodhousei (Flowerpecker Weaver Finch)
P.w.woodhousei
SE Nigeria, Cameroun, W Zaire
P.w.ansorgei
N Angola
Parmoptila jamesoni (Red-fronted Flowerpecker Weaver Finch)
P.j.rubrifrons
Ghana
P.j.jamesoni
E Zaire, W Uganda

NIGRITA
Nigrita fusconota (White-breasted Negro Finch)
N.f.uropygialis
Guinea to S Nigeria
N.f.fusconota
Fernando Po I, Gabon, Cameroun Mt to Angola, Uganda, Kenya
Nigrita bicolor (Chestnut-breasted Negro Finch)
N.b.bicolor
Guinea to Ghana
N.b.brunnescens
S Nigeria to W Uganda & N Angola
Nigrita luteifrons (Pale-fronted Negro Finch)
N.l.luteifrons
S Nigeria to N Zaire & Gabon
N.l.alexanderi
Fernando Po I
Nigrita canicapilla (Grey-crowned Negro Finch)
N.c.emilae
Guinea to Ghana
N.c.canicapilla
S Nigeria to W Zaire & Uganda
N.c.angolensis
SW Zaire, NW Angola
N.c.sparsimguttata
S Sudan, E Zaire, Uganda, NW Tanzania
N.c.schistacea
SE Sudan, Kenya, N Tanzania
N.c.diabolica
C Kenya
N.c.candida
W Tanzania

NESOCHARIS
Nesocharis shelleyi (Fernando Po Olive-back)
N.s.shelleyi
Fernando Po I, Cameroun Mt
N.s.bansoensis
SE Nigeria, Cameroun
Nesocharis ansorgei (White-collared Olive-back)
E Zaire, W Uganda
Nesocharis capistrata (Grey-headed Olive-back)
Gambia to Sudan, Uganda

PYTILIA
Pytilia phoenicoptera (Crimson-winged Pytilia) (Aurora Finch)
P.p.phoenicoptera
Senegal to Cameroun
P.p.emini
Cameroun to Uganda & S Sudan
P.p.lineata
N Ethiopia
Pytilia hypogrammica (Red-faced Pytilia)
Sierra Leone to Cameroun
Pytilia afra (Orange-winged Pytilia)
Sudan to Angola, Zambia
Pytilia melba (Green-winged Pytilia)
P.m.citerior
Senegal to Sudan
P.m.flavicaudata
Djibouti
P.m.soudanensis
E Sudan, Ethiopia, Kenya
P.m.jessei
E Ethiopia
P.m.percivali
SW Kenya, N Tanzania
P.m.belli
Uganda, E Zaire to Malawi
P.m.grotei
NE Tanzania to Mozambique
P.m.melba
Zaire & Tanzania to Namibia & Transvaal
P.m.hygrophila
N Zambia, N Malawi
P.m.thamnophilus
E Mozambique, E Natal

MANDINGOA
Mandingoa nitidula (Green-backed Twin-spot)
M.n.schlegeli
Sierra Leone to Zaire, Angola
M.n.virginiae
Fernando Po I
M.n.chubbi
S Ethiopia, S Sudan, Kenya, Tanzania, Zanzibar I
M.n.nitidula
Mozambique & Zambia to E Cape Province

CRYPTOSPIZA
Cryptospiza reichenovii (Red-faced Crimson-wing)

C.r.reichenovii
Cameroun to Uganda, N Angola
C.r.australis
S Uganda, Tanzania, Malawi, Zimbabwe, Mozambique
C.r.homogenes
E Zimbabwe
Cryptospiza salvadorii (Ethiopian Crimson-wing)
C.s.salvadorii
S Ethiopia, N Kenya
C.s.ruwenzori
E Zaire, W Uganda
C.s.kilimensis
SE Sudan, Kenya, N Tanzania
Cryptospiza jacksoni (Dusky Crimson-wing)
E Zaire, W Uganda
Cryptospiza shelleyi (Shelley's Crimson-wing)
E Zaire, W Uganda

PYRENESTES
Pyrenestes sanguineus (Crimson Seedcracker)
P.s.sanguineus
Senegal to Ivory Coast
P.s.coccineus
Sierra Leone to Gabon
Pyrenestes ostrinus (Black-bellied Seedcracker)
P.o.ostrinus
Ghana to Togo
P.o.frommi
C Togo to Cameroun, N Zaire
P.o.rothschildi
Ghana to Zaire, Angola
Pyrenestes minor (Lesser Seedcracker)
Tanzania, Malawi, Zimbabwe

SPERMOPHAGA
Spermophaga poliogenys (Grant's Bluebill)
E Zaire, W Uganda
Spermophaga haematina (Western Bluebill)
S.h.haematina
Gambia to Ghana
S.h.togoensis
Togo to SW Nigeria
S.h.pustulata
S Nigeria, Cameroun to N Zaire, N Angola
Spermophaga ruficapilla (Red-headed Bluebill)
S.r.ruficapilla
Angola to E Zaire, Uganda, S Sudan, W Kenya
S.r.kilgoris
SW Kenya
S.r.cana
Tanzania

CLYTOSPIZA
Clytospiza monteiri (Brown Twin-spot)
Cameroun to Sudan, Uganda

HYPARGOS
Hypargos margaritatus (Rosy Twin-spot)
N Natal, Mozambique
Hypargos niveoguttatus (Peter's Twin-spot)
H.n.macrospilotus
E Zaire, Kenya, Tanzania, Malawi
H.n.idius
Zambia
H.n.interior
Zimababwe
H.n.niveoguttatus
Mozambique
H.n.baddeleyi
Nacola (Mozambique)

EUSCHISTOSPIZA
Euschistospiza dybowskii (Dybowski's Dusky Twin-spot)
Sierra Leone to Sudan
Euschistospiza cinereovinacea (Dusky Twin-spot)
E.c.cinereovinacea
W Angola
E.c.graueri
E Zaire, W Tanzania

LAGONOSTICTA
Lagonosticta rara (Black-bellied Fire Finch)
L.r.forbesi
Sierra Leone to Nigeria
L.r.rara
N Cameroun to S Sudan, Uganda, Kenya
Lagonosticta rufopicta (Bar-breasted Fire Finch)
L.r.rufopicta
Senegal to N Cameroun, Central African Republic
L.r.lateritia
Sudan, NE Zaire, W Uganda
Lagonosticta nitidula (Brown Fire Finch)
L.n.nitidula
E Angola, S Zaire, N Zambia
L.n.plumbaria
S Zambia, Botswana
Lagonosticta senegala (Red-billed Fire Finch)
L.s.senegala
Senegal to Nigeria
L.s.guineensis
coast of Guinea & Sierra Leone
L.s.rhodopsis
Chad to SW Sudan
L.s.brunneiceps
Ethiopia
L.s.somaliensis
Somalia, Kenya, Tanzania
L.s.kikuyuensis
C Kenya, N Tanzania
L.s.ruberrima
Uganda, SE Zaire, Zambia, W Tanzania
L.s.rendalli
SE Zaire, S Tanzania, Sth Africa
L.s.confidens
Transvaal, Natal & Mozambique

L.s.pallidicrissa
  S Angola, N Namibia
Lagonosticta virata (Kuli Koro Fire Finch)
  Mali
Lagonosticta rubricata (African Fire Finch)
L.r.polionota
  Guinea to Nigeria
L.r.ugandae
  Cameroun to N Tanzania
L.r.congica
  Gabon to S Zaire, NW Zambia
L.r.haematocephala
  S Tanzania, Mozambique
L.r.rubricata
  S Mozambique to Cape Province
Lagonosticta landanae (Pale-billed Fire Finch)
  Cabinda, W Angola
Lagonosticta rhodopareia (Jameson's Fire Finch)
L.r.bruneli
  S Chad
L.r.rhodopareia
  Ethiopia, N Kenya
L.r.jamesoni
  S Kenya to Transvaal, Natal
L.r.taruensis
  N Zimbabwe, Mozambique to SE Kenya
L.r.ansorgei
  Cabinda, W Angola
Lagonosticta larvata (Masked Fire Finch)
  W Ethiopia, Sudan
Lagonosticta vinacea (Vinaceous Fire Finch)
L.v.vinacea
  Senegal to Guinea
L.v.togoensis
  Ghana, Togo to N Cameroun, W Sudan
L.v.nigricollis
  Central African Republic to Sudan & Uganda

URAEGINTHUS
Uraeginthus angolensis (Cordon-bleu)
U.a.angolensis
  SW Zaire, N Angola, NW Zambia
U.a.cyanopleurus
  Zimbabwe, N Botswana, W Transvaal
U.a.niassensis
  E Tanzania, SE Zaire to Zimbabwe, Transvaal
U.a.natalensis
  S Zimbabwe, Transvaal, Natal
Uraeginthus bengalus (Red-cheeked Cordon-bleu)
U.b.bengalus
  W, NC & E Africa
U.b.brunneigularis
  Kenya
U.b.littoralis
  E Kenya, Tanzania
U.b.ugogoensis
  N & W Tanzania

U.b.katangae
  S Zaire, Zambia
Uraeginthus cyanocephala (Blue-capped Cordon-bleu)
  Ethiopia to Tanzania
Uraeginthus granatina (Common Grenadier)
U.g.granatina
  S Angola to Natal
U.g.siccata
  W Angola to N Cape Province
U.g.retusa
  Mozambique
Uraeginthus ianthinogaster (Purple Grenadier)
U.i.ianthinogaster
  Somalia, N Kenya, N Uganda
U.i.hawkeri
  SE Sudan, Uganda, N Kenya
U.i.roosevelti
  Kenya
U.i.rothschildi
  Kenya

ESTRILDA
Estrilda caerulescens (Lavender Waxbill)
  Senegal to Central African Republic
Estrilda perreini (Black-tailed Waxbill)
E.p.perreini
  Gabon to N Angola & Tanzania
E.p.poliogastra
  S Tanzania to Zimbabwe, Mozambique
E.p.torrida
  Sufala (Mozambique)
E.p.incana
  Natal
Estrilda thomensis (Cinderella Waxbill)
  Sao Thome I
Estrilda melanotis (Swee Waxbill)
E.m.kilimensis
  E Zaire, Uganda to Zambia
E.m.bocagei
  W Angola
E.m.melanotis
  South Africa
Estrilda quartinia (East African Swee Waxbill)
162.1
E.q.quartinia
  Ethiopia to S & E Zambia
E.q.stuartirwini
  S Mozambique
Estrilda poliopareia (Anambra Waxbill)
  S Nigeria
Estrilda paludicola (Fawn-breasted Waxbill)
E.p.paludicola
  N Zaire, N Uganda, S Sudan
E.p.ochrogaster
  Ethiopia, SE Sudan
E.p.roseicrissa
  S Uganda, NW Tanzania
E.p.marwitzi
  W Tanzania
E.p.benguellensis
  Angola, N Zambia

E.p.ruthae
C Zaire
Estrilda melpoda (Orange-cheeked Waxbill)
E.m.melpoda
Gambia to N Zaire, N Angola, Zambia
E.m.tschadensis
Cameroun, Chad
Estrilda rhodopyga (Crimson-rumped Waxbill)
E.r.rhodopyga
Sudan, Ethiopia, N Somalia
E.r.centralis
S Ethiopia, SE Sudan, Uganda to Tanzania, Malawi
Estrilda rufibarba (Arabian Waxbill)
SW Arabia
Estrilda troglodytes (Black-rumped Waxbill)
Senegal to Ethiopia
Estrilda astrild (Common Waxbill)
E.a.kempi
Sierra Leone, Liberia
E.a.occidentalis
Fernando Po I, Cameroun to N Zaire
E.a.sousae
Sao Thome I
E.a.peasei
Ethiopia
E.a.macmillani
Sudan
E.a.adesma
Uganda, NW Tanzania
E.a.massaica
Kenya, N Tanzania
E.a.minor
E Kenya, NE Tanzania, Zanzibar
E.a.cavendishi
S Tanzania to Zambia, Transvaal, Mozambique
E.a.schoutedeni
S Zaire
E.a.ngamiensis
E Angola, Zambia, Zimbabwe
E.a.angolensis
S Zaire, W Angola
E.a.jagoensis
W Angola, Cape Verde Is
E.a.rubriventris
Gabon
E.a.damarensis
Namibia
E.a.astrild
S Botswana, W Transvaal, Orange Free State, W Cape Province
E.a.tenebridorsa
E Cape Province, SE Transvaal, Natal
Estrilda nigriloris (Black-faced Waxbill)
Zaire
Estrilda nonnula (Black-crowned Waxbill)
E.n.elizae
Fernando Po I
E.n.eisentrauti
Cameroun Mt

E.n.nonnula
E Cameroun to Sudan, Kenya, Tanzania
Estrilda atricapilla (Black-headed Waxbill)
E.a.atricapilla
S Cameroun to NE Zaire
E.a.marungensis
Marungu
E.a.avakubi
E Zaire, NE Angola
E.a.kandti
SE Zaire, Uganda, Kenya
Estrilda erythronotos (Black-cheeked Waxbill)
E.e.delamerei
Uganda, Kenya, Tanzania
E.e.soligena
Angola, Namibia to N Transvaal, N Zimbabwe
E.e.erythronotos
S Zimbabwe, Transvaal, N Cape Province
Estrilda charmosyna (Red-rumped Waxbill)
E.c.charmosyna
Somalia, S Ethiopia, S Sudan, Uganda, N Kenya
E.c.pallidior
C Kenya
E.c.kiwanukae
S Kenya, N Tanzania

AMANDAVA
Amandava amandava (Red Munia)
A.a.amandava
Pakistan, India
A.a.flavidiventris
SW China, Burma, Lesser Sunda Is
A.a.punicea
Indochina, Java, Bali
Amandava formosa (Green Munia)
C.India
Amandava subflava (Zebra Waxbill)
A.s.subflava
Senegal to Ethiopia, Uganda
A.s.niethammeri
S Zaire, Malawi to N Namibia
A.s.clarkei
Angola to Mozambique, South Africa

ORTYGOSPIZA
Ortygospiza atricollis (African Quailfinch)
O.a.atricollis
Senegal to Chad, N Zaire
O.a.ansorgei
Guinea to Ivory Coast
O.a.ugandae
S Sudan, Uganda, W Kenya
O.a.fuscocrissa
Ethiopia
O.a.muelleri
S Kenya, Tanzania, Malawi
O.a.smithersi
NE Zambia
O.a.pallida
Botswana, Zimbabwe

O.a.bradfieldi
N Namibia
O.a.digressa
S, E & N Cape Province to S Mozambique
Ortygospiza gabonensis (Red-billed Quailfinch)
O.g.gabonensis
Gabon to C Zaire
O.g.fuscata
Angola, S Zaire, Zambia
O.g.dorsostriata
E Zaire, Uganda
Ortygospiza locustella (Locust Finch)
O.l.uelensis
N Zaire
O.l.locustella
S Zaire, Zambia, Malawi, Mozambique, Zimbabwe
O.l.rendalli
Southern Africa

AEGINTHA
Aegintha temporalis (Red-browed Waxbill)
A.t.loftyi
S Australia
A.t.temporalis
Eastern Australia
A.t.minor
N Queensland

EMBLEMA
Emblema picta (Painted Finch)
C Australia
Emblema bella (Beautiful Firetail Finch)
SE Australia, Tasmania
Emblema oculata (Red-eared Firetail Finch)
SW Western Australia
Emblema guttata (Diamond Firetail Finch)
WC & SC Australia

OREOSTRUTHUS
Oreostruthus fuliginosus (Crimson-sided Mountain Finch)
O.f.fuliginosus
SE New Guinea
O.f.pallidus
W New Guinea
O.f.hagenensis
C New Guinea

NEOCHMIA
Neochmia phaeton (Crimson Finch)
N.p.evangelinae
S New Guinea
N.p.albiventer
N Queensland
N.p.phaeton
N Western Australia, Northern Territory
Neochmia ruficauda (Star Finch)
N.r.ruficauda
C Queensland

N.r.clarescens
N Queensland, Northern Territory, N Western Australia

POEPHILA
Poephila guttata (Spotted-sided Finch)(Zebra Finch)
P.g.guttata
Lesser Sunda Is
P.g.castanotis
Australia
Poephila bichenovii (Double-barred Finch)
P.b.annulosa
Northern Territory, N Western Australia
P.b.bichenovii
E Northern Territory, Queensland, N New South Wales
Poephila personata (Masked Finch)
P.p.personata
Northern Territory, W Queensland
P.p.leucotis
N Queensland
Poephila acuticauda (Long-tailed Finch)
P.a.acuticauda
Northern Australia
P.a.hecki
N Western Australia
Poephila cincta (Black-throated Finch)
P.c.nigrotecta
N Queensland
P.c.atropygialis
C Queensland
P.c.cincta
S Queensland, N New South Wales

ERYTHRURA
Erythrura hyperythra (Bamboo Parrot Finch)
E.h.brunneiventris
N Luzon, Mindoro Is
E.h.borneensis
Borneo
E.h.malayana
N Malaysia
E.h.hyperythra
W Java
E.h.intermedia
Lombok, Lesser Sunda Is
E.h.microrhyncha
Sulawesi
Erythrura prasina (Pin-tailed Parrot Finch)
E.p.prasina
S Burma, S Thailand, Malaysia, Java, Sumatra
E.p.coelica
Borneo
Erythrura viridifacies (Green-faced Parrot Finch)
Luzon I
Erythrura tricolor (Three-coloured Parrot Finch)
Timor, Wetar, Tanimbar, Damar Is

Erythrura coloria (Mount Katanglad Parrot Finch)
Mindanao I

Erythrura trichroa (Blue-faced Parrot Finch)
E.t.sanfordi
SC Sulawesi
E.t.modesta
N Moluccas
E.t.pinaiae
S Moluccas
E.t.sigillifera
NE Australia, New Guinea, New Britain, New Ireland
E.t.eichhorni
St Matthias Is, Bismarck Arch
E.t.pelewensis
Palau Is
E.t.clara
Truk, Ponapé Is
E.t.trichroa
Kusaie, Caroline Is
E.t.woodfordi
Guadalcanal I
E.t.cyanofrons
New Hebrides, Loyalty Is

Erythrura papuana (Papuan Parrot Finch)
New Guinea

Erythrura psittacea (Red-throated Parrot Finch)
New Caledonia

Erythrura pealii (Fiji Parrot Finch)
Fiji Is

Erythrura cyaneovirens (Red-headed Parrot Finch)
E.c.cyaneovirens
Samoa Is
E.c.regia
N New Hebrides
E.c.serena
S New Hebrides
E.c.efatensis
Efate I
E.c.gaughrani
Savaii I

Erythrura kleinschmidti (Pink-billed Parrot Finch)
Viti Levu I

CHLOEBIA
Chloebia gouldiae (Gouldian Finch)
Northern Australia

AIDEMOSYNE
Aidemosyne modesta (Plum-headed Finch)
W Queensland, W New South Wales

LEPIDOPYGIA
Lepidopygia nana (Bib-Finch)
Madagascar

LONCHURA
Lonchura cantans (African Silverbill)
L.c.cantans
Senegal to W Sudan

L.c.orientalis
Somalia, Ethiopia to Tanzania, S Yemen

Lonchura malabarica (Indian Silverbill)
Sauda Arabia to N India, Sri Lanka

Lonchura griseicapilla (Grey-headed Silverbill)
S Ethiopia to Tanzania

Lonchura cucullata (Bronze Mannikin)
L.c.cucullata
Senegal to Sudan, Uganda
L.c.tessellata
S Zaire to E Cape Province
L.c.scutata
Ethiopia to Angola & Cape Province

Lonchura bicolor (Black & White Mannikin)
L.b.bicolor
Guinea to Cameroun
L.b.poensis
Cameroun to Angola, Ethiopia, Kenya
L.b.stigmatophora
SW Ethiopia, Uganda
L.b.nigriceps
East Africa to Natal
L.b.minor
S Somalia
L.b.woltersi
SE Zaire, NW Zambia

Lonchura fringilloides (Magpie Mannikin)
L.f.fringilloides
Senegal to Somalia & Natal
L.f.pica
Mozambique

Lonchura striata (White-backed Munia)
L.s.acuticauda
N India, Bangladesh, Nepal, Burma
L.s.striata
S India, Sri Lanka
L.s.fumigata
Andaman Is
L.s.semistriata
Nicobar Is
L.s.subsquamicollis
S Thailand, Malaysia, Sumatra, Indochina
L.s.swinhoei
S China, Taiwan

Lonchura leucogastroides (Javanese Mannikin)
S Sumatra, Java, Bali, Lombok I

Lonchura fuscans (Dusky Mannikin)
Borneo

Lonchura molucca (Moluccan Mannikin)
Sulawesi, Lesser Sunda Is

Lonchura punctulata (Nutmeg Mannikin)
L.p.punctulata
India, S Nepal
L.p.subundulata
Bhutan, Bangladesh, Assam, W Burma
L.p.yunnanensis
SW China, NE Burma
L.p.topela
S China, Thailand, Indochina
L.p.cabanisi
Luzon, Mindoro Is

L.p.fretensis
Malaysia, Sumatra
L.p.nisoria
Java, Bali, Lombok, Sumbawa Is
L.p.sumbae
Sumba Is
L.p.blasii
Flores, Timor, Lesser Sunda Is
L.p.particeps
Sulawesi
Lonchura kelaarti (Rufous-bellied Mannikin)
L.k.vernayi
E India
L.k.jerdoni
SW India
L.k.kelaarti
Sri Lanka
Lonchura leucogastra (White-headed Munia)
L.l.leucogastra
S Thailand, Malaysia, Sumatra
L.l.everetti
Luzon, Mindoro Is
L.l.manueli
C & S Philippine Is
L.l.palawana
Palawan I, N & E Borneo
L.l.smythiesi
SW Sarawak
L.l.castanonota
S Borneo
Lonchura tristissima (Streak-headed Mannikin)
L.t.tristissima
NW New Guinea
L.t.hypomelaena
C New Guinea
L.t.calaminoros
S New Guinea
Lonchura leucosticta (White-spotted Mannikin)
S New Guinea
Lonchura quinticolor (Coloured Finch)
Timor, Lesser Sunda Is
Lonchura malacca (Chestnut Mannikin)
L.m.rubroniger
N India, E Nepal
L.m.malacca
S India, Sri Lanka
L.m.atricapilla
NE India, Bangladesh, Assam, Burma
L.m.deignani
N Thailand, Indochina
L.m.sinensis
S Thailand, Malaysia, Sumatra
L.m.bakatana
N Sumatra
L.m.formosana
N Luzon I, Taiwan
L.m.jagori
Philippine, Palawan Is, Borneo, Sulawesi
L.m.ferruginosa
Java

Lonchura maja (Pale-headed Mannikin)
S Thailand, Malaysia, Sumatra, Java, Bali
Lonchura pallida (Pallid Finch)
W Sulawesi, Lombok, Lesser Sunda Is
Lonchura grandis (Great-billed Mannikin)
L.g.grandis
SE New Guinea
L.g.ernesti
N New Guinea
L.g.destructa
N New Guinea
L.g.heurni
N New Guinea
Lonchura vana (Arfak Mannikin)
NW New Guinea
Lonchura caniceps (Grey-headed Mannikin)
L.c.caniceps
SE New Guinea
L.c.scratchleyana
SE New Guinea
L.c.kumusii
SE New Guinea
Lonchura nevermanni (White-crowned Mannikin)
S New Guinea
Lonchura spectabilis (New Britain Mannikin)
L.s.wahgiensis
E New Guinea
L.s.gajduseki
C New Guinea
L.s.mayri
N New Guinea
L.s.spectabilis
New Britain
Lonchura forbesi (New Ireland Finch)
New Ireland
Lonchura hunsteini (White-headed Finch)
L.h.hunsteini
N New Ireland
L.h.nigerrima
New Hanover
L.h.minor
Ponapé I
Lonchura flaviprymna (Yellow-tailed Mannikin)
N Australia
Lonchura castaneothorax (Chestnut-breasted Mannikin)
L.c.uropygialis
NW New Guinea
L.c.sharpii
N New Guinea
L.c.boschmai
C New Guinea
L.c.ramsayi
SE New Guinea
L.c.assimilis
Northern Territory
L.c.castaneothorax
E Queensland, E New South Wales
Lonchura stygia (Black Mannikin)
S New Guinea

Lonchura teerinki (Grand Valley Mannikin)
L.t.teerinki
  NC New Guinea
L.t.mariae
  W New Guinea
Lonchura monticola (Alpine Mannikin)
  SE New Guinea
Lonchura montana (Snow Mountain Mannikin)
  C New Guinea
Lonchura melaena (New Britain Finch)
  New Britain
Lonchura pectoralis (Pictorella Finch)
  Northern Australia

PADDA
Padda fuscata (Timor Dusky Sparrow)
  Timor I
Padda oryzivora (Java Sparrow)
  Java to Sumbawa I, Sulawesi

AMADINA
Amadina erythrocephala (Paradise Sparrow)
A.e.erythrocephala
  Angola, Sthn Africa
A.e.dissita
  E Cape Province, S Natal
Amadina fasciata (Cut-throat Weaver)
A.f.fasciata
  Senegal & N Nigeria to Sudan, Uganda
A.f.alexanderi
  Ethiopia, Somalia, Kenya, Tanzania
A.f.meridionalis
  Malawi, Zambia, Zimbabwe, Transvaal, Mozambique
A.f.contigua
  Transvaal, N Natal to S Mozambique

PHOLIDORNIS
Pholidornis rushiae (Tit-Hylia)
P.r.ussheri
  Sierra Leone to Ghana
P.r.rushiae
  S Nigeria to Angola
P.r.bedfordi
  Fernando Po I
P.r.denti
  E Zaire, Uganda

163. PLOCEIDAE (WEAVERS, SPARROWS)

VIDUINAE

VIDUA
Vidua chalybeata (Village Indigobird)
V.c.chalybeata
  Senegal to Sierra Leone
V.c.neumanni
  Mali to Sudan

V.c.ultramarina
  Ethiopia
V.c.centralis
  Kenya, Uganda, W Tanzania
V.c.amauropteryx
  Somalia to Zambia, Mozambique
V.c.okavangoensis
  W Zambia, Botswana, Angola
Vidua purpurascens (Dusky Indigobird)
  Kenya to Angola, Transvaal
Vidua funerea (Variable Indigobird)
V.f.nigerrima
  Angola, Zambia, Tanzania
V.f.codringtoni
  S Zambia, Malawi, W Zimbabwe
V.f.lusituensis
  E Zimbabwe
V.f.funerea
  C & E Sth Africa
Vidua wilsoni (Pale-winged Indigobird)
"V.f.wilsoni"
  N Nigeria to W Sudan
"V.f. camerunensis"
  Gambia to Ethiopia
"V.f.nigeriae"
  S Nigeria, Cameroun, S Sudan
  (These constitute groups rather than true sub-species)
Vidua hypocherina (Steel-blue Whydah)
  Ethiopia & Somalia to Tanzania
Vidua fischeri (Fischer's Whydah)
  Somalia to Uganda & N Tanzania
Vidua regia (Shaft-tailed Whydah)
  S Angola to S Mozambique
Vidua macroura (Pin-tailed Whydah)
  Senegal to Ethiopia & Cape Province
Vidua paradisaea (Paradise Whydah)
  E Sudan to S Angola & Natal
Vidua orientalis (Broad-tailed Paradise Whydah)
V.o.acupum
  Senegal to N Nigeria
V.o.togoensis
  Sierra Leone to Togo
V.o.orientalis
  Chad to Ethiopia
V.o.interjecta
  N Cameroun to S Sudan
V.o.obtusa
  Angola to Kenya & Mozambique

BUBALORNITHINAE

BUBALORNIS
Bubalornis albirostris (White-billed Buffalo Weaver)
B.a.albirostris
  Senegal to Ethiopia, N Uganda, Kenya
B.a.intermedius
  S Ethiopia, Somalia, Kenya, S Tanzania
Bubalornis niger (Red-billed Buffalo Weaver)
B.n.niger
  S Angola to Mozambique, Transvaal

B.n.militaris
  E Transvaal, N Natal, S Zambia, S
  Mozambique

DINEMELLIA
Dinemellia dinemelli (White-headed Buffalo Weaver)
  D.d.dinemelli
    S Sudan, S Ethiopia, Somalia, N Kenya
  D.d.boehmi
    SE Zaire, Tanzania

PASSERINAE

PLOCEPASSER
Plocepasser mahali (White-browed Sparrow Weaver)
  P.m.melanorhynchus
    S Sudan, S Ethiopia, Uganda, Kenya
  P.m.propinquatus
    S Somalia
  P.m.pectoralis
    Zambia, N Botswana to S Tanzania,
    Mozambique
  P.m.ansorgei
    S Angola, N Namibia
  P.m.terricolor
    N Namibia to NW Transvaal
  P.m.stentor
    Namibia, W Cape Province, S Botswana,
    Transvaal
  P.m.mahali
    W Orange Free State, N Cape Province
Plocepasser superciliosus (Chestnut-crowned Sparrow Weaver)
  P.s.superciliosus
    Senegal to Sudan
  P.s.brunnescens
    Central African Republic to SW Sudan,
    Uganda
Plocepasser donaldsoni (Donaldson-Smith's Sparrow Weaver)
  N Kenya
Plocepasser rufoscapulatus (Chestnut-mantled Sparrow Weaver)
  S Angola, SE Zaire to Malawi

HISTURGOPS
Histurgops ruficauda (Rufous-tailed Weaver)
  Tanzania

PSEUDONIGRITA
Pseudonigrita arnaudi (Grey-headed Social Weaver)
  P.a.arnaudi
    SW Sudan, Kenya, Uganda, N Tanzania
  P.a.australoabyssinicus
    S Ethiopia
  P.a.dorsalis
    C Tanzania
Pseudonigrita cabanisi (Black-capped Social Weaver)
  S Ethiopia, E Kenya, NE Tanzania

PHILETAIRUS
Philetairus socius (Sociable Weaver)
  P.s.geminus
    N & C Namibia
  P.s.xericus
    W Namibia
  P.s.socius
    S Namibia
  P.s.lepidus
    S Botswana, W Transvaal, N Cape Province
  P.s.eremnus
    N Cape Province

PASSER
Passer ammodendri (Saxaul Sparrow)
  P.a.korejewi
    Transcaspia, Iran
  P.a.ammodendri
    Russian Turkestan
  P.a.stoliczkae
    W China
  P.a.timidus
    S Mongolia
Passer domesticus (House Sparrow)
  P.d.domesticus
    Europe, N Asia, Americas, Sth Africa,
    Australia
  P.d.italiae
    SE France, Italy, Crete
  P.d.tingitanus
    NW Africa
  P.d.biblicus
    Asia Minor, S Arabia, Caucasus, Iran
  P.d.hufufae
    E Arabia
  P.d.niloticus
    NE Africa
  P.d.rufidorsalis
    Sudan
  P.d.indicus
    S Afghanistan, Pakistan, India, Bangladesh,
    Burma
  P.d.hyrcanus
    Transcaspia, N Iran
  P.d.bactrianus
    SC Asia
  P.d.parkini
    Himalayas
Passer hispaniolensis (Spanish Sparrow)
  P.h.hispaniolensis
    SW Europe, Nth Africa, Asia Minor
  P.h.transcaspicus
    Caucasus, Tien Shan >> S Iran, N India
Passer pyrrhonotus (Sind Jungle Sparrow)
  SE Iran, Pakistan, NW India
Passer castanopterus (Somali Sparrow)
  P.c.fulgens
    Ethiopia, N Kenya
  P.c.castanopterus
    Somalia
Passer rutilans (Cinnamon Sparrow)
  P.r.cinnamomeus
    NE Afghanistan, Himalayas, SE Tibet

P.r.intensior
Assam, N Burma, Laos, S China, N Vietnam
P.r.rutilans
China, Taiwan, Korea, N Japan
Passer flaveolus (Pegu House Sparrow)
N Burma, Thailand, Laos, S Vietnam
Passer moabiticus (Dead Sea Sparrow)
P.m.moabiticus
Jordan, Iraq, SW Iran
P.m.yatii
E Iran, W Afghanistan
Passer motitensis (Great Sparrow)
P.m.iagoensis
Cape Verde Is
P.m.cordofanicus
NW Sudan
P.m.shelleyi
S Sudan, N Uganda
P.m.hemileucus
Abd el Kuri I
P.m.subsolanus
S Zimbabwe to N Orange Free State
P.m.motitensis
Botswana, Transvaal
P.m.benguellensis
SW Africa, S Angola
Passer melanurus (Cape Sparrow)
P.m.damarensis
SW Angola, Namibia, Botswana
P.m.vicinus
N Natal, Lesotho, E Orange Free State
P.m.melanurus
Sth Africa
Passer insularis (Socotra Sparrow)
Socotra I
Passer rufocinctus (Kenya Rufous Sparrow)
Kenya, N Tanzania
Passer griseus (Grey-headed Sparrow)
P.g.griseus
Niger, Chad, Senegal to Ghana
P.g.ugandae
Ghana to Somalia, N Zaire
P.g.laeneni
E Chad
P.g.luangwae
Zambia
P.g.mosambicus
E Tanzania, Malawi, Mozambique
P.g.diffusus
Angola, N Namibia, Botswana to Natal
P.g.stygiceps
S Natal, E Cape Province
Passer swainsonii (Swainson's Sparrow)
Somalia, E & S Ethiopia
Passer gongonensis (Parrot-billed Sparrow)
Kenya, SE Tanzania
Passer suahelicus (Swahili Sparrow)
Kenya, C Tanzania
Passer simplex (Desert Sparrow)
P.s.zarudnyi
E Iran
P.s.simplex
S Sahara

P.s.saharae
W Sahara
Passer montanus (Eurasian Tree Sparrow)
P.m.montanus
Europe, W, N & NE Asia, Asia Minor
P.m.transcaucasicus
Transcaucasia, N Iran
P.m.zaissanensis
C Asia, NE Mongolia
P.m.kansuensus
NE Tsinghai, Kansu
P.m.iubilaeus
N, C & E China
P.m.dilutus
NE Iran, Pakistana, Sinkiang, Manchuria, W China
P.m.tibetanus
N Himalayas, Tibet, NW China
P.m.saturatus
Sakhalin I, S Korea, Japan, Taiwan
P.m.hepaticus
NE Assam, NW Burma
P.m.malaccensis
S Himalayas, Burma, Thailand, Indochina, Malaysia, Sumatra, Java

AURIPASSER
Auripasser luteus (Sudan Golden Sparrow)
N Nigeria, Chad, Sudan, N Ethiopia
Auripasser euchlorus (Arabian Golden Sparrow)
SW Arabia, Somalia

SORELLA
Sorella eminibey (Chestnut Sparrow)
Sudan to N Tanzania

PETRONIA
Petronia brachydactyla (Pale Rock Sparrow)
Syria, Iran >> NE Africa
Petronia xanthosterna (Yellow-spotted Rock Sparrow)
P.x.pallida
Senegal, Mauretania to S Sudan
P.x.pyrgita
Ethiopia, Somalia to NE Tanzania
P.x.transfuga
S Iraq, Iran, Afghanistan, Pakistan, NW India
P.x.xanthosterna
India
Petronia petronia (Streaked Rock Sparrow)
P.p.petronia
S Europe, Morocco, W Asia Minor
P.p.barbara
N Africa
P.p.puteicola
S Syria, Israel
P.p.exigua
Caucasus, Iraq, Iran
P.p.intermedia
Transcaspia, N Iran, C Asia, Pakistan

P.p.brevirostris
E Siberia, Mongolia, N China
Petronia superciliaris (South African Rock Sparrow)
P.s.rufitergum
SW Tanzania to Angola
P.s.boroensis
E Tanzania to E Transvaal & N Natal
P.s.flavigula
S Zambia, Zimbabwe, S Mozambique, N Transvaal
P.s.superciliaris
E Cape Province to Natal
Petronia dentata (Lesser Rock Sparrow)
P.d.dentata
Senegal to Ethiopia, SW Arabia
P.d.buchanani
S Niger

MONTIFRINGILLA
Montifringilla nivalis (White-winged Snow Finch)
M.n.nivalis
SW Europe
M.n.alpicola
Transcaucasus, Iran C Asia
M.n.kwenlunensis
S Sinkiang, W China
M.n.henrici
Tibet, W China
Montifringilla adamsi (Adams' Snow Finch)
M.a.xerophila
NW China
M.a.adamsi
Tibet, N Himalayas, Nepal
Montifringilla taczanowskii (Mandelli's Snow Finch)
Tibet, N Sikang, Tsinghai
Montifringilla davidiana (Père David's Snow Finch)
M.d.potanini
Altai, N Mongolia
M.d.davidiana
S Mongolia, NW China
Montifringilla ruficollis (Red-necked Snow Finch)
M.r.isabellina
NW China, N Tsinghai
M.r.ruficollis
Tibet, W China
Montifringilla blanfordi (Blanford's Snow Finch)
M.b.barbara
N China
M.b.ventorum
NW China, Sinkiang
M.b.blanfordi
N Himalayas, Tibet, W China
Montifringilla theresae (Meinertzhagen's Snow Finch)
Afghanistan

SPOROPIPES
Sporopipes squamifrons (Scaly Weaver)
S.s.pallidus
Mossamedes (Angola)
S.s.fuligescens
NE Cape Province, Zimbabwe, N Botswana
S.s.squamifrons
SW Angola to Transvaal, Cape Province
Sporopipes frontalis (Speckle-fronted Weaver)
S.f.frontalis
Senegal to E Ethiopia
S.f.pallidior
S Sahara
S.f.emini
S Sudan, NE Uganda, Kenya, N Tanzania

PLOCEINAE

AMBLYOSPIZA
Amblyospiza albifrons (Grosbeak Weaver)
A.a.capitalba
Sierra Leone to Nigeria
A.a.saturata
Cameroun to N Zaire
A.a.melanota
NE Zaire, Ethiopia, Uganda, NW Kenya
A.a.montana
S Kenya, Zambia, Tanzania, Zimbabwe, Malawi
A.a.unicolor
E Kenya, E Tanzania
A.a.tandae
N Angola
A.a.kasaica
S Zaire
A.a.maxima
N Botswana
A.a.woltersi
S Mozambique
A.a.albifrons
Sth Africa

PLOCEUS
Ploceus baglafecht (Baglafecht Weaver)
P.b.baglafecht
Ethiopia, S Sudan
P.b.reichenowi
Kenya, N Tanzania
P.b.stuhlmanni
E Zaire, S Uganda, W Tanzania
P.h.sharpii
SW Tanzania
P.b.nyikae
Zambia, Malawi
P.b.neumanni
Cameroun
P.b.eremobius
NE Zaire, SE Sudan
P.b.emini
S Sudan, N Uganda
Ploceus bannermani (Bannerman's Weaver)
Cameroun

Ploceus batesi (Bates's Weaver)
  Cameroun
Ploceus nigrimentum (Black-chinned Weaver)
  W Angola, S Zaire
Ploceus bertrandi (Bertrand's Weaver)
  Tanzania, Malawi, Mozambique
Ploceus pelzelni (Slender-billed Weaver)
  P.p.pelzelni
    Uganda, Kenya, Tanzania, E Zaire
  P.p.tuta
    SE Zaire
  P.p.monachus
    Ghana to Gabon & N Angola
Ploceus subpersonatus (Loanga Slender-billed Weaver)
  S Gabon
Ploceus luteolus (Little Masked Weaver)
  P.l.luteolus
    Senegal to Ethiopia
  P.l.kavirondensis
    Uganda, W Kenya, NW Tanzania
Ploceus ocularis (Spectacled Weaver)
  P.o.crocatus
    Cameroun to Ethiopia >> Tanzania & Angola
  P.o.suahelicus
    E Kenya, E Tanzania, E Zambia, Mozambique
  P.o.brevior
    N Transkei to E Transvaal
  P.o.tenuirostris
    N Botswana, N Namibia
  P.o.ocularis
    Transvaal, Natal, Cape Province
Ploceus nigricollis (Black-necked Weaver)
  P.n.brachypterus
    Senegal to Nigeria
  P.n.nigricollis
    Cameroun to Sudan, Zaire, Angola, W Kenya
  P.n.po
    Fernando Po I
  P.n.melanoxanthus
    S Ethiopia, Somalia, E Kenya, NE Tanzania
Ploceus alienus (Strange Weaver)
  E Zaire, W Uganda
Ploceus melanogaster (Black-billed Weaver)
  P.m.melanogaster
    E Nigeria, W Cameroun, Fernando Po I
  P.m.stephanophorus
    S Sudan, E Zaire, SW Uganda, NW Kenya
Ploceus capensis (Cape Weaver)
  P.c.olivaceus
    E Cape Province, Transvaal, Natal
  P.c.rubricomus
    E Transvaal, W Natal
  P.c.capensis
    W Cape Province
Ploceus temporalis (Bocage's Weaver)
  S Angola, W Zambia
Ploceus subaureus (Golden Weaver)
  P.s.aureoflavus
    E Kenya, E Tanzania, Malawi, Mozambique

P.s.tongensis
    S Mozambique
P.s.subaureus
    Natal, E Cape Province
Ploceus burnieri (Kilombero Weaver)   163.5
    EC Tanzania
Ploceus xanthops (Holub's Golden Weaver)
    Zaire & Angola to Kenya & Mozambique
Ploceus aurantius (Orange Weaver)
  P.a.aurantius
    Senegal to Cameroun, Gabon, Zaire
  P.a.rex
    Uganda, NW Tanzania
Ploceus heuglini (Heuglin's Masked Weaver)
    Senegal to NW Kenya
Ploceus bojeri (Golden Palm Weaver)
    S Somalia, Kenya
Ploceus castaneiceps (Taveta Golden Weaver)
    SE Kenya, NE Tanzania
Ploceus princeps (Principe Golden Weaver)
    Principé I
Ploceus xanthopterus (Brown-throated Golden Weaver)
  P.x.castaneigula
    N Botswana, SW Zambia
  P.x.marleyi
    Natal
  P.x.xanthopterus
    Transvaal, Natal, S Zimbabwe, Malawi, Mozambique
Ploceus castanops (Northern Brown-throated Weaver)
    E Zaire, Uganda
Ploceus galbula (Ruppell's Weaver)
    E Sudan, N Ethiopia, SW Arabia
Ploceus victoriae (Lake Victoria Weaver)
      163.1
    Entebbe (Uganda)
Ploceus taeniopterus (Northern Masked Weaver)
  P.t.furensis
    W Sudan
  P.t.taeniopterus
    SE Sudan, N Uganda
Ploceus intermedius (Lesser Masked Weaver)
  P.i.intermedius
    Sudan, Ethiopia, Somalia to E Zaire, Tanzania
  P.i.cabanisii
    SE Zaire to Botswana & Transvaal
  P.i.luebberti
    Namibia, S Angola
  P.i.beattyi
    W Angola
Ploceus velatus (African Masked Weaver)
  P.v.uluensis
    Sudan & Somalia to Tanzania
  P.v.velatus
    N Cape Province, SW Transvaal
  P.v.tahatali
    E Transvaal, Natal

P.v.caurinus
S Angola, N Namibia, Botswana
P.v.shelleyi
Malawi, Mozambique, N Natal
P.v.finschi
coast of SW Angola
P.v.nigrifrons
E Cape Province
Ploceus ruweti (Lufira Masked Weaver)
S Zaire
Ploceus katangae (Katanga Masked Weaver)
P.k.upembae
SE Zaire
P.k.katangae
NW Zambia, SE Zaire
Ploceus reichardi (Tanzanian Masked Weaver)
SW Tanzania
Ploceus vitellinus (Vitelline Masked Weaver)
P.v.vitellinus
Senegal to Chad, W Sudan
P.v.peixotoi
Sao Thome I
Ploceus spekei (Speke's Weaver)
S Ethiopia, Somalia, Kenya, N Tanzania
Ploceus spekeoides (Fox's Weaver)
Uganda
Ploceus cucullatus (Village Weaver)
P.c.cucullatus
Senegal to Cameroun, Chad, Fernando Po I
P.c.collaris
Gabon, Zaire, N Angola
P.c.bohndorffi
N Zaire, Uganda, Sudan, NW Tanzania
P.c.abyssinicus
Ethiopia
P.c.frobenii
S Zaire
P.c.graueri
E Zaire, W Tanzania
P.c.paroptus
Mozambique to E Zaire & S Somalia
P.c.dilutescens
NE Natal to SE Botswana, S Mozambique
P.c.spilonotus
S Mozambique, Natal, Transvaal, E Cape Province
Ploceus nigriceps (Layard's Black-headed Weaver)
Somalia, East Africa, N Mozambique
Ploceus grandis (Giant Weaver)
Sao Thome I
Ploceus nigerrimus (Vieillot's Black Weaver)
P.n.castaneofuscus
Sierra Leone to W Nigeria
P.n.nigerrimus
E Nigeria, Cameroun to W Kenya
Ploceus weynsi (Weyns's Weaver)
N Zaire, S Uganda, NW Tanznaia
Ploceus golandi (Clarke's Weaver)
E Kenya
Ploceus dicrocephalus (Salvadori's Weaver)
S Ethiopia, Somalia, N Kenya

Ploceus melanocephalus (Black-headed Weaver)
P.m.melanocephalus
Senegal to Benin, Niger, Chad
P.m.capitalis
Nigeria, S Chad, Central African Republic
P.m.duboisi
Zaire, N Zambia
P.m.dimidiatus
NE Sudan
P.m.fischeri
Uganda, Kenya, Tanzania
Ploceus jacksoni (Eastern Golden-backed Weaver)
S Sudan, Kenya, Uganda
Ploceus badius (Cinnamon Weaver)
P.b.badius
E Sudan
P.b.axillaris
S Sudan
Ploceus rubiginosus (Chestnut Weaver)
P.r.rubiginosus
Ethiopia, Somalia, Uganda, Kenya, N Tanzania
P.r.trothae
SW Angola, N Namibia
Ploceus aureonucha (Gold-naped Weaver)
NE Zaire
Ploceus tricolor (Yellow-mantled Weaver)
P.t.tricolor
Guinea to Cameroun, Gabon, Angola
P.t.interscapularis
Zaire, W Uganda
Ploceus albinucha (Maxwell's Black Weaver)
P.a.albinucha
Sierra Leone to Ghana
P.a.maxwelli
Fernando Po I
P.a.holomelas
E Nigeria to Gabon, Central African Republic, N Zaire
Ploceus nelicourvi (Nelicourvi Weaver)
N & E Madagascar
Ploceus hypoxanthus (Asian Golden Weaver)
P.h.hymenaicus
S Burma, Thailand, S Indochina
P.h.hypoxanthus
Sumatra, Java
Ploceus superciliosus (Compact Weaver)
Sierra Leone to Ethiopia, Uganda, Kenya & Angola
Ploceus benghalensis (Bengal Weaver)
Pakistan, N India, Nepal, Bangladesh
Ploceus manyar (Streaked Weaver)
P.m.flaviceps
Pakistan, W India, Sri Lanka
P.m.peguensis
NE India, Bangladesh, Burma
P.m.williamsoni
Thailand, Vietnam
P.m.manyar
Java, Bali
Ploceus philippinus (Baya Weaver)

P.p.philippinus
Pakistan, India, Sri Lanka
P.p.travencoreensis
SW India
P.p.burmanicus
NE India, Bangladesh, Burma
P.p.infortunatus
S Vietnam, Malaysia, Sumatra
P.p.angelorum
C Thailand
Ploceus megarhynchus (Finn's Weaver)
P.m.megarhynchus
S Himalayas
P.m.salimalii
NE India
Ploceus bicolor (Forest Weaver)
P.b.tephronotus
E Nigeria, Cameroun, Fernando Po I
P.b.analogus
S Cameroun
P.b.amaurocephalus
N Angola
P.b.mentalis
S Sudan, NE Zaire, Uganda, W Kenya
P.b.kigomaensis
S Zaire, Zambia, W Tanzania
P.b.kersteni
Somalia, E Kenya, E Tanzania
P.b.stictifrons
E Zimbabwe, SE Tanzania, Mozambique,
Malawi
P.b.sylvanus
NE Zimbabwe, W Mozambique
P.b.lebomboensis
S Mozambique, N Natal
P.b.sclateri
E Natal, SE Mozambique
P.b.bicolor
S Mozambique to E Cape Province
**Ploceus preussi (Western Golden-backed
Weaver)**
Sierra Leone to Cameroun, Central African
Republic
**Ploceus dorsomaculatus (Yellow-capped
Weaver)**
Cameroun, Central African Republic,
Congo, Zaire
**Ploceus olivaceiceps (Olive-headed Golden
Weaver)**
P.o.olivaceiceps
S Tanzania, Malawi, Mozambique
P.o.vicarius
S Mozambique
Ploceus nicolli (Usambara Weaver)
P.n.nicolli
Usambara (Tanzania)
P.n.anderseni
Morogoro (Tanzania)
Ploceus insignis (Brown-capped Weaver)
P.i.insignis
Cameroun to Sudan, Angola, Zaire, Kenya,
Tanzania

P.i.unicus
Fernando Po I
Ploceus angolensis (Bar-winged Weaver)
Angola, N Namibia, S Zaire, Zambia
Ploceus sanctaethomae (Sao Thome Weaver)
Sao Thome I

MALIMBUS
Malimbus flavipes (Yellow-legged Malimbe)
NE Zaire
Malimbus coronatus (Red-crowned Malimbe)
Cameroun
Malimbus cassini (Black-throated Malimbe)
S Cameroun, Gabon, Congo
Malimbus racheliae (Rachel's Malimbe)
E Nigeria to Gabon
Malimbus ballmani (Tai Malimbe) 163.4
Tai (Ivory Coast)
Malimbus scutatus (Red-vented Malimbe)
M.s.scutatus
Sierra Leone to Ghana
M.s.scutopartitus
S Nigeria, W Cameroun
Malimbus ibadanensis (Ibadan Malimbe)
E Nigeria
Malimbus nitens (Gray's Malimbe)
M.n.nitens
Guinea to S Nigeria
M.n.moreaui
Cameroun, Gabon, NW Zaire
M.n.microrhynchus
NE Zaire, W Uganda
Malimbus rubricollis (Red-headed Malimbe)
M.r.bartletti
Sierra Leone to Ghana
M.r.nigeriae
Benin, W Nigeria
M.r.rubricollis
E Nigeria to Sudan, Chad, Central African
Republic
M.r.rufovelatus
Fernando Po I
M.r.praedi
N Angola
**Malimbus erythrogaster (Red-bellied
Malimbe)**
E Nigeria to E Zaire
Malimbus malimbicus (Crested Malimbe)
M.m.nigrifrons
Sierra Leone to Nigeria
M.m.malimbicus
Cameroun to Uganda, S Zaire, N Angola

ANAPLECTES
Anaplectes melanotis (Red-headed Weaver)
A.m.melanotis
Senegal to Ethiopia, Uganda, W Kenya
A.m.jubaensis
S Somalia, NE Kenya
A.m.rubriceps
S Angola to Tanzania, Mozambique
A.m.gurneyi
N Namibia

QUELEA
Quelea cardinalis (Cardinal Quelea)
 Q.c.cardinalis
  S Sudan, S Ethiopia, Uganda, Kenya, NW
  Tanzania
 Q.c.rhodesiae
  Tanzania, Zambia
Quelea erythrops (Red-headed Quelea)
 Senegal to Ethiopia, Natal & Cape Province
Quelea quelea (Red-billed Quelea)
 Q.q.quelea
  Senegal to Chad, Central African Republic
 Q.q.aethiopica
  Sudan, Somalia to E Zaire, N Tanzania
 Q.q.lathamii
  Angola, S Zaire
 Q.q.spoliator
  E Cape Province to Zambia & Malawi

FOUDIA
Foudia madagascariensis (Madagascan Red Fody)
 Madagascar, Mauritius, Reunion Is
Foudia eminentissima (Mascarene Fody)
 F.e.aldabrana
  Aldabra I
 F.e.consobrina
  Great Comoro I
 F.e.anjuanensis
  Anjouan I
 F.e.eminentissima
  Moheli I
 F.e.algondae
  Mayotte I
Foudia omissa (Red Forest Fody)
 E Madagascar
Foudia rubra (Mauritius Fody)
 Mauritius I
Foudia sechellarum (Seychelles Fody)
 Seychelles Is
Foudia flavicans (Rodriguez Fody)
 Rodriguez I
Foudia sakalava (Sakalava Fody)
 F.s.sakalava
  N & NE Madagascar
 F.s.minor
  W & SW Madagascar

BRACHYCOPE
Brachycope anomala (Bob-tailed Weaver)
 SE Cameroun, Congo

EUPLECTES
Euplectes afer (Golden Bishop)
 E.a afer
  Senegal to Chad, Central African Republic
 E.a.ladoensis
  S Sudan, Uganda, N Kenya, N Tanzania
 E.a.strictus
  Ethiopia
 E.a.taha
  Sthn Africa

Euplectes diademata (Fire-fronted Bishop)
 E Kenya, NE Tanzania
Euplectes gierowii (Gierow's Bishop)
 E.g.ansorgei
  S Sudan, S Ethiopia, E Zaire, Uganda
 E.g.friederichseni
  SW Kenya, N Tanzania
 E.g.gierowii
  N Angola, SW Zaire
Euplectes nigroventris (Zanzibar Red Bishop)
 E Kenya, E Tanzania, E Mozambique,
 Zanzibar I
Euplectes hordeacea (Red-crowned Bishop)
 E.h.hordeacea
  Senegal to W Sudan, Angola, Zimbabwe
 E.h.craspedoptera
  S Sudan, SW Ethiopia, Uganda, NW Kenya
Euplectes orix (Red Bishop)
 E.o.franciscana
  Senegal to Ethiopia, Uganda, Kenya
 E.o.pusilla
  SE Ethiopia, Somalia
 E.o.nigrifrons
  E Zaire, S Kenya, Tanzania, Mozambique,
  Malawi
 E.o.sundevalli
  Transvaal, Zimbabwe, S Mozambique
 E.o.turgidus
  S Namibia, Cape Province, N Natal
 E.o.orix
  N Angola to S Mozambique & Sthn Africa
Euplectes aurea (Golden-backed Bishop)
 Sao Thome I, W Angola
Euplectes capensis (Yellow-rumped Bishop)
 E.c.phoenicomera
  SE Nigeria, Cameroun, Fernando PO i
 E.c.xanthomelas
  Sudan, Ethiopia
 E.c.crassirostris
  N & E Transvaal to E Zaire, Kenya
 E.c.approximans
  E Transvaal, Natal, E Cape Province
 E.c.capensis
  W & S Cape Province
 E.c.macrorhynchus
  NW Cape Province
Euplectes axillaris (Fan-tailed Whydah)
 E.a.bocagei
  Niger, Cameroun, Angola, S Zaire
 E.a.quanzae
  C Angola
 E.a.traversii
  N Ethiopia
 E.a.phoeniceus
  S Ethiopia, Sudan, Uganda, W Kenya, W
  Tanzania
 E.a.batesi
  Upper Volta to Upper Niger
 E.a.zanzibaricus
  Somalia, E Kenya, E Tanzania
 E.a.axillaris
  Zambia to Mozambique & E South Africa

Euplectes macrourus (Yellow-mantled Whydah)
  E.m.macrocercus
    Uganda, W Kenya
  E.m.macrourus
    West Africa to Sudan, Zaire, Angola to Mozambique
  E.m.conradsi
    NW Tanzania
  E.m.intermedius
    W Tanzania
Euplectes hartlaubi (Marsh Whydah)
  E.h.humeralis
    Cameroun to Uganda, W Kenya
  E.h.hartlaubi
    Angola, S Zaire, Zambia
Euplectes psammocromius (Mountain Marsh Whydah)
  SW Tanzania, W Malawi, NE Zambia
Euplectes albonotatus (White-winged Whydah)
  E.a.eques
    Sudan to Tanzania
  E.a.sassii
    E Zaire
  E.a.asymmetrurus
    Gabon to N Namibia
  E.a.albonotatus
    SE Zaire & Zambia to Tanzania, Natal
Euplectes ardens (Red-collared Whydah)
  E.a.concolor
    Senegal to S Sudan, Uganda, Chad
  E.a.laticauda
    SE Sudan, Ethiopia
  E.a.suahelicus
    Kenya, NE Tanzania
  E.a.ardens
    EC Sthn Africa
Euplectes progne (Long-tailed Whydah)
  E.p.delamerei
    E Kenya
  E.p.ansorgei
    E Angola, W Zambia
  E.p.progne
    E Zambia, S Africa
  E.p.definita
    W Zambia
Euplectes jacksoni (Jackson's Whydah)
  Kenya, N Tanzania

ANOMALOSPIZA
Anomalospiza imberbis (Parasitic Weaver)
  Sierra Leone to Ethiopia & Transvaal

164. STURNIDAE (STARLINGS)

STURNINAE

APLONIS
Aplonis zelandica (New Hebrides Starling)
  A.z.rufipennis
    C & N New Hebrides, Banks Is

  A.z.maxwellii
    Santa Cruz I
  A.z.zelandica
    Vanikoro I
Aplonis santovestris (Mountain Starling)
  Espiritu Santo I
Aplonis pelzelni (Ponapé Starling)
  Ponapé I
Aplonis atrifusca (Samoan Starling)
  Samoan Is
Aplonis cinerascens (Rarotonga Starling)
  Cook Is
Aplonis tabuensis (Striped Starling)
  A.t.pachyramphus
    Santa Cruz I
  A.t.tucopiae
    Tucopia I
  A.t.rotumae
    Rotuma I
  A.t.vitiensis
    Fiji Is
  A.t.manuae
    Manuan I
  A.t.tabuensis
    Tonga I
  A.t.fortunae
    Fotuna, Alofa, Uea Is
  A.t.tenebrosa
    Keppel, Boscawen Is
  A.t.nesiotes
    Niuafou I
  A.t.brunnescens
    Niue I
  A.t.tutuilae
    Tutuila I
  A.t.brevirostris
    Upolu, Savaii Is
Aplonis striata (Striated Starling)
  A.s.striata
    New Caledonia
  A.s.atronitens
    Loyalty Is
Aplonis fusca (Norfolk I Starling)
  Norfolk I
Aplonis opaca (Micronesian Starling)
  A.o.aeneus
    Takatsukasa I
  A.o.guami
    Guam I
  A.o.orii
    Palau Is
  A.o.kurodai
    Yap I
  A.o.ponapensis
    Ponapé I
  A.o.opaca
    Kusaie I
  A.o.angus
    Truk I
Aplonis cantoroides (Singing Starling)
  New Guinea, Bismarck Arch
Aplonis crassa (Tanimbar Starling)
  Tanimbar Is

Aplonis feadensis (Fead Is Starling)
A.f.feadensis
Solomon Is
A.f.heureka
Bismarck Arch
Aplonis insularis (Rennell I Starling)
Rennell I
Aplonis dichroa (San Cristobal Starling)
San Cristobal I
Aplonis grandis (Large Glossy Starling)
A.g.malaitae
Malaita I
A.g.macrura
Guadalcanal I
A.g.grandis
Bougainville, Choiseul, Ysabel Is
Aplonis mysolensis (Moluccan Starling)
A.m.mysolensis
W New Guinea islands, Moluccas
A.m.sulaensis
Sula, Banggai Is
Aplonis magna (Long-tailed Starling)
A.m.magna
Biak I
A.m.brevicauda
Numfor I
Aplonis minor (Short-tailed Starling)
A.m.minor
Lesser Sunda Is, Sulawesi
A.m.todayensis
Mindanao I
Aplonis panayensis (Philippine Glossy Starling)
A.p.affinis
Assam, W Burma, S Vietnam
A.p.strigata
S Thailand, Malaysia, Sumatra, Java, W Borneo
A.p.eustathis
E Borneo
A.p.heterochlora
Anamba, Natuna Is
A.p.tytleri
Andaman, Nicobar Is
A.p.altirostris
W Sumatran islands
A.p.leptorrhyncha
W Sumatran islands
A.p.pachistorhina
W Sumatran islands
A.p.enganensis
Enggano I
A.p.gusti
Bali
A.p.alipodis
Maratua I, E Borneo
A.p.sanghirensis
Sanghir, Talaut Is
A.p.panayensis
Sulawesi, Philippine Is
Aplonis metallica (Shining Starling)
A.m.circumscripta
Tanimbar, Damar Is

A.m.metallica
Moluccas, New Guinea, NE Queensland
A.m.nitida
Solomon Is, Bismarck Arch
A.m.purpureiceps
Admiralty Is
A.m.inornata
Biak, Numfor Is
Aplonis mystacea (Grant's Starling)
W & S New Guinea
Aplonis brunneicapilla (White-eyed Starling)
Bougainville, Rendova Is

POEOPTERA
Poeoptera kenricki (Kenrick's Starling)
P.k.bensoni
Kenya
P.k.kenricki
S Kenya, N Tanzania
Poeoptera stuhlmanni (Stuhlmann's Starling)
SW Ethiopia, W Kenya, Uganda, E Zaire
Poeoptera lugubris (Narrow-tailed Starling)
P.l.lugubris
Sierra Leone to W Uganda, N Angola
P.l.webbi
Kigezi, Uganda

GRAFISIA
Grafisia torquata (White-collared Starling)
Cameroun, N Zaire, Central African Republic

ONYCHOGNATHUS
Onychognathus walleri (Waller's Red-winged Starling)
O.w.preussi
Cameroun, Fernando Po I
O.w.elgonensis
S Sudan, Uganda, E Zaire, W Kenya
O.w.walleri
S Kenya, Tanzania, N Malawi
Onychognathus nabouroup (Pale-winged Starling)
E Angola, Namibia, Botswana, N Cape Province
Onychognathus morio (African Red-winged Starling)
O.m.modicus
Senegal, Mali, W Niger
O.m.neumanni
N Nigeria to Central African Republic & W Sudan
O.m.rueppellii
Ethiopia & Sudan to Tanzania
O.m.shelleyi
Zambia, N Zimbabwe, W Mozambique
O.m.morio
Zimbabwe, S Malawi, S Mozambique, Sthn Africa
Onychognathus blythii (Somali Chestnut-winged Starling)
Somalia, Socotra I

Onychognathus frater (Socotra Chestnut-winged Starling)
Socotra I

Onychognathus tristramii (Tristram's Starling)
Israel, Arabia

Onychognathus fulgidus (Chestnut-winged Starling)
O.f.fulgidus
Sao Thome I
O.f.hartlaubii
Guinea to W Uganda & Angola

Onychognathus tenuirostris (Slender-billed Red-winged Starling)
O.t.tenuirostris
Ethiopia, N Kenya
O.t.theresae
E Zaire to Uganda, Kenya, Tanzania, Malawi

Onychognathus albirostris (White-billed Starling)
Ethiopia

Onychognathus salvadorii (Bristle-crowned Starling)
Somalia, S Ethiopia, N Kenya

LAMPROTORNIS
Lamprotornis iris (Iris Glossy Starling)
Guinea to Ivory Coast

Lamprotornis cupreocauda (Copper-tailed Glossy Starling)
Sierra Leone to Ghana

Lamprotornis purpureiceps (Purple-headed Glossy Starling)
S Nigeria to Gabon & Uganda

Lamprotornis superbus (Superb Starling)
164.2
L.s.superbus
S Sudan, Kenya, Somalia, Ethiopia
L.s.excelsior
Tanzania

Lamprotornis corruscus (Black-bellied Glossy Starling)
L.c.corruscus
E & SE Africa
L.c.mandanus
coastline Kenya to Natal
L.c.vaughani
Pemba I

Lamprotornis purpureus (Purple Glossy Starling)
L.p.purpureus
Senegal to Nigeria, Mali
L.p.amethystinus
Nigeria & Chad to Sudan & N Kenya

Lamprotornis nitens (Red-shouldered Glossy Starling)
L.n.nitens
Gabon to Angola
L.n.phoenicopterus
SW & SC Africa
L.n.culminator
Cape Province, S Natal

Lamprotornis chalcurus (Bronze-tailed Glossy Starling)
L.c.chalcurus
Senegal to Ghana
L.c.emini
Togo to Central African Republic, NE Zaire, Uganda, Kenya

Lamprotornis chalybaeus (Greater Blue-eared Glossy Starling)
L.c.chalybaeus
Senegal to Somalia, Kenya
L.c.cyaniventris
Ethiopia, W Kenya, Uganda, E Zaire
L.c.sycobius
Tanzania, Zambia, Malawi, Mozambique
L.c.nordmanni
S Angola, Zambia, Botswana, Transvaal

Lamprotornis chloropterus (Lesser Blue-eared Glossy Starling)
L.c.chloropterus
Senegal to Ethiopia, Uganda, Kenya
L.c.cyanogenys
SE Sudan, Ethiopia, N Uganda
L.c.elisabeth
S Uganda, S Kenya, Tanzania, Zambia, Mozambique

Lamprotornis acuticaudus (Sharp-tailed Glossy Starling)
L.a.acuticaudus
Angola, Zambia, S Zaire
L.a.ecki
S Angola, N Namibia

Lamprotornis splendidus (Splendid Glossy Starling)
L.s.chrysonotis
Senegal to Sierra Leone
L.s.splendidus
Nigeria & Ethiopia to Angola, W Tanzania
L.s.lessoni
Fernando Po I
L.s.bailundensis
S Angola, S Zaire, Zambia, S Tanzania

**Lamprotornis ornatus (Principé Glossy Starling)**
Principé I

Lamprotornis australis (Burchell's Starling)
L.a.australis
E Angola to Zimbabwe
L.a.degener
NW Transvaal to S Mozambique

Lamprotornis mevesii (Meve's Starling)
L.m.chalceus
C Angola
L.m.mevesii
S Angola to S Malawi
L.m.violacior
NW Namibia, SW Angola
L.m.benguelensis
SW & W Angola

Lamprotornis purpuropterus (Ruppell's Long-tailed Glossy Starling)
L.p.aeneocephalus
E Sudan, N Ethiopia

L.p.purpuropterus
S Ethiopia, S Sudan, Uganda, Kenya, W Tanzania

Lamprotornis caudatus (Long-tailed Glossy Starling)
Senegal to Sudan

CINNYRICINCLUS
Cinnyricinclus femoralis (Abbott's Starling)
S Kenya, N Tanzania
Cinnyricinclus sharpii (Sharpe's Starling)
E Zaire to Ethiopia & Tanzania
Cinnyricinclus leucogaster (Violet Starling)
C.l.leucogaster
Senegal to Uganda, Kenya, Tanzania
C.l.arabicus
SW Arabia, NE Sudan, Somalia, N Ethiopia
C.l.friedmanni
S Ethiopia
C.l.verreauxi
Angola to Kenya & Cape Province

SPECULIPASTOR
Speculipastor bicolor (Magpie Starling)
S Ethiopia, N Kenya

NEOCICHLA
Neocichla gutturalis (White-winged Starling)
N.g.gutturalis
S Angola, W Zambia
N.g.angusta
E Zambia, Tanzania, Malawi

SPREO
Spreo fischeri (Fischer's Starling)
S Somalia, Kenya, N Tanzania
Spreo bicolor (Pied Starling)
Ethiopia, E & Sth Africa
Spreo albicapillus (White-crowned Starling)
S Ethiopia, Somalia
Spreo pulcher (Chestnut-bellied Starling)
S.p.pulcher
Senegal to S Sudan
S.p.rufiventris
Chad, Sudan, Ethiopia
Spreo hildebrandti (Hildebrandt's Starling)
S Kenya, N Tanzania
Spreo shelleyi (Shelley's Starling)
S Ethiopia, S Somalia, Kenya

COSMOPSARUS
Cosmopsarus regius (Golden-breasted Starling)
C.r.regius
S Ethiopia, S Somalia, Kenya
C.r.magnificus
E Kenya
Cosmopsarus unicolor (Ashy Starling)
S Kenya, Tanzania

SAROGLOSSA
Saroglossa aurata (Madagascar Starling)
Madagascar

Saroglossa spiloptera (Spot-winged Starling)
Himalayas, Burma, Thailand

CREATOPHORA
Creatophora cinerea (Wattled Starling)
Ethiopia to Angola & Cape Province

STURNUS
Sturnus senex (Ceylon White-headed Starling)
SW Sri Lanka
Sturnus malabaricus (Ashy-headed Starling)
S.m.blythii
SW India
S.m.malabaricus
C & E India, Assam
S.m.nemoricola
E India, Burma, Thailand, SW China, Indochina
Sturnus erythropygius (White-headed Starling)
S.e.erythropygius
Andaman Is
S.e.andamanensis
Car Nicobar I
S.e.katchalensis
Katchal I
Sturnus pagodarum (Black-headed Starling)
Afghanistan, India, Sri Lanka
Sturnus sericeus (Silky Starling)
S China
Sturnus philippensis (Violet-backed Starling)
Japan, Borneo, Philippine Is, N Sulawesi
Sturnus sturninus (Daurian Starling)
E & SE Asia
Sturnus roseus (Rose-coloured Starling)
E Europe, W & C Asia >> India
Sturnus vulgaris (Common Starling)
S.v.faroensis
Faroe Is
S.v.zetlandicus
Outer Hebrides, Shetland Is
S.v.vulgaris
N & C Europe, N Africa, N America
S.v.tauricus
SE Europe >> Iraq, W Iran
S.v.caucasicus
N Iran
S.v.purpurascens
S Russia >> Iraq, Egypt
S.v.nobilior
Transcaspia, NE Iran >> N India
S.v.poltaratskyi
C Siberia >> E Iran & E India
S.v.porphyronotus
Turkestan >> Nepal & N India
S.v.humii
W Himalayas >> N India
S.v.minor
Sind
Sturnus unicolor (Spotless Starling)
S Europe, N Africa

Sturnus cineraceus (Grey Starling)
    C Asia to Japan >> S China
Sturnus contra (Asian Pied Starling)
   S.c.contra
    N & C India
   S.c.sordidus
    N Assam
   S.c.superciliaris
    E Assam, Burma
   S.c.floweri
    S Burma, Thailand, Laos
   S.c.jalla
    Sumatra, Java, Bali
Sturnus nigricollis (Black-collared Starling)
    S China, Burma, Thailand, Malaysia,
    Indochina
Sturnus burmannicus (Jerdon's Starling)
   S.b.burmannicus
    Burma
   S.b.leucocephalus
    S Thailand, Cambodia, S Indochina
**Sturnus melanopterus (Black-winged
Starling)**
   S.m.melanopterus
    W Java
   S.m.tricolor
    E Java
   S.m.tertius
    Bali, Lombok Is
Sturnus sinensis (Chinese Starling)
    S China, N Indochina >> Malaysia

LEUCOPSAR
Leucopsar rothschildi (Rothschild's Mynah)
    Bali

ACRIDOTHERES
Acridotheres tristis (Common Mynah)
   A.t.tristis
    Afghanistan, India, Se Asia
   A.t.melanosturnus
    Sri Lanka
   A.t.tristoides
    C & N Burma, Nepal & Sth Africa (intro.)
Acridotheres ginginianus (Bank Mynah)
    Pakistan, India
Acridotheres fuscus (Indian Jungle Mynah)
   A.f.mahrattensis
    W & S India
   A.f.fuscus
    N India, Burma
   A.f.fumidus
    NE Assam
   A.f.torquatus
    N & C Malaysia
   A.f.javanicus
    Java
   A.f.cinereus
    S Sulawesi
Acridotheres grandis (Great Mynah)
    Assam, Burma, Indochina

**Acridotheres albocinctus (White-collared
Mynah)**
    E India to NW Yunnan
**Acridotheres cristatellus (Chinese Jungle
Mynah)**
   A.c.cristatellus
    C & S China, E Burma
   A.c.formosanus
    Taiwan
   A.c.brevipennis
    Hainan I, Indochina

AMPELICEPS
Ampeliceps coronatus (Gold-crested Mynah)
    Assam, Burma, Thailand, Laos

MINO
Mino anais (Golden-breasted Mynah)
   M.a.anais
    NW New Guinea
   M.a.orientalis
    N New Guinea
   M.a.robertsoni
    S New Guinea
Mino dumontii (Yellow-faced Mynah)
   M.d.dumontii
    New Guinea, Aru Is
   M.d.kreffti
    Bismarck Arch
   M.d.sanfordi
    Guadalcanal, Malaita Is

BASILORNIS
**Basilornis celebensis (Sulawesi King
Starling)**
    Sulawesi
Basilornis galeatus (Greater King Starling)
    Banggai, Sula Is
Basilornis corythaix (Seram King Starling)
    Seram I
Basilornis miranda (Mount Apo King Starling)
    Mindanao I

STREPTOCITTA
Streptocitta albicollis (Sulawesi Magpie)
   S.a.torquata
    N & E Sulawesi
   S.a.albicollis
    S & SE Sulawesi
Streptocitta albertinae (Sula Magpie)
    Sula Is

SARCOPS
Sarcops calvus (Bald Starling)
   S.c.calvus
    N Philippine Is
   S.c.melanotus
    C & SE Philippine Is
   S.c.lowii
    Sulu Is

GRACULA
Gracula ptilogenys (Ceylon Grackle)
Sri Lanka
Gracula religiosa (Southern Grackle)(Hill Mynah)
G.r.indica
SW India, Sri Lanka
G.r.peninsularis
NE India
G.r.intermedia
N India, Burma, Thailand, Indochina
G.r.andamanensis
Andaman, Nicobar Is
G.r.religiosa
Malaysia, Sumatra, Java, Bali, Borneo, Bangka I
G.r.batuensis
W Sumatran islands
G.r.robusta
Babi, Nias Is
G.r.palawanensis
Palawan I
G.r.venerata
Sumbawa I to Alor I

ENODES
Enodes erythrophris (Fiery-browed Enodes Starling)   164.1
Sulawesi

SCISSIROSTRUM
Scissirostrum dubium (Grosbeak Starling)
164.1
Sulawesi, Togian, Peleng Is

BUPHAGINAE

BUPHAGUS
Buphagus africanus (Yellow-billed Oxpecker)
B.a.africanus
Senegal & SW Ethiopia to Namibia & Natal
B.a.langi
Gabon, W Congo
Buphagus erythrorhynchus (Red-billed Oxpecker)
B.e.erythrorhynchus
Ethiopia, Sudan
B.e.archeri
N Kenya
B.e.caffer
Botswana, W Zimbabwe, W Transvaal
B.e.angolensis
S Angola to W Zambia
B.e.scotinus
S Kenya to S Mozambique

165 ORIOLIDAE (ORIOLES)

ORIOLUS
Oriolus szalayi (Brown Oriole)
New Guinea
Oriolus phaeochromus (Dusky Brown Oriole)
165.1
Halmahera I

Oriolus forsteni (Grey-collared Oriole)   165.1
Seram
Oriolus bouroensis (Black-eared Oriole)   165.1
O.b.bouroensis
Buru I
O.b.decipiens
Tanimbar Is
Oriolus melanotis (Olive Brown Oriole)   165.1
O.m.finschi
Wetar I
O.m.melanotis
Timor, Roti Is
Oriolus sagittatus (Olive-backed Oriole)
O.s.magnirostris
S New Guinea, N Queensland
O.s.affinis
N Western Australia
O.s.sagittatus
E & SE Australia
Oriolus flavocinctus (Australian Yellow Oriole)   165.1
Lesser Sunda Is, S New Guinea, N Northern Territory
Oriolus xanthonotus (Dark-throated Oriole)
O.x.xanthonotus
Malaysia, Sumatra, Java, SW Borneo
O.x.consobrinus
N, C & E Borneo
O.x.mentawi
W Sumatran islands, Siberut I
O.x.cinereogenys
Sulu Is
O.x.persuasus
Palawan I
O.x.basilanicus
Basilan, W Mindanao Is
O.x.samarensis
E Mindanao, Samar, Leyte Is
O.x.steerii
Masbate, Negros Is
O.x.assimilis
Cebu I
Oriolus albiloris (White-lored Oriole)
Bataan, Luzon Is
Oriolus isabellae (Isabella Oriole)
Bataan, Luzon Is
Oriolus oriolus (Golden Oriole)
O.o.oriolus
Europe, W & WC Asia >> E & S Africa, NW India
O.o.kundoo
C Asia, N India
Oriolus auratus (African Golden Oriole)
O.a.auratus
W & NC Africa
O.a.notatus
Eastern & Southern Africa
Oriolus chinensis (Black-naped Oriole)
O.c.tenuirostris
E Nepal, C Burma >> S Burma, Thailand
O.c.invisus
S Vietnam

O.c.diffusus
E Asia >> India, Malaysia, Indochina
O.c.andamanensis
Andaman Is
O.c.macrourus
Nicobar Is
O.c.chinensis
Philippine Is
O.c.suluensis
Sulu Is
O.c.melanisticus
Talaut I
O.c.sanghirensis
Sangihe I
O.c.formosus
Siau, Ruang, Mayu Is
O.c.frontalis
Sula, Peling, Banggai Is
O.c.mundus
Simalur I
O.c.sipora
Sipora I
O.c.richmondi
Siberut, Pagi Is
O.c.insularis
Kangean I
O.c.broderipii
Lombok I to Alor, Sumba Is
O.c.lamprochryseus
Solombo Besar I
O.c.boneratensis
Flores Sea Is
O.c.maculatus
Sumatra, Java, Borneo, Bali, Nias I
O.c.celebensis
Sulawesi
Oriolus chlorocephalus (Green-headed Oriole)
O.c.amani
Tanzania
O.c.chlorocephalus
Malawi, Mozambique
O.c.speculifer
S Mozambique
Oriolus crassirostris (Sao Thome Oriole)
Sao Thome I
Oriolus brachyrhynchus (Western Black-headed Oriole)
O.b.brachyrynchus
W Africa
O.b.laetior
WC & C Africa
Oriolus monacha (Dark-headed Oriole)
O.m.monacha
N Ethiopia
O.m.meneliki
S Ethiopia
Oriolus percivali (Montane Oriole) 165.2
E Zaire, W Uganda, W Kenya, NW Tanzania
Oriolus larvatus (African Black-headed Oriole) 165.3

O.l.angolensis
Angola, Namibia to Tanzania, Kenya
O.l.larvatus
South Africa
O.l.additus
Mozambique, SE Tanzania
Oriolus nigripennis (Black-winged Oriole)
O.n.alleni
W Africa
O.n.nigripennis
WC & C Africa
Oriolus xanthornus (Asian Black-headed Oriole)
O.x.xanthornus
N India, Thailand, Indochina
O.x.maderaspatanus
S India, Andaman Is
O.x.ceylonensis
Sri Lanka
O.x.tanakae
NE Borneo
O.x.thaiacous
S Thailand, N Malaysia
O.x.andamanensis
S Andaman Is
Oriolus hosii (Black Oriole)
Borneo
Oriolus cruentus (Crimson-breasted Oriole)
O.c.cruentus
Java
O.c.malayanus
C Malaysia
O.c.consanguineus
Sumatra
O.c.vulneratus
N Borneo
Oriolus traillii (Maroon Oriole)
O.t.traillii
Himalayas, Burma, Thailand
O.t.robinsoni
S Indochina
O.t.nigellicauda
N Vietnam, Hainan I
O.t.ardens
Taiwan
Oriolus mellianus (Stresemann's Maroon Oriole)
W China

SPHECOTHERES
Sphecotheres viridis (Figbird)
S.v.viridis
Timor, Roti Is
S.v.hypoleucus
Wetar I
S.v.vieilloti
NE Australia
S.v.salvadorii
NE Queensland, S New Guinea
S.c.flaviventris
N & NE Australia, Kai Is

166. DICRURIDAE (DRONGOS)

CHAETORHYNCHUS
Chaetorhynchus papuensis (Papuan Mountain Drongo)
New Guinea

DICRURUS
Dicrurus ludwigii (Square-tailed Drongo)
D.l.sharpei
W, WC & C Africa
D.l.saturnus
Angola
D.l.ludwigii
Eastern & Southern Africa
D.l.tephrogaster
Mozambique, E Zimbabwe, S Malawi
Dicrurus atripennis (Shining Drongo)
W Africa
Dicrurus adsimilis (Fork-tailed Drongo)
D.a.adsimilis
EC & Southern Africa
D.a.fugax
East Africa to N Natal, Zimbabwe
D.a.divaricatus
W Africa, Chad, Sudan, Ethiopia
D.a.coracinus
WC & C Africa
D.a.atactus
Upper Guinea, Nigeria
D.a.modestus
Principé I
D.a.apivorus
Namibia
Dicrurus fuscipennis (Comoro Drongo)
Great Comoro I
Dicrurus aldabranus (Aldabra Drongo)
Aldabra I
Dicrurus forficatus (Crested Drongo)
D.f.forficatus
Madagascar
D.f.potior
Anjouan I
Dicrurus waldenii (Mayotte Drongo)
Mayotte I
Dicrurus macrocercus (Black Drongo)
D.m.albirictus
SE Iran, Afghanistan, N India
D.m.macrocercus
S India
D.m.minor
Sri Lanka
D.m.cathoecus
China, N Burma, N Thailand, Laos, N Vietnam, Malaysia
D.m.thai
S Burma, S Thailand, S Vietnam
D.m.harterti
Taiwan
D.m.javanus
Java, Bali
Dicrurus leucophaeus (Ashy Drongo)
D.l.longicaudatus

E Afghanistan >> S India, Sri Lanka
D.l.hopwoodi
Sikkim, Bhutan, Assam, Burma, S China >> S Indochina
D.l.mouhoti
S Burma, N Thailand >> S Vietnam, Indochina
D.l.bondi
S Thailand, Cambodia
D.l.nigrescens
S Thailand, Malaysia
D.l.leucogenis
Manchuria, E China >> S Indochina
D.l.salangensis
SE China, S Thailand >> Hainan I, Malaysia
D.l.innexus
Hainan I
D.l.stigmatops
N Borneo
D.l.phaedrus
S Sumatra
D.l.batakensis
N Sumatra
D.l.periophthalmicus
Sipora I (Mentawei Group)
D.l.siberu
Siberut I
D.l.leucophaeus
Java, Bali, Lombok, Palawan Is
Dicrurus caerulescens (White-bellied Drongo)
D.c.caerulescens
Peninsular India
D.c.insularis
N Sri Lanka
D.c.leucopygialis
S Sri Lanka
Dicrurus annectans (Crow-billed Drongo)
Himalayas, Burma, Thailand, Malaysia, India
Dicrurus aeneus (Bronzed Drongo)
D.a.aeneus
India, Burma, S China, Thailand, Indochina
D.a.malayensis
S Malaysia, Sumatra, Borneo
D.a.braunianus
Taiwan
Dicrurus remifer (Lesser Racquet-tailed Drongo)
D.r.tectirostris
Himalayas, Burma, S China, Thailand, Indochina
D.r.remifer
Java, Sumatra
D.r.peracensis
W Laos, S Thailand, Malaysia
D.r.lefoli
S Cambodia
Dicrurus balicassius (Balicassio Drongo)
D.b.balicassius
Lubang, C & S Luzon, Mindoro Is
D.b.abraensis
N Luzon I

D.b.mirabilis
Panay, Cebu, Negros, Masbate Is
Dicrurus bracteatus (Spangled Drongo)
166.1
D.b.montanus
Sulawesi
D.b.samarensis
Samar, Leyte, Bohol Is
D.b.striatus
Mindanao, Basilan Is
D.b.morotensis
Morotai I
D.b.atrocaeruleus
Kofiau, Halmahera Is
D.b.carbonarius
New Guinea, D'Entrecasteaux Arch
D.b.bracteatus
N & E Australia >> S New Guinea
D.b.laemostictus
New Britain
D.b.meeki
Guadalcanal I
D.b.longirostris
San Cristobal I
D.b.amboinensis
S Moluccas
D.b.buruensis
Buru I
Dicrurus megarhynchus (New Ireland Drongo)
New Ireland
Dicrurus densus (Wallacean Drongo) 166.1
D.d.densus
Timor I
D.d.kuehni
Tanimbar Is
D.d.megalornis
Kai Is
D.d.sumbae
Sumba
D.d.bimaensis
Lombok to Alor, Sumbawa Is
Dicrurus sumatranus (Sumatran Drongo)
166.1
Sumatra
Dicrurus hottentottus (Hair-crested Drongo)
166.1
D.h.guillemardi
Obi Is
D.h.pectoralis
Sula Is
D.h.leucops
Sulawesi
D.h.jentincki
Bali, Kangean I
D.h.viridinitens
Mentawei Is
D.h.borneensis
N Borneo
D.h.suluensis
Sibutu I, Sulu Arch
D.h.hottentottus
India, Burma, Thailand, S Indochina

D.h.brevirostris
China, N Burma, N Laos, N Vietnam
D.h.palawanensis
Cagayan, Sulu, Palawan Is
D.h.cuyensis
Cuyo, Semirara, Philippine Is
D.h.menagei
Tablas I
Dicrurus andamanensis (Andaman Drongo)
D.a.andamanensis
S Andaman Is
D.a.dicruriformis
Gt Cocos, Table Is
Dicrurus paradiseus (Greater Racquet-tailed Drongo)
D.p.brachyphorus
Borneo
D.p.banguey
N Borneo islands
D.p.microlophus
Tioman, Anamba, N Natuna Is
D.p.platurus
S Malaysia, Sumatra, NW Sumatra Is
D.p.formosus
Java
D.p.malayensis
N Malaysia
D.p.paradiseus
S India, S Thailand, Indochina
D.p.rangoonensis
S Burma, W Thailand, C Laos
D.p.grandis
N India, N Burma, N Vietnam
D.p.johni
Hainan I
D.p.ceylonicus
Sri Lanka
D.p.lophorinus
W Sri Lanka
D.p.otiosus
Andaman Is
D.p.nicobariensis
Nicobar Is

167. CALLAEIDAE (WATTLEBIRDS)

CALLAEAS
Callaeas cinerea (Kokako)
C.c.wilsoni
North I (New Zealand)
C.c.cinerea
South I (New Zealand), Stewart I

CREADION
Creadion carunculatus (Saddleback)
C.c.rufusater
North Island (New Zealand)
C.c.carunculatus
Stewart I

168. GRALLINIDAE (MAGPIE LARKS)

GRALLININAE

GRALLINA
Grallina cyanoleuca (Magpie Lark)
Australia
Grallina bruijni (Torrent Lark)
New Guinea

CORCORACINAE

CORCORAX
Corcorax melanorhamphos (White-winged Chough)
E & SE Australia

STRUTHIDEA
Struthidea cinerea (Apostle Bird)
S.c.cinerea
Eastern Australia
S.c.dalyi
Northern Territory

169. ARTAMIDAE (WOOD SWALLOWS)

ARTAMUS
Artamus fuscus (Ashy Wood Swallow)
India to S China, Indochina
Artamus leucorhynchus (White-breasted Wood Swallow)
A.l.pelewensis
Palau Is
A.l.leucorhynchus
Philippine, Palawan Is, Borneo
A.l.amydrus
Sumatra, Bangka I, Java, Bali
A.l.humei
Andaman, Cocos Is
A.l.albiventer
Sulawesi, Lesser Sunda Is
A.l.musschenbroeki
Tanimbar Is
A.l.leucopygialis
Moluccas, Aru Is, New Guinea, N Australia
A.l.melaleucus
New Caledonia, Loyalty Is
A.l.tenuis
New Hebrides
A.l.mentalis
N Fiji Is
Artamus monachus (White-backed Wood Swallow)
Sulawesi, Sula Is
Artamus maximus (Papuan Wood Swallow)
New Guinea
Artamus insignis (Bismarck Wood Swallow)
New Britain, New Ireland
Artamus personatus (Masked Wood Swallow)
Australia
Artamus superciliosus (White-browed Wood Swallow)
SE Australia

Artamus cinereus (Black-faced Wood Swallow)
A.c.perspicillatus
Timor I
A.c.cinereus
W, C & SE Australia
A.c.hypoleucos
S New Guinea, N Queensland
A.c.normani
N Queensland
A.c.inkermani
C Queensland
Artamus cyanopterus (Dusky Wood Swallow)
A.c.cyanopterus
E & SE Australia, Tasmania
A.c.perthi
S Western Australia
Artamus minor (Little Wood Swallow)
N & C Australia

170. CRACTICIDAE (BUTCHER BIRDS)

CRACTICUS
Cracticus mentalis (Black-backed Butcher Bird)
C.m.mentalis
SE New Guinea
C.m.kempi
N Queensland
Cracticus torquatus (Grey Butcher Bird)
C.t.argenteus
Northern Territory, NW Western Australia
C.t.leucopterus
Central Australia
C.t.torquatus
E Australia
C.t.cinereus
Tasmania
Cracticus nigrogularis (Black-throated Butcher Bird)
C.n.picatus
Northern Territory, NW Western Australia
C.n.kalgoorli
Central Australia, Western Australia
C.n.nigrogularis
E & SE Australia
Cracticus cassicus (Black-headed Butcher Bird)
C.c.cassicus
New Guinea
C.c.hercules
Trobriand Is, D'Entrecasteaux Arch
Cracticus louisiadensis (White-rumped Butcher Bird)
Tagula I
Cracticus quoyi (Black Butcher Bird)
C.q.quoyi
New Guinea
C.q.spaldingi
Aru Is, Northern Territory, N Queensland
C.q.rufescens
NC Queensland

GYMNORHINA
Gymnorhina tibicen (Black-backed Magpie)
 G.t.papuana
  S New Guinea
 G.t.eylandtensis
  Northern Territory
 G.t.longirostris
  Western Australia
 G.t.finki
  Central Australia
 G.t.terraereginae
  N Northern Territory, C Queensland
 G.t.tibicen
  New South Wales, Victoria, South Australia
 G.t.leuconota
  SE South Australia, W Victoria
 G.t.dorsalis
  S Western Australia
 G.t.hypoleuca
  E Victoria, Tasmania

STREPERA
Strepera graculina (Pied Currawong)
 S.g.robinsoni
  Queensland
 S.g.graculina
  New South Wales
 S.g.ashbyi
  Victoria
 S.g.crissalis
  Lord Howe I
Strepera fuliginosa (Black Currawong)
  Tasmania
Strepera versicolor (Grey Currawong)
 S.v.versicolor
  New South Wales, E Victoria
 S.v.centralia
  N South Australia
 S.v.plumbea
  S Western Australia
 S.v.howei
  NW Victoria, E South Australia
 S.v.melanoptera
  SE South Australia, Kangaroo I
 S.v.intermedia
  S South Australia
 S.v.arguta
  Tasmania

**171. PTILONORHYNCHIDAE
(BOWERBIRDS)**

AILUROEDUS
Ailuroedus buccoides (White-eared Catbird)
 A.b.cinnamomeus
  S New Guinea
 A.b.buccoides
  W Papuan islands, NW New Guinea
 A.b.stonii
  SE New Guinea
 A.b.geislerorum
  Japen I, New Guinea

Ailuroedus crassirostris (Green Catbird)
  SE Queensland to N Victoria
Ailuroedus melanotis (Spotted Catbird)
 A.m.maculosus
  N Queensland
 A.m.melanotis
  Aru Is, S New Guinea
 A.m.melanocephalus
  SE New Guinea
 A.m.facialis
  WC New Guinea
 A.m.guttaticollis
  N New Guinea
 A.m.astigmaticus
  E New Guinea
 A.m.jobiensis
  WC New Guinea
 A.m.arfakianus
  NW New Guinea
 A.m.misoliensis
  Misol I

SCENOPOEETES
**Scenopoeetes dentirostris (Tooth-billed
Catbird)**
  NE Queensland

ARCHBOLDIA
**Archboldia papuensis (Archbold's
Bowerbird)**
 A.p.papuensis
  WC New Guinea
 A.p.sanfordi
  EC New Guinea

AMBLYORNIS
**Amblyornis inornatus (Vogelkop Gardener
Bowerbird)**
  NW New Guinea
**Amblyornis macgregoriae (Macgregor's
Gardener Bowerbird)**
 A.m.mayri
  WC New Guinea
 A.m.amati
  Adelbert Mts
 A.m.macgregoriae
  EC New Guinea
 A.m.germanus
  E New Guinea
 A.m.kombok
  EC New Guinea
 A.m.nubicola
  SE New Guinea
**Amblyornis subalaris (Striped Gardener
Bowerbird)**
  SE New Guinea
**Amblyornis flavifrons (Yellow-fronted
Gardener Bowerbird)**
  W New Guinea

PRIONODURA
**Prionodura newtoniana (Newton's Golden
Bowerbird)**
  NE Queensland

SERICULUS
Sericulus aureus (Flamed Bowerbird)
  S.a.aureus
    N & W New Guinea
  S.a.ardens
    S New Guinea
Sericulus bakeri (Adelbert Bowerbird)
    NE New Guinea
Sericulus chrysocephalus (Regent Bowerbird)
  S.c.chrysocephalus
    NE New South Wales
  S.c.rothschildi
    C & S Queensland

PTILONORHYNCHUS
Ptilonorhynchus violaceus (Satin Bowerbird)
  P.v.violaceus
    SE Queensland to Victoria
  P.v.minor
    NE Queensland

CHLAMYDERA
Chlamydera maculata (Spotted Bowerbird)
  C.m.maculata
    EC Australia
  C.m.guttata
    WC Australia
Chlamydera nuchalis (Great Grey Bowerbird)
  C.n.oweni
    N Western Australia
  C.n.nuchalis
    N Western Australia to NW Queensland
  C.n.yorki
    N Queensland
  C.n.orientalis
    NW Queensland
Chlamydera lauterbachi (Lauterbach's Bowerbird)
  C.l.lauterbachi
    NC New Guinea
  C.l.uniformis
    C New Guinea
Chlamydera cerviniventris (Fawn-breasted Bowerbird)
    E New Guinea, N Queensland

172. PARADISAEIDAE (BIRDS OF PARADISE)

CNEMOPHILINAE

LORIA
Loria loriae (Loria's Bird of Paradise)
  L.l.inexpectata
    WC New Guinea
  L.l.loriae
    SE New Guinea
  L.l.amethystina
    EC New Guinea

LOBOPARADISEA
Loboparadisea sericea (Wattle-billed Bird of Paradise)
  L.s.sericea
    C New Guinea
  L.s.aurora
    E New Guinea

CNEMOPHILUS
Cnemophilus macgregorii (Sickle-crested Bird of Paradise)
  C.m.sanguineus
    EC New Guinea
  C.m.macgregorii
    SE New Guinea

PARADISAEINAE

MACGREGORIA
Macgregoria pulchra (Macgregor's Bird of Paradise)
  M.p.pulchra
    SE New Guinea
  M.p.carolinae
    WC New Guinea

LYCOCORAX
Lycocorax pyrrhopterus (Paradise Crow)
  L.p.obiensis
    Obi Is
  L.p.pyrrhopterus
    Batjan, Halmahera Is
  L.p.morotensis
    Morotai, Rau Is

MANUCODIA
Manucodia ater (Glossy-mantled Manucode)
  M.a.ater
    C & W New Guinea
  M.a.subalter
    Aru Is, SE New Guinea
  M.a.alter
    Tagula I
Manucodia jobiensis (Jobi Manucode)
  M.j.jobiensis
    Japen I
  M.j.rubiensis
    N & W New Guinea
Manucodia chalybatus (Crinkle-collared Manucode)
    Misol I, all New Guinea except mountains
Manucodia comrii (Curl-crested Manucode)
  M.c.comrii
    Ferguson, Goodenough, Normanby Is
  M.c.trobriandi
    Trobriand Is

PHONYGAMMUS
Phonygammus keraudrenii (Trumpet Bird)
  P.k.keraudrenii
    NW New Guinea
  P.k.adelberti
    N New Guinea

P.k.neumanni
  NC New Guinea
P.k.mayri
  NE New Guinea
P.k.jamesii
  Aru Is, S New Guinea
P.k.purpureoviolaceus
  SE New Guinea
P.k.hunsteini
  D'Entrecasteaux Arch
P.k.gouldii
  N Queensland

PTILORIS
Ptiloris paradiseus (Paradise Riflebird)
  SE Queensland, NE New South Wales
Ptiloris victoriae (Queen Victoria Riflebird)
  NE Queensland
Ptiloris magnificus (Magnificent Riflebird)
P.m.intercedens
  E New Guinea
P.m.magnificus
  S & W New Guinea
P.m.alberti
  N Queensland

SEMIOPTERA
Semioptera wallacei (Wallace's Standardwing)
S.w.halmaherae
  Halmahera I
S.w.wallacei
  Batjan I

SELEUCIDIS
Seleucidis melanoleuca (Twelve-wired Bird of Paradise)
S.m.melanoleuca
  Salawati I, coast of New Guinea
S.m.auripennis
  N New Guinea

PARADIGALLA
Paradigalla carunculata (Long-tailed Paradigalla)
P.c.carunculata
  Arfak Mts (New Guinea)
P.c.intermedia
  WC New Guinea
Paradigalla brevicauda (Short-tailed Paradigalla)
  C New Guinea

DREPANORNIS
Drepanornis albertisii (Black-billed Sicklebill)
D.a.albertisii
  NW New Guinea
D.a.cervinicauda
  C New Guinea
D.a.geisleri
  E New Guinea
Drepanornis bruijnii (Pale-billed Sicklebill)
  NW New Guinea

EPIMACHUS
Epimachus fastuosus (Black Sicklebill)
E.f.fastuosus
  NW New Guinea
E.f.atratus
  WC New Guinea
E.f.ultimus
  Mt Menawa (N New Guinea)
E.f.stresemanni
  EC New Guinea
Epimachus meyeri (Brown Sicklebill)
E.m.megarhynchus
  Weyland Mts (New Guinea)
E.m.albicans
  C New Guinea
E.m.bloodi
  EC New Guinea
E.m.meyeri
  SE New Guinea

ASTRAPIA
Astrapia nigra (Arfak Bird of Paradise)
  Arfak Mts (New Guinea)
Astrapia splendidissima (Splendid Bird of Paradise)
A.s.helios
  NW New Guinea
A.s.splendidissima
  Weyland Mts (New Guinea)
A.s.elliottsmithi
  WC New Guinea
Astrapia mayeri (Ribbon-tailed Bird of Paradise)
  EC New Guinea
Astrapia stephaniae (Princess Stephanie's Bird of Paradise)
A.s.feminina
  EC New Guinea
A.s.ducalis
  E New Guinea
A.s.stephaniae
  SE New Guinea
Astrapia rothschildi (Huon Bird of Paradise)
  E New Guinea

LOPHORINA
Lophorina superba (Superb Bird of Paradise)
L.s.superba
  Arfak Mts (NW New Guinea)
L.s.niedda
  Mt Wondiwoi (W New Guinea)
L.s.feminina
  WC New Guinea
L.s.pseudoparotia
  EC New Guinea
L.s.latipennis
  E New Guinea
L.s.connectens
  E New Guinea
L.s.minor
  SE New Guinea
L.s.sphinx
  SE New Guinea

PAROTIA
Parotia sefilata (Arfak Parotia)
Arfak Mts (New Guinea)
Parotia carolae (Queen Carola's Parotia)
P.c.clelandiae
Victor Emmanuel Mts (N New Guinea)
P.c.meeki
Nassau, Oranje Mts (New Guinea)
P.c.carolae
Weyland Mts (C New Guinea)
P.c.chalcothorax
Idenburg river (C New Guinea)
P.c.berlepschi
Van Rees Mts (C New Guinea) ?
P.c.chrysenia
Bismarck Mts (C New Guinea)
Parotia lawesii (Lawes' Parotia)
P.l.lawesii
C & SE New Guinea
P.l.helenae
SE New Guinea   172.1
Parotia wahnesi (Wahnes' Parotia)
E New Guinea

PTERIDOPHORA
Pteridophora alberti (King of Saxony Bird of Paradise)
P.a.alberti
C New Guinea
P.a.hallstromi
EC New Guinea
P.a.buergersi
EC New Guinea

CICINNURUS
Cicinnurus regius (King Bird of Paradise)
C.r.regius
Aru Is
C.r.rex
W New Guinea islands, New Guinea
C.r.coccineifrons
Japen I
C.r.similis
N New Guinea
C.r.cryptorhynchus
NW New Guinea
C.r.gymnorhynchus
NE New Guinea

DIPHYLLODES
Diphyllodes magnificus (Magnificent Bird of Paradise)
D.m.magnificus
NW New Guinea, Salawati I
D.m.intermedius
Weyland Mts (C New Guinea)
D.m.chrysopterus
Japen I, N New Guinea
D.m.hunsteini
E New Guinea
Diphyllodes respublica (Wilson's Bird of Paradise)
Waigeu, Batanta Is

PARADISAEA
Paradisaea apoda (Greater Bird of Paradise)
P.a.apoda
Aru Is, (Lt Tobago I, introd.)
P.a.novaeguineae
S New Guinea
Paradisaea raggiana (Raggiana Bird of Paradise)
P.r.augustaevictoriae
NE New Guinea
P.r.intermedia
E New Guinea
P.r.granti
E New Guinea
P.r.salvadorii
S New Guinea
P.r.raggiana
SE New Guinea
Paradisaea minor (Lesser Bird of Paradise)
P.m.minor
NW & W New Guinea
P.m.finschi
NC New Guinea
P.m.jobiensis
Japen I
P.m.pulchra
Misol I
Paradisaea decora (Goldie's Bird of Paradise)
Fergusson, Normanby Is
Paradisaea rubra (Red Bird of Paradise)
Batanta, Waigeu, Saonek Is
Paradisaea guilielmi (Emperor of Germany Bird of Paradise)
E New Guinea
Paradisaea rudolphi (Blue Bird of Paradise)
P.r.ampla
Hertzog Mts (W New Guinea)
P.r.margaritae
WC New Guinea
P.r.rudolphi
E New Guinea

173. CORVIDAE (CROWS, JAYS)

PLATYLOPHUS
Platylophus galericulatus (Crested Shrike-Jay)
P.g.ardesiacus
S Thailand, Malaysia
P.g.coronatus
Borneo, Sumatra
P.g.galericulatus
Java

PLATYSMURUS
Platysmurus leucopterus (White-winged Magpie)
P.l.leucopterus
Malaysia, Sumatra
P.l.aterrimus
Borneo

GYMNORHINUS
Gymnorhinus cyanocephala (Pinyon Jay)
  W USA, NW Mexico

CYANOCITTA
Cyanocitta cristata (Blue Jay)
  C.c.bromia
    S Canada, C USA >> SE USA
  C.c.cristata
    EC & SE USA
  C.c.semplei
    S Florida
  C.c.cyanotephra
    SC USA
Cyanocitta stelleri (Steller's Jay)
  C.s.stelleri
    W Canada, NW USA
  C.s.carlottae
    Queen Charlotte Is
  C.s.annectens
    W Canada, WC USA
  C.s.frontalis
    W USA
  C.s.carbonacea
    W California
  C.s.macrolopha
    C & S USA, N Mexico
  C.s.diademata
    NC Mexico
  C.s.coronata
    SC Mexico
  C.s.purpurea
    SW Mexico
  C.s.azteca
    C Mexico
  C.s.teotepecencis
    S Mexico
  C.s.ridgwayi
    S Mexico to El Salvador
  C.s.suavis
    Honduras, Nicaragua

APHELOCOMA
Aphelocoma coerulescens (Scrub Jay)
  A.c.immanis
    W Oregon
  A.c.caurina
    W USA
  A.c.oocleptica
    W USA
  A.c.californica
    W California
  A.c.cana
    California
  A.c.obscura
    N Baja California
  A.c.cactophila
    C Baja California
  A.c.hypoleuca
    C & S Baja California
  A.c.insularis
    Santa Cruz I

  A.c.nevadae
    WC USA, N Mexico
  A.c.woodhouseii
    WC & SC USA
  A.c.texana
    WC Texas
  A.c.grisea
    NW Mexico
  A.c.cyanotis
    EC Mexico
  A.c.sumichrasti
    SC Mexico
  A.c.remota
    SW Mexico
  A.c.coerulescens
    S Florida
Aphelocoma ultramarina (Mexican Jay)
  A.u.arizonae
    SW USA, NW Mexico
  A.u.wollweberi
    W Mexico
  A.u.gracilis
    WC Mexico
  A.u.couchii
    S Texas, NE Mexico
  A.u.potosina
    EC Mexico
  A.u.ultramarina
    SC Mexico
  A.u.colimae
    SW Mexico
Aphelocoma unicolor (Unicoloured Jay)
  A.u.guerrerensis
    C Mexico
  A.u.oaxacae
    S Mexico
  A.u.concolor
    SE Mexico
  A.u.unicolor
    SE Mexico
  A.u.griscomi
    El Salvador, W Honduras

CYANOLYCA
Cyanolyca viridicyana (White-collared Jay)
  C.v.joylaea
    NC Peru
  C.v.cyanolaema
    SE Peru
  C.v.viridicyana
    W Bolivia
Cyanolyca armillata (Collared Jay)
  C.a.meridana
    NW Venezuela
  C.a.armillata
    E Colombia, W Venezuela
  C.a.quindiuna
    S Colombia, N Ecuador
Cyanolyca turcosa (Turquoise Jay)
    S Colombia, N Peru
Cyanolyca pulchra (Beautiful Jay)
    SW Colombia, W Ecuador
Cyanolyca cucullata (Azure-hooded Jay)

C.c.mitrata
 E & S Mexico, Guatemala
C.c.guatemalae
 Chiapas, SE Mexico
C.c.hondurensis
 W Honduras
C.c.cucullata
 Costa Rica, W Panama
Cyanolyca pumilo (Black-throated Jay)
 S Mexico to Honduras
Cyanolyca nana (Dwarf Jay)
 S Mexico
Cyanolyca mirabilis (White-throated Jay)
 SW Mexico
Cyanolyca argentigula (Silvery-throated Jay)
C.a.albior
 Costa Rica
C.a.argentigula
 S Costa Rica

CISSILOPHA
Cissilopha melanocyanea (Bushy-crested Jay)
C.m.melanocyanea
 Guatemala to Honduras
C.m.chavezi
 S Honduras, N Nicaragua
Cissilopha sanblasiana (San Blas Jay)
C.s.nelsoni
 SW Mexico
C.s.sanblasiana
 SW Mexico
Cissilopha yucatanica (Yucatan Jay)
C.y.yucatanica
 SE Mexico, Guatemala
C.y.rivularis
 SE Mexico
Cissilopha beecheii (Purplish-backed Jay)
 NW Mexico

CYANOCORAX
Cyanocorax caeruleus (Azure Jay)
 SE Brazil to N Argentina
Cyanocorax cyanomelas (Purplish Jay)
 SE Peru to N Argentina
Cyanocorax violaceus (Violaceous Jay)
C.v.pallidus
 N Venezuela
C.v.violaceus
 N Sth America
Cyanocorax cristatellus (Curl-crested Jay)
 C & E Brazil
Cyanocorax heilprini (Azure-naped Jay)
 S Venezuela, NW Brazil
Cyanocorax cayanus (Cayenne Jay)
 SE Venezuela, the Guianas, N Brazil
Cyanocorax affinis (Black-chested Jay)
C.a.zeledoni
 S Costa Rica, Panama
C.a.affinis
 N Colombia, NW Venezuela

Cyanocorax chrysops (Plush-crested Jay)
C.c.diesingii
 N Brazil
C.c.chrysops
 E Bolivia to SE Brazil
C.c.tucumanus
 NW Argentina
Cyanocorax cyanopogon (White-naped Jay)
 E Brazil
Cyanocorax mystacalis (White-tailed Jay)
 SW Ecuador, NW Peru
Cyanocorax dickeyi (Tufted Jay)
 W Mexico
Cyanocorax yncas (Green Jay)
C.y.glaucescens
 NE Mexico
C.y.speciosus
 W Mexico
C.y.vividus
 SW Mexico
C.y.luxuosus
 E Mexico
C.y.centralis
 SE Mexico, Guatemala, Honduras
C.y.maya
 SE Mexico
C.y.cozumelae
 Cozumel I
C.y.galeatus
 C Colombia
C.y.cyanodorsalis
 E Colombia
C.y.andicolus
 NW Venezuela
C.y.guatimalensis
 N Venezuela
C.y.yncas
 SW Colombia, Ecuador, Peru, N Bolivia
C.y.longirostris
 N Peru

PSILORHINUS
Psilorhinus morio (Brown Jay)
P.m.palliatus
 NE & C Mexico
P.m.morio
 SE Mexico
P.m.cyanogenys
 SE Mexico, Central America
P.m.mexicanus
 E Mexico
P.m.vociferus
 SE Mexico

CALOCITTA
Calocitta formosa (White-throated Magpie-Jay)
C.f.formosa
 SW Mexico
C.f.azurea
 SE Mexico, Guatemala

C.f.pompata
  S Mexico to Costa Rica
Calocitta colliei (Collie's Magpie-Jay)
  W Mexico

Garrulus glandarius (Jay)
  G.g.rufitergum
    S Scotland, England, N France
  G.g.hibernicus
    N Scotland, Ireland
  G.g.glandarius
    N & C Europe
  G.g.fasciatus
    Iberia
  G.g.ichnusae
    Sardinia
  G.g.corsicanus
    Corsica
  G.g.albipectus
    Italy
  G.g.cretorum
    Greece, Crete
  G.g.glaszneri
    Cyprus
  G.g.hansguentheri
    Istanbul
  G.g.cervicalis
    E Algeria, Tunisia
  G.g.whitakeri
    N Morocco, W Algeria
  G.g.minor
    NW Africa
  G.g.atricapillus
    Iraq, W Iran
  G.g.rhodius
    Rhodes
  G.g.krynicki
    Turkey, Caucasus
  G.g.iphigenia
    Crimea
  G.g.hyrcanus
    N Iran
  G.g.suianae
    Kurdistan, NE Iran
  G.g.severzowii
    Scandinavia, S Russia
  G.g.brandtii
    NE Russia, C & NE Asia
  G.g.kansuensis
    W China
  G.g.pekingensis
    N China, NW Manchuria
  G.g.sinensis
    SW China, NE Burma
  G.g.taivanus
    Taiwan
  G.g.leucotis
    E Burma, Thailand, Indochina
  G.g.oatesi
    C Burma
  G.g.haringtoni
    SC Burma

G.g.interstinctus
  E Himalayas, SE Tibet
G.g.persaturatus
  Assam
G.g.bispecularis
  W Himalayas
G.g.japonicus
  N Japan
G.g.tokugawae
  Sado I
G.g.hiugaensis
  S Japan
G.g.orii
  Yakushima Is
G.g.namiyei
  Tsushima I
Garrulus lanceolatus (Lanceolated Jay)
  W Himalayas, N India
Garrulus lidthi (Purple Jay)
  N Riukiu Is

Perisoreus canadensis (Grey Jay)
  P.c.pacificus
    NW Alaska
  P.c.canadensis
    C Canada, N USA
  P.c.nigricapillus
    NE Canada
  P.c.arcus
    SW Canada
  P.c.albescens
    W Canada, NW USA
  P.c.bicolor
    SW Canada, NW USA
  P.c.capitalis
    WC & SC USA
  P.c.griseus
    SW Canada, NW USA
  P.c.obscurus
    NW USA
Perisoreus infaustus (Siberian Jay)
  P.i.infaustus
    Lapland
  P.i.ostjakorum
    NW Siberia
  P.i.yakutensis
    N & NE Asia
  P.i.ruthenus
    C Russia, C Scandinavia
  P.i.opicus
    NC Asia
  P.i.rogosowi
    C Siberia
  P.i.sibericus
    Outer Mongolia
  P.i.varnak
    N Manchuria
  P.i.sakhalinensis
    N Sakhalin I
  P.i.maritimus
    Lower Amur river

Perisoreus internigrans (Szechwan Grey Jay)
W China

UROCISSA
Urocissa ornata (Ceylon Blue Magpie)
Sri Lanka
Urocissa caerulea (Formosan Blue Magpie)
Taiwan
Urocissa flavirostris (Yellow-billed Blue Magpie)
U.f.cucullata
W Himalayas
U.f.flavirostris
E Himalayas, N Burma
U.f.schaferi
W Burma
U.f.robini
N Vietnam
Urocissa erythrorhyncha (Red-billed Blue Magpie)
U.e.brevivexilla
N China
U.e.erythrorhyncha
C & S China, N Vietnam
U.e.alticola
SW China, NE Burma
U.e.occipitalis
Himalayas
U.e.magnirostris
Assam to Indochina
Urocissa whiteheadi (White-winged Magpie)
U.w.whiteheadi
Hainan I
U.w.xanthomelana
C Laos, N Vietnam

CISSA
Cissa chinensis (Green Magpie)
C.c.chinensis
Himalayas, N Indochina
C.c.robinsoni
Malaysia
C.c.klossi
C Indochina
C.c.margaritae
S Vietnam
C.c.minor
Sumatra, NW Borneo
Cissa hypoleuca (Eastern Green Magpie)
C.h.jini
SE China
C.h.concolor
N Vietnam
C.h.chauleti
C Vietnam
C.h.hypoleuca
E Thailand, S Indochina
C.h.katsumatae
Hainan I
Cissa thalassina (Short-tailed Green Magpie)
C.t.thalassina
Java

C.t.jeffreyi
NW Borneo

CYANOPICA
Cyanopica cyana (Azure-winged Magpie)
C.c.cooki
W Spain, Portugal
C.c.cyana
C & EC Asia
C.c.pallescens
NE Asia
C.c.koreensis
Korea
C.c.stegmanni
Manchuria
C.c.swinhoei
E China
C.c.interposita
N China
C.c.kansuensis
W China
C.c.japonica
Japan

DENDROCITTA
Dendrocitta vagabunda (Indian Tree Pie)
D.v.pallida
W Himalayas, NW India
D.v.vagabunda
E Himalayas, NE India
D.v.parvula
SW India
D.v.vernayi
SE India
D.v.sclateri
W Burma
D.v.kinneari
S Burma, NW Thailand
D.v.saturatior
S Thailand
D.v.sakeratensis
E Thailand, Indochina
Dendrocitta occipitalis (Malaysian Tree Pie)
D.o.occipitalis
Sumatra
D.o.cinerascens
Borneo
Dendrocitta formosae (Himalayan Tree Pie)
D.f.occidentalis
W Himalayas
D.f.himalayensis
E Himalayas, Burma, N Laos
D.f.sarkari
E India
D.f.assimilis
S Burma, Thailand, Andaman Is
D.f.sinica
E & S China, N Vietnam
D.f.sapiens
W China
D.f.formosae
Taiwan

D.f.insulae
Hainan I
Dendrocitta leucogastra (Southern Tree Pie)
S India
Dendrocitta frontalis (Black-browed Tree Pie)
Himalayas to N Vietnam
Dendrocitta baileyi (Andaman Tree Pie)
Andaman Is

CRYPSIRINA
Crypsirina temia (Black Racquet-tailed Tree Pie)
S Burma to Indochina & Java
Crypsirina cucullata (Hooded Racquet-tailed Tree Pie)
N & C Burma

TEMNURUS
Temnurus temnurus (Notch-tailed Tree Pie)
N Vietnam, Hainan I

PICA
Pica pica (Black-billed Magpie)
P.p.fennorum
N Scandinavia, W Russia
P.p.pica
British Isles, C & E Europe
P.p.galliae
W Europe
P.p.melanotos
Spain, Portugal
P.p.mauretanica
NW Africa
P.p.asirensis
SW Arabia
P.p.bactriana
C Russia to N India
P.p.hemileucoptera
W & S Siberia, WC Asia
P.p.leucoptera
EC Asia
P.p.camtschatica
NE Asia
P.p.sericea
S China, Burma, Indochina
P.p.bottanensis
N Himalayas, Tibet
P.p.hudsonia
W Canada, W USA
Pica nuttalli (Yellow-billed Magpie)
W California

ZAVATTARIORNIS
Zavattariornis stresemanni (Stresemann's Bush Crow)
S Ethiopia

PODOCES
Podoces hendersoni (Henderson's Ground Jay)
C Asia, W China
Podoces biddulphi (Biddulph's Ground Jay)
W Sinkiang

Podoces panderi (Pander's Ground Jay)
S Russia
Podoces pleskei (Pleske's Ground Jay)
E Iran

PSEUDOPODOCES
Pseudopodoces humilis (Hume's Ground Chough)
Tsinghai, W China

NUCIFRAGA
Nucifraga columbiana (Clark's Nutcracker)
SW Canada, W USA
Nucifraga caryocatactes (Spotted Nutcracker)
N.c.caryocatactes
N & E Europe >> S Russia
N.c.macrorhynchos
N & NE Asia >> N Iran & N China
N.c.rothschildi
Russia, Turkestan
N.c.japonica
N Japan
N.c.owstoni
Taiwan
N.c.interdicta
N China
N.c.multipunctata
Pakistan, NW India
N.c.hemispila
W Himalayas
N.c.macella
E Himalayas, Burma, W China
N.c.yunnanensis
SW China

PYRRHOCORAX
Pyrrhocorax pyrrhocorax (Red-billed Chough)
P.p.pyrrhocorax
England, Ireland
P.p.erythrorhamphus
W Europe
P.p.barbarus
Canary Is, NW Africa
P.p.baileyi
N Ethiopia
P.p.docilis
E Europe to Arabia, Iran
P.p.centralis
C Asia, Pakistan, NW India
P.p.himalayanus
N India, Himalayas, W China
P.p.brachypus
N China, NE Asia
Pyrrhocorax graculus (Alpine Chough)
P.g.graculus
Europe, N Africa, Caucasus
P.g.digitatus
Iran, C Asia, Himalayas

PTILOSTOMUS
Ptilostomus afer (Piapiac)
Senegal to Ethiopia, Uganda

CORVUS

Corvus monedula (Jackdaw)
  C.m.monedula
    Scandinavia
  C.m.spermologus
    W & C Europe
  C.m.soemmerringii
    E Europe, N & C Asia >> Iran, W India
  C.m.cirtensis
    N Africa

Corvus dauuricus (Daurian Jackdaw)
  C & NE Asia >> SE China & Japan

Corvus splendens (House Crow)
  C.s.zugmayeri
    Baluchistan, NW India
  C.s.splendens
    India
  C.s.protegatus
    Sri Lanka, Malaysia
  C.s.maledivicus
    Laccadive, Maldive Is
  C.s.insolens
    S Burma, SW Thailand, W Yunnan

Corvus moneduloides (New Caledonian Crow)
  New Caledonia, Loyalty Is

Corvus unicolor (Banggai Crow)   173.1
  Banggai I

Corvus enca (Slender-billed Crow)
  C.e.compilator
    Malaysia, Sumatra, Borneo
  C.e.enca
    Java, Bali, Montawi Is
  C.e.celebensis
    Sulawesi
  C.e.mangoli
    Sula Arch
  C.e.violaceus
    Seram I
  C.e.pusillus
    Balabac, Palawan, Mindoro Is
  C.e.sierramadrensis
    NE Luzon I
  C.e.samarensis
    Samar, Mindanao Is

Corvus typicus (Piping Crow)   173.1
  C & S Sulawesi, Muna I

Corvus florensis (Flores Crow)
  Flores I

Corvus kubaryi (Marianas Crow)
  Guam, Rota Is

Corvus validus (Moluccan Crow)
  N Moluccas

Corvus woodfordi (White-billed Crow)
  C.w.meeki
    Bougainville I
  C.w.woodfordi
    Guadalcanal I
  C.w.vegetus
    Choiseul, Ysabel Is

Corvus fuscicapillus (Brown-headed Crow)
  C.f.fuscicapillus
    Aru Is, New Guinea

  C.f.megarhynchus
    Waigeu, Geimen Is

Corvus tristis (Grey Crow)
  New Guinea, D'Entrecasteaux Arch

Corvus capensis (Black Crow)
  Eastern & Southern Africa

Corvus frugilegus (Rook)
  C.f.frugilegus
    Europe, W & C Asia >> N Africa & NW India
  C.f.pastinator
    E Asia >> Japan & SE China

Corvus brachyrhynchos (American Crow)
  C.b.hesperis
    W Canada, W USA
  C.b.brachyrhynchos
    C & E Canada, C & NE USA >> E USA
  C.b.paulus
    E & SE USA
  C.b.pascuus
    S Florida

Corvus caurinus (Northwestern Crow)
  W Canada, NW USA

Corvus imparatus (Tamaulipas Crow)
  N Mexico

Corvus sinaloae (Sinaloa Crow)
  NW Mexico

Corvus ossifragus (Fish Crow)
  E USA

Corvus palmarum (Palm Crow)
  C.p.minutus
    Cuba
  C.p.palmarum
    Hispaniola

Corvus jamaicensis (Jamaican Crow)
  Jamaica

Corvus nasicus (Cuban Crow)
  Cuba, Grand Caicos I

Corvus leucognaphalus (White-necked Crow)
  Hispaniola, Puerto Rico

Corvus corone (Carrion Crow)
  C.c.corone
    W Europe >> N Africa
  C.c.cornix
    N & E Europe
  C.c.sardonius
    S & SE Europe, Asia Minor
  C.c.sharpii
    Siberia, Iraq, Iran to Turkestan, NW India
  C.c.capellanus
    S Iraq, SW Iran
  C.c.orientalis
    E Asia, Japan >> NW India, S China

Corvus macrorhynchos (Jungle Crow)
  C.m.japonensis
    Sakhalin I, Japan
  C.m.conectens
    C & S Riukiu Is
  C.m.osai
    S Riukiu Is
  C.m.mandschuricus
    NE Asia

C.m.colonorum
  China, N Indochina
C.m.hainanus
  Hainan I
C.m.mengtszensis
  SW China
C.m.tibetosinensis
  E Himalayas, N Burma, W China
C.m.intermedius
  W Himalayas, NW India
C.m.culminatus
  S India, Sri Lanka
C.m.levaillantii
  NE India, Burma, Thailand
C.m.macrorhynchos
  Malaysia, S Indochina, Sunda Is
C.m.philippinus
  Philippine Is
Corvus orru (Australian Crow)
C.o.orru
  Moluccas, New Guinea
C.o.insularis
  New Britain, New Ireland, New Hanover
C.o.latirostris
  Tanimbar, Babar Is
C.o.ceciliae
  Australia
Corvus bennetti (Little Crow)
  W & C Australia
Corvus coronoides (Australian Raven)
C.c.coronoides
  E, S & SW Australia
C.c.boreus
  New South Wales, Victoria
Corvus tasmanicus (Forest Raven)
C.t.novaanglica
  NE New South Wales
C.t.tasmanicus
  Tasmania
Corvus mellori (Little Raven)
  SE Australia

Corvus torquatus (Collared Crow)
  E & C China, N Vietnam
Corvus albus (Pied Crow)
  W, C, E & Sthn Africa, Madagascar
Corvus hawaiiensis (Hawaiian Crow)   173.2
  Hawaii Is
Corvus cryptoleucus (White-necked Raven)
C.c.cryptoleucus
  SW USA, N Mexico
C.c.reai
  Arizona
Corvus ruficollis (Brown-necked Raven)
C.r.ruficollis
  N Africa to Pakistan
C.r.edithae
  Somalia
Corvus corax (Common Raven)
C.c.principalis
  Alaska, Canada, N USA
C.c.sinuatus
  WC USA, Central America
C.c.varius
  Iceland, Faroe Is
C.c.corax
  Europe, W Asia
C.c.subcorax
  SE Europe, Asia Minor to Pakistan
C.c.tingitanus
  N Africa
C.c.tibetanus
  C Asia, Himalayas
C.c.kamtschaticus
  NE Asia, N Japan
Corvus rhipidurus (Fan-tailed Raven)
  NE Africa to Syria, Arabia
Corvus albicollis (African White-necked Raven)
  E & S Africa
Corvus crassirostris (Thick-billed Raven)
  Ethiopia, E Sudan

INDEX OF LATIN NAMES

INDEX OF ENGLISH NAMES